D1031713

STANDARD CANTILEVER

RETAINING WALLS

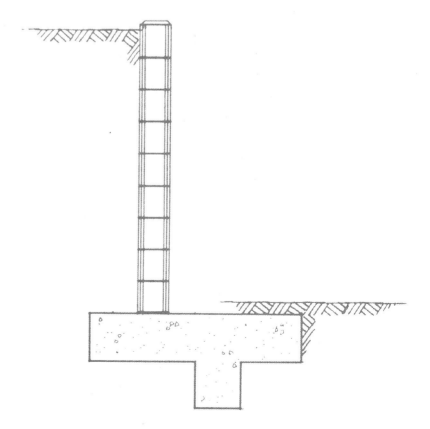

McGRAW-HILL BOOK COMPANY

New York
St. Louis
San Francisco
Auckland
Bogota
Dusseldorf
Johannesburg
London
Madrid
Mexico
Montreal
New Delhi
Panama
Paris
Sao Paulo
Singapore
Sydney
Tokyo
Toronto

STANDARD
RETAINING

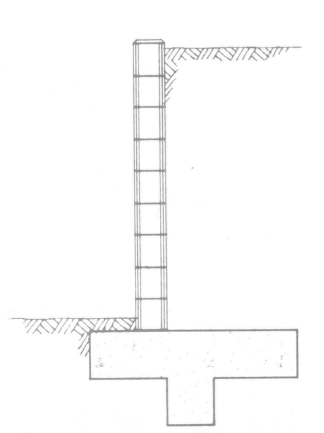

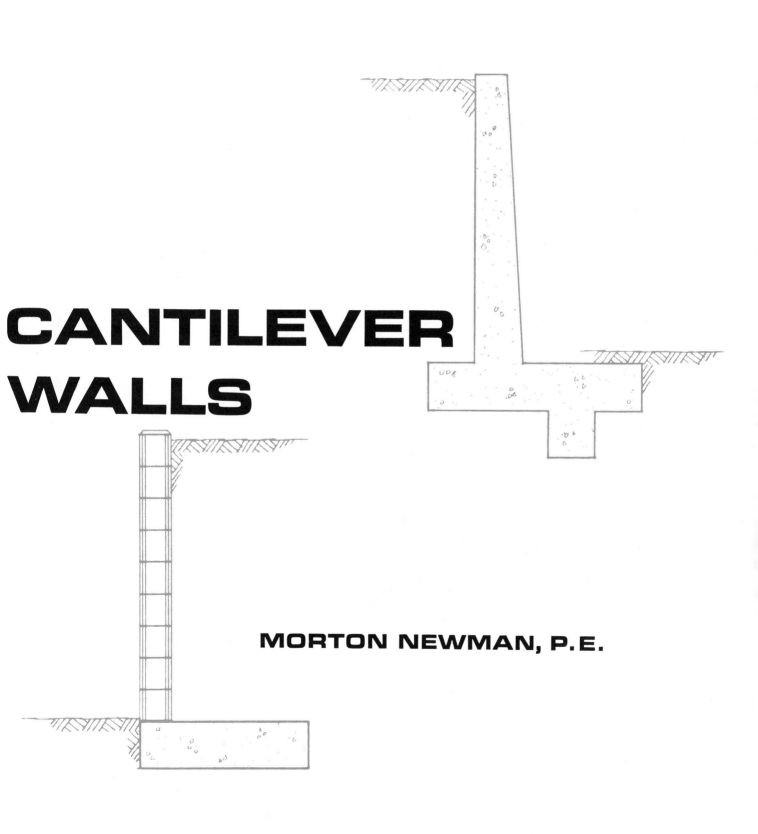

CANTILEVER
WALLS

MORTON NEWMAN, P.E.

Library of Congress Cataloging in Publication Data

Newman, Morton.
   Standard cantilever retaining walls.

   Bibliography: p.
   Includes index.
   1. Retaining walls. I. Title.
TA770.N48   624'.16   76-23250
ISBN 0-07-046347-6

1234567890 HDHD 785432109876

TO FAY AND SUZANNE

Contents

Contents

Preface

The purpose of this book is to coordinate the many building code requirements for the design and the construction of concrete and masonry cantilever retaining walls, and to compile and present this information as a standard reference handbook. The increasing costs of construction labor and materials, along with the more rigid requirements of current building codes, require that retaining walls be more accurately designed and constructed. The trial and error procedure used for originating a structural design of a retaining wall is time consuming and will not give a sufficient degree of accuracy to achieve the optimum use of construction materials. Design and construction methods, the degree of accuracy in design and the efficient use of construction materials are directly related to the economics of a structure.

There are a variety of types of retaining walls used on construction projects. This book is concerned only with the cantilever-type of retaining wall. The other types of walls are used when special job site conditions prevail to require a particular method of construction, such as a gravity retaining wall or a counterfort retaining wall. The cantilever retaining wall is the most commonly used wall because it can resist relatively large vertical and lateral loads for a variety of soil conditions. Also, the wall is constructed with standard building materials. The four basic structural requirements for cantilever retaining walls are material shear and bending capacity, foundation soil pressure, the ability of the wall to resist overturning, and the capacity of the soil to prevent lateral sliding. The structural design procedure must consider the bearing capacity of the foundation soil, the magnitude of the lateral force on the stem of the wall that results from the retained soil, the possible application of a surcharge or an axial load, the location of the resultant of all of the forces and loads on the wall, and the kinds of construction materials used. All of these interacting design criteria and considerations and their relevant building code requirements were used to develop a computer program capable of producing the structural design of either a concrete or masonry cantilever retaining wall for fourteen different stem loading conditions, and for stem heights ranging from 3 feet to 12 feet. The magnitude and type of load of the fourteen different stem loading conditions are given in Chapter 1. The loads were selected as those most usually encountered in the design of this type of retaining wall. These loads fall within one of three general categories: axial loads that are applied to the top of the wall stem and coincide with the stem vertical centerline; surcharge loads which are uniformly distributed loads applied to the flat surface of the retained soil; and loads caused by the angle of slope of the retained soil. The wall stem height of the computer program was limited to a maximum of twelve feet to maintain the scope of this book to walls within a medium height range.

The amount of information that this book endeavors to present within a usable reference format made it necessary to develop a system of organization which consists of nine individual chapters. Each chapter, except for Chapter 1, deals with a specific shape of retaining wall. Chapter 1 gives a general discussion on retaining walls and treats such related subjects as soil mechanics, materials specifications, building code requirements, design and construction methods, descriptions of the various load conditions, and a description of the nomenclature of the different types and parts of cantilever retaining walls. Chapter 2 through Chapter 9 are each concerned with a particular retaining wall shape or cross section configuration. The title of each of these chapters describes the shape of the wall, whether the wall stem is constructed with concrete or masonry, and the placement of the retained soil in relation to the wall foundation. Each of these retaining wall chapters consists of a description of the wall and a discussion of the application of the retaining wall that is designated by the chapter title, three design examples to demonstrate the derivation of the tabulated design data, and a series of drawings with design data tables for each loading condition. The drawings are arranged in order of ascending stem heights of 1 foot increments. This method of presentation is given for each retaining wall chapter and for each of the fourteen loading conditions.

Eight different retaining wall shapes with fourteen load conditions and for ten stem heights required a total of 1,120 retaining wall drawings and 112 design data tables. There are several instances in which a wall shape and reinforcement are the same for different load conditions, particularly for lower stem wall heights. These walls are placed in their respective section of the chapter so that they can be consecutively reference numbered within the system of organization. The chapters of this book and the wall loads are numbered and arranged so that any wall can be located or identified by a corresponding reference number. For example, a wall designated as 4.5.12 can be identified by the first number which indicates Chapter 4, "Property Line Retaining Wall—Concrete Stem—Soil Not Over Footing;" the second number, 5, indicates that the wall is designed for load condition No. 5, that is, an 800 lbs./ft. axial stem load; and the last number, 12, indicates the stem height in feet, measured from the top of the wall foundation. Thus, by this method, any retaining wall in this book can be identified and located by a set of three numbers, each separated by a decimal point, and each number corresponding respectively to the chapter number, the load condition number, and the stem height in feet.

Retaining walls that are designed to sustain either axial or surcharge loads represent their maximum design case. However, the source of an axial or surcharge load may not be constant or they may be an intermittently applied load. For this reason, these same shaped cross sections are designed without the additional load. The tables of design data for an unloaded condition of a retaining wall can be found in Appendix A. The numbers of the tables in the appendix correspond to the table number of the particular wall for its maximum design load. For example, Table 4.8 in the chapter gives the design data for a retaining wall with a 200 lbs./sq/ft surcharge load and Table 4.8a in the appendix gives the design data for the same wall without the surcharge load.

The design data and construction information that is presented in this book was developed and produced by a Fortran Computer program which was run on a Control Data 6600. It should be noted that computer output data, however accurate, should not be viewed as a total alternative to proven engineering experience and judgment. The tabulated data and the drawings in this book can be used as an aid in making design calculations, or as a guide in selecting a particular retaining wall for a specific job site requirement. With these points in mind, the author must caution the reader that the retaining walls in the following chapters cannot be used arbitrarily. The reader must know the bearing capacity and type of soil at the job site, the strength of the construction materials employed, and the physical characteristics of the soils that are to be retained. The design of cantilever retaining walls requires the technology of soil mechanics, strength of materials, and structural mechanics. Each of these subjects encompasses a whole field of scientific study, therefore it was necessary to limit the scope of this book to the extent that they are treated only to the degree that they relate to applied design criteria and accepted methods of construction for retaining walls.

The design criteria and material specifications used in preparing this book are in accordance with the requirements of The American Concrete Institute Building Code, The International Conference of Building Official Uniform Building Code, The Los Angeles City Building Code, The Masonry Institute of America, The Building Code of the City of New York, and The American Society for Testing and Materials.

The author would like to acknowledge the able assistance of Mr. Gary M. Meadville who produced the drawings, Mr. James E. Amrhein of the Masonry Institute of America and Mr. James P. Cassidy of Computer Design Service for their technical consultation, and Mr. James W. Meadville who developed the computer program that produced the design data.

Morton Newman

Symbols and Notations

a the coefficient used in the calculation to determine the area of reinforcing steel

A a foundation dimension (see the wall table diagrams)

A_g the gross cross section area

A_s the cross section area of reinforcement

A_v the cross section area of shear reinforcement

Adjust   adjustable

Alt.   alternate

A.C.I.   American Concrete Institute

A.S.T.M.   American Society for Testing and Materials

Arch.   architect

B footing key width dimension (see the wall table diagrams)

b width of a rectangular cross section, in.

Blk.   block

Bott.   bottom

Bldg.   building

C footing dimension (see the wall table diagrams)

Calcs.   calculations

Cem.   cement

C.L.   centerline

C.E.   Civ Engineer

Clr.   clear

C.M.   center of mass

Conc.   concrete

Cont.   continuous

Cu.   cubic

D depth of footing key (see the wall table diagrams)

d the effective depth of a flexural member measured from the compression face to the center of the reinforcement

Dbl.   double

Defl.   deflection

Depr.   depress

Det.   detail

Diag.   diagonal

Dia.   diameter

Dim.   dimension

D.L.   dead load

Drwg.   drawing

E the eccentricity of the resultant of forces, measured as the horizontal distance from the center of the wall foundation, ft.

Ea.   each

E_c the modulus of elasticity of concrete, p.s.i.

E_m	the modulus of elasticity of masonry, p.s.i.
E_s	the modulus of elasticity of reinforcing steel, p.s.i.
E.F.P.	equivalent fluid pressure, p.c.f.
Engr.	engineer
Eq.	equation
Equip.	equipment
Exist.	existing
Exp.	exposed
Ext.	exterior
F	the lateral force caused by the E.F.P., lbs.
F_p	the force of the passive soil pressure, lbs.
f_a	the calculated axial compression unit stress, p.s.i.
f_b	the calculated flexural unit stress, p.s.i.
f_c	the calculated compressive stress in the extreme fiber of a concrete member acting in flexure, p.s.i.
f_m	the allowable compressive stress in the extreme fiber of a masonry member acting in flexure, p.s.i.
f_s	the tensile stress of the reinforcement, p.s.i.
f_t	the calculated tensile stress in the extreme fiber of a concrete member acting in flexure, p.s.i.
f_y	the yield stress of reinforcing steel, p.s.i.
Fr	the coefficient of frictional resistance to lateral sliding
Ft.	foot or feet
Ft. K	foot kips
Ftg.	footing
Fdn.	foundation
Grt.	grout
H	the stem height dimension of a retaining wall measured from the top of the wall foundation, ft.
H_1	a concrete stem height dimension of a masonry retaining wall, ft. (see the wall table diagrams)
H_2	the stem height dimension of a 12-in. wide concrete block of a masonry retaining wall, ft. (see the wall table diagrams)
H_3	the stem height dimension of an 8-in. wide concrete block of a masonry retaining wall, ft. (see the wall table diagrams)
Hk.	reinforcing bar hook
Horiz.	horizontal
I	moment of inertia
In.	inch or inches
In. Lbs.	inch pounds
Incl.	inclusive
Insp.	inspection
Int.	interior
I.C.B.O.	International Conference of Building Officials
j	the ratio of the distance (jd) between the resultant tension and compression on the cross section of a concrete or masonry member acting in flexure
Jnt.	joint
K	kips = 1000 lbs.
K	$\frac{1}{2}f_c jk$
k	ratio of the distance between the extreme fiber and the neutral axis to the effective depth or to the total depth
L	the length of a retaining wall foundation (see the wall table diagrams)

Lat.	lateral
Ld.	load
Lbs.	pounds
L.L.	live load
Long.	longitudinal
M	bending moment: ft. lbs., in. lbs., ft. K, in. K
M_e	the moment caused by the eccentricity of forces and loads on the wall foundation, ft. lbs.
M_o	the overturning moment of the wall, ft. lbs.
M_r	the resiting moment of the wall, ft. lbs.
Mas.	masonry
Max.	maximum
Memb.	membrane
Min.	minimum
n	the ratio of the modulus of elasticity of the reinforcing steel (E_s) to that of concrete (E_c) or masonry (E_m)
Nat.	natural
No.	number or #
o.c.	on center
Opng.	opening
Opp.	opposite
O.T.R.	the overturning ratio = M_r/M_o
P_A	active soil pressure, lbs.[3]
P_p	passive soil pressure, lbs.[3]
p	percent of reinforcement steel per cross section area
p.c.f.	pounds per cubic foot
p.l.f.	pounds per linear foot
p.s.f.	pounds per square foot
p.s.i.	pounds per square inch
P	an axial load applied at the top of the wall stem and acting through the centerline of the stem, lbs./ft.
Press.	pressure
R	the resultant of the summation of forces and loads on a retaining wall section
Rect.	rectangular
Reinf.	reinforcing or reinforcement
Reqd.	required
RS	reference standard (number) from the New York City Building Code
S	section modulus, in.[3] or ft.[3]
Sect.	section
SP	soil pressure
SP_h	the soil pressure at the heel of the retaining wall foundation
SP_t	the soil pressure at the toe of the retaining wall foundation
Spcg.	spacing
Spec.	specification
Stb.	setback (see the wall table diagrams)
Std.	standard
Stgr.	stagger
Stl.	steel
Stirr.	stirrup
Sq.	square
Sq. Ft.	square foot or square feet

Sq. In.   square inch or square inches

Sym.   symmetrical

T retaining wall foundation thickness, ft. or in. (see the wall table diagrams)

Thk.   thick or thickness

Thru   through

U retaining wall foundation undercut dimension, ft. or in. (see the wall table diagrams)

u the unit bond stress for reinforcing bars, p.s.i.

U.B.C.   Uniform Building Code

Unif.   uniform

v the unit shear stress, p.s.i.

Vert.   vertical

W the total weight of all vertical forces and loads acting on a retaining wall section, lbs.

W_b the width of the retaining wall stem at its base, in.

W_t the width of the retaining wall stem at the top, in.

WWF   welded wire fabric

$\bar{X}$ the horizontal dimension location on the length of the retaining wall foundation of the resultant of all forces and loads acting on a retaining wall section

β beta—the angle of the slope of the backfill soil with respect to the horizontal plane

γ gamma—the density of the soil, p.c.f.

δ delta—the angle of wall friction; the vertical component of the active or passive force

ϕ phi—the angle of internal friction of the soil

ω omega—the angle between a vertical plane and the face of the wall stem

List of Tables and Figures

Appendixes

CHAPTER 1

Cantilever Retaining Wall Design and Construction

A retaining wall is a structure that is designed and constructed to provide the lateral resistance necessary to prevent horizontal movement of the vertical plane of a mass of soil. Retaining walls act as an opposing element to the tendency of the soil to displace laterally to form a natural slope. The wall must be able to resist the lateral force of the movement of the soil and to maintain it in a vertical and static condition (Ref. 4). There are several types of retaining walls that can perform this function. The general shape of each of these walls determines the mechanics of reaction to the possible soil movement. The type of retaining wall that is selected for a project depends on the job site conditions, the materials that are available, the soil bearing capacity at the job site, the economics of construction, and the height and physical characteristics of the soil that is to be retained.

Figure 1.1 shows the cross sections for five types of retaining walls that are presently used on general construction projects (Ref. 11). In Fig. 1.1a, a cross section of a gravity-type retaining wall is shown. Its name is derived from the fact that the lateral force of the retained soil is resisted by the weight of the wall and by the resultant frictional resistance and passive soil pressure at its base. This type of wall is rarely used to retain high embankments and is uneconomical since it does not make the most efficient use of concrete. Large rocks and stones are usually added to reduce the volume of concrete used and to increase its weight. Gravity retaining walls are seldom constructed with masonry.

1

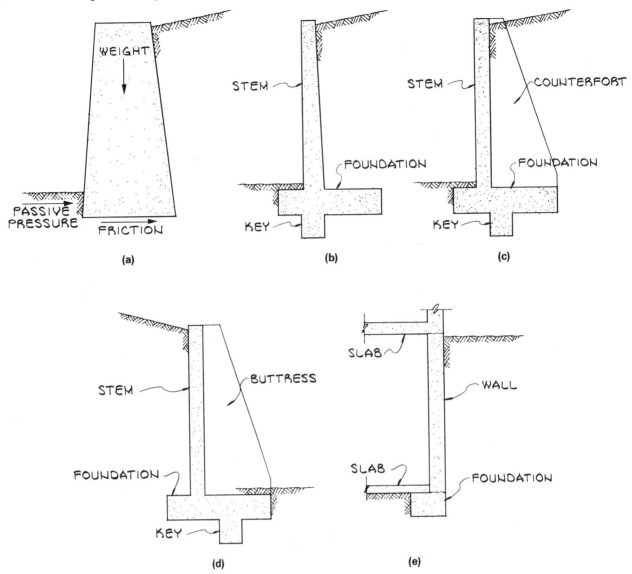

Fig. 1.1

Figure 1.1*b* is a cross section of a cantilever retaining wall. There are several variations of the shape of this type of retaining wall which will be discussed later in this chapter. The wall that is shown in Fig. 1.1*b* is classified as an "inverted T-shaped" wall. However, many cantilever retaining walls are "L-shaped." The use of an inverted T- or an L-shaped wall depends on job-site limiting dimensions and restrictions. The cantilever retaining wall makes the most efficient use of construction materials when it is compared to the other types of retaining walls. The wall stem and foundation react as a structural unit to externally applied forces and each component part of the retaining wall works as a cantilevered slab. The cantilever type of retaining wall is the most commonly used wall because it is economical to construct and versatile in its application for a variety of loads and stem heights.

In Fig. 1.1*c*, a cross section of counterfort retaining wall is shown. Actually, this type of retaining wall is a variation of the cantilever retaining wall. The primary difference between the two types of wall is that the stem and the footing of the counterfort wall work as one-way slabs instead of cantilevered slabs. The slabs span between resisting vertical counterfort walls which are placed normal to the interior surface of the stem wall. The lateral force of the retained soil is transferred to the counterfort wall through a connection acting in tension.

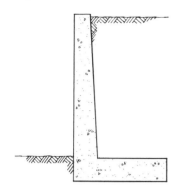

Property Line Retaining Wall—Concrete Stem—Soil Over Footing Chapter 2

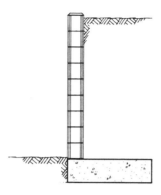

Property Line Retaining Wall—Masonry Stem—Soil Over Footing Chapter 3

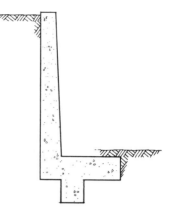

Property Line Retaining Wall—Concrete Stem—Soil Not Over Footing Chapter 4

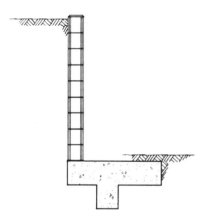

Property Line Retaining Wall—Masonry Stem—Soil Not Over Footing Chapter 5

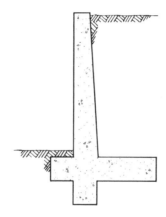

Undercut Footing Retaining Wall—Concrete Stem—Soil at Toe Side Chapter 6

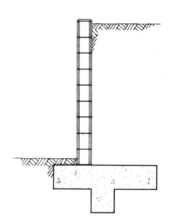

Undercut Footing Retaining Wall—Masonry Stem—Soil at Toe Side Chapter 7

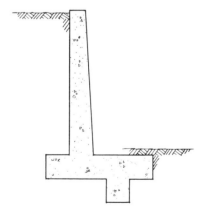

General Retaining Wall—Concrete Stem—Soil at Heel Side Chapter 8

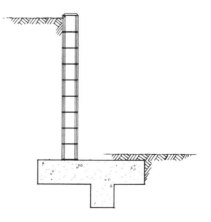

General Retaining Wall—Masonry Stem—Soil at Heel Side Chapter 9

Chapter Cross Section Shapes of Retaining Walls

Fig. 1.2

Figure 1.1d shows a cross section of a buttress-type retaining wall. This type of retaining wall is basically the same as a counterfort retaining wall except that the one-way slabs of the stem and footing span between opposing vertical buttress walls. The lateral force of the retained soil in this case is transferred to the buttress wall through a connection acting in compression. On a comparable basis, that is, all site and height requirements being equal, a counterfort or a buttress retaining wall is not as economical to construct as a cantilever retaining wall. However, a counterfort or a buttress retaining wall is generally used when the embankment to be retained is relatively high compared to a limited space that may be available for the foundation.

Figure 1.1e is classified as a basement wall (Ref. 12). Basement walls can be constructed with either concrete or masonry. The choice of construction materials relies on such structural criteria as the height of the wall, the amount of soil to be retained, and the possible application of an axial or surcharge load. The wall is designed as a one-way slab that spans from the foundation to a resisting concrete slab or floor at the top. The structural design of this type of wall must also consider the combination of the axial and bending stresses at points of maximum stress on its cross section. The concrete slab or floor at the top of the wall and the foundation must be capable of resisting the horizontal reactions of the force of the retained soil. Therefore, it is important that the backfill soil to be retained is not placed until after the wall and the concrete slab have attained sufficient working strength, or the floor at the top of the wall is completed; otherwise the wall must be laterally braced. If these conditions are not met, the wall will fail since it is not designed to act as a cantilevered structure.

As previously stated, the general cross-section configurations of cantilever retaining walls are either inverted T-shaped or L-shaped. These two shapes can be further classified as retaining walls constructed with either a concrete or a masonry stem wall. Another consideration in the classification of cantilever retaining walls is that the required use of a wall places the soil to be retained either against the exterior surface or against the interior surface of the stem wall. Figure 1.2 is a graphic illustration of the above statements on the wall shapes, the types of construction materials used, and the placement of the retained soil. It can be seen that the three major classification considerations, each with two alternatives, will produce eight variations of types of cantilever retaining wall.

Each of the cross sections that are shown in Fig. 1.2 is designated with the related chapter number and title for the design and construction of that particular retaining wall. The chapter titles describe the wall and the placement of the retained soil. Also, the terms "inverted T-shaped" and "L-shaped" are replaced by the titles "General Retaining Walls," "Property Line Retaining Walls," and "Undercut Footing Retaining Walls." This was done so that the chapter titles can correspond to the accepted terminology in the field. It should be noted that the undercut footing retaining walls are the same shape as general retaining walls. These shapes are the most basic and economical for retaining walls because they permit the foundation toe to extend past the exterior face of the stem wall. This gains the most efficient use of the construction materials and produces a more even distribution of soil pressure along the length of the wall foundation. The property line retaining wall, although not as economical as the general retaining wall, is used when the exterior face of the stem wall is located directly adjacent to a property line or to some restricting construction line that does not allow the toe of the foundation to extend beyond the exterior face of the stem wall. This shape of wall has an advantage over the general retaining wall when the soil that it retains is placed against the interior face of the stem and over the footing heel. The weight of the soil increases the wall resistance to overturning and the friction resistance to lateral sliding. The least efficient use of the property line retaining wall occurs when the retained soil is placed against the exterior face of the stem. In this case, the lateral force of the retained soil is resisted only by the weight of the wall, except for a small vertical force component that is caused by the soil reaction against the stem-wall surface.

The structural design of a cantilever retaining wall is performed for the total wall acting as a monolithic structure and for the function of each of its component parts (Ref. 13). Figure 1.3 shows typical cross sections for cantilever retaining walls with either a concrete or a masonry stem. The dimension notations used in Figs. 1.3a and 1.3b are the same as those used in the design data tables presented in this book. Each wall is composed of a stem wall, a foundation or footing, and a foundation key. The

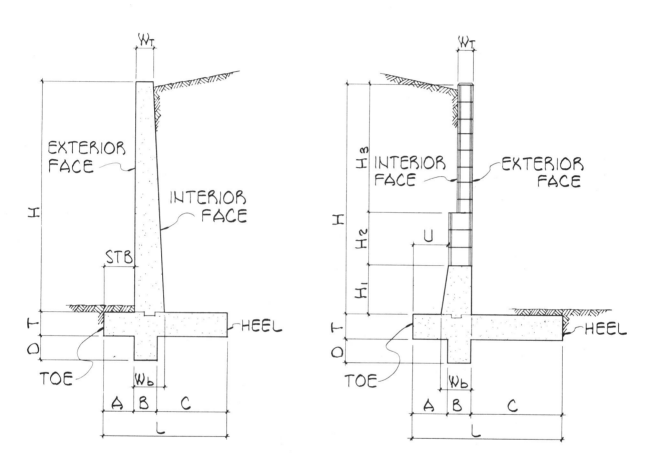

Retaining Wall Dimension Designations
Fig. 1.3

total height dimension of the stem wall is designated as H. The top and the bottom width dimensions of the stem wall are respectively designated as W_t and W_b. In Fig. 1.3b the height dimension H of the stem wall consists of the sum of H_1, H_2, and H_3. H_1 is the dimension of the height of the poured concrete part of the stem wall, H_2 is the dimension of the height of the 12-inch wide concrete block, and H_3 is the dimension of the height of the 8-inch wide concrete block.

The length and the thickness dimensions of the foundation are respectively designated as L and T. The edge of the foundation closest to the stem wall is referred to as the footing "toe." In the case of a property line retaining wall, the exterior face of the stem wall is directly in line with the toe edge of the foundation. The opposite end of the foundation length is referred to as the footing "heel." These two terms are used to delineate the placement of the retained soil in relation to the stem wall and the footing. The surface of the stem wall that is adjacent to the retained soil is considered the interior surface, while the surface that is opposite to the retained soil is considered the exterior surface. The foundation soil bearing pressure varies from a maximum to a minimum numerical value from one end of the footing to the other. (See the design data tables in the following chapters.)

SP_t and SP_h indicate the soil pressure on the foundation toe or heel (Ref. 13). The foundation length L consists of the sum of the dimension designations A, B, and C. The dimension A is the distance from the edge of the foundation to the face of the foundation key; the B dimension is the width of the foundation key; and the C dimension is the distance from the interior side of the foundation key to the end of the foundation. The depth of the foundation key is given by the dimension designated as D. The design data table diagrams in the following chapters show that not all retaining walls require a foundation key. This requirement is determined by structural design calculations. A foundation key must be used when the friction resistance between the bottom of the footing and the soil is not sufficient to restrain the wall from sliding (Ref. 17).

The function of the key is to increase the vertical plane of the foundation depth to obtain a larger surface normal to the soil passive pressure. The passive pressure compression of the soil on this surface adds to the resistance against lateral movement of the wall. Although the foundation key is a small percent of the total area of a cross section of a retaining wall, its weight and location are included in the calculations of the total weight, W, and the overturning resisting moment, M_r. The retaining walls presented in this book are designed by a computer program which produced data to a high degree of numerical accuracy. However, numerical accuracy should not be considered as an alternative to engineering judgment and experience. Presented in this book are instances of retaining wall designs in which the friction resistance F_r is within, but very close to, the allowable design criteria (Ref. 16). In these cases, a foundation key is added to the wall cross section. The designs for each shape of wall and for any case of loading maintain a foundation length within 5 percent of the design criteria.

The undercut footing retaining wall is an example of a general retaining wall shape with the retained soil placed against the exterior face of the stem wall. The term "undercut" is used because the embankment that is to be retained must be excavated to accommodate the footing toe of the wall. The distance the face of the stem is set back from the edge of the footing toe is given by the dimension designated as U. This dimension is used only for undercut retaining walls. In the case of a general retaining wall, the stem is set back from the edge of the footing toe by the dimension designation stb. The magnitude and direction of the force resulting from the retained soil is determined by the height of the stem and by the placement of the soil in relation to the toe or the heel of the footing (Ref. 13). All of the dimensions in Figs. 1.3a and 1.3b are functions of the stem height, the relative placement of the soil, plus the addition of any externally applied axial or surcharge load.

The lateral deformation or movement of the mass of a retained soil produces a force which acts normal to the vertical plane of a retaining wall. This force is called the "active soil pressure." The horizontal component of the active soil pressure is resisted by friction on the horizontal plane of the foundation and by compression of the soil against the vertical plane of the foundation. The compressive reaction of the soil is called the "passive soil pressure." The active soil pressure exerted by the retained soil depends on the cohesion of the soil, the internal friction of the granules, the degree of compaction, the moisture content, and the wall resistance to the lateral strain of the soil (Ref. 37). It is not possible to derive an accurate numerical value for active soil pressure due to the variable and indeterminate physical characteristics of soils. However, semiempirical formulas are used to estimate active and passive soil pressures for different soil classifications, shapes of retaining walls, and slopes of retained soils. These semiempirical formulas give approximate values; however, they are the accepted design criteria used throughout the engineering field. Experience has proven these criteria to be valid for the safe and economical design of retaining walls up to 20 feet high.

The two theories used to derive the formulas for active and passive soil pressures of a retained soil are the Coulomb theory (Ref. 10), and the Rankine theory (Ref. 21). Coulomb published his work on this subject in France in 1776. The Coulomb analysis is based on the concept that a lateral force is exerted on the stem of the wall by the movement of a wedge shape of the retained soil. The sides of the wedge are defined by the slope of the backfill soil, the interior face of the stem wall, and the plane of failure of the soil that permits the soil to move. This theory uses two assumptions: first, that the plane of failure is a straight line; second, that the force on the wall acts in a known direction. The weight of the wedge of soil, the application of any surcharge loads, the angle of internal friction of the soil, and the cohesion of the soil have varying effects on these two assumptions. A change of the angle of the plane of failure will produce a corresponding change in the direction and magnitude of the active force.

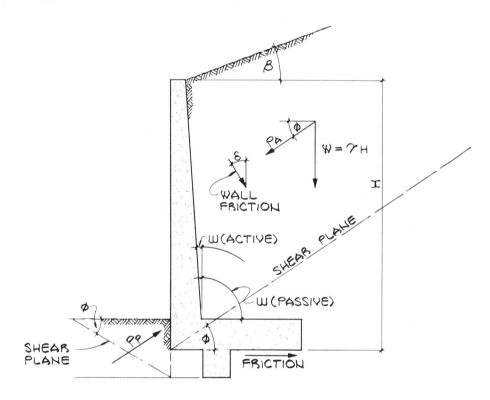

Equation Notation
Fig. 1.4

The variables involved in this theory are expressed in Eq. 1.1 which is Coulomb's formula (Ref. 15) for the resultant active pressure against the stem of a retaining wall.

$$P_A = \tfrac{1}{2}\gamma H^2 \; \frac{\cos^2(\phi - \omega)}{\cos^2\omega \cos(\delta + \omega)\left[1 + \sqrt{\dfrac{\sin(\delta + \phi)\sin(\phi - \beta)}{\cos(\delta + \omega)\cos(\omega - \beta)}}\,\right]^2} \tag{1.1}$$

Figure 1.4 gives a graphic illustration of the equation. The cohesive quality of soils is uncertain, and Eq. 1.1 makes no direct provision for this variable (Ref. 37). However, since soil cohesion will reduce the magnitude of the active pressure, its omission makes the results of the equation conservative for design. The angle of internal friction, ϕ, is the only soil characteristic variable in the equation. The other variables relate to the shape of the wall and the slope of the backfill. The lateral movement of the wedge of soil that exerts the active force is the result of a shear failure of the soil along an assumed plane whose angle of inclination is approximately equal to the angle of internal friction of the soil (Ref. 24). The mechanics of this statement can be generally demonstrated by a vector diagram of a body resisting sliding on an inclined plane. The coefficient of friction that prevents the shear failure of the soil consists of the cohesion in the soil and the friction between the soil granules. For practical purposes, it is usually assumed that the angle of internal friction varies between 30 and 40 degrees, depending upon the soil classification (Ref. 25). However, the most commonly used value is $33°40'0''$. The second assumption in the derivation of the Coulomb equation states that the active force acts in a known direction. The angle of the plane of failure, or the angle of internal friction, defines the angle of the active force. The vector diagram of the force P_A shown in Fig. 1.4 shows a vertical component force that acts at an angle equal to $\tan^{-1} \frac{1}{3}$ ($18.43495°$) (Ref. 17).

The calculation in Ex. 1.1 demonstrates the use of the Coulomb equation to determine the active soil pressure. Case 1 of this example gives a value of 25.898 p.c.f. which is conservative when compared to the value that is required in Table 23-E of App. B (Ref. 17). Cases 2, 3, and 4 of Ex. 1.1 are performed to show that a plus or minus variation of 10 percent in the value of the angle of internal friction will produce an average active soil pressure within 4 percent of the value specified in Table 23-E of App. B. Case 5 of Ex. 1.1 indicates the effect the angle of the surface of the stem wall has on the value of the active soil pressure. It should be noted that the value of $33°40'0''$ is a valid assumption for most soil classifications. However, the Coulomb equation should be used with experienced judgment and with a knowledge of the soils at the job site.

The Rankine theory of active and passive soil pressures was first published in London in 1856 (Ref. 21). Rankine assumed that the active pressure exerted on the vertical plane of a retaining wall is the result of the lateral strain of the soil mass caused by vertical pressure on a horizontal plane, and that the wall stem has no effect on the shear stress of the soil (Ref. 24). The vertical pressure on a horizontal plane of the soil is calculated by the density of the soil multiplied by the depth of the plane plus any surcharge loads on the surface of the embankment. Eq. 1.2 is the Rankine formula (Ref. 24) for active soil pressure.

$$P_A = \frac{1}{2}\gamma H^2 \left(\cos \beta \; \frac{\cos \beta - \sqrt{\cos^2 \beta - \cos^2 \phi}}{\cos \beta + \sqrt{\cos^2 \beta - \cos^2 \phi}} \right) \tag{1.2}$$

Figure 1.4 gives a graphic illustration of the equation.

A comparison of the Coulomb equation (1.1) with the Rankine equation (1.2) indicates that Rankine made no provision for the slope of the face of the stem wall or for the vertical component of the active force. The two variables in the Rankine equation are the slope of the backfill of the soil and the angle of internal friction. The calculation in Ex. 1.2 demonstrates the use of the Rankine equation. Case 1 of this example shows that the active soil pressure value calculated is approximately equal to the building code value specified in Table 23-E in App. B (Ref. 17). The average of Cases 2, 3, and 4 of Ex. 1.2 is calculated to show that a plus or minus variation of 10 percent in the angle of internal friction will produce an average active soil pressure within 1.5 percent of the value specified in Table 23-E. The value of the angle of internal friction can be determined either by shear tests on a soil sample or from known values for specific soil classifications. As a general rule, a stress-strain curve for a dense cohesionless soil rises faster and higher than one for loose cohesionless soils. When the soil is well compacted in place, the value of ϕ increases. It can be seen in the calculation examples of the Coulomb equation and the Rankine equation that an inverse relationship exists between the value of the internal friction and the active soil pressure (Ref. 37).

| EXAMPLE 1.1 – ACTIVE SOIL PRESSURE | SHT. 1 OF 3 |

TO CALCULATE THE ACTIVE SOIL PRESSURE BY THE COULOMB EQUATION:

$$P_A = \tfrac{1}{2}\, \gamma\, H^2 \, \frac{\cos^2(\phi - \omega)}{\cos^2 \omega \cos(\delta + \omega)\left[1 + \sqrt{\dfrac{\sin(\delta + \phi)\sin(\phi - \beta)}{\cos(\delta + \omega)\cos(\omega - \beta)}}\right]^2}$$

δ = ANGLE OF WALL FRICTION (INCLINATION OF P_A)

ω = ANGLE BETWEEN VERT. PLANE & BACK OF WALL

β = ANGLE OF BACKFILL SLOPE

H = HEIGHT OF VERT. PLANE OF PRESSURE SURFACE

γ = DENSITY OF SOIL = 100 p.c.f.

ϕ = SOIL ANGLE OF INTERNAL FRICTION

W_e = EQUIVALENT UNIT PRESSURE = p.c.f.

$$P_A = \tfrac{1}{2}\, W_e\, H^2$$

CASE 1 – NO BACKFILL SLOPE

GIVEN:

$\delta = \text{TAN}^{-1}\,\tfrac{1}{3} = 18.43495°$, $\omega = 0$, $\beta = 0$, $\phi = 33°\text{-}40'\text{-}0''$

REDUCED EQ.:
$$W_e = \frac{\gamma \cos^2 \phi}{\cos \delta \left[1 + \sqrt{\dfrac{\sin(\delta + \phi)\sin \phi}{\cos \delta}}\,\right]^2}$$

$$W_e = \frac{100 \cos^2(33.667)}{\cos(18.43495)\left[1 + \sqrt{\dfrac{\sin(18.43495 + 33.667)\sin(33.667)}{\cos(18.43495)}}\,\right]^2}$$

$$W_e = \frac{100 \times 0.83227^2}{0.94868\left[1 + \sqrt{\dfrac{0.7891 \times 0.55436}{0.94868}}\,\right]^2}$$

$$\underline{W_e = 25.898 \text{ p.c.f.}}$$

EXAMPLE 1.1 - ACTIVE SOIL PRESSURE	SHT. 2 OF 3

CASE 2 - BACKFILL SLOPE = 2 TO 1 (β = 26°-33'-54")

GIVEN: $\delta = 18.43495°$, $\omega = 0$, $\beta = 26.5650°$, $\phi = 33.6667°$

REDUCED EQ.: $W_e = \dfrac{\gamma \cos^2 \phi}{\cos \delta \left[1 + \sqrt{\dfrac{\sin(\delta + \phi)\sin(\phi - \beta)}{\cos \delta \cos \beta}} \right]^2}$

$W_e = \dfrac{100 \cos^2(33.667)}{\cos(18.43495)\left[1 + \sqrt{\dfrac{\sin(18.43495 + 33.667)\sin(33.667 - 26.5650)}{\cos(18.43495)\cos(26.5650)}} \right]^2}$

$W_e = \dfrac{100 \times 0.83227^2}{0.94868\left[1 + \sqrt{\dfrac{0.7891 \times 0.1236}{0.94868 \times 0.89443}} \right]^2}$

$\underline{W_e = 40.721 \text{ p.c.f.}}$

CASE 3 - BACKFILL SLOPE = 2 TO 1 (β = 26°-33'-54")

GIVEN: $\delta = 18.43495°$, $\omega = 0$, $\beta = 26.5650$, $\phi = 37.0°$

REDUCED EQ.: $W_e = \dfrac{\gamma \cos^2 \phi}{\cos \delta \left[1 + \sqrt{\dfrac{\sin(\delta + \phi)\sin(\phi - \beta)}{\cos \delta \cos \beta}} \right]^2}$

$W_e = \dfrac{100 \cos^2 37.00}{\cos(18.43495)\left[1 + \sqrt{\dfrac{\sin(18.43495 + 37.00)\sin(37.00 - 26.565)}{\cos(18.43495)\cos(26.565)}} \right]^2}$

$W_e = \dfrac{100 \times 0.79864^2}{0.94868\left[1 + \sqrt{\dfrac{0.8325 \times .1812}{0.94868 \times 0.89443}} \right]^2}$

$\underline{W_e = 33.266 \text{ p.c.f.}}$

| EXAMPLE 1.1 - ACTIVE SOIL PRESSURE | SHT. 3 of 3 |

CASE 4 - BACKFILL SLOPE = 2 to 1

GIVEN: $\delta = 18.43495°$, $\omega = 0$, $\beta = 26.5650°$, $\phi = 30.50°$

REDUCED EQ.: $W_e = \dfrac{\gamma \cos^2 \phi}{\cos \delta \left[1 + \sqrt{\dfrac{\sin(\delta + \phi) \sin(\phi - \beta)}{\cos \delta \cos \beta}} \right]^2}$

$W_e = \dfrac{100 \cos^2(30.50)}{\cos(18.43495)\left[1 + \sqrt{\dfrac{\sin(18.43495 + 30.50)\sin(30.50 - 26.565)}{\cos(18.43495)\cos(30.50)}}\right]^2}$

$W_e = \dfrac{100 \times 0.86163^2}{0.94868\left[1 + \sqrt{\dfrac{0.75396 \times 0.06862}{0.94868 \times 0.86163}}\right]^2}$

$\underline{W_e = 49.958 \text{ p.c.f.}}$

AVERAGE OF CASE 2, 3 & 4 $= \dfrac{40.721 + 33.266 + 49.958}{3} = 41.315$ p.c.f.

NOTE THAT 41.315 IS 4% LESS THAN BLDG. CODE VALUE

CASE 5 - BACKFILL SLOPE 2 to 1, SLOPE STEM WALL

GIVEN: $\delta = 18.43495$, $\omega = 2.50°$, $\beta = 26.5650°$, $\phi = 33.667°$

USE EQ. ON SHT. 1

$W_e = \dfrac{100 \cos^2(33.667 - 2.50)}{\cos^2(2.50)\cos(18.43495 + 2.50)\left[1 + \sqrt{\dfrac{\sin(18.43495 + 33.667)\sin(33.667 - 26.565)}{\cos(18.43495 + 2.50)\cos(2.50 - 26.565)}}\right]^2}$

$W_e = \dfrac{100 \times 0.85565^2}{0.9905^2 \times 0.93398\left[1 + \sqrt{\dfrac{0.7910 \times 0.12363}{0.93398 \times 0.91308}}\right]^2}$

$\underline{W_e = 44.588 \text{ p.c.f.} \quad +9.5\% \text{ CASE 2}}$

The derivations of the Coulomb and Rankine equations are based on the operation of a conjugate relationship between the vertical soil pressure and the active soil pressure (Ref. 26). The vertical pressure of the retained soil is directly proportional to the resultant active soil pressure. The retaining wall, acting as a unit structure, must be restrained against sliding and rotation that result from the force of the active soil pressure. The lateral movement of the wall is resisted at the foundation by passive soil pressure on its vertical plane and by friction on its horizontal plane. As the wall tries to move laterally it compresses the soil against the face of the foundation and induces a state of passive soil pressure. The intensity of the compression causes the soil to fail along a sloped shear plane and thus permits it to expand vertically. The angle of the plane of failure and the vertical expansion result from the lateral deformation of the compressed soil.

The unit compressive pressure normal to the vertical plane of the soil is analogous to the vertical pressure that produces active soil pressure. Passive soil pressure differs from active soil pressure in that, although in each case a conjugate relationship exists between the lateral and vertical pressures, in the passive case the relationship is inversely proportional. Conversely stated, the passive soil pressure is inversely proportional to the resultant vertical pressure. The slope of the shear plane depends on the same factors that influence the plane of failure for the active soil pressure: that is, soil cohesion, friction, moisture content, and the degree of soil compaction. The passive soil pressure reacts to oppose the force of the active soil pressure. The angle of the plane of failure of the soil for each type of pressure defines the directions of the forces. Figure 1.4 shows the passive force acting at an angle of opposite sign to the angle of the active force. This change in sign is reflected in the equations for passive soil pressure. Equation 1.3 (Ref. 24) is the Coulomb equation for passive soil pressure.

$$P_P = \frac{1}{2}\gamma H^2 \left[\frac{\csc \omega \sin (\omega + \phi)}{\sqrt{\sin (\omega + \delta)} - \sqrt{\dfrac{\sin (\phi - \delta) \sin (\phi + \beta)}{\sin (\omega - \beta)}}} \right]^2 \tag{1.3}$$

The calculation in Ex. 1.3 demonstrates the use of this equation. The angle δ is set equal to $0°$. This value can vary and may approximate the value of ϕ. However, δ should not exceed $\frac{3}{4}\phi$ to obtain conservative values for passive soil pressures. Also, in Eq. 1.3 the angle ω is measured as the angle between a horizontal line and the plane of the stem wall (*see* Fig. 1.4). Equation 1.4 (Ref. 37) is the Rankine equation for passive soil pressure.

$$P_P = \frac{1}{2}\gamma H^2 \left(\cos \beta \; \frac{\cos \beta + \sqrt{\cos^2 \beta - \cos^2 \phi}}{\cos \beta - \sqrt{\cos^2 \beta - \cos^2 \phi}} \right) \tag{1.4}$$

A comparison of Eq. 1.2 with this equation indicates that this equation gives the inverse of the active pressure. The change of sign of the angle of direction of the active force reverses the signs in the equations. The calculation in Ex. 1.4 similarly demonstrates the use of Eq. 1.4.

The restraint of a retaining wall against lateral sliding depends on the physical characteristics of the soil at the wall foundation. Movement of the wall is resisted by passive soil pressure and by friction. Tables 28-B (Ref. 16) and 29-B (Ref. 29) of App. B specify allowable building code values of passive soil pressures and friction coefficients. It is important that the calculation design values of these soil properties accurately reflect the capacity of the soil at the construction site. The use of an unrealistically high passive pressure or friction coefficient may allow the wall to slide, whereas low or conservative values will increase the cost of the wall. The retaining walls presented in the following chapters of this book are designed by using a passive soil pressure of 300 p.c.f. and a friction coefficient of 0.40. These values were

EXAMPLE 1.2 — ACTIVE SOIL PRESSURE	SHT. 1 OF 2

TO CALCULATE THE ACTIVE SOIL PRESSURE BY THE RANKINE EQUATION:

$$P_A = \frac{1}{2}\,\gamma\,H^2\left[\cos\beta\,\frac{\cos\beta - \sqrt{\cos^2\beta - \cos^2\phi}}{\cos\beta + \sqrt{\cos^2\beta - \cos^2\phi}}\right]$$

β = ANGLE OF BACKFILL SLOPE
H = HEIGHT OF VERT. PLANE OF PRESSURE SURFACE
γ = DENSITY OF SOIL = 100 p.c.f.
ϕ = SOIL ANGLE OF INTERNAL FRICTION
W_e = EQUIVALENT UNIT PRESSURE

$$P_A = \frac{1}{2}\,W_e\,H^2$$

CASE 1 — NO BACKFILL SLOPE

GIVEN: $\beta = 0°$, $\phi = 33°-40'-0''$

$$W_e = \gamma\,\cos\beta\,\frac{\cos\beta - \sqrt{\cos^2\beta - \cos^2\phi}}{\cos\beta + \sqrt{\cos^2\beta - \cos^2\phi}}$$

$$W_e = 100 \times 1.0\,\frac{1 - \sqrt{1 - 0.83227^2}}{1 + \sqrt{1 - 0.83227^2}} = 30.825 \text{ p.c.f.}$$

CASE 2 — BACKFILL SLOPE = 2 TO 1 ($\beta = 26°-33'-54''$)

GIVEN: $\beta = 26.565°$, $\phi = 33.6677°$

$$W_e = 100\,\cos(26.5650)\left[\frac{\cos(26.5650) - \sqrt{\cos^2(26.565) - \cos^2(33.6677)}}{\cos(26.5650) - \sqrt{\cos^2(26.565) - \cos^2(33.6677)}}\right]$$

$$W_e = 100 \times 0.89443 \times \frac{0.89443 - \sqrt{0.89443^2 - 0.83227^2}}{0.89443 + \sqrt{0.89443^2 - 0.83227^2}}$$

$$\underline{\underline{W_e = 41.486 \text{ p.c.f.}}}$$

| EXAMPLE 1.2 – ACTIVE SOIL PRESSURE | SHT. 2 OF 2 |

CASE 3 – BACKFILL SLOPE = 2 TO 1 ($\beta = 26°-33'-54''$)

GIVEN: $\beta = 26.565°$, $\phi = 37.0°$

$$W_e = 100 \cos(26.565) \left[\frac{\cos(26.565) - \sqrt{\cos^2(26.565) - \cos^2(37.0)}}{\cos(26.565) + \sqrt{\cos^2(26.565) - \cos^2(37.0)}} \right]$$

$$W_e = 100 \times 0.89443 \left[\frac{0.89443 - \sqrt{0.89443^2 - 0.79864^2}}{0.89443 + \sqrt{0.89443^2 - 0.79864^2}} \right]$$

$\underline{W_e = 33.906 \text{ p.c.f.}}$

CASE 4 – BACKFILL SLOPE = 2 TO 1 ($\beta = 26°-33'-54''$)

GIVEN: $\beta = 26.565°$, $\phi = 30.50°$

$$W_e = 100 \cos(26.565) \left[\frac{\cos(26.565) - \sqrt{\cos^2(26.565) - \cos^2(30.50)}}{\cos(26.565) + \sqrt{\cos^2(26.565) - \cos^2(30.50)}} \right]$$

$$W_e = 100 \times 0.89443 \left[\frac{0.89443 - \sqrt{0.89443^2 - 0.86163^2}}{0.89443 + \sqrt{0.89443^2 - 0.86163^2}} \right]$$

$\underline{W_e = 51.598 \text{ p.c.f.}}$

AVERAGE OF CASES 2,3 & 4 $= \frac{41.486 + 33.906 + 51.598}{3} = 42.330$

NOTE THAT 42.330 p.c.f. IS 1.5% LESS THAN BUILDING CODE.

| EXAMPLE 1.3 — PASSIVE SOIL PRESSURE | SHT. 1 OF 1 |

TO CALCULATE THE PASSIVE SOIL PRESSURE BY THE COULOMB EQUATION:

$$P_P = \frac{1}{2}\, \gamma\, H^2 \left[\frac{CSC\ \omega\ SIN\,(\omega + \phi)}{\sqrt{SIN\,(\omega + \delta)} - \sqrt{\dfrac{SIN\,(\phi - \delta)\ SIN\,(\phi + \beta)}{SIN\,(\omega - \beta)}}} \right]^2$$

δ = ANGLE OF WALL FRICTION 0° (ASSUMED)

ω = ANGLE BETWEEN HORIZ. PLANE & BACK OF WALL

β = ANGLE OF BACKFILL SLOPE

H = HEIGHT OF VERT. PLANE OF PRESSURE SURFACE

γ = DENSITY OF SOIL = 100 p.c.f.

ϕ = SOIL ANGLE OF INTERNAL FRICTION

W_e = EQUIVALENT UNIT PRESSURE

$$P_P = \frac{1}{2}\, W_e\, H^2$$

CASE 1 - NO BACKFILL SLOPE

GIVEN:

$\delta = 0°$, $\omega = 90°$, $\beta = 0°$, $\phi = 33°\text{-}40'\text{-}0''$

$$W_e = 100 \left[\frac{CSC\,(90)\ SIN\,(90 + 33.667)}{\sqrt{SIN\,(90)} - \sqrt{\dfrac{SIN\,(33.667)\ SIN\,(33.667 + 0)}{SIN\,(90 - 0)}}} \right]^2$$

$$W_e = 100 \left[\frac{1.0 \times 0.83227}{1 - \sqrt{\dfrac{0.55436^2}{1}}} \right]^2$$

$$\underline{W_e = 348.78\ p.c.f.}$$

EXAMPLE 1.4 — PASSIVE SOIL PRESSURE | SHT. 1 of 2

TO CALCULATE THE PASSIVE SOIL PRESSURE BY THE RANKINE EQUATION:

$$P_p = \tfrac{1}{2}\,\gamma\,H^2\left[\cos\beta\;\frac{\cos\beta + \sqrt{\cos^2\beta - \cos^2\phi}}{\cos\beta - \sqrt{\cos^2\beta - \cos^2\phi}}\right]$$

β = ANGLE OF BACKFILL SLOPE
H = HEIGHT OF VERT. PLANE OF PRESSURE SURFACE
γ = DENSITY OF SOIL = 100 p.c.f.
ϕ = SOIL ANGLE OF INTERNAL FRICTION
W_e = EQUIVALENT UNIT PRESSURE

$$P_p = \tfrac{1}{2}\,W_e\,H^2$$

CASE 1 — NO BACKFILL SLOPE

GIVEN: $\beta = 0°$, $\phi = 33°\text{-}40'\text{-}0''$

REDUCED EQ.: $W_e = 100\left[\dfrac{1 + \sqrt{1 - \cos^2\phi}}{1 - \sqrt{1 - \cos^2\phi}}\right]$

$W_e = 100\left[\dfrac{1 + \sqrt{1 - \cos^2(33.667)}}{1 - \sqrt{1 - \cos^2(33.667)}}\right]$

$W_e = 100\left[\dfrac{1 + \sqrt{1 - 0.69268}}{1 - \sqrt{1 - 0.69268}}\right]$

$\underline{W_e = 348.78 \text{ p.c.f.}}$

EXAMPLE 1.4 – PASSIVE SOIL PRESSURE	SHT. 2 OF 2

CASE 2 – BACKFILL SLOPE = 2 TO 1 ($\beta = 26°-33'-54"$)

GIVEN: $\beta = 26.565°$, $\phi = 33.667°$

$$We = 100 \cos(26.565)\left[\frac{\cos(26.565)+\sqrt{\cos^2(26.565)-\cos^2(33.667)}}{\cos(26.565)-\sqrt{\cos^2(26.565)-\cos^2(33.667)}}\right]$$

$$We = 100 \times 0.89443\left[\frac{0.89443+\sqrt{0.89443^2-0.83227^2}}{0.89443-\sqrt{0.89443^2-0.83227^2}}\right]$$

$\underline{We = 192.837 \text{ p.c.f.}}$

CASE 3 – BACKFILL SLOPE = 2 TO 1 ($\beta = 26°-33'-54"$)

GIVEN: $\beta = 26.565°$, $\phi = 37.0°$

$$We = 100 \cos(26.565)\left[\frac{\cos(26.565)+\sqrt{\cos^2(26.565)-\cos^2(37.0)}}{\cos(26.565)-\sqrt{\cos^2(26.565)-\cos^2(37.0)}}\right]$$

$$We = 100 \times 0.89443\left[\frac{0.89443+\sqrt{0.89443^2-0.79864^2}}{0.89443-\sqrt{0.89443^2-0.79864^2}}\right]$$

$\underline{We = 235.950 \text{ p.c.f.}}$

CASE 4 – BACKFILL SLOPE = 2 TO 1 ($\beta = 26°-33'-54"$)

GIVEN: $\beta = 26.565°$, $\phi = 30.5°$

$$We = 100 \cos(26.565)\left[\frac{\cos(26.565)+\sqrt{\cos^2(26.565)-\cos^2(30.5)}}{\cos(26.565)-\sqrt{\cos^2(26.565)-\cos^2(30.5)}}\right]$$

$$We = 100 \times 0.89443\left[\frac{0.89443+\sqrt{0.89443^2-0.86163^2}}{0.89443-\sqrt{0.89443^2-0.86163^2}}\right]$$

$\underline{We = 155.045 \text{ p.c.f.}}$

AVERAGE OF CASE 2, 3 & 4 = $\dfrac{192.837 + 235.950 + 155.045}{3} = 194.61 \text{ p.c.f.}$

selected because they are generally used and are relatively realistic for most conditions. They are therefore reliable values for preventing lateral sliding of the wall.

Rational design requires a close approximation to natural conditions. Therefore, the engineer should endeavor to substantiate the use of assumed design values. This can be achieved through the combination of experience, judgment, research, and observation. The design examples presented in each of the following chapters demonstrate the calculations necessary to determine wall resistance to lateral sliding. When the calculated friction coefficient exceeds the allowable value of 0.40, a key is added to the wall foundation. The combination of the passive soil pressure on the key and the frictional resistance should be greater than the horizontal active force on the wall (Ref. 17).

The equations used in Exs. 1.1 and 1.2 give active soil pressures designated as W_e in terms of pounds per cubic foot. It is assumed that the active pressure diagram will approximate a triangular configuration. This shape of diagram is based on the assumption that the soil acts against the wall as an equivalent fluid pressure with a density equal to W_e. The assumption is valid for retaining walls up to approximately 15 feet in height. The shape of the pressure diagram also depends on the soil cohesion, compaction, moisture content, and the angle of internal friction. Based upon a triangular-shaped pressure diagram, the total magnitude of P_A can be calculated as the area of a triangle with an altitude equal to the height, H, and a base equal to $W_e \times H$. The horizontal line of application of P_A occurs at a dimension equal to $\frac{1}{3}$ of H above the triangle base. The angle of the slope of an embankment that is laterally supported by a retaining wall is directly related to the magnitude of the active soil pressure. The slope angle is measured as the angle between a horizontal line and the surface of the soil. When the surface of the slope of the retained soil is irregular, the angle is measured between a horizontal line and a line from the face of the wall connecting the highest point on the slope that is within a horizontal distance equal to the stem height (Ref. 17). The angle of the slope can be expressed in degrees or radians, but it is usually expressed in a ratio of horizontal distance to vertical rise. Table 23-E in App. B (Ref. 17) specifies values of equivalent fluid pressures for various surface slopes of retained soils. Table 1.1 is an extended form of the building code table. The surface slope ratios are converted to degrees, minutes, and seconds. Figure 1.5 is a graph of the equivalent fluid pressures obtained by calculation of the Coulomb equation, the Rankine equation, and the values specified in Table 1.1. The equations do not give accurate results for slopes greater than the angle ϕ. The curves are extrapolated beyond this point. It can be seen that the building code values for the equivalent fluid pressure are consistently conservative, particularly for walls that retain high backfill slopes. The higher specified values for equivalent fluid pressure will compensate for a variation of the assumed triangular-shaped pressure diagram. The design data tables and the drawings of the retaining walls presented in the following chapters of this book are based upon the equivalent fluid pressure values specified in Table 1.1.

TABLE 1.1 Equivalent Fluid Pressures for Retained Slopes

Surface slope of retained soil, horizontal to vertical	Surface slope angle			Equivalent fluid pressure, p.c.f.
Level	0°	0'	0"	30
5 to 1	11°	18'	36"	32
4 to 1	14°	02'	10"	35
3 to 1	18°	26'	06"	38
2 to 1	26°	33'	54"	43
1½ to 1	33°	41'	24"	55
1 to 1	45°	0'	0"	80

The design examples at the beginning of each chapter demonstrate the assumed triangular-shaped pressure diagram using an equivalent fluid pressure for the active soil pressure on the wall. The level surface of a retained soil is often required to sustain a superimposed surcharge load. Applied surcharge loads act normal to the surface of the soil and originate from dead loads, live loads, or from a combination of both. Three cases of surcharge load are used for the design of the retaining walls in this book: 50 p.s.f., 100 p.s.f., and 200 p.s.f. These values represent the uniformly distributed surcharge loads most often encountered and are compatible with building code requirements (Ref. 30). The total wall and each component part are designed to support the applied surcharge. The addition of a uniformly applied load to the soil surface is converted to an equivalent increment of height of retained soil. The ratio of the surcharge load to the weight of the soil (100 p.c.f.) equals the increase in the height to be added to the total height of the soil pressure diagram. Design examples of retaining walls are presented in each .chapter to demonstrate the method of calculation to convert surcharge loads to an equivalent height for the active pressure diagram (Ref. 18). It should be noted that the total height of the soil pressure diagram is measured from the bottom of the footing to the top of the wall stem plus the equivalent increment of height derived from the presence of a surcharge load.

Example 1.5 is included in this text to show the method of calculation for a special case of surcharge loading on a retaining wall (Ref. 18). The load originates from a continuous or isolated footing that is supported by the retained soil. The proximity of the applied load is within a distance from the retaining wall so that the increased vertical bearing pressure of the footing will result in a lateral load to the stem wall. Active soil pressure derived from this type of loading is applied to the retaining wall for each linear foot of the length of the continuous or isolated footing. The lateral load will diminish uniformly to zero within a distance from the end of the footing equal to its distance from the face of the retaining wall.

The sloped backfill and surcharge loads on a retained soil induce the active soil pressure on the wall. The weight of the soil and the superimposed surcharge transmit proportional horizontal forces to the retaining wall through the compression of the soil. Another type of load that retaining walls are often required to support is an axial load applied to the top of the wall stem. This type of load does not affect the active soil pressure since it is applied directly to the structure. Just as in the case of surcharge loads, an axial load may be composed of a combination of dead load and live load. The load acts vertically through the centerline of the stem wall. The retaining walls in the following chapters present four cases of axial loading: 200 lbs./ft., 400 lbs./ft., 600 lbs./ft., and 800 lbs./ft. These loads represent only an intermediate range of values. Walls with higher axial loads should be investigated by design calculations for soil bearing pressure on the foundation and for overturning resistance. A retaining wall carrying an axial load requires that the stem wall be designed for the combined stresses of bending and compression. This is particularly important for walls with masonry stems due to the relatively low allowable stress that is permitted without continuous field inspection of the construction (Ref. 31). In general, the effect of an axial load on a retaining wall is to increase the soil bearing pressure on the foundation, the wall resistance to overturning, and the frictional resistance to sliding (Ref. 5).

Figure 1.2 shows the shapes of the various retaining walls presented in the following chapters. The designated chapter numbers correspond to the chapters in the book that deal with the design and construction of that particular shape of wall. Each of the shapes of retaining wall is designed for the cases of loadings previously described: that is, for backfill slopes, surcharge loads, and axial loads. Figure 1.6 specifies the fourteen different cases of loads for which each shape of retaining wall is designed. These loadings and wall designs are illustrated in the following chapters. The equivalent fluid pressures specified in Table 1.1 are given for the walls that retain sloped backfills of soil. Each load condition in Fig. 1.6 is marked with a reference number which corresponds to the section of the chapter that uses that particular design load. The retaining walls in each section of each chapter range in stem height from 3 feet to 12 feet and are arranged in order of ascending heights of one foot increments. There are instances in which a wall shape and reinforcement are the same for different load conditions, particularly for walls of low stem heights. These walls are described in their respective section of each chapter so they can be consecutively reference-numbered. The cross-section drawing of each retaining wall represents the shape and reinforcement of a one-foot-long wall. The section number of a wall consists of three designation numbers, each of which is separated by a decimal point: the first number refers to the chapter number which also describes the shape and use of the wall (Fig. 1.2); the second number indicates the design load condition that is specified in Fig. 1.6; and the third number gives the stem height of the wall, measured from the top of the foundation to the top of the wall. This system of organization of designation numbers makes it convenient to refer to any retaining wall cross section in this book by a series of preassigned numbers. For example, a retaining wall with section number 6.12.10 can be identified by the first number which indicates Chapter 6, "Undercut Footing Retaining Walls—Concrete Stem" (see Fig. 1.2); the second number, 12, refers to load condition number 12 shown in Fig. 1.6, which specifies a backfill slope of 2 to 1 that requires a design equivalent fluid pressure of 43 p.c.f.; and last number, 10, gives the stem height equal to 10 feet.

Retaining walls required to sustain externally applied loads such as surcharge loads or axial loads represent their maximum design case. The source of these applied loads may not be constant, or the loads may occur only intermittently. Axial loads are composed of a combination of dead load and live load and in many instances the surcharge may represent a specific building code requirement (Ref. 18). All of the alternative conditions of design on the total wall and on each of its component parts should be investigated to determine the maximum design condition. The removal of an applied load can alter the soil bearing pressure diagram on the foundation, decrease the friction resistance to sliding, and decrease the wall resistance to overturning. For these reasons, the same shape and size of wall are also designed without the added axial or surcharge load. Design data tables for the unloaded condition can be found in App. A. The numbers of the tables in the appendix correspond to the table in a chapter for the maximum design case. For example, Table 6.4 in the chapter gives the design data for the retaining wall with a 600 lbs./ft. axial load, and Table 6.4a in the appendix gives the design data for the same shape and size of wall without the axial load.

The reliability of a retaining wall to perform its function is determined by the results of structural calculations. The critical requirements of stability are wall overturning resistance, foundation soil bearing capacity, structural strength of the wall, and the foundation restraint against lateral sliding. Design examples are given in each chapter to demonstrate the calculation procedures performed to satisfy these requirements. The first sheet of an example shows a free body diagram of all of the forces, loads, and reactions on a retaining wall. A state of structural stability requires that the sum of the vertical and horizontal forces and loads equals zero. The structure will not move in translation or rotation when its free body diagram of loads and forces are in a static state of equilibrium. Figure 1.7 is a graphic illustration of all of the forces, loads, and reactions on two different shapes of retaining walls for an assumed one-foot-long wall. The weight of each part of the wall, the weight of the soil, and the surcharge or axial load are designated by W with a respective numerical subscript. The area of the triangular active pressure diagram is equal to F. Fr is the designation for the coefficient of friction. The area of the triangular diagram of passive soil pressure is equal to P_P. R represents the vertical resultant of all of the forces and loads acting on the wall and is located at a distance $\overline{X}$ from the point about

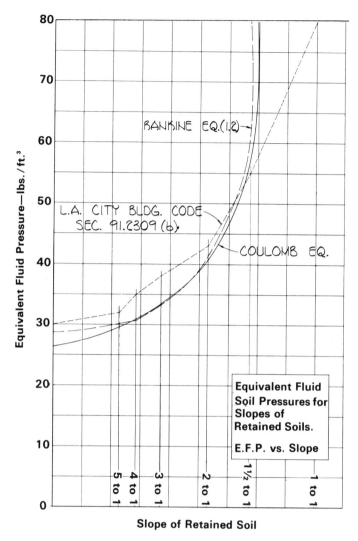

Fig. 1.5

which the moment of resistance is calculated. SP_t and SP_h are the calculated soil bearing pressures at the toe and heel of the foundation. The points about which the overturning moment and the resisting moment are calculated are respectively designated as M_O and M_R. Figure 1.7a shows that the retained soil has a sloped backfill. The weight of the soil segment designated as W_6 is calculated from the area of the triangle defined by a horizontal line at the top of the wall, the slope of the backfill, and a line projected vertically from the heel of the foundation. The balance of the weight of the soil, W_7, on this wall is calculated from the area defined by the horizontal lines at the top and the bottom of the stem, a line projected vertically from the heel of the foundation, and the interior face of the stem wall. W_5, shown in Fig. 1.7b, represents either a surcharge or an axial load. In the case of a surcharge loading condition, the height of the triangular diagram of active soil pressure is increased to an equivalent height to convert the surcharge load to active soil pressure. The load W_8 in Fig. 1.7a and W_4 in Fig. 1.7b represent the vertical component of the active force on the surface of the stem wall. This load is calculated as one-third of the active force on the stem wall, using a height of active pressure diagram equal to the height of the soil contact on the stem (Ref. 17). The density of the materials used in the design examples of each chapter is as follows: soil = 100 p.c.f.; concrete = 150 p.c.f.; 8-inch-wide concrete block with the cells filled = 92 p.s.f.; 12-inch-wide concrete block with the cells filled = 140 p.s.f. (Ref. 1).

| EXAMPLE 1.5 — SURCHARGE LOADS | SHT. 1 OF 2 |

TO CALCULATE THE SURCHARGE LOAD DUE
TO A CONTINUOUS OR ISOLATED FOOTING
REF. SEC. 91.2309(c) OF L.A. CITY BLDG. CODE

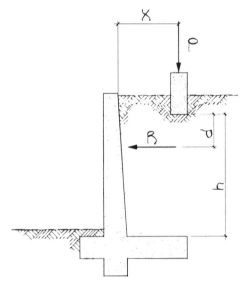
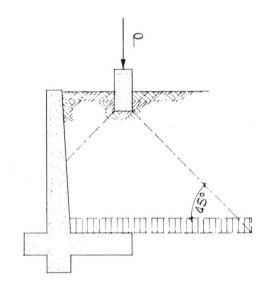

LATERAL RESULTANT   VERTICAL PRESSURE

$$R = \frac{0.30\, Ph^2}{X^2 + h^2} \quad , \quad d = X\left[\left(\frac{X^2}{h^2} + 1\right) TAN^{-1}\frac{h}{X} - \left(\frac{X}{h}\right)\right]$$

R = RESULTANT LATERAL FORCE — LB./FT.

P = LOAD OF CONT. FTG. OR ISOLATED FTG. — LBS./FT.

X = DISTANCE FROM BACK OF WALL TO ℄ OF
   LOAD ON FTG. — FT.

d = DEPTH OF LATERAL RESUTANT BELOW POINT
   OF APPLICATION OF FTG. LOAD — FT.

$\left(TAN^{-1}\frac{h}{X}\right)$ = THE ANGLE IN RADIANS WHOSE TAN.
   IS EQUAL TO $\frac{h}{X}$

EXAMPLE 1.5 — SURCHARGE LOADS | SHT. 2 OF 2

CASE 1 — CONTINUOUS FOOTING LOAD

GIVEN: $P = 1600 \,^{\#}/_{FT.}$, $X = 3'\text{-}6''$, $h = 5'\text{-}0''$

$$R = \frac{0.30 \times 1600 \times 5.0^2}{3.5^2 \times 5.0^2} = 322.15 \,^{\#}/_{FT.}$$

$$\frac{h}{X} = \frac{5.0}{3.5} = 1.42857 , \quad TAN^{-1} 1.42857 = 55°\text{-}00'\text{-}29''$$

$$55°\text{-}00'\text{-}29'' = 0.96 \text{ RADIANS}$$

$$d = 3.5 \left[\left(\frac{3.5^2}{5.0^2} + 1 \right) 0.96 - \left(\frac{3.5}{5.0} \right) \right] = \underline{2.556 \text{ FT.}}$$

CASE 2 — SPREAD FOOTING LOAD

GIVEN: SPREAD FTG. 2'-6" SQ.
SOIL BEARING PRESSURE = 2500 p.s.f.
$P = 2.50 \times 2500 = 6250 \,^{\#}/_{FT.}$
$X = 4.0'$, $h = 7'\text{-}6''$

$$R = \frac{0.30 \times 6250 \times 7.5^2}{4.0^2 + 7.5^2} = 1459.78 \,^{\#}/_{FT.}$$

NOTE — CODE ALLOWS "R" REDUCTION TO $\frac{1}{6}$ FOR SPREAD FTGS. LESS THAN 3'-0"

$$\therefore R = \frac{1459.78}{6} = \underline{243.30 \,^{\#}/_{FT.}}$$

$$\frac{h}{X} = \frac{7.5}{4.0} = 1.875 , \quad TAN^{-1} 1.875 = 61°\text{-}55'\text{-}39''$$

$$61°\text{-}55'\text{-}39'' = 1.0743 \text{ RADIANS}$$

$$d = 4.0 \left[\left(\frac{4.0^2}{7.5^2} + 1 \right) 1.0743 - \left(\frac{4.0}{7.5} \right) \right] = \underline{3.386 \text{ FT.}}$$

The first criteria for the stability of a retaining wall is its capacity to resist overturning rotation. The rotational moments that act on a retaining wall are the product of the force or load and the distance to its line of action relative to a point of rotation. Overturning rotation of a retaining wall originates from the force exerted by the active soil pressure. The lateral force F in Fig. 1.7 is located at a distance above the bottom of the wall foundation equal to one-third of the height of the active pressure diagram. M_O in the design data tables in the following chapters is the designation for the numerical value of the overturning moment given in ft. lbs. Rotational stability of the wall is achieved by a counteracting moment composed of the product of the vertical loads and their respective distances from the point designated as M_R in Fig. 1.7. M_R in the design data tables is the designation for the numerical value of the resisting moment given in ft. lbs. The ratio of the magnitude of the resisting moment and the overturning moment is called the "overturning ratio" and is designated as O.T.R. The structural calculations for the resisting and overturning moments are performed using certain valid assumptions which may vary sufficiently to significantly alter the design results. The density of the retained soil, the shape of the active soil pressure diagram, the weights of the construction materials, and the intensity of axial or surcharge loads have an effect on the accuracy of the calculations. Therefore, the resisting moment should have a factor of safety at least $1\frac{1}{2}$ times greater than the overturning moment (Ref. 5). Resistance to overturning is verified in each wall design by the calculation of the O.T.R. which must be equal to or greater than 1.50 (Ref. 5).

The soil that supports the foundation of a retaining wall, or any building, is a basic structural material of the construction system. The capacity of the soil beneath a foundation must be capable of sustaining the superstructure without damage from differential settlement or structural deformation. However, the soil that exists at a job site cannot be expected to perform to the same degree of predictability as other structural materials such as concrete, masonry, or steel. The mechanical properties of the various classifications of soil depend on the compaction, chemical composition, angle of internal friction, cohesion, moisture content, plasticity, and compressibility of the soil (Ref. 27). The types of soil found in different geographical regions may be traced to its geological history and local topography. Deposits of soil resulting from prehistoric phenomena, such as weathering, flooding, or glacial movement, create a condition of several strata or a mixture of different types of soil within a particular locality.

Soils that are transported and placed either by natural or mechanical means are referred to as fill soil, as opposed to soils existing in their natural undisturbed states. The value of this type of material for foundation support depends on the angle of internal friction and the degree of compaction in place. Generally, fill soil is not as reliable as natural soils. It is not the purpose of this book to expound on the subjects of soil mechanics (Ref. 28) or the types and classifications of soils and their structural behavior (Ref. 14). It is important that the engineer be aware of the soil conditions at a job site and have enough knowledge of these subjects to anticipate possible problems. The engineer should visit the site to inspect the actual conditions to make a valid estimate of the capacity of the soil. Experience and judgment are valuable tools for this purpose; however, there are times when observation alone will not render all of the information required to make a rational decision. This is particularly true in situations where the substrata soils may be weaker than the surface materials. The degree of soil investigation depends on the interrelationship of reliability of the soils, the possible reduction of construction costs that may result from a more accurate evaluation of the foundation soils, reliability versus the use of the structure, and the time and costs required to make the investigation. Field borings, taking samples, and laboratory tests are relatively inexpensive when they are considered a one-payment insurance policy that will last for the life of the structure.

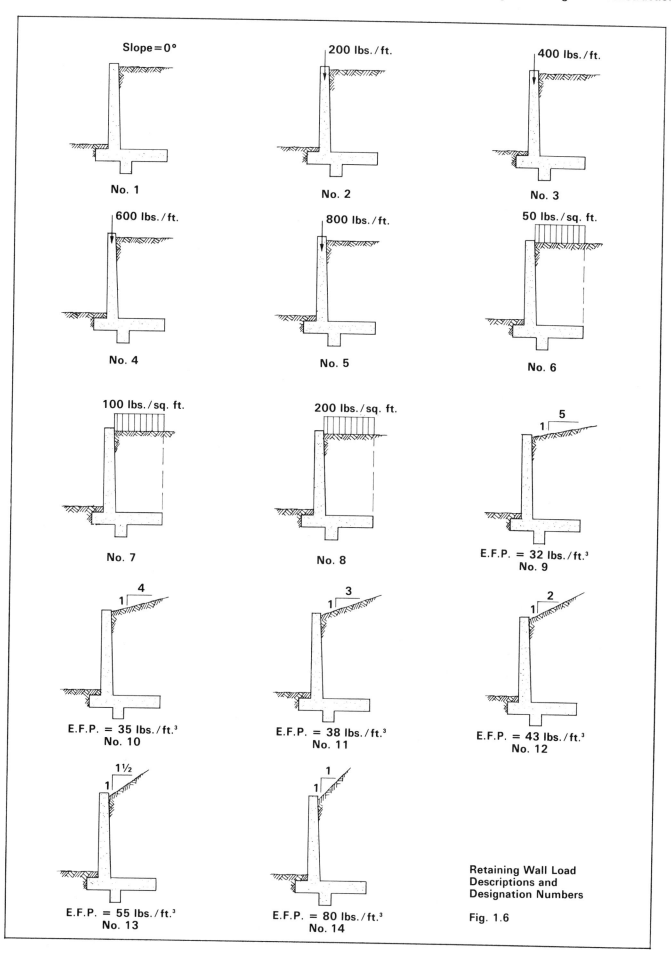

Slope = 0°

No. 1

200 lbs./ft.

No. 2

400 lbs./ft.

No. 3

600 lbs./ft.

No. 4

800 lbs./ft.

No. 5

50 lbs./sq. ft.

No. 6

100 lbs./sq. ft.

No. 7

200 lbs./sq. ft.

No. 8

5
1

E.F.P. = 32 lbs./ft.³
No. 9

4
1

E.F.P. = 35 lbs./ft.³
No. 10

3
1

E.F.P. = 38 lbs./ft.³
No. 11

2
1

E.F.P. = 43 lbs./ft.³
No. 12

1½
1

E.F.P. = 55 lbs./ft.³
No. 13

1
1

E.F.P. = 80 lbs./ft.³
No. 14

Retaining Wall Load
Descriptions and
Designation Numbers

Fig. 1.6

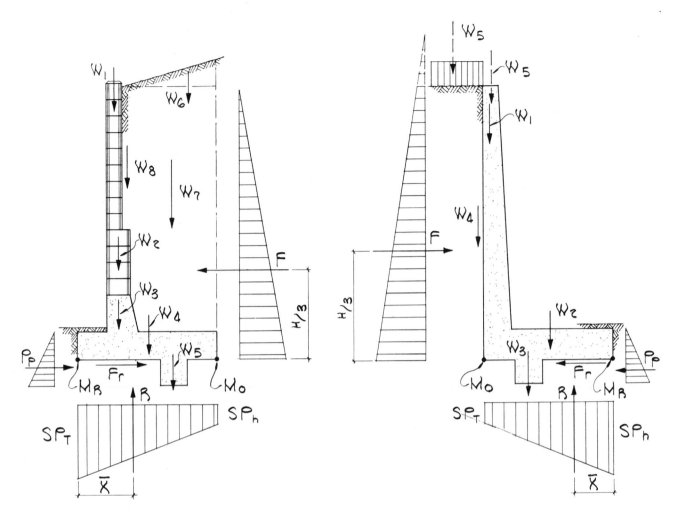

Retaining Wall Load Free Body Diagrams
Fig. 1.7

Information obtained by soil investigation should be commensurate with the reliability requirements of the structure. The amount of load that is placed on the foundation of a medium-height cantilever retaining wall is substantially less than that placed on a foundation wall that supports a heavy superstructure. The average of the values of the maximum and minimum soil bearing pressures on the foundation of the retaining walls in this book are within the range of values permitted by most building codes for specific soil classifications (Ref. 6). These maximum allowable soil bearing capacities should be applied to a retaining wall design only when the nature of the existing foundation soil is reasonably assured. Loose compacted soils, mixtures of clays, or expansive soils may yield unacceptable vertical displacement of the wall as a unit or produce large or extensive cracks from load consolidation. The trapezoidal-shaped pressure diagram shown in Fig. 1.7 is not the actual soil pressure beneath the foundation of the retaining wall. The assumed straight-line relationship between the maximum and minimum soil pressure is modified by the mechanical properties of the soil and the flexural deflection of the cantilevered foundation. The profile of the soil pressure diagram actually approximates the shape of a partial ellipse with the horizontal plane of the foundation as an increment length of the major axis. The numerical values in the design data tables designated as SP_t and SP_h are calculated unit soil bearing pressures.

Example 1.6 demonstrates the method of calculating the values of SP_t and SP_h for a one-foot-long retaining wall. The soil bearing pressure at each end of the foundation consists of the algebraic sum of the unit pressures resulting from vertical load and eccentric moment. Diagram (a) in the example shows that the vertical load produces a uniformly distributed unit pressure equal to W divided by the foundation area. The algebraic signs assume pressures acting downward as plus and pressures acting upward as minus. Diagram (b) in Ex. 1.6 is the resultant unit pressure of the eccentricity of the forces and loads of the wall acting on the foundation. The horizontal dimension to the location of the resultant of forces and loads is designated as $\overline{X}$. This dimension is calculated by dividing the net moment of the wall, M_R minus M_O, by the vertical load W. The value of the eccentricity e is equal to the distance between $\overline{X}$ and the center of the length of the foundation. The eccentric moment M_e on the retaining wall is calculated as the product of the vertical load W and the eccentricity e (Ref. 7). The soil pressure resulting from the eccentric moment is determined by dividing M_e by the section modulus S of the one-foot-wide length of the foundation. The algebraic addition of these two unit pressures is shown in diagram (c) of the example. It can be seen that the pressure is maximum at the edge of the foundation closest to the location of the resultant of the forces: The minimum soil bearing pressure should not be less than zero to prevent a reversal of the direction of bending in the foundation. The location of the resultant of the forces on the retaining wall should be within the middle one-third of the length of the wall foundation to obtain a minimum soil pressure equal to or greater than zero (Ref. 37). The criteria for the retaining walls in the following chapters also limits the maximum unit soil bearing pressure to 4000 p.s.f. The design data tables give the calculated maximum and minimum unit soil pressures for each loading condition. The engineer should verify that the calculated soil pressure is compatible with the actual soil conditions on which the retaining wall is supported. The diagram (d) in Example 1.6 is the probable profile of the unit soil pressure beneath the wall foundation.

The forces resulting from the active and bearing soil pressures on a retaining wall are resisted by cantilever bending of the wall stem and the foundation slab. The resistance capacity of each of the parts of a wall is calculated by the equations for concrete and masonry design given in App. C. A detailed design procedure is demonstrated in the design examples in each chapter for three different cases of wall loading conditions. In all of these design examples the stem wall and the foundation are checked for shear stress and flexural reinforcement. Retaining walls with masonry stems are also checked for the flexural compression stress. The allowable stress criteria for concrete and masonry are specified in Tables C1 and C2 in App. C. It is important to note that all of the retaining walls presented in this book are designed for the following material specifications: concrete f'_c = 2000 p.s.i. (Ref. 19); masonry f'_m = 1500 p.s.i. (Ref. 33); reinforcing steel = 20,000 p.s.i. (Ref. 3); grout f'_c = 2000 p.s.i. (Ref. 2); mortar (type S) = 1800 p.s.i. (Ref. 32).

The design of the concrete stem walls and the foundation slabs are performed using the design coefficients for concrete of f'_c = 2000 p.s.i. The amount of reinforcement that is required by using this value is conservative. A higher strength of concrete will increase the design coefficient a and thus reduce the calculated required reinforcing steel. As an example, using the equations for reinforced concrete in App. C, if f'_c = 3000 p.s.i., n = 9.2, j = 0.864, and a = 1.29 (Ref. 22). The value of a for the higher strength concrete represents an approximate 14 percent reduction of the reinforcing steel when compared with that required for concrete with an f'_c = 2000 p.s.i. The strength of concrete that is used for the construction of a retaining wall depends primarily on the magnitude of the stresses it will be required to withstand. The efficient use of construction materials also requires a comparison of the costs of field inspections and laboratory tests (Ref. 34) with a possible reduction of reinforcement. The area of reinforcing steel shown in the drawings in this book are determined for concrete with an f'_c = 2000 p.s.i. because it meets the stress requirements of design and the economic factors for construction of medium-height retaining walls. All of the information that is presented in the drawings and the design data tables is for a retaining wall length of one foot. Shear reinforcement is not usually used in retaining walls; however, the unit shear stress is calculated for each wall at points of maximum

Cantilever Retaining Wall Design and Construction

EXAMPLE 1.6 – UNIT SOIL BEARING PRESSURE | SHT. 1 OF 1

TO CALCULATE THE UNIT SOIL BEARING PRESSURE BENEATH WALL FOUNDATION

GIVEN: $W = 8318.6^{\#}$, $M_O = 6962'^{\#}$, $M_R = 25211'^{\#}$
FTG. $L = 6'\text{-}6''$, WIDTH $= 1'\text{-}0''$ (SEE DES. EX. 3.7.9)

$A = 1.0 \times 6.5 = 6.50$ SQ. FT.

$S = \dfrac{bL^2}{6} = \dfrac{1.0 \times 6.5^2}{6} = 7.04'^3$

a) $\dfrac{W}{A} = \dfrac{8318.6}{6.50} = 1279.8$ p.s.f.

b) NET $M = 25211 - 6962 = 18249'^{\#}$

$\bar{X} = \dfrac{18249}{8318.6} = 2.194'$

$e = \dfrac{6.50}{2} - 2.194 = 1.056'$

$Me = 1.056 \times 8318.6 = 8784.4'^{\#}$

$\dfrac{Me}{S} = \dfrac{8784.4}{7.04} = 1247.8$ p.s.f.

c) $SP_T = 1279.8 + 1247.8 = 2527.6$ p.s.f.

$SP_H = 1279.8 - 1247.8 = 32.0$ p.s.f.

$+\dfrac{W}{A}$

(a)

$+\dfrac{Me}{S}$

$-\dfrac{Me}{S}$

(b)

$SP_T = \dfrac{P}{A} + \dfrac{Me}{S}$

$SP_h = \dfrac{P}{A} - \dfrac{Me}{S}$

$\bar{X}$ e

$L/2$

(c)

NAT. GRADE

PROFILE OF SOIL PRESSURE

(d)

stress to determine that it is within the allowable stress specified in Table C1 of App. C.

The design equations given in App. C for reinforced concrete and masonry are based on the following assumptions: a straight line relationship exists between the working stress of the tension and the compression; the materials are individually elastic; the concrete or the masonry is not capable of resisting tension stress; and the tension stress of a member is resisted by steel reinforcement. Also, the reinforcing steel and the concrete are bonded together so that the materials will react as a monolithic structure. In the case of reinforced masonry, the masonry units are bonded by the mortar in the joints and the grout in the cells of the hollow units. The equation notations illustrated in the stress diagram of Fig. C1(a) indicates that the unit compression stress of the concrete or masonry is proportional to the unit tension stress of the reinforcement. This proportional relationship is designated as n. Table C1 gives a value of $n = 11$ for concrete with $f_c' = 2000$ p.s.i.; however the value of n for reinforced masonry specified in Table C2 depends on whether special field inspection of the construction process is required. The primary difference between the design of reinforced concrete and reinforced grouted masonry depends on the modulus of elasticity E of each material. The numerical value of n is derived from the ratio of E_s to E_c in Eq. C.1 and the ratio of E_s to E_m in Eq. C.14.

The retaining wall designs in this book use solid grouted hollow masonry units with $f_m' = 1500$ p.s.i. (Ref. 1). Table C2 in App. C specifies values for $E_m = 1,500,000$ p.s.i. with continuous field inspection. These values indicate that the quality of workmanship is an important factor in determining the strength of reinforced grouted masonry. It should be noted that for an ultimate compressive strength of $f_m' = 1500$ p.s.i. the specified allowable stresses are doubled when the work is performed under continuous field inspection. The Building Code of the City of New York does not make an allowance for uninspected reinforced grouted masonry (Ref. 8); however, the specified allowable stresses in the Uniform Building Code (Ref. 35) and the Los Angeles City Building Code (Ref. 35) depend on the requirement of continuous field inspection of the construction by a registered deputy masonry inspector.

The hollow masonry units used for the retaining walls in this book are commonly referred to as concrete blocks. In the following chapters, the stems of the retaining walls constructed with concrete block masonry react to the force exerted by the active soil pressure as a cantilevered slab. A state of optimum flexural design occurs when the reinforcing steel is stressed to its maximum allowable tension stress and the concrete block is stressed to its maximum allowable compression stress. This state of stress on the cross section of a stem wall is a balanced design and makes the most efficient use of the materials. If the amount of reinforcing steel provided in a wall is more than that required to resist the tensile stress, then compressive stress will not reach its maximum allowable value. Conversely stated, if the amount of reinforcing steel that is provided is not sufficient to resist the tensile stress, the compressive stress will be exceeded. The allowable flexural compressive stress for masonry with $f_m' = 1500$ p.s.i. is specified in Table C2 of App. C.

The design examples of the retaining walls with masonry stems demonstrate the design procedure to determine the amount of reinforcing steel required to approximate a balanced design condition. In general, the design method involves calculating the area of reinforcing steel by use of Eq. C.18 and checking it against the compressive stress of the masonry by use of Eq. C.23. The depth of the cross section in Eq. C.18 used to calculate the area of reinforcing steel depends on the width of the concrete block. The standard widths of the concrete blocks that are generally used are 8 and 12 inches (Ref. 9). The two values specified for allowable compressive stress of masonry, depending on field inspection, and the two widths of concrete blocks produce four design alternatives. The designs of the masonry stem walls in this book considered the alternatives in the following sequence: (1) 8-inch concrete block without field inspection, (2) 8-inch concrete block with field inspection, (3) 12-inch concrete block without field inspection, and (4) 12-inch concrete block with field inspection. There are instances in which a high bending moment on a stem wall will produce a compressive stress which exceeds the allowable value for 12-inch-wide concrete block with continuous field inspection.

Although the magnitude of the bending moment on the stem decreases rapidly with the height of the retained soil, a reinforced concrete stub wall is provided to resist the maximum moment that occurs at the base of the stem. The higher allowable flexural compression stress of the concrete and the greater depth of the cross section increase the moment resistive capacity of the wall at the bottom of the stem. Figure 1.3b shows an example of such a concrete stub wall. The height of the stub, H_1, is the dimension above the top of the foundation at which 12-inch-wide reinforced concrete block is capable of resisting the bending moment. The moment resistance of reinforced grouted masonry depends on the compressive capacity of the concrete block, which in turn depends on the continuous field inspection of the construction. The design data tables for masonry walls in the following chapters indicate the requirement of field inspection by a "yes" or "no" statement in the last line of the data.

The design equations in App. C indicate that reinforced grouted masonry is designed by the same basic methods and principles used to design reinforced concrete. The bonding together of the composite materials of a masonry wall, that is, masonry, mortar, grout, and reinforcing steel, permits them to react as an analogous homogeneous monolithic material. The area of the reinforcing steel can be transformed into an equivalent area of concrete or grout by multiplying $n \times f_s'$ for the purpose of design or analysis. The concrete blocks are bonded together by the horizontal and vertical mortar joints. Horizontal mortar joints, or bed joints, are made with a full cover of mortar over the exterior face shell and cross web of the block. Vertical joints, or head joints, are made by placing mortar between the ends of the blocks for a width not less than the thickness of the block face shell. Any mortar that may overflow into the interior cells of the blocks or onto the exterior surfaces as a result of positioning the concrete block should be removed. Mortar joints should be $\frac{3}{8}$-inch thick and flush with the surface of the block. The design depth of a masonry wall is decreased when mortar joints are made less than the width of the masonry units. Raked or otherwise depressed mortar joints can be used; however, they reduce the structural capacity of the wall. The strengths and mixtures of the various types of mortars are given in App. B. Mortar should be freshly prepared and uniformly mixed (Ref. 32); also, it is advisable to substitute high early strength cement, Type III Portland Cement, in lieu of Type I Portland Cement, when the mortar is to be used in cold weather.

Hollow unit masonry walls are constructed by placing the concrete blocks in successive vertical courses so that the hollow cells align to form a vertical unobstructed core within the wall. Alignment of the cells can be accomplished by either a common bond or a stacked bond arrangement of placing the blocks. Common bond walls are constructed by lapping the vertical courses by one half of a block length so that the vertical head joint is staggered by 8 inches. Stack bond walls are constructed by placing the successive vertical courses directly over each other so that the vertical head joint forms a straight line. It should be noted that stack bond construction requires that the blocks be tied together with continuous reinforcement or that open end masonry units be used to attain a bond between the blocks at the head joints. The common bond type of wall is the more usual method of construction.

The retaining walls in this book are designed for the stresses specified for hollow concrete masonry units with all vertical cells filled solid with a grout having an $f_c' = 2000$ p.s.i. (Ref. 35) and with flush mortar joints. The method of placing and the space clearance determine the volume proportions of a concrete grout mixture.

Transporting grout into place in the vertical cells of a wall require a fluid consistency without segregation of the aggregates of the mixture. Pouring grout into a space with the least dimension, less than 3 inches, requires a mixture by volumes of 1 part Portland Cement to $2\frac{1}{4}$ to 3 parts of damp loose sand. Pea gravel aggregate may be added to the grout when it is poured into a space with the least dimension equal to or greater than 3 inches. The mixture by volumes then becomes 1 part Portland Cement, 2 to 3 parts damp loose sand, and 2 parts pea gravel. When grout is placed by pumping it into the cells, the mixture by volume should be seven sacks of Portland Cement per each cubic yard with either sand or pea gravel aggregate.

The condition of balanced design of a cross section of concrete or masonry acting in flexure requires that the working stresses of the materials be within a compatible range of each other. It would serve no purpose in design or construction to use high strength reinforcing steel for concrete or grouted masonry with relatively lower allowable stresses. The flexural compressive stress would limit the full utilization of the reinforcement. It was determined that reinforced concrete with $f'_c = 2000$ p.s.i. and reinforced grouted masonry with $f'_m = 1500$ p.s.i. are normally commensurate for the requirements of medium-height retaining walls. Therefore, intermediate grade, deformed reinforcing bars (Ref. 3) with $f'_s = 20,000$ p.s.i. are used in the design of the retaining walls in this book. Each drawing of a retaining wall section gives the size and spacing of the reinforcement required for the particular stem height and loading condition. Table C3 gives the sizes, cross section area, and perimeter dimension for each reinforcing bar designation number. Table C4 gives the cross section areas and perimeter dimensions of reinforcing bars for various spacings within sections 1 foot wide (Ref. 22). Figure C1a indicates that the location d of the reinforcement of a section depends on the amount of required clear protection of concrete or masonry. Tables C5 and C6 give the minimum clear coverage and spacing of reinforcing bars in concrete. Table C7 gives the minimum clear coverage and spacing of reinforcing bars in grouted masonry. It is important that the reinforcing bars be accurately placed and secured prior to placing the concrete or grout (Ref. 20).

The clear dimension for protection of the reinforcement is not specifically called out on the drawings in each of the chapters; however, these dimensions can be found in Tables C5, C5a, and C6. The spacing of the principal reinforcement in concrete foundations and stem walls should not exceed three times the wall or slab thickness or be greater than 18 inches apart. Section 2607(n) of the U.B.C. (Uniform Building Code) in App. B specifies the minimum requirements for shrinkage and temperature reinforcement in concrete. The drawings of the retaining walls show the shrinkage and temperature reinforcement placed perpendicular to the principal reinforcement at the interior face of the stem. Many building codes in regions with a wide range of ambient temperature require that shrinkage and temperature reinforcement be placed at both faces of the stem wall. The spacing of reinforcing bars in concrete block walls is controlled by the 8-inch modular spacing of the vertical grout cells and the horizontal mortar joints. Section 2418(j)3 of the U.B.C. in App. B gives the minimum spacing and temperature requirements for solid grouted hollow unit masonry.

When a reinforced concrete or masonry member reacts to an imposed bending moment, the resultant tension force exerted on the reinforcing bars produces a stress between the surface of the bar and the surrounding concrete. The transfer of the tension of the reinforcing bar to the concrete depends on the capacity of the two materials to bond together. Reinforcing bars are manufactured with a raised rib surface pattern, called "deformations," to increase the area for bonding and to provide a degree of mechanical restraint against slipping. The amount of surface area required to provide sufficient bond resistance to hold a reinforcing bar to the concrete can be calculated by Eq. C.13. Table C8 gives the unit bond stress values for the various reinforcing bar sizes for concrete with $f'_c = 2000$ p.s.i. and Table C2 gives the unit bond stress for reinforced solid grouted masonry for $f'_m = 1500$ p.s.i. The length of a reinforcing bar required to develop the bond resistance can be calculated by Eq. C.27. Table C9 lists the required length of reinforcement to develop the bond anchorage for the various sizes of bars. In cases where there is not enough space in the member to extend the length of the bar to develop the required bond anchorage, the bars are bent in either a 90 degree bend or a 180 degree hook. Table C10 gives the tension capacity of hooks and bends for anchorage of the various sizes of reinforcing bars. Table C10*a* indicates the required length of the end of the bend equal to 12 times the diameter of the bar. Also, this length is required to extend reinforcing bars 12 diameters beyond the point at which it is no longer needed to resist stress (Ref. 36). There are several drawings in this book which show the foundation reinforcement hooked at the end where the bending and shear are maximum. This was done because the foundation did not have sufficient depth to accommodate the required 12-bar-diameter length of a bend as indicated in Table C10*a*.

The component parts of a retaining wall are designed and connected to permit the wall to act as a monolithic unit. The most important connection of the wall is between the stem wall and the foundation. This connection is critical since it is the point of maximum bending and shear of the stem. The transfer of the wall bending into the foundation is accomplished through a tension lap splice of foundation dowels to the vertical reinforcement of the wall. The size and space of the dowels are the same as the wall vertical bars. The length of the lap of the bar in the wall should be sufficient to develop a bond resistance equal to the tension of the reinforcing bar. This length can be calculated by Eq. C.27; however, building codes usually require that the lapped bar be extended 12-bar diameters past the point where is no longer needed or the minimum length of lap not be less than 12 inches. Table C11 gives the minimum lengths of reinforcing bars for lapped tension splices. The spacing between a pair of lapped spliced bars should not exceed 20 percent of the required lap length or be more than 6 inches apart. Shear resistance across the plane of contact between the stem wall and the top of the foundation can be achieved through the bonding of the materials. There are instances in which the magnitude of the shear will require a shear key between the two elements of the wall, particularly in walls with high active soil pressures or surcharge loads on high stem walls. A convenient method of constructing a shear key at the top of the foundation is by forming a depressed slot with a nominal 2- X 6-inch wood plank.

The force exerted by active soil pressure on the stem wall depends on the density of the retained soil. The retaining walls in this book are designed for a soil density equal to 100 p.c.f. This value can be substantially increased by the presence of accumulated water in the retained embankment. In regions where there is high rainfall, or a high level water table strata, along with a condition of alternate freezing and thawing, the density of the retained soil will vary unless some means of removing the water is provided. Expansion of freezing water that may be permitted to remain immediately behind the stem of a retaining wall can exert lateral forces much greater than the designed capacity of the wall. Also, the foundation of the retaining wall should be placed at a level below the natural grade that is below the local frost level to prevent vertical heaving or displacement caused by seasonal volume changes of the bearing soil. The source of water behind the stem wall can be traced to percolation of rain water on the surface of the retained embankment or to capillary action of trapped water within the soil. It is difficult to determine the exact amount of water that may accumulate at the wall stem; however, because the water can cause serious damage to the wall, it should be removed. The most common method of draining water from retained soil is to place a series of weep holes at the base of the stem with a 1-foot square layer of loose gravel at the interior surface of the stem behind the weep holes. The weep holes should be 4-inch diameter galvanized iron pipe and should be placed with a slight slope down toward the exterior surface of the wall.

The size and spacing of weep holes depends on the local climate and soil conditions. Quite often, the exterior face of a stem wall will be the interior face of a building, and water will not be permitted to drain through the wall. Drainage for this condition is provided by placing a perforated metal or clay pipe within the loose gravel pocket behind the wall, and by sloping the pipe to a subgrade sump to collect the water. The volume of the gravel pocket has internal voids to allow the water to accumulate. Retaining walls constructed with concrete blocks often provide for water drainage by omitting the alternate vertical mortar joint between the blocks at the first course. It should be noted that the weep holes are not specifically called out on the drawings in this book. It remains for the engineer's experience and judgment to determine the size and spacing of weep holes. In any case, except for extreme arid climate conditions, weep holes should be provided.

The thermal expansion and shrinkage of long lengths of retaining walls can cause cracks on the surface of the wall if there is no provision made for longitudinal movement. The length of the wall will expand and contract, depending on the climate conditions. Vertical construction joints should be provided at intervals of approximately 60 feet and at points where the wall may be restrained by abutments or wing walls. The expansion joints should pass through the full vertical plane of the cross section of the wall and be filled with a compressible bituminous material or a manufactured rubber water stop gasket. The width of the joint should be sufficient to allow the horizontal movement of the length of the wall.

CHAPTER **2**

Property Line Retaining Walls
Concrete Stem
Soil over the Footing

The design data and drawings presented in this chapter are concerned with property line concrete retaining walls. The retained soil is placed over the footing of the wall. The chapter consists of three design examples of retaining walls with various types of loading, a design data table for each of the fourteen loading conditions specified in Fig. 1.6, and a series of corresponding drawings following each design data table. The retaining wall drawings and the design data in this chapter are for a wall length of 1 foot.

The retaining walls in this chapter are designed using the following criteria:

Weight of soil = 100 p.c.f.   Weight of concrete = 150 p.c.f.
Concrete f'_c = 2000 p.s.i.   Reinforcing steel f'_s = 20,000 p.s.i.
Maximum allowable soil pressure = 4000 p.s.f.
Minimum allowable soil pressure = 0 p.s.f.
$\overline{X} \geqslant L/3$ Minimum O.T.R. = 1.50
Coefficient of soil friction = 0.40
Passive soil pressure = 300 p.c.f.

See Table C5 for reinforcement cover. Weep holes are not shown on the drawings.

2.0 Design Example—Property Line Retaining Wall—Concrete Stem—Soil over Footing

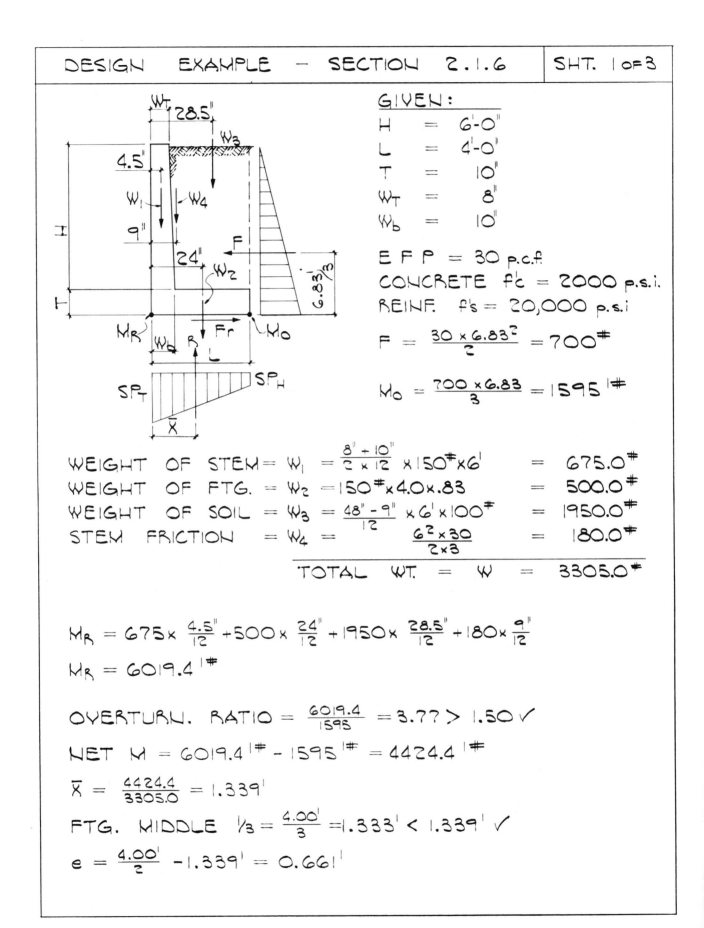

GIVEN:

$H = 6'\text{-}0''$

$L = 4'\text{-}0''$

$T = 10''$

$W_T = 8''$

$W_b = 10''$

$EFP = 30 \text{ p.c.f.}$

$CONCRETE\ f'_c = 2000\ \text{p.s.i.}$

$REINF.\ f'_s = 20,000\ \text{p.s.i}$

$F = \dfrac{30 \times 6.83^2}{2} = 700^{\#}$

$M_O = \dfrac{700 \times 6.83}{3} = 1595^{'\#}$

$\text{WEIGHT OF STEM} = W_1 = \dfrac{8''+10''}{2 \times 12} \times 150^{\#} \times 6' = 675.0^{\#}$

$\text{WEIGHT OF FTG.} = W_2 = 150^{\#} \times 4.0 \times .83 = 500.0^{\#}$

$\text{WEIGHT OF SOIL} = W_3 = \dfrac{48''-9''}{12} \times 6' \times 100^{\#} = 1950.0^{\#}$

$\text{STEM FRICTION} = W_4 = \dfrac{6^2 \times 30}{2 \times 3} = 180.0^{\#}$

$$\text{TOTAL WT.} = W = 3305.0^{\#}$$

$M_R = 675 \times \dfrac{4.5''}{12} + 500 \times \dfrac{24''}{12} + 1950 \times \dfrac{28.5''}{12} + 180 \times \dfrac{9''}{12}$

$M_R = 6019.4^{'\#}$

$\text{OVERTURN. RATIO} = \dfrac{6019.4}{1595} = 3.77 > 1.50 \checkmark$

$\text{NET M} = 6019.4^{'\#} - 1595^{'\#} = 4424.4^{'\#}$

$\bar{X} = \dfrac{4424.4}{3305.0} = 1.339'$

$\text{FTG. MIDDLE } \frac{1}{3} = \dfrac{4.00'}{3} = 1.333' < 1.339' \checkmark$

$e = \dfrac{4.00'}{2} - 1.339' = 0.661'$

DESIGN EXAMPLE — SECTION 2.1.6 | SHT. 2 OF 3

FOOTING DESIGN:

$e = 0.661'$ $W = 3305.0^{\#}$

$M_e = 0.661' \times 3305^{\#} = 2185'^{\#}$

FTG. AREA $= 4.00'^2$

SECTION MODULUS $= \dfrac{1' \times 4.00^2}{6} = 2.667'^3$

SOIL PRESSURE $= \dfrac{3305^{\#}}{4.00} \pm \dfrac{2185'^{\#}}{2.667} = 826.25 \pm 819.4$

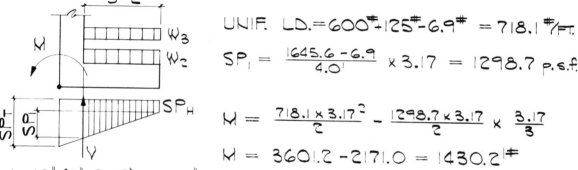

$\underline{SP_T = 1645.6 \text{ p.s.f.}}$ 　　　$\underline{SP_H = 6.9 \text{ p.s.f.}}$

UNIF. LD. $= 600^{\#} + 125^{\#} - 6.9^{\#} = 718.1^{\#}/\text{FT.}$

$SP_1 = \dfrac{1645.6 - 6.9}{4.0'} \times 3.17 = 1298.7 \text{ p.s.f.}$

$M = \dfrac{718.1 \times 3.17^2}{2} - \dfrac{1298.7 \times 3.17}{2} \times \dfrac{3.17}{3}$

$M = 3601.2 - 2171.0 = 1430.2'^{\#}$

$V = 3.17' \times 718.1^{\#} - \dfrac{1298.7 \times 3.17'}{2}$

$V = 2274.2 - 2056.5 = 217.7^{\#}$

$\nu = \dfrac{217.7}{12 \times .89 \times 7.75} = 2.63 \text{ p.s.i} < 54 \text{ p.s.i} \checkmark$

$d = 10'' - (2'' + .5 \times .5) = 7.75''$

$a = 1.13$

$A_s = \dfrac{1.430'^K}{1.13 \times 7.75} = 0.16 \text{ SQ. IN./FT.}$

USE #4 @ 9"o.c.

MIN. $A_s = 10 \times 48 \times .0012 = 0.576$

USE LONG. REINF. $= 3 - \#4's$

DESIGN EXAMPLE — SECTION 2.1.6	SHT. 3 OF 3

STEM DESIGN AT TOP OF FTG. :

$$V = \frac{30 \times 6^2}{2} = 540.0^{\#}$$

$$M = \frac{540^{\#} \times 6'}{3} = 1080'^{\#}; \quad d = 10'' - (2 + .5 \times .5) = 7.75''; \quad a = 1.13$$

$$v = \frac{540^{\#}}{12 \times .89 \times 7.75} = 6.52 \, p.s.i < 54.0 \, p.s.i$$

$$A_s = \frac{1.08}{1.13 \times 7.75} = 0.1233 \, sq. \, in.; \quad MIN. \, A_s = 10 \times 12 \times .0012 = 0.144 \, sq. in.$$

USE   #5 @ 16" o.c. VERT.
        #4 @ 24" o.c. HORIZ.

TOTAL   REINF. = 0.15 + 0.10 = 0.25 sq. in.

SLIDING :

$$W = 3305.0^{\#} \qquad F = 700.0^{\#}$$

$$F_r = \frac{700.0}{3305.0} = 0.212 < 0.40 \checkmark$$

```
┌────────────────────────────────────────────────────────────┐
│ DESIGN   EXAMPLE  -  SECTION  2.6.6 │ SHT. 1 OF 4 │
└────────────────────────────────────────────────────────────┘
```

GIVEN:

$$H = 6'\text{-}0''$$
$$L = 4'\text{-}4''$$
$$T = 10''$$
$$W_T = 8''$$
$$W_b = 10''$$

SURCHARGE $= 50$ psf

EFP $= 30$ pcf

CONCRETE $f'_c = 2000$ psi

REINF. $f'_s = 20{,}000$ psi

WEIGHT OF SOIL $= 100$ pcf

EQUIV. HT. $= 6.0' + 0.83' + \dfrac{50}{100} = 7.33'$

$$F = \frac{30 \times 7.33^2}{2} = 807^{\#}$$

$$M_o = \frac{807 \times 7.33'}{3} = 1972.0'^{\#}$$

WEIGHT OF STEM $= W_1 = \dfrac{8'' + 10''}{2 \times 12} \times 150 \times 6.0' = 675.0^{\#}$

WEIGHT OF FTG. $= W_2 = 150^{\#} \times 4.33 \times .83 = 541.6^{\#}$

WEIGHT OF SOIL $= W_3 = \dfrac{52'' - 9''}{12} \times 6.0' \times 100 = 2150.0^{\#}$

STEM FRICTION $= W_4 = \dfrac{6.5^2 \times 30}{3 \times 2} = 211.3$

SURCHARGE LD. $= W_5 = 3.58 \times 50 = 179.1^{\#}$

TOTAL WT. $= W = 3757.0^{\#}$

$M_R = 675 \times \dfrac{4.5''}{12} + 541.6 \times \dfrac{26''}{12} + 2150 \times \dfrac{30.5''}{12} + 211.3 \times \dfrac{9''}{12} + 179.1 \times \dfrac{30.5''}{12}$

$M_R = 7505.0'^{\#}$

OVERTURN. RATIO $= \dfrac{7505}{1972} = 3.806 > 1.50 \checkmark$

NET M $= 7505 - 1972 = 5533'^{\#}$

$\bar{X} = \dfrac{5533}{3757} = 1.473'$

FTG. MIDDLE $\frac{1}{3} = \dfrac{4.33}{3} = 1.44 < 1.473 \checkmark$

$e = \dfrac{4.33'}{2} - 1.473' = 0.694'$

DESIGN EXAMPLE — SECTION 2.6.6.	SHT. 2 OF 4

FOOTING   DESIGN :

$e = 0.694'$ $W = 3757.^{\#}$

$M_e = 3757 \times .694 = 2607^{'\#}$

FTG.   AREA $= 4.33$ SQ. FT.

FTG.   SECT.   MODULUS $= \dfrac{1 \times 4.33^2}{6} = 3.125^{'3}$

SOIL   PRESSURE $= \dfrac{3757}{4.33} \pm \dfrac{2607}{3.125} = 867.0 \pm 834.2$

$\underline{\underline{SP_T = 1701. \; psf}}$ $\underline{\underline{SP_H = 33. \; psf}}$

UNIF.   LD. $= 600 + 50 + 125 - 33 = 742^{\#}/\text{FT.}$

$SP_1 = \dfrac{1701 - 33}{4.33} \times 3.50 = 1347. \; psf$

$-M = \dfrac{742 \times 3.50^2}{2} - \dfrac{1347 \times 3.50^2}{2 \times 3}$

$-M = 4544.7 - 2750.1 = 1794.6^{'\#}$

$V = 3.50 \times 742 - \dfrac{1347 \times 3.50}{2}$

$V = 239.8^{\#}$

$d = 10'' - (2 + .5 \times .5) = 7.75''$; $a = 1.13$; $j = 0.89$

$\upsilon = \dfrac{239.8}{12 \times .89 \times 7.75} = 3.0 \; psi < 54 \; psi \; \checkmark$

$A_s = \dfrac{1.794}{1.13 \times 7.75} = 0.205 \; \text{SQ.IH.}/\text{FT.}$

$\underline{\underline{\text{USE} \#5 @ 16'' \text{ o.c.} = 0.23 \; \text{SQ.IH.}/\text{FT.}}}$

LONG. REINF. :

$A_s = 0.0012 \times 52 \times 10 = 0.62 \; \text{SQ. IN.}$

$\underline{\underline{\text{USE   LONG. REINF.} = 4 - \#4}}$

| DESIGN EXAMPLE – SECTION 2.6.6. | SHT. 3 OF 4 |

STEM DESIGN AT TOP OF FTG. :

$EFP = 30$ p.c.f.

EQUIV. STEM HT. $= 6.0' + \frac{50}{100} = 6.50'$

$V = \frac{30 \times 6.5^2}{2} = 633.75^{\#}$

$M = \frac{633.75 \times 6.5}{3} = 1373.1'^{\#}$

$d = 10'' - (2 + .5 \times .5) = 7.75''$; $a = 1.13$; $j = 0.89$

$v = \frac{633.75}{12 \times .89 \times 7.75} = 7.6$ psi < 54 psi $\checkmark$

$A_s = \frac{1.373}{1.13 \times 7.75} = 0.156$ SQ. IN./FT. __USE #5 @ 16" o.c.__

LONG. REINF. :

$A_s = 10 \times 12 \times .0012 = 0.144$ SQ. IN./FT.

__USE # 4 @ 22" o.c. HORIZ.__

SLIDING :

$W = 3757.0^{\#}$ $F = 807.0^{\#}$

$Fr = \frac{807.0}{3757.0} = 0.215 < 0.40 \checkmark$

DESIGN EXAMPLE — SECTION 2.6.6	SHT. 4 OF 4

CHECK WALL WITHOUT 50 p.s.f. SURCHARGE LOAD

EQUIV. HT. $= 6.0 + 0.83 = 6.83'$;   E F P $= 30$ p.c.f.

$F = \dfrac{30 \times 6.83^2}{2} = 700^{\#}$; $M_0 = \dfrac{700 \times 6.83}{3} = 1595^{'\#}$

$W = 3757 - 211.3 + \dfrac{6.0^2 \times 30}{2 \times 3} - 179.1 = 3547^{\#}$

$M_R = 7505 - 211.3 \times \dfrac{9''}{12} + 180 \times \dfrac{9''}{12} - 179.1 \times \dfrac{30.5}{12} = 7026^{'\#}$

OVERTURN. RATIO $= \dfrac{7026}{1595} = 4.405 > 1.50$ ✓

NET M $= 7026 - 1595 = 5431^{'\#}$

$\bar{X} = \dfrac{5431}{3547} = 1.531' > 1.44'$ ✓

$e = \dfrac{4.33'}{2} - 1.531 = 0.635'$

$M_e = 3547 \times .635 = 2254^{'\#}$

SOIL PRESSURE $= \dfrac{3547}{4.33} \pm \dfrac{2254}{3.125} = 819.2 \pm 721.3$

$\underline{SP_T = 1540.5 \text{ p.s.f}}$ $\underline{SP_H = 97.9 \text{ p.s.f}}$

CHECK SLIDING:

$F_r = \dfrac{700}{3547} = 0.197 < 0.40$ ✓

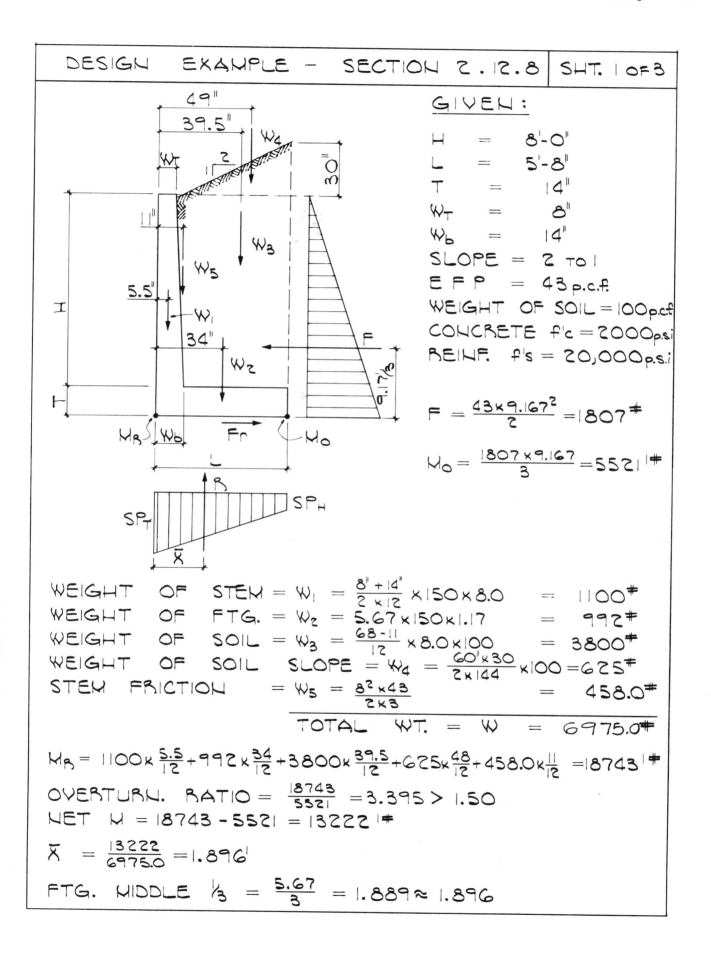

DESIGN EXAMPLE – SECTION 2.12.8 | SHT. 1 OF 3

GIVEN:

$$H = 8'-0''$$
$$L = 5'-8''$$
$$T = 14''$$
$$W_T = 8''$$
$$W_b = 14''$$
SLOPE = 2 TO 1
E F P = 43 p.c.f.
WEIGHT OF SOIL = 100 p.c.f.
CONCRETE f'c = 2000 p.s.i
REINF. f's = 20,000 p.s.i

$$F = \frac{43 \times 9.167^2}{2} = 1807^{\#}$$

$$M_O = \frac{1807 \times 9.167}{3} = 5521^{'\#}$$

WEIGHT OF STEM = W_1 = $\frac{8'' + 14''}{2 \times 12} \times 150 \times 8.0$ = $1100^{\#}$
WEIGHT OF FTG. = W_2 = $5.67 \times 150 \times 1.17$ = $992^{\#}$
WEIGHT OF SOIL = W_3 = $\frac{68-11}{12} \times 8.0 \times 100$ = $3800^{\#}$
WEIGHT OF SOIL SLOPE = W_4 = $\frac{60'' \times 30}{2 \times 144} \times 100$ = $625^{\#}$
STEM FRICTION = W_5 = $\frac{8^2 \times 43}{2 \times 3}$ = $458.0^{\#}$

TOTAL WT. = W = $6975.0^{\#}$

$$M_R = 1100 \times \frac{5.5}{12} + 992 \times \frac{34}{12} + 3800 \times \frac{39.5}{12} + 625 \times \frac{48}{12} + 458.0 \times \frac{11}{12} = 18743^{'\#}$$

OVERTURN. RATIO = $\frac{18743}{5521}$ = 3.395 > 1.50

NET M = 18743 - 5521 = 13222$^{'\#}$

$$\bar{X} = \frac{13222}{6975.0} = 1.896'$$

FTG. MIDDLE $\frac{1}{3}$ = $\frac{5.67}{3}$ = 1.889 ≈ 1.896

43

DESIGN   EXAMPLE - SECTION 2.12.8 | SHT. 2 of 3

FOOTING   DESIGN :

$\bar{X} = 1.896'$

$e = \dfrac{5.67}{2} - 1.896 = 0.938'$ $W = 6975.0^{\#}$

$M_e = 6975 \times .938 = 6539'^{\#}$

FTG. AREA = 5.67 SQ. FT.

FTG. SECT. MODULUS $= \dfrac{1 \times 5.67^2}{6} = 5.35'^3$

SOIL PRESSURE $= \dfrac{6975}{5.67} \pm \dfrac{6539}{5.35} = 1230. \pm 1222.2$

$\underline{SP_T = 2454.2 \text{ p.s.f.}}$ $\underline{SP_H = 7.8 \text{ p.s.f.}}$

UNIF. LD. $= 175 + 800 - 7.8 = 967.2^{\#}/\text{FT.}$

$SP'_T = \dfrac{2454.2 - 7.8}{5.67} \times 4.50 = 1941.6 \text{ p.s.f.}$

$M = \dfrac{967.2 \times 4.5^2}{2} + \dfrac{625 \times 2 \times 4.50}{3} - \dfrac{1941.6 \times 4.50^2}{2 \times 3}$

$M = 5115'^{\#}$

$V = 4.50 \times 967.2 + 625 - \dfrac{1941.6 \times 4.50}{2}$

$V = 608.8^{\#}$

$d = 14 - (2 + .5 \times .5) = 11.75''$; $a = 1.13$; $j = 0.89$

$v = \dfrac{608.8}{12 \times .89 \times 11.75} = 4.85 \text{ p.s.i.} < 54 \text{ p.s.i.}$

$A_s = \dfrac{5.115}{1.13 \times 11.75} = 0.385 \text{ SQ. IN./FT.}$

$\underline{USE \quad \#5 @ 9'' \text{ O.C.}}$

LONG. REINF. :

$A_s = 0.0015 \times 12 \times 5.67 \times 11.75 = 1.20 \text{ SQ. IN.}$

$\underline{USE \quad 4 - \#5\text{'s}}$

DESIGN EXAMPLE — SECTION 2.12.8	SHT. 3 OF 3

STEM DESIGN AT TOP OF FTG. :

$EFP = 43$ p.c.f. $HT. = 8'-0''$

$V = \dfrac{43 \times 8^2}{2} = 1367^{\#}$; $M = \dfrac{1367 \times 8}{3} = 3669.3'^{\#}$

$d = 14 - (2 + .5 \times .5) = 11.68''$; $a = 1.13$; $j = 0.89$

$v = \dfrac{1367}{12 \times .89 \times 11.75} = 10.9$ p.s.i. < 54.0 p.s.i. ✓

$As = \dfrac{3.669}{1.13 \times 11.75} = 0.276$ SQ. IN./FT.

<u>USE #6 @ 18'' o.c.</u>

STEM DESIGN AT H = 4'-0'' (MID HEIGHT)

$V = \dfrac{43 \times 4^2}{2} = 344^{\#}$; $M = \dfrac{344 \times 4}{3} = 458.6'^{\#}$

$d = 11 - (2 + .5 \times .5) = 8.75''$; $a = 1.13$; $j = 0.89$

$v = \dfrac{344}{12 \times .89 \times 8.75} = 3.70$ p.s.i

$As = \dfrac{0.458}{1.13 \times 8.75} = 0.046$ SQ. IN./FT.

<u>USE #5 @ 18'' o.c.</u>

LONG. REINF. :

$As = 0.0015 \times 8 \times 12 \times 11 = 1.58$ SQ. IN.

<u>USE #5 @ 18'' o.c.</u>

RESISTANCE TO SLIDING:

$Fr = \dfrac{1807}{7014.6} = 0.257 < 0.40$ ✓

TABLE 2.1	PROPERTY LINE RETAINING WALLS – CONCRETE STEM – SOIL OVER FOOTING
	SLOPE = 0 to 1 SURCHARGE = 0 lbs./sq. ft. AXIAL = 0 lbs./ft.

SOIL PRESSURE DIAGRAM

Stem Height	3'- 0''	4'- 0''	5'- 0''	6'- 0''	7'- 0''	8'- 0''	9'- 0''	10'- 0''	11'- 0''	12'- 0''
Ft. L ft.	2.167	2.833	3.500	4.000	4.667	5.333	6.000	6.667	7.167	7.833
Ft. T ft.	0.667	0.667	0.833	0.833	0.833	1.000	1.000	1.167	1.167	1.333
Wt ft.	0.667	0.667	0.667	0.667	0.667	0.667	0.667	0.667	0.667	0.667
Wb ft.	0.667	0.667	0.833	0.833	0.833	1.000	1.000	1.167	1.167	1.333
W lbs.	1012	1630	2500	3305	4357	5720	7080	8792	10247	12287
F lbs.	202	327	510	700	920	1215	1500	1870	2220	2667
Mo ft.-lbs.	246	508	992	1595	2403	3645	5000	6962	9005	11852
Mr ft.-lbs.	1002	2105	3992	6019	9266	13917	19394	26780	33528	43973
O.T.R.	4.066	4.142	4.022	3.773	3.855	3.818	3.879	3.846	3.723	3.710
$\overline{X}$ ft.	0.747	0.979	1.200	1.339	1.575	1.796	2.033	2.254	2.393	2.614
e ft.	0.336	0.437	0.550	0.661	0.759	0.871	0.967	1.079	1.190	1.302
Me ft.-lbs.	340	713	1375	2186	3305	4982	6846	9488	12194	16002
S ft.^{3}	0.782	1.338	2.042	2.667	3.630	4.741	6.000	7.407	8.560	10.227
S.P.t p.s.f.	902	1108	1388	1646	1844	2123	2321	2600	2854	3133
S.P.h p.s.f.	32	43	41	6	23	22	39	38	5	4
Friction	0.199	0.200	0.204	0.212	0.211	0.212	0.212	0.213	0.217	0.217

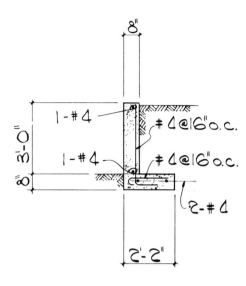

E.F.P. = **30** p.c.f.
Section 2.1.3

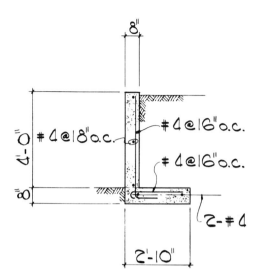

E.F.P. = **30** p.c.f.
Section 2.1.4

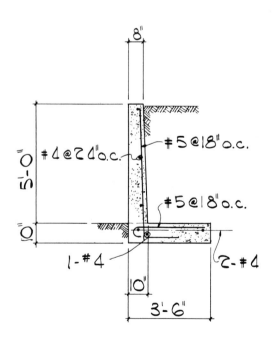

E.F.P. = **30** p.c.f.
Section 2.1.5

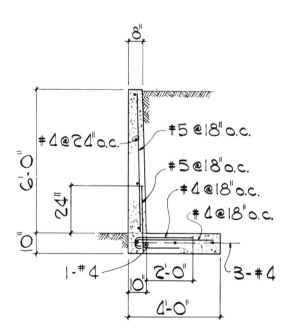

E.F.P. = **30** p.c.f.
Section 2.1.6

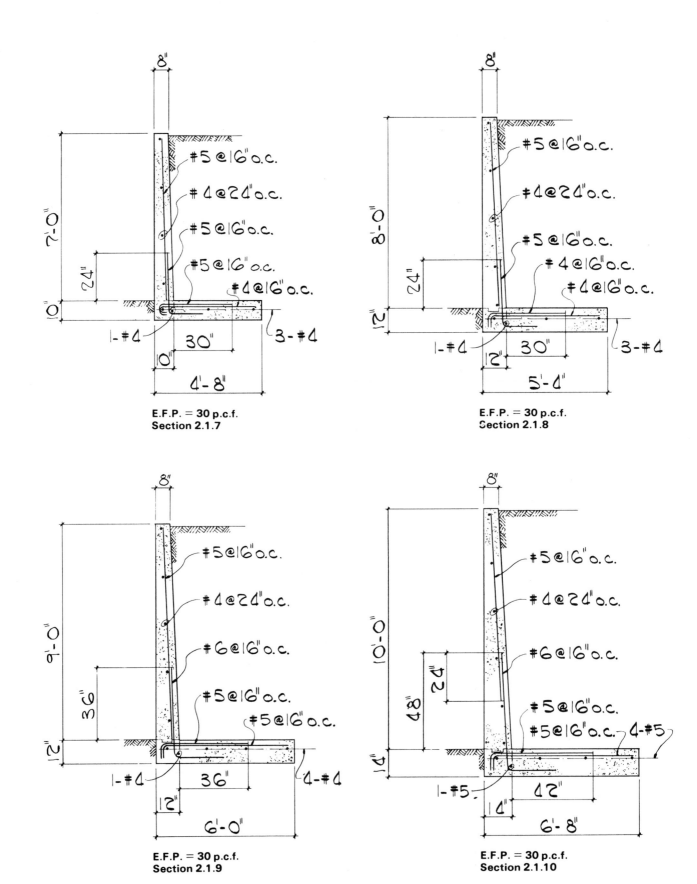

8"

#5 @ 16" o.c.

#4 @ 24" o.c.

#5 @ 16" o.c.

#5 @ 16" o.c.

#4 @ 16" o.c.

7'-0"

24"

1-#4

30"

3-#4

4'-8"

E.F.P. = 30 p.c.f.
Section 2.1.7

8"

#5 @ 16" o.c.

#4 @ 24" o.c.

#5 @ 16" o.c.

#4 @ 16" o.c.

#4 @ 16" o.c.

8'-0"

24"

12"

1-#4

12"

30"

3-#4

5'-4"

E.F.P. = 30 p.c.f.
Section 2.1.8

8"

#5 @ 16" o.c.

#4 @ 24" o.c.

#6 @ 16" o.c.

#5 @ 16" o.c.

#5 @ 16" o.c.

9'-0"

36"

12"

1-#4

36"

4-#4

12"

6'-0"

E.F.P. = 30 p.c.f.
Section 2.1.9

8"

#5 @ 16" o.c.

#4 @ 24" o.c.

#6 @ 16" o.c.

#5 @ 16" o.c.

#5 @ 16" o.c.

4-#5

10'-0"

48"

24"

14"

1-#5

14"

42"

6'-8"

E.F.P. = 30 p.c.f.
Section 2.1.10

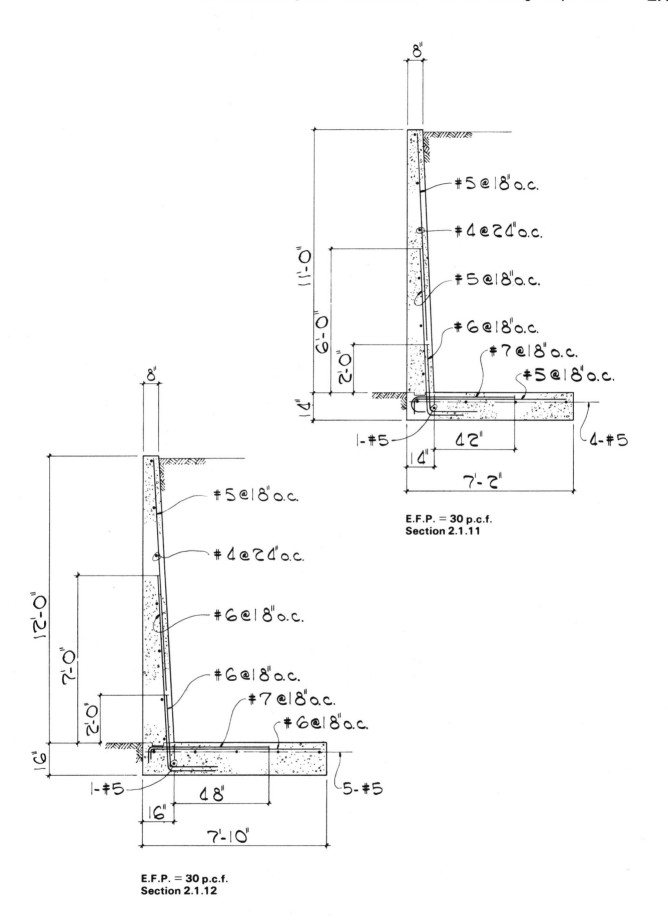

8"

#5@18" o.c.

#4@24" o.c.

#5@18" o.c.

#6@18" o.c.

#7@18" o.c.

#5@18" o.c.

11'-0"

9'-0"

2'-0"

1'-4"

1-#5

42"

4"

7'-2"

4-#5

E.F.P. = 30 p.c.f.
Section 2.1.11

8"

#5@18" o.c.

#4@24" o.c.

#6@18" o.c.

#6@18" o.c.

#7@18" o.c.

#6@18" o.c.

12'-0"

7'-0"

2'-0"

6"

1-#5

48"

16"

7'-10"

5-#5

E.F.P. = 30 p.c.f.
Section 2.1.12

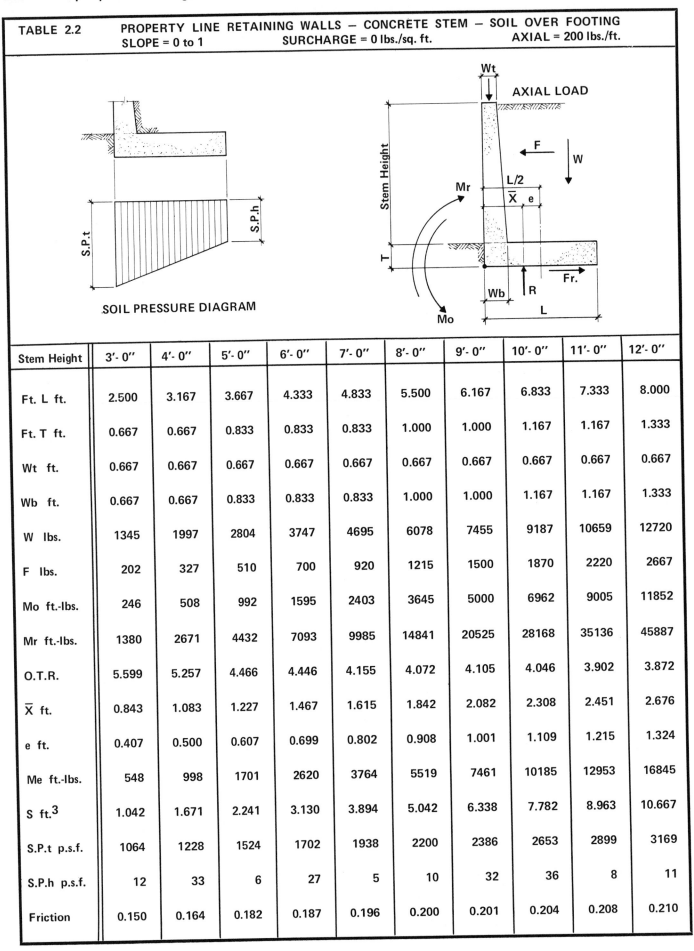

Stem Height	3'- 0''	4'- 0''	5'- 0''	6'- 0''	7'- 0''	8'- 0''	9'- 0''	10'- 0''	11'- 0''	12'- 0''
Ft. L ft.	2.500	3.167	3.667	4.333	4.833	5.500	6.167	6.833	7.333	8.000
Ft. T ft.	0.667	0.667	0.833	0.833	0.833	1.000	1.000	1.167	1.167	1.333
Wt ft.	0.667	0.667	0.667	0.667	0.667	0.667	0.667	0.667	0.667	0.667
Wb ft.	0.667	0.667	0.833	0.833	0.833	1.000	1.000	1.167	1.167	1.333
W lbs.	1345	1997	2804	3747	4695	6078	7455	9187	10659	12720
F lbs.	202	327	510	700	920	1215	1500	1870	2220	2667
Mo ft.-lbs.	246	508	992	1595	2403	3645	5000	6962	9005	11852
Mr ft.-lbs.	1380	2671	4432	7093	9985	14841	20525	28168	35136	45887
O.T.R.	5.599	5.257	4.466	4.446	4.155	4.072	4.105	4.046	3.902	3.872
$\overline{X}$ ft.	0.843	1.083	1.227	1.467	1.615	1.842	2.082	2.308	2.451	2.676
e ft.	0.407	0.500	0.607	0.699	0.802	0.908	1.001	1.109	1.215	1.324
Me ft.-lbs.	548	998	1701	2620	3764	5519	7461	10185	12953	16845
S ft.^{3}	1.042	1.671	2.241	3.130	3.894	5.042	6.338	7.782	8.963	10.667
S.P.t p.s.f.	1064	1228	1524	1702	1938	2200	2386	2653	2899	3169
S.P.h p.s.f.	12	33	6	27	5	10	32	36	8	11
Friction	0.150	0.164	0.182	0.187	0.196	0.200	0.201	0.204	0.208	0.210

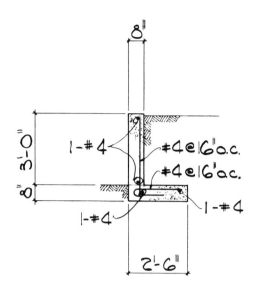

Axial load = 200 lbs./ft.
E.F.P. = 30 p.c.f.
Section 2.2.3

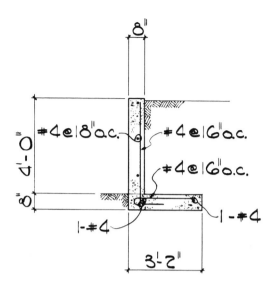

Axial load = 200 lbs./ft.
E.F.P. = 30 p.c.f.
Section 2.2.4

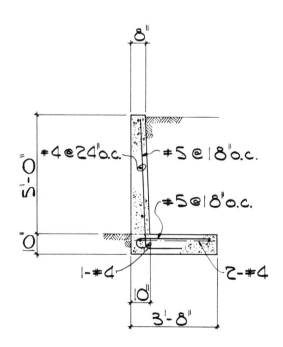

Axial load = 200 lbs./ft.
E.F.P. = 30 p.c.f.
Section 2.2.5

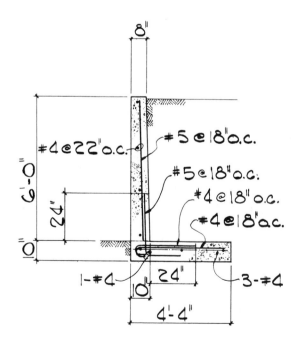

Axial load = 200 lbs./ft.
E.F.P. = 30 p.c.f.
Section 2.2.6

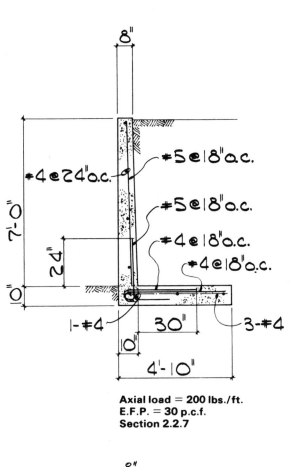

8"

#5 @ 18"o.c.

#4 @ 24"o.c.

#5 @ 18"o.c.

#4 @ 18"o.c.

#4 @ 18"o.c.

7'-0"

24"

10"

1-#4

30"

3-#4

10"

4'-10"

Axial load = 200 lbs./ft.
E.F.P. = 30 p.c.f.
Section 2.2.7

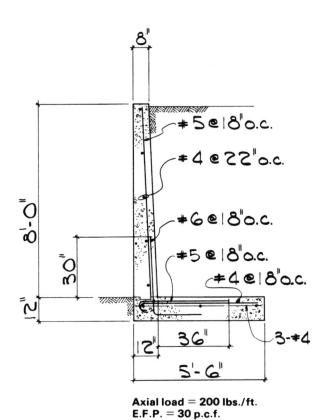

8"

#5 @ 18"o.c.

#4 @ 22"o.c.

#6 @ 18"o.c.

#5 @ 18"o.c.

#4 @ 18"o.c.

8'-0"

30"

12"

12"

36"

3-#4

5'-6"

Axial load = 200 lbs./ft.
E.F.P. = 30 p.c.f.
Section 2.2.8

8"

#5 @ 16"o.c.

#4 @ 20"o.c.

#6 @ 16"o.c.

#6 @ 16"o.c.

#4 @ 16"o.c.

9'-0"

36"

12"

36"

12"

4-#4

6'-2"

Axial load = 200 lbs./ft.
E.F.P. = 30 p.c.f.
Section 2.2.9

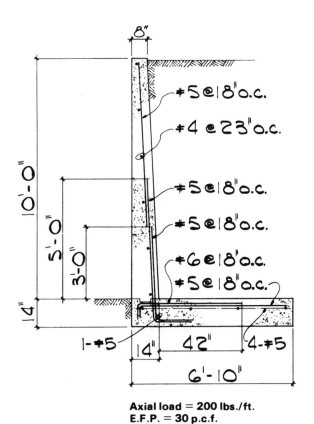

8"

#5 @ 18"o.c.

#4 @ 23"o.c.

#5 @ 18"o.c.

#5 @ 18"o.c.

#6 @ 18"o.c.

#5 @ 18"o.c.

10'-0"

5'-0"

3'-0"

14"

1-#5

14"

42"

4-#5

6'-10"

Axial load = 200 lbs./ft.
E.F.P. = 30 p.c.f.
Section 2.2.10

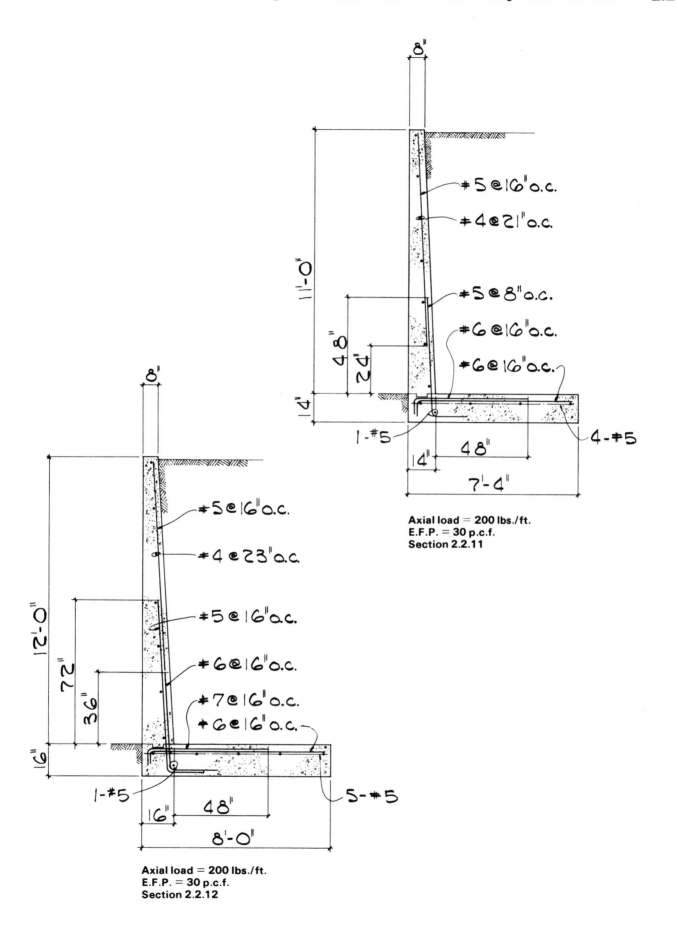

8"

11'-0"

#5 @ 16" o.c.

#4 @ 21" o.c.

#5 @ 8" o.c.

#6 @ 16" o.c.

#6 @ 16" o.c.

48"

24"

4"

1-#5

4-#5

4"

48"

7'-4"

Axial load = 200 lbs./ft.
E.F.P. = 30 p.c.f.
Section 2.2.11

8"

12'-0"

#5 @ 16" o.c.

#4 @ 23" o.c.

#5 @ 16" o.c.

#6 @ 16" o.c.

#7 @ 16" o.c.

#6 @ 16" o.c.

72"

36"

6"

1-#5

5-#5

16"

48"

8'-0"

Axial load = 200 lbs./ft.
E.F.P. = 30 p.c.f.
Section 2.2.12

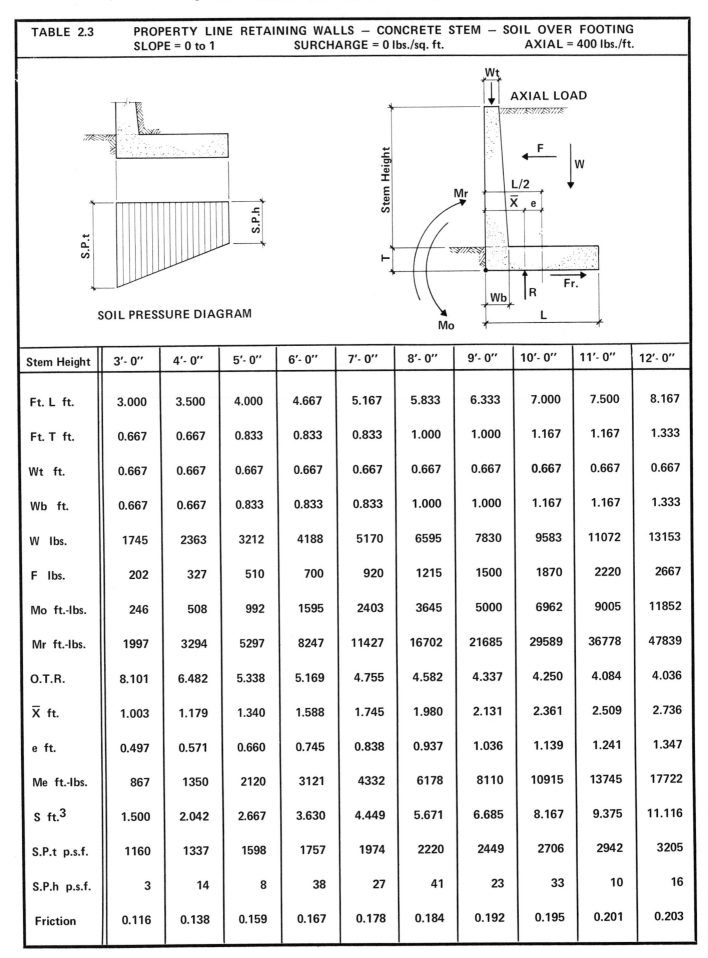

TABLE 2.3	PROPERTY LINE RETAINING WALLS — CONCRETE STEM — SOIL OVER FOOTING
	SLOPE = 0 to 1 SURCHARGE = 0 lbs./sq. ft. AXIAL = 400 lbs./ft.

SOIL PRESSURE DIAGRAM

Stem Height	3'- 0"	4'- 0"	5'- 0"	6'- 0"	7'- 0"	8'- 0"	9'- 0"	10'- 0"	11'- 0"	12'- 0"
Ft. L ft.	3.000	3.500	4.000	4.667	5.167	5.833	6.333	7.000	7.500	8.167
Ft. T ft.	0.667	0.667	0.833	0.833	0.833	1.000	1.000	1.167	1.167	1.333
Wt ft.	0.667	0.667	0.667	0.667	0.667	0.667	0.667	0.667	0.667	0.667
Wb ft.	0.667	0.667	0.833	0.833	0.833	1.000	1.000	1.167	1.167	1.333
W lbs.	1745	2363	3212	4188	5170	6595	7830	9583	11072	13153
F lbs.	202	327	510	700	920	1215	1500	1870	2220	2667
Mo ft.-lbs.	246	508	992	1595	2403	3645	5000	6962	9005	11852
Mr ft.-lbs.	1997	3294	5297	8247	11427	16702	21685	29589	36778	47839
O.T.R.	8.101	6.482	5.338	5.169	4.755	4.582	4.337	4.250	4.084	4.036
$\overline{X}$ ft.	1.003	1.179	1.340	1.588	1.745	1.980	2.131	2.361	2.509	2.736
e ft.	0.497	0.571	0.660	0.745	0.838	0.937	1.036	1.139	1.241	1.347
Me ft.-lbs.	867	1350	2120	3121	4332	6178	8110	10915	13745	17722
S ft.^{3}	1.500	2.042	2.667	3.630	4.449	5.671	6.685	8.167	9.375	11.116
S.P.t p.s.f.	1160	1337	1598	1757	1974	2220	2449	2706	2942	3205
S.P.h p.s.f.	3	14	8	38	27	41	23	33	10	16
Friction	0.116	0.138	0.159	0.167	0.178	0.184	0.192	0.195	0.201	0.203

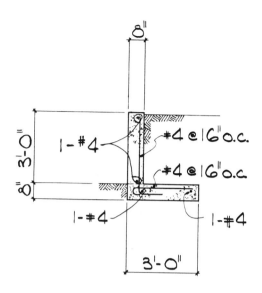

Axial load = 400 lbs./ft.
E.F.P. = 30 p.c.f.
Section 2.3.3

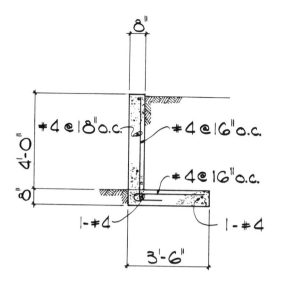

Axial load = 400 lbs./ft.
E.F.P. = 30 p.c.f.
Section 2.3.4

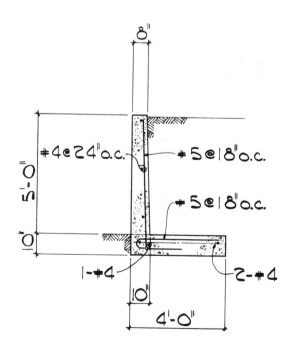

Axial load = 400 lbs./ft.
E.F.P. = 30 p.c.f.
Section 2.3.5

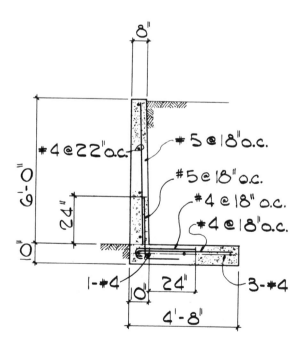

Axial load = 400 lbs./ft.
E.F.P. = 30 p.c.f.
Section 2.3.6

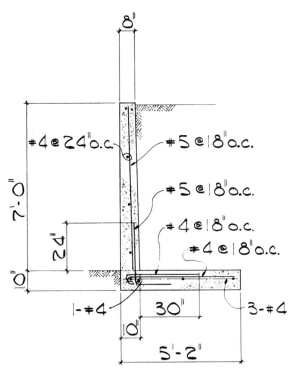

Axial load = 400 lbs./ft.
E.F.P. = 30 p.c.f.
Section 2.3.7

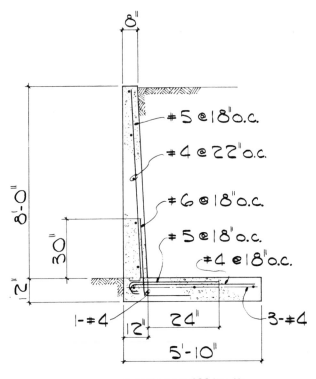

Axial load = 400 lbs./ft.
E.F.P. = 30 p.c.f.
Section 2.3.8

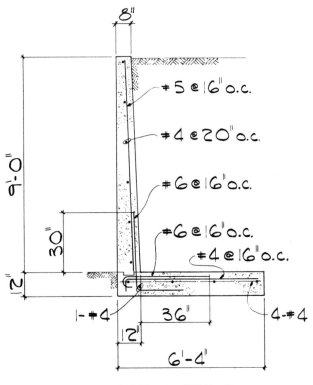

Axial load = 400 lbs./ft.
E.F.P. = 30 p.c.f.
Section 2.3.9

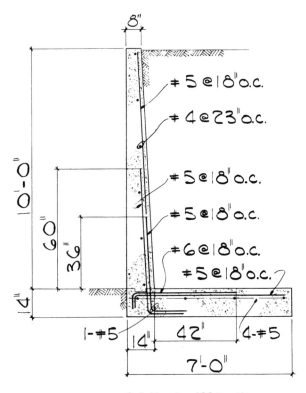

Axial load = 400 lbs./ft.
E.F.P. = 30 p.c.f.
Section 2.3.10

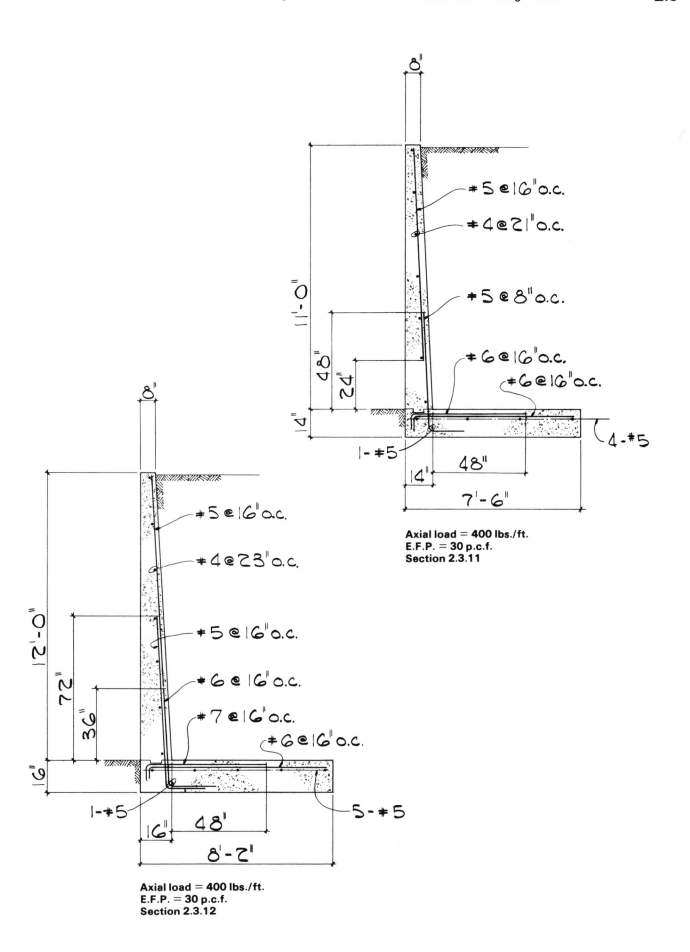

Axial load = **400 lbs./ft.**
E.F.P. = **30 p.c.f.**
Section **2.3.11**

Axial load = **400 lbs./ft.**
E.F.P. = **30 p.c.f.**
Section **2.3.12**

TABLE 2.4 PROPERTY LINE RETAINING WALLS — CONCRETE STEM — SOIL OVER FOOTING
SLOPE = 0 to 1   SURCHARGE = 0 lbs./sq. ft.   AXIAL = 600 lbs./ft.

SOIL PRESSURE DIAGRAM

AXIAL LOAD

Stem Height	3'- 0''	4'- 0''	5'- 0''	6'- 0''	7'- 0''	8'- 0''	9'- 0''	10'- 0''	11'- 0''	12'- 0''
Ft. L ft.	3.667	4.000	4.333	4.833	5.333	6.000	6.500	7.167	7.667	8.333
Ft. T ft.	0.667	0.667	0.833	0.833	0.833	1.000	1.000	1.167	1.167	1.333
Wt ft.	0.667	0.667	0.667	0.667	0.667	0.667	0.667	0.667	0.667	0.667
Wb ft.	0.667	0.667	0.833	0.833	0.833	1.000	1.000	1.167	1.167	1.333
W lbs.	2212	2813	3621	4509	5507	6953	8205	9979	11484	13587
F lbs.	202	327	510	700	920	1215	1500	1870	2220	2667
Mo ft.-lbs.	246	508	992	1595	2403	3645	5000	6962	9005	11852
Mr ft.-lbs.	2952	4298	6232	8888	12216	17706	22875	31043	38456	49831
O.T.R.	11.977	8.458	6.279	5.571	5.083	4.857	4.575	4.459	4.271	4.204
$\overline{X}$ ft.	1.223	1.347	1.447	1.617	1.782	2.022	2.179	2.413	2.565	2.795
e ft.	0.610	0.653	0.720	0.799	0.885	0.978	1.071	1.170	1.269	1.371
Me ft.-lbs.	1349	1837	2605	3605	4874	6799	8791	11678	14571	18632
S ft.^{3}	2.241	2.667	3.130	3.894	4.741	6.000	7.042	8.560	9.796	11.574
S.P.t p.s.f.	1205	1392	1668	1859	2061	2292	2511	2757	2985	3240
S.P.h p.s.f.	1	14	3	7	4	26	14	28	11	21
Friction	0.091	0.116	0.141	0.155	0.167	0.175	0.183	0.187	0.193	0.196

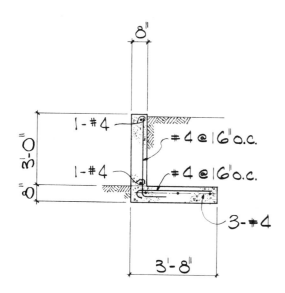

Axial load = 600 lbs./ft.
E.F.P. = 30 p.c.f.
Section 2.4.3

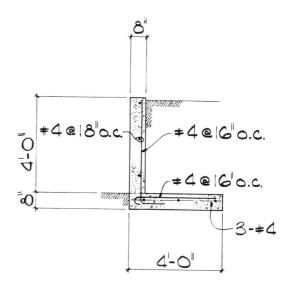

Axial load = 600 lbs./ft.
E.F.P. = 30 p.c.f.
Section 2.4.4

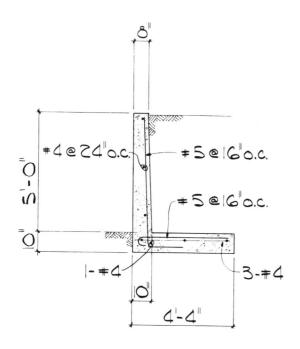

Axial load = 600 lbs./ft.
E.F.P. = 30 p.c.f.
Section 2.4.5

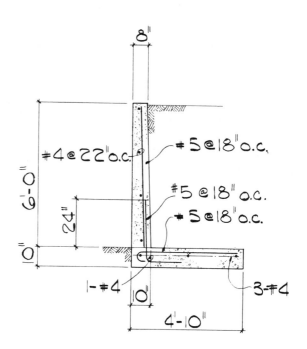

Axial load = 600 lbs./ft.
E.F.P. = 30 p.c.f.
Section 2.4.6

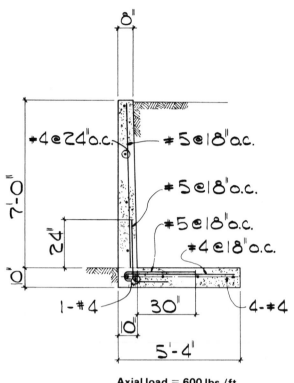

Axial load = 600 lbs./ft.
E.F.P. = 30 p.c.f.
Section 2.4.7

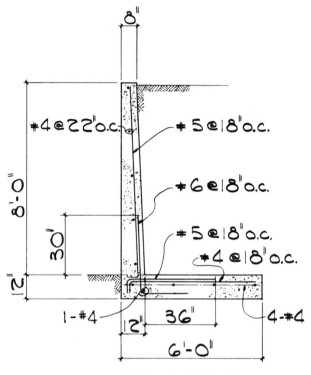

Axial load = 600 lbs./ft.
E.F.P. = 30 p.c.f.
Section 2.4.8

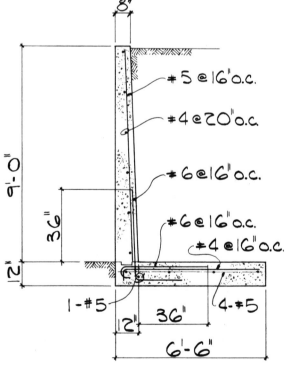

Axial load = 600 lbs./ft.
E.F.P. = 30 p.c.f.
Section 2.4.9

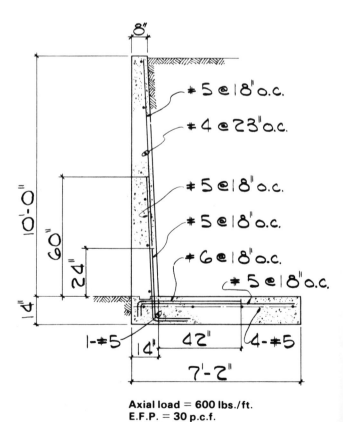

Axial load = 600 lbs./ft.
E.F.P. = 30 p.c.f.
Section 2.4.10

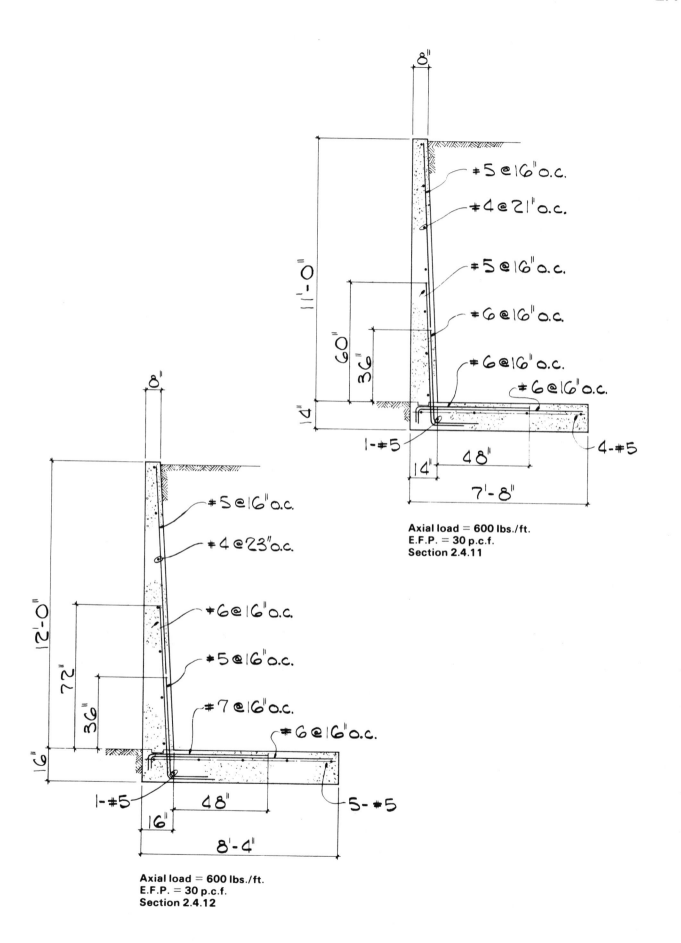

Axial load = 600 lbs./ft.
E.F.P. = 30 p.c.f.
Section 2.4.11

Axial load = 600 lbs./ft.
E.F.P. = 30 p.c.f.
Section 2.4.12

TABLE 2.5	PROPERTY LINE RETAINING WALLS – CONCRETE STEM – SOIL OVER FOOTING
	SLOPE = 0 to 1 SURCHARGE = 0 lbs./sq. ft. AXIAL = 800 lbs./ft.

SOIL PRESSURE DIAGRAM

AXIAL LOAD

Stem Height	3'- 0"	4'- 0"	5'- 0"	6'- 0"	7'- 0"	8'- 0"	9'- 0"	10'- 0"	11'- 0"	12'- 0"
Ft. L ft.	4.500	4.500	4.833	5.167	5.667	6.167	6.667	7.333	7.833	8.500
Ft. T ft.	0.667	0.667	0.833	0.833	0.833	1.000	1.000	1.167	1.167	1.333
Wt ft.	0.667	0.667	0.667	0.667	0.667	0.667	0.667	0.667	0.667	0.667
Wb ft.	0.667	0.667	0.833	0.833	0.833	1.000	1.000	1.167	1.167	1.333
W lbs.	2745	3263	4133	4951	5982	7312	8580	10375	11897	14020
F lbs.	202	327	510	700	920	1215	1500	1870	2220	2667
Mo ft.-lbs.	246	508	992	1595	2403	3645	5000	6962	9005	11852
Mr ft.-lbs.	4380	5427	7731	10163	13795	18735	24094	32530	40170	51862
O.T.R.	17.770	10.680	7.790	6.370	5.740	5.140	4.819	4.672	4.461	4.376
$\overline{X}$ ft.	1.506	1.507	1.630	1.730	1.904	2.064	2.225	2.464	2.620	2.854
e ft.	0.744	0.743	0.786	0.853	0.929	1.019	1.108	1.202	1.297	1.396
Me ft.-lbs.	2043	2424	3250	4222	5559	7454	9506	12474	15430	19575
S ft.^{3}	3.375	3.375	3.894	4.449	5.352	6.338	7.407	8.963	10.227	12.042
S.P.t p.s.f.	1215	1443	1690	1907	2094	2362	2570	2807	3028	3275
S.P.h p.s.f.	5	7	20	9	17	10	4	23	10	24
Friction	0.073	0.100	0.123	0.141	0.154	0.166	0.175	0.180	0.187	0.190

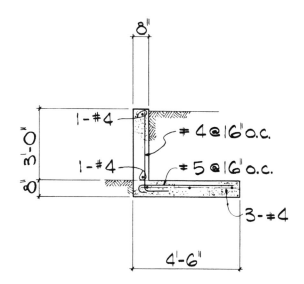

Axial load = 800 lbs./ft.
E.F.P. = 30 p.c.f.
Section 2.5.3

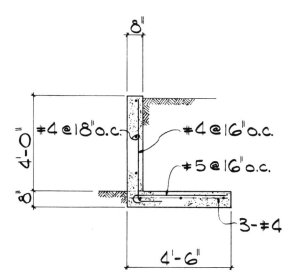

Axial load = 800 lbs./ft.
E.F.P. = 30 p.c.f.
Section 2.5.4

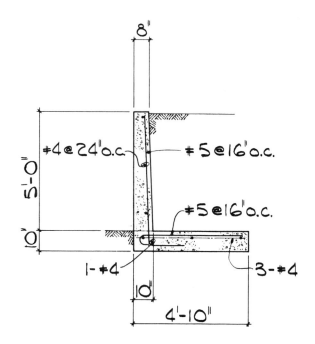

Axial load = 800 lbs./ft.
E.F.P. = 30 p.c.f.
Section 2.5.5

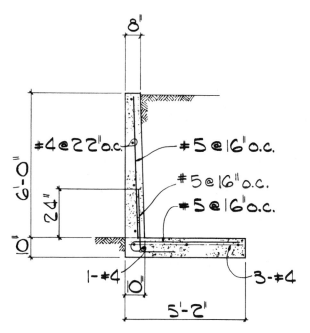

Axial load = 800 lbs./ft.
E.F.P. = 30 p.c.f.
Section 2.5.6

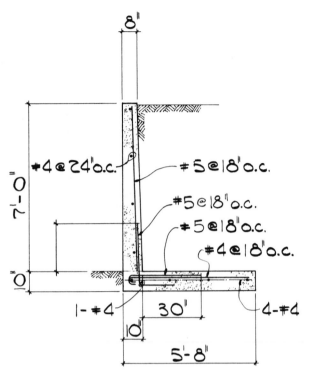

Axial load = 800 lbs./ft.
E.F.P. = 30 p.c.f.
Section 2.5.7

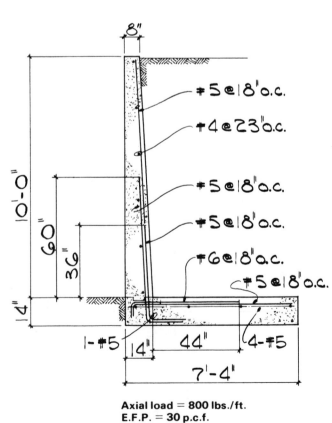

Axial load = 800 lbs./ft.
E.F.P. = 30 p.c.f.
Section 2.5.8

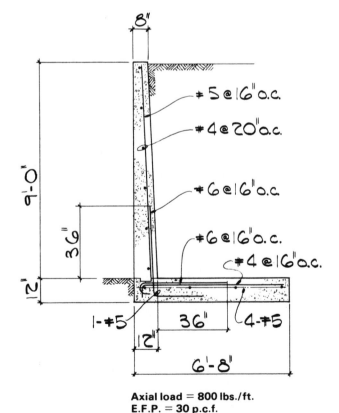

Axial load = 800 lbs./ft.
E.F.P. = 30 p.c.f.
Section 2.5.9

Axial load = 800 lbs./ft.
E.F.P. = 30 p.c.f.
Section 2.5.10

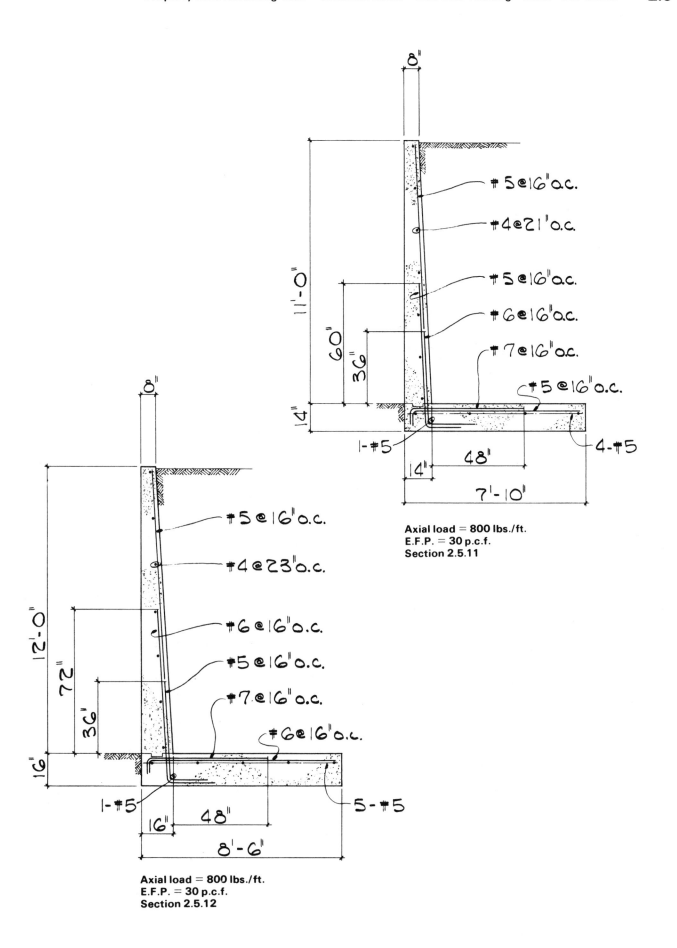

#5@16"o.c.

#4@21"o.c.

#5@16"o.c.

#6@16"o.c.

#7@16"o.c.

#5@16"o.c.

8"

11'-0"

60"

36"

4"

1-#5

4"

48"

4-#5

14"

7'-10"

Axial load = 800 lbs./ft.
E.F.P. = 30 p.c.f.
Section 2.5.11

#5@16"o.c.

#4@23"o.c.

#6@16"o.c.

#5@16"o.c.

#7@16"o.c.

#6@16"o.c.

8"

12'-0"

72"

36"

16"

1-#5

5-#5

16"

48"

8'-6"

Axial load = 800 lbs./ft.
E.F.P. = 30 p.c.f.
Section 2.5.12

TABLE 2.6   PROPERTY LINE RETAINING WALLS – CONCRETE STEM – SOIL OVER FOOTING
SLOPE = 0 to 1   SURCHARGE = 50 lbs./sq. ft.   AXIAL = 0 lbs./ft.

SOIL PRESSURE DIAGRAM

Stem Height	3'- 0''	4'- 0''	5'- 0''	6'- 0''	7'- 0''	8'- 0''	9'- 0''	10'- 0''	11'- 0''	12'- 0''
Ft. L ft.	2.333	3.000	3.667	4.333	5.000	5.667	6.333	6.833	7.500	8.167
Ft. T ft.	0.667	0.833	0.833	0.833	1.000	1.000	1.167	1.167	1.333	1.333
Wt ft.	0.667	0.667	0.667	0.667	0.667	0.667	0.667	0.667	0.667	0.667
Wb ft.	0.667	0.833	0.833	0.833	1.000	1.000	1.167	1.167	1.333	1.333
W lbs.	1178	1939	2776	3757	5031	6320	7943	9335	11286	13173
F lbs.	260	427	602	807	1084	1354	1707	2042	2470	2870
Mo ft.-lbs.	362	759	1270	1972	3071	4287	6068	7940	10568	13236
Mr ft.-lbs.	1288	2706	4707	7505	11589	16478	23144	29295	38880	49410
O.T.R.	3.561	3.567	3.706	3.806	3.774	3.844	3.814	3.690	3.679	3.733
$\overline{X}$ ft.	0.786	1.004	1.238	1.473	1.693	1.929	2.150	2.288	2.509	2.746
e ft.	0.380	0.496	0.595	0.694	0.807	0.904	1.017	1.129	1.241	1.337
Me ft.-lbs.	448	961	1653	2607	4060	5714	8077	10538	14011	17615
S ft.^{3}	0.907	1.500	2.241	3.130	4.167	5.352	6.685	7.782	9.375	11.116
S.P.t p.s.f.	998	1287	1495	1700	1981	2183	2462	2720	2999	3198
S.P.h p.s.f.	11	6	20	34	32	48	46	12	10	28
Friction	0.221	0.220	0.217	0.215	0.215	0.214	0.215	0.219	0.219	0.218

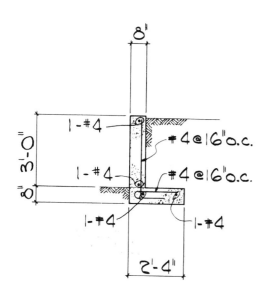

Surcharge load = 50 p.s.f.
E.F.P. = 30 p.c.f.
Section 2.6.3

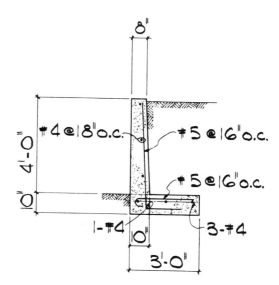

Surcharge load = 50 p.s.f.
E.F.P. = 30 p.c.f.
Section 2.6.4

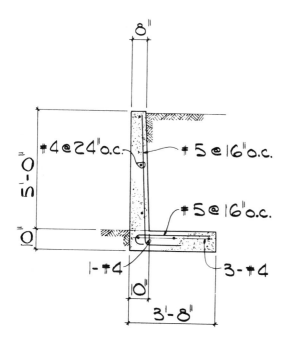

Surcharge load = 50 p.s.f.
E.F.P. = 30 p.c.f.
Section 2.6.5

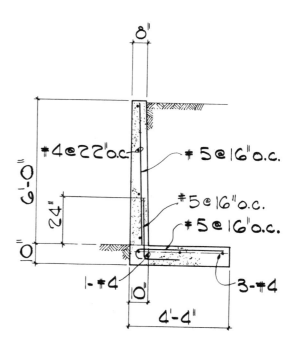

Surcharge load = 50 p.s.f.
E.F.P. = 30 p.c.f.
Section 2.6.6

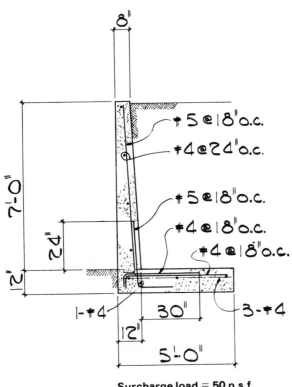

Surcharge load = 50 p.s.f.
E.F.P. = 30 p.c.f.
Section 2.6.7

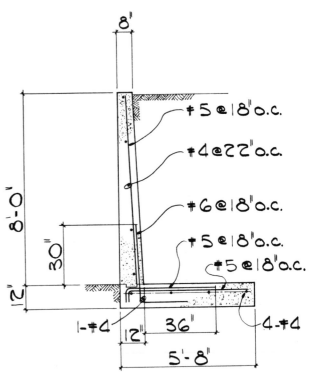

Surcharge load = 50 p.s.f.
E.F.P. = 30 p.c.f.
Section 2.6.8

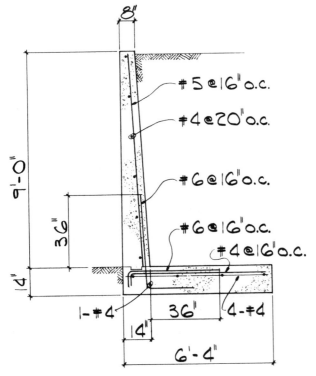

Surcharge load = 50 p.s.f.
E.F.P. = 30 p.c.f.
Section 2.6.9

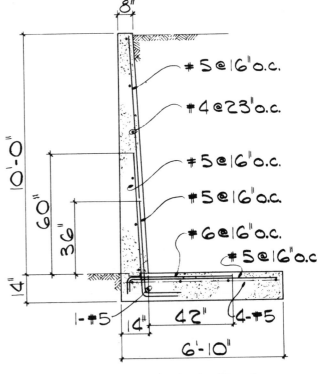

Surcharge load = 50 p.s.f.
E.F.P. = 30 p.c.f.
Section 2.6.10

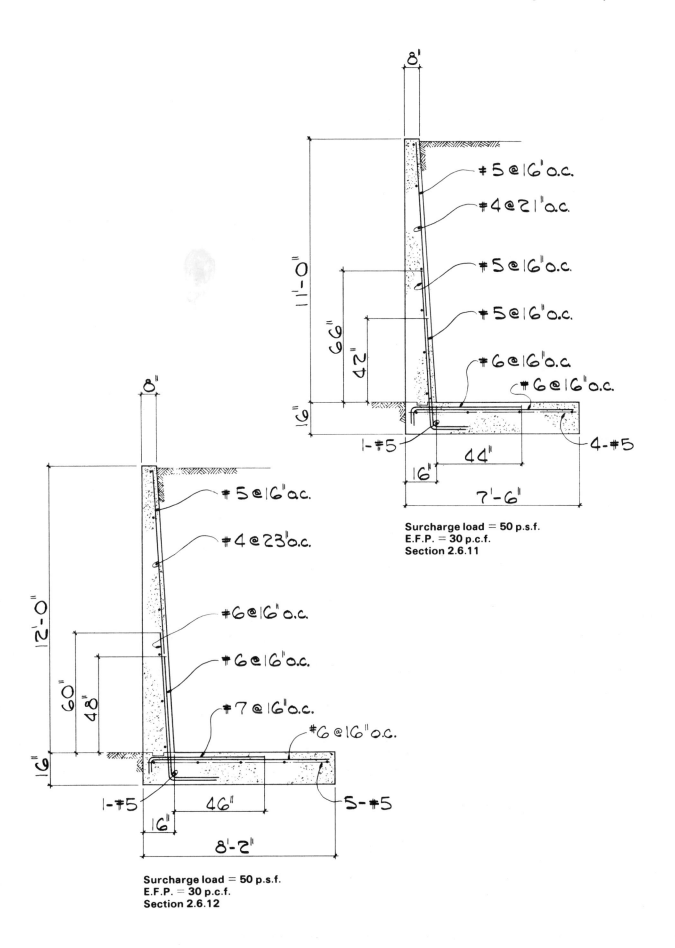

8"

#5 @ 16' o.c.

#4 @ 21" o.c.

#5 @ 16" o.c.

#5 @ 16" o.c.

#6 @ 16" o.c.

#6 @ 16" o.c.

11'-0"

66"

42"

6"

1-#5

16"

44"

4-#5

7'-6"

Surcharge load = 50 p.s.f.
E.F.P. = 30 p.c.f.
Section 2.6.11

8"

#5 @ 16" o.c.

#4 @ 23" o.c.

#6 @ 16" o.c.

#6 @ 16" o.c.

#7 @ 16" o.c.

#6 @ 16" o.c.

12'-0"

60"

48"

6"

1-#5

16"

46"

5-#5

8'-2"

Surcharge load = 50 p.s.f.
E.F.P. = 30 p.c.f.
Section 2.6.12

TABLE 2.7	PROPERTY LINE RETAINING WALLS — CONCRETE STEM — SOIL OVER FOOTING
	SLOPE = 0 to 1 SURCHARGE = 100 lbs./sq. ft. AXIAL = 0 lbs./ft.

SOIL PRESSURE DIAGRAM

Stem Height	3'- 0''	4'- 0''	5'- 0''	6'- 0''	7'- 0''	8'- 0''	9'- 0''	10'- 0''	11'- 0''	12'- 0''
Ft. L ft.	2.667	3.333	4.000	4.500	5.167	5.833	6.500	7.167	7.833	8.500
Ft. T ft.	0.667	0.833	0.833	0.833	1.000	1.000	1.167	1.167	1.333	1.333
Wt ft.	0.667	0.667	0.667	0.667	0.667	0.667	0.667	0.667	0.667	0.667
Wb ft.	0.667	0.833	0.833	0.833	1.000	1.000	1.167	1.167	1.333	1.333
W lbs.	1447	2283	3192	4107	5437	6780	8458	10109	12137	14095
F lbs.	327	510	700	920	1215	1500	1870	2220	2667	3082
Mo ft.-lbs.	508	992	1595	2403	3645	5000	6962	9005	11852	14724
Mr ft.-lbs.	1842	3594	5977	8593	13033	18306	25427	33465	43898	55282
O.T.R.	3.625	3.621	3.747	3.576	3.576	3.661	3.652	3.716	3.704	3.755
$\overline{X}$ ft.	0.922	1.139	1.373	1.507	1.727	1.963	2.183	2.420	2.640	2.878
e ft.	0.411	0.527	0.627	0.743	0.856	0.954	1.067	1.164	1.276	1.372
Me ft.-lbs.	595	1204	2003	3052	4656	6469	9024	11764	15489	19345
S ft.^{3}	1.185	1.852	2.667	3.375	4.449	5.671	7.042	8.560	10.227	12.042
S.P.t p.s.f.	1044	1335	1549	1817	2099	2303	2583	2785	3064	3265
S.P.h p.s.f.	41	35	47	8	6	22	20	36	35	52
Friction	0.226	0.224	0.219	0.224	0.223	0.221	0.221	0.220	0.220	0.219

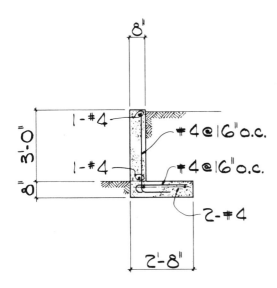

Surcharge load = 100 p.s.f.
E.F.P. = 30 p.c.f.
Section 2.7.3

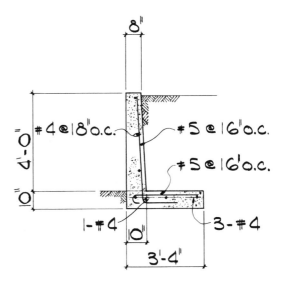

Surcharge load = 100 p.s.f.
E.F.P. = 30 p.c.f.
Section 2.7.4

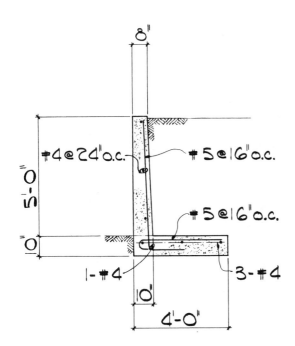

Surcharge load = 100 p.s.f.
E.F.P. = 30 p.c.f.
Section 2.7.5

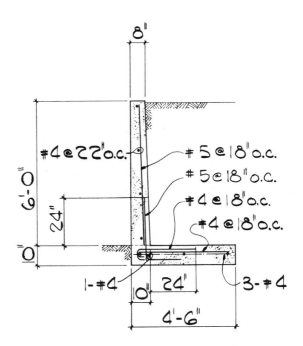

Surcharge load = 100 p.s.f.
E.F.P. = 30 p.c.f.
Section 2.7.6

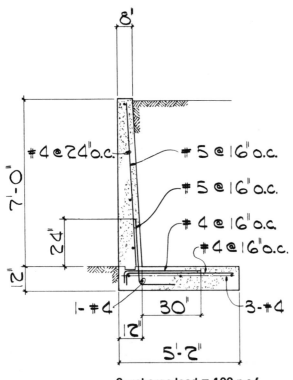

Surcharge load = 100 p.s.f.
E.F.P. = 30 p.c.f.
Section 2.7.7

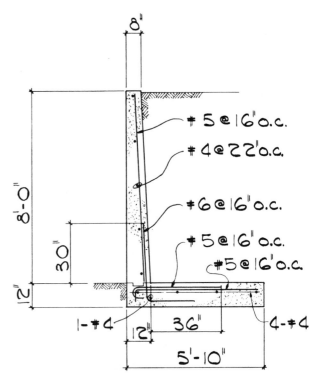

Surcharge load = 100 p.s.f.
E.F.P. = 30 p.c.f.
Section 2.7.8

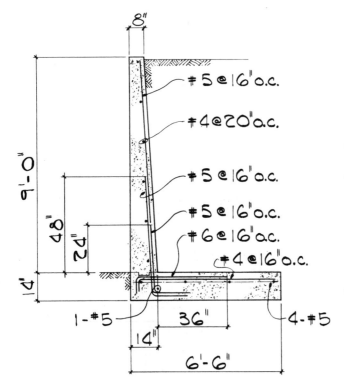

Surcharge load = 100 p.s.f.
E.F.P. = 30 p.c.f.
Section 2.7.9

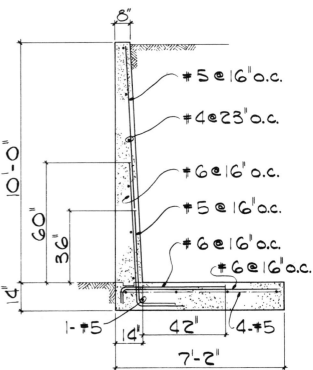

Surcharge load = 100 p.s.f.
E.F.P. = 30 p.c.f.
Section 2.7.10

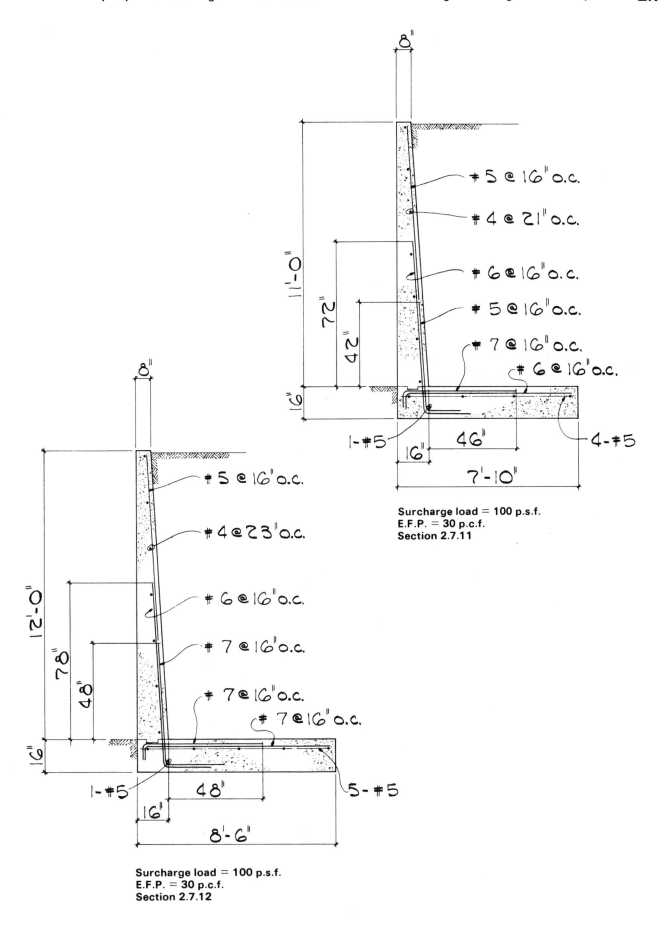

8"

#5 @ 16" o.c.

#4 @ 21" o.c.

#6 @ 16" o.c.

#5 @ 16" o.c.

#7 @ 16" o.c.

#6 @ 16" o.c.

11'-0"

72"

42"

6"

1-#5

16"   46"   4-#5

7'-10"

Surcharge load = 100 p.s.f.
E.F.P. = 30 p.c.f.
Section 2.7.11

8"

#5 @ 16" o.c.

#4 @ 23" o.c.

#6 @ 16" o.c.

#7 @ 16" o.c.

#7 @ 16" o.c.

#7 @ 16" o.c.

12'-0"

78"

48"

6"

1-#5

16"   48"   5-#5

8'-6"

Surcharge load = 100 p.s.f.
E.F.P. = 30 p.c.f.
Section 2.7.12

TABLE 2.8	PROPERTY LINE RETAINING WALLS — CONCRETE STEM — SOIL OVER FOOTING
	SLOPE = 0 to 1 SURCHARGE = 200 lbs./sq. ft. AXIAL = 0 lbs./ft.

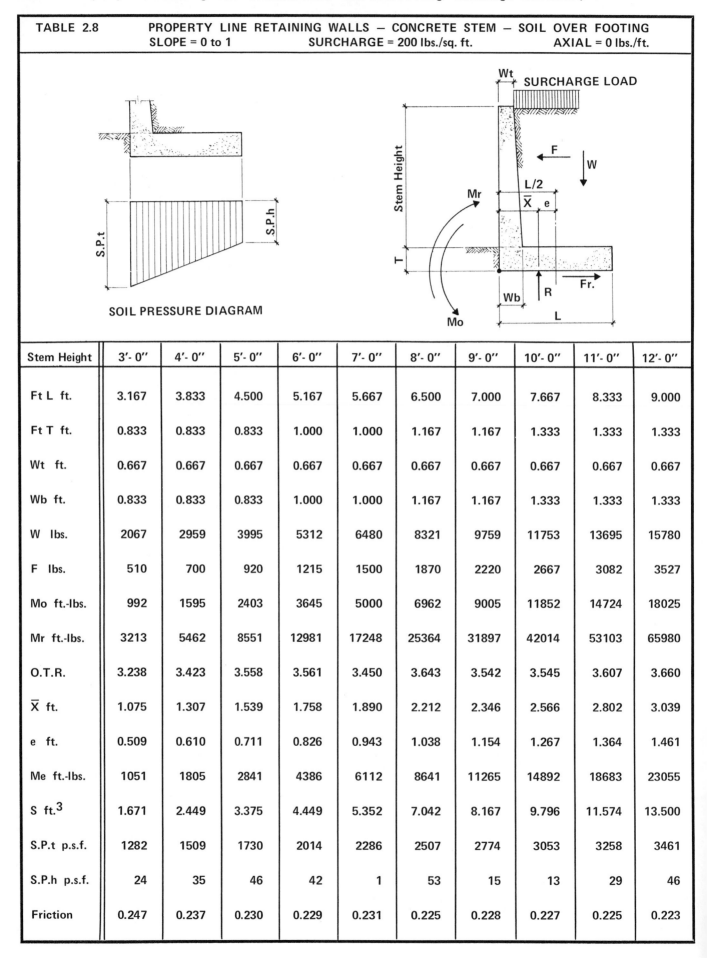

SOIL PRESSURE DIAGRAM

SURCHARGE LOAD

Stem Height	3'- 0"	4'- 0"	5'- 0"	6'- 0"	7'- 0"	8'- 0"	9'- 0"	10'- 0"	11'- 0"	12'- 0"
Ft L ft.	3.167	3.833	4.500	5.167	5.667	6.500	7.000	7.667	8.333	9.000
Ft T ft.	0.833	0.833	0.833	1.000	1.000	1.167	1.167	1.333	1.333	1.333
Wt ft.	0.667	0.667	0.667	0.667	0.667	0.667	0.667	0.667	0.667	0.667
Wb ft.	0.833	0.833	0.833	1.000	1.000	1.167	1.167	1.333	1.333	1.333
W lbs.	2067	2959	3995	5312	6480	8321	9759	11753	13695	15780
F lbs.	510	700	920	1215	1500	1870	2220	2667	3082	3527
Mo ft.-lbs.	992	1595	2403	3645	5000	6962	9005	11852	14724	18025
Mr ft.-lbs.	3213	5462	8551	12981	17248	25364	31897	42014	53103	65980
O.T.R.	3.238	3.423	3.558	3.561	3.450	3.643	3.542	3.545	3.607	3.660
$\overline{X}$ ft.	1.075	1.307	1.539	1.758	1.890	2.212	2.346	2.566	2.802	3.039
e ft.	0.509	0.610	0.711	0.826	0.943	1.038	1.154	1.267	1.364	1.461
Me ft.-lbs.	1051	1805	2841	4386	6112	8641	11265	14892	18683	23055
S ft.^{3}	1.671	2.449	3.375	4.449	5.352	7.042	8.167	9.796	11.574	13.500
S.P.t p.s.f.	1282	1509	1730	2014	2286	2507	2774	3053	3258	3461
S.P.h p.s.f.	24	35	46	42	1	53	15	13	29	46
Friction	0.247	0.237	0.230	0.229	0.231	0.225	0.228	0.227	0.225	0.223

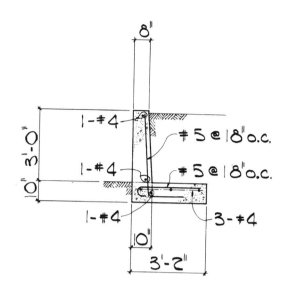

Surcharge load = 200 p.s.f.
E.F.P. = 30 p.c.f.
Section 2.8.3

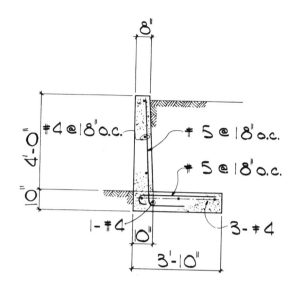

Surcharge load = 200 p.s.f.
E.F.P. = 30 p.c.f.
Section 2.8.4

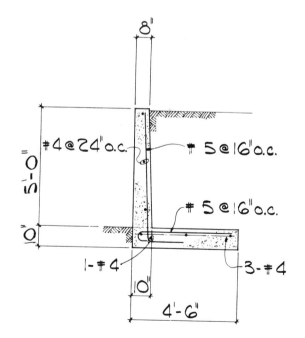

Surcharge load = 200 p.s.f.
E.F.P. = 30 p.c.f.
Section 2.8.5

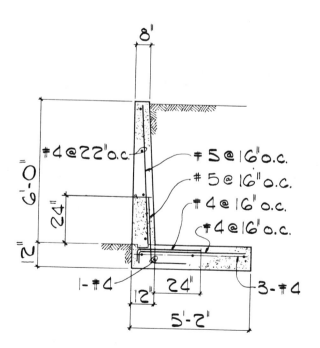

Surcharge load = 200 p.s.f.
E.F.P. = 30 p.c.f.
Section 2.8.6

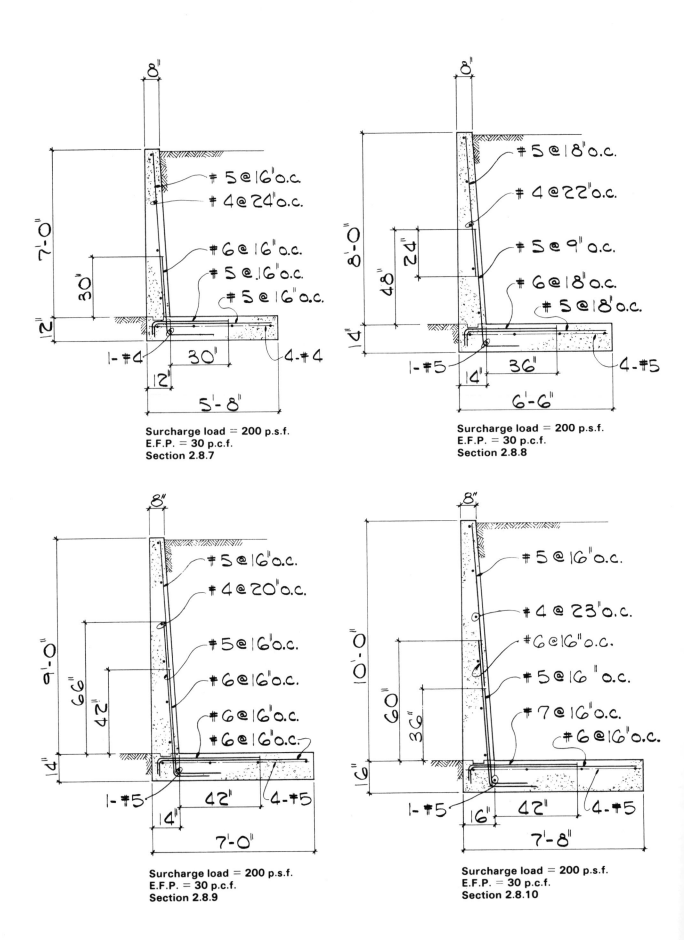

Surcharge load = 200 p.s.f.
E.F.P. = 30 p.c.f.
Section 2.8.7

Surcharge load = 200 p.s.f.
E.F.P. = 30 p.c.f.
Section 2.8.8

Surcharge load = 200 p.s.f.
E.F.P. = 30 p.c.f.
Section 2.8.9

Surcharge load = 200 p.s.f.
E.F.P. = 30 p.c.f.
Section 2.8.10

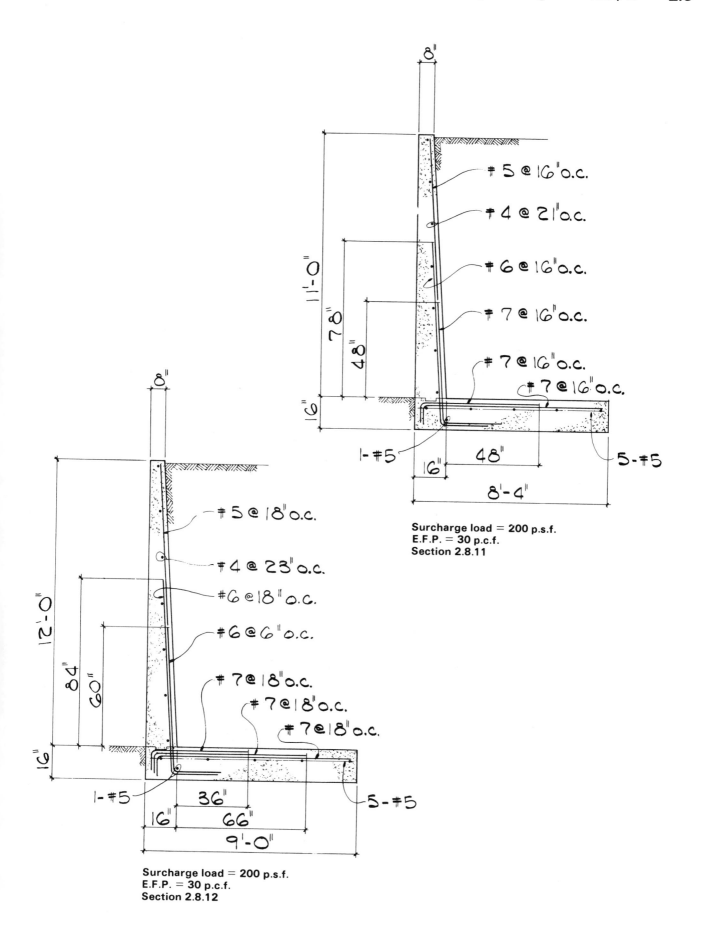

#5 @ 16" o.c.

#4 @ 21" o.c.

#6 @ 16" o.c.

#7 @ 16" o.c.

#7 @ 16" o.c.

#7 @ 16" o.c.

1-#5

5-#5

11'-0"

78"

48"

6"

16"

48"

8'-4"

Surcharge load = 200 p.s.f.
E.F.P. = 30 p.c.f.
Section 2.8.11

#5 @ 18" o.c.

#4 @ 23" o.c.

#6 @ 18" o.c.

#6 @ 6" o.c.

#7 @ 18" o.c.

#7 @ 18" o.c.

#7 @ 18" o.c.

1-#5

5-#5

12'-0"

84"

60"

6"

16"

36"

66"

9'-0"

Surcharge load = 200 p.s.f.
E.F.P. = 30 p.c.f.
Section 2.8.12

TABLE 2.9	PROPERTY LINE RETAINING WALLS — CONCRETE STEM — SOIL OVER FOOTING
	SLOPE = 5 to 1 SURCHARGE = 0 lbs./sq. ft. AXIAL = 0 lbs./ft.

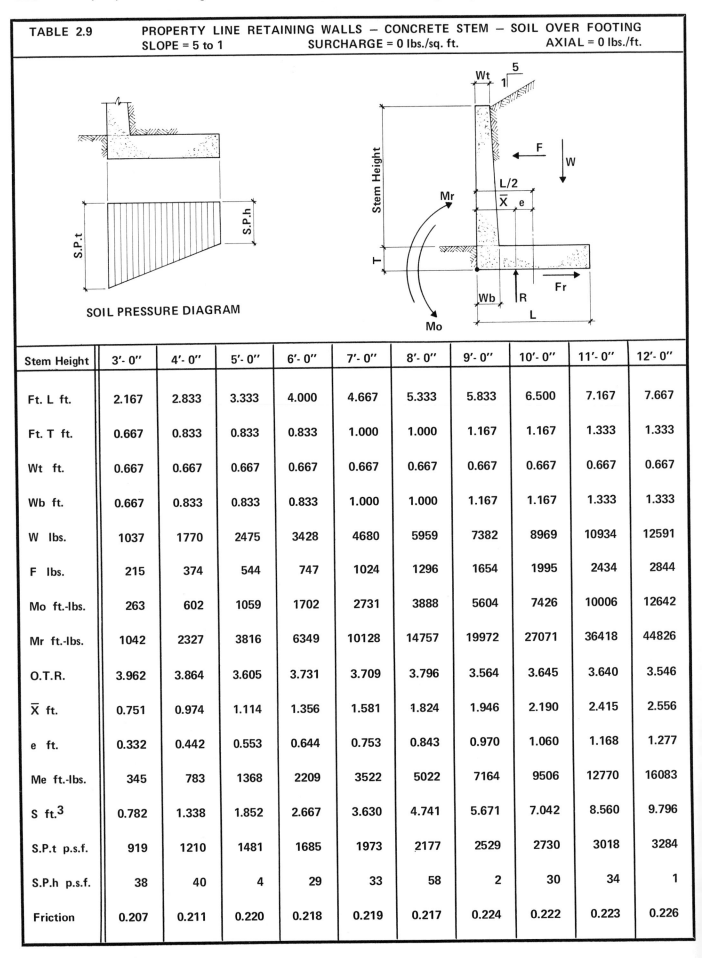

SOIL PRESSURE DIAGRAM

Stem Height	3'- 0"	4'- 0"	5'- 0"	6'- 0"	7'- 0"	8'- 0"	9'- 0"	10'- 0"	11'- 0"	12'- 0"
Ft. L ft.	2.167	2.833	3.333	4.000	4.667	5.333	5.833	6.500	7.167	7.667
Ft. T ft.	0.667	0.833	0.833	0.833	1.000	1.000	1.167	1.167	1.333	1.333
Wt ft.	0.667	0.667	0.667	0.667	0.667	0.667	0.667	0.667	0.667	0.667
Wb ft.	0.667	0.833	0.833	0.833	1.000	1.000	1.167	1.167	1.333	1.333
W lbs.	1037	1770	2475	3428	4680	5959	7382	8969	10934	12591
F lbs.	215	374	544	747	1024	1296	1654	1995	2434	2844
Mo ft.-lbs.	263	602	1059	1702	2731	3888	5604	7426	10006	12642
Mr ft.-lbs.	1042	2327	3816	6349	10128	14757	19972	27071	36418	44826
O.T.R.	3.962	3.864	3.605	3.731	3.709	3.796	3.564	3.645	3.640	3.546
$\overline{X}$ ft.	0.751	0.974	1.114	1.356	1.581	1.824	1.946	2.190	2.415	2.556
e ft.	0.332	0.442	0.553	0.644	0.753	0.843	0.970	1.060	1.168	1.277
Me ft.-lbs.	345	783	1368	2209	3522	5022	7164	9506	12770	16083
S ft.^{3}	0.782	1.338	1.852	2.667	3.630	4.741	5.671	7.042	8.560	9.796
S.P.t p.s.f.	919	1210	1481	1685	1973	2177	2529	2730	3018	3284
S.P.h p.s.f.	38	40	4	29	33	58	2	30	34	1
Friction	0.207	0.211	0.220	0.218	0.219	0.217	0.224	0.222	0.223	0.226

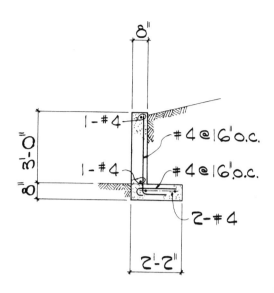

Soil slope = 5 to 1
E.F.P. = 32 p.c.f.
Section 2.9.3

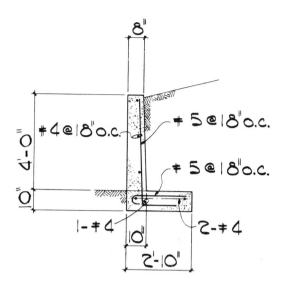

Soil slope = 5 to 1
E.F.P. = 32 p.c.f.
Section 2.9.4

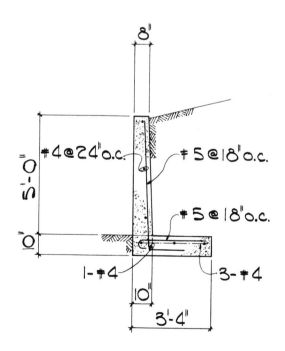

Soil slope = 5 to 1
E.F.P. = 32 p.c.f.
Section 2.9.5

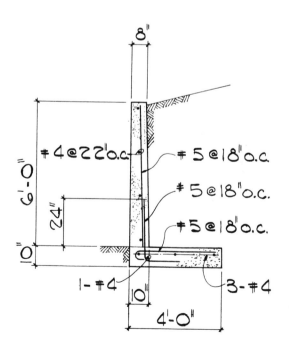

Soil slope = 5 to 1
E.F.P. = 32 p.c.f.
Section 2.9.6

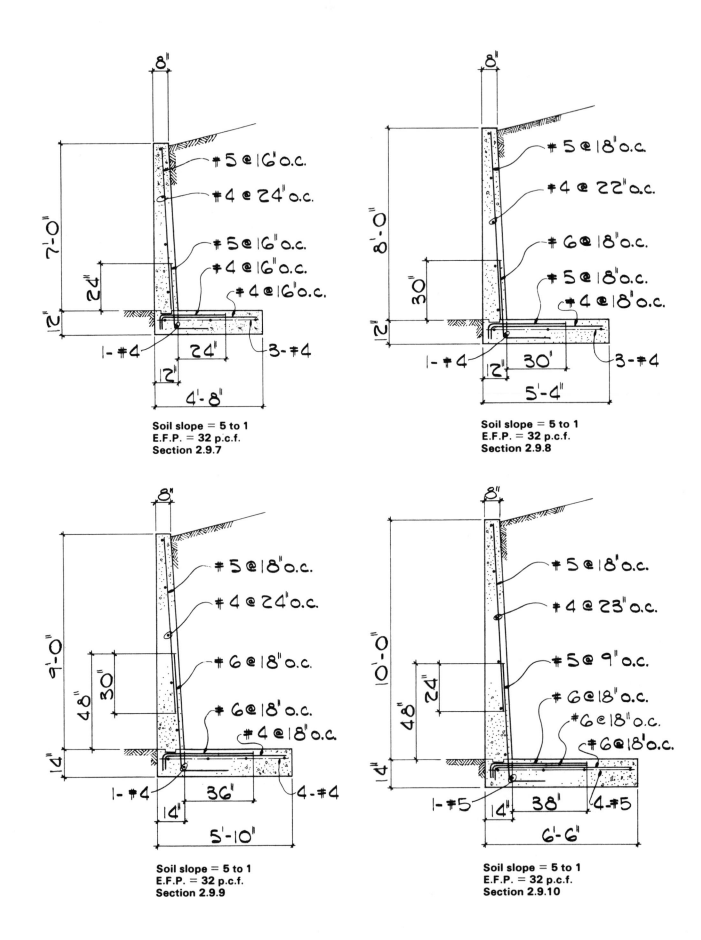

Soil slope = 5 to 1
E.F.P. = 32 p.c.f.
Section 2.9.7

Soil slope = 5 to 1
E.F.P. = 32 p.c.f.
Section 2.9.8

Soil slope = 5 to 1
E.F.P. = 32 p.c.f.
Section 2.9.9

Soil slope = 5 to 1
E.F.P. = 32 p.c.f.
Section 2.9.10

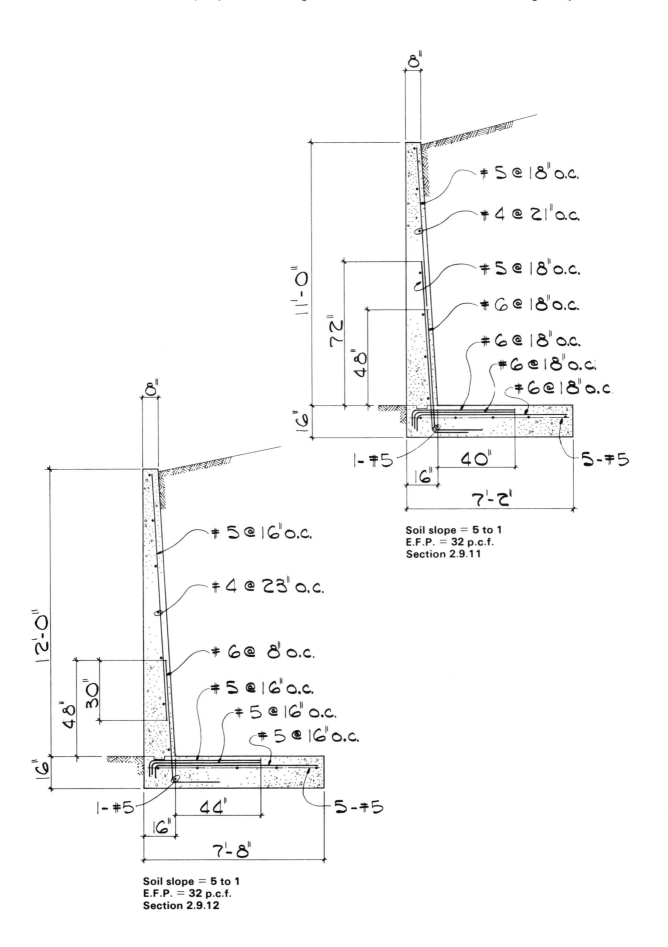

#5 @ 18" o.c.

#4 @ 21" o.c.

#5 @ 18" o.c.

#6 @ 18" o.c.

#6 @ 18" o.c.

#6 @ 18" o.c.

#6 @ 18" o.c.

8"

11'-0"

72"

48"

6"

1-#5

40"

16"

5-#5

7'-2"

Soil slope = 5 to 1
E.F.P. = 32 p.c.f.
Section 2.9.11

#5 @ 16" o.c.

#4 @ 23" o.c.

#6 @ 8" o.c.

#5 @ 16" o.c.

#5 @ 16" o.c.

#5 @ 16" o.c.

8"

12'-0"

48"

30"

6"

1-#5

44"

16"

5-#5

7'-8"

Soil slope = 5 to 1
E.F.P. = 32 p.c.f.
Section 2.9.12

TABLE 2.10	PROPERTY LINE RETAINING WALLS — CONCRETE STEM — SOIL OVER FOOTING
	SLOPE = 4 to 1 SURCHARGE = 0 lbs./sq. ft. AXIAL = 0 lbs./ft.

SOIL PRESSURE DIAGRAM

Stem Height	3'- 0"	4'- 0"	5'- 0"	6'- 0"	7'- 0"	8'- 0"	9'- 0"	10'- 0"	11'- 0"	12'- 0"
Ft. L ft.	2.167	2.833	3.500	4.167	4.833	5.500	6.167	6.667	7.333	8.000
Ft. T ft.	0.667	0.833	0.833	0.833	1.000	1.000	1.167	1.167	1.333	1.333
Wt ft.	0.667	0.667	0.667	0.667	0.667	0.667	0.667	0.667	0.667	0.667
Wb ft.	0.667	0.833	0.833	0.833	1.000	1.000	1.167	1.167	1.333	1.333
W lbs.	1047	1790	2621	3609	4903	6224	7892	9325	11345	13312
F lbs.	235	409	595	817	1120	1417	1809	2182	2662	3111
Mo ft.-lbs.	288	659	1158	1861	2987	4252	6130	8122	10944	13827
Mr ft.-lbs.	1054	2357	4264	6995	11036	15954	22701	28956	38776	49675
O.T.R.	3.666	3.579	3.683	3.758	3.695	3.752	3.703	3.565	3.543	3.593
$\overline{X}$ ft.	0.732	0.949	1.185	1.422	1.642	1.880	2.100	2.234	2.453	2.693
e ft.	0.351	0.467	0.565	0.661	0.775	0.870	0.984	1.099	1.213	1.307
Me ft.-lbs.	368	836	1481	2385	3800	5413	7764	10250	13765	17401
S ft.^{3}	0.782	1.338	2.042	2.894	3.894	5.042	6.338	7.407	8.963	10.667
S.P.t p.s.f.	954	1257	1474	1691	1990	2205	2505	2782	3083	3295
S.P.h p.s.f.	13	6	24	42	39	58	55	15	11	33
Friction	0.225	0.228	0.227	0.226	0.228	0.228	0.229	0.234	0.235	0.234

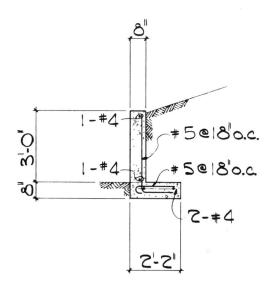

Soil slope = 4 to 1
E.F.P. = 35 p.c.f.
Section 2.10.3

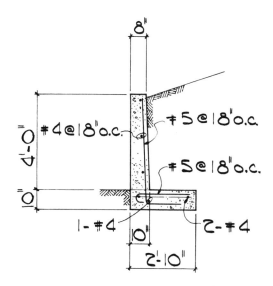

Soil slope = 4 to 1
E.F.P. = 35 p.c.f.
Section 2.10.4

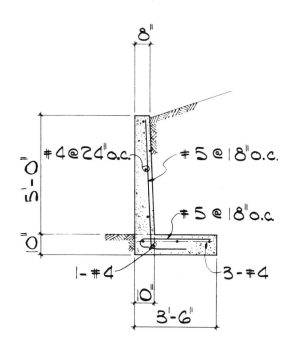

Soil slope = 4 to 1
E.F.P. = 35 p.c.f.
Section 2.10.5

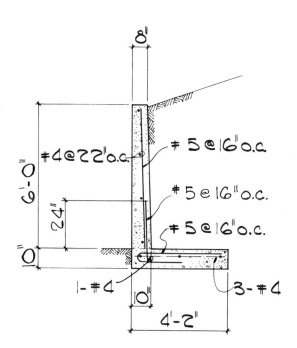

Soil slope = 4 to 1
E.F.P. = 35 p.c.f.
Section 2.10.6

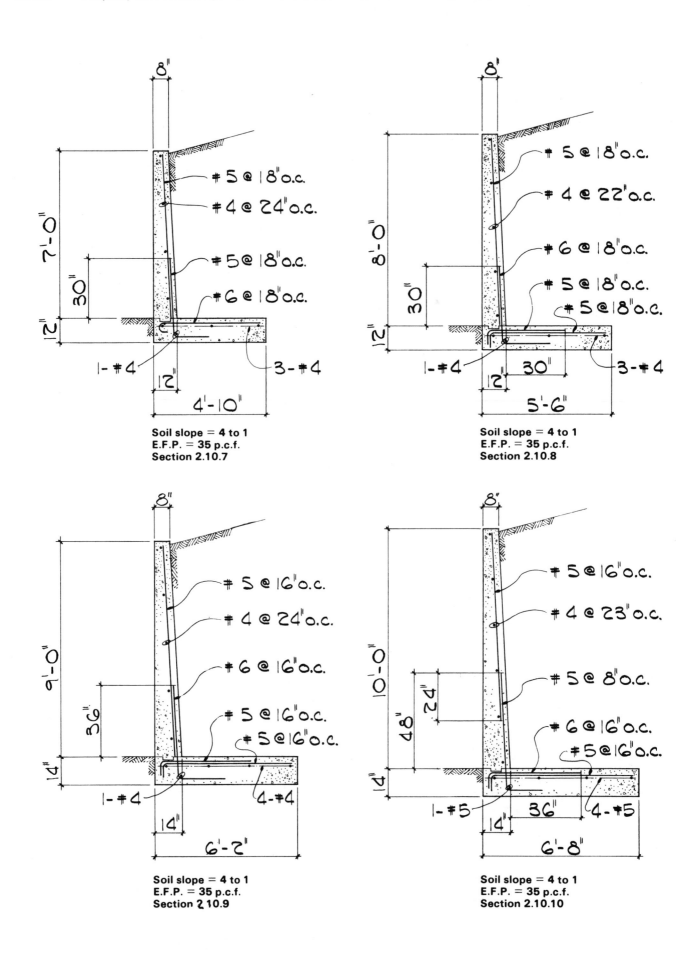

Section 2.10.7

8"

7'-0"

30"

12"

#5 @ 18" o.c.
#4 @ 24" o.c.
#5 @ 18" o.c.
#6 @ 18" o.c.

1-#4

3-#4

12"

4'-10"

Soil slope = 4 to 1
E.F.P. = 35 p.c.f.
Section 2.10.7

Section 2.10.8

8"

8'-0"

30"

12"

#5 @ 18" o.c.
#4 @ 22" o.c.
#6 @ 18" o.c.
#5 @ 18" o.c.
#5 @ 18" o.c.

1-#4

30"

3-#4

12"

5'-6"

Soil slope = 4 to 1
E.F.P. = 35 p.c.f.
Section 2.10.8

Section 2.10.9

8"

9'-0"

36"

14"

#5 @ 16" o.c.
#4 @ 24" o.c.
#6 @ 16" o.c.
#5 @ 16" o.c.
#5 @ 16" o.c.

1-#4

4-#4

14"

6'-2"

Soil slope = 4 to 1
E.F.P. = 35 p.c.f.
Section 2.10.9

Section 2.10.10

8"

10'-0"

48"

24"

14"

#5 @ 16" o.c.
#4 @ 23" o.c.
#5 @ 8" o.c.
#6 @ 16" o.c.
#5 @ 16" o.c.

1-#5

36"

4-#5

14"

6'-8"

Soil slope = 4 to 1
E.F.P. = 35 p.c.f.
Section 2.10.10

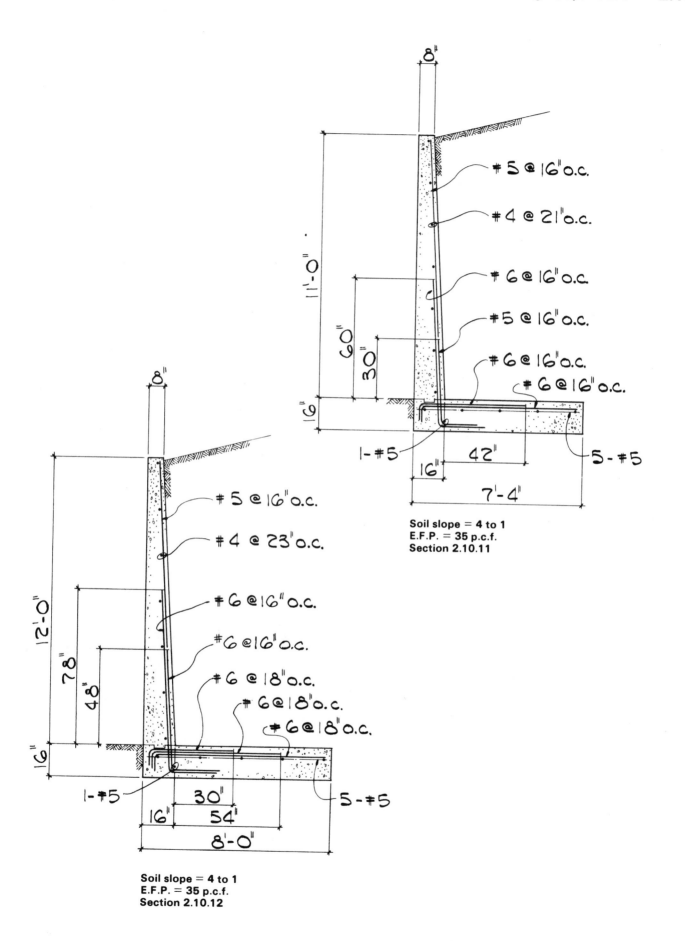

#5 @ 16" o.c.

#4 @ 21" o.c.

#6 @ 16" o.c.

#5 @ 16" o.c.

#6 @ 16" o.c.

#6 @ 16" o.c.

11'-0"

60"

30"

6"

1-#5

42"

16"

5-#5

7'-4"

Soil slope = 4 to 1
E.F.P. = 35 p.c.f.
Section 2.10.11

#5 @ 16" o.c.

#4 @ 23" o.c.

#6 @ 16" o.c.

#6 @ 16" o.c.

#6 @ 18" o.c.

#6 @ 18" o.c.

#6 @ 18" o.c.

12'-0"

78"

48"

6"

1-#5

30"

16"

54"

5-#5

8'-0"

Soil slope = 4 to 1
E.F.P. = 35 p.c.f.
Section 2.10.12

TABLE 2.11	PROPERTY LINE RETAINING WALLS — CONCRETE STEM — SOIL OVER FOOTING
	SLOPE = 3 to 1 SURCHARGE = 0 lbs./sq. ft. AXIAL = 0 lbs./ft.

SOIL PRESSURE DIAGRAM

Stem Height	3'- 0"	4'- 0"	5'- 0"	6'- 0"	7'- 0"	8'- 0"	9'- 0"	10'- 0"	11'- 0"	12'- 0"
Ft. L ft.	2.333	3.000	3.667	4.167	4.833	5.500	6.167	6.833	7.500	8.167
Ft. T ft.	0.667	0.833	0.833	0.833	1.000	1.000	1.167	1.167	1.333	1.333
Wt ft.	0.667	0.667	0.667	0.667	0.667	0.667	0.667	0.667	0.667	0.667
Wb ft.	0.667	0.833	0.833	0.833	1.000	1.000	1.167	1.167	1.333	1.333
W lbs.	1137	1917	2787	3678	5000	6353	8059	9755	11845	13883
F lbs.	255	444	647	887	1216	1539	1964	2369	2890	3378
Mo ft.-lbs.	312	715	1257	2021	3243	4617	6655	8819	11882	15012
Mr ft.-lbs.	1243	2696	4790	7161	11305	16360	23284	31252	41668	53211
O.T.R.	3.980	3.771	3.811	3.544	3.486	3.543	3.499	3.544	3.507	3.544
$\overline{X}$ ft.	0.818	1.033	1.268	1.398	1.613	1.848	2.063	2.300	2.515	2.751
e ft.	0.348	0.467	0.566	0.686	0.804	0.902	1.020	1.117	1.235	1.332
Me ft.-lbs.	396	894	1577	2522	4020	5728	8219	10895	14631	18490
S ft.^{3}	0.907	1.500	2.241	2.894	3.894	5.042	6.338	7.782	9.375	11.116
S.P.t p.s.f.	923	1235	1464	1754	2067	2291	2604	2827	3140	3363
S.P.h p.s.f.	51	43	56	11	2	19	10	28	19	37
Friction	0.225	0.232	0.232	0.241	0.243	0.242	0.244	0.243	0.244	0.243

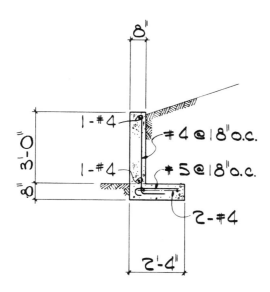

Soil slope = 3 to 1
E.F.P. = 38 p.c.f.
Section 2.11.3

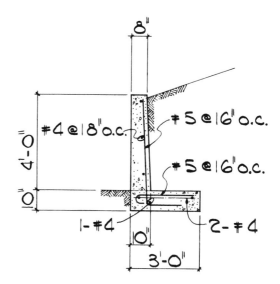

Soil slope = 3 to 1
E.F.P. = 38 p.c.f.
Section 2.11.4

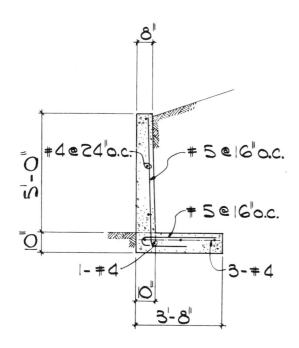

Soil slope = 3 to 1
E.F.P. = 38 p.c.f.
Section 2.11.5

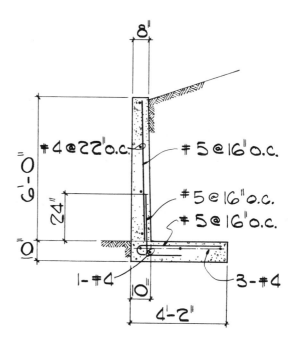

Soil slope = 3 to 1
E.F.P. = 38 p.c.f.
Section 2.11.6

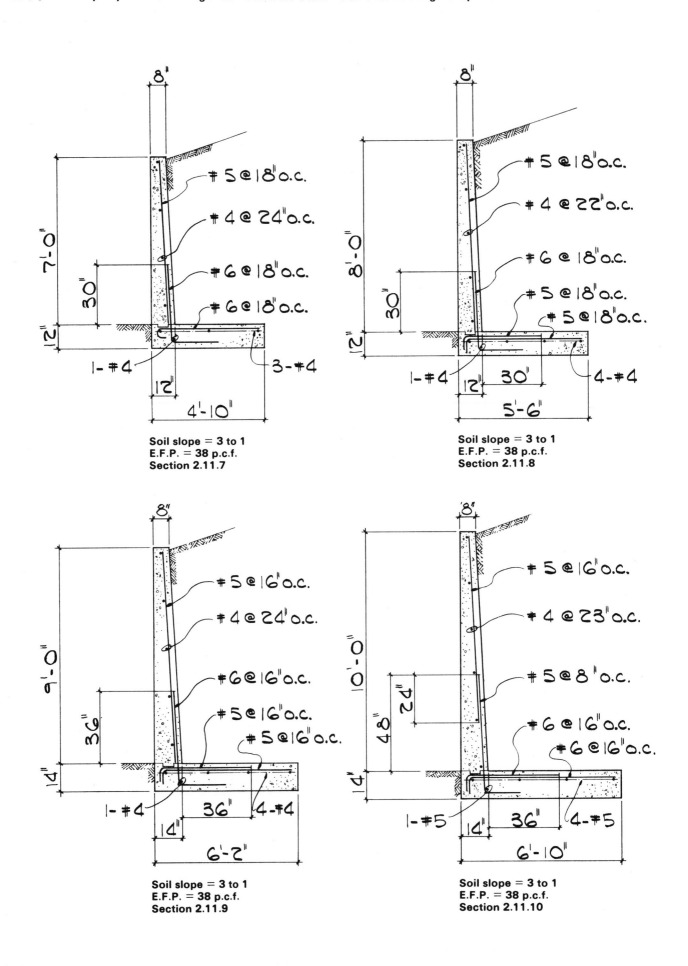

Soil slope = 3 to 1
E.F.P. = 38 p.c.f.
Section 2.11.7

Soil slope = 3 to 1
E.F.P. = 38 p.c.f.
Section 2.11.8

Soil slope = 3 to 1
E.F.P. = 38 p.c.f.
Section 2.11.9

Soil slope = 3 to 1
E.F.P. = 38 p.c.f.
Section 2.11.10

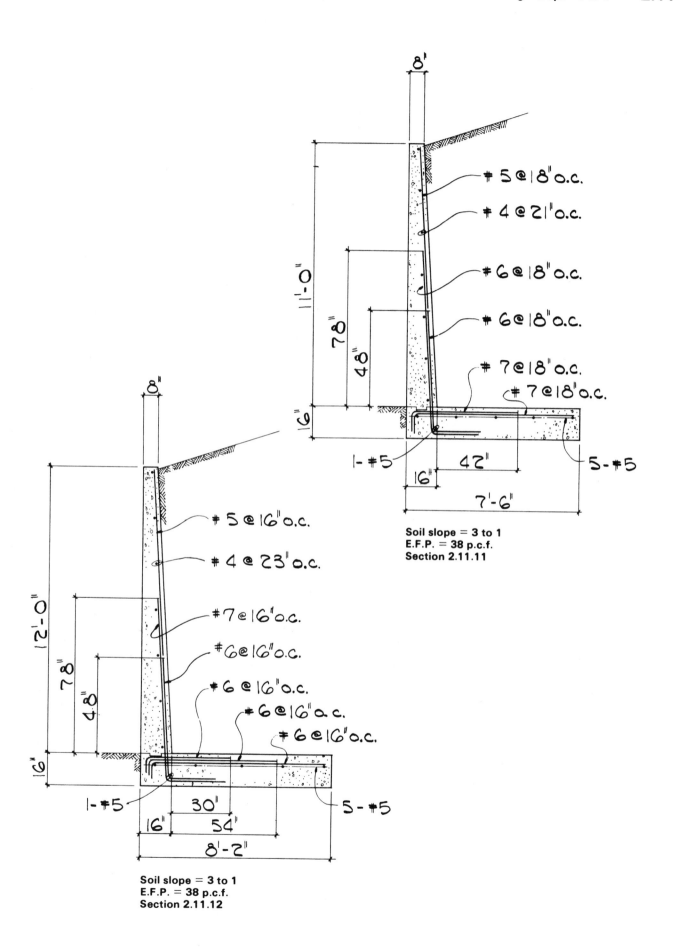

8"

#5 @ 18" o.c.

#4 @ 21" o.c.

#6 @ 18" o.c.

#6 @ 18" o.c.

#7 @ 18" o.c.

#7 @ 18" o.c.

11'-0"

78"

48"

6"

1-#5

42"

5-#5

16"

7'-6"

Soil slope = 3 to 1
E.F.P. = 38 p.c.f.
Section 2.11.11

8"

#5 @ 16" o.c.

#4 @ 23" o.c.

#7 @ 16" o.c.

#6 @ 16" o.c.

#6 @ 16" o.c.

#6 @ 16" o.c.

#6 @ 16" o.c.

12'-0"

78"

48"

6"

1-#5

30"

5-#5

16"

54"

8'-2"

Soil slope = 3 to 1
E.F.P. = 38 p.c.f.
Section 2.11.12

TABLE 2.12	PROPERTY LINE RETAINING WALLS − CONCRETE STEM − SOIL OVER FOOTING
	SLOPE = 2 to 1 SURCHARGE = 0 lbs./sq. ft. AXIAL = 0 lbs./ft.

SOIL PRESSURE DIAGRAM

Stem Height	3'- 0"	4'- 0"	5'- 0"	6'- 0"	7'- 0"	8'- 0"	9'- 0"	10'- 0"	11'- 0"	12'- 0"
Ft. L ft.	2.333	3.000	3.667	4.333	5.000	5.667	6.333	7.000	7.667	8.333
Ft. T ft.	0.667	0.833	0.833	0.833	1.000	1.167	1.167	1.333	1.333	1.333
Wt ft.	0.667	0.667	0.667	0.667	0.667	0.667	0.667	0.667	0.667	0.667
Wb ft.	0.667	0.833	0.833	0.833	1.000	1.167	1.167	1.333	1.333	1.333
W lbs.	1167	1976	2883	3961	5362	6975	8604	10619	12609	14768
F lbs.	289	502	732	1004	1376	1807	2222	2762	3270	3822
Mo ft.-lbs.	353	809	1423	2287	3669	5520	7531	10433	13445	16988
Mr ft.-lbs.	1289	2807	5006	8130	12708	18743	25849	35269	45881	58433
O.T.R.	3.648	3.469	3.519	3.556	3.463	3.395	3.432	3.381	3.413	3.440
$\overline{X}$ ft.	0.801	1.011	1.243	1.475	1.686	1.896	2.129	2.339	2.572	2.806
e ft.	0.365	0.489	0.590	0.691	0.814	0.938	1.038	1.161	1.261	1.360
Me ft.-lbs.	426	966	1703	2738	4367	6541	8929	12331	15898	20088
S ft.^{3}	0.907	1.500	2.241	3.130	4.167	5.352	6.685	8.167	9.796	11.574
S.P.t p.s.f.	970	1302	1546	1789	2120	2453	2694	3027	3267	3508
S.P.h p.s.f.	30	15	27	39	24	9	23	7	22	37
Friction	0.248	0.254	0.254	0.253	0.257	0.259	0.258	0.260	0.259	0.259

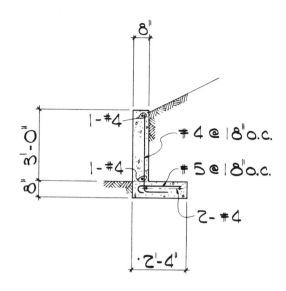

Soil slope = 2 to 1
E.F.P. = 43 p.c.f.
Section 2.12.3

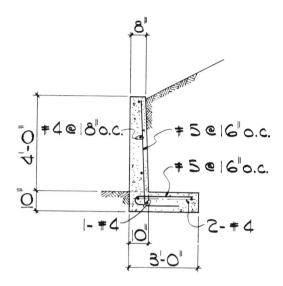

Soil slope = 2 to 1
E.F.P. = 43 p.c.f.
Section 2.12.4

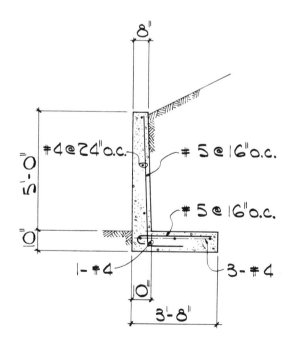

Soil slope = 2 to 1
E.F.P. = 43 p.c.f.
Section 2.12.5

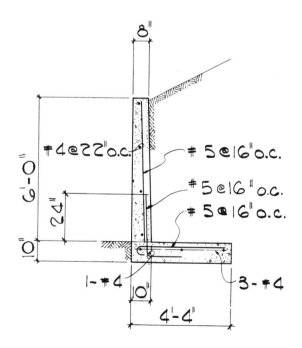

Soil slope = 2 to 1
E.F.P. = 43 p.c.f.
Section 2.12.6

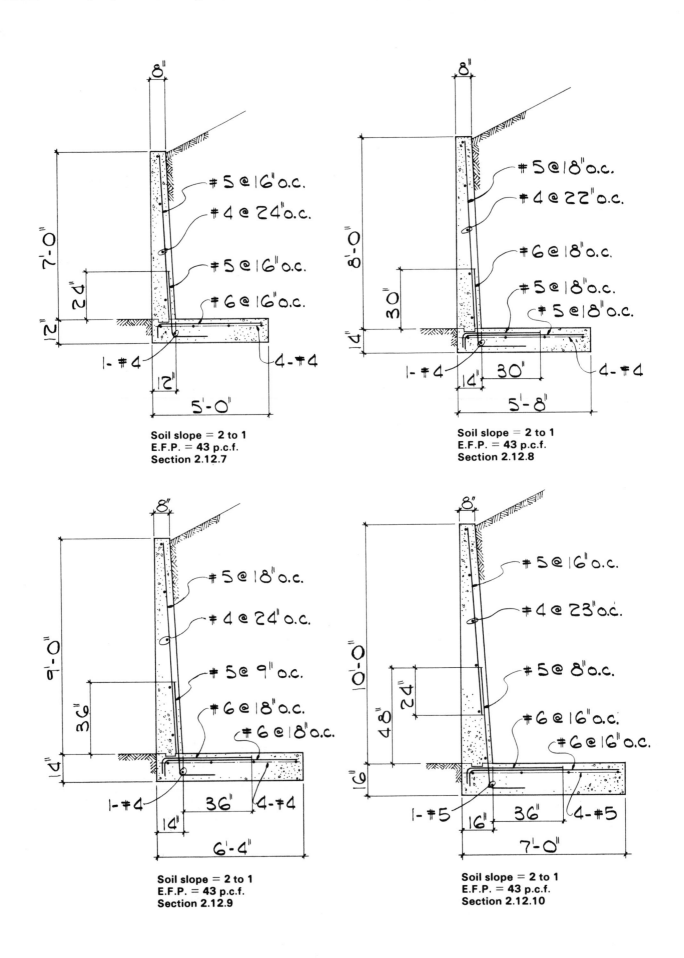

Soil slope = 2 to 1
E.F.P. = 43 p.c.f.
Section 2.12.7

Soil slope = 2 to 1
E.F.P. = 43 p.c.f.
Section 2.12.8

Soil slope = 2 to 1
E.F.P. = 43 p.c.f.
Section 2.12.9

Soil slope = 2 to 1
E.F.P. = 43 p.c.f.
Section 2.12.10

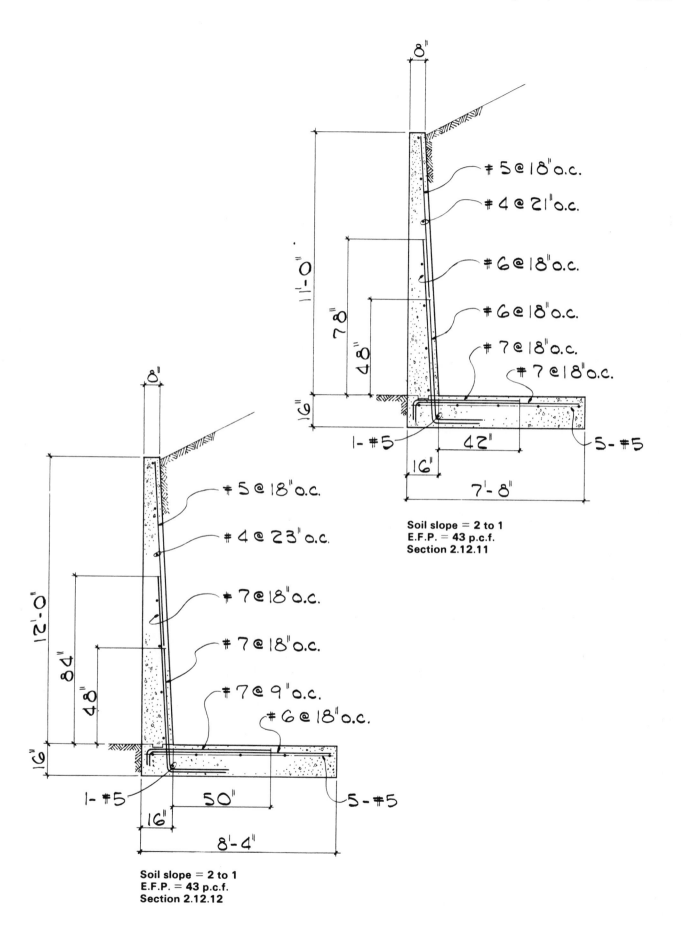

Soil slope = 2 to 1
E.F.P. = 43 p.c.f.
Section 2.12.11

Soil slope = 2 to 1
E.F.P. = 43 p.c.f.
Section 2.12.12

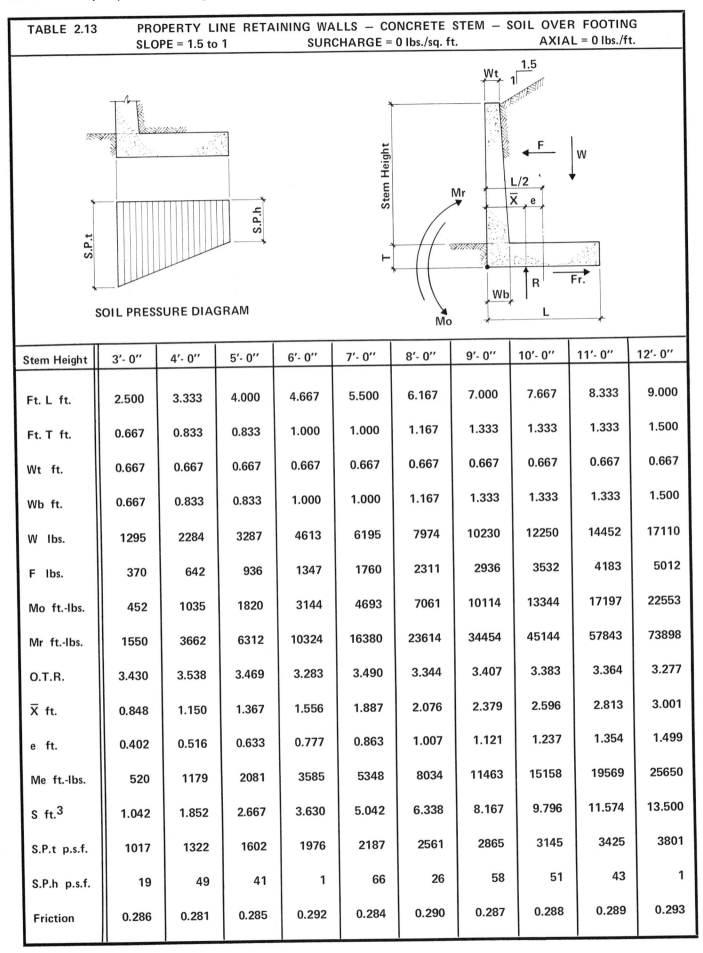

TABLE 2.13   PROPERTY LINE RETAINING WALLS – CONCRETE STEM – SOIL OVER FOOTING
SLOPE = 1.5 to 1   SURCHARGE = 0 lbs./sq. ft.   AXIAL = 0 lbs./ft.

SOIL PRESSURE DIAGRAM

Stem Height	3'- 0"	4'- 0"	5'- 0"	6'- 0"	7'- 0"	8'- 0"	9'- 0"	10'- 0"	11'- 0"	12'- 0"
Ft. L ft.	2.500	3.333	4.000	4.667	5.500	6.167	7.000	7.667	8.333	9.000
Ft. T ft.	0.667	0.833	0.833	1.000	1.000	1.167	1.333	1.333	1.333	1.500
Wt ft.	0.667	0.667	0.667	0.667	0.667	0.667	0.667	0.667	0.667	0.667
Wb ft.	0.667	0.833	0.833	1.000	1.000	1.167	1.333	1.333	1.333	1.500
W lbs.	1295	2284	3287	4613	6195	7974	10230	12250	14452	17110
F lbs.	370	642	936	1347	1760	2311	2936	3532	4183	5012
Mo ft.-lbs.	452	1035	1820	3144	4693	7061	10114	13344	17197	22553
Mr ft.-lbs.	1550	3662	6312	10324	16380	23614	34454	45144	57843	73898
O.T.R.	3.430	3.538	3.469	3.283	3.490	3.344	3.407	3.383	3.364	3.277
$\overline{X}$ ft.	0.848	1.150	1.367	1.556	1.887	2.076	2.379	2.596	2.813	3.001
e ft.	0.402	0.516	0.633	0.777	0.863	1.007	1.121	1.237	1.354	1.499
Me ft.-lbs.	520	1179	2081	3585	5348	8034	11463	15158	19569	25650
S ft.^{3}	1.042	1.852	2.667	3.630	5.042	6.338	8.167	9.796	11.574	13.500
S.P.t p.s.f.	1017	1322	1602	1976	2187	2561	2865	3145	3425	3801
S.P.h p.s.f.	19	49	41	1	66	26	58	51	43	1
Friction	0.286	0.281	0.285	0.292	0.284	0.290	0.287	0.288	0.289	0.293

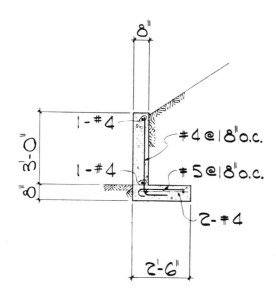

Soil slope = 1½ to 1
E.F.P. = 55 p.c.f.
Section 2.13.3

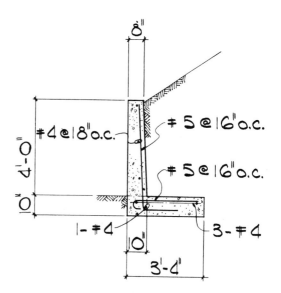

Soil slope = 1½ to 1
E.F.P. = 55 p.c.f.
Section 2.13.4

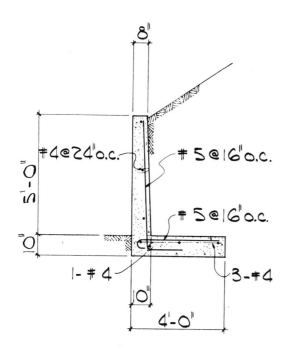

Soil slope = 1½ to 1
E.F.P. = 55 p.c.f.
Section 2.13.5

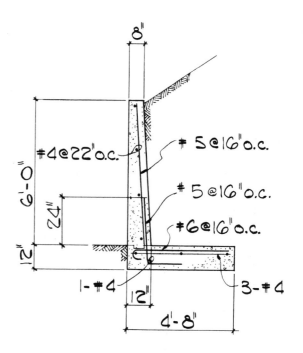

Soil slope = 1½ to 1
E.F.P. = 55 p.c.f.
Section 2.13.6

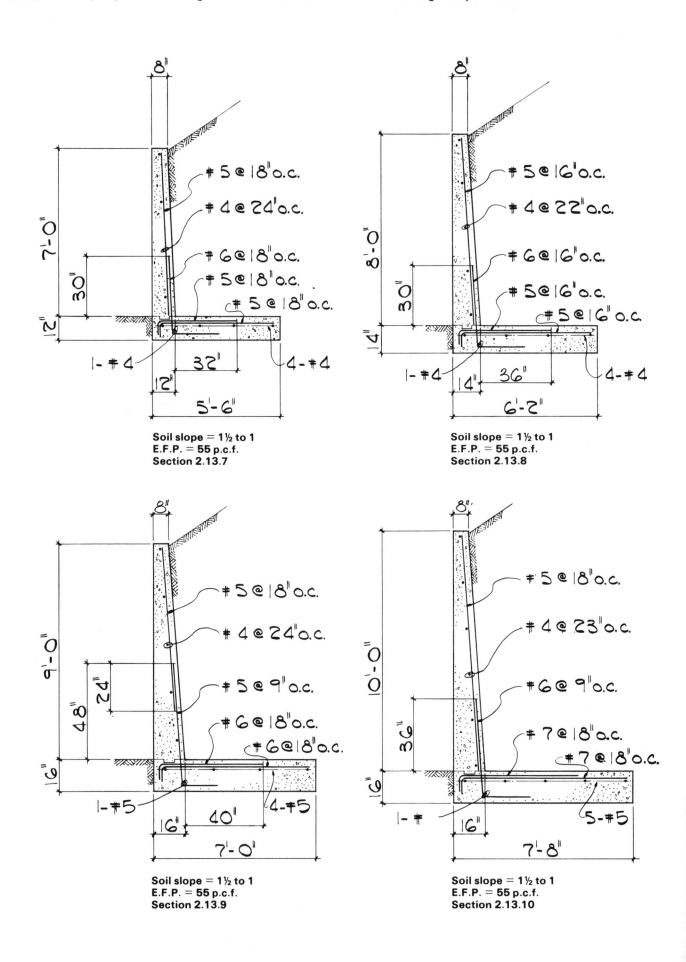

Soil slope = 1½ to 1
E.F.P. = 55 p.c.f.
Section 2.13.7

Soil slope = 1½ to 1
E.F.P. = 55 p.c.f.
Section 2.13.8

Soil slope = 1½ to 1
E.F.P. = 55 p.c.f.
Section 2.13.9

Soil slope = 1½ to 1
E.F.P. = 55 p.c.f.
Section 2.13.10

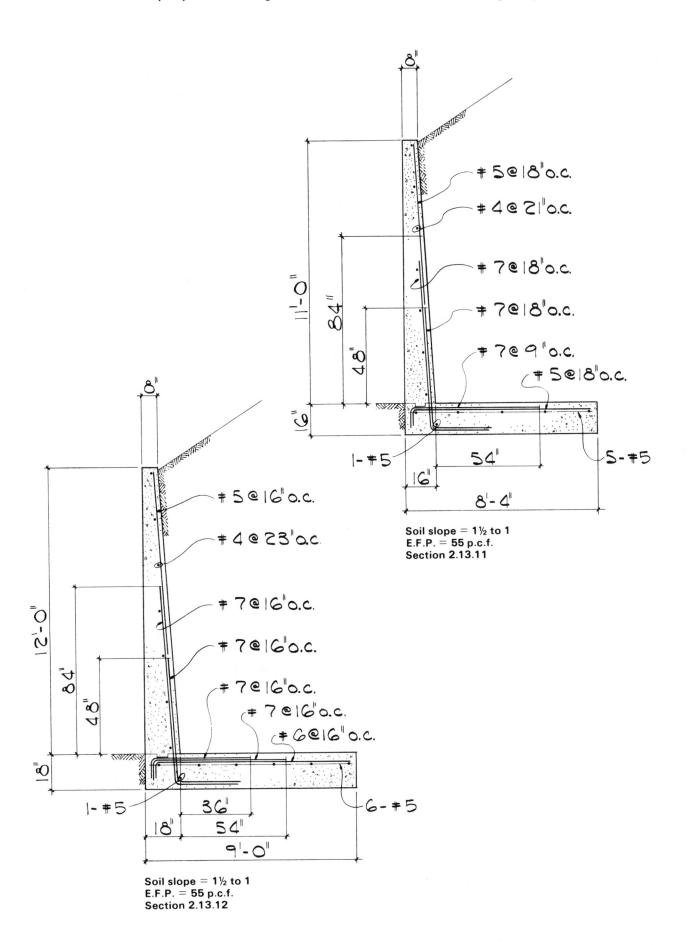

#5@18"o.c.

#4@21"o.c.

#7@18"o.c.

#7@18"o.c.

#7@9"o.c.

#5@18"o.c.

8"

11'-0"

84"

48"

6"

1-#5

54"

16"

5-#5

8'-4"

Soil slope = 1½ to 1
E.F.P. = 55 p.c.f.
Section 2.13.11

8"

#5@16"o.c.

#4@23"o.c.

#7@16"o.c.

#7@16"o.c.

#7@16"o.c.

#7@16"o.c.

#6@16"o.c.

12'-0"

84"

48"

18"

1-#5

36"

54"

18"

6-#5

9'-0"

Soil slope = 1½ to 1
E.F.P. = 55 p.c.f.
Section 2.13.12

TABLE 2.14	**PROPERTY LINE RETAINING WALLS — CONCRETE STEM — SOIL OVER FOOTING**									
	SLOPE = 1 to 1 SURCHARGE =0 lbs./sq. ft. AXIAL = 0 lbs./ft.									

SOIL PRESSURE DIAGRAM

Stem Height	3'- 0''	4'- 0''	5'- 0''	6'- 0''	7'- 0''	8'- 0''	9'- 0''	10'- 0''	11'- 0''	12'- 0''
Ft. L ft.	2.833	3.667	4.500	5.333	6.167	7.000	7.667	8.500	9.333	10.500
Ft. T ft.	0.833	0.833	0.833	1.000	1.167	1.333	1.333	1.333	1.500	1.500
Wt ft.	0.667	0.667	0.667	0.667	0.667	0.667	0.667	0.667	0.667	0.667
Wb ft.	0.833	0.833	0.833	1.000	1.167	1.333	1.333	1.333	1.500	1.500
W lbs.	1671	2738	4068	5819	7882	10259	12413	15101	18331	22367
F lbs.	588	934	1361	1960	2668	3484	4271	5138	6250	7290
Mo ft.-lbs.	751	1505	2647	4573	7262	10840	14712	19409	26042	32805
Mr ft.-lbs.	2334	4945	9016	15284	23937	35358	46699	63001	83984	115903
O.T.R.	3.107	3.285	3.407	3.342	3.296	3.262	3.174	3.246	3.225	3.533
$\overline{X}$ ft.	0.947	1.256	1.566	1.841	2.115	2.390	2.577	2.887	3.161	3.715
e ft.	0.470	0.577	0.684	0.826	0.968	1.110	1.256	1.363	1.506	1.535
Me ft.-lbs.	785	1580	2784	4806	7629	11388	15597	20590	27604	34330
S ft.^{3}	1.338	2.241	3.375	4.741	6.338	8.167	9.796	12.042	14.519	18.375
S.P.t p.s.f.	1177	1452	1729	2105	2482	2860	3211	3486	3865	3999
S.P.h p.s.f.	3	42	79	77	74	71	27	67	63	262
Friction	0.352	0.341	0.335	0.337	0.338	0.340	0.344	0.340	0.341	0.326

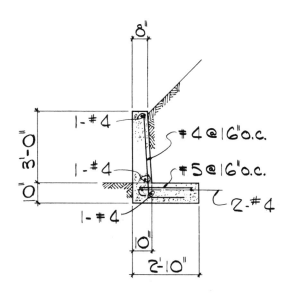

Soil slope = 1 to 1
E.F.P. = 80 p.c.f.
Section 2.14.3

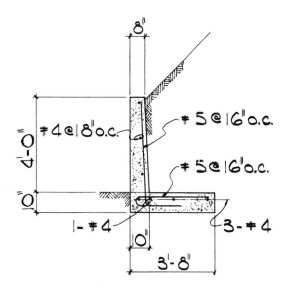

Soil slope = 1 to 1
E.F.P. = 80 p.c.f.
Section 2.14.4

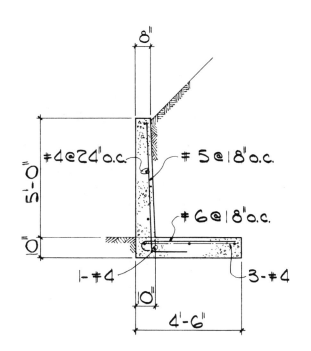

Soil slope = 1 to 1
E.F.P. = 80 p.c.f.
Section 2.14.5

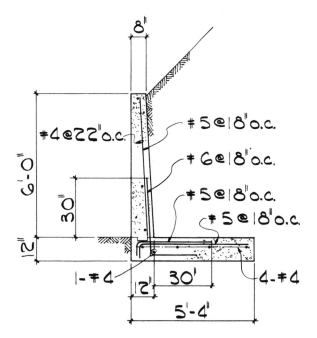

Soil slope = 1 to 1
E.F.P. = 80 p.c.f.
Section 2.14.6

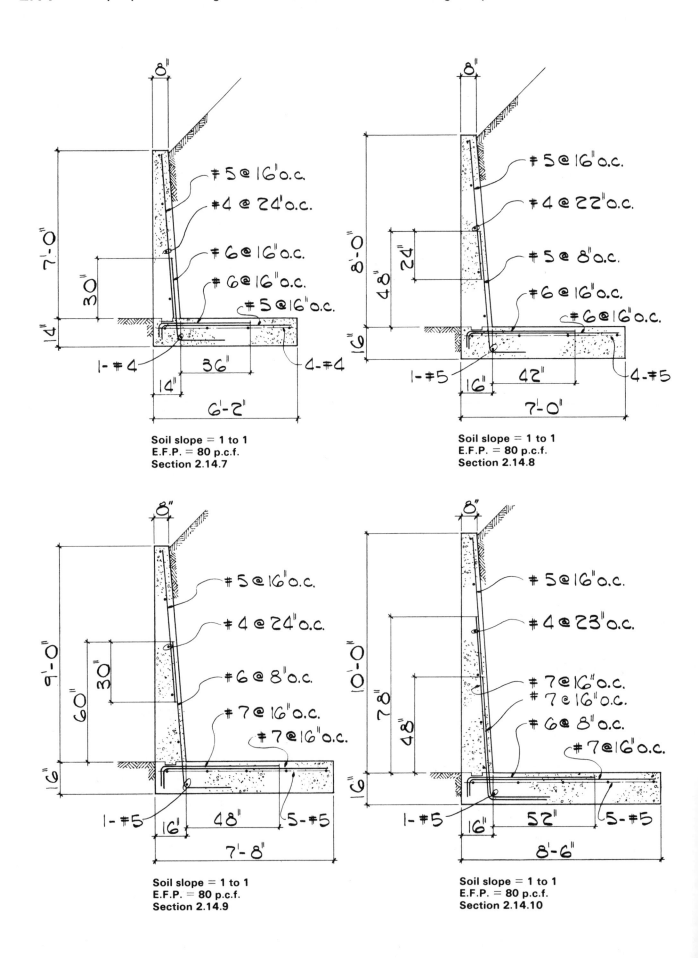

Soil slope = 1 to 1
E.F.P. = 80 p.c.f.
Section 2.14.7

Soil slope = 1 to 1
E.F.P. = 80 p.c.f.
Section 2.14.8

Soil slope = 1 to 1
E.F.P. = 80 p.c.f.
Section 2.14.9

Soil slope = 1 to 1
E.F.P. = 80 p.c.f.
Section 2.14.10

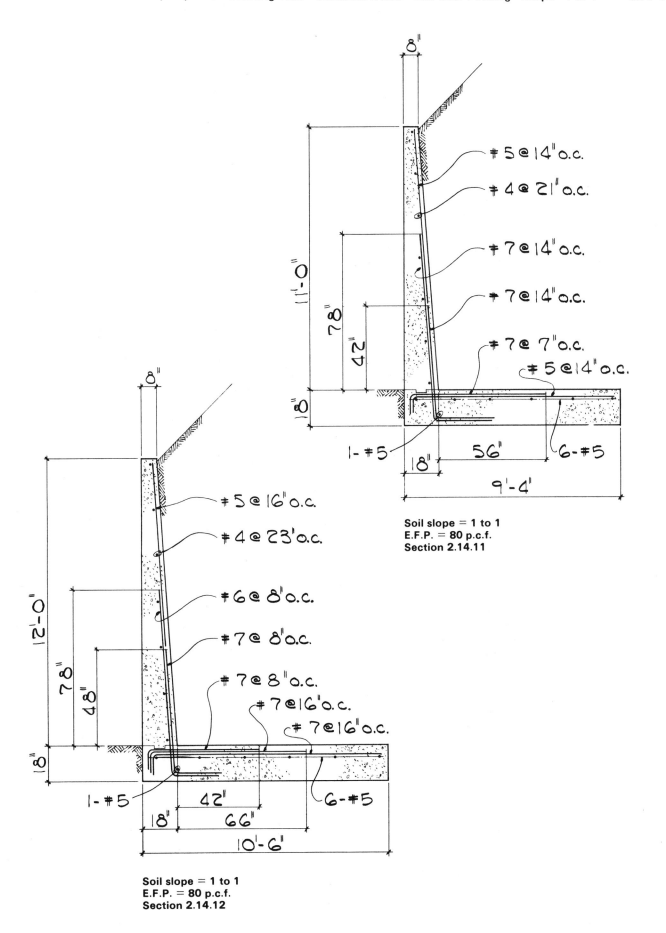

8"

11'-0"

78"

42"

8"

#5 @ 14" o.c.

#4 @ 21" o.c.

#7 @ 14" o.c.

#7 @ 14" o.c.

#7 @ 7" o.c.

#5 @ 14" o.c.

1 - #5

18"

56"

6 - #5

9'-4"

Soil slope = 1 to 1
E.F.P. = 80 p.c.f.
Section 2.14.11

8"

12'-0"

78"

48"

18"

#5 @ 16" o.c.

#4 @ 23' o.c.

#6 @ 8' o.c.

#7 @ 8" o.c.

#7 @ 8" o.c.

#7 @ 16" o.c.

#7 @ 16" o.c.

1 - #5

42"

66"

6 - #5

18"

10'-6"

Soil slope = 1 to 1
E.F.P. = 80 p.c.f.
Section 2.14.12

CHAPTER 3

Property Line Retaining Walls
Masonry Stem
Soil over the Footing

The design data and drawings presented in this chapter are concerned with property line retaining walls constructed with concrete masonry stem walls. The retained soil is placed over the footing of the wall. The chapter consists of three design examples of retaining walls with various types of loading, a design data table for each of the fourteen loading conditions specified in Fig. 1.6, and a series of corresponding drawings following each design data table. The retaining wall drawings and the design data in this chapter are for a wall length of 1 foot.

The retaining walls in this chapter are designed using the following criteria:

Weight of soil = 100 p.c.f.   Weight of concrete = 150 p.c.f.
Concrete f'_c = 2000 p.s.i.   Reinforcing steel f'_c = 20,000 p.s.i.
Hollow masonry units, grouted solid, grade N, f'_m = 1500 p.s.i.
Field inspection of masonry required (see the design data table)
Weight of 8-inch concrete block with all cells filled = 92 p.s.f.
Weight of 12-inch concrete block with all cells filled = 140 p.s.f.
Grout f'_c = 2000 p.s.i.   Mortar (Type S) = 1800 p.s.i.
Maximum allowable soil pressure = 4000 p.s.f.
Minimum allowable soil pressure = 0 p.s.f.
$\overline{X} \geqslant L/3$ Minimum O.T.R. = 1.50
Coefficient of soil friction = 0.40
Passive soil pressure = 300 p.c.f.

See Tables C5 and C6 for reinforcement protection. Weep holes are not shown on the drawings.

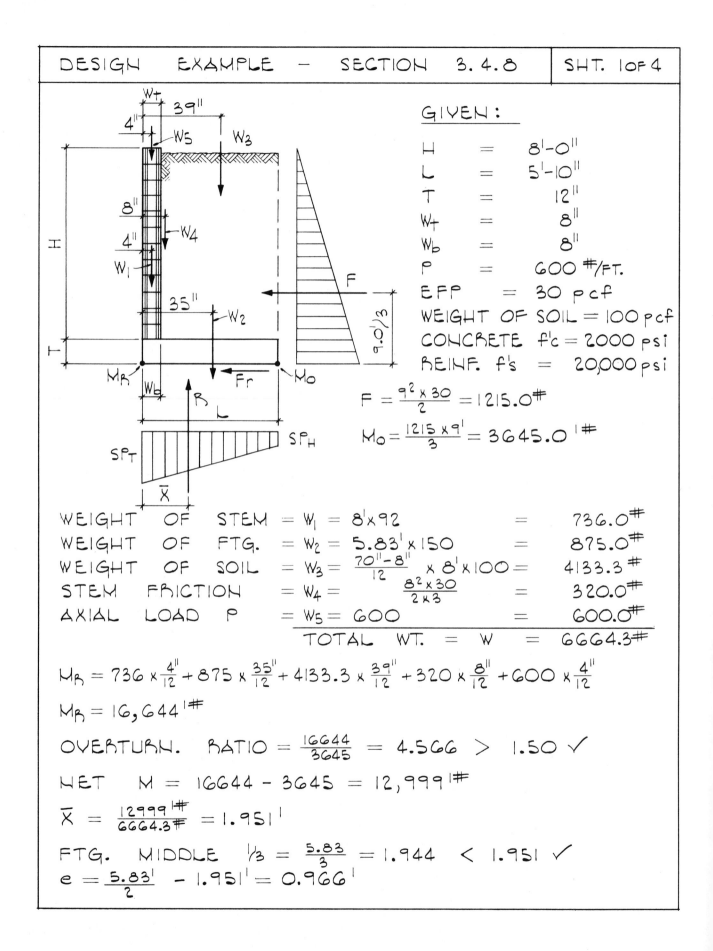

DESIGN EXAMPLE — SECTION 3.4.8 | SHT. 1 OF 4

GIVEN:

$$H = 8'\text{-}0''$$
$$L = 5'\text{-}10''$$
$$T = 12''$$
$$W_t = 8''$$
$$W_b = 8''$$
$$P = 600\ \#/\text{FT.}$$
$$EFP = 30\ pcf$$
$$\text{WEIGHT OF SOIL} = 100\ pcf$$
$$\text{CONCRETE } f'_c = 2000\ psi$$
$$\text{REINF. } f'_s = 20{,}000\ psi$$

$$F = \frac{9^2 \times 30}{2} = 1215.0\#$$

$$M_O = \frac{1215 \times 9'}{3} = 3645.0^{'\#}$$

WEIGHT OF STEM	$= W_1 =$	$8' \times 92$	$= 736.0\#$
WEIGHT OF FTG.	$= W_2 =$	$5.83' \times 150$	$= 875.0\#$
WEIGHT OF SOIL	$= W_3 =$	$\frac{70''-8''}{12} \times 8' \times 100$	$= 4133.3\#$
STEM FRICTION	$= W_4 =$	$\frac{8^2 \times 30}{2 \times 3}$	$= 320.0\#$
AXIAL LOAD P	$= W_5 =$	600	$= 600.0\#$

$$\text{TOTAL WT.} = W = 6664.3\#$$

$$M_R = 736 \times \frac{4''}{12} + 875 \times \frac{35''}{12} + 4133.3 \times \frac{39''}{12} + 320 \times \frac{8''}{12} + 600 \times \frac{4''}{12}$$

$$M_R = 16{,}644^{'\#}$$

$$\text{OVERTURN. RATIO} = \frac{16644}{3645} = 4.566 > 1.50 \checkmark$$

$$\text{NET } M = 16644 - 3645 = 12{,}999^{'\#}$$

$$\bar{X} = \frac{12999^{'\#}}{6664.3\#} = 1.951'$$

$$\text{FTG. MIDDLE } \frac{1}{3} = \frac{5.83}{3} = 1.944 < 1.951 \checkmark$$

$$e = \frac{5.83'}{2} - 1.951' = 0.966'$$

DESIGN EXAMPLE − SECTION 3.4.8	SHT. 2 OF 4

FOOTING   DESIGN :

$e = 0.966'$ $W = 6664.3^{\#}$

$M_e = 0.966 \times 6664.3 = 6438^{'\#}$

FTG. AREA = 5.83 SQ. FT.

FTG. SECT. MODULUS $= \frac{1 \times 5.83^2}{6} = 5.67^{'3}$

SOIL   PRESSURE $= \frac{6664.3}{5.83} \pm \frac{6438}{5.67} = 1143.1 \pm 1135.5$

$\underline{SP_T = 2278.6 \text{ psf}}$ $\underline{SP_H = 7.6 \text{ psf}}$

UNIF. LD. $= 800 + 150 - 7.6 = 942.4^{\#}/\text{FT.}$

W_2 $SP_1 = \frac{2278.6 - 7.6}{5.83} \times 5.17 = 2012 \text{ psf}$

W_3

$-M_1 = \frac{942.4 \times 5.17^2}{2} - \frac{2012 \times 5.17}{2} \times \frac{5.17}{3}$

$-M_1 = 12,595 - 8952 = 3643^{'\#}$

SP_H $V_1 = 942.4 \times 5.17 - \frac{2012 \times 5.17}{2}$

$V_1 = 329^{\#}$

$d = 12'' - (2 + .5 \times .5) = 9.75''$

$a = 1.13$; $j = 0.89$

$\nu = \frac{329}{12 \times .89 \times 9.75} = 3.20 \text{ psi} < 54 \text{ psi} \checkmark$

$A_s = \frac{3.643}{1.13 \times 9.75} = 0.33 \text{ SQ. IN./FT.}$

MIN. $A_s = 0.0012 \times 70 \times 12 = 1.00 \text{ SQ. IN.}$

$\underline{\text{USE } \#5 @ 16'' \text{ o.c. } \& \ \#4 @ 16'' \text{ o.c. } ; \quad 4 - \#4 \text{ LONG.}}$

$-M_2$ AT $2'-6''$ FROM END OF FTG.

$SP_2 = \frac{2278.6 - 7.6}{5.83} \times 2.50 = 974 \text{ psf}$

$-M = \frac{942.4 \times 2.5^2}{2} - \frac{974 \times 2.50}{2} \times \frac{2.5}{3} = 1930^{'\#}$

$A_s = \frac{1.93}{1.13 \times 9.75} = 0.175 \text{ SQ. IN./FT.}$ $\underline{\text{USE } \#5 @ 16'' \text{ o.c.}}$

DESIGN EXAMPLE — SECTION 3.4.8	SHT. 3 OF 4

STEM DESIGN AT TOP OF FTG.:

V & M AT BASE OF STEM ; H = 8'-0" ; EFP = 30 p.c.f.

$$V = \frac{30 \times 8^2}{2} = 960^\# \quad ; \quad M = \frac{960 \times 8}{3} = 2560'^\#$$

$$d = 8'' - (2'' + .5 \times .5) = 5.75'' \qquad a = 1.13 \; ; \; j = 0.89$$

$$\nu = \frac{960}{12 \times .85 \times 5.75} = 15.60 \; psi \approx 15.0 \, psi$$

$$A_s = \frac{2.56}{1.13 \times 5.75} = 0.394 \; sq.in./ft. \quad ; \quad TRY \; \#5 @ 8'' o.c. = 0.47 \, sq.in.$$

$$P = \frac{0.47}{12 \times 5.75} = 0.0068 \; ; \; np = 20 \times 0.0068 = 0.136 \; ; \; \frac{2}{kj} = 5.73$$

$$fm = \frac{2560 \times 12 \times 5.73}{12 \times 5.75^2} = 443.6 \, psi < 500 \, psi \; \checkmark$$

$$fa = \frac{600 + 736}{12 \times 8} = 13.9 \; psi \qquad \underline{USE \; \#5 @ 8'' \, o.c. \; VERT.}$$

$$COMBINED \; STEM \; STRESS = \frac{443.6}{500} + \frac{13.9}{270} = 0.94 < 1.0 \; \checkmark$$

STEM DESIGN AT 3'-0" ABOVE TOP OF FTG.:

$$V = \frac{30 \times 5^2}{2} = 375^\# \quad ; \quad M = \frac{375 \times 5}{3} = 625'^\#$$

$$A_s = \frac{0.625}{1.13 \times 5.75} = 0.096 \; sq.in./ft. \quad TRY \#5 @ 16'' o.c. \quad A_s \; 0.23 sq.in$$

$$P = \frac{0.23}{12 \times 8} = 0.0033 \quad ; \quad np = 20 \times 0.0033 = 0.066$$

$$2/kj = 7.34 \quad ; \quad fm = \frac{625 \times 12 \times 7.34}{12 \times 5.75^2} = 138.7 psi < 500 \, psi \checkmark$$

$$\underline{USE \; \#5 @ 16'' \, o.c. \; AT \; 5'-0'' \; FROM \; TOP \; OF \; STEM}$$

HORIZ. STEEL = 0.0007 × 8 × 12 × 8 = 0.537 sq. in.
$$\underline{USE \; 2 - \#4 \; TOP \; \& \; BOTT \; \& \; \#4 @ 48'' o.c. \; HORIZ.}$$

$$F_R = \frac{1215}{6664.3} = 0.182 < 0.40 \; \checkmark \quad NO \; KEY \; REQ'D.$$

DESIGN EXAMPLE – SECTION 3.4.8	SHT. 4 OF 4

CHECK WALL WITHOUT 600 #/FT. AXIAL LOAD

$F = 1215^{\#}$; $M_o = 3645^{'\#}$; $EFP = 30$ p.c.f.

$W = 6664.3 - 600 = 6064.3^{\#}$

$M_R = 16644 - 600 \times \frac{4}{12} = 16444^{'\#}$

NET $M = 16444 - 3645 = 12799^{'\#}$

$\bar{X} = \frac{12799}{6064.3} = 2.111' > 1.50' \checkmark$

$e = \frac{5.83}{2} - 2.111 = 0.806'$

$M_e = 6064.3 \times .806 = 4888^{'\#}$

SOIL PRESSURE $= \frac{6064.3}{5.83} \pm \frac{4888}{5.67} = 1040 \pm 862$

$\underline{SP_T = 1902}$ p.s.f. $\underline{SP_H = 178}$ p.s.f.

CHECK SLIDING:

$F_r = \frac{1215}{6064.3} = 0.200 < 0.40$

NO KEY REQ'D.

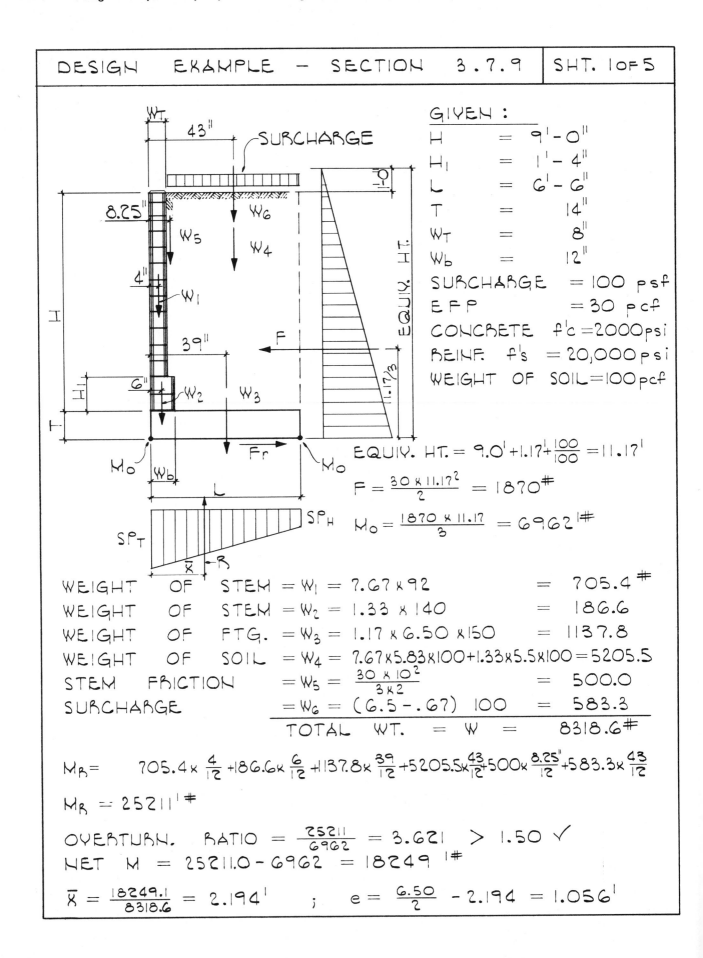

GIVEN:
$H = 9'-0''$
$H_1 = 1'-4''$
$L = 6'-6''$
$T = 14''$
$W_T = 8''$
$W_b = 12''$
SURCHARGE $= 100$ psf
EFP $= 30$ pcf
CONCRETE $f'_c = 2000$ psi
REINF. $f'_s = 20,000$ psi
WEIGHT OF SOIL $= 100$ pcf

EQUIV. HT. $= 9.0' + 1.17 + \frac{100}{100} = 11.17'$

$F = \frac{30 \times 11.17^2}{2} = 1870^{\#}$

$M_0 = \frac{1870 \times 11.17}{3} = 6962'^{\#}$

WEIGHT OF STEM $= W_1 = 7.67 \times 92 = 705.4^{\#}$
WEIGHT OF STEM $= W_2 = 1.33 \times 140 = 186.6$
WEIGHT OF FTG. $= W_3 = 1.17 \times 6.50 \times 150 = 1137.8$
WEIGHT OF SOIL $= W_4 = 7.67 \times 5.83 \times 100 + 1.33 \times 5.5 \times 100 = 5205.5$
STEM FRICTION $= W_5 = \frac{30 \times 10^2}{3 \times 2} = 500.0$
SURCHARGE $= W_6 = (6.5 - .67) 100 = 583.3$
TOTAL WT. $= W = 8318.6^{\#}$

$M_R = 705.4 \times \frac{4}{12} + 186.6 \times \frac{6}{12} + 1137.8 \times \frac{39}{12} + 5205.5 \times \frac{43}{12} + 500 \times \frac{8.25''}{12} + 583.3 \times \frac{43}{12}$

$M_R = 25211'^{\#}$

OVERTURN. RATIO $= \frac{25211}{6962} = 3.621 > 1.50$ ✓

NET M $= 25211.0 - 6962 = 18249'^{\#}$

$\bar{x} = \frac{18249.1}{8318.6} = 2.194'$; $e = \frac{6.50}{2} - 2.194 = 1.056'$

DESIGN EXAMPLE – SECTION 3.7.9	SHT. 2 OF 5

FOOTING   DESIGN :

$e = 1.056'$ $M_e = 8318.6 \times 1.056 = 8784.4'^{\#}$

FTG.   AREA = 6.50 SQ. FT.

FTG.   SECT. MODULUS = $\dfrac{1 \times 6.50^2}{6} = 7.04'^3$

SOIL   PRESSURE = $\dfrac{8318.6}{6.50} \pm \dfrac{8784.4}{7.04} = 1279.8 \pm 1247.8$

$SP_T = 2527.6$ psf $SP_H = 32.0$ psf

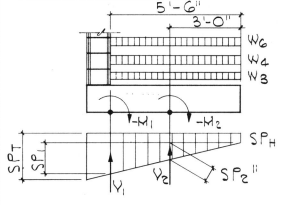

UNIF. LD. = $175 + 900 + 100 - 32 = 1143.0^{\#}/$FT.

$SP_1 = \dfrac{2527.6 - 32}{6.5} \times 5.50 = 2111.7$ psf

$-M_1 = \dfrac{1143 \times 5.5^2}{2} - \dfrac{2111.7 \times 5.5^2}{2 \times 3} = 6641'^{\#}$

$V_1 = 5.5 \times 1143 - \dfrac{2111.7 \times 5.5}{2} = 479.3^{\#}$

$d = 14 - (2 + .5 \times .62) = 11.7"$

$a = 1.13$; $j = 0.89$

$v = \dfrac{479.3}{12 \times .89 \times 11.7} = 3.83$ psi

$A_s = \dfrac{6.64}{1.13 \times 11.7} = 0.502$ SQ. IN.)FT.

USE # 5 @ 8" O.C.

$-M_2 :$ $SP_2 = \dfrac{2527.6 - 32.0}{6.50} \times 3.0 = 1151.8$ psf

$M_2 = \dfrac{1143 \times 3^2}{2} - \dfrac{1151.8 \times 3^2}{2 \times 3} = 3415.8'^{\#}$

$A_s = \dfrac{3.42}{1.13 \times 11.7} = 0.258$ SQ. IN.)FT.

CUT #5 @ 8" O.C. AT 3'-6" FROM FACE OF STEM

MIN. $A_s = 0.0012 \times 14 \times 78 = 1.31$ SQ. IN.

LONG. REINF. = 4-#5's

DESIGN EXAMPLE — SECTION 3.7.9	SHT. 3 OF 5

STEM DESIGN AT TOP OF FTG. :

EQUIV. STEM HT. $= 9.0' + \dfrac{100}{100} = 10.0'$

$V = \dfrac{30 \times 10^2}{2} = 1500^{\#}$; $M = \dfrac{1500 \times 10}{3} = 5000'^{\#}$

$d = 12 - (2 + .5 \times .5) = 9.75''$; $a = 1.13$; $j = 0.89$

$v = \dfrac{1500}{12 \times .89 \times 9.75} = 14.4 \, psi < 30 \, psi$

$A_s = \dfrac{5.0}{1.13 \times 9.75} = 0.454 \, sq. in./ft.$; TRY #5 @ 8" o.c.

$A_s = 0.47 \, sq. in.$; $P = \dfrac{0.47}{12 \times 9.75} = 0.0040$; $np = 20 \times .0040 = 0.080$

$^2/_{kj} = 6.85$ $fm = \dfrac{12 \times 5000 \times 6.85}{12 \times 9.75^2} = 360.3 < 500 \, psi$

USE #5 @ 8" o.c. CONT. INSPECTION REQ'D.

STEM DESIGN AT 16" ABOVE FTG. :

EQUIV. HT. $= 9.0' - 1.33' + \dfrac{100}{100} = 8.67'$

$V = \dfrac{30 \times 8.67^2}{2} = 1127.5^{\#}$ $M = \dfrac{1127.5 \times 8.67}{3} = 3257.4'^{\#}$

$d = 8 - (2 + .5 \times .5) = 5.75$; $a = 1.13$; $j = 0.89$; 8" BLOCK

$v = \dfrac{1127.5}{12 \times .89 \times 5.75} = 18.4 \, psi < 30 \, psi$

$A_s = \dfrac{3.257}{1.13 \times 5.75} = 0.50 \, sq.in./ft.$ TRY #6 @ 8" o.c. $A_s = 0.66 \, sq. in.$

$P = \dfrac{0.66}{12 \times 5.75} = .0096$; $np = 20 \times .0096 = 0.191$; $^2/_{kj} = 5.17$

$fm = \dfrac{12 \times 3257.4 \times 5.17}{12 \times 5.75^2} = 509.4 \, psi \approx 500 \, psi$ (+1.8%)

USE #6 @ 8" o.c. - CONT. INSPECTION REQ'D.

COMBINED STRESS OF 8" STEM :

VERT. LD. $= 7.67 \times 92 = 705.4^{\#}$ $fa = \dfrac{705.4}{12 \times 5.75} = 10.2 \, psi$

COMBINED STRESS $= \dfrac{10.2}{270} + \dfrac{509.4}{500} = 1.05 \approx 1.00 \checkmark$

DESIGN EXAMPLE — SECTION 3.7.9	SHT. 4 OF 5

STEM DESIGN AT 4'-0" ABOVE FTG.:

EQUIV. HT. $= 9.0' - 4.0' + \dfrac{100}{100} = 6.0'$

$V = \dfrac{30 \times 6^2}{2} = 540^{\#}$; $M = \dfrac{540 \times 6}{3} = 1080'^{\#}$

$\nu = \dfrac{540}{12 \times .89 \times 5.75} = 8.80 \, psi < 30 \, psi$

$A_s = \dfrac{1.08}{1.13 \times 5.75} = 0.166 \, sq.in./ft.$ TRY #5 @ 16" o.c.

$A_s = 0.23 \, sq.in.$; $P = \dfrac{0.23}{12 \times 5.75} = 0.0033$; $np = 20 \times 0.0033 = 0.066$

$^2/_{kj} = 7.34$ $fm = \dfrac{12 \times 1080 \times 7.34}{12 \times 5.75^2} = 239.8 \, psi < 500 \, psi \checkmark$

USE #5 @ 16" o.c.

STEM DESIGN AT 6'-0" ABOVE FTG.:

EQUIV. HT. $= 9.0' - 6.0' + \dfrac{100}{100} = 4.0'$

$V = \dfrac{30 \times 4^2}{2} = 240^{\#}$; $M = \dfrac{240 \times 4}{3} = 320'^{\#}$

TRY #5 @ 32" o.c. A_s 0.115 sq.in./ft.

$P = \dfrac{.115}{12 \times 5.75} = 0.0016$ $np = 20 \times .0016 = 0.032$

$^2/_{kj} = 9.70$; $fm = \dfrac{12 \times 320 \times 9.70}{12 \times 5.75^2} = 93.9 \, psi < 500 \, psi \checkmark$

USE #5 @ 32" o.c.

HORIZ. REINF. $= 0.0007 \times 8 \times 12 \times 7.67 = 0.515 \, sq.in.$

2 - #5 AT BOTT. , 2 - #4 AT TOP & 1 - #4 MID. HEIGHT

RESISTANCE TO SLIDING

$F_R = \dfrac{1870}{8318.6} = 0.225 < 0.40 \checkmark$

DESIGN EXAMPLE — SECTION 3.7.9	SHT. 5 of 5

CHECK WALL DESIGN WITHOUT 100 p.s.f. SURCHARGE LD.

EQUIV. HT. $= 9.0' + 1.17' = 10.17'$; E F P $= 30$ p.c.f.

$F = \dfrac{30 \times 10.17^2}{2} = 1550^{\#}$; $M_o = \dfrac{1550 \times 10.17}{3} = 5254'^{\#}$

$W = 8318.6 - 500 + \dfrac{30 \times 9^2}{2 \times 3} - 583.3 = 7640^{\#}$

$M_R = 25211 - 500 \times \dfrac{8.25}{12} + 405 \times \dfrac{8.25}{12} - 583.3 \times \dfrac{43}{12}$

$M_R = 23088'^{\#}$

OVERTURN. RATIO $= \dfrac{23088}{5254} = 4.394 > 1.50$

NET M $= 23088 - 5254 = 17834'^{\#}$

$\bar{X} = \dfrac{17834}{7640} = 2.334' > 2.17'$ ✓

$e = \dfrac{6.5}{2} - 2.334 = 0.916'$

$M_e = 7640 \times .916 = 6998'^{\#}$

SOIL PRESSURE $= \dfrac{7640}{6.50} \pm \dfrac{6998}{7.04} = 1175.4 \pm 994$

$\underline{SP_T = 2169.4\ p.s.f.}$ $\underline{SP_H = 181.4\ p.s.f.}$

CHECK FTG. DESIGN (SEE CALC. SHT. No. 2)

UNIF. LD. $= 175 + 900 - 181.4 = 893.6^{\#}/_{FT.}$

$SP_1 = \dfrac{2169.4 - 181.4}{6.5} \times 5.50 = 1682$ p.s.f.

$-M_1 = \dfrac{893.6 \times 5.5^2}{2} - \dfrac{1682 \times 5.5^2}{2 \times 3} = 5035.6'^{\#} < 6641'^{\#}$ ✓

CHECK SLIDING :

$F_r = \dfrac{1550}{7640} = 0.203 < 0.40$

NO KEY REQ'D.

DESIGN EXAMPLE — SECTION 3.13.5	SHT. 1 OF 3

GIVEN:

$$H = 5'\text{-}0''$$
$$L = 4'\text{-}0''$$
$$T = 10''$$
$$W_t = 8''$$
$$W_b = 8''$$
$$SLOPE = 1\tfrac{1}{2} \text{ TO } 1$$
$$EFP = 55 \text{ pcf}$$
$$\text{WEIGHT OF SOIL} = 100 \text{ pcf}$$
$$\text{CONCRETE } f'_c = 2000 \text{ psi}$$
$$\text{REINF. } f'_s = 20{,}000 \text{ psi}$$

$$F = \frac{55 \times 5.83^2}{2} = 936^\#$$

$$M_O = \frac{936 \times 5.83}{3} = 1820^{'\#}$$

WEIGHT OF STEM $= W_1 = 5 \times 92 \qquad\qquad = 460.0^\#$

WEIGHT OF FTG. $= W_2 = 4.0 \times 150 \times 0.83 = 500.0^\#$

WEIGHT OF SOIL $= W_3 = \dfrac{48'' - 8''}{12} \times 5.0 \times 100 = 1666.7^\#$

WEIGHT OF SOIL — SLOPE $= W_4 = \dfrac{40'' \times 26.67''}{2 \times 144} \times 100 = 370.4^\#$

STEM FRICTION $= W_5 = \dfrac{5^2 \times 55}{2 \times 3} \qquad\qquad = 229.2^\#$

$$\text{TOTAL WT.} = W = 3226.3^\#$$

$$M_B = 460 \times \frac{4}{12} + 500 \times \frac{24}{12} + 1666.7 \times \frac{28}{12} + 370.4 \times \frac{34.67}{12} + 229.2 \times \frac{8}{12}$$

$$M_B = 6265^{'\#}$$

OVERTURN. RATIO $= \dfrac{6265}{1820} = 3.442 > 1.50 \checkmark$

NET M $= 6265 - 1820 = 4445^{'\#}$

$$\bar{X} = \frac{4445}{3226.3} = 1.378'$$

FTG. MIDDLE $\frac{1}{3} = \dfrac{4.0}{3} = 1.333' < 1.378'$

$$e = \frac{4.0}{2} - 1.378 = 0.622'$$

DESIGN   EXAMPLE   —   SECTION   3.13.5 | SHT. 2 OF 3

FOOTING   DESIGN :

$e = 0.622'$ $W = 3226.3$

$Me = 3226.3 \times .622 = 2006.8^{'\#}$

FTG. AREA = 4.0 SQ. FT.

FTG. SECT. MODULUS $= \frac{1 \times 4^2}{6} = 2.66^{3}$

SOIL PRESSURE $= \frac{3226.3}{4.0} + \frac{2006.8}{2.66} = 806.6 \pm 754.4$

$\underline{SP_T = 1561 \text{ psf}}$ $\underline{SP_H = 52.2 \text{ psf}}$

UNIF. LD. $= 125 + 500 - 52.2 = 572.8^{\#}/FT.$

$SP_1 = \frac{1561 - 52.2}{4.0} \times 3.33 = 1257.3 \text{ psf}$

$-M = \frac{572.8 \times 3.33^2}{2} + 370.4 \times \frac{2 \times 3.33}{3} - \frac{1257.3 \times 3.33^2}{2 \times 3}$

$-M = 1670^{'\#}$

$Y = 3.33 \times 572.8 + 370.4 - \frac{1257.3 \times 3.33}{2}$

$Y = 184.4^{\#}$

$d = 10 - (2 + .5 \times .5) = 7.75''$; $a = 1.13$; $j = 0.89$

$\nu = \frac{184.4}{12 \times .89 \times 7.75} = 2.2 \text{ p.s.i} < 54 \text{ p.s.i} \checkmark$

$A_s = \frac{1.670}{1.13 \times 7.75} = 0.19 \text{ SQ. IN./FT.}$

USE   #5 @ 16" o.c.

LONG. REINF. :

$A_s = 0.0012 \times 12 \times 4.0 \times 10 = 0.576 \text{ SQ. IN.}$

USE   3 - #4's

DESIGN EXAMPLE — SECTION 3.13.5	SHT. 3 of 3

STEM   DESIGN :

HT. = 5.0'   EFP = 55 pcf

$V = \dfrac{55 \times 5.0^2}{2} = 687.5^{\#}$ $M = \dfrac{687.5 \times 5.0}{3} = 1145.8'^{\#}$

$d = 8 - (2 + .5 \times .5) = 5.75''$; $a = 1.13$; $j = 0.89$

$v = \dfrac{687.5}{12 \times .89 \times 5.75} = 11.2$ psi < 15 psi ✓—NO INSPECT. REQ'D.

$A_s = \dfrac{1.145}{1.13 \times 5.75} = 0.176$ SQ.IN./FT. ; TRY #5 @ 16" o.c. $A_s = 0.23$ SQ.IN.

$P = \dfrac{0.23}{12 \times 5.75} = 0.0033$; $np = 40 \times .0033 = 0.132$; $^2/_{kj} = 5.79$

$fm = \dfrac{12 \times 1145.8 \times 5.79}{12 \times 5.75^2} = 200.7$ psi < 250 psi ✓

USE # 5 @ 16" o.c. — NO INSPECTION REQ'D

LONG REINF. :

$A_s = 0.0007 \times 5 \times 12 \times 8 = 0.336$ SQ. IN.

USE 2 - #4 TOP & BOTT. OF STEM

RESISTANCE TO SLIDING

$F_R = \dfrac{936}{3226.3} = 0.29 < 0.40$ ✓

TABLE 3.1	PROPERTY LINE RETAINING WALLS – MASONRY STEM – SOIL OVER FOOTING
	SLOPE = 0 to 1 SURCHARGE = 0 lbs./sq. ft. AXIAL = 0 lbs./ft.

SOIL PRESSURE DIAGRAM

H1 = Concrete Stem
H2 = 12″ Concrete Block
H3 = 8″ Concrete Block

Stem Height	3'- 0″	4'- 0″	5'- 0″	6'- 0″	7'- 0″	8'- 0″	9'- 0″	10'- 0″	11'- 0″	12'- 0″
Ft. L ft.	2.167	2.667	3.333	4.000	4.667	5.333	5.833	6.500	7.167	7.833
Ft. T ft.	0.667	0.667	0.833	0.833	0.833	1.000	1.000	1.167	1.167	1.333
8 in. blk. ft.	3.000	4.000	5.000	6.000	7.000	8.000	7.000	8.000	8.333	8.667
12 in. blk. ft.	—	—	—	—	—	—	2.000	2.000	2.667	2.667
Conc. Stem	—	—	—	—	—	—	—	—	—	0.667
Wb ft.	—	—	—	—	—	—	—	—	—	1.333
W lbs.	988	1515	2335	3232	4272	5589	6768	8410	10060	12052
F lbs.	202	327	510	700	920	1215	1500	1870	2220	2667
Mo ft.-lbs.	246	508	992	1595	2403	3645	5000	6962	9005	11852
Mr ft.-lbs.	994	1865	3598	5971	9206	13792	18227	25274	33314	43646
O.T.R.	4.034	3.670	3.625	3.742	3.830	3.784	3.645	3.630	3.699	3.683
$\overline{X}$ ft.	0.757	0.896	1.116	1.354	1.592	1.815	1.954	2.177	2.416	2.638
e ft.	0.326	0.438	0.551	0.646	0.741	0.851	0.962	1.073	1.167	1.279
Me ft.-lbs.	322	663	1286	2089	3166	4758	6513	9021	11741	15408
S ft.^{3}	0.782	1.185	1.852	2.667	3.630	4.741	5.671	7.042	8.560	10.227
S.P.t p.s.f.	868	1127	1395	1591	1788	2052	2309	2575	2775	3045
S.P.h p.s.f.	44	9	6	25	43	44	12	13	32	32
Friction	0.204	0.216	0.219	0.217	0.215	0.217	0.222	0.222	0.221	0.221
Inspct.	NO	NO	NO	NO	YES	YES	YES	YES	YES	YES

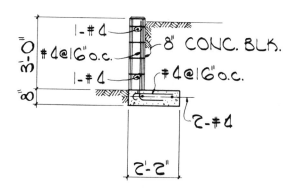

E.F.P. = 30 p.c.f.
Section 3.1.3

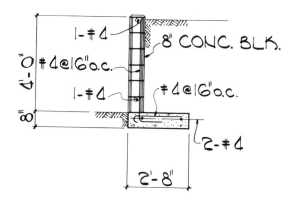

E.F.P. = 30 p.c.f.
Section 3.1.4

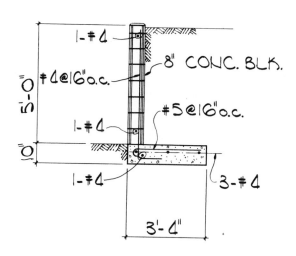

E.F.P. = 30 p.c.f.
Section 3.1.5

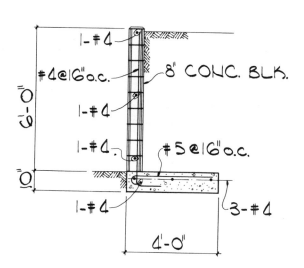

E.F.P. = 30 p.c.f.
Section 3.1.6

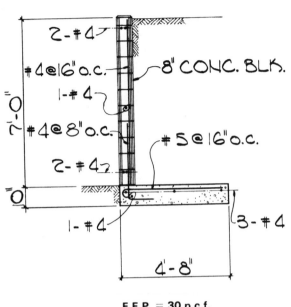

2-#4
#4@16"o.c.
1-#4
#4@8"o.c.
2-#4
7'-0"
8" CONC. BLK.
#5@16"o.c.
1-#4
3-#4
4'-8"

E.F.P. = 30 p.c.f.
Section 3.1.7

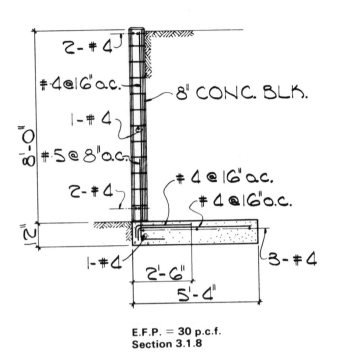

2-#4
#4@16"o.c.
1-#4
#5@8"o.c.
2-#4
8'-0"
8" CONC. BLK.
#4@16"o.c.
#4@16"o.c.
1-#4
3-#4
2'-6"
5'-4"

E.F.P. = 30 p.c.f.
Section 3.1.8

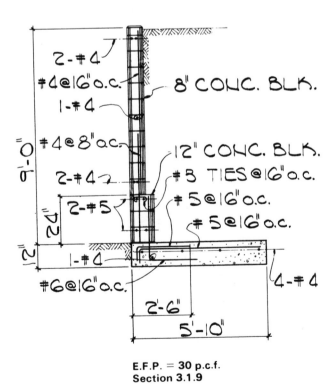

2-#4
#4@16"o.c.
1-#4
#4@8"o.c.
2-#4
9'-0"
8" CONC. BLK.
12" CONC. BLK.
#3 TIES@16"o.c.
#5@16"o.c.
#5@16"o.c.
2-#5
24"
2"
1-#4
#6@16"o.c.
2-#6
4-#4
2'-6"
5'-10"

E.F.P. = 30 p.c.f.
Section 3.1.9

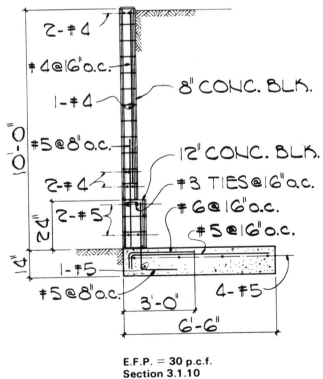

2-#4
#4@16"o.c.
1-#4
#5@8"o.c.
2-#4
10'-0"
8" CONC. BLK.
12" CONC. BLK.
#3 TIES@16"o.c.
#6@16"o.c.
#5@16"o.c.
2-#5
24"
4"
1-#5
#5@8"o.c.
4-#5
3'-0"
6'-6"

E.F.P. = 30 p.c.f.
Section 3.1.10

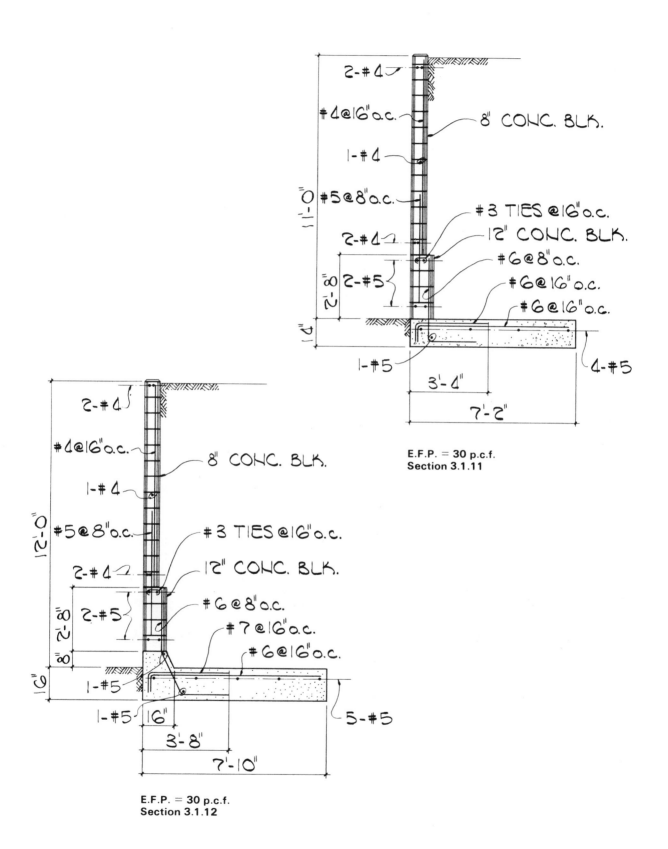

2-#4
#4@16"o.c.
1-#4
#5@8"o.c.
2-#4
2-#5
1'-0"
8"
4"
8" CONC. BLK.
#3 TIES @16"o.c.
12" CONC. BLK.
#6@8"o.c.
#6@16"o.c.
#6@16"o.c.
1-#5
4-#5
3'-4"
7'-2"

E.F.P. = 30 p.c.f.
Section 3.1.11

2-#4
#4@16"o.c.
1-#4
#5@8"o.c.
2-#4
2-#5
12'-0"
2'-8"
8"
6"
8" CONC. BLK.
#3 TIES @16"o.c.
12" CONC. BLK.
#6@8"o.c.
#7@16"o.c.
#6@16"o.c.
1-#5
1-#5
16"
3'-8"
7'-10"
5-#5

E.F.P. = 30 p.c.f.
Section 3.1.12

TABLE 3.2	PROPERTY LINE RETAINING WALLS – MASONRY STEM – SOIL OVER FOOTING
	SLOPE = 0 to 1 · · · SURCHARGE = 0 lbs./sq. ft. · · · AXIAL = 200 lbs./ft.

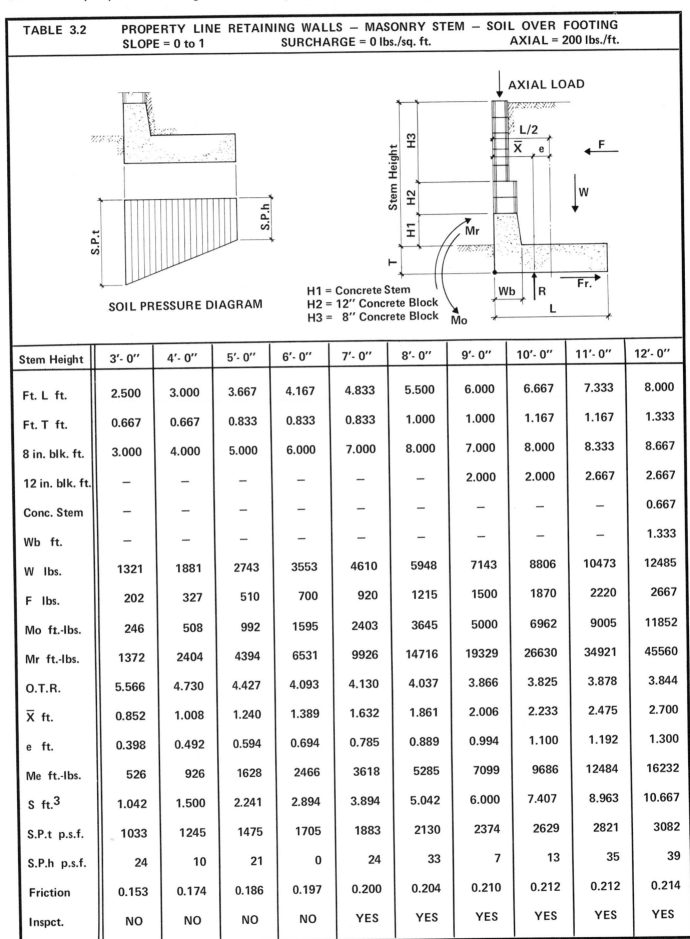

SOIL PRESSURE DIAGRAM

H1 = Concrete Stem
H2 = 12″ Concrete Block
H3 = 8″ Concrete Block

Stem Height	3'- 0''	4'- 0''	5'- 0''	6'- 0''	7'- 0''	8'- 0''	9'- 0''	10'- 0''	11'- 0''	12'- 0''
Ft. L ft.	2.500	3.000	3.667	4.167	4.833	5.500	6.000	6.667	7.333	8.000
Ft. T ft.	0.667	0.667	0.833	0.833	0.833	1.000	1.000	1.167	1.167	1.333
8 in. blk. ft.	3.000	4.000	5.000	6.000	7.000	8.000	7.000	8.000	8.333	8.667
12 in. blk. ft.	–	–	–	–	–	–	2.000	2.000	2.667	2.667
Conc. Stem	–	–	–	–	–	–	–	–	–	0.667
Wb ft.	–	–	–	–	–	–	–	–	–	1.333
W lbs.	1321	1881	2743	3553	4610	5948	7143	8806	10473	12485
F lbs.	202	327	510	700	920	1215	1500	1870	2220	2667
Mo ft.-lbs.	246	508	992	1595	2403	3645	5000	6962	9005	11852
Mr ft.-lbs.	1372	2404	4394	6531	9926	14716	19329	26630	34921	45560
O.T.R.	5.566	4.730	4.427	4.093	4.130	4.037	3.866	3.825	3.878	3.844
$\bar{X}$ ft.	0.852	1.008	1.240	1.389	1.632	1.861	2.006	2.233	2.475	2.700
e ft.	0.398	0.492	0.594	0.694	0.785	0.889	0.994	1.100	1.192	1.300
Me ft.-lbs.	526	926	1628	2466	3618	5285	7099	9686	12484	16232
S ft.^{3}	1.042	1.500	2.241	2.894	3.894	5.042	6.000	7.407	8.963	10.667
S.P.t p.s.f.	1033	1245	1475	1705	1883	2130	2374	2629	2821	3082
S.P.h p.s.f.	24	10	21	0	24	33	7	13	35	39
Friction	0.153	0.174	0.186	0.197	0.200	0.204	0.210	0.212	0.212	0.214
Inspct.	NO	NO	NO	NO	YES	YES	YES	YES	YES	YES

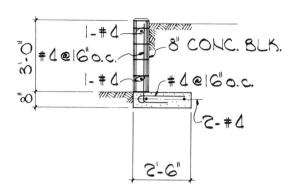

Axial load = 200 lbs./ft.
E.F.P. = 30 p.c.f.
Section 3.2.3

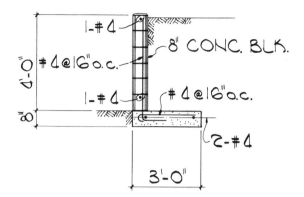

Axial load = 200 lbs./ft.
E.F.P. = 30 p.c.f.
Section 3.2.4

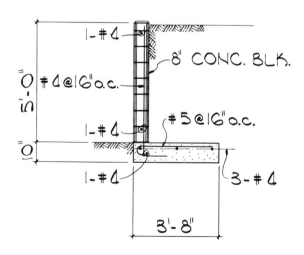

Axial load = 200 lbs./ft.
E.F.P. = 30 p.c.f.
Section 3.2.5

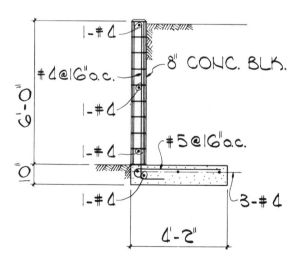

Axial load = 200 lbs./ft.
E.F.P. = 30 p.c.f.
Section 3.2.6

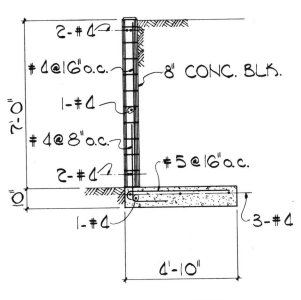

8" CONC. BLK.

2-#4

#4@16"o.c.

1-#4

#4@8"o.c.

2-#4

#5@16"o.c.

1-#4

7'-0"

0"

3-#4

4'-10"

Axial load = 200 lbs./ft.
E.F.P. = 32 p.c.f.
Section 3.2.7

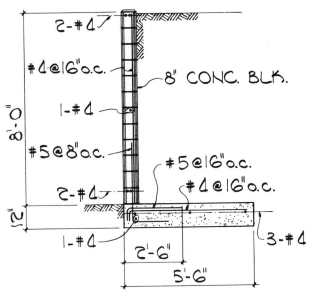

8" CONC. BLK.

2-#4

#4@16"o.c.

1-#4

#5@8"o.c.

2-#4

#5@16"o.c.

#4@16"o.c.

1-#4

8'-0"

2"

3-#4

2'-6"

5'-6"

Axial load = 200 lbs./ft.
E.F.P. = 32 p.c.f.
Section 3.2.8

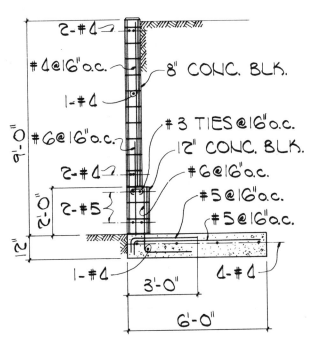

8" CONC. BLK.

2-#4

#4@16"o.c.

1-#4

#6@16"o.c.

2-#4

2-#5

#3 TIES @16"o.c.

12" CONC. BLK.

#6@16"o.c.

#5@16"o.c.

#5@16"o.c.

1-#4

9'-0"

2'-0"

2"

3'-0"

4-#4

6'-0"

Axial load = 200 lbs./ft.
E.F.P. = 30 p.c.f.
Section 3.2.9

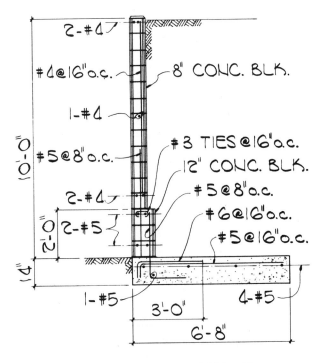

8" CONC. BLK.

2-#4

#4@16"o.c.

1-#4

#5@8"o.c.

2-#4

2-#5

#3 TIES @16"o.c.

12" CONC. BLK.

#5@8"o.c.

#6@16"o.c.

#5@16"o.c.

1-#5

10'-0"

2'-0"

1"

3'-0"

4-#5

6'-8"

Axial load = 200 lbs./ft.
E.F.P. = 30 p.c.f.
Section 3.2.10

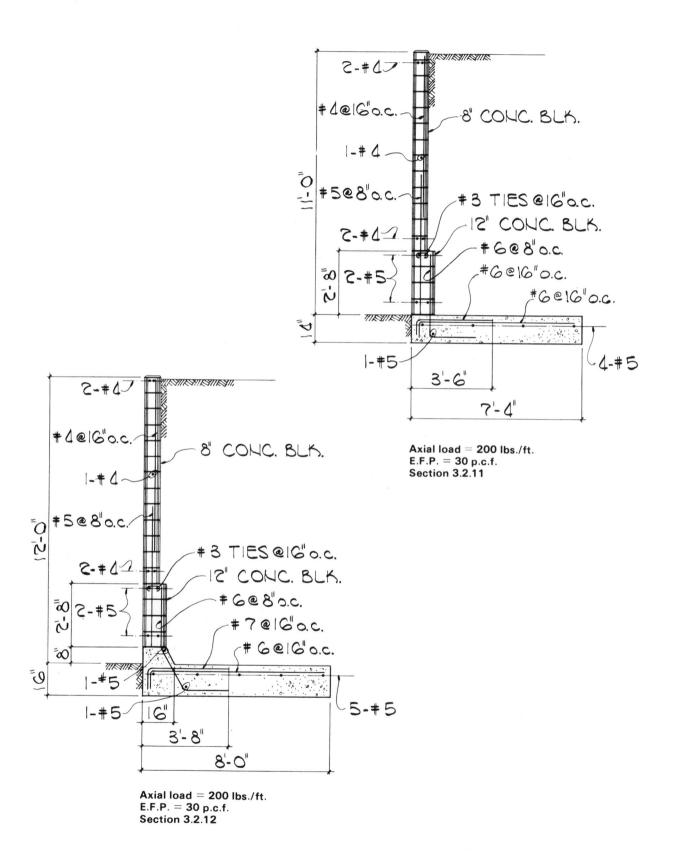

Axial load = 200 lbs./ft.
E.F.P. = 30 p.c.f.
Section 3.2.11

Axial load = 200 lbs./ft.
E.F.P. = 30 p.c.f.
Section 3.2.12

TABLE 3.3	PROPERTY LINE RETAINING WALLS – MASONRY STEM – SOIL OVER FOOTING
	SLOPE = 0 to 1 SURCHARGE = 0 lbs./sq. ft. AXIAL = 400 lbs./ft.

SOIL PRESSURE DIAGRAM

AXIAL LOAD

H1 = Concrete Stem
H2 = 12″ Concrete Block
H3 = 8″ Concrete Block

Stem Height	3'- 0''	4'- 0''	5'- 0''	6'- 0''	7'- 0''	8'- 0''	9'- 0''	10'- 0''	11'- 0''	12'- 0''
Ft. L ft.	3.000	3.500	4.000	4.500	5.000	5.667	6.167	6.833	7.500	8.167
Ft. T ft.	0.667	0.667	0.833	0.833	0.833	1.000	1.000	1.167	1.167	1.333
8 in. blk. ft.	3.000	4.000	5.000	6.000	7.000	8.000	7.000	8.000	8.333	8.667
12 in. blk. ft.	—	—	—	—	—	—	2.000	2.000	2.667	2.667
Conc. Stem	—	—	—	—	—	—	—	—	—	0.667
Wb ft.	—	—	—	—	—	—	—	—	—	1.333
W lbs.	1721	2331	3152	3994	4947	6306	7518	9202	10885	12918
F lbs.	202	327	510	700	920	1215	1500	1870	2220	2667
Mo ft.-lbs.	246	508	992	1595	2403	3645	5000	6962	9005	11852
Mr ft.-lbs.	1989	3283	5259	7645	10668	15667	20460	28019	36564	47513
O.T.R.	8.068	6.461	5.299	4.792	4.439	4.298	4.092	4.024	4.060	4.009
$\overline{X}$ ft.	1.012	1.190	1.354	1.514	1.671	1.906	2.056	2.288	2.532	2.760
e ft.	0.488	0.560	0.646	0.736	0.829	0.927	1.027	1.128	1.218	1.323
Me ft.-lbs.	839	1305	2037	2938	4103	5845	7720	10384	13261	17089
S ft.^{3}	1.500	2.042	2.667	3.375	4.167	5.352	6.338	7.782	9.375	11.116
S.P.t p.s.f.	1133	1305	1552	1758	1974	2205	2437	2681	2866	3119
S.P.h p.s.f.	14	27	24	17	5	21	1	12	37	44
Friction	0.117	0.140	0.162	0.175	0.186	0.193	0.200	0.203	0.204	0.206
Inspct.	NO	NO	NO	NO	YES	YES	YES	YES	YES	YES

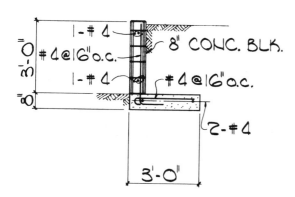

Axial load = 400 lbs./ft.
E.F.P. = 30 p.c.f.
Section 3.3.3

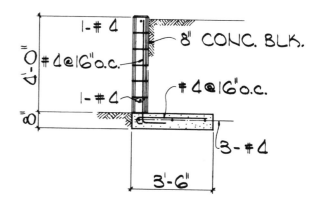

Axial load = 400 lbs./ft.
E.F.P. = 30 p.c.f.
Section 3.3.4

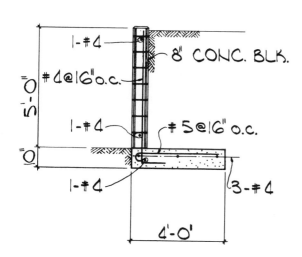

Axial load = 400 lbs./ft.
E.F.P. = 30 p.c.f.
Section 3.3.5

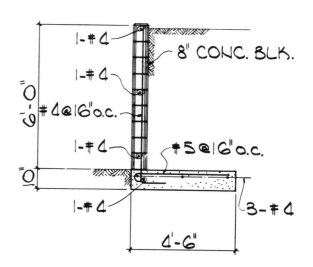

Axial load = 400 lbs./ft.
E.F.P. = 30 p.c.f.
Section 3.3.6

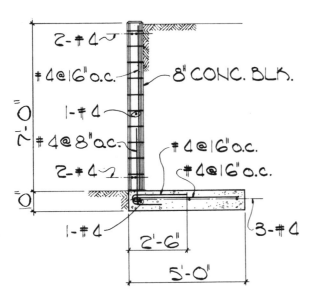

Axial load = 400 lbs./ft.
E.F.P. = 30 p.c.f.
Section 3.3.7

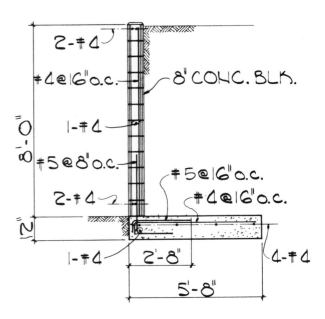

Axial load = 400 lbs./ft.
E.F.P. = 30 p.c.f.
Section 3.3.8

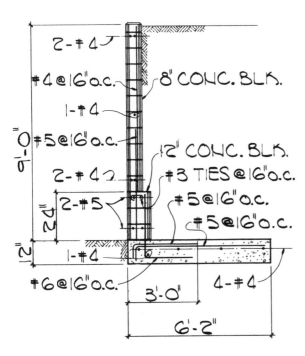

Axial load = 400 lbs./ft.
E.F.P. = 30 p.c.f.
Section 3.3.9

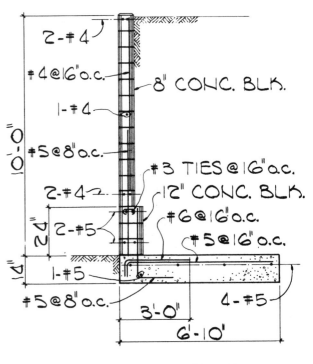

Axial load = 400 lbs./ft.
E.F.P. = 30 p.c.f.
Section 3.3.10

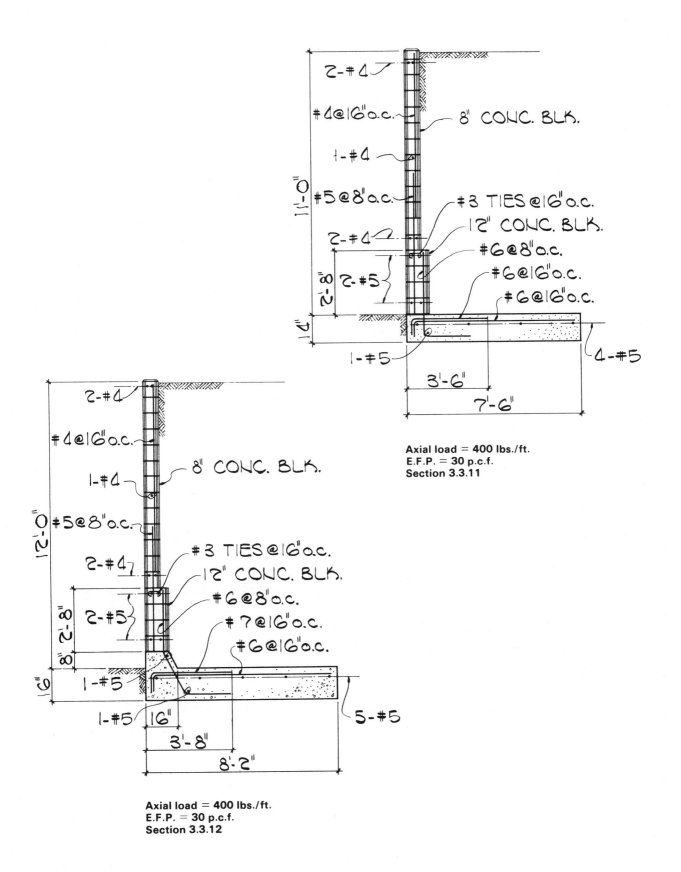

2-#4

#4@16"o.c.

8" CONC. BLK.

1-#4

1'-0"

#5@8"o.c.

#3 TIES@16"o.c.

12" CONC. BLK.

2-#4

#6@8"o.c.

2'-8"

2-#5

#6@16"o.c.

2'-2"

#6@16"o.c.

1'-4"

1-#5

4-#5

3'-6"

7'-6"

Axial load = 400 lbs./ft.
E.F.P. = 30 p.c.f.
Section 3.3.11

2-#4

#4@16"o.c.

1-#4

8" CONC. BLK.

12'-0"

#5@8"o.c.

2-#4

#3 TIES@16"o.c.

12" CONC. BLK.

2'-8"

2-#5

#6@8"o.c.

#7@16"o.c.

8"

2'-2"

#6@16"o.c.

1-#5

6"

1-#5   16"

5-#5

3'-8"

8'-2"

Axial load = 400 lbs./ft.
E.F.P. = 30 p.c.f.
Section 3.3.12

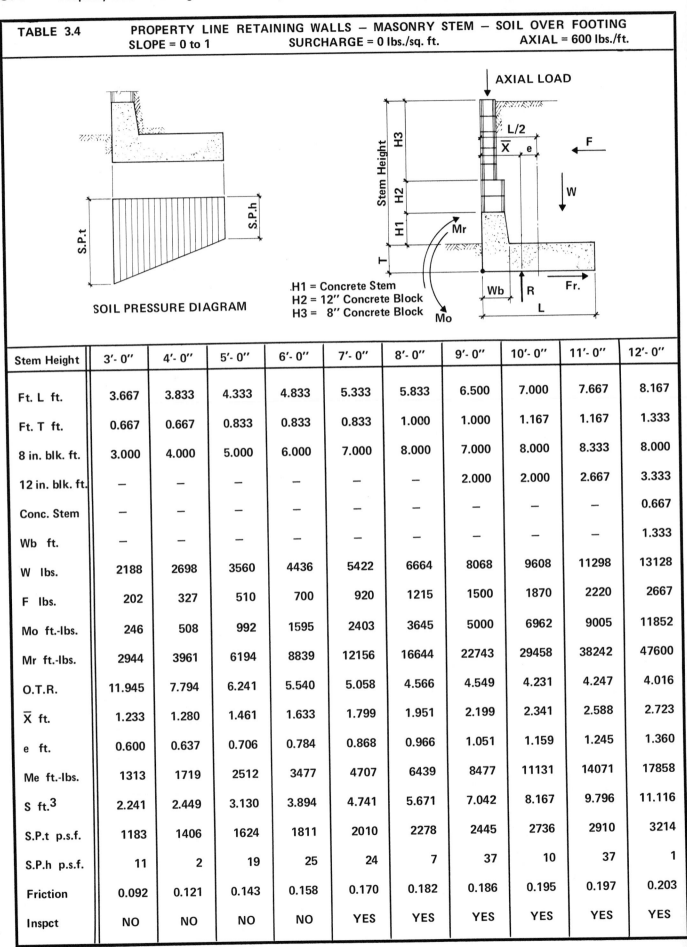

TABLE 3.4	PROPERTY LINE RETAINING WALLS — MASONRY STEM — SOIL OVER FOOTING
	SLOPE = 0 to 1 SURCHARGE = 0 lbs./sq. ft. AXIAL = 600 lbs./ft.

SOIL PRESSURE DIAGRAM

AXIAL LOAD

.H1 = Concrete Stem
H2 = 12" Concrete Block
H3 = 8" Concrete Block

Stem Height	3'- 0"	4'- 0"	5'- 0"	6'- 0"	7'- 0"	8'- 0"	9'- 0"	10'- 0"	11'- 0"	12'- 0"
Ft. L ft.	3.667	3.833	4.333	4.833	5.333	5.833	6.500	7.000	7.667	8.167
Ft. T ft.	0.667	0.667	0.833	0.833	0.833	1.000	1.000	1.167	1.167	1.333
8 in. blk. ft.	3.000	4.000	5.000	6.000	7.000	8.000	7.000	8.000	8.333	8.000
12 in. blk. ft.	—	—	—	—	—	—	2.000	2.000	2.667	3.333
Conc. Stem	—	—	—	—	—	—	—	—	—	0.667
Wb ft.	—	—	—	—	—	—	—	—	—	1.333
W lbs.	2188	2698	3560	4436	5422	6664	8068	9608	11298	13128
F lbs.	202	327	510	700	920	1215	1500	1870	2220	2667
Mo ft.-lbs.	246	508	992	1595	2403	3645	5000	6962	9005	11852
Mr ft.-lbs.	2944	3961	6194	8839	12156	16644	22743	29458	38242	47600
O.T.R.	11.945	7.794	6.241	5.540	5.058	4.566	4.549	4.231	4.247	4.016
$\overline{X}$ ft.	1.233	1.280	1.461	1.633	1.799	1.951	2.199	2.341	2.588	2.723
e ft.	0.600	0.637	0.706	0.784	0.868	0.966	1.051	1.159	1.245	1.360
Me ft.-lbs.	1313	1719	2512	3477	4707	6439	8477	11131	14071	17858
S ft.^{3}	2.241	2.449	3.130	3.894	4.741	5.671	7.042	8.167	9.796	11.116
S.P.t p.s.f.	1183	1406	1624	1811	2010	2278	2445	2736	2910	3214
S.P.h p.s.f.	11	2	19	25	24	7	37	10	37	1
Friction	0.092	0.121	0.143	0.158	0.170	0.182	0.186	0.195	0.197	0.203
Inspct	NO	NO	NO	NO	YES	YES	YES	YES	YES	YES

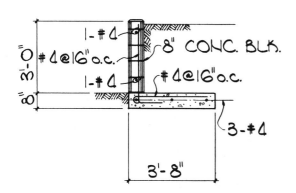

Axial load = 600 lbs./ft.
E.F.P. = 30 p.c.f.
Section 3.4.3

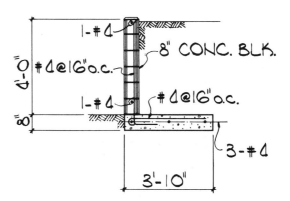

Axial load = 600 lbs./ft.
E.F.P. = 30 p.c.f.
Section 3.4.4

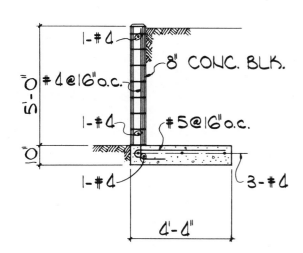

Axial load = 600 lbs./ft.
E.F.P. = 30 p.c.f.
Section 3.4.5

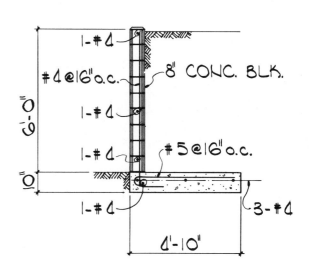

Axial load = 600 lbs./ft.
E.F.P. = 30 p.c.f.
Section 3.4.6

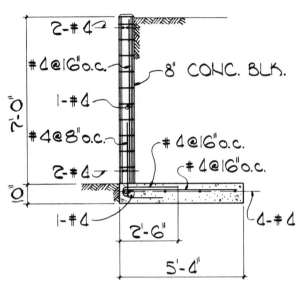

2-#4
#4@16"o.c.
1-#4
#4@8"o.c.
2-#4
8" CONC. BLK.
#4@16"o.c.
#4@16"o.c.
7'-0"
0"
1-#4
4-#4
2'-6"
5'-4"

Axial load = 600 lbs./ft.
E.F.P. = 30 p.c.f.
Section 3.4.7

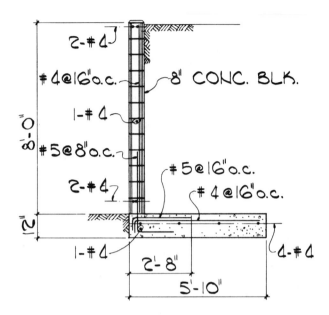

2-#4
#4@16"o.c.
1-#4
#5@8"o.c.
2-#4
8" CONC. BLK.
#5@16"o.c.
#4@16"o.c.
8'-0"
2"
1-#4
4-#4
2'-8"
5'-10"

Axial load = 600 lbs./ft.
E.F.P. = 30 p.c.f.
Section 3.4.8

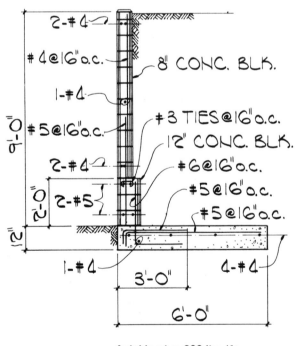

2-#4
#4@16"o.c.
1-#4
#5@16"o.c.
2-#4
2-#5
8" CONC. BLK.
#3 TIES@16"o.c.
12" CONC. BLK.
#6@16"o.c.
#5@16"o.c.
#5@16"o.c.
9'-0"
2'-0"
2"
1-#4
4-#4
3'-0"
6'-0"

Axial load = 600 lbs./ft.
E.F.P. = 30 p.c.f.
Section 3.4.9

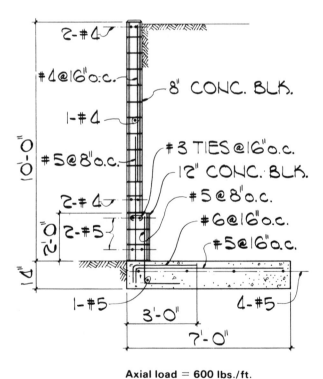

2-#4
#4@16"o.c.
1-#4
#5@8"o.c.
2-#4
2-#5
8" CONC. BLK.
#3 TIES@16"o.c.
12" CONC. BLK.
#5@8"o.c.
#6@16"o.c.
#5@16"o.c.
10'-0"
2'-0"
2"
4"
1-#5
4-#5
3'-0"
7'-0"

Axial load = 600 lbs./ft.
E.F.P. = 30 p.c.f.
Section 3.4.10

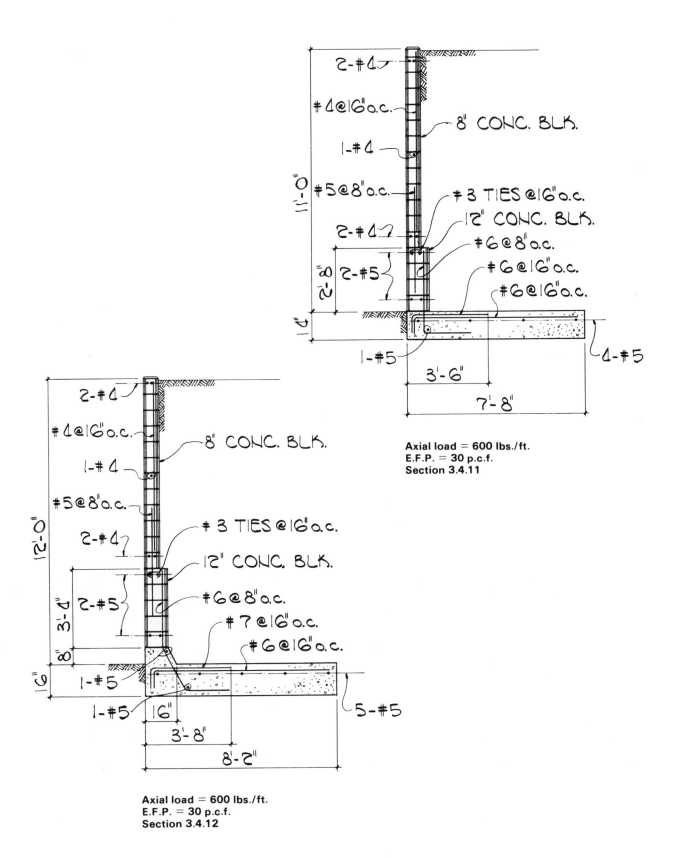

2-#4

#4@16" o.c.

1-#4

#5@8" o.c.

2-#4

2-#5

1'-0"

8" o.c.

4"

8" CONC. BLK.

#3 TIES @16" o.c.

12" CONC. BLK.

#6@8" o.c.

#6@16" o.c.

#6@16" o.c.

1-#5

4-#5

3'-6"

7'-8"

Axial load = 600 lbs./ft.
E.F.P. = 30 p.c.f.
Section 3.4.11

2-#4

#4@16" o.c.

1-#4

#5@8" o.c.

2-#4

2-#5

12'-0"

3'-4"

8"

6"

8" CONC. BLK.

#3 TIES @16" o.c.

12" CONC. BLK.

#6@8" o.c.

#7@16" o.c.

#6@16" o.c.

1-#5

1-#5

16"

5-#5

3'-8"

8'-2"

Axial load = 600 lbs./ft.
E.F.P. = 30 p.c.f.
Section 3.4.12

TABLE 3.5	PROPERTY LINE RETAINING WALLS — MASONRY STEM — SOIL OVER FOOTING
	SLOPE = 0 to 1 SURCHARGE = 0 lbs./sq. ft. AXIAL = 800 lbs./ft.

SOIL PRESSURE DIAGRAM

AXIAL LOAD

H1 = Concrete Stem
H2 = 12″ Concrete Block
H3 = 8″ Concrete Block

Stem Height	3′- 0″	4′- 0″	5′- 0″	6′- 0″	7′- 0″	8′- 0″	9′- 0″	10′- 0″	11′- 0″	12′- 0″
Ft. L ft.	4.500	4.500	4.667	5.167	5.500	6.167	6.667	7.167	7.833	8.333
Ft. T ft.	0.667	0.667	0.833	0.833	0.833	1.000	1.000	1.167	1.167	1.333
8 in. blk. ft.	3.000	4.000	5.000	6.000	7.000	8.000	7.000	8.000	8.333	8.000
12 in. blk. ft.	–	–	–	–	–	–	2.000	2.000	2.667	3.333
Conc. Stem	–	–	–	–	–	–	–	–	–	0.667
Wb ft.	–	–	–	–	–	–	–	–	–	1.333
W lbs.	2721	3231	3968	4878	5760	7181	8443	10003	11710	13562
F lbs.	202	327	510	700	920	1215	1500	1870	2220	2667
Mo ft.-lbs.	246	508	992	1595	2403	3645	5000	6962	9005	11852
Mr ft.-lbs.	4372	5416	7198	10114	12967	18611	23962	30912	39955	49592
O.T.R.	17.738	10.659	7.252	6.340	5.396	5.106	4.792	4.440	4.437	4.184
$\overline{X}$ ft.	1.516	1.519	1.564	1.746	1.834	2.084	2.246	2.394	2.643	2.783
e ft.	0.734	0.731	0.770	0.837	0.916	0.999	1.087	1.189	1.274	1.384
Me ft.-lbs.	1997	2362	3054	4082	5276	7176	9180	11896	14915	18766
S ft.^{3}	3.375	3.375	3.630	4.449	5.042	6.338	7.407	8.560	10.227	11.574
S.P.t p.s.f.	1196	1418	1692	1862	2094	2297	2506	2786	2953	3249
S.P.h p.s.f.	13	18	9	27	1	32	27	6	37	6
Friction	0.074	0.101	0.129	0.144	0.160	0.169	0.178	0.187	0.190	0.197
Inspct.	NO	NO	NO	NO	YES	YES	YES	YES	YES	YES

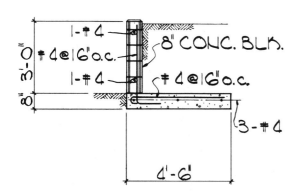

Axial load = 800 lbs./ft.
E.F.P. = 30 p.c.f.
Section 3.5.3

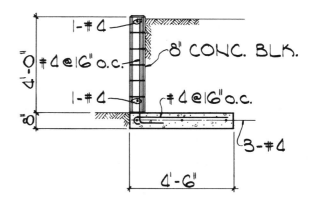

Axial load = 800 lbs./ft.
E.F.P. = 30 p.c.f.
Section 3.5.4

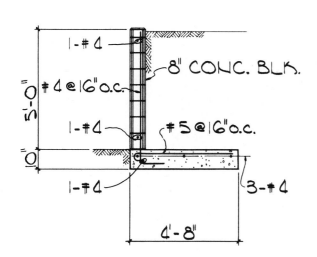

Axial load = 800 lbs./ft.
E.F.P. = 30 p.c.f.
Section 3.5.5

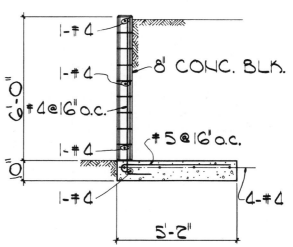

Axial load = 800 lbs./ft.
E.F.P. = 30 p.c.f.
Section 3.5.6

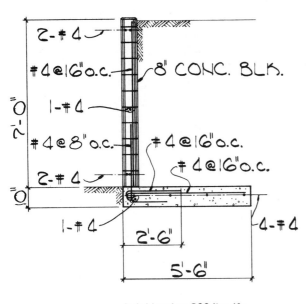

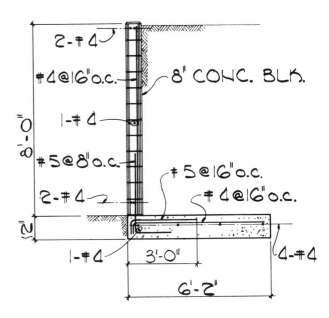

Axial load = 800 lbs./ft.
E.F.P. = 30 p.c.f.
Section 3.5.7

Axial load = 800 lbs./ft.
E.F.P. = 30 p.c.f.
Section 3.5.8

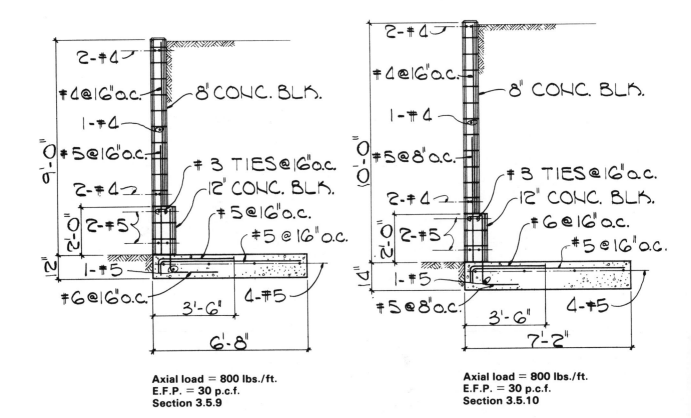

Axial load = 800 lbs./ft.
E.F.P. = 30 p.c.f.
Section 3.5.9

Axial load = 800 lbs./ft.
E.F.P. = 30 p.c.f.
Section 3.5.10

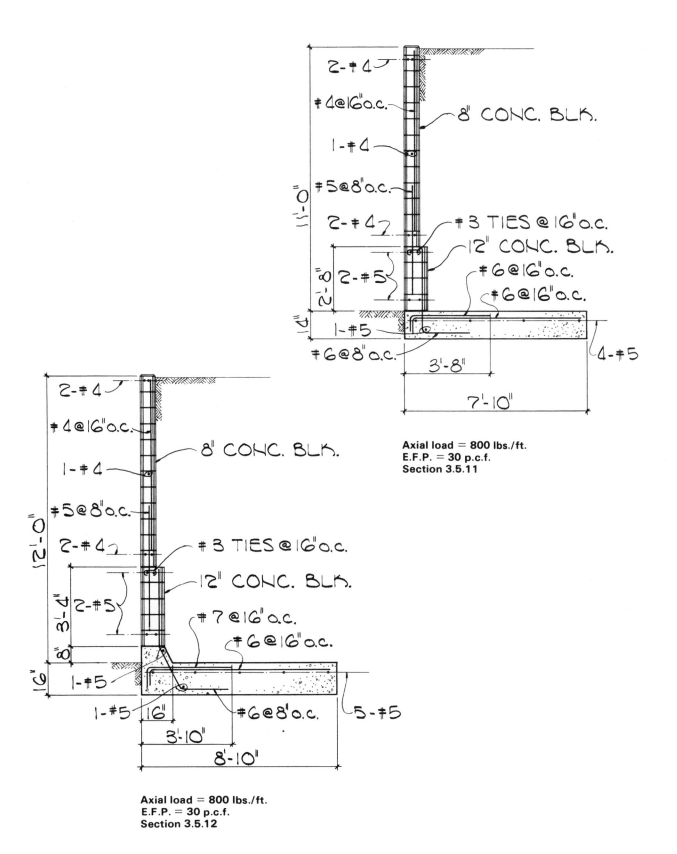

2-#4
#4@16"o.c.
8" CONC. BLK.
1-#4
#5@8"o.c.
2-#4
#3 TIES @ 16"o.c.
12" CONC. BLK.
2-#5
#6@16"o.c.
#6@16"o.c.
1-#5
#6@8"o.c.
4-#5
11'-0"
2'-8"
4"
3'-8"
7'-10"

Axial load = 800 lbs./ft.
E.F.P. = 30 p.c.f.
Section 3.5.11

2-#4
#4@16"o.c.
8" CONC. BLK.
1-#4
#5@8"o.c.
2-#4
#3 TIES @ 16"o.c.
12" CONC. BLK.
2-#5
#7@16"o.c.
#6@16"o.c.
1-#5
1-#5
16"
#6@8"o.c.
5-#5
12'-0"
3'-4"
8"
16"
3'-10"
8'-10"

Axial load = 800 lbs./ft.
E.F.P. = 30 p.c.f.
Section 3.5.12

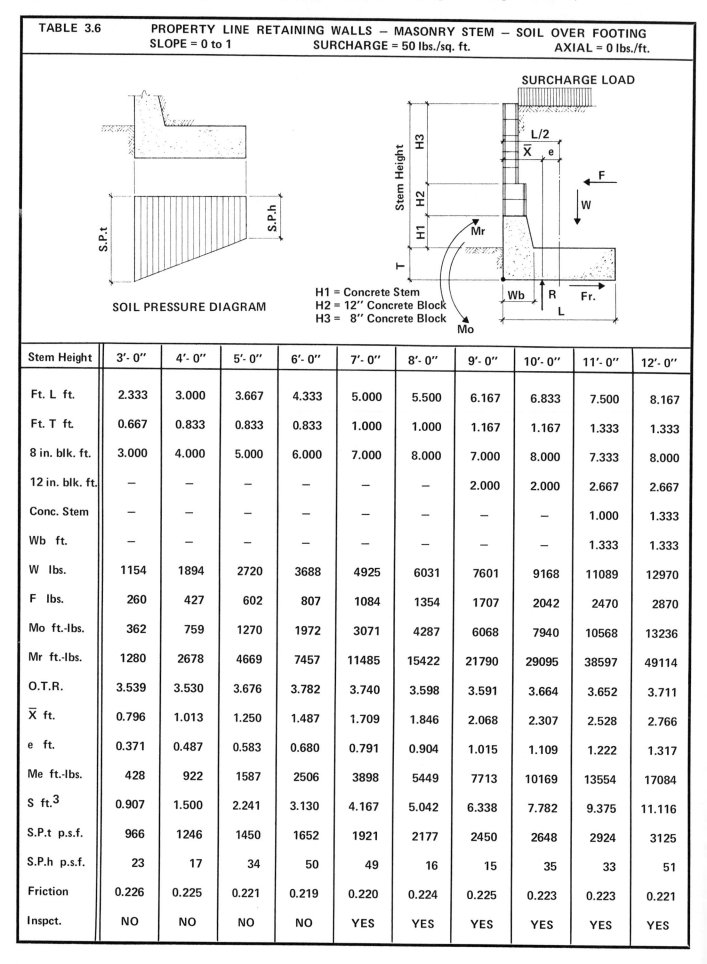

TABLE 3.6	PROPERTY LINE RETAINING WALLS — MASONRY STEM — SOIL OVER FOOTING
SLOPE = 0 to 1	SURCHARGE = 50 lbs./sq. ft. AXIAL = 0 lbs./ft.

H1 = Concrete Stem
H2 = 12″ Concrete Block
H3 = 8″ Concrete Block

SOIL PRESSURE DIAGRAM

SURCHARGE LOAD

Stem Height	3′- 0″	4′- 0″	5′- 0″	6′- 0″	7′- 0″	8′- 0″	9′- 0″	10′- 0″	11′- 0″	12′- 0″
Ft. L ft.	2.333	3.000	3.667	4.333	5.000	5.500	6.167	6.833	7.500	8.167
Ft. T ft.	0.667	0.833	0.833	0.833	1.000	1.000	1.167	1.167	1.333	1.333
8 in. blk. ft.	3.000	4.000	5.000	6.000	7.000	8.000	7.000	8.000	7.333	8.000
12 in. blk. ft.	–	–	–	–	–	–	2.000	2.000	2.667	2.667
Conc. Stem	–	–	–	–	–	–	–	–	1.000	1.333
Wb ft.	–	–	–	–	–	–	–	–	1.333	1.333
W lbs.	1154	1894	2720	3688	4925	6031	7601	9168	11089	12970
F lbs.	260	427	602	807	1084	1354	1707	2042	2470	2870
Mo ft.-lbs.	362	759	1270	1972	3071	4287	6068	7940	10568	13236
Mr ft.-lbs.	1280	2678	4669	7457	11485	15422	21790	29095	38597	49114
O.T.R.	3.539	3.530	3.676	3.782	3.740	3.598	3.591	3.664	3.652	3.711
$\overline{X}$ ft.	0.796	1.013	1.250	1.487	1.709	1.846	2.068	2.307	2.528	2.766
e ft.	0.371	0.487	0.583	0.680	0.791	0.904	1.015	1.109	1.222	1.317
Me ft.-lbs.	428	922	1587	2506	3898	5449	7713	10169	13554	17084
S ft.^{3}	0.907	1.500	2.241	3.130	4.167	5.042	6.338	7.782	9.375	11.116
S.P.t p.s.f.	966	1246	1450	1652	1921	2177	2450	2648	2924	3125
S.P.h p.s.f.	23	17	34	50	49	16	15	35	33	51
Friction	0.226	0.225	0.221	0.219	0.220	0.224	0.225	0.223	0.223	0.221
Inspct.	NO	NO	NO	NO	YES	YES	YES	YES	YES	YES

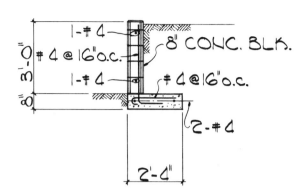

Surcharge load = 50 p.s.f.
E.F.P. = 30 p.c.f.
Section 3.6.3

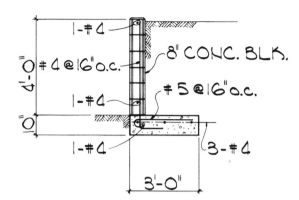

Surcharge load = 50 p.s.f.
E.F.P. = 30 p.c.f.
Section 3.6.4

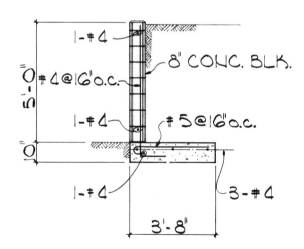

Surcharge load = 50 p.s.f.
E.F.P. = 30 p.c.f.
Section 3.6.5

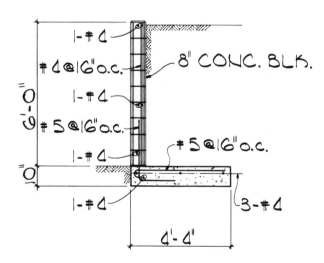

Surcharge load = 50 p.s.f.
E.F.P. = 30 p.c.f.
Section 3.6.6

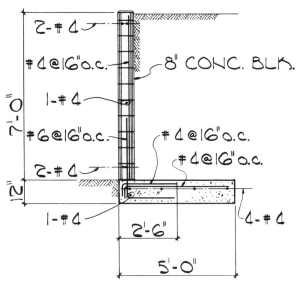

8" CONC. BLK.

2-#4
#4@16"o.c.
1-#4
#6@16"o.c.
2-#4
7'-0"
#4 @16"o.c.
#4@16"o.c.
1-#4
2"
4-#4
2'-6"
5'-0"

Surcharge load = 50 p.s.f.
E.F.P. = 30 p.c.f.
Section 3.6.7

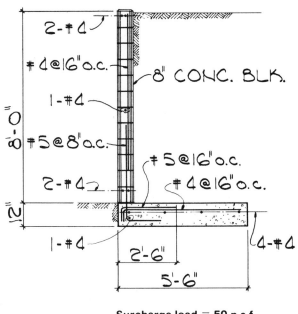

8" CONC. BLK.

2-#4
#4@16"o.c.
1-#4
#5@8"o.c.
2-#4
8'-0"
#5@16"o.c.
#4@16"o.c.
1-#4
2"
4-#4
2'-6"
5'-6"

Surcharge load = 50 p.s.f.
E.F.P. = 30 p.c.f.
Section 3.6.8

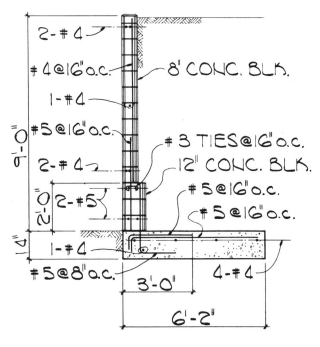

8" CONC. BLK.

2-#4
#4@16"o.c.
1-#4
#5@16"o.c.
2-#4
2-#5
9'-0"
2'-0"
#3 TIES@16"o.c.
12" CONC. BLK.
#5@16"o.c.
#5@16"o.c.
1-#4
#5@8"o.c.
1"
3'-0"
4-#4
6'-2"

Surcharge load = 50 p.s.f.
E.F.P. = 30 p.c.f.
Section 3.6.9

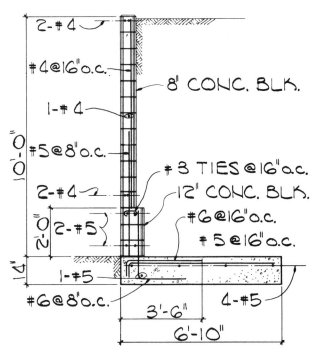

8" CONC. BLK.

2-#4
#4@16"o.c.
1-#4
#5@8"o.c.
2-#4
2-#5
10'-0"
2'-0"
#3 TIES @16"o.c.
12" CONC. BLK.
#6@16"o.c.
#5@16"o.c.
1-#5
#6@8"o.c.
1"
3'-6"
4-#5
6'-10"

Surcharge load = 50 p.s.f.
E.F.P. = 30 p.c.f.
Section 3.6.10

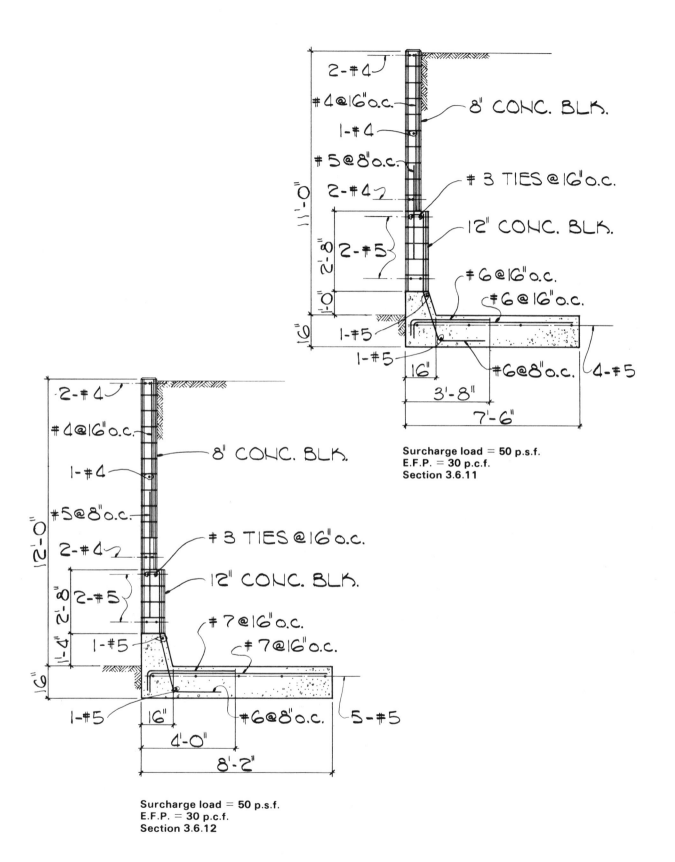

Section 3.6.11

2-#4

#4@16" o.c.

1-#4

#5@8" o.c.

2-#4

2-#5

8" CONC. BLK.

#3 TIES @16" o.c.

12" CONC. BLK.

#6@16" o.c.

#6@16" o.c.

1-#5

1-#5

11'-0"

2'-8"

2'-0"

6"

16"

3'-8"

7'-6"

#6@8" o.c.

4-#5

Surcharge load = 50 p.s.f.
E.F.P. = 30 p.c.f.
Section 3.6.11

2-#4

#4@16" o.c.

1-#4

#5@8" o.c.

2-#4

2-#5

8" CONC. BLK.

#3 TIES @16" o.c.

12" CONC. BLK.

#7@16" o.c.

#7@16" o.c.

1-#5

1-#5

12'-0"

2'-8"

1'-4"

6"

16"

4'-0"

8'-2"

#6@8" o.c.

5-#5

Surcharge load = 50 p.s.f.
E.F.P. = 30 p.c.f.
Section 3.6.12

TABLE 3.7	PROPERTY LINE RETAINING WALLS — MASONRY STEM — SOIL OVER FOOTING
	SLOPE = 0 to 1 SURCHARGE = 100 lbs./sq. ft. AXIAL = 0 lbs./ft.

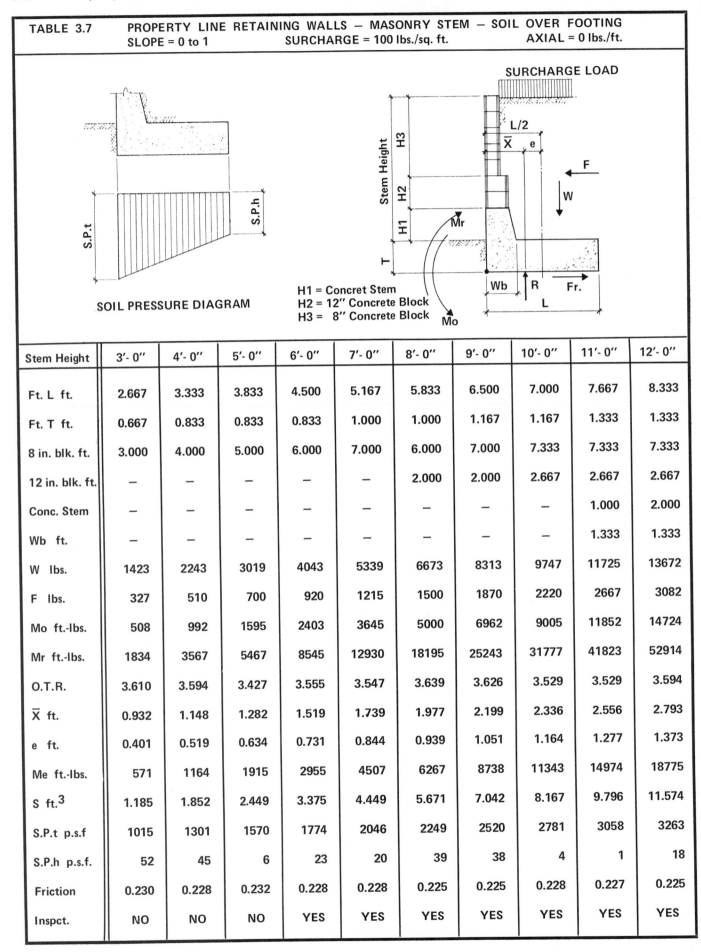

SOIL PRESSURE DIAGRAM

SURCHARGE LOAD

H1 = Concret Stem
H2 = 12" Concrete Block
H3 = 8" Concrete Block

Stem Height	3'- 0"	4'- 0"	5'- 0"	6'- 0"	7'- 0"	8'- 0"	9'- 0"	10'- 0"	11'- 0"	12'- 0"
Ft. L ft.	2.667	3.333	3.833	4.500	5.167	5.833	6.500	7.000	7.667	8.333
Ft. T ft.	0.667	0.833	0.833	0.833	1.000	1.000	1.167	1.167	1.333	1.333
8 in. blk. ft.	3.000	4.000	5.000	6.000	7.000	6.000	7.000	7.333	7.333	7.333
12 in. blk. ft.	—	—	—	—	—	2.000	2.000	2.667	2.667	2.667
Conc. Stem	—	—	—	—	—	—	—	—	1.000	2.000
Wb ft.	—	—	—	—	—	—	—	—	1.333	1.333
W lbs.	1423	2243	3019	4043	5339	6673	8313	9747	11725	13672
F lbs.	327	510	700	920	1215	1500	1870	2220	2667	3082
Mo ft.-lbs.	508	992	1595	2403	3645	5000	6962	9005	11852	14724
Mr ft.-lbs.	1834	3567	5467	8545	12930	18195	25243	31777	41823	52914
O.T.R.	3.610	3.594	3.427	3.555	3.547	3.639	3.626	3.529	3.529	3.594
$\overline{X}$ ft.	0.932	1.148	1.282	1.519	1.739	1.977	2.199	2.336	2.556	2.793
e ft.	0.401	0.519	0.634	0.731	0.844	0.939	1.051	1.164	1.277	1.373
Me ft.-lbs.	571	1164	1915	2955	4507	6267	8738	11343	14974	18775
S ft.^{3}	1.185	1.852	2.449	3.375	4.449	5.671	7.042	8.167	9.796	11.574
S.P.t p.s.f	1015	1301	1570	1774	2046	2249	2520	2781	3058	3263
S.P.h p.s.f.	52	45	6	23	20	39	38	4	1	18
Friction	0.230	0.228	0.232	0.228	0.228	0.225	0.225	0.228	0.227	0.225
Inspct.	NO	NO	NO	YES	YES	YES	YES	YES	YES	YES

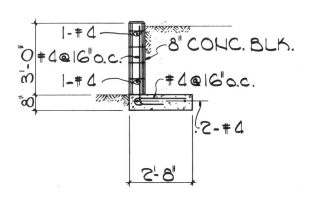

3'-0"

8"

1-#4

#4@16"o.c.

1-#4

8" CONC. BLK.

#4@16"o.c.

2-#4

2'-8"

Surcharge load = **100 p.s.f.**
E.F.P. = **30 p.c.f.**
Section **3.7.3**

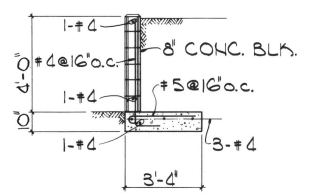

4'-0"

0"

1-#4

#4@16"o.c.

1-#4

1-#4

8" CONC. BLK.

#5@16"o.c.

3-#4

3'-4"

Surcharge load = **100 p.s.f.**
E.F.P. = **30 p.c.f.**
Section **3.7.4**

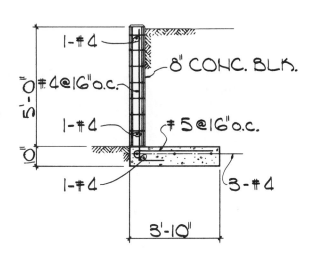

5'-0"

0"

1-#4

#4@16"o.c.

1-#4

1-#4

8" CONC. BLK.

#5@16"o.c.

3-#4

3'-10"

Surcharge load = **100 p.s.f.**
E.F.P. = **30 p.c.f.**
Section **3.7.5**

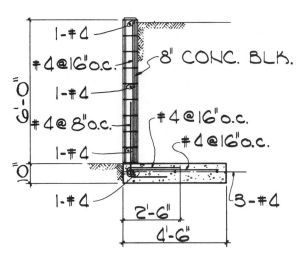

6'-0"

0"

1-#4

#4@16"o.c.

1-#4

#4@8"o.c.

1-#4

1-#4

8" CONC. BLK.

#4@16"o.c.

#4@16"o.c.

3-#4

2'-6"

4'-6"

Surcharge load = **100 p.s.f.**
E.F.P. = **30 p.c.f.**
Section **3.7.6**

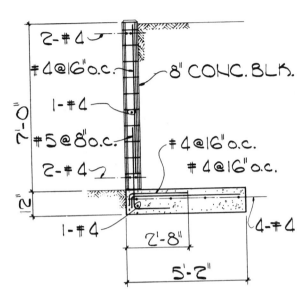

2-#4

#4@16"o.c.

1-#4

8" CONC. BLK.

7'-0"

#5@8"o.c.

2-#4

#4@16"o.c.
#4@16"o.c.

2"

1-#4

2'-8"

4-#4

5'-2"

Surcharge load = 100 p.s.f.
E.F.P. = 30 p.c.f.
Section 3.7.7

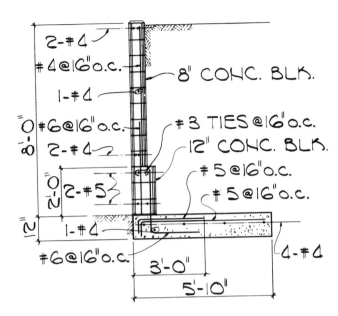

2-#4

#4@16"o.c.

1-#4

8" CONC. BLK.

8'-0"

#6@16"o.c.

2-#4

#3 TIES @16"o.c.
12" CONC. BLK.

#5 @16"o.c.
#5 @16"o.c.

2'-0"

2-#5

2"

1-#4

#6@16"o.c.

3'-0"

4-#4

5'-10"

Surcharge load = 100 p.s.f.
E.F.P. = 30 p.c.f.
Section 3.7.8

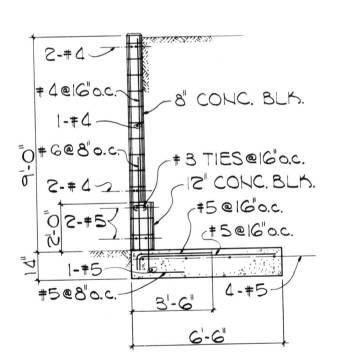

2-#4

#4@16"o.c.

1-#4

8" CONC. BLK.

9'-0"

#6@8"o.c.

2-#4

#3 TIES @16"o.c.
12" CONC. BLK.

#5 @16"o.c.
#5 @16"o.c.

2'-0"

2-#5

2"

4"

1-#5

#5@8"o.c.

3'-6"

4-#5

6'-6"

Surcharge load = 100 p.s.f.
E.F.P. = 30 p.c.f.
Section 3.7.9

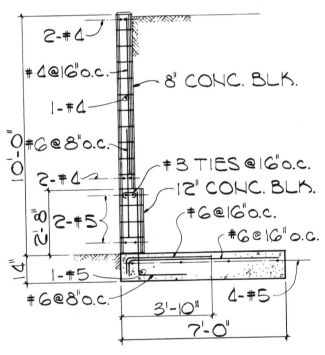

2-#4

#4@16"o.c.

1-#4

8" CONC. BLK.

10'-0"

#6@8"o.c.

2-#4

#3 TIES @16"o.c.
12" CONC. BLK.

#6@16"o.c.
#6 @16"o.c.

2'-8"

2-#5

2"

4"

1-#5

#6@8"o.c.

3'-10"

4-#5

7'-0"

Surcharge load = 100 p.s.f.
E.F.P. = 30 p.c.f.
Section 3.7.10

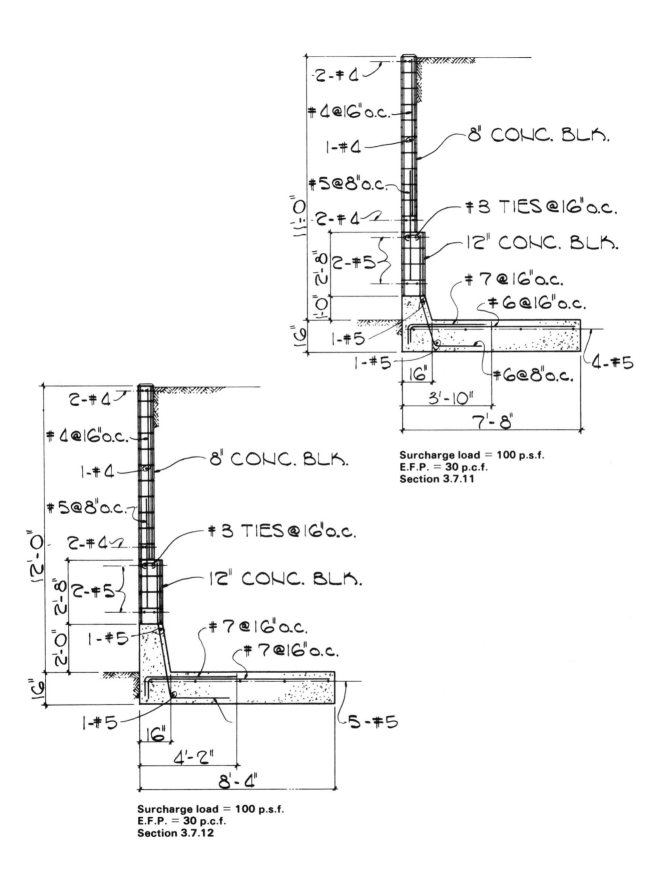

2-#4

#4@16"o.c.

1-#4

#5@8"o.c.

2-#4

2-#5

1-#5

1-#5

8" CONC. BLK.

#3 TIES@16"o.c.

12" CONC. BLK.

#7@16"o.c.

#6@16"o.c.

4-#5

#6@8"o.c.

11'-0"

2'-8"

1'-0"

9"

16"

3'-10"

7'-8"

Surcharge load = 100 p.s.f.
E.F.P. = 30 p.c.f.
Section 3.7.11

2-#4

#4@16"o.c.

1-#4

#5@8"o.c.

2-#4

2-#5

1-#5

1-#5

8" CONC. BLK.

#3 TIES@16"o.c.

12" CONC. BLK.

#7@16"o.c.

#7@16"o.c.

5-#5

12'-0"

2'-8"

2'-0"

9"

16"

4'-2"

8'-4"

Surcharge load = 100 p.s.f.
E.F.P. = 30 p.c.f.
Section 3.7.12

TABLE 3.8	PROPERTY LINE RETAINING WALLS — MASONRY STEM — SOIL OVER FOOTING
	SLOPE = 0 to 1 SURCHARGE = 200 lbs./sq. ft. AXIAL = 0 lbs./ft.

SOIL PRESSURE DIAGRAM

SURCHARGE LOAD

H1 = Concrete Stem
H2 = 12″ Concrete Block
H3 = 8″ Concrete Block

Stem Height	3'-0"	4'-0"	5'-0"	6'-0"	7'-0"	8'-0"	9'-0"	10'-0"	11'-0"	12'-0"
Ft. L ft.	3.167	3.833	4.500	5.167	5.667	6.333	7.000	7.667	8.333	8.833
Ft. T ft.	0.833	0.833	0.833	1.000	1.000	1.167	1.167	1.333	1.333	1.333
8 in. blk. ft.	3.000	4.000	5.000	6.000	5.000	6.000	6.333	6.667	6.667	6.667
12 in. blk. ft.	—	—	—	—	2.000	2.000	2.667	2.667	2.667	2.667
Conc. Stem	—	—	—	—	—	—	—	0.667	1.667	2.667
Wb ft.	—	—	—	—	—	—	—	1.333	1.333	1.333
W lbs.	2047	2927	3951	5247	6402	8019	9644	11610	13553	15374
F lbs.	510	700	920	1215	1500	1870	2220	2667	3082	3527
Mo ft.-lbs.	992	1595	2403	3645	5000	6962	9005	11852	14724	18025
Mr ft.-lbs.	3198	5436	8514	12899	17158	23954	31744	41774	52865	63368
O.T.R.	3.222	3.407	3.543	3.539	3.432	3.441	3.525	3.525	3.590	3.516
$\overline{X}$ ft.	1.077	1.312	1.547	1.764	1.899	2.119	2.358	2.577	2.814	2.949
e ft.	0.506	0.605	0.703	0.820	0.934	1.048	1.142	1.256	1.352	1.467
Me ft.-lbs.	1035	1770	2778	4300	5982	8403	11015	14583	18330	22559
S ft.^{3}	1.671	2.449	3.375	4.449	5.352	6.685	8.167	9.796	11.574	13.005
S.P.t p.s.f.	1266	1486	1701	1982	2248	2523	2726	3003	3210	3475
S.P.h p.s.f.	27	41	55	49	12	9	29	26	43	6
Friction	0.249	0.239	0.233	0.232	0.234	0.233	0.230	0.230	0.227	0.229
Inspct.	NO	NO	YES	YES	YES	YES	YES	YES	YES	YES

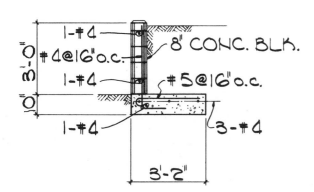

Surcharge load = 200 p.s.f.
E.F.P. = 30 p.c.f.
Section 3.8.3

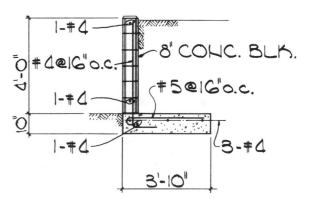

Surcharge load = 200 p.s.f.
E.F.P. = 30 p.c.f.
Section 3.8.4

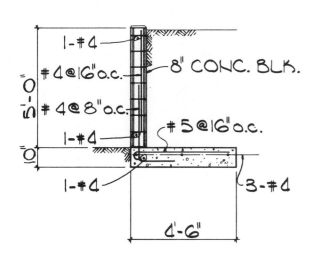

Surcharge load = 200 p.s.f.
E.F.P. = 30 p.c.f.
Section 3.8.5

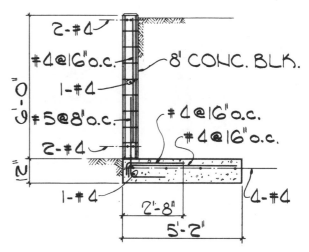

Surcharge load = 200 p.s.f.
E.F.P. = 30 p.c.f.
Section 3.8.6

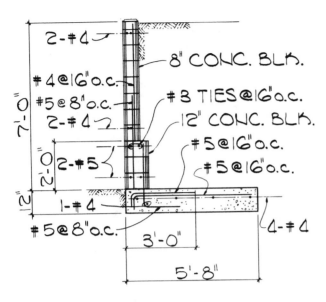

2-#4

#4@16"o.c.

#5@8"o.c.

2-#4

7'-0"

2'-0"

2-#5

4"

1-#4

#5@8"o.c.

8" CONC. BLK.

#3 TIES@16"o.c.

12" CONC. BLK.

#5@16"o.c.

#5@16"o.c.

4-#4

3'-0"

5'-8"

Surcharge load = 200 p.s.f.
E.F.P. = 30 p.c.f.
Section 3.8.7

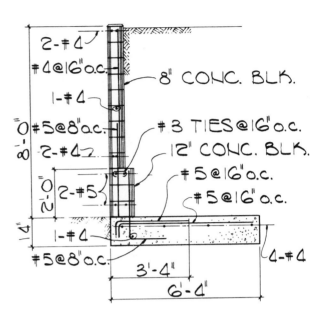

2-#4

#4@16"o.c.

1-#4

#5@8"o.c.

2-#4

8'-0"

2'-0"

2-#5

4"

1-#4

#5@8"o.c.

8" CONC. BLK.

#3 TIES@16"o.c.

12" CONC. BLK.

#5@16"o.c.

#5@16"o.c.

4-#4

3'-4"

6'-4"

Surcharge load = 200 p.s.f.
E.F.P. = 30 p.c.f.
Section 3.8.8

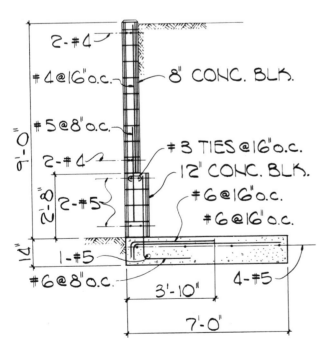

2-#4

#4@16"o.c.

#5@8"o.c.

2-#4

9'-0"

2'-8"

2-#5

4"

1-#5

#6@8"o.c.

8" CONC. BLK.

#3 TIES@16"o.c.

12" CONC. BLK.

#6@16"o.c.

#6@16"o.c.

4-#5

3'-10"

7'-0"

Surcharge load = 200 p.s.f.
E.F.P. = 30 p.c.f.
Section 3.8.9

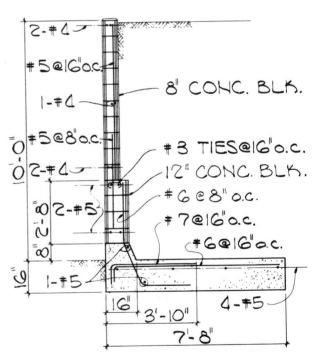

2-#4

#5@16"o.c.

1-#4

#5@8"o.c.

2-#4

10'-0"

2'-0"

2-#5

8"

1-#5

8" CONC. BLK.

#3 TIES@16"o.c.

12" CONC. BLK.

#6@8"o.c.

#7@16"o.c.

#6@16"o.c.

4-#5

16"

3'-10"

7'-8"

Surcharge load = 200 p.s.f.
E.F.P. = 30 p.c.f.
Section 3.8.10

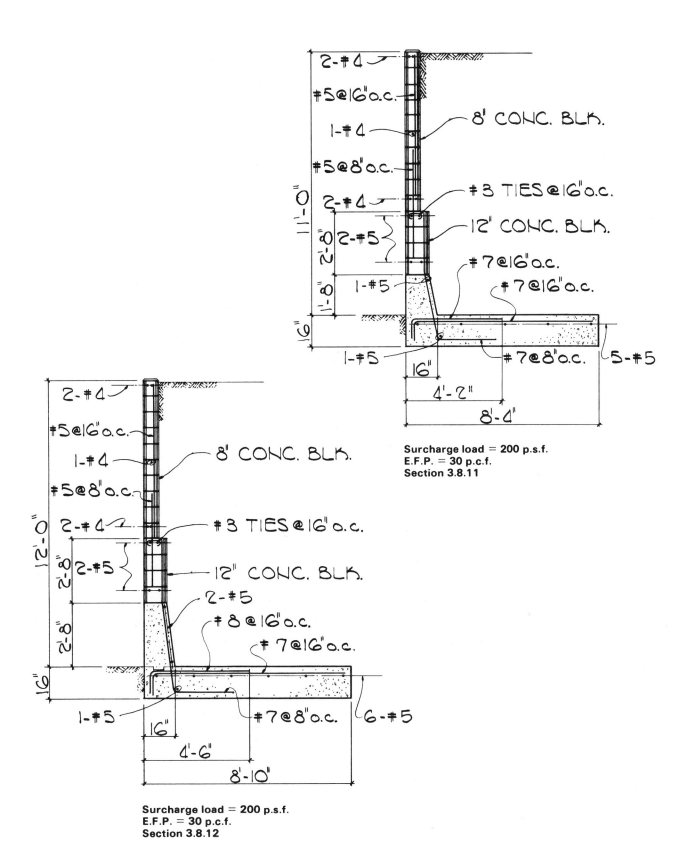

2-#4
#5@16"o.c.
1-#4
#5@8"o.c.
2-#4
2-#5
1-#5
1-#5

8" CONC. BLK.
#3 TIES @16"o.c.
12" CONC. BLK.
#7@16"o.c.
#7@16"o.c.
#7@8"o.c. **5-#5**

11'-0"   2'-8"   1'-8"   6"

16"
4'-2"
8'-4"

Surcharge load = 200 p.s.f.
E.F.P. = 30 p.c.f.
Section 3.8.11

2-#4
#5@16"o.c.
1-#4
#5@8"o.c.
2-#4
2-#5
2-#5
1-#5

8" CONC. BLK.
#3 TIES @16"o.c.
12" CONC. BLK.
#8 @16"o.c.
#7@16"o.c.
#7@8"o.c. **6-#5**

12'-0"   2'-8"   2'-8"   16"

16"
4'-6"
8'-10"

Surcharge load = 200 p.s.f.
E.F.P. = 30 p.c.f.
Section 3.8.12

TABLE 3.9	PROPERTY LINE RETAINING WALLS — MASONRY STEM — SOIL OVER FOOTING
	SLOPE = 5 to 1 SURCHARGE = 0 lbs./sq. ft. AXIAL = 0 lbs./ft.

SOIL PRESSURE DIAGRAM

H1 = Concrete Stem
H2 = 12″ Concrete Block
H3 = 8″ Concrete Block

Stem Height	3'- 0″	4'- 0″	5'- 0″	6'- 0″	7'- 0″	8'- 0″	9'- 0″	10'- 0″	11'- 0″	12'- 0″
Ft. L ft.	2.167	2.833	3.333	4.000	4.667	5.167	5.833	6.500	7.167	7.667
Ft. T ft.	0.667	0.833	0.833	0.833	1.000	1.000	1.167	1.167	1.333	1.333
8 in. blk. ft.	3.000	4.000	5.000	6.000	7.000	8.000	7.000	8.000	8.333	8.000
12 in. blk. ft.	—	—	—	—	—	—	2.000	2.000	2.667	2.667
Conc. Stem	—	—	—	—	—	—	—	—	—	1.333
Wb ft.	—	—	—	—	—	—	—	—	—	1.333
W lbs.	1013	1721	2414	3355	4565	5655	7208	8794	10702	12378
F lbs.	215	374	544	747	1024	1296	1654	1995	2434	2844
Mo ft.-lbs.	263	602	1059	1702	2731	3888	5604	7426	10006	12642
Mr ft.-lbs.	1034	2297	3777	6300	10022	13717	19768	26867	36098	44523
O.T.R.	3.932	3.814	3.568	3.702	3.670	3.528	3.527	3.618	3.608	3.522
$\overline{X}$ ft.	0.761	0.985	1.126	1.370	1.597	1.738	1.965	2.211	2.438	2.576
e ft.	0.323	0.432	0.541	0.630	0.736	0.845	0.952	1.039	1.145	1.258
Me ft.-lbs.	327	743	1306	2112	3361	4779	6858	9139	12257	15569
S ft.^{3}	0.782	1.338	1.852	2.667	3.630	4.449	5.671	7.042	8.560	9.796
S.P.t p.s.f.	885	1163	1429	1631	1904	2169	2445	2651	2925	3204
S.P.h p.s.f.	50	52	19	47	52	20	26	55	61	25
Friction	0.212	0.217	0.225	0.223	0.223	0.229	0.229	0.227	0.227	0.230
Inspct.	NO	NO	NO	NO	YES	YES	YES	YES	YES	YES

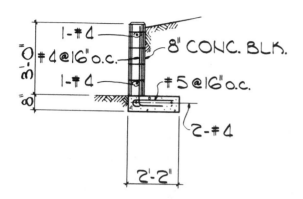

Soil slope = **5 to 1**
E.F.P. = **32** p.c.f.
Section **3.9.3**

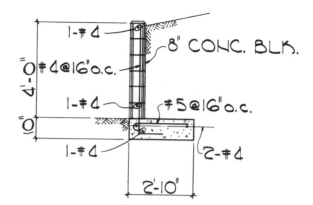

Soil slope = **5 to 1**
E.F.P. = **32** p.c.f.
Section **3.9.4**

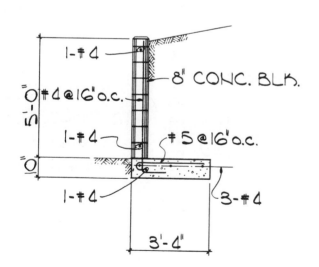

Soil slope = **5 to 1**
E.F.P. = **32** p.c.f.
Section **3.9.5**

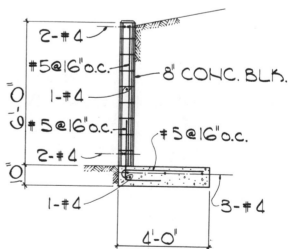

Soil slope = **5 to 1**
E.F.P. = **32** p.c.f.
Section **3.9.6**

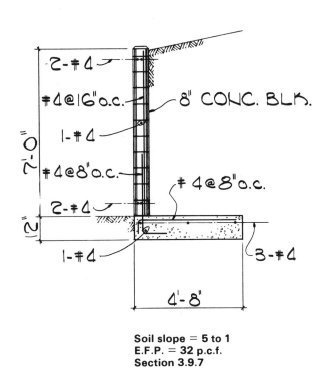

Soil slope = 5 to 1
E.F.P. = 32 p.c.f.
Section 3.9.7

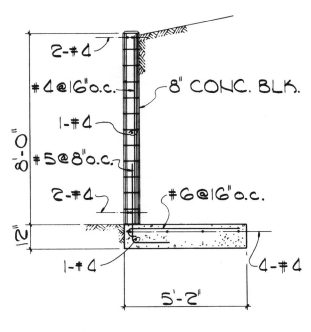

Soil slope = 5 to 1
E.F.P. = 32 p.c.f.
Section 3.9.8

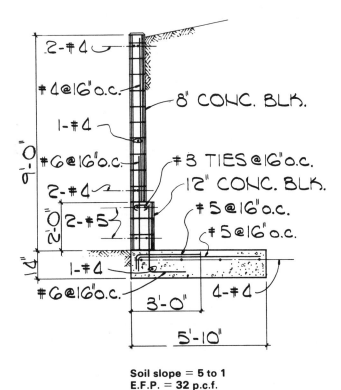

Soil slope = 5 to 1
E.F.P. = 32 p.c.f.
Section 3.9.9

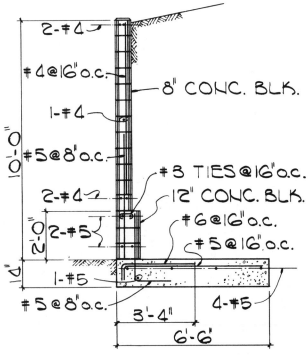

Soil slope = 5 to 1
E.F.P. = 32 p.c.f.
Section 3.9.10

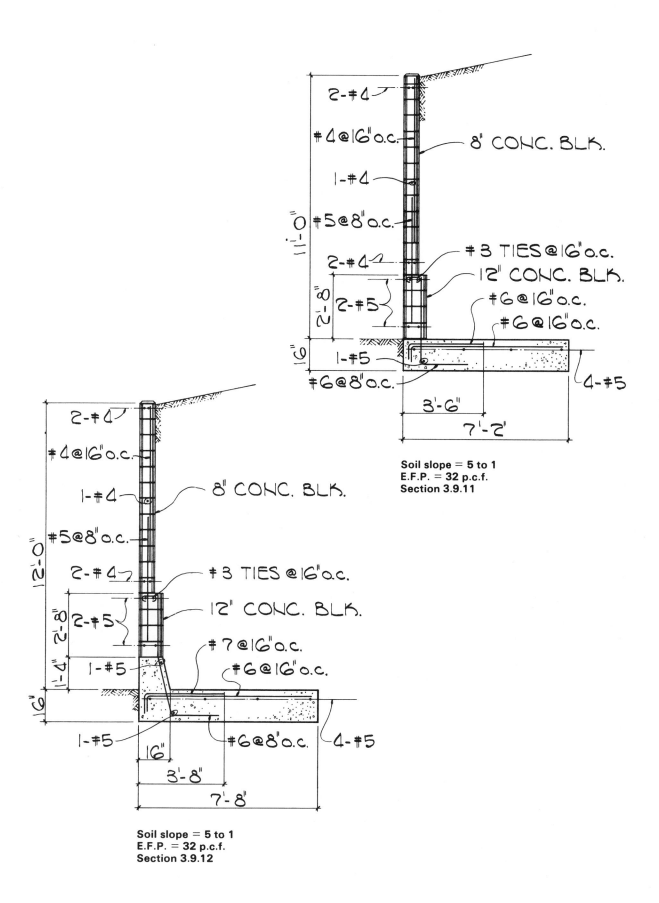

8" CONC. BLK.

2-#4
#4@16"o.c.
1-#4
#5@8"o.c.
2-#4
2-#5

#3 TIES@16"o.c.
12" CONC. BLK.
#6@16"o.c.
#6@16"o.c.

1-#5
#6@8"o.c.
4-#5

11'-0"
2'-8"
6"

3'-6"
7'-2"

Soil slope = 5 to 1
E.F.P. = 32 p.c.f.
Section 3.9.11

2-#4
#4@16"o.c.
1-#4
#5@8"o.c.
2-#4
2-#5
1-#5

8" CONC. BLK.

#3 TIES@16"o.c.
12" CONC. BLK.
#7@16"o.c.
#6@16"o.c.

1-#5
#6@8"o.c.
4-#5

12'-0"
2'-8"
1'-4"
6"

16"
3'-8"
7'-8"

Soil slope = 5 to 1
E.F.P. = 32 p.c.f.
Section 3.9.12

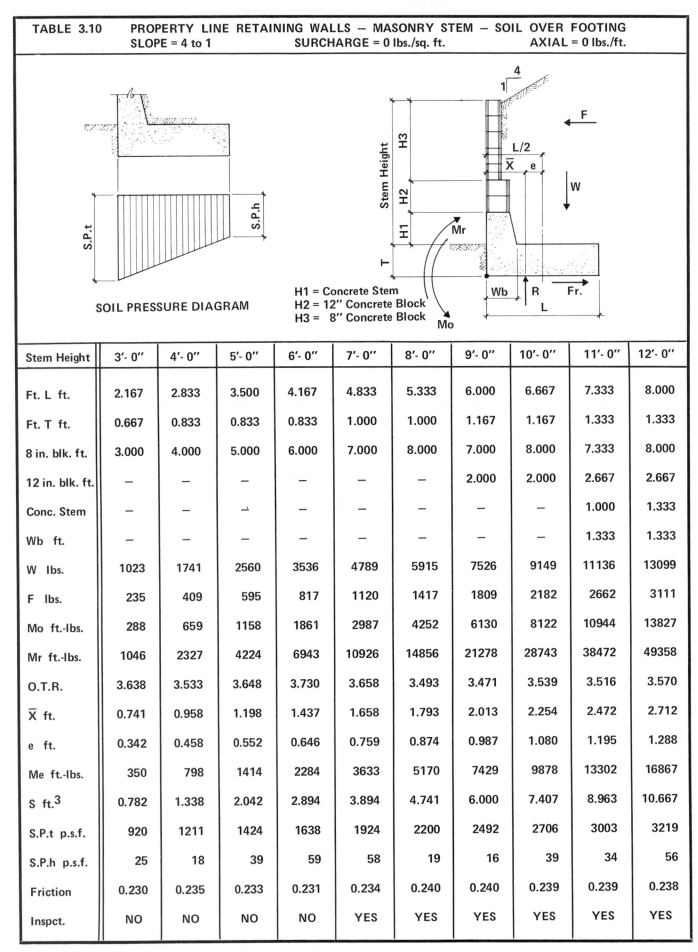

TABLE 3.10   PROPERTY LINE RETAINING WALLS — MASONRY STEM — SOIL OVER FOOTING
SLOPE = 4 to 1   SURCHARGE = 0 lbs./sq. ft.   AXIAL = 0 lbs./ft.

H1 = Concrete Stem
H2 = 12″ Concrete Block
H3 = 8″ Concrete Block

SOIL PRESSURE DIAGRAM

Stem Height	3′- 0″	4′- 0″	5′- 0″	6′- 0″	7′- 0″	8′- 0″	9′- 0″	10′- 0″	11′- 0″	12′- 0″
Ft. L ft.	2.167	2.833	3.500	4.167	4.833	5.333	6.000	6.667	7.333	8.000
Ft. T ft.	0.667	0.833	0.833	0.833	1.000	1.000	1.167	1.167	1.333	1.333
8 in. blk. ft.	3.000	4.000	5.000	6.000	7.000	8.000	7.000	8.000	7.333	8.000
12 in. blk. ft.	—	—	—	—	—	—	2.000	2.000	2.667	2.667
Conc. Stem	—	—	—	—	—	—	—	—	1.000	1.333
Wb ft.	—	—	—	—	—	—	—	—	1.333	1.333
W lbs.	1023	1741	2560	3536	4789	5915	7526	9149	11136	13099
F lbs.	235	409	595	817	1120	1417	1809	2182	2662	3111
Mo ft.-lbs.	288	659	1158	1861	2987	4252	6130	8122	10944	13827
Mr ft.-lbs.	1046	2327	4224	6943	10926	14856	21278	28743	38472	49358
O.T.R.	3.638	3.533	3.648	3.730	3.658	3.493	3.471	3.539	3.516	3.570
$\overline{X}$ ft.	0.741	0.958	1.198	1.437	1.658	1.793	2.013	2.254	2.472	2.712
e ft.	0.342	0.458	0.552	0.646	0.759	0.874	0.987	1.080	1.195	1.288
Me ft.-lbs.	350	798	1414	2284	3633	5170	7429	9878	13302	16867
S ft.^{3}	0.782	1.338	2.042	2.894	3.894	4.741	6.000	7.407	8.963	10.667
S.P.t p.s.f.	920	1211	1424	1638	1924	2200	2492	2706	3003	3219
S.P.h p.s.f.	25	18	39	59	58	19	16	39	34	56
Friction	0.230	0.235	0.233	0.231	0.234	0.240	0.240	0.239	0.239	0.238
Inspct.	NO	NO	NO	NO	YES	YES	YES	YES	YES	YES

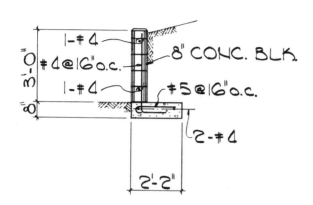

Soil slope = **4 to 1**
E.F.P. = **35** p.c.f.
Section **3.10.3**

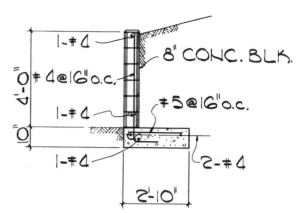

Soil slope = **4 to 1**
E.F.P. = **35** p.c.f.
Section **3.10.4**

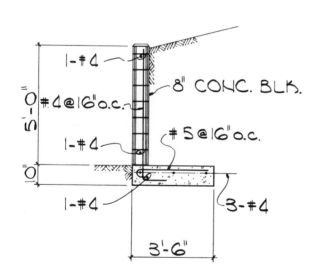

Soil slope = **4 to 1**
E.F.P. = **35** p.c.f
Section **3.10.5**

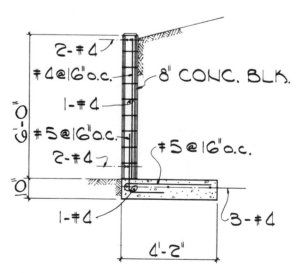

Soil slope = **4 to 1**
E.F.P. = **35** p.c.f.
Section **3.10.6**

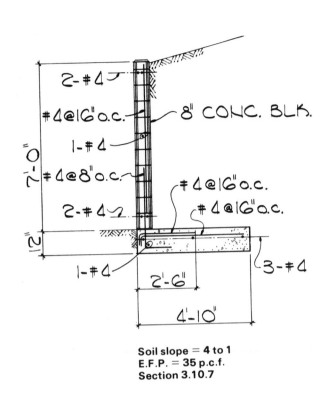

2-#4

#4@16"o.c.

1-#4

#4@8"o.c.

2-#4

7'-0"

2"

8" CONC. BLK.

#4@16"o.c.

#4@16"o.c.

1-#4

2-#4

3-#4

2'-6"

4'-10"

Soil slope = **4 to 1**
E.F.P. = **35** p.c.f.
Section 3.10.7

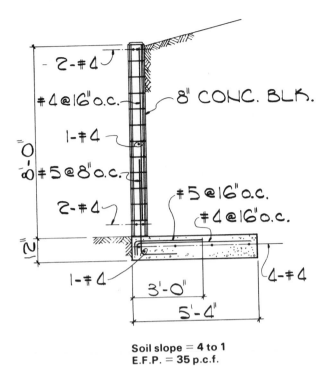

2-#4

#4@16"o.c.

1-#4

#5@8"o.c.

2-#4

8'-0"

2"

8" CONC. BLK.

#5@16"o.c.

#4@16"o.c.

1-#4

4-#4

3'-0"

5'-4"

Soil slope = **4 to 1**
E.F.P. = **35** p.c.f.
Section 3.10.8

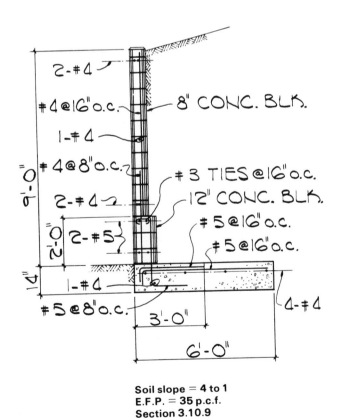

2-#4

#4@16"o.c.

1-#4

#4@8"o.c.

2-#4

9'-0"

2'-0"

2"

4"

2-#5

8" CONC. BLK.

#3 TIES @16"o.c.

12" CONC. BLK.

#5@16"o.c.

#5@16"o.c.

1-#4

#5@8"o.c.

3'-0"

4-#4

6'-0"

Soil slope = **4 to 1**
E.F.P. = **35** p.c.f.
Section 3.10.9

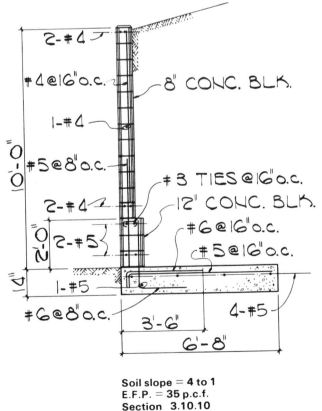

2-#4

#4@16"o.c.

1-#4

#5@8"o.c.

2-#4

10'-0"

2'-0"

2"

4"

2-#5

8" CONC. BLK.

#3 TIES @16"o.c.

12" CONC. BLK.

#6@16"o.c.

#5@16"o.c.

1-#5

#6@8"o.c.

3'-6"

4-#5

6'-8"

Soil slope = **4 to 1**
E.F.P. = **35** p.c.f.
Section 3.10.10

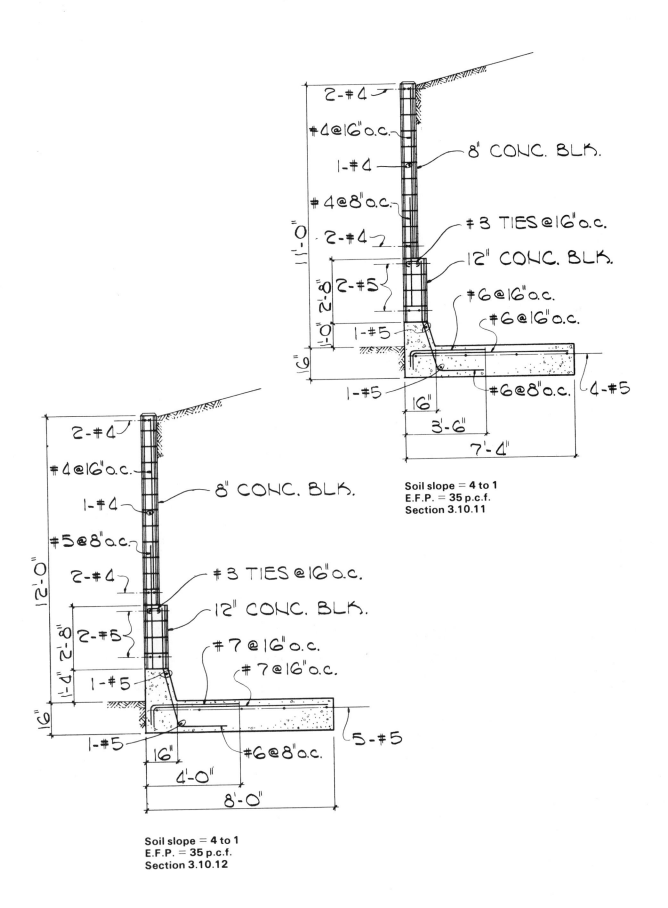

2-#4

#4@16"o.c.

1-#4

8" CONC. BLK.

#4@8"o.c.

2-#4

#3 TIES @16"o.c.

12" CONC. BLK.

2-#5

#6@16"o.c.

#6@16"o.c.

1-#5

1-0"

2'-8"

1'-0"

6"

1-#5

#6@8"o.c.

4-#5

16"

3'-6"

7'-4"

Soil slope = 4 to 1
E.F.P. = 35 p.c.f.
Section 3.10.11

2-#4

#4@16"o.c.

1-#4

8" CONC. BLK.

#5@8"o.c.

2-#4

#3 TIES @16"o.c.

12" CONC. BLK.

2-#5

#7 @16"o.c.

#7@16"o.c.

1-#5

1-#5

5-#5

16"

#6@8"o.c.

4'-0"

8'-0"

12'-0"

2'-8"

1'-4"

16"

Soil slope = 4 to 1
E.F.P. = 35 p.c.f.
Section 3.10.12

TABLE 3.11   PROPERTY LINE RETAINING WALLS — MASONRY STEM — SOIL OVER FOOTING
SLOPE = 3 to 1   SURCHARGE = 0 lbs./sq. ft.   AXIAL = 0 lbs./ft.

SOIL PRESSURE DIAGRAM

H1 = Concrete Stem
H2 = 12″ Concrete Block
H3 = 8″ Concrete Block

Stem Height	3′- 0″	4′- 0″	5′- 0″	6′- 0″	7′- 0″	8′- 0″	9′- 0″	10′- 0″	11′- 0″	12′- 0″
Ft. L ft.	2.167	3.000	3.500	4.167	4.833	5.500	6.167	6.833	7.500	8.167
Ft. T ft.	0.667	0.833	0.833	0.833	1.000	1.000	1.167	1.167	1.333	1.333
8 in. blk. ft.	3.000	4.000	5.000	6.000	7.000	8.000	7.000	8.000	8.000	8.000
12 in. blk. ft.	—	—	—	—	—	—	2.000	2.000	2.000	2.000
Conc. Stem	—	—	—	—	—	—	—	—	1.000	2.000
Wb ft.	—	—	—	—	—	—	—	—	1.333	1.333
W lbs.	1037	1868	2606	3605	4885	6222	7894	9579	11636	13682
F lbs.	255	444	647	887	1216	1539	1964	2369	2890	3378
Mo ft.-lbs.	312	715	1257	2021	3243	4617	6655	8819	11882	15012
Mr ft.-lbs.	1065	2665	4318	7109	11191	16221	23081	31029	41351	52894
O.T.R.	3.410	3.727	3.435	3.518	3.451	3.513	3.468	3.519	3.480	3.523
$\overline{X}$ ft.	0.726	1.044	1.174	1.411	1.627	1.865	2.081	2.319	2.533	2.769
e ft.	0.358	0.456	0.576	0.672	0.790	0.885	1.003	1.098	1.217	1.315
Me ft.-lbs.	371	852	1500	2423	3858	5508	7914	10518	14164	17987
S ft.^{3}	0.782	1.500	2.042	2.894	3.894	5.042	6.338	7.782	9.375	11.116
S.P.t p.s.f.	953	1191	1480	1702	2002	2224	2529	2753	3062	3294
S.P.h p.s.f.	4	55	10	28	20	39	31	50	41	57
Friction	0.246	0.238	0.248	0.246	0.249	0.247	0.249	0.247	0.248	0.247
Inspct.	NO	NO	NO	NO	YES	YES	YES	YES	YES	YES

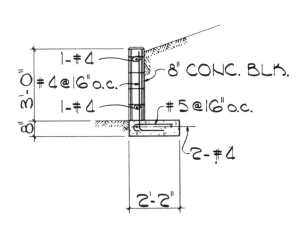

Soil slope = 3 to 1
E.F.P. = 38 p.c.f.
Section 3.11.3

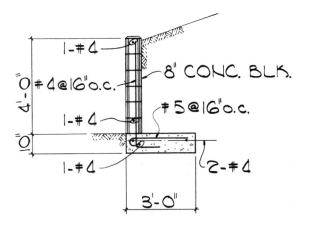

Soil slope = 3 to 1
E.F.P. = 38 p.c.f.
Section 3.11.4

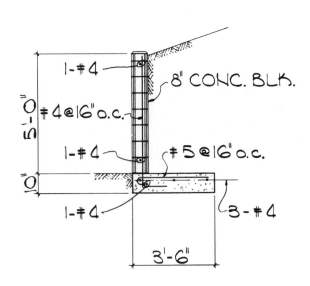

Soil slope = 3 to 1
E.F.P. = 38 p.c.f.
Section 3.11.5

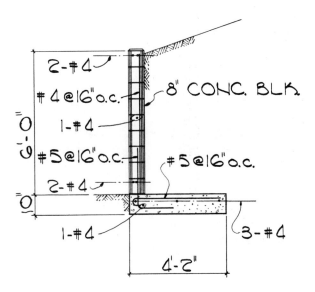

Soil slope = 3 to 1
E.F.P. = 38 p.c.f.
Section 3.11.6

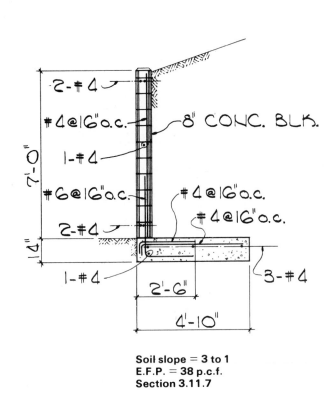

Soil slope = 3 to 1
E.F.P. = 38 p.c.f.
Section 3.11.7

8" CONC. BLK.

2-#4
#4@16" o.c.
1-#4
#6@16" o.c.
2-#4
#4@16" o.c.
#4@16" o.c.
1-#4
3-#4
7'-0"
4"
2'-6"
4'-10"

Soil slope = 3 to 1
E.F.P. = 38 p.c.f.
Section 3.11.8

2-#4
#4@16" o.c.
1-#4
#5@8" o.c.
2-#4
1-#4
8" CONC. BLK.
#6@16" o.c.
#4@16" o.c.
4-#4
8'-0"
2"
3'-0"
5'-6"

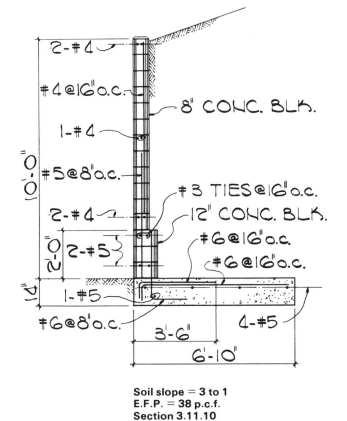

Soil slope = 3 to 1
E.F.P. = 38 p.c.f.
Section 3.11.9

2-#4
#4@16" o.c.
1-#4
#6@16" o.c.
2-#4
2-#5
1-#4
#5@8" o.c.
8" CONC. BLK.
#3 TIES @16" o.c.
12" CONC. BLK.
#5@16" o.c.
#5@16" o.c.
4-#4
9'-0"
2'-0"
4"
3'-0"
6'-2"

Soil slope = 3 to 1
E.F.P. = 38 p.c.f.
Section 3.11.10

2-#4
#4@16" o.c.
1-#4
#5@8" o.c.
2-#4
2-#5
1-#5
#6@8" o.c.
8" CONC. BLK.
#3 TIES @16" o.c.
12" CONC. BLK.
#6@16" o.c.
#6@16" o.c.
4-#5
10'-0"
2'-0"
4"
3'-6"
6'-10"

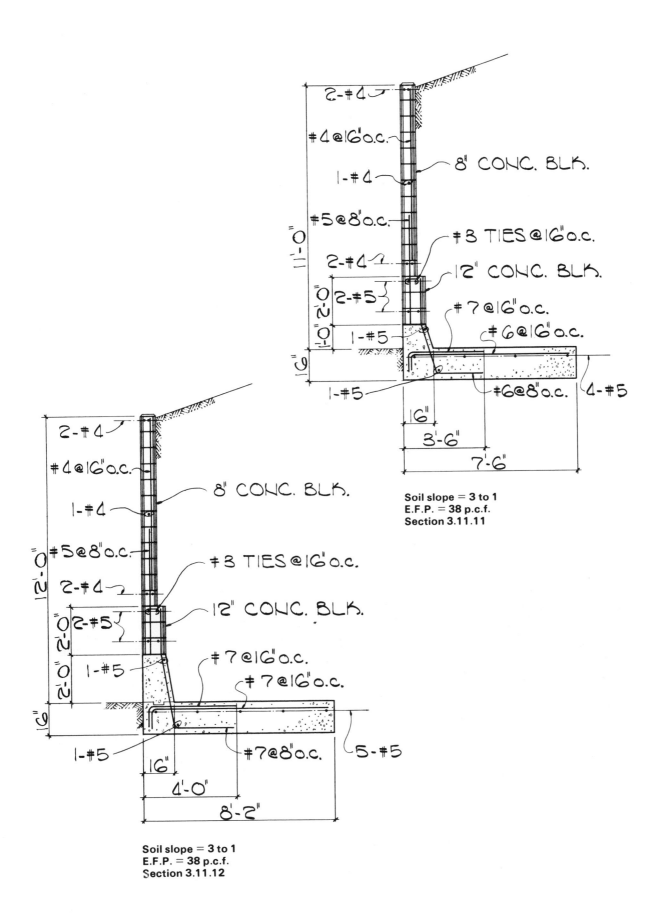

2-#4

#4@16"o.c.

1-#4

8" CONC. BLK.

#5@8"o.c.

#3 TIES@16"o.c.

2-#4

12" CONC. BLK.

2-#5

#7@16"o.c.

1-#5

#6@16"o.c.

1-#5

#6@8"o.c.

4-#5

11'-0"

2'-0"

1'-0"

6"

6"

3'-6"

7'-6"

Soil slope = 3 to 1
E.F.P. = 38 p.c.f.
Section 3.11.11

2-#4

#4@16"o.c.

1-#4

8" CONC. BLK.

#5@8"o.c.

#3 TIES@16"o.c.

2-#4

12" CONC. BLK.

2-#5

#7@16"o.c.

1-#5

#7@16"o.c.

1-#5

#7@8"o.c.

5-#5

12'-0"

2'-0"

2'-0"

6"

16"

4'-0"

8'-2"

Soil slope = 3 to 1
E.F.P. = 38 p.c.f.
Section 3.11.12

TABLE 3.12   PROPERTY LINE RETAINING WALLS – MASONRY STEM – SOIL OVER FOOTING
SLOPE = 2 to 1   SURCHARGE = 0 lbs./sq. ft.   AXIAL = 0 lbs./ft.

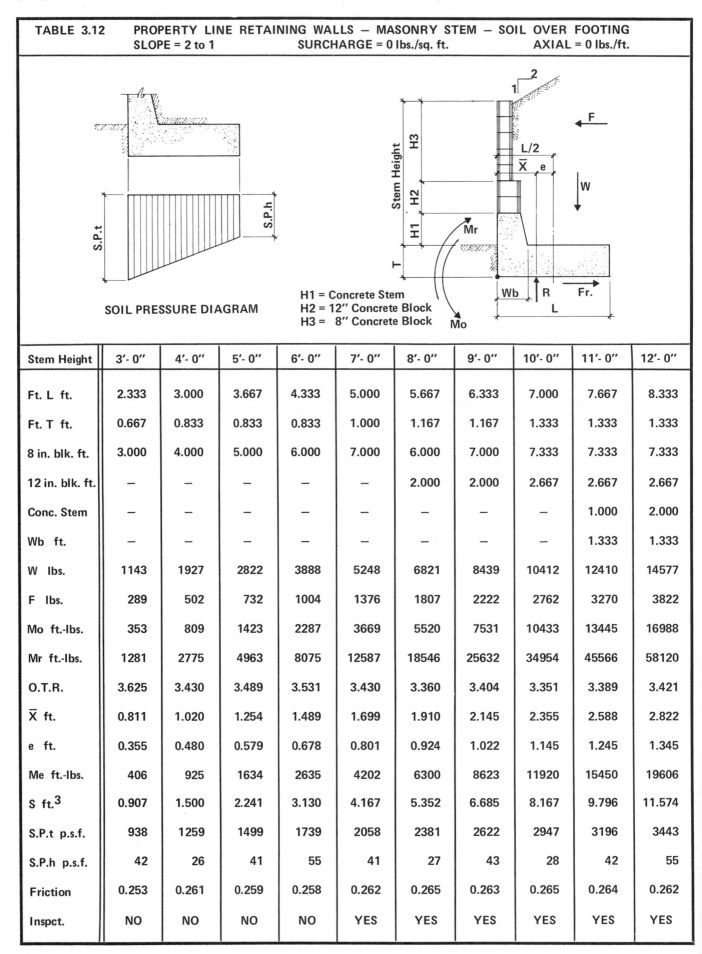

SOIL PRESSURE DIAGRAM

H1 = Concrete Stem
H2 = 12″ Concrete Block
H3 = 8″ Concrete Block

Stem Height	3′- 0″	4′- 0″	5′- 0″	6′- 0″	7′- 0″	8′- 0″	9′- 0″	10′- 0″	11′- 0″	12′- 0″
Ft. L ft.	2.333	3.000	3.667	4.333	5.000	5.667	6.333	7.000	7.667	8.333
Ft. T ft.	0.667	0.833	0.833	0.833	1.000	1.167	1.167	1.333	1.333	1.333
8 in. blk. ft.	3.000	4.000	5.000	6.000	7.000	6.000	7.000	7.333	7.333	7.333
12 in. blk. ft.	–	–	–	–	–	2.000	2.000	2.667	2.667	2.667
Conc. Stem	–	–	–	–	–	–	–	–	1.000	2.000
Wb ft.	–	–	–	–	–	–	–	–	1.333	1.333
W lbs.	1143	1927	2822	3888	5248	6821	8439	10412	12410	14577
F lbs.	289	502	732	1004	1376	1807	2222	2762	3270	3822
Mo ft.-lbs.	353	809	1423	2287	3669	5520	7531	10433	13445	16988
Mr ft.-lbs.	1281	2775	4963	8075	12587	18546	25632	34954	45566	58120
O.T.R.	3.625	3.430	3.489	3.531	3.430	3.360	3.404	3.351	3.389	3.421
$\overline{X}$ ft.	0.811	1.020	1.254	1.489	1.699	1.910	2.145	2.355	2.588	2.822
e ft.	0.355	0.480	0.579	0.678	0.801	0.924	1.022	1.145	1.245	1.345
Me ft.-lbs.	406	925	1634	2635	4202	6300	8623	11920	15450	19606
S ft.^{3}	0.907	1.500	2.241	3.130	4.167	5.352	6.685	8.167	9.796	11.574
S.P.t p.s.f.	938	1259	1499	1739	2058	2381	2622	2947	3196	3443
S.P.h p.s.f.	42	26	41	55	41	27	43	28	42	55
Friction	0.253	0.261	0.259	0.258	0.262	0.265	0.263	0.265	0.264	0.262
Inspct.	NO	NO	NO	NO	YES	YES	YES	YES	YES	YES

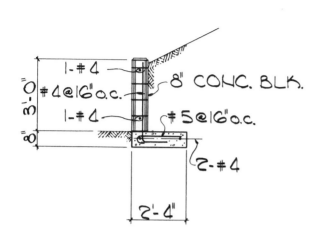

Soil slope = 2 to 1
E.F.P. = 43 p.c.f.
Section 3.12.3

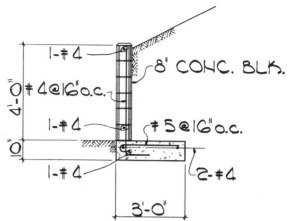

Soil slope = 2 to 1
E.F.P. = 43 p.c.f.
Section 3.12.4

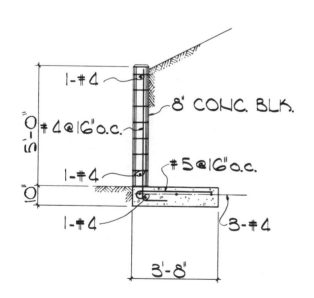

Soil slope = 2 to 1
E.F.P. = 43 p.c.f.
Section 3.12.5

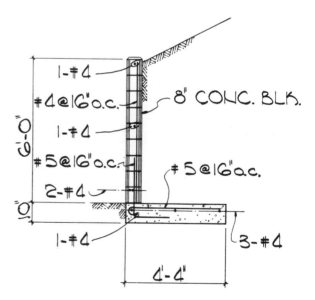

Soil slope = 2 to 1
E.F.P. = 43 p.c.f.
Section 3.12.6

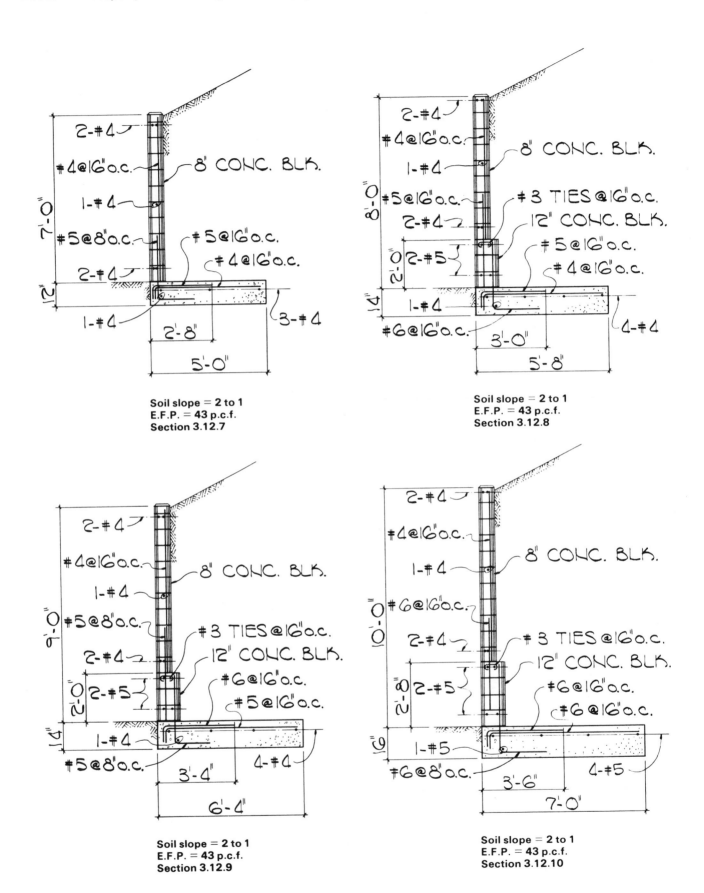

Soil slope = **2 to 1**
E.F.P. = **43 p.c.f.**
Section **3.12.7**

Soil slope = **2 to 1**
E.F.P. = **43 p.c.f.**
Section **3.12.8**

Soil slope = **2 to 1**
E.F.P. = **43 p.c.f.**
Section **3.12.9**

Soil slope = **2 to 1**
E.F.P. = **43 p.c.f.**
Section **3.12.10**

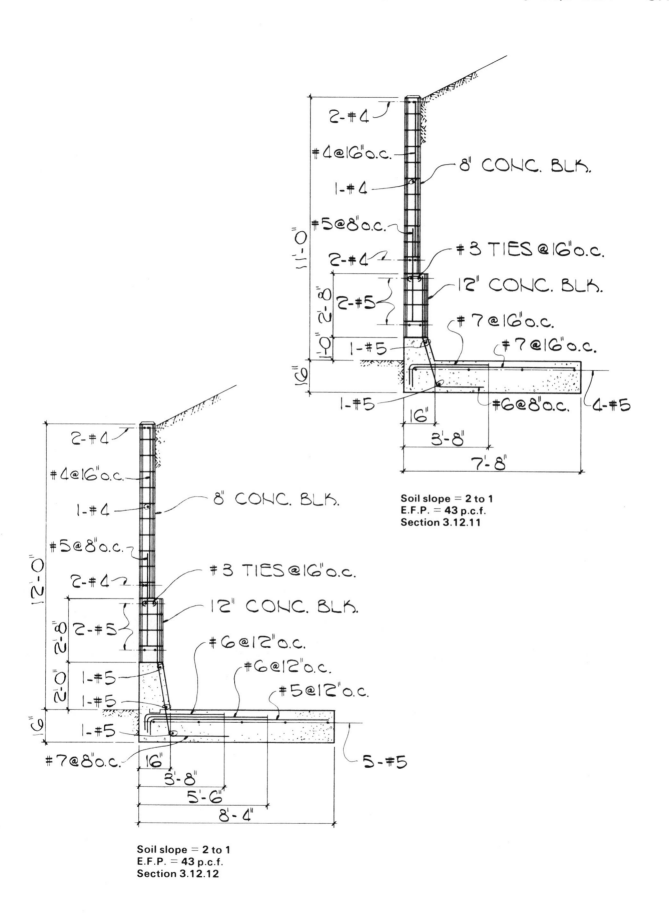

2-#4

#4@16"o.c.

1-#4

8" CONC. BLK.

#5@8"o.c.

2-#4

#3 TIES @16"o.c.

12" CONC. BLK.

2-#5

#7@16"o.c.

1-#5

#7@16"o.c.

1-#5

#6@8"o.c.

4-#5

16"

3'-8"

7'-8"

1'-0"

2'-8"

9"

6"

Soil slope = 2 to 1
E.F.P. = 43 p.c.f.
Section 3.12.11

2-#4

#4@16"o.c.

1-#4

8" CONC. BLK.

#5@8"o.c.

2-#4

#3 TIES @16"o.c.

12" CONC. BLK.

2-#5

#6@12"o.c.

1-#5

#6@12"o.c.

1-#5

#5@12"o.c.

1-#5

#7@8"o.c.

5-#5

16"

3'-8"

5'-6"

8'-4"

12'-0"

2'-8"

2'-0"

9"

Soil slope = 2 to 1
E.F.P. = 43 p.c.f.
Section 3.12.12

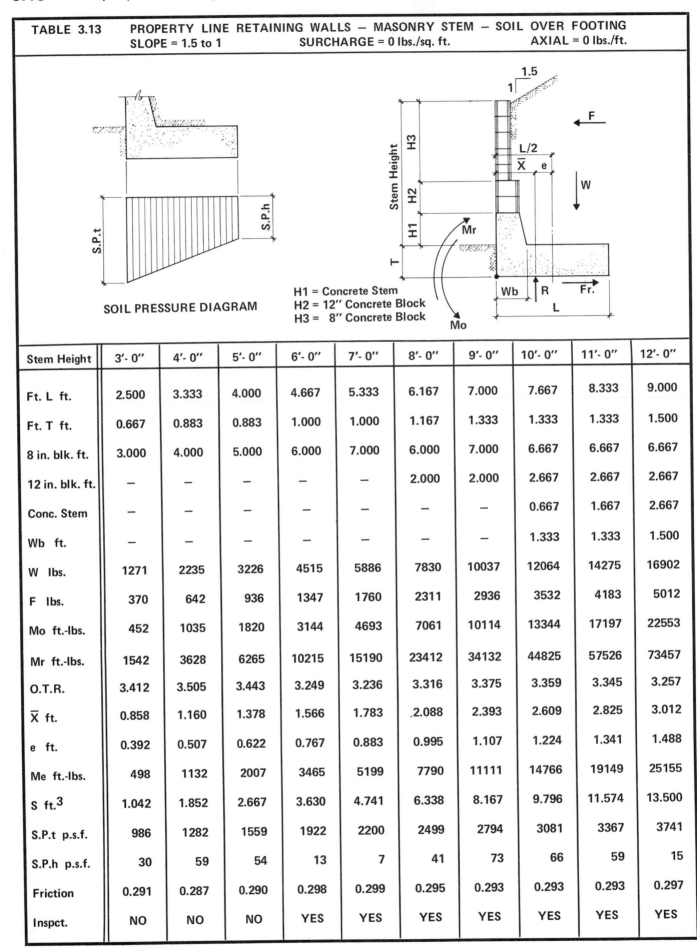

TABLE 3.13 PROPERTY LINE RETAINING WALLS — MASONRY STEM — SOIL OVER FOOTING
SLOPE = 1.5 to 1   SURCHARGE = 0 lbs./sq. ft.   AXIAL = 0 lbs./ft.

H1 = Concrete Stem
H2 = 12" Concrete Block
H3 = 8" Concrete Block

SOIL PRESSURE DIAGRAM

Stem Height	3'- 0"	4'- 0"	5'- 0"	6'- 0"	7'- 0"	8'- 0"	9'- 0"	10'- 0"	11'- 0"	12'- 0"
Ft. L ft.	2.500	3.333	4.000	4.667	5.333	6.167	7.000	7.667	8.333	9.000
Ft. T ft.	0.667	0.883	0.883	1.000	1.000	1.167	1.333	1.333	1.333	1.500
8 in. blk. ft.	3.000	4.000	5.000	6.000	7.000	6.000	7.000	6.667	6.667	6.667
12 in. blk. ft.	—	—	—	—	—	2.000	2.000	2.667	2.667	2.667
Conc. Stem	—	—	—	—	—	—	—	0.667	1.667	2.667
Wb ft.	—	—	—	—	—	—	—	1.333	1.333	1.500
W lbs.	1271	2235	3226	4515	5886	7830	10037	12064	14275	16902
F lbs.	370	642	936	1347	1760	2311	2936	3532	4183	5012
Mo ft.-lbs.	452	1035	1820	3144	4693	7061	10114	13344	17197	22553
Mr ft.-lbs.	1542	3628	6265	10215	15190	23412	34132	44825	57526	73457
O.T.R.	3.412	3.505	3.443	3.249	3.236	3.316	3.375	3.359	3.345	3.257
$\overline{X}$ ft.	0.858	1.160	1.378	1.566	1.783	2.088	2.393	2.609	2.825	3.012
e ft.	0.392	0.507	0.622	0.767	0.883	0.995	1.107	1.224	1.341	1.488
Me ft.-lbs.	498	1132	2007	3465	5199	7790	11111	14766	19149	25155
S ft.^{3}	1.042	1.852	2.667	3.630	4.741	6.338	8.167	9.796	11.574	13.500
S.P.t p.s.f.	986	1282	1559	1922	2200	2499	2794	3081	3367	3741
S.P.h p.s.f.	30	59	54	13	7	41	73	66	59	15
Friction	0.291	0.287	0.290	0.298	0.299	0.295	0.293	0.293	0.293	0.297
Inspct.	NO	NO	NO	YES	YES	YES	YES	YES	YES	YES

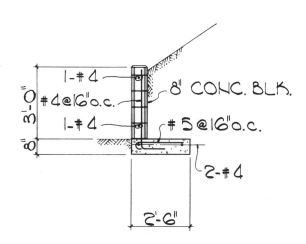

Soil slope = 1½ to 1
E.F.P. = 55 p.c.f.
Section 3.13.3

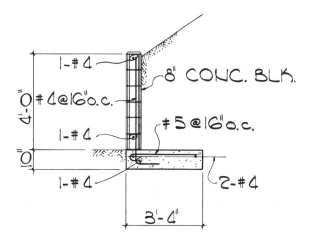

Soil slope = 1½ to 1
E.F.P. = 55 p.c.f.
Section 3.13.4

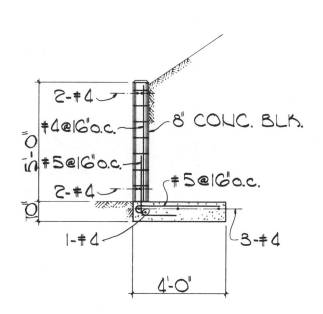

Soil slope = 1½ to 1
E.F.P. = 55 p.c.f.
Section 3.13.5

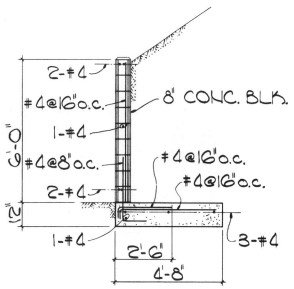

Soil slope = 1½ to 1
E.F.P. = 55 p.c.f.
Section 3.13.6

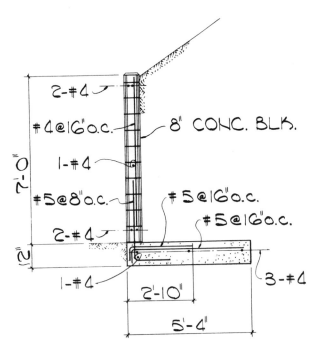

2-#4
#4@16"o.c.
1-#4
#5@8"o.c.
2-#4
1-#4

8" CONC. BLK.

#5@16"o.c.
#5@16"o.c.

3-#4

7'-0"
2"
2'-10"
5'-4"

Soil slope = 1½ to 1
E.F.P. = 55 p.c.f.
Section 3.13.7

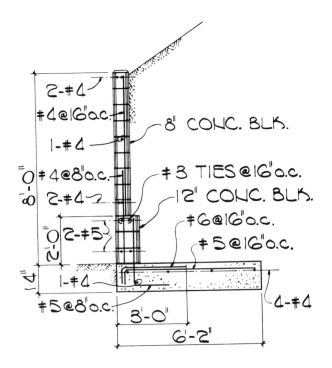

2-#4
#4@16"o.c.
1-#4
#4@8"o.c.
2-#4
2-#5
1-#4
#5@8"o.c.

8" CONC. BLK.
#3 TIES@16"o.c.
12" CONC. BLK.
#6@16"o.c.
#5@16"o.c.

4-#4

8'-0"
2'-0"
4"
3'-0"
6'-2"

Soil slope = 1½ to 1
E.F.P. = 55 p.c.f.
Section 3.13.8

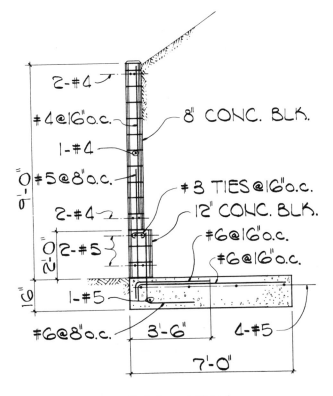

2-#4
#4@16"o.c.
1-#4
#5@8"o.c.
2-#4
2-#5
1-#5
#6@8"o.c.

8" CONC. BLK.

#3 TIES@16"o.c.
12" CONC. BLK.
#6@16"o.c.
#6@16"o.c.

4-#5

9'-0"
2'-0"
6"
3'-6"
7'-0"

Soil slope = 1½ to 1
E.F.P. = 55 p.c.f.
Section 3.13.9

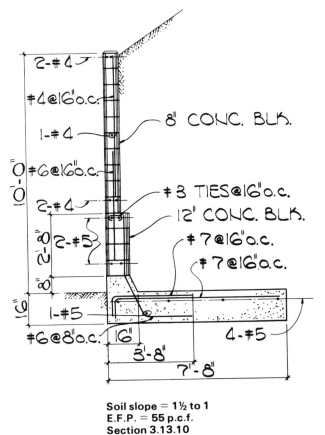

2-#4
#4@16"o.c.
1-#4
#6@16"o.c.
2-#4
2-#5
1-#5
#6@8"o.c.

8" CONC. BLK.

#3 TIES@16"o.c.
12" CONC. BLK.
#7@16"o.c.
#7@16"o.c.

4-#5

10'-0"
2'-0"
8"
6"
16"
3'-8"
7'-8"

Soil slope = 1½ to 1
E.F.P. = 55 p.c.f.
Section 3.13.10

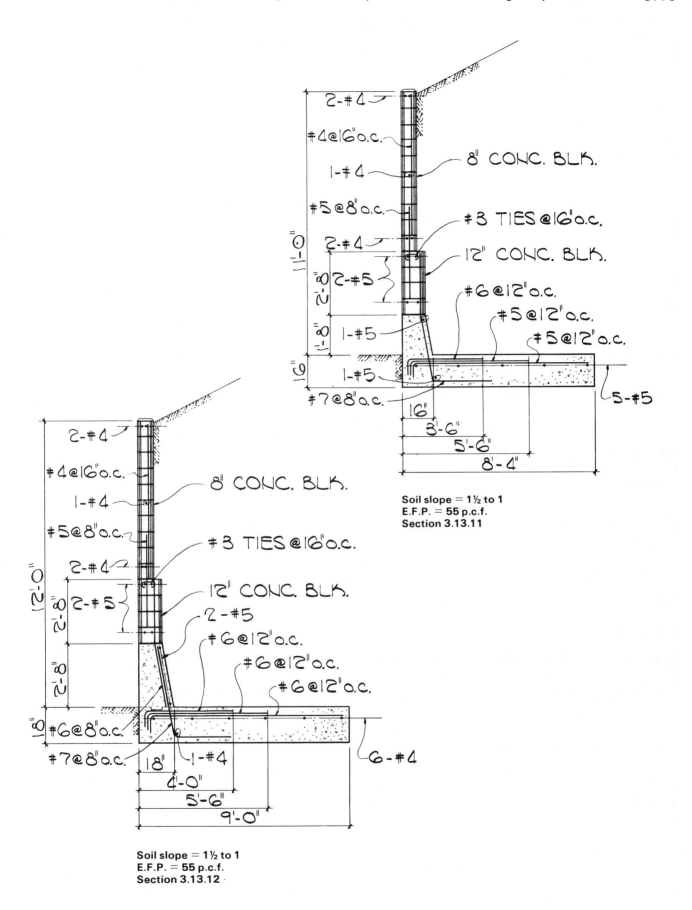

2-#4
#4@16"o.c.
1-#4
#5@8"o.c.
2-#4
2-#5
1-#5
#7@8"o.c.

8" CONC. BLK.
#3 TIES @16"o.c.
12" CONC. BLK.
#6@12"o.c.
#5@12"o.c.
#5@12"o.c.
5-#5

11'-0"
2'-0"
1'-0"
6"
16"
3'-6"
5'-6"
8'-4"

Soil slope = 1½ to 1
E.F.P. = 55 p.c.f.
Section 3.13.11

2-#4
#4@16"o.c.
1-#4
#5@8"o.c.
2-#4
2-#5
2-#5
#6@12"o.c.
#6@12"o.c.
#6@12"o.c.
#6@8"o.c.
#7@8"o.c.
1-#4
6-#4

8" CONC. BLK.
#3 TIES @16"o.c.
12" CONC. BLK.

12'-0"
2'-0"
2'-0"
18"
18"
4'-0"
5'-6"
9'-0"

Soil slope = 1½ to 1
E.F.P. = 55 p.c.f.
Section 3.13.12

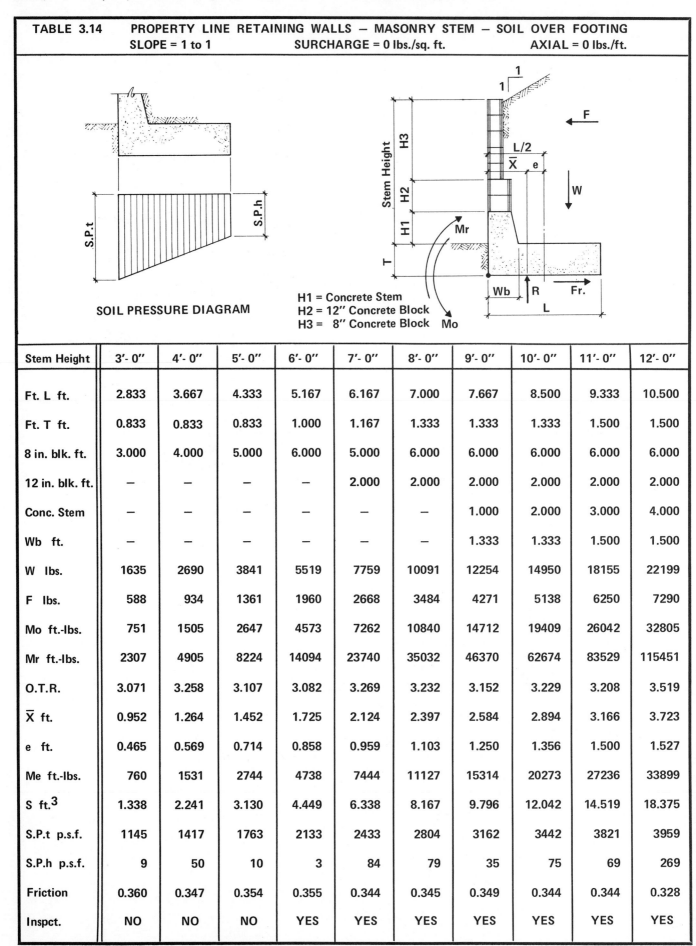

TABLE 3.14   PROPERTY LINE RETAINING WALLS — MASONRY STEM — SOIL OVER FOOTING

SLOPE = 1 to 1   SURCHARGE = 0 lbs./sq. ft.   AXIAL = 0 lbs./ft.

SOIL PRESSURE DIAGRAM

H1 = Concrete Stem
H2 = 12″ Concrete Block
H3 = 8″ Concrete Block

Stem Height	3'- 0″	4'- 0″	5'- 0″	6'- 0″	7'- 0″	8'- 0″	9'- 0″	10'- 0″	11'- 0″	12'- 0″
Ft. L ft.	2.833	3.667	4.333	5.167	6.167	7.000	7.667	8.500	9.333	10.500
Ft. T ft.	0.833	0.833	0.833	1.000	1.167	1.333	1.333	1.333	1.500	1.500
8 in. blk. ft.	3.000	4.000	5.000	6.000	5.000	6.000	6.000	6.000	6.000	6.000
12 in. blk. ft.	—	—	—	—	2.000	2.000	2.000	2.000	2.000	2.000
Conc. Stem	—	—	—	—	—	—	1.000	2.000	3.000	4.000
Wb ft.	—	—	—	—	—	—	1.333	1.333	1.500	1.500
W lbs.	1635	2690	3841	5519	7759	10091	12254	14950	18155	22199
F lbs.	588	934	1361	1960	2668	3484	4271	5138	6250	7290
Mo ft.-lbs.	751	1505	2647	4573	7262	10840	14712	19409	26042	32805
Mr ft.-lbs.	2307	4905	8224	14094	23740	35032	46370	62674	83529	115451
O.T.R.	3.071	3.258	3.107	3.082	3.269	3.232	3.152	3.229	3.208	3.519
$\overline{X}$ ft.	0.952	1.264	1.452	1.725	2.124	2.397	2.584	2.894	3.166	3.723
e ft.	0.465	0.569	0.714	0.858	0.959	1.103	1.250	1.356	1.500	1.527
Me ft.-lbs.	760	1531	2744	4738	7444	11127	15314	20273	27236	33899
S ft.^{3}	1.338	2.241	3.130	4.449	6.338	8.167	9.796	12.042	14.519	18.375
S.P.t p.s.f.	1145	1417	1763	2133	2433	2804	3162	3442	3821	3959
S.P.h p.s.f.	9	50	10	3	84	79	35	75	69	269
Friction	0.360	0.347	0.354	0.355	0.344	0.345	0.349	0.344	0.344	0.328
Inspct.	NO	NO	NO	YES	YES	YES	YES	YES	YES	YES

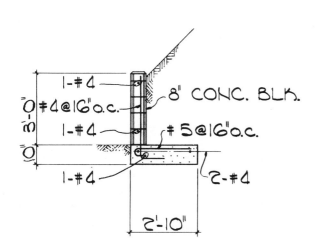

Soil slope = 1 to 1
E.F.P. = 80 p.c.f.
Section 3.14.3

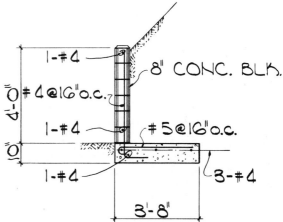

Soil slope = 1 to 1
E.F.P. = 80 p.c.f.
Section 3.14.4

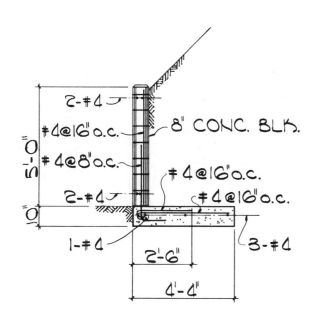

Soil slope = 1 to 1
E.F.P. = 80 p.c.f.
Section 3.14.5

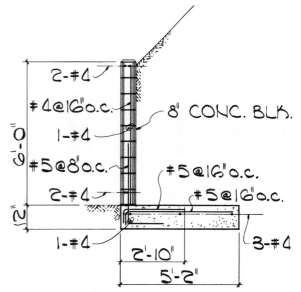

Soil slope = 1 to 1
E.F.P. = 80 p.c.f.
Section 3.14.6

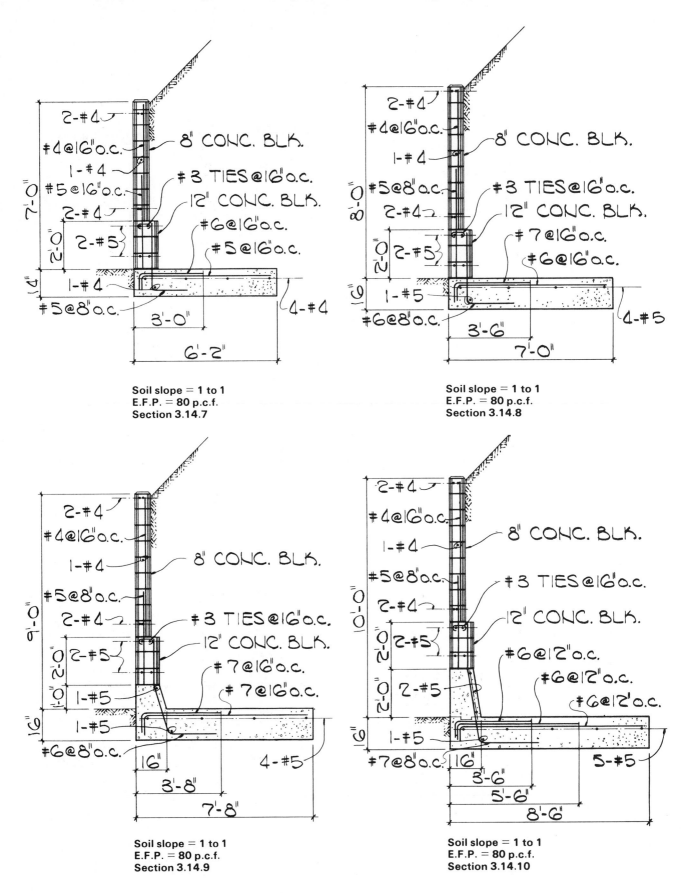

Soil slope = 1 to 1
E.F.P. = 80 p.c.f.
Section 3.14.7

Soil slope = 1 to 1
E.F.P. = 80 p.c.f.
Section 3.14.8

Soil slope = 1 to 1
E.F.P. = 80 p.c.f.
Section 3.14.9

Soil slope = 1 to 1
E.F.P. = 80 p.c.f.
Section 3.14.10

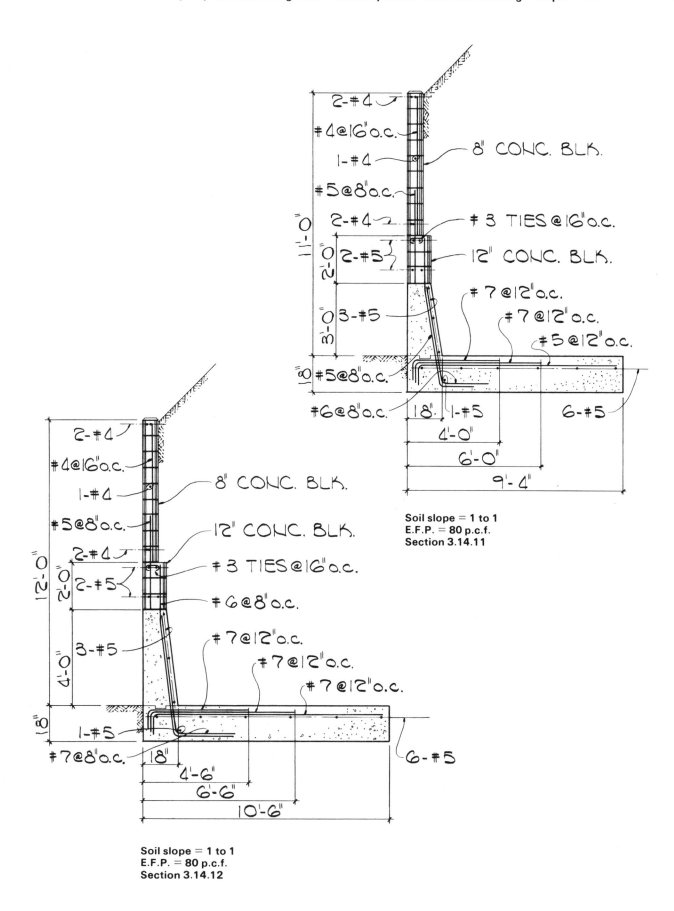

Soil slope = 1 to 1
E.F.P. = 80 p.c.f.
Section 3.14.11

Soil slope = 1 to 1
E.F.P. = 80 p.c.f.
Section 3.14.12

CHAPTER 4

Property Line Retaining Walls
Concrete Stem
Soil Not over the Footing

The design data and drawings presented in this chapter are concerned with property line concrete retaining walls. The retained soil is not placed over the wall footing, but at the toe side of the stem wall. The chapter consists of three design examples of retaining walls with various types of loading, a design data table for each of the fourteen loading conditions specified in Fig. 1.6, and a series of corresponding drawings following each design data table. The retaining wall drawings and the design data in this chapter are for a wall length of 1 foot.

The retaining walls in this chapter are designed using the following criteria:

Weight of soil = 100 p.c.f.   Weight of concrete = 150 p.c.f.
Concrete $f_c' = 2000$ p.s.i.   Reinforcing steel $f_s' = 20,000$ p.s.i.
Maximum allowable soil pressure = 4000 p.s.f.
Minimum allowable soil pressure = 0 p.s.f.
$\overline{X} \geqslant L/3$ Minimum O.T.R. = 1.50
Coefficient of soil friction = 0.40
Passive soil pressure = 300 p.c.f.

See Table C5 for reinforcement protection. Weep holes are not shown on the drawings.

DESIGN EXAMPLE – SECTION 4.7.6 | SHT. 1 OF 4

GIVEN:

$$H = 6'\text{-}0''$$
$$L = 3'\text{-}6''$$
$$T = 10''$$
$$W_T = 8''$$
$$W_b = 10''$$
$$A = 22''$$
$$B = 10''$$
$$C = 10''$$
$$D = 10''$$

SURCHARGE = 100 psf
EFP = 30 pcf
WEIGHT OF SOIL = 100 pcf
PASSIVE SOIL PRESSURE = 300 pcf
CONCRETE f'_c = 2000 psi
REINF. f'_s = 20,000 psi
EQUIV. HT. = $6.0 + 0.83' + \dfrac{100}{100} = 7.83'$

$$F = \frac{30 \times 7.83^2}{2} = 920^{\#}$$

$$M_O = \frac{920 \times 7.83}{3} = 2403^{'\#}$$

WEIGHT OF STEM = $W_1 = \dfrac{8'' + 10''}{2 \times 12} \times 6.0' \times 150 = 675.0^{\#}$

WEIGHT OF FTG. = $W_2 = 3.5 \times 0.83 \times 150 = 437.5$

WEIGHT OF KEY = $W_3 = 0.83^2 \times 150 = 104.2$

STEM FRICTION = $W_4 = 920/3 = 306.7$
_____
TOTAL WT. = W = 1523.4#

$$M_R = 675 \times \frac{37.5}{12} + 437.5 \times \frac{21}{12} + 104.2 \times \frac{27}{12} + 306.7 \times \frac{42}{12}$$

$$M_R = 4183.0^{'\#}$$

OVERTURN. RATIO = $\dfrac{4183}{2403} = 1.741 > 1.50$ ✓

NET M = $4183 - 2403 = 1780^{'\#}$

$$\bar{X} = \frac{1780}{1523.4} = 1.168'$$

FTG. MIDDLE $\frac{1}{3}$ = $\dfrac{3.50}{3} = 1.167 \approx 1.168$ ✓

DESIGN　EXAMPLE　—　SECTION　4.7.6.　| SHT. 2 OF 4

FOOTING　DESIGN:

$W = 1523.4^{\#}$; $\bar{X} = 1.168'$; $e = \dfrac{3.50}{2} - 1.168 = 0.582'$

$M_e = 1523.4 \times 0.582 = 886.6'^{\#}$

FTG. AREA = 3.50 SQ. FT.

FTG. SECT. MODULUS $= \dfrac{1 \times 3.5^2}{6} = 2.042'^3$

SOIL PRESSURE $= \dfrac{1523.4}{3.5} + \dfrac{886.6}{2.042} = 435.3 \pm 434.3$

$\underline{SP_H = 869.6 \text{ psf}}$　　　　　$\underline{SP_T = 1.0 \text{ psf}}$

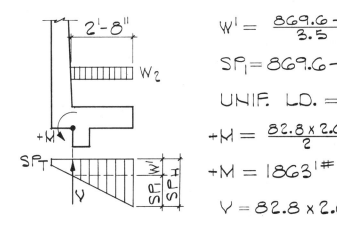

$W' = \dfrac{869.6 - 1}{3.5} \times 0.83 + 1.0 = 207.8 \text{ psf}$

$SP_1 = 869.6 - 207.8 = 661.8 \text{ psf}$

UNIF. LD. $= 207.8 - 125 = 82.8^{\#}/\text{FT.}$

$+M = \dfrac{82.8 \times 2.67^2}{2} + \dfrac{661.8 \times 2.67^2 \times 2}{2 \times 3}$

$+M = 1863'^{\#}$

$Y = 82.8 \times 2.67 + \dfrac{661.8 \times 2.67}{2}$

$Y = 1103.3^{\#}$

$d = 10 - (3 + .5 \times .62) = 6.70''$

$a = 1.13$; $j = 0.89$

$v = \dfrac{1103.3}{12 \times .89 \times 6.68} = 15.5 \text{ psi}$

$A_s = \dfrac{1.863}{1.13 \times 6.70} = 0.246 \text{ SQ. IN./FT.}$

$\underline{\text{USE } \# 5 @ 16'' \text{ O.C.}}$

LONG. REINF. $= 0.0012 \times 12 \times 3.50 \times 10 = 0.50 \text{ SQ. IN.}$

$\underline{\text{USE } 3 - \#4's}$

DESIGN EXAMPLE — SECTION 4.7.6 | SHT. 3 OF 4

STEM DESIGN AT TOP OF FTG. :

EQUIVALENT HT. $= 6.0' + 1.0' = 7.00'$; EFP $= 30$ p.c.f.

$$V = \frac{30 \times 7^2}{2} = 735^{\#} \quad ; \quad M = \frac{735 \times 7}{3} = 1715^{'\#}$$

$$d = 10'' - (2 + .5 \times .62) = 6.70'' \quad ; \quad a = 1.13 \quad ; \quad j = 0.89$$

$$\nu = \frac{735}{12 \times .89 \times 6.70} = 10.3 \text{ psi} \checkmark \quad ; \quad A_s = \frac{1.715}{1.13 \times 6.70} = 0.227 \text{ sq.in/ft.}$$

USE # 5 @ 16" o.c. · VERT.

HORIZ. REINF. $= 0.0012 \times 6 \times 12 \times 9 = 0.777$ SQ.IN.

USE # 4 @ 22" o.c. HORIZ.

SLIDING :

$$Fr = \frac{920}{1523.4} = 0.604 > 0.40$$

KEY REQUIRED

PASSIVE PRESSURE $= 300$ pcf

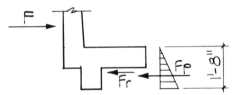

RESISTANCE TO SLIDING $= 1523.4 \times 0.4 + \frac{300 \times 1.67^2}{2} = 1026.2^{\#} > 920^{\#} \checkmark$

DESIGN EXAMPLE - SECTION 4.7.6	SHT. 4 OF 4

CHECK WALL WITHOUT 100 p.s.f. SURCHARGE LD.

EQUIV. HT. = 6.0' + 0.83' = 6.83' ; EFP = 30 p.c.f.

$$Y = \frac{30 \times 6.83^2}{2} = 700^{\#} \quad ; \quad M_o = \frac{30 \times 6.83}{3} = 1595'^{\#}$$

$$W = 1523.4 - 306.7 + \frac{700}{3} = 1450^{\#}$$

$$M_R = 4183.0 - 306.7 \times \frac{42}{12} + 233.3 \times \frac{42}{12} = 3927'^{\#}$$

OVERTURN. RATIO = $\frac{3927}{1595}$ = 2.462 > 1.50 ✓

NET M = 3927 - 1595 = 2332'^{\#}

$$\bar{X} = \frac{2332}{1450} = 1.608' > 1.167' \checkmark$$

$$e = \frac{3.50}{2} - 1.608 = 0.142$$

$$M_e = 1450 \times .142 = 207'^{\#}$$

SOIL PRESSURE = $\frac{1450}{3.5} \pm \frac{207}{2.042}$ = 414.3 ± 101.3

$\underline{SP_T = 515.6 \text{ p.s.f.}}$ $\qquad$ $\underline{SP_H = 313 \text{ p.s.f.}}$

CHECK SLIDING:

$$F_r = \frac{700}{1450} = 0.483 > 0.40$$

A KEY IS REQ'D.

4.0　Design Example—Property Line Retaining Wall—Concrete Stem—Soil Not over Footing

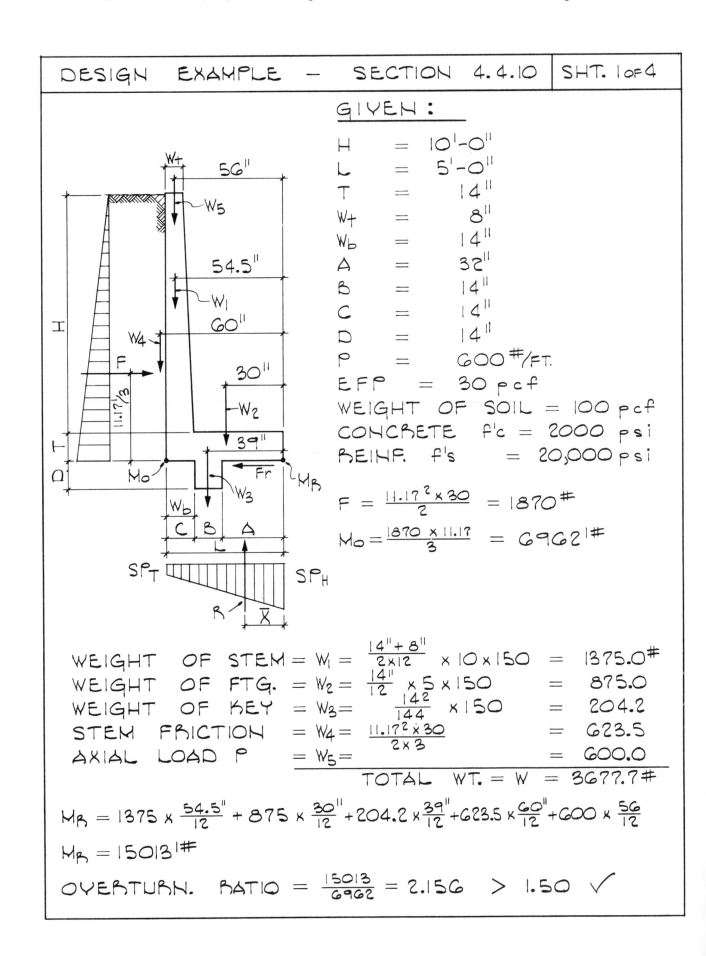

DESIGN EXAMPLE — SECTION 4.4.10 | SHT. 1 OF 4

GIVEN:

$H = 10'\text{-}0''$
$L = 5'\text{-}0''$
$T = 14''$
$W_t = 8''$
$W_b = 14''$
$A = 32''$
$B = 14''$
$C = 14''$
$D = 14''$
$P = 600\ \#/FT.$
$EFP = 30\ pcf$
WEIGHT OF SOIL $= 100\ pcf$
CONCRETE $f'_c = 2000\ psi$
REINF. $f's = 20,000\ psi$

$F = \dfrac{11.17^2 \times 30}{2} = 1870\ \#$

$M_o = \dfrac{1870 \times 11.17}{3} = 6962\ '\#$

WEIGHT OF STEM $= W_1 = \dfrac{14''+8''}{2\times12} \times 10 \times 150 = 1375.0\ \#$
WEIGHT OF FTG. $= W_2 = \dfrac{14''}{12} \times 5 \times 150 = 875.0$
WEIGHT OF KEY $= W_3 = \dfrac{14^2}{144} \times 150 = 204.2$
STEM FRICTION $= W_4 = \dfrac{11.17^2 \times 30}{2\times3} = 623.5$
AXIAL LOAD P $= W_5 = 600.0$

TOTAL WT. $= W = 3677.7\ \#$

$M_R = 1375 \times \dfrac{54.5''}{12} + 875 \times \dfrac{30''}{12} + 204.2 \times \dfrac{39''}{12} + 623.5 \times \dfrac{60''}{12} + 600 \times \dfrac{56}{12}$

$M_R = 15013\ '\#$

OVERTURN. RATIO $= \dfrac{15013}{6962} = 2.156 > 1.50 ✓$

DESIGN   EXAMPLE — SECTION 4.4.10 | SHT. 2 OF 4

$M_R = 15013^{1\#}$ $M_o = 6962^{1\#}$ $W = 3677.7^{\#}$

NET M $= 15013 - 6962 = 8151^{1\#}$

$\bar{X} = \dfrac{8151^{1\#}}{3677.7^{\#}} = 2.189^{1}$

FTG. MIDDLE $\frac{1}{3} = \dfrac{5.00}{3} = 1.667^{1} < 2.189^{1}$ ✓

$e = \dfrac{5.00}{2} - 2.189 = 0.311^{1}$

FOOTING   DESIGN :

$M_e = 3677.7 \times 0.311 = 1143.8^{1\#}$; $W = 3677.7^{\#}$

FTG.   AREA = 5.00 SQ. FT.

FTG.   SECT.   MODULUS $= \dfrac{1 \times 5^2}{6} = 4.167^{13}$

SOIL   PRESSURE $= \dfrac{3677.7}{5.0} \pm \dfrac{1143.8}{4.167} = 735.5 \pm 274.5$

$\underline{SP_T = 461 \; psf}$ $\underline{SP_H = 1010 \; psf}$

$W^{1} = \dfrac{1010 - 461}{5.0} \times 1.17 + 461 = 589 \, psf$

$SP_1 = 1010 - 589 = 421 \, psf$

UNIF. LD. $= 589 - 175 = 414^{\#}/FT.$

$+M = \dfrac{414 \times 3.83^2}{2} + \dfrac{421 \times 3.83^2 \times 2}{2 \times 3}$

$+M = 5104^{1\#}$

$V = 414 \times 3.83 + \dfrac{421 \times 3.83}{2} = 2392^{\#}$

$\nu = \dfrac{2392}{12 \times .89 \times 10.70} = 21\,p.s.i < 54\,p.s.i$

$d = 14 - (3 + .5 \times .62) = 10.70^{11}$

$a = 1.13$; $j = 0.89$

$A_s = \dfrac{5.104}{1.13 \times 10.70} = .422 \; SQ. IN./FT.$

USE # 5 @ 8'' o.c.

LONG.   REINF. $= 0.0012 \times 12 \times 5.0 \times 14$ $= 1.00$ SQ. IN.
USE 5 - # 4's

DESIGN EXAMPLE – SECTION 4.4.10	SHT. 3 OF 4

STEM DESIGN AT TOP OF FTG.:

$$V = \frac{30 \times 10^2}{2} = 1500^{\#} \quad ; \quad M = \frac{1500 \times 10}{3} = 5000^{'\#}$$

$$d = 14'' - (2 + .5 \times .62) = 11.70'' \qquad a = 1.13 \quad ; \quad j = 0.89$$

$$v = \frac{1500}{12 \times .89 \times 11.70} = 12.0 \text{ psi} \checkmark \quad ; \quad A_S = \frac{5.0}{1.13 \times 11.70} = 0.378^{SQ.IN}/_{FT}$$

USE #5 @ 8" o.c. VERT.

STEM DESIGN AT MID STEM HEIGHT

$$V = \frac{30 \times 5^2}{2} = 375^{\#} \quad ; \quad M = \frac{375 \times 5}{3} = 625^{'\#}$$

$$d = 8 + \frac{14'' - 8''}{10} \times 5 - (2 + .5 \times .62) = 8.70'' \quad ; \quad a = 1.13 \quad ; \quad j = 0.89$$

$$v = \frac{375}{12 \times .89 \times 8.70} = 4.0 \text{ psi} \checkmark \quad ; \quad A_S = \frac{.625}{1.13 \times 8.70} = .063 \text{ SQ.IN./FT.}$$

MIN. $A_S = 12 \times 11 \times .0020 = 0.246$ SQ. IN./FT.

USE #5 @ 18" o.c. VERT.

STEM HORIZ. REINF.

$A_S = 0.0012 \times 10 \times 12 \times 11 = 1.58$ SQ. IN.

USE #4 @ 18" o.c. HORIZ.

SLIDING :   KEY DEPTH + FTG. THICKNESS = 2.33'

$$F_r = \frac{1870}{3677.7} = 0.508 > 0.40 \text{ (ALLOWABLE)}$$

$$\text{PASSIVE SOIL RESISTANCE} = \frac{2.33^2 \times 300}{2} = 816.4^{\#}$$

SLIDING RESISTANCE $= 0.40 \times 3677.7 + 816.4 = 2287.5^{\#} > 1870^{\#}$

| DESIGN EXAMPLE — SECTION 4.4.10 | SHT. 4 OF 4 |

CHECK WALL WITHOUT 600 #/FT. AXIAL LD.

$F = 1870^{\#}$; $M_0 = 6962^{'\#}$; E F P = 30 p.c.f.

$W = 3677.7 - 600 = 3077.7^{\#}$

$M_R = 1501.3 - 600 \times \dfrac{56}{12} = 12213^{'\#}$

OVERTURN. RATIO $= \dfrac{12213}{6962} = 1.754 > 1.50$ ✓

NET $M = 12213 - 6962 = 5251^{'\#}$

$\bar{X} = \dfrac{5251}{3077.7} = 1.706' > 1.667'$ ✓

$e = \dfrac{5.0}{2} - 1.706 = 0.794'$

$M_e = 3077.7 \times .794 = 2443^{'\#}$

SOIL PRESSURE $= \dfrac{3077.7}{5.0} \pm \dfrac{2443}{4.167} = 615.5 \pm 586.5$

$\underline{SP_T = 29 \text{ p.s.f.}}$ $\underline{SP_H = 1202 \text{ p.s.f.}}$

CHECK SLIDING :

$F_r = \dfrac{1870}{3077.7} = 0.607 > 0.40$

A KEY IS REQ'D.

RESISTANCE TO SLIDING $= 0.40 \times 3077.7 + 816.4 = 2047^{\#} > 1870^{\#}$ ✓

DESIGN EXAMPLE — SECTION 4.11.8 | SHT. 1 OF 4

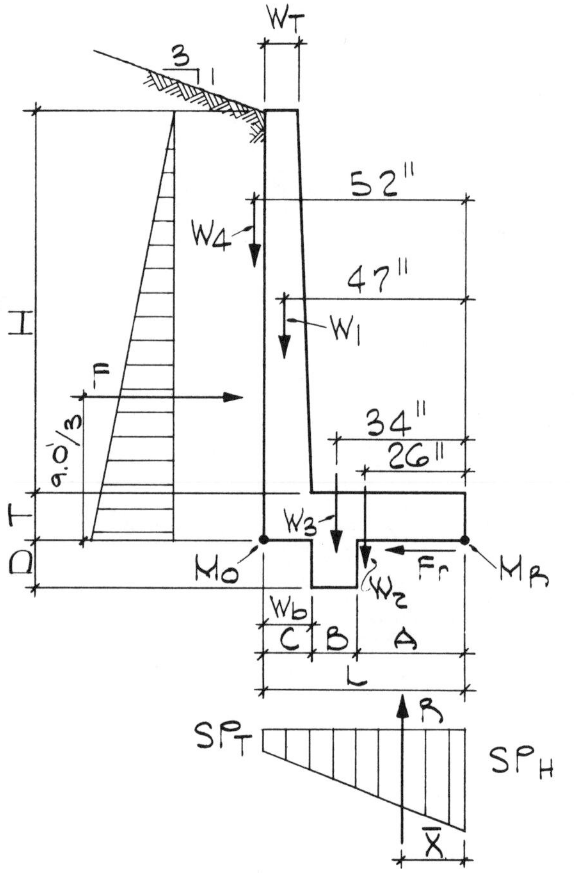

GIVEN :

H	=	8'-0"
L	=	4'-4"
T	=	12"
W_T	=	8"
W_b	=	12"
A	=	28"
B	=	12"
C	=	12"
D	=	12"

SOIL SLOPE = 3 TO 1
EFP = 38 pcf
WEIGHT OF SOIL = 100 pcf
PASSIVE SOIL PRESSURE = 300 pcf
CONCRETE f'c = 2000 psi
REINF. f's = 20,000 psi

$$F = \frac{38 \times 9^2}{2} = 1539^{\#}$$

$$M_o = \frac{1539 \times 9}{3} = 4617^{'\#}$$

WEIGHT OF STEM = W_1 = 150 × 0.83 × 8.0 = 1000$^{\#}$
WEIGHT OF FTG. = W_2 = 150 × 4.33 = 650
WEIGHT OF KEY = W_3 = 1^2 × 150 = 150
STEM FRICTION = W_4 = $\frac{1539}{3}$ = 513

TOTAL WT. = W = 2313$^{\#}$

$$M_R = 1000 \times \frac{47}{12} + 650 \times \frac{26}{12} + 150 \times \frac{34}{12} + 513 \times \frac{52}{12}$$

$$M_R = 7973^{'\#}$$

OVERTURN. RATIO = $\frac{7973}{4617}$ = 1.727 > 1.50 ✓

NET M = 7973 − 4617 = 3356$^{'\#}$

$\bar{X}$ = $\frac{3356}{2313}$ = 1.451' ; FTG. MIDDLE ⅓ = $\frac{4.33}{3}$ = 1.444 ✓

DESIGN EXAMPLE — SECTION 4.11.8	SHT. 2 OF 4

FOOTING DESIGN :

$W = 2313^{\#}$; $\bar{X} = 1.45'$; $e = \dfrac{4.33}{2} - 1.451 = 0.714'$

$M_e = 2313 \times 0.714 = 1651.5'^{\#}$

FTG. AREA = 4.33 SQ. FT.

FTG. SECT. MODULUS $= \dfrac{1 \times 4.33^2}{6} = 3.13'^3$

SOIL PRESSURE $= \dfrac{2313}{4.33} \pm \dfrac{1651.5}{3.13} = 534 \pm 528$

$\underline{SP_H = 1062 \ psf}$ $\underline{SP_T = 6.0 \ psf}$

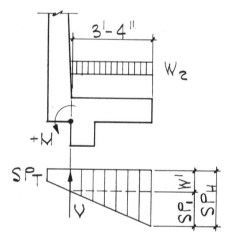

$W' = \dfrac{1062 - 6}{4.33} = \times 1.0 + 6.0 = 250 \ psf$

$SP_1 = 1062 - 250.0 = 812.0 \ psf$

UNIF. LD. $= 250 - 150 = 100 \ ^{\#}/_{FT.}$

$+M = \dfrac{100 \times 3.33^2}{2} + \dfrac{812 \times 3.33^2 \times 2}{2 \times 3}$

$+M = 3555'^{\#}$

$V = 100 \times 3.33 + \dfrac{812 \times 3.33}{2}$

$V = 1685^{\#}$

$a = 1.13$; $j = 0.89$

$d = 12'' - (3 + .5 \times .62) = 8.70''$

$v = \dfrac{1685}{12 \times .89 \times 8.70} = 18.1 \ psi \quad \checkmark$

$A_s = \dfrac{3.55}{1.13 \times 8.70} = 0.361 \ SQ. \ IN./FT.$

$\underline{USE \ \#5 @ 9'' \ o.c.}$

LONG. REINF. $= A_s = 0.0012 \times 12 \times 4.33 \times 12 = 0.75 \ SQ. IN.$

$\underline{USE \ 4 - \#4's}$

DESIGN   EXAMPLE — SECTION 4.11.8 | SHT. 3 OF 4

STEM DESIGN AT TOP OF FTG.:

$$V = \frac{38 \times 8^2}{2} = 1216\# \quad ; \quad M = \frac{1216 \times 8}{3} = 3242'\#$$

$$d = 12'' - (2 + .5 \times .62) = 9.70'' \quad ; \quad a = 1.13 \quad ; \quad j = 0.89$$

$$\nu = \frac{1216}{12 \times .89 \times 9.70} = 11.7 \text{ psi} \checkmark \; ; \; A_s = \frac{3.242}{1.13 \times 9.70} = 0.295 \text{ sq.in./ft}$$

USE # 4 @ 9'' o.c. VERT.

STEM DESIGN AT 3'-0'' ABOVE FTG.:

$$H = 5'-0'' \quad ; \quad M = \frac{38 \times 5^2 \times 5}{2 \times 3} = 791.7'\#$$

$$d = 8'' + \frac{12-8}{8} \times 5 - (2 + .5 \times .62) = 8.20''$$

$$A_s = \frac{0.7917}{1.13 \times 8.20} = 0.0855 \text{ sq. in./ft}$$

USE # 4 @ 16'' o.c.

MIN. VERT. $A_s = 0.0020 \times 10 \times 12 = 0.24$ sq.in./ft $\checkmark$

HORIZ. REINF. $= 0.0012 \times 8 \times 12 \times 10 = 1.15$ sq.in./ft

USE # 4 @ 18'' o.c. HORIZ.

SLIDING :

KEY DEPTH + FTG. THICKNESS = 2.0'

RESISTANCE TO SLIDING $= 2313 \times 0.40 + \frac{300 \times 2^2}{2} = 1525 \approx 1539 \checkmark$

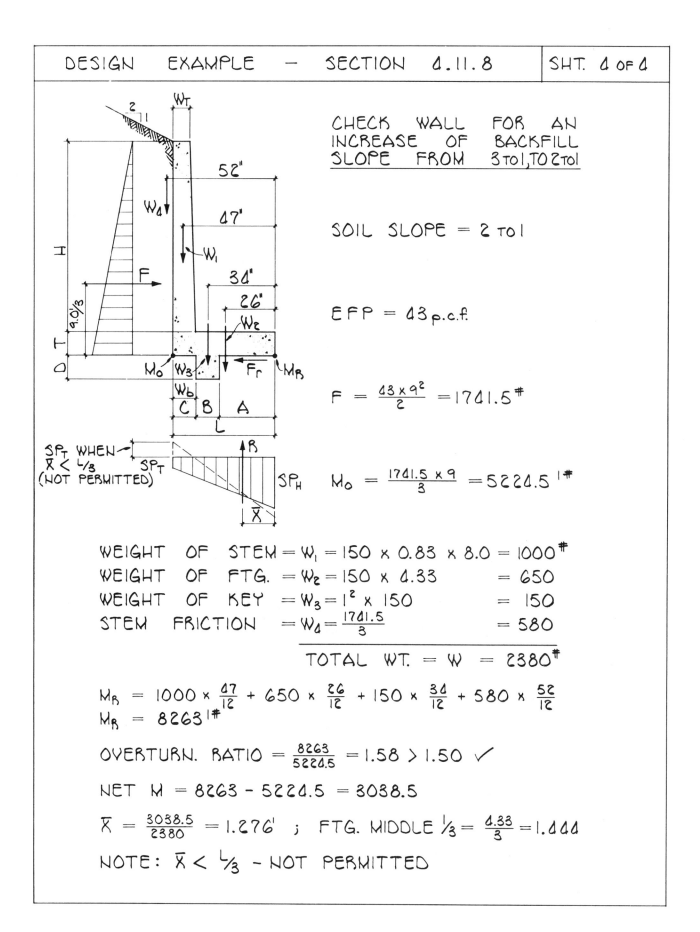

DESIGN EXAMPLE — SECTION 4.11.8 | SHT. 4 OF 4

CHECK WALL FOR AN INCREASE OF BACKFILL SLOPE FROM 3 TO 1, TO 2 TO 1

SOIL SLOPE = 2 TO 1

EFP = 43 p.c.f.

$$F = \frac{43 \times 9^2}{2} = 1741.5^{\#}$$

$$M_O = \frac{1741.5 \times 9}{3} = 5224.5^{\,\prime\#}$$

WEIGHT OF STEM = W_1 = 150 × 0.83 × 8.0 = 1000$^{\#}$
WEIGHT OF FTG. = W_2 = 150 × 4.33 = 650
WEIGHT OF KEY = W_3 = 1^2 × 150 = 150
STEM FRICTION = W_4 = $\frac{1741.5}{3}$ = 580
_____
TOTAL WT. = W = 2380$^{\#}$

$M_R = 1000 \times \frac{47}{12} + 650 \times \frac{26}{12} + 150 \times \frac{34}{12} + 580 \times \frac{52}{12}$
$M_R = 8263^{\,\prime\#}$

OVERTURN. RATIO = $\frac{8263}{5224.5}$ = 1.58 > 1.50 ✓

NET M = 8263 − 5224.5 = 3038.5

$\bar{X} = \frac{3038.5}{2380} = 1.276'$; FTG. MIDDLE ⅓ = $\frac{4.33}{3} = 1.444$

NOTE: $\bar{X} < \frac{L}{3}$ — NOT PERMITTED

TABLE 4.1	PROPERTY LINE RETAINING WALLS – CONCRETE STEM – SOIL NOT OVER FOOTING
SLOPE = 0 to 1	SURCHARGE = 0 lbs./sq. ft. AXIAL = 0 lbs./ft.

SOIL PRESSURE DIAGRAM

Stem Height	3'- 0''	4'- 0''	5'- 0''	6'- 0''	7'- 0''	8'- 0''	9'- 0''	10'- 0''	11'- 0''	12'- 0''
Ft. L ft.	1.667	1.833	2.167	2.667	3.333	3.833	4.500	5.000	5.667	6.167
Ft. T ft.	0.667	0.667	0.833	0.833	0.833	1.000	1.000	1.167	1.167	1.333
Wt ft.	0.667	0.667	0.667	0.667	0.667	0.667	0.667	0.667	0.667	0.667
Wb ft.	0.667	0.667	0.833	0.833	0.833	1.000	1.000	1.167	1.167	1.333
A ft.	0.500	0.667	0.833	1.333	1.667	2.000	2.500	2.667	3.167	3.333
B ft.	0.667	0.667	0.833	0.833	0.833	1.000	1.000	1.167	1.167	1.333
C ft.	0.500	0.500	0.500	0.500	0.833	0.833	1.000	1.167	1.333	1.500
D ft.	0.667	0.667	0.833	0.833	0.833	1.000	1.000	1.167	1.167	1.333
W lbs.	601	759	1108	1346	1615	2130	2450	3078	3448	4189
F lbs.	202	327	510	700	920	1215	1500	1870	2220	2667
Mo ft.-lbs.	246	508	992	1595	2403	3645	5000	6962	9005	11852
Mr ft.-lbs.	706	1034	1800	2796	4264	6446	8812	12213	15647	20551
O.T.R.	2.866	2.036	1.814	1.753	1.774	1.769	1.763	1.754	1.738	1.734
$\overline{X}$ ft.	0.766	0.693	0.729	0.892	1.152	1.315	1.556	1.706	1.926	2.077
e ft.	0.067	0.223	0.354	0.441	0.515	0.602	0.694	0.794	0.907	1.007
Me ft.-lbs.	40	169	392	594	831	1281	1700	2443	3129	4217
S ft.^{3}	0.463	0.560	0.782	1.185	1.852	2.449	3.375	4.167	5.352	6.338
S.P.h p.s.f.	448	716	1013	1006	933	1079	1048	1202	1193	1345
S.P.t p.s.f.	273	111	10	4	36	32	41	29	24	14
Friction	0.336	0.430	0.461	0.520	0.570	0.570	0.612	0.608	0.644	0.637

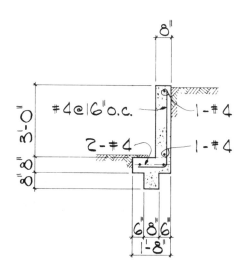

E.F.P. = 30 p.c.f.
Section 4.1.3

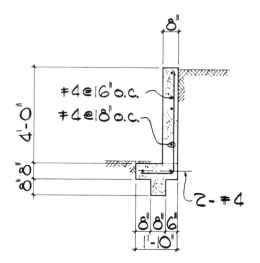

E.F.P. = 30 p.c.f.
Section 4.1.4

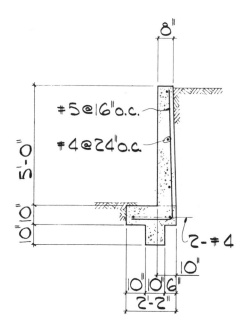

E.F.P. = 30 p.c.f.
Section 4.1.5

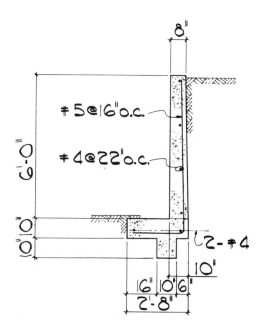

E.F.P. = 30 p.c.f.
Section 4.1.6

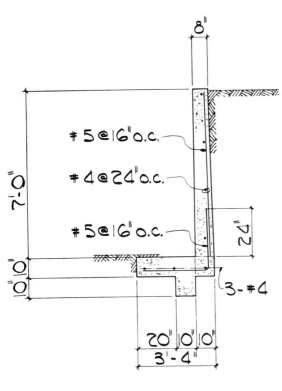

#5@16"o.c.

#4@24"o.c.

#5@16"o.c.

7'-0"

24"

3-#4

20" 10" 10"
3'-4"

E.F.P. = 30 p.c.f.
Section 4.1.7

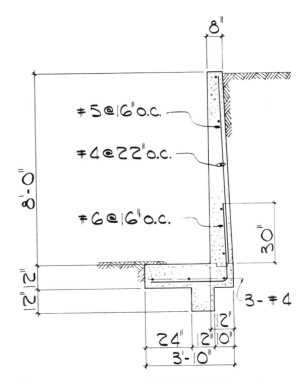

#5@16"o.c.

#4@22"o.c.

#6@16"o.c.

8'-0"

30"

3-#4

24" 12" 2"
3'-10"

E.F.P. = 30 p.c.f.
Section 4.1.8

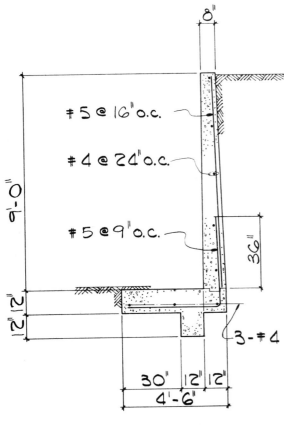

#5@16"o.c.

#4@24"o.c.

#5@9"o.c.

9'-0"

36"

3-#4

30" 12" 12"
4'-6"

E.F.P. = 30 p.c.f.
Section 4.1.9

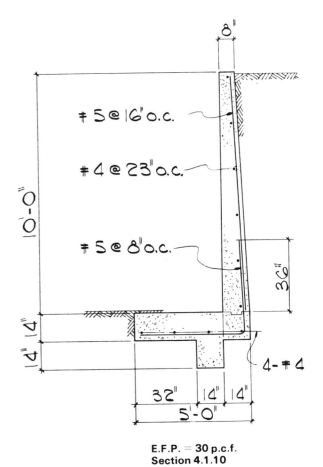

#5@16"o.c.

#4@23"o.c.

#5@8"o.c.

10'-0"

36"

4-#4

32" 14" 14"
5'-0"

E.F.P. = 30 p.c.f.
Section 4.1.10

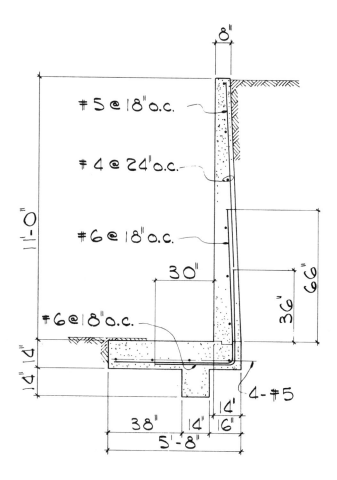

E.F.P. = 30 p.c.f.
Section 4.1.11

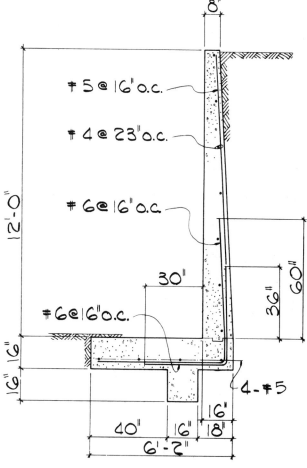

E.F.P. = 30 p.c.f.
Section 4.1.12

TABLE 4.2	PROPERTY LINE RETAINING WALLS – CONCRETE STEM – SOIL NOT OVER FOOTING
	SLOPE = 0 to 1 SURCHARGE = 0 lbs./sq. ft. AXIAL = 200 lbs./ft.

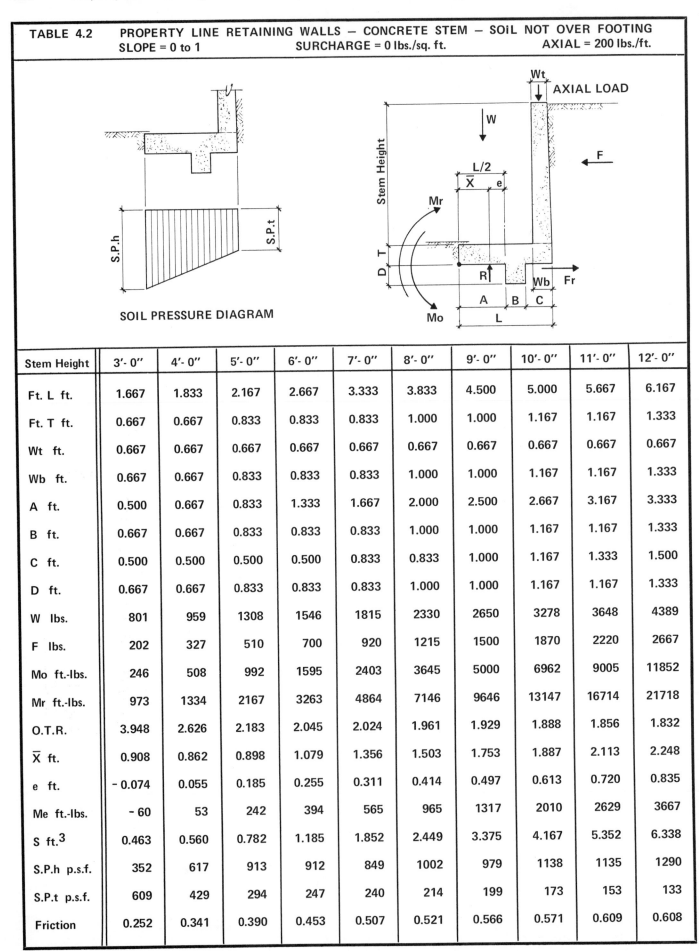

SOIL PRESSURE DIAGRAM

Stem Height	3'- 0''	4'- 0''	5'- 0''	6'- 0''	7'- 0''	8'- 0''	9'- 0''	10'- 0''	11'- 0''	12'- 0''
Ft. L ft.	1.667	1.833	2.167	2.667	3.333	3.833	4.500	5.000	5.667	6.167
Ft. T ft.	0.667	0.667	0.833	0.833	0.833	1.000	1.000	1.167	1.167	1.333
Wt ft.	0.667	0.667	0.667	0.667	0.667	0.667	0.667	0.667	0.667	0.667
Wb ft.	0.667	0.667	0.833	0.833	0.833	1.000	1.000	1.167	1.167	1.333
A ft.	0.500	0.667	0.833	1.333	1.667	2.000	2.500	2.667	3.167	3.333
B ft.	0.667	0.667	0.833	0.833	0.833	1.000	1.000	1.167	1.167	1.333
C ft.	0.500	0.500	0.500	0.500	0.833	0.833	1.000	1.167	1.333	1.500
D ft.	0.667	0.667	0.833	0.833	0.833	1.000	1.000	1.167	1.167	1.333
W lbs.	801	959	1308	1546	1815	2330	2650	3278	3648	4389
F lbs.	202	327	510	700	920	1215	1500	1870	2220	2667
Mo ft.-lbs.	246	508	992	1595	2403	3645	5000	6962	9005	11852
Mr ft.-lbs.	973	1334	2167	3263	4864	7146	9646	13147	16714	21718
O.T.R.	3.948	2.626	2.183	2.045	2.024	1.961	1.929	1.888	1.856	1.832
$\overline{X}$ ft.	0.908	0.862	0.898	1.079	1.356	1.503	1.753	1.887	2.113	2.248
e ft.	- 0.074	0.055	0.185	0.255	0.311	0.414	0.497	0.613	0.720	0.835
Me ft.-lbs.	- 60	53	242	394	565	965	1317	2010	2629	3667
S ft.^{3}	0.463	0.560	0.782	1.185	1.852	2.449	3.375	4.167	5.352	6.338
S.P.h p.s.f.	352	617	913	912	849	1002	979	1138	1135	1290
S.P.t p.s.f.	609	429	294	247	240	214	199	173	153	133
Friction	0.252	0.341	0.390	0.453	0.507	0.521	0.566	0.571	0.609	0.608

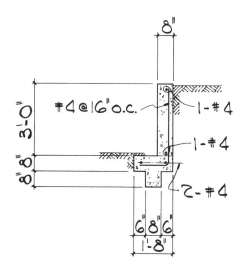

Axial load = 200 lbs./ft.
E.F.P. = 30 p.c.f.
Section 4.2.3

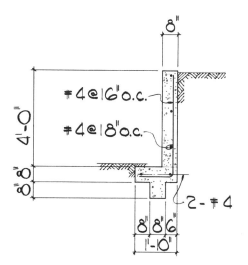

Axial load = 200 lbs./ft.
E.F.P. = 30 p.c.f.
Section 4.2.4

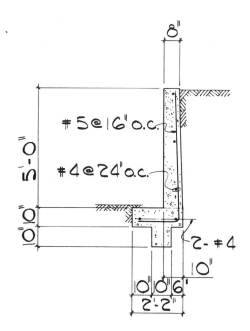

Axial load = 200 lbs./ft.
E.F.P. = 30 p.c.f.
Section 4.2.5

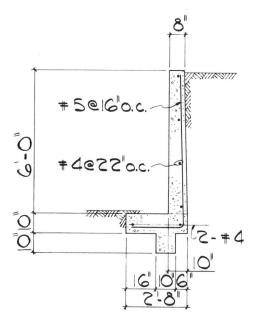

Axial load = 200 lbs./ft.
E.F.P. = 30 p.c.f.
Section 4.2.6

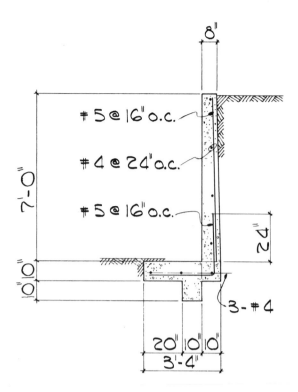

Axial load = 200 lbs./ft.
E.F.P. = 30 p.c.f.
Section 4.2.7

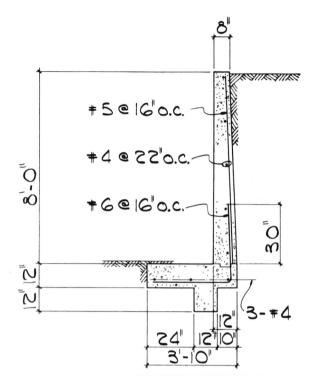

Axial load = 200 lbs./ft.
E.F.P. = 30 p.c.f.
Section 4.2.8

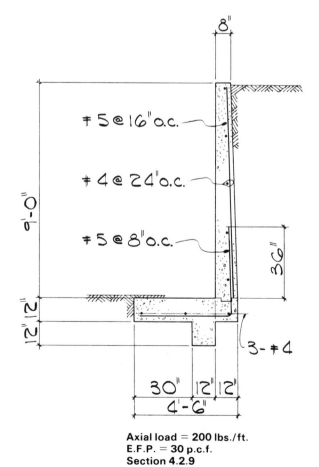

Axial load = 200 lbs./ft.
E.F.P. = 30 p.c.f.
Section 4.2.9

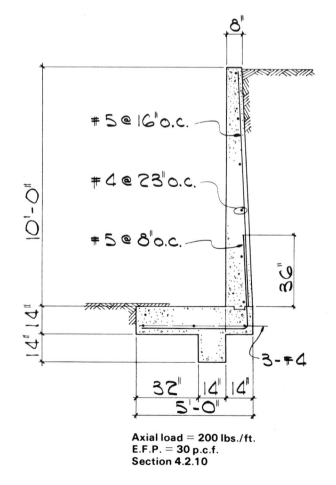

Axial load = 200 lbs./ft.
E.F.P. = 30 p.c.f.
Section 4.2.10

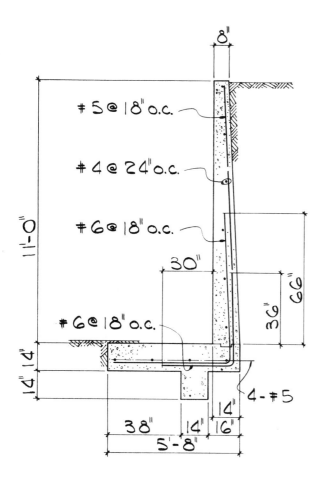

#5 @ 18" o.c.

#4 @ 24" o.c.

#6 @ 18" o.c.

#6 @ 18" o.c.

11'-0"

30"

66"

36"

14"

14"

4-#5

38"

14"

14"

16"

5'-8"

Axial load = 200 lbs./ft.
E.F.P. = 30 p.c.f.
Section 4.2.11

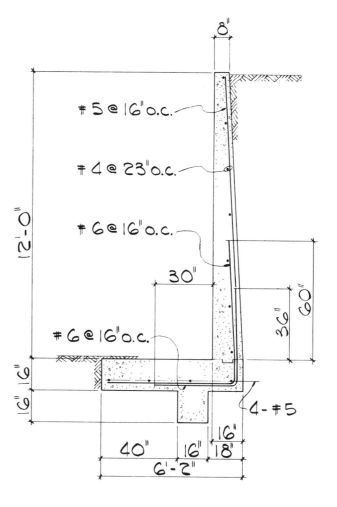

#5 @ 16" o.c.

#4 @ 23" o.c.

#6 @ 16" o.c.

#6 @ 16" o.c.

12'-0"

30"

60"

36"

16"

16"

4-#5

40"

16"

16"

18"

6'-2"

Axial load = 200 lbs./ft.
E.F.P. = 30 p.c.f.
Section 4.2.12

TABLE 4.3   PROPERTY LINE RETAINING WALLS — CONCRETE STEM — SOIL NOT OVER FOOTING
SLOPE = 0 to 1   SURCHARGE = 0 lbs./sq. ft.   AXIAL = 400 lbs./ft.

SOIL PRESSURE DIAGRAM

Stem Height	3'- 0"	4'- 0"	5'- 0"	6'- 0"	7'- 0"	8'- 0"	9'- 0"	10'- 0"	11'- 0"	12'- 0"
Ft. L ft.	1.667	1.833	2.167	2.667	3.333	3.833	4.500	5.000	5.667	6.167
Ft. T ft.	0.667	0.667	0.833	0.833	0.833	1.000	1.000	1.167	1.167	1.333
Wt ft.	0.667	0.667	0.667	0.667	0.667	0.667	0.667	0.667	0.667	0.667
Wb ft.	0.667	0.667	0.833	0.833	0.833	1.000	1.000	1.167	1.167	1.333
A ft.	0.500	0.667	0.833	1.333	1.667	2.000	2.500	2.667	3.167	3.333
B ft.	0.667	0.667	0.833	0.833	0.833	1.000	1.000	1.167	1.167	1.333
C ft.	0.500	0.500	0.500	0.500	0.833	0.833	1.000	1.167	1.333	1.500
D ft.	0.667	0.667	0.833	0.833	0.833	1.000	1.000	1.167	1.167	1.333
W lbs.	1001	1159	1508	1746	2015	2530	2850	3478	3848	4589
F lbs.	202	327	510	700	920	1215	1500	1870	2220	2667
Mo ft.-lbs.	246	508	992	1595	2403	3645	5000	6962	9005	11852
Mr ft.-lbs.	1240	1634	2533	3730	5464	7846	10479	14080	17780	22884
O.T.R.	5.030	3.216	2.553	2.338	2.273	2.153	2.096	2.022	1.974	1.931
$\overline{X}$ ft.	0.993	0.972	1.022	1.222	1.519	1.661	1.923	2.047	2.280	2.404
e ft.	-0.159	-0.055	0.061	0.111	0.148	0.256	0.327	0.453	0.553	0.679
Me ft.-lbs.	-160	-64	92	194	298	648	933	1576	2129	3117
S ft.^{3}	0.463	0.560	0.782	1.185	1.852	2.449	3.375	4.167	5.352	6.338
S.P.h p.s.f.	256	518	814	818	765	925	910	1074	1077	1236
S.P.t p.s.f.	945	746	578	491	444	395	357	317	281	252
Friction	0.202	0.282	0.339	0.401	0.457	0.480	0.526	0.538	0.577	0.581

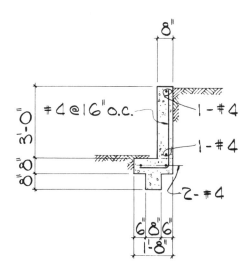

Axial load = 400 lbs./ft.
E.F.P. = 30 p.c.f.
Section 4.3.3

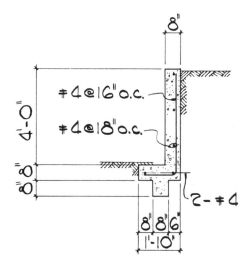

Axial load = 400 lbs./ft.
E.F.P. = 30 p.c.f.
Section 4.3.4

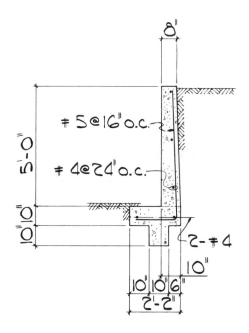

Axial load = 400 lbs./ft.
E.F.P. = 30 p.c.f.
Section 4.3.5

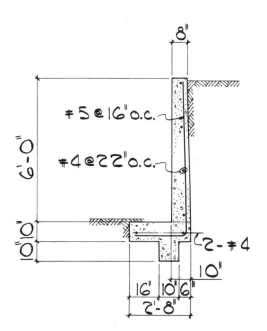

Axial load = 400 lbs./ft.
E.F.P. = 30 p.c.f.
Section 4.3.6

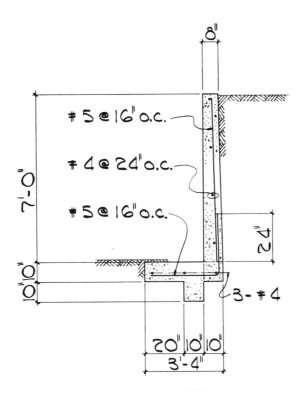

Axial load = 400 lbs./ft.
E.F.P. = 30 p.c.f.
Section 4.3.7

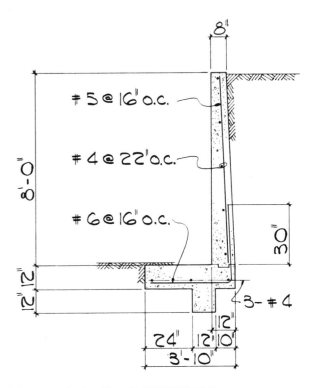

Axial load = 400 lbs./ft.
E.F.P. = 30 p.c.f.
Section 4.3.8

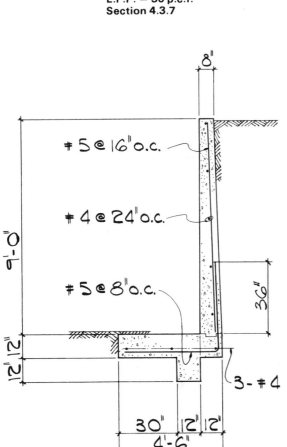

Axial load = 400 lbs./ft.
E.F.P. = 30 p.c.f.
Section 4.3.9

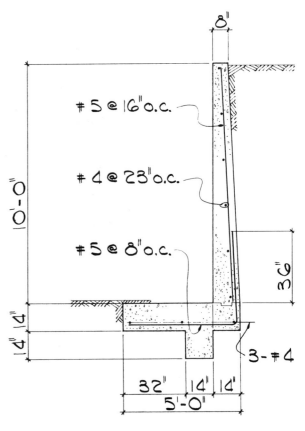

Axial load = 400 lbs./ft.
E.F.P. = 30 p.c.f.
Section 4.3.10

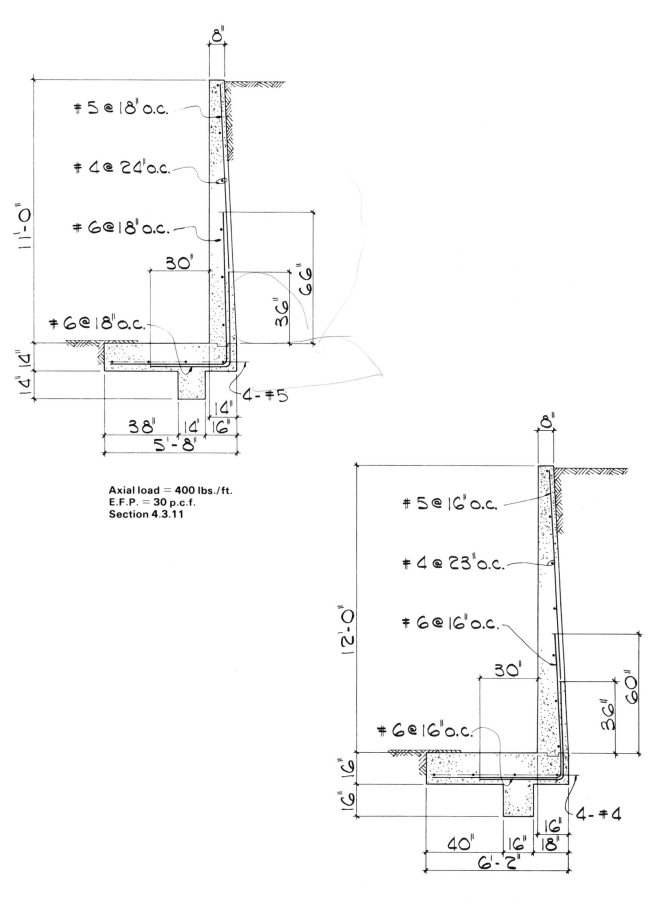

Axial load = 400 lbs./ft.
E.F.P. = 30 p.c.f.
Section 4.3.11

#5 @ 18" o.c.
#4 @ 24" o.c.
#6 @ 18" o.c.
#6 @ 18" o.c.
4-#5
8"
11'-0"
14"
14"
30"
36"
66"
38"
14"
14"
16"
5'-8"

#5 @ 16" o.c.
#4 @ 23" o.c.
#6 @ 16" o.c.
#6 @ 16" o.c.
4-#4
8"
12'-0"
16"
16"
16"
30"
36"
60"
40"
16"
16"
18"
6'-2"

Axial load = 400 lbs./ft.
E.F.P. = 30 p.c.f.
Section 4.3.12

TABLE 4.4	PROPERTY LINE RETAINING WALLS — CONCRETE STEM — SOIL NOT OVER FOOTING
SLOPE = 0 to 1	SURCHARGE = 0 lbs./sq. ft. AXIAL = 600 lbs./ft.

SOIL PRESSURE DIAGRAM

Stem Height	3'- 0''	4'- 0''	5'- 0''	6'- 0''	7'- 0''	8'- 0''	9'- 0''	10'- 0''	11'- 0''	12'- 0''
Ft. L ft.	1.667	1.833	2.167	2.667	3.333	3.833	4.500	5.000	5.667	6.167
Ft. T ft.	0.667	0.667	0.833	0.833	0.833	1.000	1.000	1.167	1.167	1.333
Wt ft.	0.667	0.667	0.667	0.667	0.667	0.667	0.667	0.667	0.667	0.667
Wb ft.	0.667	0.667	0.833	0.833	0.833	1.000	1.000	1.167	1.167	1.333
A ft.	0.500	0.667	0.833	1.333	1.667	2.000	2.500	2.667	3.167	3.333
B ft.	0.667	0.667	0.833	0.833	0.833	1.000	1.000	1.167	1.167	1.333
C ft.	0.500	0.500	0.500	0.500	0.833	0.833	1.000	1.167	1.333	1.500
D ft.	0.667	0.667	0.833	0.833	0.833	1.000	1.000	1.167	1.167	1.333
W lbs.	1201	1359	1708	1946	2215	2730	3050	3678	4048	4789
F lbs.	202	327	510	700	920	1215	1500	1870	2220	2667
Mo ft.-lbs.	246	508	992	1595	2403	3645	5000	6962	9005	11852
Mr ft.-lbs.	1506	1934	2900	4196	6064	8546	11312	15013	18847	24051
O.T.R.	6.112	3.807	2.922	2.630	2.523	2.345	2.263	2.156	2.093	2.029
$\overline{X}$ ft.	1.050	1.050	1.117	1.337	1.653	1.795	2.070	2.189	2.431	2.547
e ft.	-0.216	-0.133	-0.034	-0.003	0.014	0.121	0.180	0.311	0.402	0.536
Me ft.-lbs.	-260	-181	-58	-6	31	331	550	1143	1629	2567
S ft.^{3}	0.463	0.560	0.782	1.185	1.852	2.449	3.375	4.167	5.352	6.338
S.P.h p.s.f.	160	419	714	725	681	847	841	1010	1019	1182
S.P.t p.s.f.	1281	1064	862	735	648	577	515	461	410	372
Friction	0.168	0.240	0.299	0.360	0.416	0.445	0.492	0.509	0.548	0.557

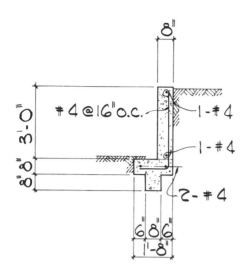

Axial load = 600 lbs./ft.
E.F.P. = 30 p.c.f.
Section 4.4.3

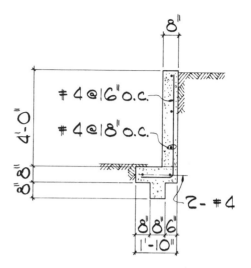

Axial load = 600 lbs./ft.
E.F.P. = 30 p.c.f.
Section 4.4.4

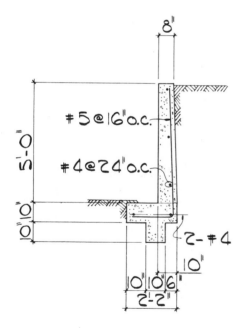

Axial load = 600 lbs./ft.
E.F.P. = 30 p.c.f.
Section 4.4.5

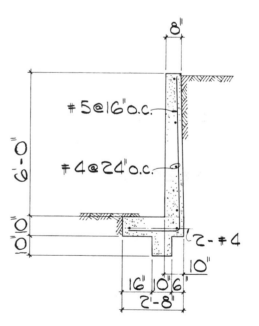

Axial load = 600 lbs./ft.
E.F.P. = 30 p.c.f.
Section 4.4.6

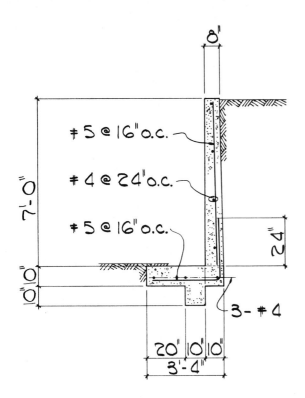

Axial load = 600 lbs./ft.
E.F.P. = 30 p.c.f.
Section 4.4.7

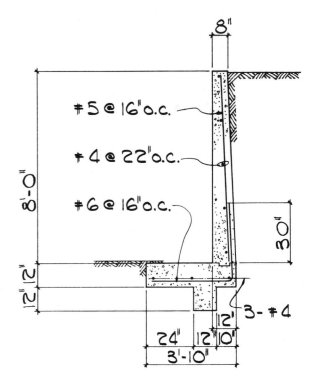

Axial load = 600 lbs./ft.
E.F.P. = 30 p.c.f.
Section 4.4.8

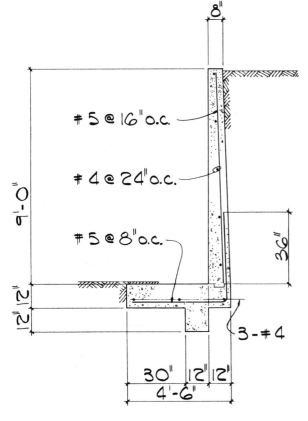

Axial load = 600 lbs./ft.
E.F.P. = 30 p.c.f.
Section 4.4.9

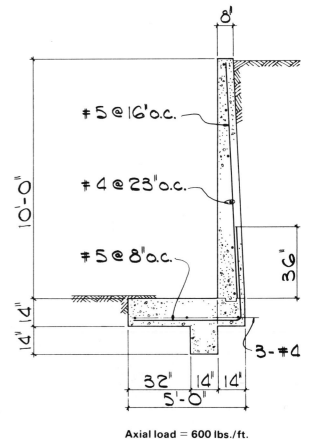

Axial load = 600 lbs./ft.
E.F.P. = 30 p.c.f.
Section 4.4.10

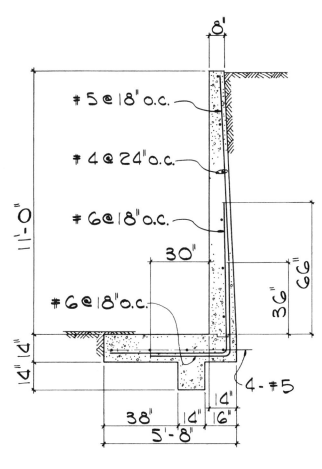

5 @ 18" o.c.

4 @ 24" o.c.

6 @ 18" o.c.

6 @ 18" o.c.

8'

11'-0"

30"

66"

36"

4"

4-#5

4"

38"   4"   6"

5'-8"

Axial load = 600 lbs./ft.
E.F.P. = 30 p.c.f.
Section 4.4.11

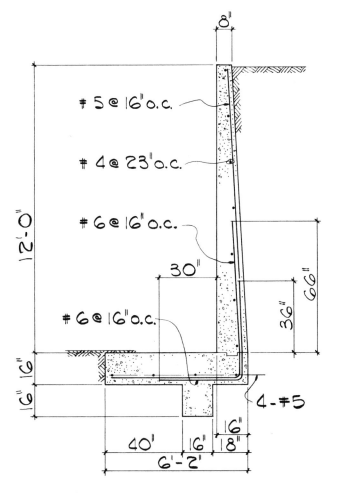

5 @ 16" o.c.

4 @ 23" o.c.

6 @ 16" o.c.

6 @ 16" o.c.

8"

12'-0"

30"

66"

36"

6"

6"

4-#5

40"   6"   6"

18"

6'-2"

Axial load = 600 lbs./ft.
E.F.P. = 30 p.c.f.
Section 4.4.12

TABLE 4.5   PROPERTY LINE RETAINING WALLS — CONCRETE STEM — SOIL NOT OVER FOOTING
SLOPE = 0 to 1   SURCHARGE = 0 lbs./sq. ft.   AXIAL = 800 lbs./ft.

SOIL PRESSURE DIAGRAM

Stem Height	3'- 0"	4'- 0"	5'- 0"	6'- 0"	7'- 0"	8'- 0"	9'- 0"	10'- 0"	11'- 0"	12'- 0"
Ft. L ft.	1.667	1.833	2.167	2.667	3.333	3.833	4.500	5.000	5.667	6.167
Ft. T ft.	0.667	0.667	0.833	0.833	0.833	1.000	1.000	1.167	1.167	1.333
Wt ft.	0.667	0.667	0.667	0.667	0.667	0.667	0.667	0.667	0.667	0.667
Wb ft.	0.667	0.667	0.833	0.833	0.833	1.000	1.000	1.167	1.167	1.333
A ft.	0.500	0.667	0.833	1.167	1.667	2.000	2.500	2.667	3.167	3.333
B ft.	0.667	0.667	0.833	0.833	0.833	1.000	1.000	1.167	1.167	1.333
C ft.	0.500	0.500	0.500	0.500	0.833	0.833	1.000	1.167	1.333	1.500
D ft.	0.667	0.667	0.833	0.833	0.833	1.000	1.000	1.167	1.167	1.333
W lbs.	1401	1559	1908	2146	2415	2930	3250	3878	4248	4989
F lbs.	202	327	510	700	920	1215	1500	1870	2220	2667
Mo ft.-lbs.	246	508	992	1595	2403	3645	5000	6962	9005	11852
Mr ft.-lbs.	1773	2234	3267	4663	6664	9246	12146	15947	19914	25218
O.T.R.	7.194	4.397	3.291	2.923	2.773	2.537	2.429	2.290	2.211	2.128
$\overline{X}$ ft.	1.090	1.107	1.192	1.429	1.764	1.912	2.199	2.317	2.568	2.679
e ft.	-0.257	-0.191	-0.109	-0.096	-0.097	0.005	0.051	0.183	0.266	0.404
Me ft.-lbs.	-360	-297	-208	-206	-235	15	167	710	1129	2017
S ft.^{3}	0.463	0.560	0.782	1.185	1.852	2.449	3.375	4.167	5.352	6.338
S.P.h p.s.f.	64	320	615	631	597	770	772	946	961	1127
S.P.t p.s.f.	1617	1381	1146	979	852	758	673	605	539	491
Friction	0.144	0.210	0.268	0.326	0.381	0.415	0.462	0.482	0.523	0.535

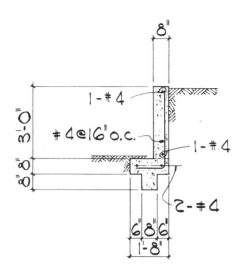

Axial load = 800 lbs./ft.
E.F.P. = 30 p.c.f.
Section 4.5.3

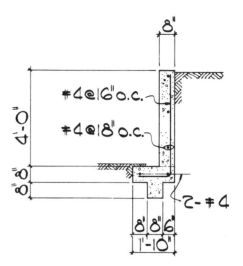

Axial load = 800 lbs./ft.
E.F.P. = 30 p.c.f.
Section 4.5.4

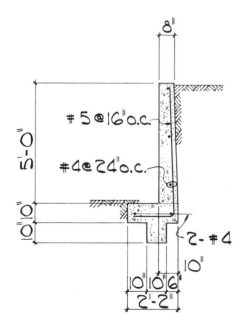

Axial load = 800 lbs./ft.
E.F.P. = 30 p.c.f.
Section 4.5.5

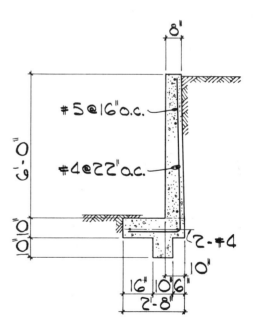

Axial load = 800 lbs./ft.
E.F.P. = 30 p.c.f.
Section 4.5.6

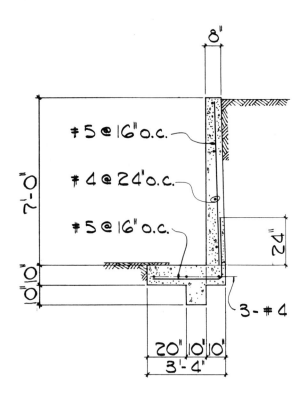

Axial load = 800 lbs./ft.
E.F.P. = 30 p.c.f.
Section 4.5.7

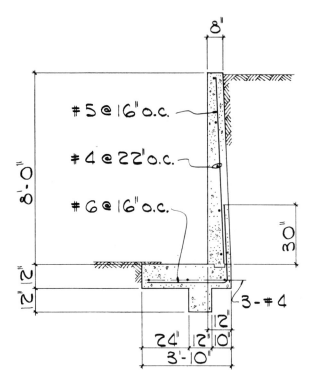

Axial load = 800 lbs./ft.
E.F.P. = 30 p.c.f.
Section 4.5.8

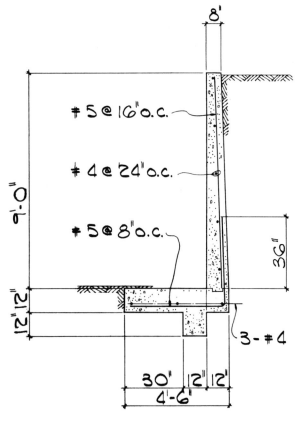

Axial load = 800 lbs./ft.
E.F.P. = 30 p.c.f.
Section 4.5.9

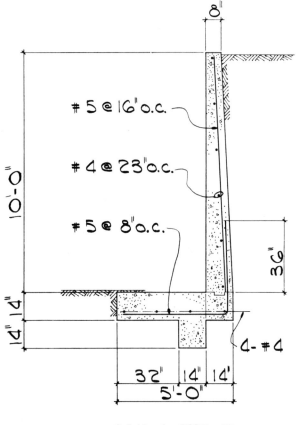

Axial load = 800 lbs./ft.
E.F.P. = 30 p.c.f.
Section 4.5.10

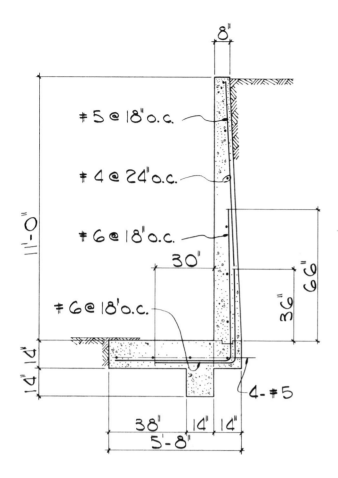

#5 @ 18" o.c.

#4 @ 24" o.c.

#6 @ 18" o.c.

#6 @ 18" o.c.

11'-0"

4-#5

30"

36"

66"

14"

14"

38" 14" 14"

5'-8"

8"

Axial load = 800 lbs./ft.
E.F.P. = 30 p.c.f.
Section 4.5.11

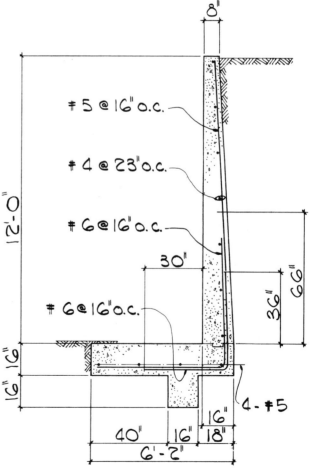

#5 @ 16" o.c.

#4 @ 23" o.c.

#6 @ 16" o.c.

#6 @ 16" o.c.

12'-0"

4-#5

30"

36"

66"

16"

16"

16"

18"

40" 16"

6'-2"

8"

Axial load = 800 lbs./ft.
E.F.P. = 30 p.c.f.
Section 4.5.12

TABLE 4.6	PROPERTY LINE RETAINING WALLS — CONCRETE STEM — SOIL NOT OVER FOOTING
SLOPE = 0 to 1	SURCHARGE = 50 lbs./sq. ft. AXIAL = 0 lbs./ft.

SOIL PRESSURE DIAGRAM

Stem Height	3'- 0''	4'- 0''	5'- 0''	6'- 0''	7'- 0''	8'- 0''	9'- 0''	10'- 0''	11'- 0''	12'- 0''
Ft. L ft.	1.667	2.167	2.667	3.167	3.667	4.333	4.833	5.500	6.000	6.667
Ft. T ft.	0.667	0.833	0.833	0.833	1.000	1.000	1.167	1.167	1.333	1.333
Wt ft.	0.667	0.667	0.667	0.667	0.667	0.667	0.667	0.667	0.667	0.667
Wb ft.	0.667	0.833	0.833	0.833	1.000	1.000	1.167	1.167	1.333	1.333
A ft.	0.500	0.833	1.333	1.667	1.833	2.333	2.500	3.000	3.333	3.667
B ft.	0.667	0.833	0.833	0.833	1.000	1.000	1.167	1.167	1.333	1.333
C ft.	0.500	0.500	0.500	0.667	0.833	1.000	1.167	1.333	1.333	1.667
D ft.	0.667	0.833	0.833	0.833	1.000	1.000	1.167	1.167	1.333	1.333
W lbs.	620	967	1201	1444	1936	2251	2856	3222	3940	4357
F lbs.	260	427	602	807	1084	1354	1707	2042	2470	2870
Mo ft.-lbs.	362	759	1270	1972	3071	4287	6068	7940	10568	13236
Mr ft.-lbs.	739	1538	2451	3580	5527	7705	10837	14054	18682	23079
O.T.R.	2.044	2.028	1.929	1.815	1.800	1.797	1.786	1.770	1.768	1.744
$\overline{X}$ ft.	0.609	0.806	0.983	1.113	1.268	1.519	1.670	1.897	2.059	2.259
e ft.	0.225	0.277	0.350	0.470	0.565	0.648	0.747	0.853	0.941	1.074
Me ft.-lbs.	139	268	420	678	1094	1459	2134	2747	3706	4680
S ft.^{3}	0.463	0.782	1.185	1.671	2.241	3.130	3.894	5.042	6.000	7.407
S.P.h p.s.f.	673	789	805	862	1016	986	1139	1131	1274	1285
S.P.t p.s.f.	71	103	96	50	40	53	43	41	39	22
Friction	0.420	0.441	0.501	0.559	0.560	0.601	0.597	0.634	0.627	0.659

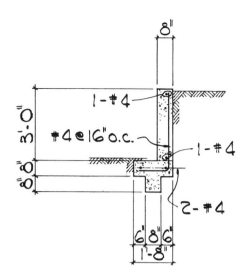

Surcharge load = **50 p.s.f.**
E.F.P. = **30 p.c.f.**
Section **4.6.3**

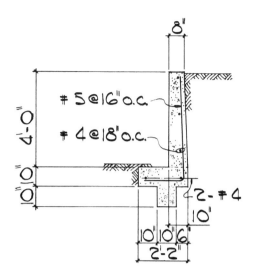

Surcharge load = **50 p.s.f.**
E.F.P. = **30 p.c.f.**
Section **4.6.4**

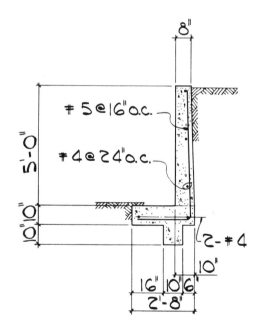

Surcharge load = **50 p.s.f.**
E.F.P. = **30 p.c.f.**
Section **4.6.5**

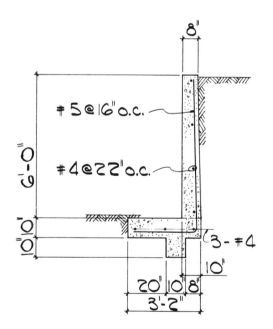

Surcharge load = **50 p.s.f.**
E.F.P. = **30 p.c.f.**
Section **4.6.6**

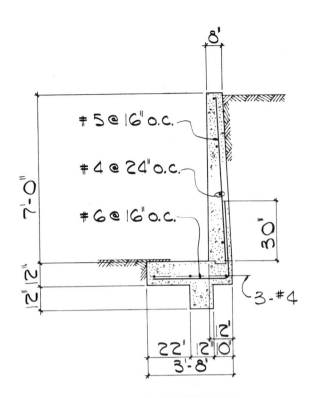

Surcharge load = 50 p.s.f.
E.F.P. = 30 p.c.f.
Section 4.6.7

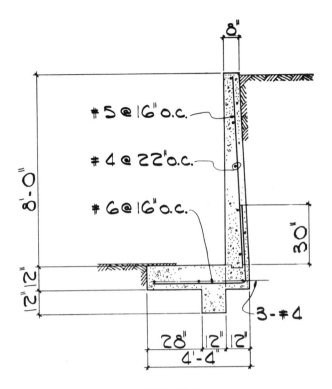

Surcharge load = 50 p.s.f.
E.F.P. = 30 p.c.f.
Section 4.6.8

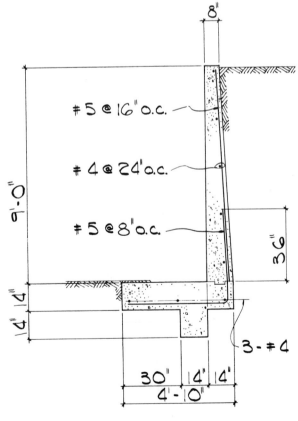

Surcharge load = 50 p.s.f.
E.F.P. = 30 p.c.f.
Section 4.6.9

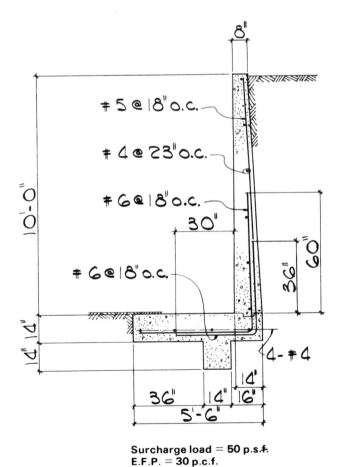

Surcharge load = 50 p.s.f.
E.F.P. = 30 p.c.f.
Section 4.6.10

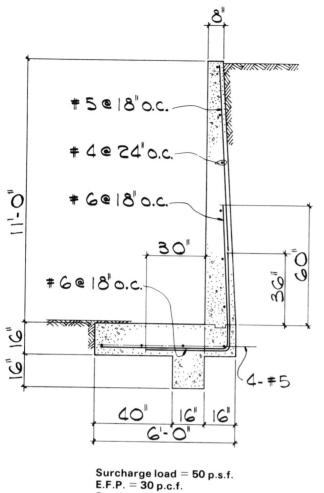

#5 @ 18" o.c.

#4 @ 24" o.c.

#6 @ 18" o.c.

#6 @ 18" o.c.

8"

11'-0"

30"

60"

36"

4-#5

16"

16"

16"

40"   16"   16"

6'-0"

Surcharge load = 50 p.s.f.
E.F.P. = 30 p.c.f.
Section 4.6.11

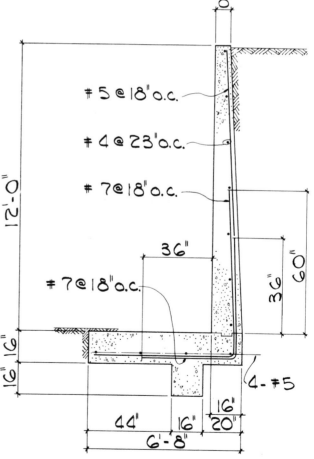

8"

#5 @ 18" o.c.

#4 @ 23" o.c.

#7 @ 18" o.c.

#7 @ 18" o.c.

12'-0"

36"

60"

36"

16"

16"

16"

16"

4-#5

44"   16"   20"

6'-8"

Surcharge load = 50 p.s.f.
E.F.P. = 30 p.c.f.
Section 4.6.12

TABLE 4.7   PROPERTY LINE RETAINING WALLS – CONCRETE STEM – SOIL NOT OVER FOOTING
SLOPE = 0 to 1   SURCHARGE = 100 lbs./sq. ft.   AXIAL = 0 lbs./ft.

SOIL PRESSURE DIAGRAM

Stem Height	3'- 0''	4'- 0''	5'- 0''	6'- 0''	7'- 0''	8'- 0''	9'- 0''	10'- 0''	11'- 0''	12'- 0''
Ft. L ft.	2.000	2.500	3.000	3.500	4.000	4.667	5.167	6.000	6.500	7.167
Ft. T ft.	0.667	0.833	0.833	0.833	1.000	1.000	1.167	1.167	1.333	1.333
Wt ft.	0.667	0.667	0.667	0.667	0.667	0.667	0.667	0.667	0.667	0.667
Wb ft.	0.667	0.833	0.833	0.833	1.000	1.000	1.167	1.167	1.333	1.333
A ft.	0.833	1.167	1.500	1.833	2.167	2.500	2.833	3.333	3.667	4.000
B ft.	0.667	0.833	0.833	0.833	1.000	1.000	1.167	1.167	1.333	1.333
C ft.	0.500	0.500	0.667	0.833	0.833	1.167	1.167	1.500	1.500	1.833
D ft.	0.667	0.833	0.833	0.833	1.000	1.000	1.167	1.167	1.333	1.333
W lbs.	676	1037	1275	1523	2030	2350	2969	3369	4106	4527
F lbs.	327	510	700	920	1215	1500	1870	2220	2667	3082
Mo ft.-lbs.	508	992	1595	2403	3645	5000	6962	9005	11852	14724
Mr ft.-lbs.	996	1937	2939	4183	6355	8667	12081	16010	21058	25742
O.T.R.	1.959	1.952	1.842	1.741	1.744	1.733	1.735	1.778	1.777	1.748
$\overline{X}$ ft.	0.721	0.911	1.054	1.168	1.335	1.560	1.724	2.079	2.242	2.434
e ft.	0.279	0.339	0.446	0.582	0.665	0.773	0.859	0.921	1.008	1.149
Me ft.-lbs.	188	351	569	886	1350	1817	2552	3103	4137	5204
S ft.^{3}	0.667	1.042	1.500	2.042	2.667	3.630	4.449	6.000	7.042	8.560
S.P.h p.s.f.	620	752	804	869	1014	1004	1148	1079	1219	1240
S.P.t p.s.f.	56	77	46	1	1	3	1	44	44	24
Friction	0.484	0.492	0.549	0.604	0.599	0.638	0.630	0.659	0.650	0.681

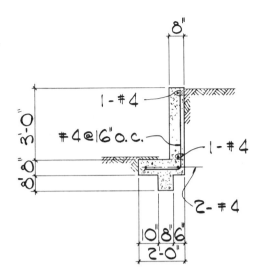

Surcharge load = **100 p.s.f.**
E.F.P. = **30 p.c.f.**
Section **4.7.3**

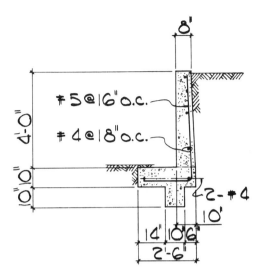

Surcharge load = **100 p.s.f.**
E.F.P. = **30 p.c.f.**
Section **4.7.4**

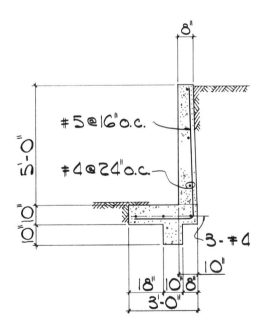

Surcharge load = **100 p.s.f.**
E.F.P. = **30 p.c.f.**
Section **4.7.5**

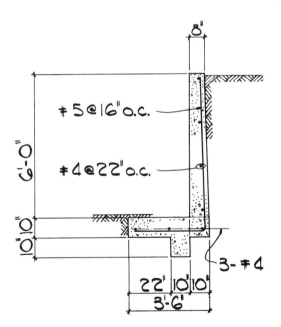

Surcharge load = **100 p.s.f.**
E.F.P. = **30 p.c.f.**
Section **4.7.6**

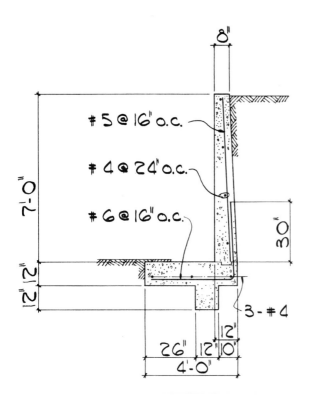

#5 @ 16" o.c.

#4 @ 24" o.c.

#6 @ 16" o.c.

8"

7'-0"

30"

3 - #4

12" 12"

26" 12" 10"

4'-0"

Surcharge load = 100 p.s.f.
E.F.P. = 30 p.c.f.
Section 4.7.7

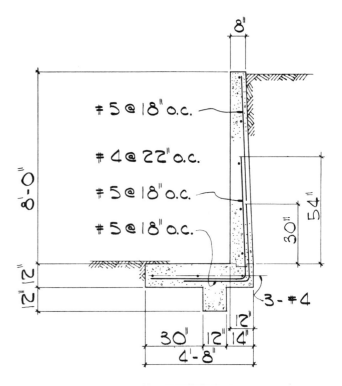

#5 @ 18" o.c.

#4 @ 22" o.c.

#5 @ 18" o.c.

#5 @ 18" o.c.

8"

8'-0"

30"   54"

3 - #4

12" 12"

30" 12" 14"

4'-8"

Surcharge load = 100 p.s.f.
E.F.P. = 30 p.c.f.
Section 4.7.8

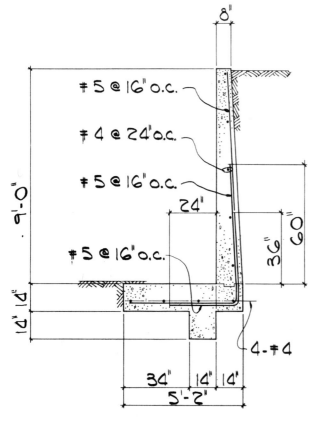

#5 @ 16" o.c.

#4 @ 24" o.c.

#5 @ 16" o.c.

#5 @ 16" o.c.

8"

9'-0"

24"

36"   60"

4-#4

14" 14"

34" 14" 14"

5'-2"

Surcharge load = 100 p.s.f.
E.F.P. = 30 p.c.f.
Section 4.7.9

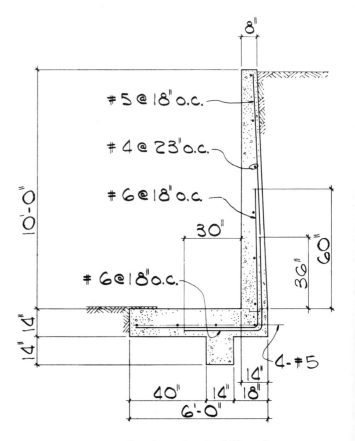

#5 @ 18" o.c.

#4 @ 23" o.c.

#6 @ 18" o.c.

#6 @ 18" o.c.

8"

10'-0"

30"

36"   60"

4-#5

14" 14"

40" 14" 18"

6'-0"

Surcharge load = 100 p.s.f.
E.F.P. = 30 p.c.f.
Section 4.7.10

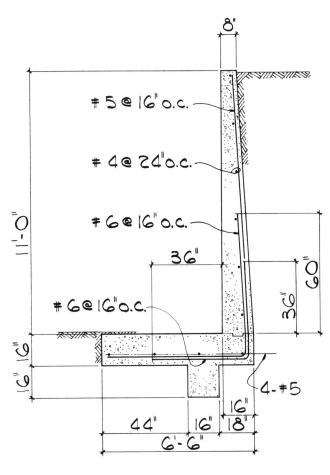

Surcharge load = 100 p.s.f.
E.F.P. = 30 p.c.f.
Section 4.7.11

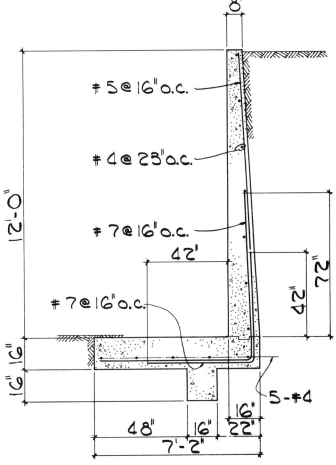

Surcharge load = 100 p.s.f.
E.F.P. = 30 p.c.f.
Section 4.7.12

TABLE 4.8	PROPERTY LINE RETAINING WALLS – CONCRETE STEM – SOIL NOT OVER FOOTING
	SLOPE = 0 to 1 SURCHARGE = 200 lbs./sq. ft. AXIAL = 0 lbs./ft.

SOIL PRESSURE DIAGRAM

Stem Height	3'- 0"	4'- 0"	5'- 0"	6'- 0"	7'- 0"	8'- 0"	9'- 0"	10'- 0"	11'- 0"	12'- 0"
Ft. L ft.	2.667	3.333	3.833	4.333	5.000	5.500	6.167	6.667	7.500	8.167
Ft. T ft.	0.833	0.833	0.833	1.000	1.000	1.167	1.167	1.333	1.333	1.333
Wt ft.	0.667	0.667	0.667	0.667	0.667	0.667	0.667	0.667	0.667	0.667
Wb ft.	0.833	0.833	0.833	1.000	1.000	1.167	1.167	1.333	1.333	1.333
A ft.	1.333	1.667	2.000	2.333	2.833	3.000	3.500	3.667	4.333	4.667
B ft.	0.833	0.833	0.833	1.000	1.000	1.167	1.167	1.333	1.333	1.333
C ft.	0.500	0.833	1.000	1.000	1.167	1.333	1.500	1.667	1.833	2.167
D ft.	0.833	0.833	0.833	1.000	1.000	1.167	1.167	1.333	1.333	1.333
W lbs.	945	1204	1453	1955	2275	2890	3261	3989	4444	4876
F lbs.	510	700	920	1215	1500	1870	2220	2667	3082	3527
Mo ft.-lbs.	992	1595	2403	3645	5000	6962	9005	11852	14724	18025
Mr ft.-lbs.	1854	3021	4292	6526	8885	12353	15789	20776	26212	31492
O.T.R.	1.868	1.894	1.786	1.790	1.777	1.774	1.753	1.753	1.780	1.747
$\overline{X}$ ft.	0.911	1.184	1.300	1.474	1.708	1.865	2.080	2.237	2.585	2.762
e ft.	0.422	0.483	0.617	0.693	0.792	0.885	1.003	1.096	1.165	1.321
Me ft.-lbs.	399	582	896	1355	1802	2557	3270	4372	5176	6442
S ft.^{3}	1.185	1.852	2.449	3.130	4.167	5.042	6.338	7.407	9.375	11.116
S.P.h p.s.f.	691	675	745	884	887	1033	1045	1189	1145	1177
S.P.t p.s.f.	18	47	13	18	23	18	13	8	40	17
Friction	0.540	0.582	0.634	0.621	0.659	0.647	0.681	0.669	0.693	0.723

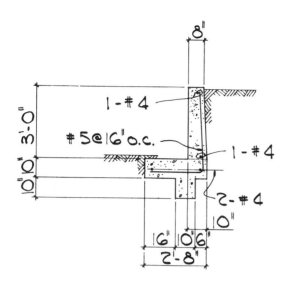

Surcharge load = 200 p.s.f.
E.F.P. = 30 p.c.f.
Section 4.8.3

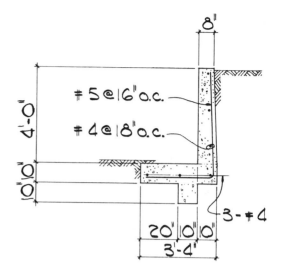

Surcharge load = 200 p.s.f.
E.F.P. = 30 p.c.f.
Section 4.8.4

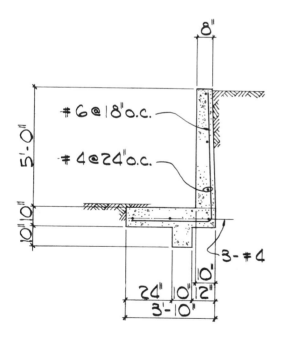

Surcharge load = 200 p.s.f.
E.F.P. = 30 p.c.f.
Section 4.8.5

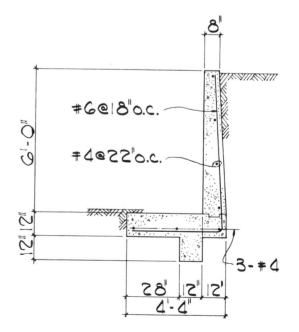

Surcharge load = 200 p.s.f.
E.F.P. = 30 p.c.f.
Section 4.8.6

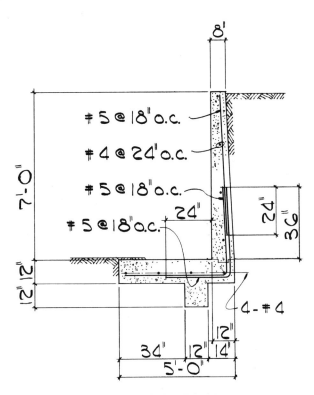

Surcharge load = **200 p.s.f.**
E.F.P. = **30 p.c.f.**
Section **4.8.7**

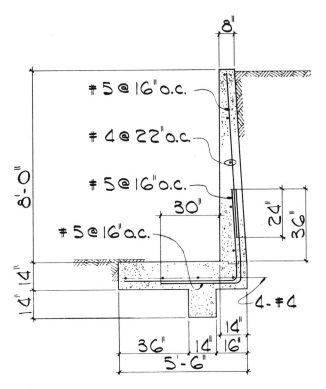

Surcharge load = **200 p.s.f.**
E.F.P. = **30 p.c.f.**
Section **4.8.8**

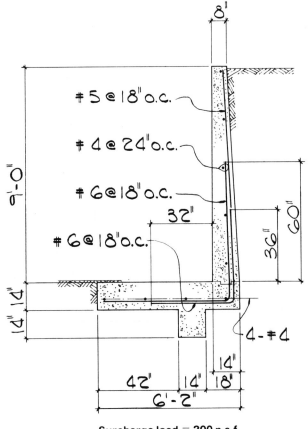

Surcharge load = **200 p.s.f.**
E.F.P. = **30 p.c.f.**
Section **4.8.9**

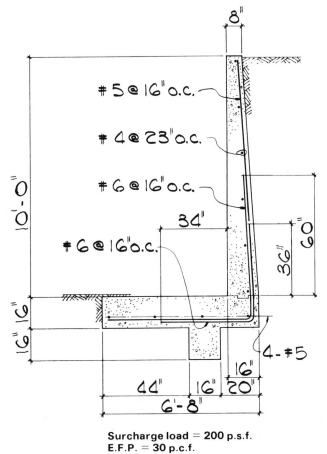

Surcharge load = **200 p.s.f.**
E.F.P. = **30 p.c.f.**
Section **4.8.10**

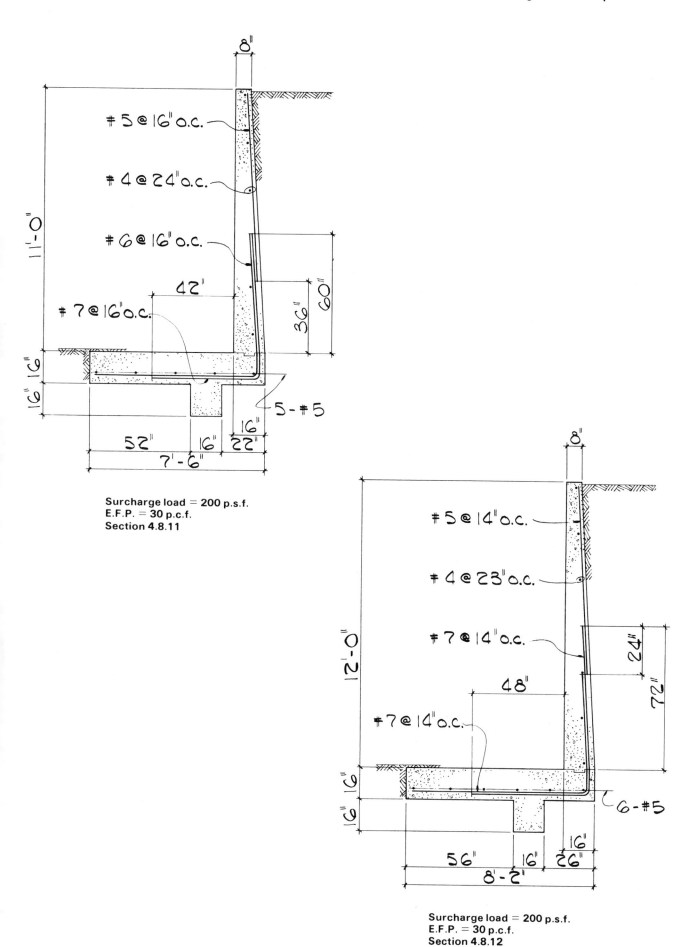

Surcharge load = 200 p.s.f.
E.F.P. = 30 p.c.f.
Section 4.8.11

Surcharge load = 200 p.s.f.
E.F.P. = 30 p.c.f.
Section 4.8.12

TABLE 4.9   PROPERTY LINE RETAINING WALLS − CONCRETE STEM − SOIL NOT OVER FOOTING
SLOPE = 5 to 1   SURCHARGE = 0 lbs./sq. ft.   AXIAL = 0 lbs./ft.

SOIL PRESSURE DIAGRAM

Stem Height	3'- 0"	4'- 0"	5'- 0"	6'- 0"	7'- 0"	8'- 0"	9'- 0"	10'- 0"	11'- 0"	12'- 0"
Ft. L ft.	1.667	1.833	2.333	2.833	3.333	4.000	4.500	5.167	5.667	6.500
Ft. T ft.	0.667	0.833	0.833	0.833	1.000	1.000	1.167	1.167	1.333	1.333
Wt ft.	0.667	0.667	0.667	0.667	0.667	0.667	0.667	0.667	0.667	0.667
Wb ft.	0.667	0.833	0.833	0.833	1.000	1.000	1.167	1.167	1.333	1.333
A ft.	0.500	0.500	1.000	1.333	1.667	2.167	2.333	2.833	3.000	3.667
B ft.	0.667	0.833	0.833	0.833	1.000	1.000	1.167	1.167	1.333	1.333
C ft.	0.500	0.500	0.500	0.667	0.667	0.833	1.000	1.167	1.333	1.500
D ft.	0.667	0.833	0.833	0.833	1.000	1.000	1.167	1.167	1.333	1.333
W lbs.	605	908	1140	1382	1866	2182	2780	3148	3861	4315
F lbs.	215	374	544	747	1024	1296	1654	1995	1434	2844
Mo ft.-lbs.	263	602	1059	1702	2731	3888	5604	7426	10006	12642
Mr ft.-lbs.	714	1190	2013	3049	4848	6911	9850	12943	17311	22344
O.T.R.	2.716	1.976	1.901	1.792	1.775	1.778	1.757	1.743	1.730	1.767
$\overline{X}$ ft.	0.745	0.648	0.837	0.975	1.135	1.386	1.527	1.752	1.892	2.248
e ft.	0.088	0.269	0.329	0.442	0.532	0.614	0.723	0.831	0.941	1.002
Me ft.-lbs.	53	244	376	611	993	1341	2011	2616	3635	4322
S ft.^{3}	0.463	0.560	0.907	1.338	1.852	2.667	3.375	4.449	5.352	7.042
S.P.h p.s.f.	478	931	902	945	1096	1048	1214	1197	1361	1278
S.P.t p.s.f.	248	59	75	31	24	43	22	21	2	50
Friction	0.356	0.412	0.478	0.540	0.549	0.594	0.595	0.634	0.630	0.659

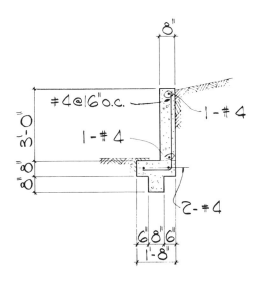

Soil slope = 5 to 1
E.F.P. = 32 p.c.f.
Section 4.9.3

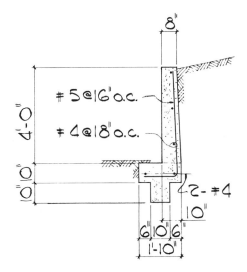

Soil slope = 5 to 1
E.F.P. = 32 p.c.f.
Section 4.9.4

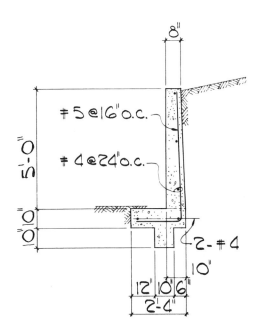

Soil slope = 5 to 1
E.F.P. = 32 p.c.f.
Section 4.9.5

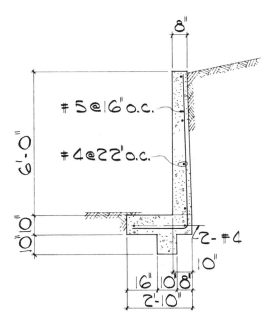

Soil slope = 5 to 1
E.F.P. = 32 p.c.f.
Section 4.9.6

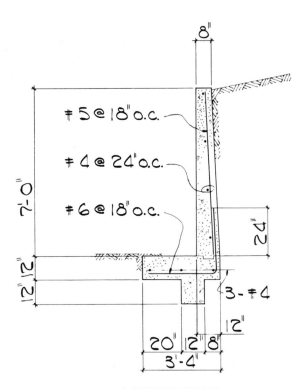

Soil slope = 5 to 1
E.F.P. = 32 p.c.f.
Section 4.9.7

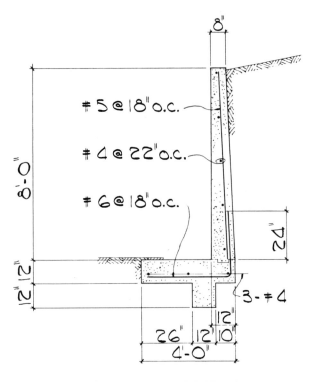

Soil slope = 5 to 1
E.F.P. = 32 p.c.f.
Section 4.9.8

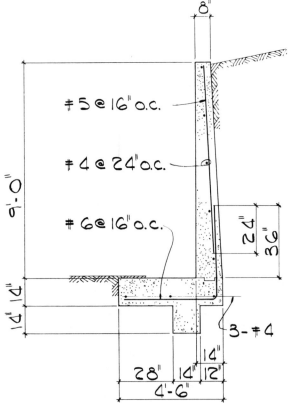

Soil slope = 5 to 1
E.F.P. = 32 p.c.f.
Section 4.9.9

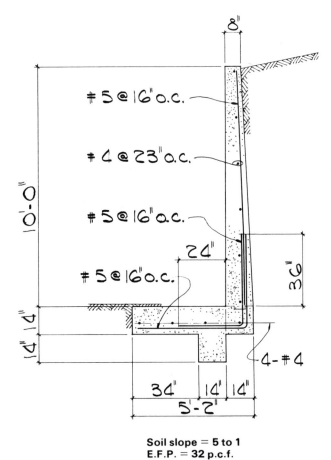

Soil slope = 5 to 1
E.F.P. = 32 p.c.f.
Section 4.9.10

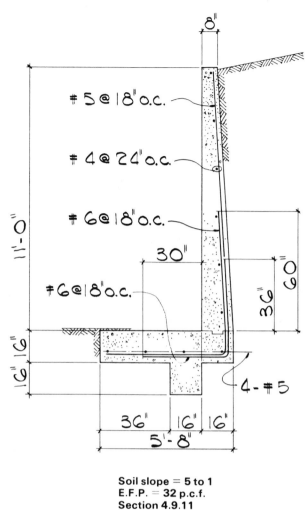

Soil slope = **5 to 1**
E.F.P. = **32 p.c.f.**
Section **4.9.11**

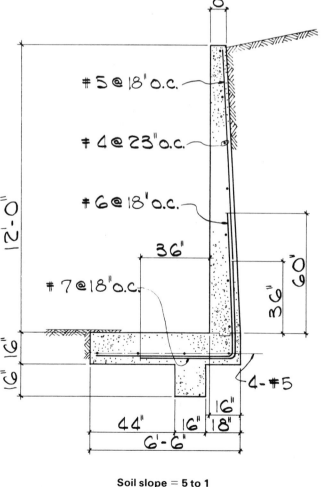

Soil slope = **5 to 1**
E.F.P. = **32 p.c.f.**
Section **4.9.12**

TABLE 4.10   PROPERTY LINE RETAINING WALLS — CONCRETE STEM — SOIL NOT OVER FOOTING
SLOPE = 4 to 1   SURCHARGE = 0 lbs./sq. ft.   AXIAL = 0 lbs./ft.

SOIL PRESSURE DIAGRAM

Stem Height	3'-0"	4'-0"	5'-0"	6'-0"	7'-0"	8'-0"	9'-0"	10'-0"	11'-0"	12'-0"
Ft. L ft.	1.667	2.000	2.500	3.000	3.500	4.167	4.833	5.500	6.000	6.833
Ft. T ft.	0.667	0.833	0.833	0.833	1.000	1.000	1.167	1.167	1.333	1.333
Wt ft.	0.667	0.667	0.667	0.667	0.667	0.667	0.667	0.667	0.667	0.667
Wb ft.	0.667	0.833	0.833	0.833	1.000	1.000	1.167	1.167	1.333	1.333
A ft.	0.500	0.667	1.167	1.500	1.833	2.167	2.500	3.000	3.333	3.833
B ft.	0.667	0.833	0.833	0.833	1.000	1.000	1.167	1.167	1.333	1.333
C ft.	0.500	0.500	0.500	0.667	0.667	1.000	1.167	1.333	1.333	1.667
D ft.	0.667	0.833	0.833	0.833	1.000	1.000	1.167	1.167	1.333	1.333
W lbs.	612	940	1178	1427	1923	2247	2890	3269	4004	4470
F lbs.	235	409	595	817	1120	1417	1809	2182	2662	3111
Mo ft.-lbs.	288	659	1158	1861	2987	4252	6130	8122	10944	13827
Mr ft.-lbs.	725	1367	2247	3351	5273	7421	11002	14311	19066	24356
O.T.R.	2.522	2.075	1.941	1.800	1.766	1.745	1.795	1.762	1.742	1.761
$\overline{X}$ ft.	0.715	0.753	0.925	1.044	1.189	1.410	1.686	1.893	2.028	2.355
e ft.	0.118	0.247	0.325	0.456	0.561	0.674	0.731	0.857	0.972	1.061
Me ft.-lbs.	72	232	383	650	1079	1514	2113	2801	3890	4745
S ft,^{3}	0.463	0.667	1.042	1.500	2.042	2.894	3.894	5.042	6.000	7.782
S.P.h p.s.f.	523	819	839	909	1078	1063	1141	1150	1316	1264
S.P.t p.s.f.	211	122	104	42	21	16	55	39	19	44
Friction	0.385	0.435	0.506	0.573	0.582	0.631	0.626	0.668	0.665	0.696

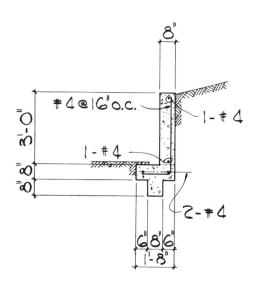

Soil slope = **4** to **1**
E.F.P. = **35** p.c.f.
Section **4.10.3**

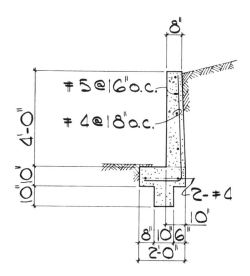

Soil slope = **4** to **1**
E.F.P. = **35** p.c.f.
Section **4.10.4**

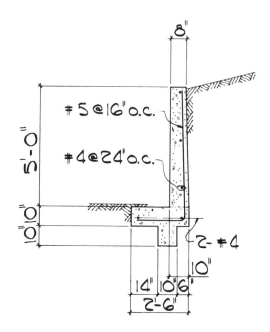

Soil slope = **4** to **1**
E.F.P. = **35** p.c.f.
Section **4.10.5**

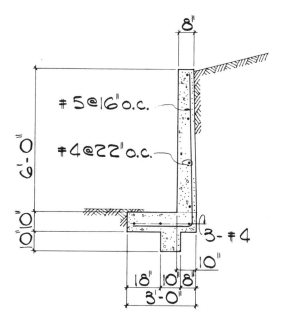

Soil slope = **4** to **1**
E.F.P. = **35** p.c.f.
Section **4.10.6**

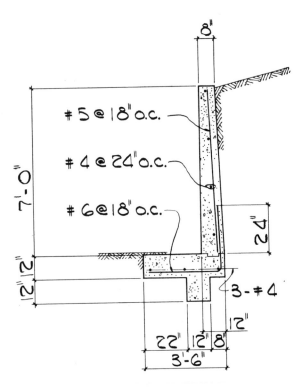

#5 @ 18" o.c.

#4 @ 24" o.c.

#6 @ 18" o.c.

7'-0"

2'-1"

24"

3-#4

22" 12" 8"

12"

3'-6"

Soil slope = **4 to 1**
E.F.P. = **35 p.c.f.**
Section 4.10.7

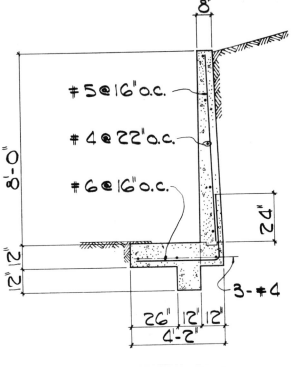

#5 @ 16" o.c.

#4 @ 22" o.c.

#6 @ 16" o.c.

8'-0"

2'-1"

24"

3-#4

26" 12" 12"

4'-2"

Soil slope = **4 to 1**
E.F.P. = **35 p.c.f.**
Section 4.10.8

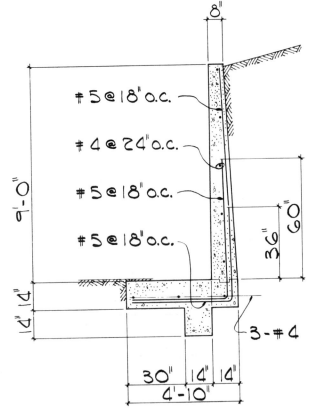

#5 @ 18" o.c.

#4 @ 24" o.c.

#5 @ 18" o.c.

#5 @ 18" o.c.

9'-0"

4'-4"

36"

60"

3-#4

30" 14" 14"

4'-10"

Soil slope = **4 to 1**
E.F.P. = **35 p.c.f.**
Section 4.10.9

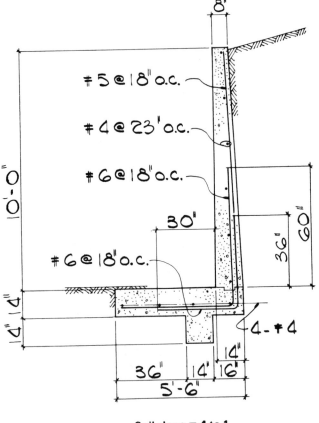

#5 @ 18" o.c.

#4 @ 23" o.c.

#6 @ 18" o.c.

#6 @ 18" o.c.

10'-0"

4'-4"

30"

36"

60"

4-#4

36" 14" 16"

5'-6"

Soil slope = **4 to 1**
E.F.P. = **35 p.c.f.**
Section 4.10.10

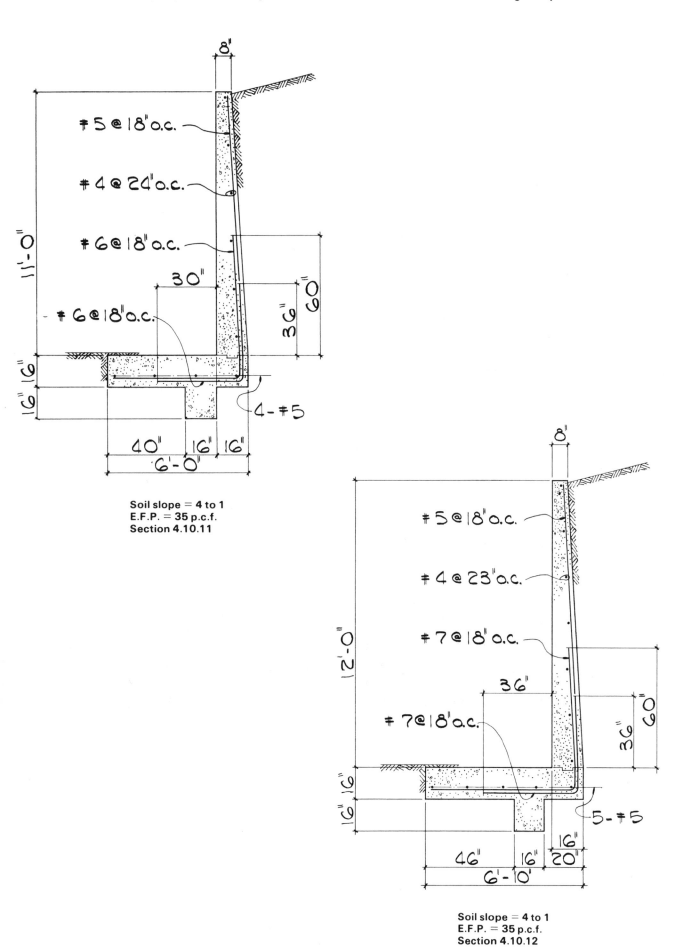

#5 @ 18" o.c.

#4 @ 24" o.c.

#6 @ 18" o.c.

#6 @ 18" o.c.

8"

11'-0"

30"

60"

36"

6"

6"

6"

4-#5

40"   16"   16"

6'-0"

Soil slope = 4 to 1
E.F.P. = 35 p.c.f.
Section 4.10.11

#5 @ 18" o.c.

#4 @ 23" o.c.

#7 @ 18" o.c.

#7 @ 18" o.c.

8"

12'-0"

36"

60"

36"

6"

6"

6"

5-#5

46"   16"   20"

16"

6'-10"

Soil slope = 4 to 1
E.F.P. = 35 p.c.f.
Section 4.10.12

TABLE 4.11	PROPERTY LINE RETAINING WALLS — CONCRETE STEM — SOIL NOT OVER FOOTING
SLOPE = 3 to 1	SURCHARGE = 0 lbs./sq. ft.　　　AXIAL = 0 lbs./ft.

SOIL PRESSURE DIAGRAM

Stem Height	3'- 0"	4'- 0"	5'- 0"	6'- 0"	7'- 0"	8'- 0"	9'- 0"	10'- 0"	11'- 0"	12'- 0"
Ft. L ft.	1.667	2.000	2.500	3.167	3.667	4.333	5.000	5.667	6.333	7.167
Ft. T ft.	0.778	0.833	0.833	0.833	1.000	1.000	1.167	1.167	1.333	1.333
Wt ft.	0.667	0.667	0.667	0.667	0.667	0.667	0.667	0.667	0.667	0.667
Wb ft.	0.667	0.833	0.833	0.833	1.000	1.000	1.167	1.167	1.333	1.333
A ft.	0.500	0.667	1.167	1.667	1.833	2.333	2.667	3.167	3.500	4.000
B ft.	0.667	0.833	0.833	0.833	1.000	1.000	1.167	1.167	1.333	1.333
C ft.	0.500	0.500	0.500	0.667	0.833	1.000	1.167	1.333	1.500	1.833
D ft.	0.667	0.833	0.833	0.833	1.000	1.000	1.167	1.167	1.333	1.333
W lbs.	618	952	1195	1471	1980	2313	2971	3361	4147	4626
F lbs.	255	444	647	887	1216	1539	1964	2369	2890	3378
Mo ft.-lbs.	312	715	1257	2021	3243	4617	6655	8819	11882	15012
Mr ft.-lbs.	736	1390	2290	3665	5688	7973	11744	15212	20849	26450
O.T.R.	2.359	1.944	1.821	1.813	1.754	1.727	1.765	1.725	1.755	1.762
$\overline{X}$ ft.	0.686	0.709	0.864	1.118	1.235	1.451	1.713	1.902	2.162	2.472
e ft.	0.148	0.291	0.386	0.466	0.598	0.716	0.787	0.931	1.004	1.111
Me ft.-lbs.	91	277	461	685	1185	1655	2339	3128	4164	5139
S ft.^{3}	0.463	0.667	1.042	1.671	2.241	3.130	4.167	5.352	6.685	8.560
S.P.h p.s.f.	568	892	920	874	1069	1063	1156	1178	1278	1246
S.P.t p.s.f.	174	60	35	55	11	5	33	9	32	45
Friction	0.413	0.466	0.541	0.603	0.614	0.665	0.661	0.705	0.697	0.730

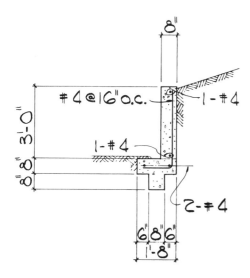

Soil slope = 3 to 1
E.F.P. = 38 p.c.f.
Section 4.11.3

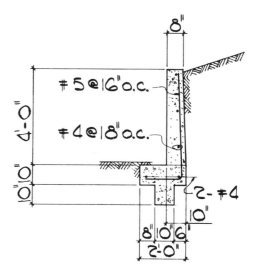

Soil slope = 3 to 1
E.F.P. = 38 p.c.f.
Section 4.11.4

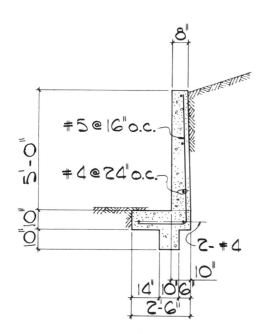

Soil slope = 3 to 1
E.F.P. = 38 p.c.f.
Section 4.11.5

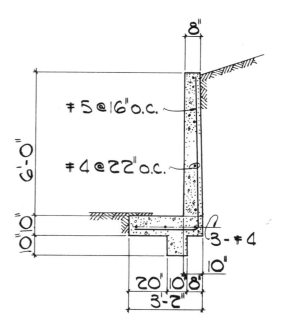

Soil slope = 3 to 1
E.F.P. = 38 p.c.f.
Section 4.11.6

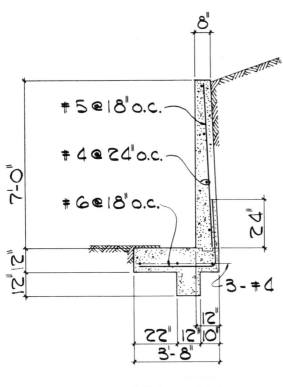

Soil slope = 3 to 1
E.F.P. = 38 p.c.f.
Section 4.11.7

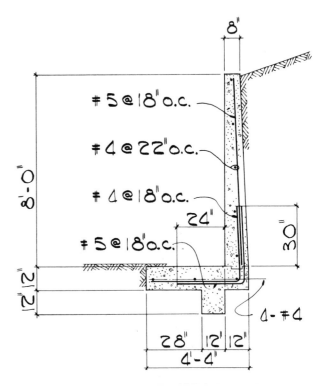

Soil slope = 3 to 1
E.F.P. = 38 p.c.f.
Section 4.11.8

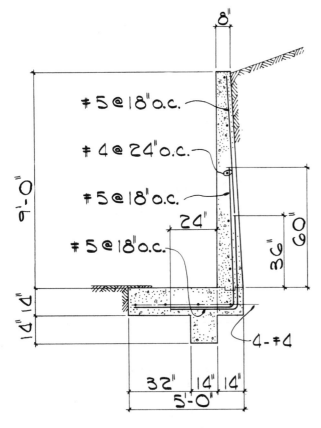

Soil slope = 3 to 1
E.F.P. = 38 p.c.f.
Section 4.11.9

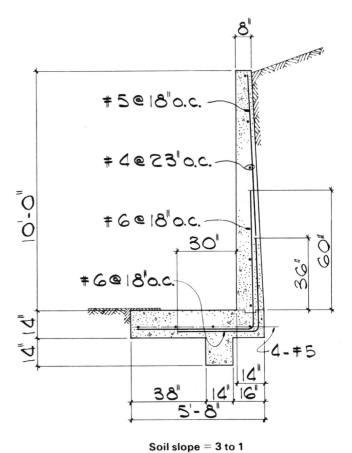

Soil slope = 3 to 1
E.F.P. = 38 p.c.f.
Section 4.11.10

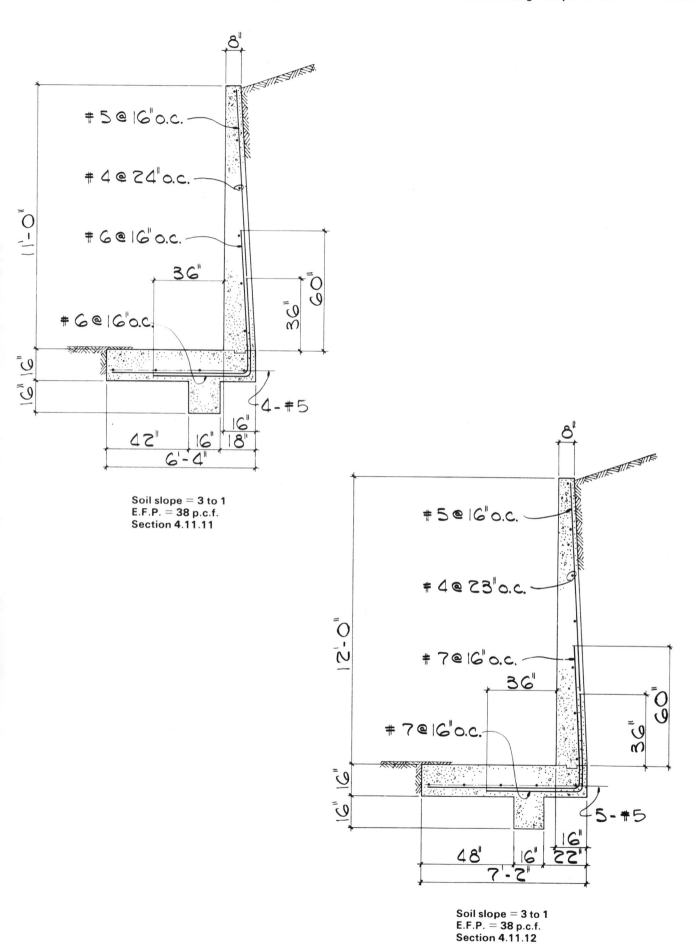

#5 @ 16" o.c.

#4 @ 24" o.c.

#6 @ 16" o.c.

#6 @ 16" o.c.

8"

11'-0"

36"

36"

60"

4-#5

16"

16"

16"

42"

16"

18"

6'-4"

Soil slope = 3 to 1
E.F.P. = 38 p.c.f.
Section 4.11.11

#5 @ 16" o.c.

#4 @ 23" o.c.

#7 @ 16" o.c.

#7 @ 16" o.c.

8"

12'-0"

36"

36"

60"

5-#5

16"

16"

16"

48"

16"

22"

7'-2"

Soil slope = 3 to 1
E.F.P. = 38 p.c.f.
Section 4.11.12

TABLE 4.12   PROPERTY LINE RETAINING WALLS – CONCRETE STEM – SOIL NOT OVER FOOTING
SLOPE = 2 to 1   SURCHARGE = 0 lbs./sq. ft.   AXIAL = 0 lbs./ft.

SOIL PRESSURE DIAGRAM

Stem Height	3'- 0''	4'- 0''	5'- 0''	6'- 0''	7'- 0''	8'- 0''	9'- 0''	10'- 0''	11'- 0''	12'- 0''
Ft. L ft.	1.667	2.167	2.667	3.333	4.000	4.667	5.333	6.000	6.833	7.500
Ft. T ft.	0.667	0.833	0.833	0.833	1.000	1.167	1.167	1.333	1.333	1.333
Wt ft.	0.667	0.667	0.667	0.667	0.667	0.667	0.667	0.667	0.667	0.667
Wb ft.	0.667	0.833	0.833	0.833	1.000	1.167	1.167	1.333	1.333	1.333
A ft.	0.500	0.833	1.333	1.667	2.167	2.500	2.833	3.333	3.833	4.333
B ft.	0.667	0.833	0.833	0.833	1.000	1.167	1.167	1.333	1.333	1.333
C ft.	0.500	0.500	0.500	0.833	0.833	1.000	1.333	1.333	1.667	1.833
D ft.	0.667	0.833	0.833	0.833	1.000	1.167	1.167	1.333	1.333	1.333
W lbs.	630	992	1244	1530	2084	2723	3116	3887	4373	4841
F lbs.	289	502	732	1004	1376	1807	2222	2762	3270	3822
Mo ft.-lbs.	353	809	1423	2287	3669	5520	7531	10433	13445	16988
Mr ft.-lbs.	755	1593	2566	4024	6570	9974	13170	18440	23769	29114
O.T.R.	2.137	1.968	1.804	1.760	1.791	1.807	1.749	1.768	1.768	1.714
$\overline{X}$ ft.	0.638	0.789	0.919	1.135	1.392	1.636	1.810	2.060	2.361	2.505
e ft.	0.195	0.294	0.414	0.532	0.608	0.698	0.857	0.940	1.056	1.245
Me ft.-lbs.	123	292	515	814	1267	1899	2670	3654	4619	6027
S ft.^{3}	0.463	0.782	1.185	1.852	2.667	3.630	4.741	6.000	7.782	9.375
S.P.h p.s.f.	643	831	901	899	996	1107	1147	1257	1234	1288
S.P.t p.s.f.	112	85	32	20	46	60	21	39	47	3
Friction	0.459	0.506	0.588	0.656	0.660	0.663	0.713	0.710	0.748	0.790

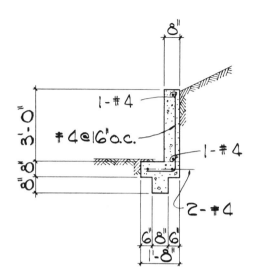

Soil slope = 2 to 1
E.F.P. = 43 p.c.f.
Section 4.12.3

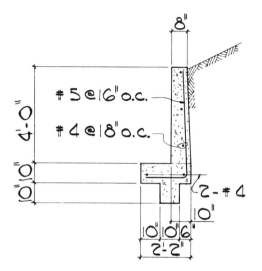

Soil slope = 2 to 1
E.F.P. = 43 p.c.f.
Section 4.12.4

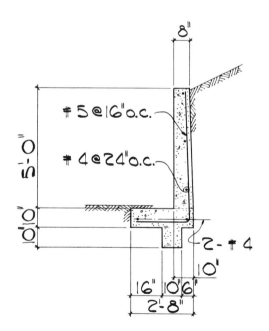

Soil slope = 2 to 1
E.F.P. = 43 p.c.f.
Section 4.12.5

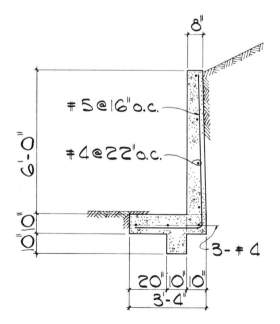

Soil slope = 2 to 1
E.F.P. = 43 p.c.f.
Section 4.12.6

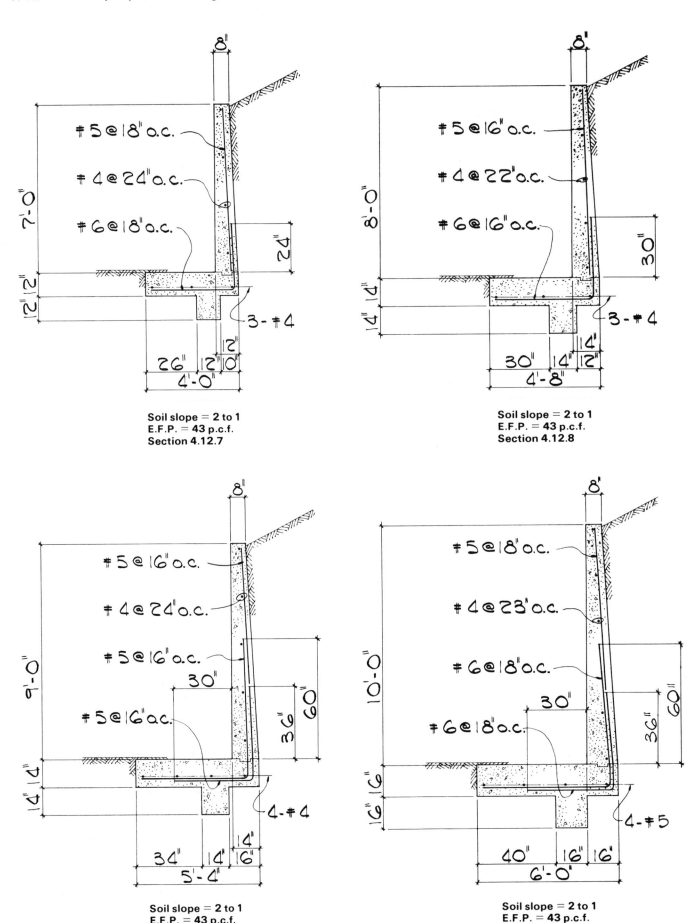

Soil slope = **2 to 1**
E.F.P. = **43 p.c.f.**
Section **4.12.7**

Soil slope = **2 to 1**
E.F.P. = **43 p.c.f.**
Section **4.12.8**

Soil slope = **2 to 1**
E.F.P. = **43 p.c.f.**
Section **4.12.9**

Soil slope = **2 to 1**
E.F.P. = **43 p.c.f.**
Section **4.12.10**

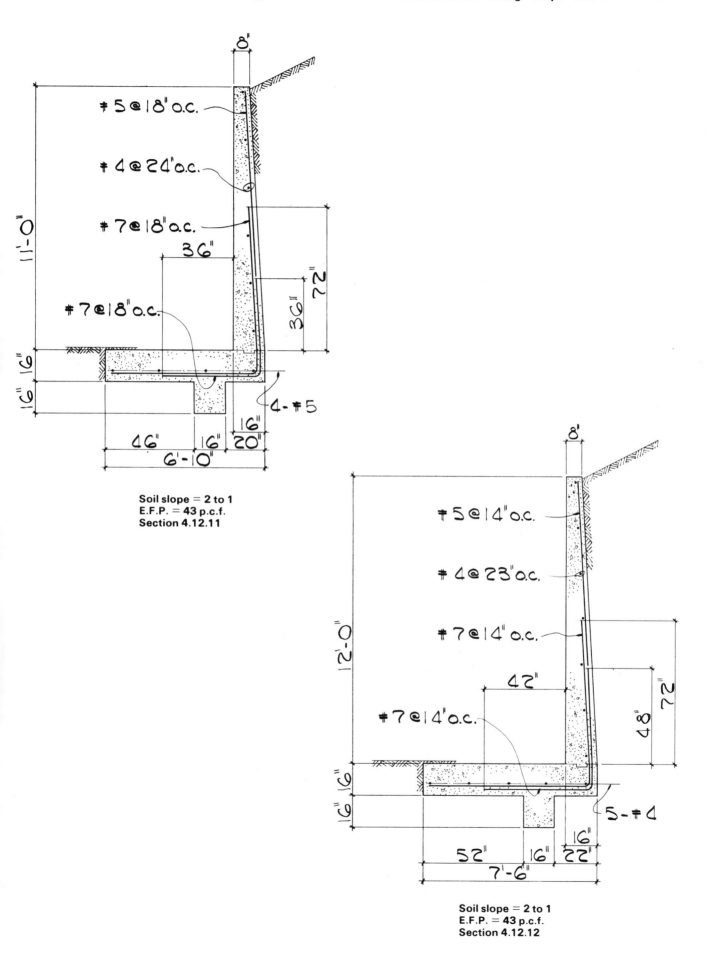

#5@18" o.c.

#4@24" o.c.

#7@18" o.c.

8"

36"

72"

36"

#7@18" o.c.

6"

6"

4-#5

46"

16"

16"

20"

6'-0"

11'-0"

Soil slope = 2 to 1
E.F.P. = 43 p.c.f.
Section 4.12.11

#5@14" o.c.

#4@23" o.c.

#7@14" o.c.

8"

42"

72"

48"

#7@14" o.c.

6"

6"

5-#4

52"

16"

16"

22"

7'-6"

12'-0"

Soil slope = 2 to 1
E.F.P. = 43 p.c.f.
Section 4.12.12

TABLE 4.13 PROPERTY LINE RETAINING WALLS — CONCRETE STEM — SOIL NOT OVER FOOTING
SLOPE = 1.5 to 1   SURCHARGE = 0 lbs./sq. ft.   AXIAL = 0 lbs./ft.

SOIL PRESSURE DIAGRAM

Stem Height	3'- 0"	4'- 0"	5'- 0"	6'- 0"	7'- 0"	8'- 0"	9'- 0"	10'- 0"	11'- 0"	12'- 0"
Ft. L ft.	1.833	2.333	3.000	3.833	4.500	5.167	6.000	6.833	7.667	8.500
Ft. T ft.	0.667	0.833	0.833	1.000	1.000	1.167	1.333	1.333	1.333	1.500
Wt ft.	0.667	0.667	0.667	0.667	0.667	0.667	0.667	0.667	0.667	0.667
Wb ft.	0.667	0.833	0.833	1.000	1.000	1.167	1.333	1.333	1.333	1.500
A ft.	0.667	1.000	1.500	2.000	2.500	2.833	3.333	3.833	4.333	4.833
B ft.	0.667	0.833	0.833	1.000	1.000	1.167	1.333	1.333	1.333	1.500
C ft.	0.500	0.500	0.667	0.833	1.000	1.167	1.333	1.667	2.000	2.167
D ft.	0.667	0.833	0.833	1.000	1.000	1.167	1.333	1.333	1.333	1.500
W lbs.	673	1060	1354	1924	2287	2979	3795	4311	4844	5871
F lbs.	370	642	936	1347	1760	2311	2936	3532	4183	5012
Mo ft.-lbs.	452	1035	1820	3144	4693	7061	10114	13344	17197	22553
Mr ft.-lbs.	911	1869	3174	5761	8182	12192	17964	23415	29726	39732
O.T.R.	2.015	1.806	1.745	1.832	1.743	1.727	1.776	1.755	1.729	1.762
$\overline{X}$ ft.	0.681	0.787	1.001	1.360	1.526	1.723	2.068	2.336	2.586	2.926
e ft.	0.235	0.380	0.499	0.556	0.724	0.861	0.932	1.080	1.247	1.324
Me ft.-lbs.	158	403	675	1071	1657	2563	3536	4657	6041	7772
S ft.^{3}	0.560	0.907	1.500	2.449	3.375	4.449	6.000	7.782	9.796	12.042
S.P.h p.s.f.	650	898	901	939	999	1153	1222	1229	1249	1336
S.P.t p.s.f.	85	10	1	65	17	0	43	32	15	45
Friction	0.549	0.606	0.691	0.700	0.770	0.776	0.774	0.819	0.863	0.854

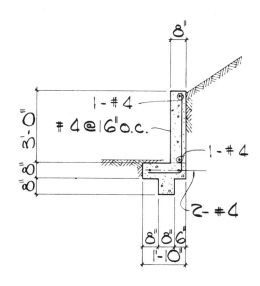

Soil slope = 1½ to 1
E.F.P. = 55 p.c.f.
Section **4.13.3**

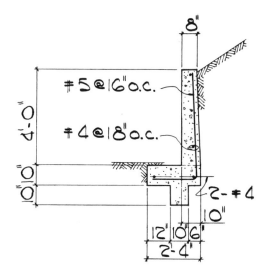

Soil slope = 1½ to 1
E.F.P. = 55 p.c.f.
Section **4.13.4**

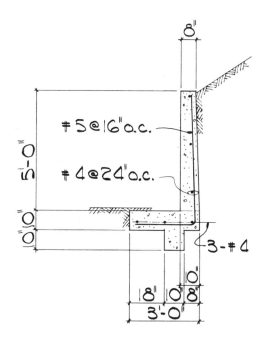

Soil slope = 1½ to 1
E.F.P. = 55 p.c.f.
Section **4.13.5**

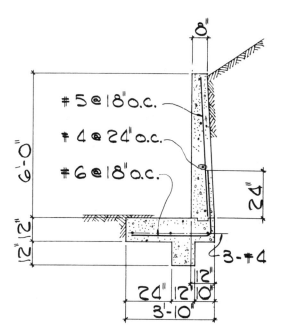

Soil slope = 1½ to 1
E.F.P. = 55 p.c.f.
Section **4.13.6**

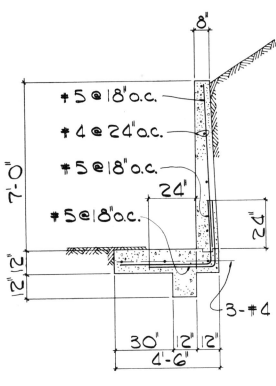

#5 @ 18" o.c.

#4 @ 24" o.c.

#5 @ 18" o.c.

#5 @ 18" o.c.

8"

7'-0"

2'-2"

24"

24"

3-#4

30" 12" 12"

4'-6"

Soil slope = 1½ to 1
E.F.P. = 55 p.c.f.
Section 4.13.7

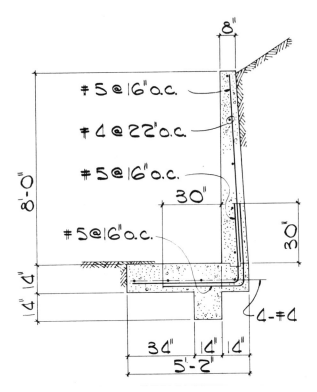

#5 @ 16" o.c.

#4 @ 22" o.c.

#5 @ 16" o.c.

#5 @ 16" o.c.

8"

8'-0"

4'-4"

30"

30"

4-#4

34" 14" 14"

5'-2"

Soil slope = 1½ to 1
E.F.P. = 55 p.c.f.
Section 4.13.8

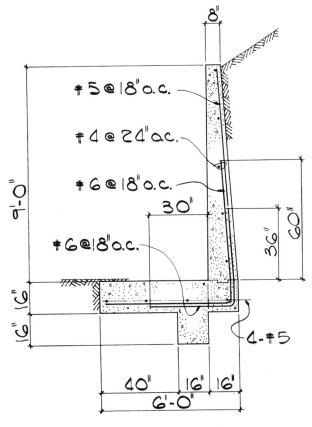

#5 @ 18" o.c.

#4 @ 24" o.c.

#6 @ 18" o.c.

#6 @ 18" o.c.

8"

9'-0"

6"

6"

30"

36"

60"

4-#5

40" 16" 16"

6'-0"

Soil slope = 1½ to 1
E.F.P. = 55 p.c.f.
Section 4.13.9

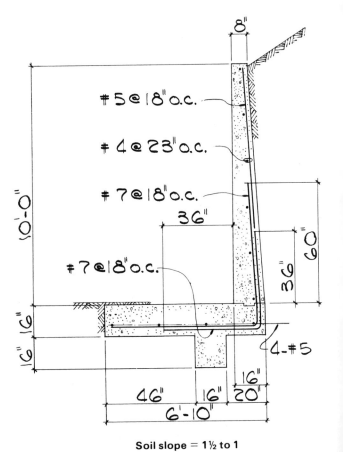

#5 @ 18" o.c.

#4 @ 23" o.c.

#7 @ 18" o.c.

#7 @ 18" o.c.

8"

10'-0"

6"

6"

36"

36"

60"

4-#5

16"

46" 16" 20"

6'-10"

Soil slope = 1½ to 1
E.F.P. = 55 p.c.f.
Section 4.13.10

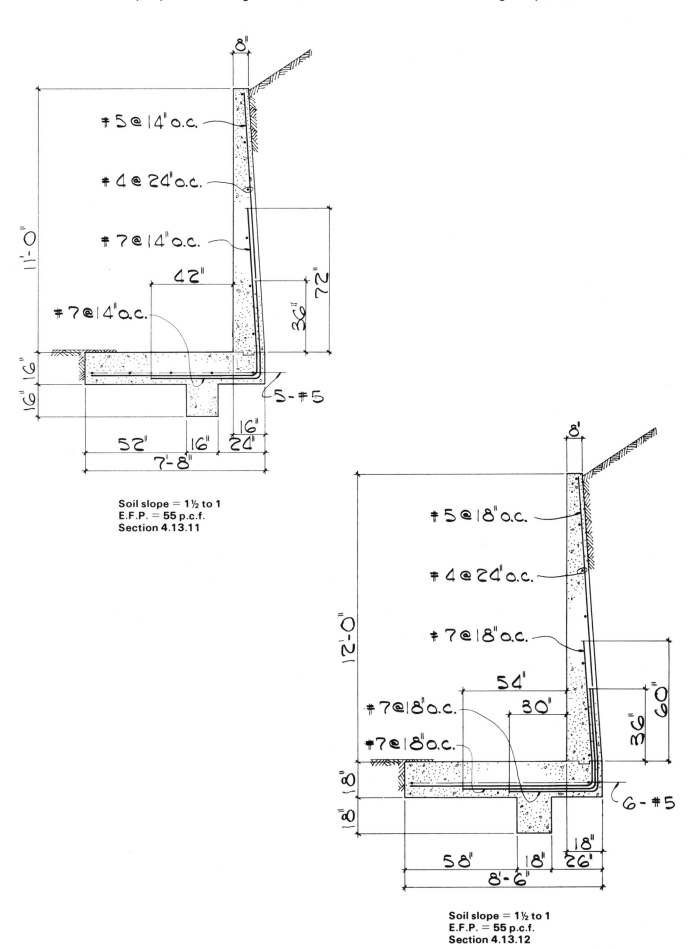

#5 @ 14" o.c.

#4 @ 24" o.c.

#7 @ 14" o.c.

#7 @ 14" o.c.

11'-0"

42"

72"

36"

5-#5

6"

16"

52"   16"   24"

16"

7'-8"

Soil slope = 1½ to 1
E.F.P. = 55 p.c.f.
Section 4.13.11

#5 @ 18" o.c.

#4 @ 24" o.c.

#7 @ 18" o.c.

#7 @ 18" o.c.

#7 @ 18" o.c.

12'-0"

54"

30"

36"

60"

6-#5

8"

18"

58"   18"   26"

18"

8'-6"

Soil slope = 1½ to 1
E.F.P. = 55 p.c.f.
Section 4.13.12

TABLE 4.14   PROPERTY LINE RETAINING WALLS – CONCRETE STEM – SOIL NOT OVER FOOTING½
SLOPE = 1 to 1   SURCHARGE = 0 lbs./sq. ft.   AXIAL = 0 lbs./ft.

SOIL PRESSURE DIAGRAM

Stem Height	3'- 0''	4'- 0''	5'- 0''	6'- 0''	7'- 0''	8'- 0''	9'- 0''	10'- 0''	11'- 0''	12'- 0''
Ft. L ft.	2.167	2.833	3.667	4.500	5.333	6.167	7.167	8.167	9.000	10.000
Ft. T ft.	0.833	0.833	0.833	1.000	1.167	1.333	1.333	1.333	1.500	1.500
Wt ft.	0.667	0.667	0.667	0.667	0.667	0.667	0.667	0.667	0.667	0.667
Wb ft.	0.833	0.833	0.833	1.000	1.167	1.333	1.333	1.333	1.500	1.500
A ft.	0.833	1.333	2.000	2.500	2.833	3.333	4.000	4.667	5.167	5.833
B ft.	0.833	0.833	0.833	1.000	1.167	1.333	1.333	1.333	1.500	1.500
C ft.	0.500	0.667	0.833	1.000	1.333	1.500	1.833	2.167	2.333	2.667
D ft.	0.833	0.833	0.833	1.000	1.167	1.333	1.333	1.333	1.500	1.500
W lbs.	908	1220	1579	2228	2989	3861	4474	5113	6233	6967
F lbs.	588	934	1361	1960	2668	3484	4271	5138	6250	7290
Mo ft.-lbs.	751	1505	2647	4573	7262	10840	14712	19409	26042	32805
Mr ft.-lbs.	1453	2673	4607	7971	12621	18832	25584	33578	44979	56216
O.T.R.	1.934	1.775	1.741	1.743	1.738	1.737	1.739	1.730	1.727	1.714
$\overline{X}$ ft.	0.772	0.957	1.242	1.525	1.793	2.070	2.430	2.771	3.038	3.360
e ft.	0.311	0.460	0.591	0.725	0.874	1.014	1.153	1.312	1.462	1.640
Me ft.-lbs.	282	561	934	1616	2612	3915	5159	6708	9113	11427
S ft.^{3}	0.782	1.338	2.241	3.375	4.741	6.338	8.560	11.116	13.500	16.667
S.P.h p.s.f.	780	850	847	974	1112	1244	1227	1229	1368	1382
S.P.t p.s.f.	58	11	14	16	9	9	22	23	18	11
Friction	0.647	0.766	0.862	0.880	0.892	0.902	0.955	1.005	1.003	1.046

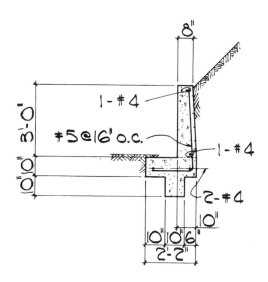

1 - #4

#5@16" o.c.

1 - #4

2 - #4

3'-0"

8"

2'-2"

Soil slope = 1 to 1
E.F.P. = 80 p.c.f.
Section 4.14.3

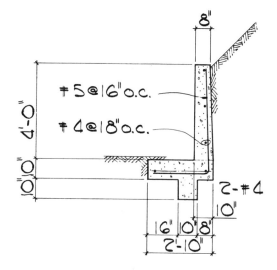

#5@16" o.c.

#4@18" o.c.

2 - #4

4'-0"

8"

2'-10"

Soil slope = 1 to 1
E.F.P. = 80 p.c.f.
Section 4.14.4

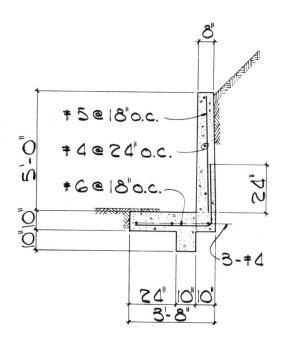

#5@18" o.c.

#4@24" o.c.

#6@18" o.c.

24"

3 - #4

5'-0"

8"

3'-8"

Soil slope = 1 to 1
E.F.P. = 80 p.c.f.
Section 4.14.5

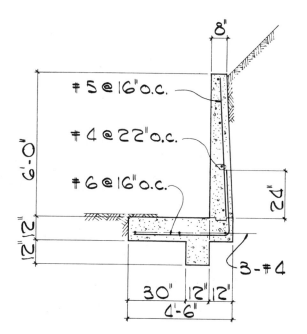

#5@16" o.c.

#4@22" o.c.

#6@16" o.c.

24"

3 - #4

6'-0"

8"

4'-6"

Soil slope = 1 to 1
E.F.P. = 80 p.c.f.
Section 4.14.6

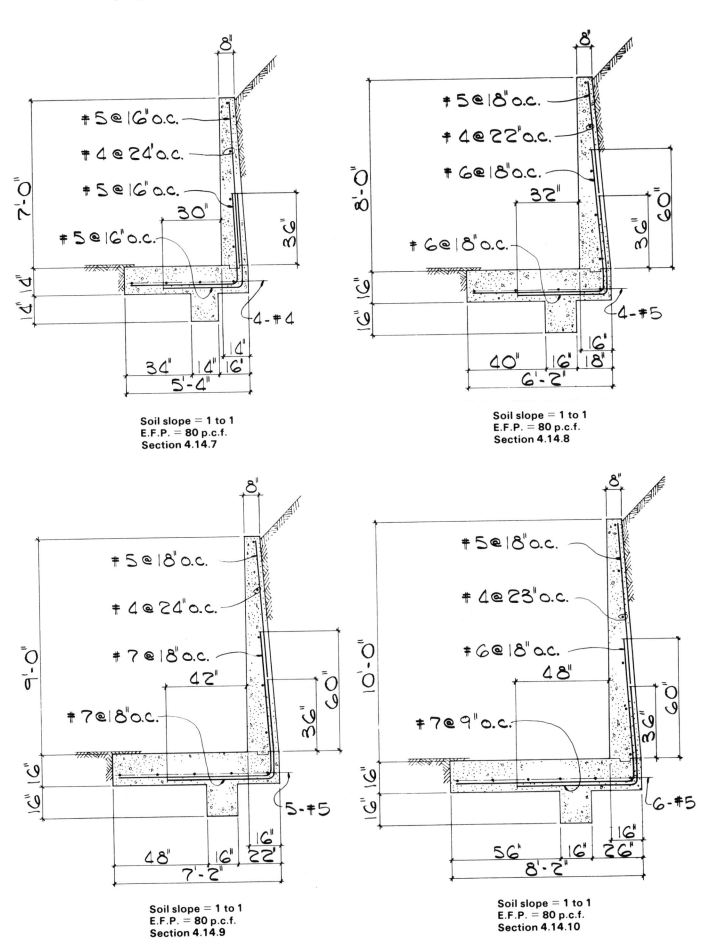

Soil slope = 1 to 1
E.F.P. = 80 p.c.f.
Section 4.14.7

Soil slope = 1 to 1
E.F.P. = 80 p.c.f.
Section 4.14.8

Soil slope = 1 to 1
E.F.P. = 80 p.c.f.
Section 4.14.9

Soil slope = 1 to 1
E.F.P. = 80 p.c.f.
Section 4.14.10

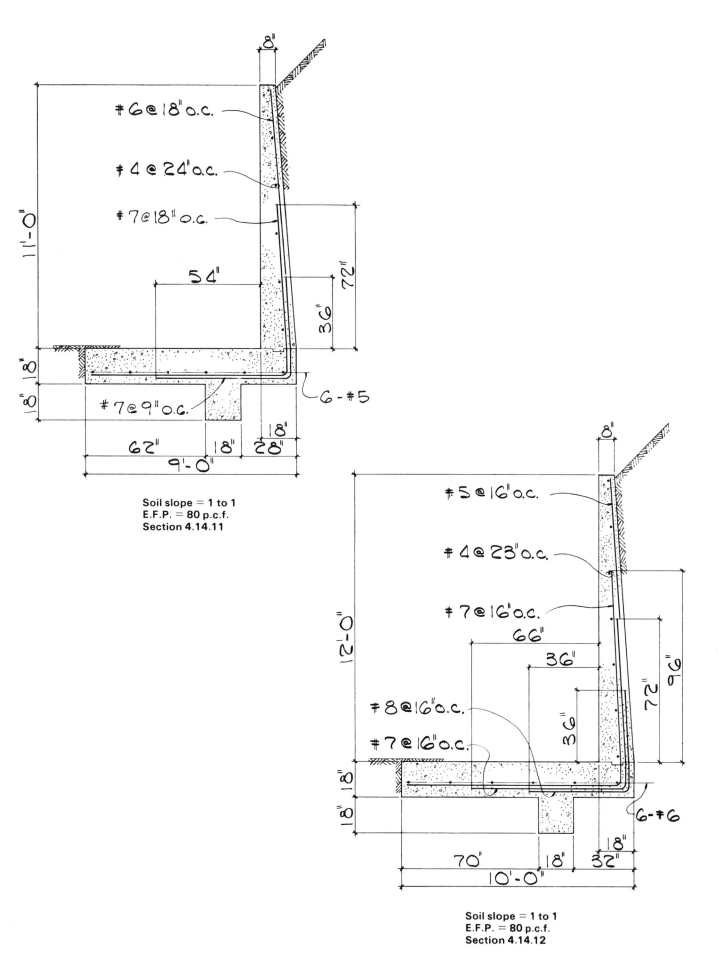

#6 @ 18" o.c.

#4 @ 24" o.c.

#7 @ 18" o.c.

54"

72"

36"

11'-0"

18"

18"

18"

#7 @ 9" o.c.

6 - #5

62"

18"

28"

9'-0"

Soil slope = 1 to 1
E.F.P. = 80 p.c.f.
Section 4.14.11

8"

#5 @ 16" o.c.

#4 @ 23" o.c.

#7 @ 16" o.c.

66"

36"

72"

96"

36"

12'-0"

#8 @ 16" o.c.

#7 @ 16" o.c.

18"

18"

6 - #6

70"

18"

32"

10'-0"

Soil slope = 1 to 1
E.F.P. = 80 p.c.f.
Section 4.14.12

CHAPTER 5

Property Line Retaining Walls
Masonry Stem
Soil Not over the Footing

The design data and drawings presented in this chapter are concerned with property line retaining walls constructed with concrete masonry stem walls. The retained soil is not placed over the wall footing, but at the toe side of the stem wall. The chapter consists of three design examples of retaining walls with various types of loading, a design data table for each of the fourteen loading conditions specified in Fig. 1.6, and a series of corresponding drawings following each design data table. The retaining wall drawings and the design data in this chapter are for a wall length of 1 foot.

The retaining walls in this chapter are designed using the following criteria:

Weight of soil = 100 p.c.f.   Weight of concrete = 150 p.c.f.
Concrete $f_c' = 2000$ p.s.i.   Reinforcing steel $f_c' = 20,000$ p.s.i.
Hollow masonry units, grouted soild, grade N, $f_m' = 1500$ p.s.i.
Field inspection of masonry required (see the design data table)
Weight of 8-inch concrete block with all cells filled = 92 p.s.f.
Weight of 12-inch concrete block with all cells filled = 140 p.s.f.
Grout $f_c' = 2000$ p.s.i.   Mortar (type S) = 1800 p.s.i.
Maximum allowable soil pressure = 4000 p.s.f.
Minimum allowable soil pressure = 0 p.s.f.
$\overline{X} \geqslant L/3$ Minimum O.T.R. = 1.50
Coefficient of soil friction = 0.40
Passive soil pressure = 300 p.c.f.

See Tables C5 and C6 for reinforcement protection. Weep holes are not shown on the drawings.

DESIGN EXAMPLE – SECTION 5.5.8 | SHT. 1 OF 4

GIVEN:

$$H = 8'-0''$$
$$L = 4'-4''$$
$$T = 12''$$
$$W_t = 8''$$
$$W_b = 8''$$
$$A = 28''$$
$$B = 12''$$
$$C = 12''$$
$$D = 12''$$
$$P = 800 \#/FT.$$

$$EFP = 30 \text{ pcf}$$

WEIGHT OF SOIL = 100 pcf

CONCRETE $f'_c = 2000$ psi

REINF. $f'_s = 20,000$ psi

PASSIVE SOIL PRESS. = 300 p.c.f.

$$F = \frac{30 \times 9^2}{2} = 1215 \#$$

$$M_o = \frac{1215 \times 9}{3} = 3645'\#$$

WEIGHT OF STEM	$= W_1 = 8 \times 92$	$= 736\#$
WEIGHT OF FTG.	$= W_2 = 4.33 \times 150$	$= 650$
WEIGHT OF KEY	$= W_3 = 1 \times 150$	$= 150$
STEM FRICTION	$= W_4 = 1215/3$	$= 405$
AXIAL LOAD	$= W_5 =$	$= 800$

TOTAL WT. = W = 2741#

$$M_R = 736 \times \frac{48}{12} + 650 \times \frac{26}{12} + 150 \times \frac{34}{12} + 405 \times \frac{52}{12} + 800 \times \frac{48}{12}$$

$$M_R = 9732'\#$$

OVERTURN. RATIO $= \frac{9732}{3645} = 2.67 > 1.50$

NET M = 9732 - 3645 = 6087'#

$$\bar{X} = \frac{6087}{2741} = 2.221'$$

FTG. MIDDLE $\frac{1}{3} = \frac{4.33}{3} = 1.444 < 2.221'$ ✓

| DESIGN EXAMPLE — SECTION 5.5.8 | SHT. 2 OF 4 |

FOOTING DESIGN :

$W = 2741^{\#}$; $\bar{X} = 2.221'$; $e = \dfrac{4.33}{2} - 2.221 = -0.054'$

$M_e = 2741 \times .054 = -148'^{\#}$

FTG. AREA = 4.33 SQ. FT.

FTG. SECT. MODULUS $= \dfrac{1 \times 4.33^2}{6} = 3.13'^3$

SOIL PRESSURE $= \dfrac{2741}{4.33} \mp \dfrac{148}{3.13} = 632.6 \mp 47.4$

$\underline{SP_T = 585.2 \ psf} \qquad\qquad \underline{SP_H = 680 \ psf}$

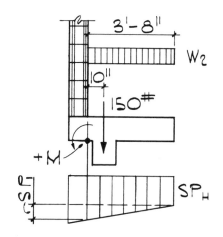

UNIF. LD $= 585.2 - 150 = 435.2^{\#}/_{FT.}$

$SP_1 = \dfrac{680 - 585.2}{4.33} \times 3.67 = 80.3 \ psf$

$+M = \dfrac{435.2 \times 3.67^2}{2} + \dfrac{80.3 \times 3.67^2}{2 \times 3} - 0.83 \times 150$

$+M = 2981'^{\#}$

$V = 435.2 \times 3.67 + \dfrac{80.3 \times 3.67}{2} - 150$

$V = 1593^{\#}$

$d = 12 - (3 + .50 \times .62) = 8.70''$

$a = 1.13$; $j = 0.89$

$v = \dfrac{1593}{12 \times .89 \times 8.70} = 17.1 \ psi$

$A_s = \dfrac{2.981}{1.13 \times 8.70} = 0.303 \ SQ.IN./FT.$

USE #4 @ 8" o.c.

LONG. REINF. $= 0.0012 \times 12 \times 4.33 \times 12 = 0.748''$

USE 4 - #4's

| DESIGN   EXAMPLE – SECTION 5.5.8. | SHT. 3 OF 4 |

STEM DESIGN AT TOP OF FTG.

$$V = \frac{30 \times 8^2}{2} = 960^{\#} \quad ; \quad M = \frac{960 \times 8}{3} = 2560^{\#}$$

$$d = 8'' - (2 + .5 \times .62) = 5.70'' \quad ; \quad a = 1.13 \quad ; \quad j = 0.89$$

$$\nu = \frac{960}{12 \times .89 \times 5.70} = 15.8 < 27 \, psi \qquad INSPECT. \; REQ'D$$

$$A_S = \frac{2.56}{1.13 \times 5.70} = 0.397 \, sq.in./ft. \qquad TRY \; \#5 @ 8'' o.c.$$

$$A_S = 0.47 \, sq.in. \; ; \; p = \frac{0.47}{12 \times 8} = 0.0048 \; ; \; np = 20 \times .0048 = .096$$

$$2/kj = 6.43 \; ; \quad fm = \frac{12 \times 2560 \times 4.94}{12 \times 5.70^2} = 506.6 \, psi \approx 500 \, psi$$

USE #5 @ 8'' o.c. – CONT. INSPECTION REQ'D.

$$AXIAL \; fa = \frac{800 + 736}{12 \times 8} = 16.0 \, psi < 270 \, psi$$

$$COMBINED \quad STRESS = \frac{506.6}{500} + \frac{16.0}{270} = 1.07 \approx 1.0 \checkmark$$

STEM DESIGN AT 3'-4'' ABOVE FTG. :

$$H = 8.0' - 3.33' = 4.67' \; ; \quad V = \frac{30 \times 4.67^2}{2} = 326.7^{\#}$$

$$M = \frac{326.7 \times 4.67}{3} = 508.6'^{\#} \; ; \quad A_S = \frac{0.508}{1.13 \times 5.70} = 0.079 \, sq.in./ft.$$

$$TRY \; \#4 @ 16'' o.c. \qquad A_S = 0.15 \, sq.in. \; ; \; p = \frac{0.15}{12 \times 5.7} = 0.0022$$

$$np = 20 \times .0022 = 0.0440 \; ; \; 2/kj = 8.54$$

$$fm = \frac{12 \times 508.6 \times 8.54}{12 \times 5.70^2} = 133.7 \, psi < 500 \, psi \qquad \underline{USE \; \#4 @ 16'' o.c.}$$

$$HORIZ. \; REINF. = .0007 \times 8 \times 12 \times 8 = 0.537 \, sq.in.$$

USE 4 – #4's

SLIDING :   KEY DEPTH + FTG. THICKNESS = 2.0'

$$RESISTANCE \; TO \; SLIDING = 2741 \times 0.41 + \frac{300 \times 2^2}{2} \; 1696^{\#} > 1215^{\#} \checkmark$$

| DESIGN   EXAMPLE — SECTION 5.5.8 | SHT. 4 OF 4 |

CHECK WALL WITHOUT 800#/FT. AXIAL LD.

$F = 1215^\#$; $M_o = 3645^{'\#}$

$W = 2741 - 800 = 1941^\#$

$M_R = 9732 - 800 \times \dfrac{48}{12} = 6532^{'\#}$

OVERTURN. RATIO $= \dfrac{6532}{3645} = 1.792 > 1.50$ ✓

NET $M = 6532 - 3645 = 2887^{'\#}$

$\bar{X} = \dfrac{2887}{1941} = 1.487' > 1.444'$ ✓

$e = \dfrac{4.33}{2} - 1.487 = 0.679'$

$M_e = 1941 \times .679 = 1318^{'\#}$

SOIL PRESSURE $= \dfrac{1941}{4.33} \pm \dfrac{1318}{3.13} = 448 \pm 421$

$\underline{SP_T = 27 \text{ p.s.f}}$ $\underline{SP_H = 869 \text{ p.s.f}}$

CHECK SLIDING :

$F_r = \dfrac{1215}{1941} = 0.626 > 0.40$

A KEY IS PROVIDED - SEE CALC. SHT. No. 3

DESIGN EXAMPLE — SECTION 5.8.6 | SHT. 1 of 5

GIVEN:

$$H = 6'-0''$$
$$L = 4'-10''$$
$$T = 12''$$
$$A = 32''$$
$$B = 12''$$
$$C = 14''$$
$$D = 12''$$
$$W_T = 8''$$
$$W_b = 8''$$

SURCHARGE = 200 psf
EFP = 30 pcf
CONCRETE f'_c = 2000 psi
REINF. f'_s = 20,000 psi
WEIGHT OF SOIL = 100 pcf

EQUIV. HT. $= 1.0 + 6.0 + \dfrac{200}{100} = 9.0'$

$$F = \frac{30 \times 9^2}{2} = 1215^{\#}$$

$$M_o = \frac{1215 \times 9}{3} = 3645^{\#}$$

WEIGHT OF STEM $= W_1 = 6 \times 92 = 552^{\#}$
WEIGHT OF FTG. $= W_2 = 4.83 \times 150 = 725$
WEIGHT OF KEY $= W_3 = 1 \times 150 = 150$
STEM FRICTION $= W_4 = \dfrac{1215}{3} = 405$

TOTAL WT. $= W = 1832^{\#}$

$$M_R = 552 \times \frac{54}{12} + 725 \times \frac{29}{12} + 150 \times \frac{38}{12} + 405 \times \frac{58}{12}$$

$$M_R = 6669^{'\#}$$

OVERTURN. RATIO $= \dfrac{6669}{3645} = 1.83 < 1.50$ ✓

NET M $= 6669 - 3645 = 3024^{'\#}$

$$\bar{X} = \frac{3024}{1832} = 1.65'$$

FTG. MIDDLE $\frac{1}{3} = \dfrac{4.83}{3} = 1.61 < 1.65$ ✓

DESIGN   EXAMPLE - SECTION 5.8.6 | SHT. 2 OF 5

FOOTING   DESIGN :

$W = 1832^{\#}$; $\bar{X} = 1.65'$; $e = \dfrac{4.83}{2} - 1.65 = 0.766'$

$Me = 1832 \times .766 = 1404'^{\#}$

FTG. AREA $= 4.83$ SQ. FT.

FTG. SECT. MODULUS $= \dfrac{1 \times 4.83^2}{6} = 3.894'^3$

SOIL PRESSURE $= \dfrac{1832}{4.83} \pm \dfrac{1404}{3.894} = 379 \pm 361$

$\underline{SP_T = 18 \ psf}$ $\underline{SP_H = 740 \ psf}$

$W' = \dfrac{740 - 18}{4.83} \times .67 + 18 = 118 \ psf$

$SP_1 = 740 - 118 = 622 \ psf$

UNIF. LD. $= 150 - 118 = 32^{\#}/FT.$

$+M = \dfrac{622 \times 4.17^2 \times 2}{2 \times 3} - \dfrac{32 \times 4.17^2}{2}$

$+M = 3322'^{\#}$

$Y = \dfrac{622 \times 4.17}{2} - 32 \times 4.17$

$Y = 1163.4^{\#}$

$d = 12'' - (3 + .5 \times .62) = 8.70''$; $a = 1.13$; $j = 0.89$

$\nu = \dfrac{1163.4}{12 \times .89 \times 8.7} = 12.5 \ psi < 54 \ psi$ ✓

$A_s = \dfrac{3.322}{1.13 \times 8.70} = 0.34$ SQ. IN./FT.

USE #5 @ 16'' o.c. & #4 @ 16'' o.c.

LONG. REINF. :
  $A_s = .0012 \times 12 \times 4.833 \times 12 = 0.835$ SQ.IN.

USE 4 - #4

DESIGN   EXAMPLE - SECTION   5.8.6 | SHT. 3 OF 5

FOOTING   DESIGN (CONT.)

MOMENT & SHEAR AT 2'-0" FROM HEEL OF FTG.

$$W' = \frac{740-18}{4.833} \times 2.83 + 18 = 441.2 \text{ psf}$$

$$SP_1 = 740 - 441.2 = 298.8 \text{ psf}$$

$$\text{UNIF. LD.} = 441.2 - 150 = 291.2 \text{ \#/FT.}$$

$$+M = \frac{298.8 \times 2^2 \times 2}{2 \times 3} + \frac{291.2 \times 2^2}{2}$$

$$+M = 980.8 \text{ '}\#$$

$$A_s = \frac{.980}{1.133 \times 8.70} = 0.10 \text{ SQ. IN./FT.}$$

<u>USE #4 @ 16" O.C.</u>

STEM   DESIGN   AT   TOP   OF   FTG. :

EQUIV. HT. $= 6.0' + \frac{200}{100} = 8.0'$

$$V = \frac{30 \times 8^2}{2} = 960 \text{ \#} \quad ; \quad M = \frac{960 \times 8}{3} = 2560 \text{ '}\#$$

$$d = 8 - (2 + .5 \times .62) = 5.70" \quad ; \quad a = 1.13 \quad ; \quad j = 0.89$$

$$v = \frac{960}{12 \times .89 \times 5.70} = 15.8 \text{ psi} \approx 15.0 \text{ psi} \checkmark$$

$$A_s = \frac{2.560}{1.13 \times 5.70} = 0.39 \text{ SQ. IN./FT.} \quad \text{TRY } \#5 @ 8" \text{ O.C.} \quad ; \quad A_s = 0.47 \text{ SQ. IN.}$$

$$P = \frac{0.47}{12 \times 5.70} = .0068 \quad ; \quad np = 20 \times .0068 = 0.136 \quad ; \quad {}^2/kj = 5.73$$

$$fm = \frac{12 \times 2560 \times 5.73}{12 \times 5.70^2} = 451.5 \text{ psi} < 500 \text{ psi}$$

<u>USE #5 @ 8" O.C.   CONT. INSPECTION REQ'D.</u>

DESIGN EXAMPLE — SECTION 5.8.6	SHT. 4 OF 5

STEM DESIGN AT 2'-0" ABOVE FTG. :

EQUIV. HT. $= 6.0' - 2.0' + \frac{200}{100} = 6.0'$

$V = \frac{30 \times 6^2}{2} = 540^{\#}$; $M = \frac{540 \times 6}{3} = 1080'^{\#}$

$\nu = \frac{540}{12 \times .89 \times 5.70} = 8.9 \, psi < 15 \, psi$

$A_s = \frac{1.08}{1.13 \times 5.70} = 0.167 \, sq.in./ft.$; TRY #5 @ 16" o.c. ; $A_s = 0.23 \, sq.in.$

$P = \frac{0.23}{12 \times 5.70} = 0.00336$; $np = 20 \times .00336 = 0.067$; $^2\!/kj = 7.30$

$f_m = \frac{12 \times 1080 \times 7.30}{12 \times 5.70^2} = 242.7 \, psi < 500 \, psi$

USE #5 @ 16" o.c. - CONT. INSPECTION REQ'D.

HORIZ. REINF. : $A_s = 0.0012 \times 6 \times 8 \times 12 = 0.69$ USE 5 - #4's

SLIDING :

$F_r = \frac{1215}{1832} = 0.663 > 0.40$

A KEY IS REQUIRED

PASSIVE PRESSURE $= 300 \, pcf$

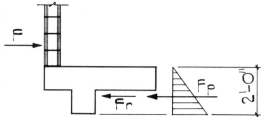

RESISTANCE TO SLIDING $= 1832 \times .4 + \frac{300 \times 2^2}{2} = 1332.8^{\#} < 1215^{\#} \checkmark$

KEY $\nu = \frac{1832}{12^2} = 12.7 \, p.s.i < 54 \, p.s.i$

DESIGN EXAMPLE — SECTION 5.8.6 | SHT. 5 of 5

CHECK WALL WITHOUT 200 p.s.f. SURCHARGE LD.

EQUIV. HT. $= 6.0' + 1.0' = 7.0'$; EFP $= 30$ p.c.f.

$V = \dfrac{30 \times 7^2}{2} = 735^{\#}$; $M_o = \dfrac{735 \times 7}{3} = 1715'^{\#}$

$W = 1832 - 405 + \dfrac{30 \times 7^2}{2 \times 3} = 1672^{\#}$

$M_R = 6669 - \dfrac{405 \times 58}{12} + \dfrac{245 \times 58}{12} = 5895'^{\#}$

OVERTURN. RATIO $= \dfrac{5895}{1715} = 3.437' < 1.50 \checkmark$

NET M $= 5895 - 1715 = 4180'^{\#}$

$\bar{X} = \dfrac{4180}{1672} = 2.50' > 1.61' \checkmark$

$e = \dfrac{4.83}{2} - 2.50 = -0.085'$

$M_e = 1672 \times .085 = 142.1'^{\#}$

SOIL PRESSURE $= \dfrac{1672}{4.83} \mp \dfrac{142.1}{3.894} = 346.2 \mp 36.5$

$SP_T = 382.7$ p.s.f. $SP_H = 309.7$ p.s.f.

CHECK SLIDING :

$F_r = \dfrac{735}{1672} = 0.44 > 0.40$

KEY IS REQ'D. — SEE CALC. SHT. No. 4

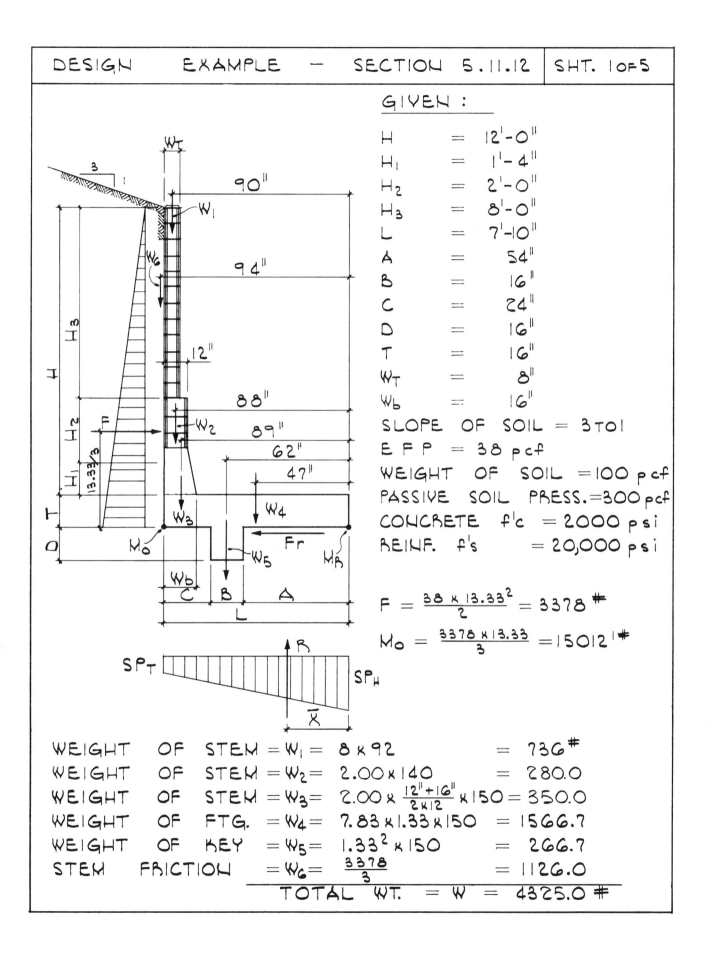

GIVEN:

$H = 12'\text{-}0''$

$H_1 = 1'\text{-}4''$

$H_2 = 2'\text{-}0''$

$H_3 = 8'\text{-}0''$

$L = 7'\text{-}10''$

$A = 54''$

$B = 16''$

$C = 24''$

$D = 16''$

$T = 16''$

$W_T = 8''$

$W_b = 16''$

SLOPE OF SOIL = 3 TO 1

EFP = 38 pcf

WEIGHT OF SOIL = 100 pcf

PASSIVE SOIL PRESS. = 300 pcf

CONCRETE f'_c = 2000 psi

REINF. f'_s = 20,000 psi

$$F = \frac{38 \times 13.33^2}{2} = 3378^{\#}$$

$$M_O = \frac{3378 \times 13.33}{3} = 15012'^{\#}$$

WEIGHT OF STEM = W_1 = 8 × 92 = 736$^{\#}$

WEIGHT OF STEM = W_2 = 2.00 × 140 = 280.0

WEIGHT OF STEM = W_3 = 2.00 × $\frac{12'' + 16''}{2 \times 12}$ × 150 = 350.0

WEIGHT OF FTG. = W_4 = 7.83 × 1.33 × 150 = 1566.7

WEIGHT OF KEY = W_5 = 1.33^2 × 150 = 266.7

STEM FRICTION = W_6 = $\frac{3378}{3}$ = 1126.0

TOTAL WT. = W = 4325.0$^{\#}$

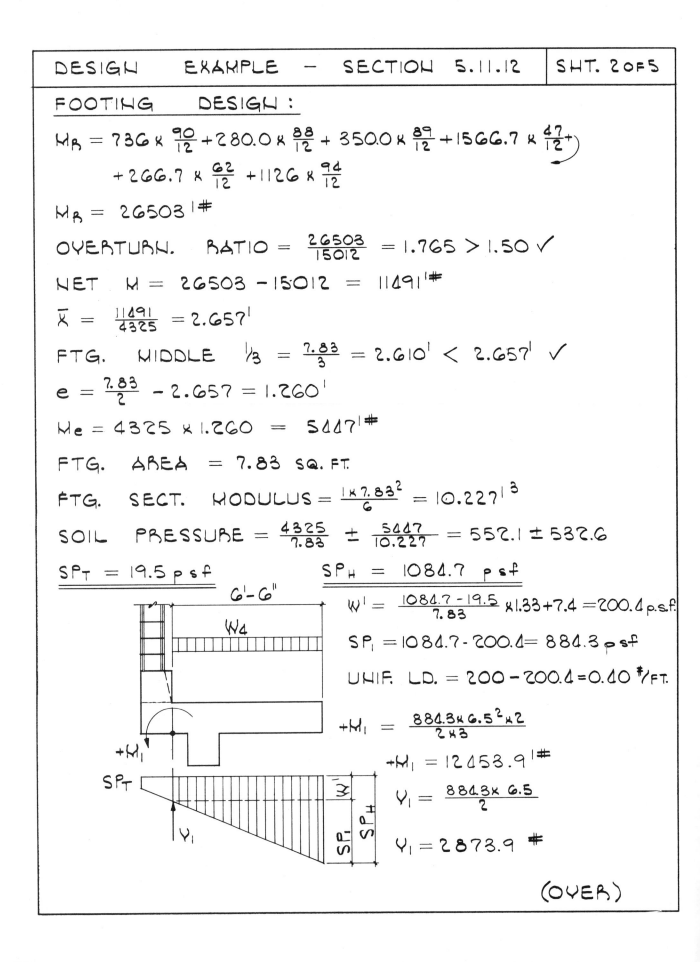

DESIGN EXAMPLE — SECTION 5.11.12 | SHT. 2 of 5

FOOTING DESIGN :

$M_R = 736 \times \frac{90}{12} + 280.0 \times \frac{88}{12} + 350.0 \times \frac{89}{12} + 1566.7 \times \frac{47}{12} +)$

$\qquad + 266.7 \times \frac{62}{12} + 1126 \times \frac{94}{12}$

$M_R = 26503^{'\#}$

OVERTURN. RATIO $= \frac{26503}{15012} = 1.765 > 1.50$ ✓

NET M $= 26503 - 15012 = 11491^{'\#}$

$\bar{X} = \frac{11491}{4325} = 2.657'$

FTG. MIDDLE $\frac{1}{3} = \frac{7.83}{3} = 2.610' < 2.657'$ ✓

$e = \frac{7.83}{2} - 2.657 = 1.260'$

$M_e = 4325 \times 1.260 = 5447^{'\#}$

FTG. AREA $= 7.83$ SQ. FT.

FTG. SECT. MODULUS $= \frac{1 \times 7.83^2}{6} = 10.227^{'3}$

SOIL PRESSURE $= \frac{4325}{7.83} \pm \frac{5447}{10.227} = 552.1 \pm 532.6$

$\underline{SP_T = 19.5 \, psf} \qquad \underline{SP_H = 1084.7 \, psf}$

6'-6"

W_4

$W' = \frac{1084.7 - 19.5}{7.83} \times 1.33 + 7.4 = 200.4 \, p.s.f.$

$SP_1 = 1084.7 - 200.4 = 884.3 \, psf$

$+M_1$

UNIF. LD. $= 200 - 200.4 = 0.40 \, \#/FT.$

$+M_1 = \frac{884.3 \times 6.5^2 \times 2}{2 \times 3}$

SP_T

$+M_1 = 12453.9^{'\#}$

$Y_1 = \frac{884.3 \times 6.5}{2}$

Y_1

$Y_1 = 2873.9 \, \#$

(OVER)

DESIGN EXAMPLE — SECTION 5.11.12	SHT. 3 OF 5

FOOTING DESIGN (CONT.)

$+M_1 = 12453.9'^{\#}$; $V_1 = 2873.9^{\#}$

$d = 16-(3+.5\times.62) = 12.7''$; $a = 1.13$; $j = 0.89$

$v = \dfrac{2873.9}{12\times.89\times12.7} = 21.2\,psi < 54\,psi$ ✓

$A_s = \dfrac{12.453}{1.13\times12.7} = 0.867\ sq.\ in./FT.$

USE #7 @ 8" o.c. & #5 @ 16" o.c.

$+M_A$: $L = 5'-0''$

$W' = \dfrac{1084.7-19.4}{7.83}\times2.83+19.4 = 404.3\,psf$

$SP_1 = 1084.7-404.3 = 680.4\ psf$

UNIF. LD. $= 200-404.3 = -204.3^{\#}/FT.$

$+M_A = \dfrac{680.4\times5^2\times2}{2\times3} + \dfrac{204.3\times5^2}{2} = 8223.7'^{\#}$

$+M_B$: $L = 4'-0''$

$W' = \dfrac{1084.7-19.4}{7.83}\times3.83+19.4 = 540.7\,psf$

$SP_1 = 1084.7-540.7 = 544.0\ psf$

UNIF. LD. $= 200-544.7 = -344.7\ ^{\#}/FT.$

$+M_B = \dfrac{544.0\times4^2\times2}{2\times3} + \dfrac{344.7\times4^2}{2} = 5659.0'^{\#}$

$+M_C$: $L = 3'-0''$

$W' = \dfrac{1084.7-19.4}{7.83}\times4.83+19.4 = 676.5\,psf$

$SP_1 = 1084.7-676.5 = 408.2\ p.s.f.$

UNIF. LD. $= 200-676.5 = -467.5^{\#}/FT.$

$M_C = \dfrac{408.2\times3^2\times2}{3\times2} + \dfrac{467.5\times3^2}{2} = 3228.4^{\#}$

TRY #7 @ 16" o.c. ; $A_s = 0.45\ sq.\ in./FT.$

$M = 1.13\times12.7\times.45 = 6.45'^k$ USE #7 @ 16" o.c. FROM PNT. "C"

LONG. REINF. $= 0.0012\times7.83\times12\times16 = 1.80\ sq.\ in.$

LONG. REINF. $= 5-\#5$

255

DESIGN EXAMPLE — SECTION 5.11.12 | SHT. 4 OF 5

STEM DESIGN AT TOP OF FTG.: EFP = 38 p.c.f.

$H = 12'-0''$; $V = \frac{38 \times 12^2}{2} = 2736^\#$; $M = \frac{2736 \times 12}{3} = 10944'^\#$

$d = 16'' - (2 + .5 \times .62) = 13.7''$; $a = 1.13$; $j = 0.89$

$v = \frac{2736}{12 \times .89 \times 13.7} = 18.7 \, psi < 54 \, psi$ ✓

$A_s = \frac{10.94}{1.13 \times 13.7} = 0.71 \, sq.in.$ USE #7 @ 8'' o.c.

STEM DESIGN AT 2'-0'' ABOVE FTG. :

$H = 12.0' - 2.00' = 10.0'$; EFP = 38 p.c.f.
$V = \frac{38 \times 10.00^2}{2} = 1900.0^\#$; $M = \frac{1900.0 \times 10.0}{3} = 6333.0'^\#$

$d = 12 - (2 + .5 \times .62) = 9.70''$; $a = 1.13$; $j = 0.89$

$v = \frac{1900.8}{12 \times .89 \times 9.7} = 18.3 \, psi < 30 \, psi - INSPECTION \, REQ'D.$

$A_s = \frac{6.33}{1.13 \times 9.70} = 0.58 \, sq.in./ft.$; TRY #5 @ 8'' o.c. & #4 @ 16'' o.c.

$A_s = 0.62 \, sq.in.$; $p = \frac{0.62}{12 \times 9.7} = 0.00532$; $np = 20 \times .00567 = 0.107$

$2/kj = 6.20$ $fm = \frac{12 \times 6333.0 \times 6.20}{12 \times 9.7^2} = 417.3 \, psi < 500 \, psi$

USE #5 @ 8'' o.c. & #4 @ 16'' o.c. INSPECTION REQ'D.

STEM DESIGN AT 4'-0'' ABOVE FTG.: $H = 8.0'$

$V = \frac{38 \times 8.0^2}{2} = 1216^\#$; $M = \frac{1216 \times 8.0}{3} = 3242.7'^\#$

$d = 8 - (2 + .5 \times .62) = 5.70''$; $a = 1.13$; $j = 0.89$

$v = \frac{1216}{12 \times .89 \times 5.7} = 11.7 \, psi < 30 \, psi$ ✓ INSPECTION REQ'D.

$A_s = \frac{3.24}{1.13 \times 5.7} = 0.50 \, sq.in./ft.$; TRY #5 @ 8'' o.c. & #4 @ 16'' o.c.

$A_s = 0.62$; $p = \frac{0.62}{12 \times 5.7} = 0.00906$; $np = 20 \times .00965 = 0.181$

$2/kj = 5.26$; $fm = \frac{12 \times 3242.7 \times 5.26}{12 \times 5.7^2} = 525 \, psi \approx 500 \, psi$ ✓

USE #5 @ 8'' o.c. & #4 @ 16'' o.c. INSPECTION REQ'D.

DESIGN EXAMPLE – SECTION 5.11.12	SHT. 5 OF 5

STEM   DESIGN   AT 6'-0"   ABOVE   FTG.: $H = 6.0'$

$$V = \frac{38 \times 6.0^2}{2} = 684^{\#} \quad ; \quad M = \frac{648 \times 6.0}{3} = 1368'^{\#} \quad ; \quad d = 5.70''$$

$$v = \frac{684}{12 \times .89 \times 5.7} = 11.23 \, psi < 30 \, psi \checkmark$$

$$A_s = \frac{1.368}{1.13 \times 5.7} = 0.21 \, SQ.\,IN./FT. \qquad TRY \quad \#4 @ 16'' \, o.c.$$

$$A_s = 0.15 \, SQ.\,IN. \quad ; \quad p = \frac{.15}{12 \times 5.70} = 0.00220 ; \quad np = 20 \times .0036 = 0.044$$

$$^2/_{kj} = 8.54 \quad ; \quad fm = \frac{12 \times 1368 \times 8.54}{12 \times 5.70^2} = 359.6 \, psi < 500 \, psi \checkmark$$

USE   #5 @ 16" o.c. – INSPECTION   REQ'D

LONG.   REINF.

STEM – $H_1 = 0.007 \times 16 \times 16 = 0.179$ SQ.IN.   USE 1–#5 TOP & BOTT.
STEM – $H_2 = 0.007 \times 32 \times 12 = 0.268$ SQ.IN.   USE 2–#5 TOP & BOTT.
STEM – $H_3 = 0.007 \times 96 \times 8 = 0.537$ SQ.IN.   USE 2–#4 TOP & BOTT.
                                                      & 1–#4 AT MID. HT.

SLIDING :

$$Fr = \frac{3378^{\#}}{4302^{\#}} = 0.785 > 0.40$$

KEY   REQUIRED

RESISTANCE TO SLIDING $= 4302 \times .40 + \frac{300 \times 2.67^2}{2} = 2787.5^{\#}$
INCREASE KEY DEPTH TO 2'-0"
RESISTANCE TO SLIDING $= 4302 \times .4 + \frac{300 \times 3.33^2}{2} = 3387^{\#} > 3378^{\#}$
ALLOW. $v = 54 \, psi$

KEY $v = \frac{3378}{12 \times 16} = 17.6 \, psi \checkmark$

MAKE   KEY 16" WIDE × 24" DEEP

TABLE 5.1 PROPERTY LINE RETAINING WALLS – MASONRY STEM – SOIL NOT OVER FOOTING
SLOPE = 0 to 1   SURCHARGE = 0 lbs./sq. ft.   AXIAL = 0 lbs./ft.

SOIL PRESSURE DIAGRAM

H1 = Concrete Stem
H2 = 12″ Concrete Block
H3 = 8″ Concrete

Stem Height	3'- 0''	4'- 0''	5'- 0''	6'- 0''	7'- 0''	8'- 0''	9'- 0''	10'- 0''	11'- 0''	12'- 0''
Ft. L ft.	1.667	1.833	2.333	3.000	3.667	4.333	5.000	5.667	6.333	7.000
Ft. T ft.	0.667	0.667	0.833	0.833	0.833	1.000	1.000	1.167	1.167	1.333
A ft.	0.500	0.667	1.000	1.500	2.000	2.333	2.833	3.167	3.500	4.000
B ft.	0.667	0.667	0.833	0.833	0.833	1.000	1.000	1.167	1.167	1.333
C ft.	0.500	0.500	0.500	0.667	0.833	1.000	1.167	1.333	1.667	1.667
D ft.	0.667	0.667	0.833	0.833	0.833	1.000	1.000	1.167	1.167	1.333
8 in. blk. ft.	3.000	4.000	5.000	6.000	7.000	8.000	7.000	8.000	8.333	8.667
12 in. blk. ft.	—	—	—	—	—	—	2.000	2.000	2.667	2.667
Conc. Stem	—	—	—	—	—	—	—	—	—	0.667
Wb ft.	—	—	—	—	—	—	—	—	—	1.333
W lbs.	577	727	1026	1265	1513	1941	2260	2803	3193	3843
F lbs.	202	327	510	700	920	1215	1500	1870	2220	2667
Mo ft.-lbs.	246	508	992	1595	2403	3645	5000	6962	9005	11852
Mr ft.-lbs.	674	986	1805	2935	4364	6532	8878	12334	15825	20883
O.T.R.	2.736	1.941	1.819	1.839	1.816	1.792	1.776	1.772	1.757	1.762
$\overline{X}$ ft.	0.742	0.658	0.792	1.059	1.295	1.488	1.716	1.916	2.136	2.350
e ft.	0.091	0.259	0.375	0.441	0.538	0.679	0.784	0.917	1.031	1.150
Me ft.-lbs.	52	188	385	558	814	1318	1772	2570	3290	4419
S ft.^{3}	0.463	0.560	0.907	1.500	2.241	3.130	4.167	5.352	6.685	8.167
S.P.h p.s.f.	459	732	864	793	776	869	877	975	996	1090
S.P.t p.s.f.	233	61	16	50	49	27	27	14	12	8
Friction	0.350	0.449	0.497	0.554	0.608	0.626	0.664	0.667	0.695	0.694
Inspct.	NO	NO	NO	NO	YES	YES	YES	YES	YES	YES

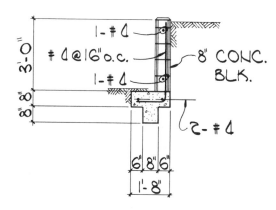

E.F.P. = 30 p.c.f.
Section 5.1.3

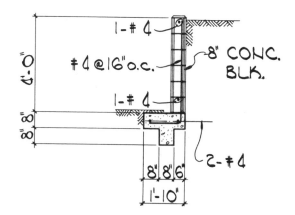

E.F.P. = 30 p.c.f.
Section 5.1.4

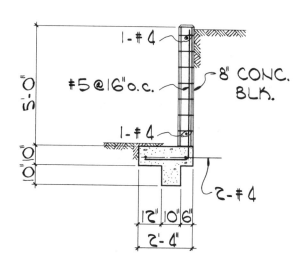

E.F.P. = 30 p.c.f.
Section 5.1.5

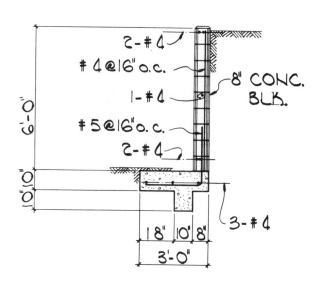

E.F.P. = 30 p.c.f.
Section 5.1.6

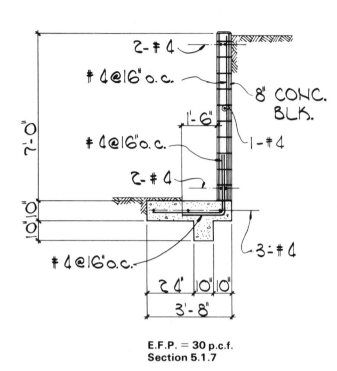

2-#4

#4@16" o.c.

8" CONC. BLK.

#4@16" o.c.

1-#4

2-#4

7'-0"

1'-6"

1'-0"

#4@16" o.c.

3-#4

24"   10"   10"

3'-8"

E.F.P. = 30 p.c.f.
Section 5.1.7

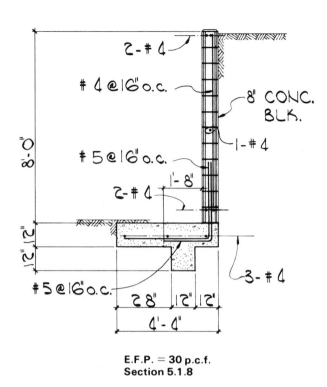

2-#4

#4@16" o.c.

8" CONC. BLK.

#5@16" o.c.

1-#4

1'-8"

2-#4

8'-0"

2"

12"

#5@16" o.c.

3-#4

28"   12"   12"

4'-4"

E.F.P. = 30 p.c.f.
Section 5.1.8

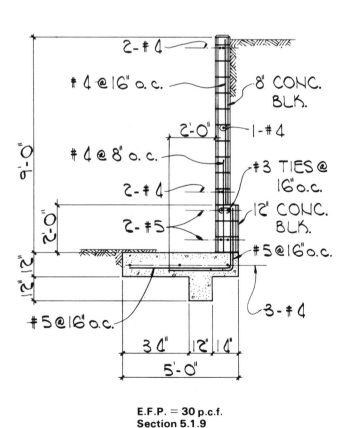

2-#4

#4@16" o.c.

8" CONC. BLK.

2'-0"

1-#4

#4@8" o.c.

#3 TIES @ 16"o.c.

2-#4

12" CONC. BLK.

2-#5

#5@16" o.c.

9'-0"

2'-0"

12"

#5@16" o.c.

3-#4

30"   12"   18"

5'-0"

E.F.P. = 30 p.c.f.
Section 5.1.9

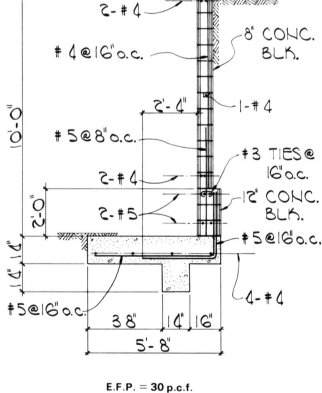

2-#4

#4@16" o.c.

8" CONC. BLK.

2'-4"

1-#4

#5@8" o.c.

#3 TIES @ 16"o.c.

2-#4

12" CONC. BLK.

2-#5

#5@16" o.c.

10'-0"

2'-0"

14"

#5@16" o.c.

4-#4

38"   14"   16"

5'-8"

E.F.P. = 30 p.c.f.
Section 5.1.10

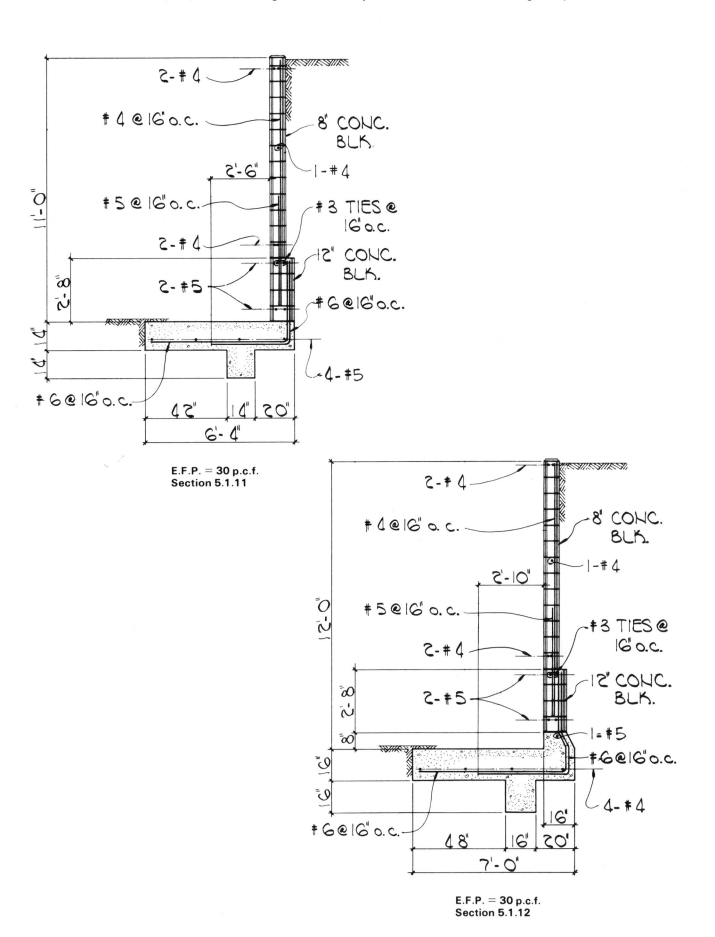

2 - # 4

4 @ 16" o.c.

8" CONC. BLK.

1 - # 4

2'-6"

5 @ 16" o.c.

3 TIES @ 16" o.c.

2 - # 4

12" CONC. BLK.

2 - # 5

6 @ 16" o.c.

11'-0"

2'-8"

4"

4"

4 - # 5

6 @ 16" o.c.

42"   14"   20"

6'-4"

E.F.P. = 30 p.c.f.
Section 5.1.11

2 - # 4

4 @ 16" o.c.

8" CONC. BLK.

1 - # 4

2'-10"

5 @ 16" o.c.

3 TIES @ 16" o.c.

2 - # 4

12" CONC. BLK.

2 - # 5

1 - # 5

6 @ 16" o.c.

12'-0"

2'-8"

8"

6"

6"

4 - # 4

6 @ 16" o.c.

16"

48"   16"   20"

7'-0"

E.F.P. = 30 p.c.f.
Section 5.1.12

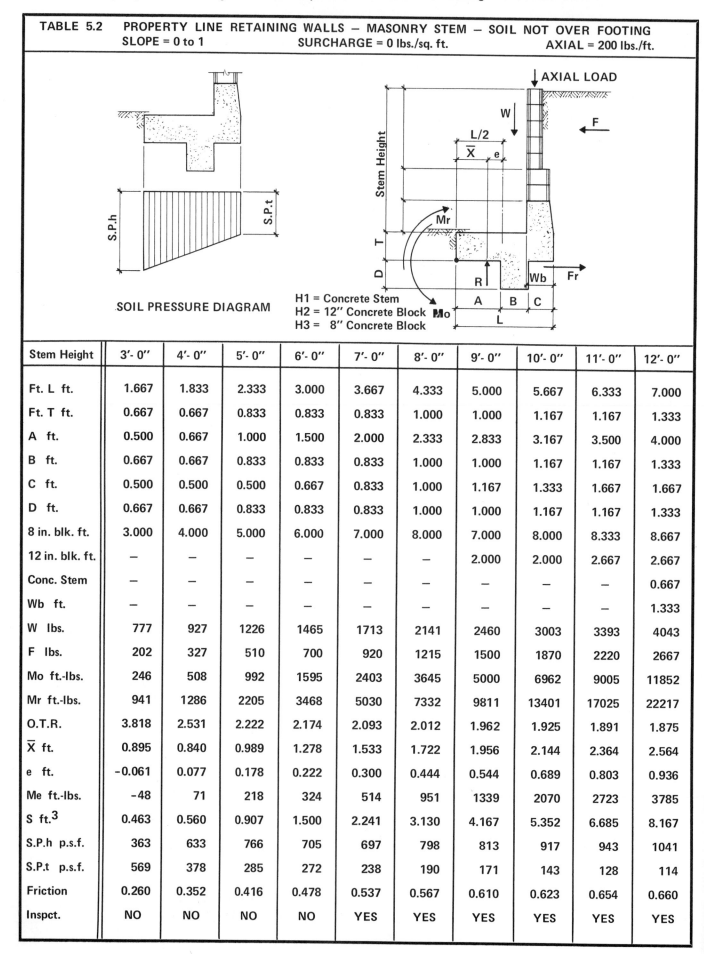

TABLE 5.2   PROPERTY LINE RETAINING WALLS – MASONRY STEM – SOIL NOT OVER FOOTING

SLOPE = 0 to 1   SURCHARGE = 0 lbs./sq. ft.   AXIAL = 200 lbs./ft.

SOIL PRESSURE DIAGRAM

H1 = Concrete Stem
H2 = 12" Concrete Block
H3 = 8" Concrete Block

Stem Height	3'- 0"	4'- 0"	5'- 0"	6'- 0"	7'- 0"	8'- 0"	9'- 0"	10'- 0"	11'- 0"	12'- 0"
Ft. L ft.	1.667	1.833	2.333	3.000	3.667	4.333	5.000	5.667	6.333	7.000
Ft. T ft.	0.667	0.667	0.833	0.833	0.833	1.000	1.000	1.167	1.167	1.333
A ft.	0.500	0.667	1.000	1.500	2.000	2.333	2.833	3.167	3.500	4.000
B ft.	0.667	0.667	0.833	0.833	0.833	1.000	1.000	1.167	1.167	1.333
C ft.	0.500	0.500	0.500	0.667	0.833	1.000	1.167	1.333	1.667	1.667
D ft.	0.667	0.667	0.833	0.833	0.833	1.000	1.000	1.167	1.167	1.333
8 in. blk. ft.	3.000	4.000	5.000	6.000	7.000	8.000	7.000	8.000	8.333	8.667
12 in. blk. ft.	—	—	—	—	—	—	2.000	2.000	2.667	2.667
Conc. Stem	—	—	—	—	—	—	—	—	—	0.667
Wb ft.	—	—	—	—	—	—	—	—	—	1.333
W lbs.	777	927	1226	1465	1713	2141	2460	3003	3393	4043
F lbs.	202	327	510	700	920	1215	1500	1870	2220	2667
Mo ft.-lbs.	246	508	992	1595	2403	3645	5000	6962	9005	11852
Mr ft.-lbs.	941	1286	2205	3468	5030	7332	9811	13401	17025	22217
O.T.R.	3.818	2.531	2.222	2.174	2.093	2.012	1.962	1.925	1.891	1.875
$\overline{X}$ ft.	0.895	0.840	0.989	1.278	1.533	1.722	1.956	2.144	2.364	2.564
e ft.	-0.061	0.077	0.178	0.222	0.300	0.444	0.544	0.689	0.803	0.936
Me ft.-lbs.	-48	71	218	324	514	951	1339	2070	2723	3785
S ft.^{3}	0.463	0.560	0.907	1.500	2.241	3.130	4.167	5.352	6.685	8.167
S.P.h p.s.f.	363	633	766	705	697	798	813	917	943	1041
S.P.t p.s.f.	569	378	285	272	238	190	171	143	128	114
Friction	0.260	0.352	0.416	0.478	0.537	0.567	0.610	0.623	0.654	0.660
Inspct.	NO	NO	NO	NO	YES	YES	YES	YES	YES	YES

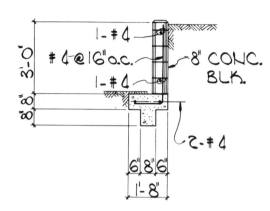

1-#4
#4 @ 16" o.c.
1-#4
8" CONC. BLK.
2-#4
3'-0"
8"
8"
6" 8" 6"
1'-8"

Axial load = 200 lbs./ft.
E.F.P. = 30 p.c.f.
Section 5.2.3

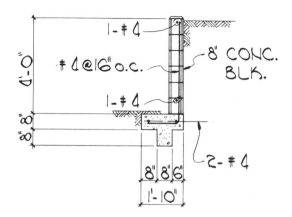

1-#4
#4 @ 16" o.c.
1-#4
8" CONC. BLK.
2-#4
4'-0"
8"
8"
8" 8" 6"
1'-10"

Axial load = 200 lbs./ft.
E.F.P. = 30 p.c.f.
Section 5.2.4

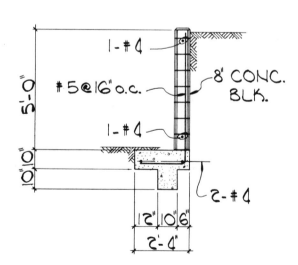

1-#4
#5 @ 16" o.c.
1-#4
8" CONC. BLK.
2-#4
5'-0"
10"
10"
12" 10" 6"
2'-4"

Axial load = 200 lbs./ft.
E.F.P. = 30 p.c.f.
Section 5.2.5

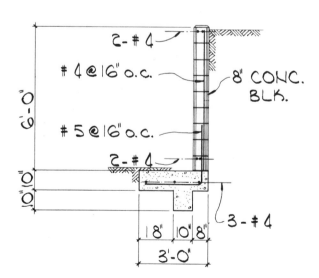

2-#4
#4 @ 16" o.c.
#5 @ 16" o.c.
2-#4
8" CONC. BLK.
3-#4
6'-0"
10"
10"
18" 10" 8"
3'-0"

Axial load = 200 lbs./ft.
E.F.P. = 30 p.c.f.
Section 5.2.6

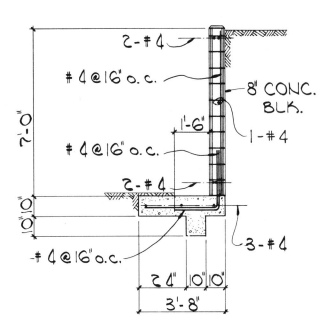

Axial load = 200 lbs./ft.
E.F.P. = 30 p.c.f.
Section 5.2.7

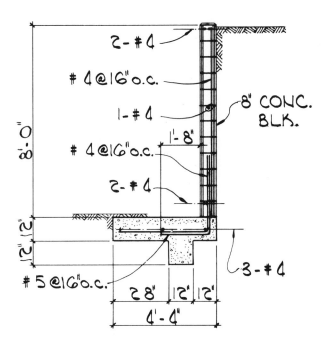

Axial load = 200 lbs./ft.
E.F.P. = 30 p.c.f.
Section 5.2.8

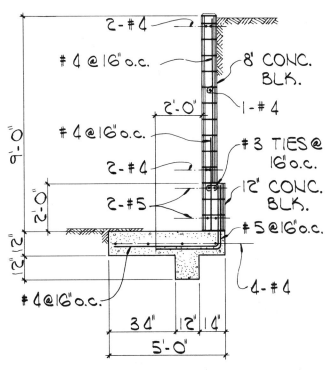

Axial load = 200 lbs./ft.
E.F.P. = 30 p.c.f.
Section 5.2.9

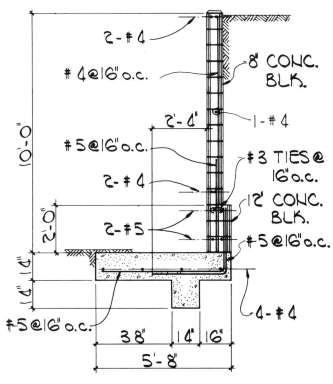

Axial load = 200 lbs./ft.
E.F.P. = 30 p.c.f.
Section 5.2.10

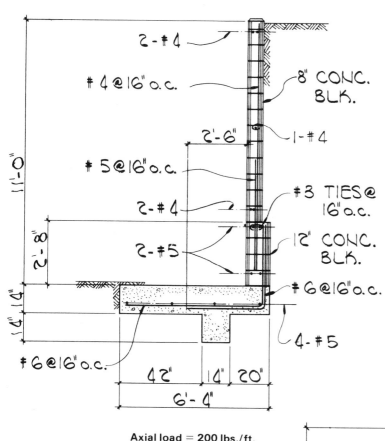

2 - #4

#4 @ 16" o.c.

8" CONC. BLK.

1 - #4

2'-6"

#5 @ 16" o.c.

2 - #4

#3 TIES @ 16" o.c.

2 - #5

12" CONC. BLK.

#6 @ 16" o.c.

4 - #5

#6 @ 16" o.c.

11'-0"

2'-8"

4"

42"   14"   20"

6'-4"

Axial load = 200 lbs./ft.
E.F.P. = 30 p.c.f.
Section 5.2.11

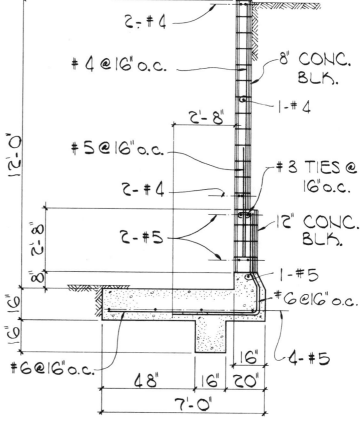

2 - #4

#4 @ 16" o.c.

8" CONC. BLK.

1 - #4

2'-8"

#5 @ 16" o.c.

2 - #4

#3 TIES @ 16" o.c.

2 - #5

12" CONC. BLK.

1 - #5

#6 @ 16" o.c.

4 - #5

#6 @ 16" o.c.

12'-0"

2'-8"

8"

16"

6"

48"   16"   20"

16"

7'-0"

Axial load = 200 lbs./ft.
E.F.P. = 30 p.c.f.
Section 5.2.12

TABLE 5.3   PROPERTY LINE RETAINING WALLS – MASONRY STEM – SOIL NOT OVER FOOTING

SLOPE = 0 to 1   SURCHARGE = 0 lbs./sq. ft.   AXIAL = 400 lbs./ft.

SOIL PRESSURE DIAGRAM

H1 = Concrete Stem
H2 = 12″ Concrete Block
H3 = 8″ Concrete Block

Stem Height	3'- 0″	4'- 0″	5'- 0″	6'- 0″	7'- 0″	8'- 0″	9'- 0″	10'- 0″	11'- 0″	12'- 0″
Ft. L ft.	1.667	1.833	2.333	3.000	3.667	4.333	5.000	5.667	6.333	7.000
Ft. T ft.	0.667	0.667	0.833	0.833	0.833	1.000	1.000	1.167	1.167	1.333
A ft.	0.500	0.667	1.000	1.500	2.000	2.333	2.833	3.167	3.500	4.000
B ft.	0.667	0.667	0.833	0.833	0.833	1.000	1.000	1.167	1.167	1.333
C ft.	0.500	0.500	0.500	0.667	0.833	1.000	1.167	1.333	1.667	1.667
D ft.	0.667	0.667	0.833	0.833	0.833	1.000	1.000	1.167	1.167	1.333
8 in. blk. ft.	3.000	4.000	5.000	6.000	7.000	8.000	7.000	8.000	8.333	8.667
12 in. blk. ft.	—	—	—	—	—	—	2.000	2.000	2.667	2.667
Conc. Stem	—	—	—	—	—	—	—	—	—	0.667
Wb ft.	—	—	—	—	—	—	—	—	—	1.333
W lbs.	977	1127	1426	1665	1913	2341	2660	3203	3593	4243
F lbs.	202	327	510	700	920	1215	1500	1870	2220	2667
Mo ft.-lbs.	246	508	992	1595	2403	3645	5000	6962	9005	11852
Mr. ft.-lbs.	1208	1586	2605	4001	5697	8132	10744	14468	18225	23550
O.T.R.	4.900	3.122	2.625	2.508	2.370	2.231	2.149	2.078	2.024	1.987
$\overline{X}$ ft.	0.984	0.957	1.131	1.445	1.721	1.917	2.160	2.343	2.566	2.757
e ft.	-0.151	-0.040	0.036	0.055	0.112	0.250	0.340	0.490	0.600	0.743
Me ft.-lbs.	-148	-45	51	91	214	585	906	1570	2157	3152
S ft.^{3}	0.463	0.560	0.907	1.500	2.241	3.130	4.167	5.352	6.685	8.167
S.P.h p.s.f.	267	534	668	616	617	727	749	859	890	992
S.P.t p.s.f.	905	695	555	494	426	353	315	272	245	220
Friction	0.207	0.290	0.358	0.421	0.481	0.519	0.564	0.584	0.618	0.629
Inspct.	NO	NO	NO	NO	YES	YES	YES	YES	YES	YES

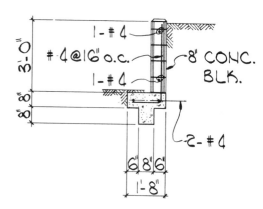

Axial load = 400 lbs./ft.
E.F.P. = 30 p.c.f.
Section 5.3.3

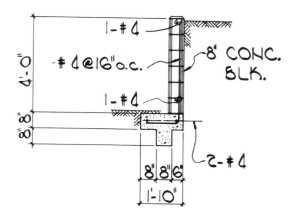

Axial load = 400 lbs./ft.
E.F.P. = 30 p.c.f.
Section 5.3.4

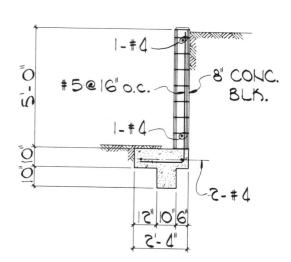

Axial load = 400 lbs./ft.
E.F.P. = 30 p.c.f.
Section 5.3.5

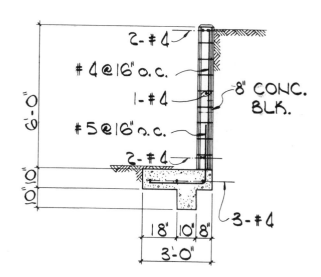

Axial load = 400 lbs./ft.
E.F.P. = 30 p.c.f.
Section 5.3.6

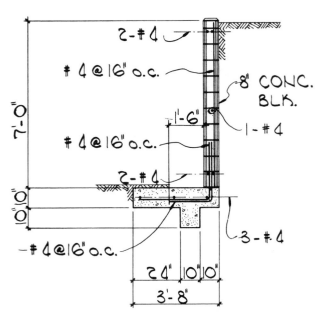

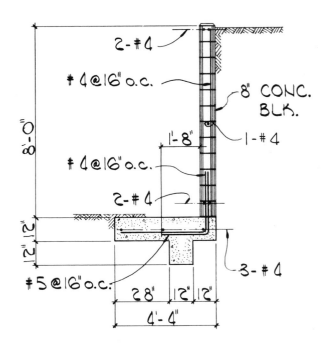

Axial load = 400 lbs./ft.
E.F.P. = 30 p.c.f.
Section 5.3.7

Axial load = 400 lbs./ft.
E.F.P. = 30 p.c.f.
Section 5.3.8

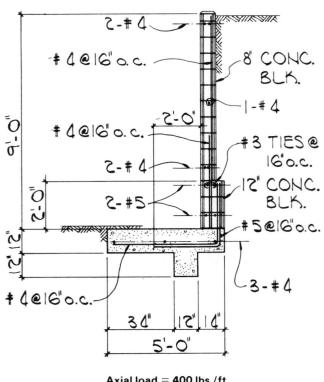

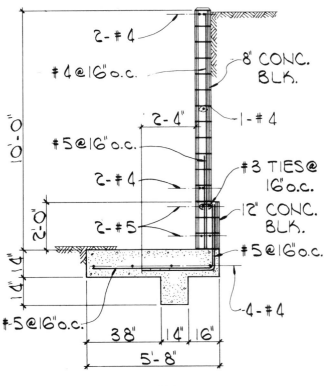

Axial load = 400 lbs./ft.
E.F.P. = 30 p.c.f.
Section 5.3.9

Axial load = 400 lbs./ft.
E.F.P. = 30 p.c.f.
Section 5.3.10

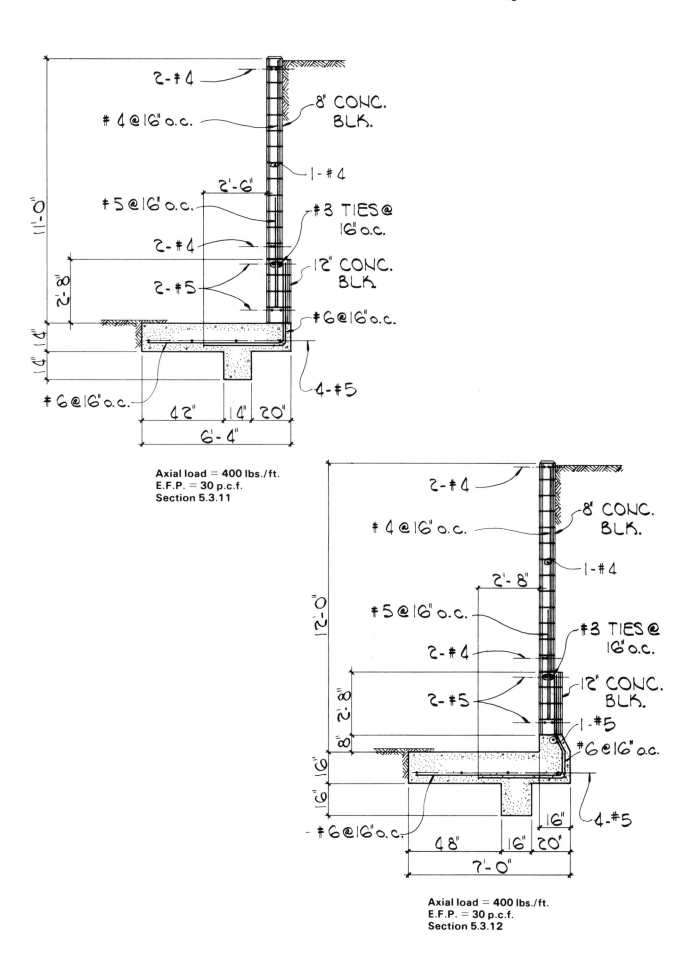

2-#4

8" CONC. BLK.

#4 @16" o.c.

1-#4

#5 @16" o.c.

2'-6"

#3 TIES @ 16" o.c.

2-#4

12" CONC. BLK

2-#5

#6@16" o.c.

4-#5

11'-0"

2'-8"

1'-4"

#6@16" o.c.

4'2"

4"

20"

6'-4"

Axial load = 400 lbs./ft.
E.F.P. = 30 p.c.f.
Section 5.3.11

2-#4

8" CONC. BLK.

#4 @16" o.c.

1-#4

#5 @16" o.c.

2'-8"

#3 TIES @ 16" o.c.

2-#4

12" CONC. BLK.

2-#5

1-#5

#6@16" o.c.

4-#5

12'-0"

2'-8"

8"

6"

6"

#6@16" o.c.

48"

16"

16"

20"

7'-0"

Axial load = 400 lbs./ft.
E.F.P. = 30 p.c.f.
Section 5.3.12

TABLE 5.4   PROPERTY LINE RETAINING WALLS – MASONRY STEM – SOIL NOT OVER FOOTING
SLOPE = 0 to 1   SURCHARGE = 0 lbs./sq. ft.   AXIAL = 600 lbs./ft.

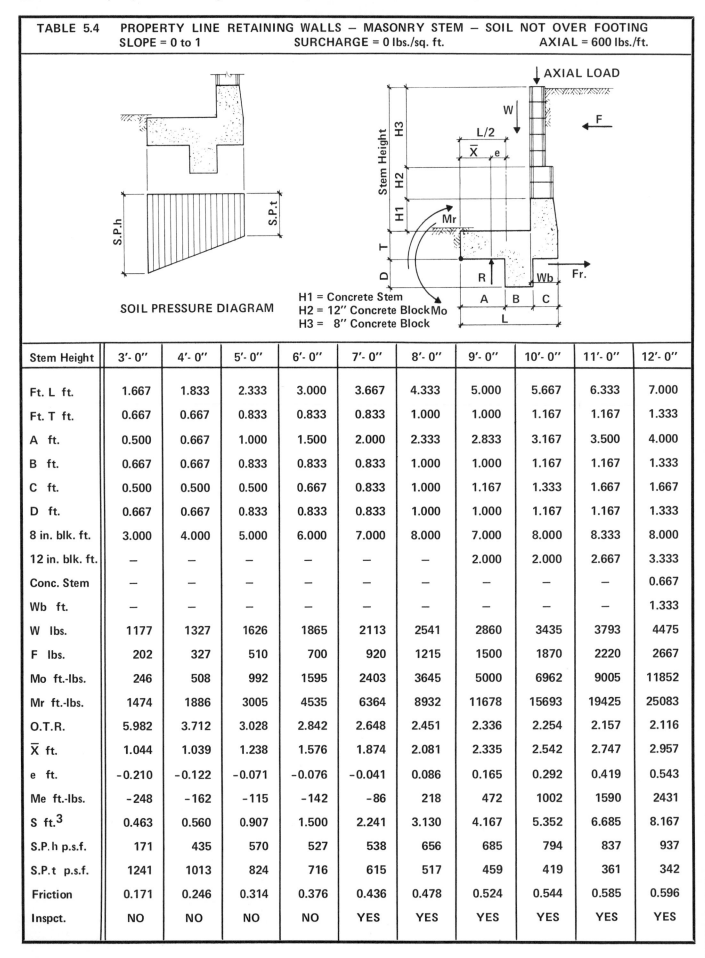

SOIL PRESSURE DIAGRAM

H1 = Concrete Stem
H2 = 12″ Concrete Block
H3 = 8″ Concrete Block

Stem Height	3′- 0″	4′- 0″	5′- 0″	6′- 0″	7′- 0″	8′- 0″	9′- 0″	10′- 0″	11′- 0″	12′- 0″
Ft. L ft.	1.667	1.833	2.333	3.000	3.667	4.333	5.000	5.667	6.333	7.000
Ft. T ft.	0.667	0.667	0.833	0.833	0.833	1.000	1.000	1.167	1.167	1.333
A ft.	0.500	0.667	1.000	1.500	2.000	2.333	2.833	3.167	3.500	4.000
B ft.	0.667	0.667	0.833	0.833	0.833	1.000	1.000	1.167	1.167	1.333
C ft.	0.500	0.500	0.500	0.667	0.833	1.000	1.167	1.333	1.667	1.667
D ft.	0.667	0.667	0.833	0.833	0.833	1.000	1.000	1.167	1.167	1.333
8 in. blk. ft.	3.000	4.000	5.000	6.000	7.000	8.000	7.000	8.000	8.333	8.000
12 in. blk. ft.	–	–	–	–	–	–	2.000	2.000	2.667	3.333
Conc. Stem	–	–	–	–	–	–	–	–	–	0.667
Wb ft.	–	–	–	–	–	–	–	–	–	1.333
W lbs.	1177	1327	1626	1865	2113	2541	2860	3435	3793	4475
F lbs.	202	327	510	700	920	1215	1500	1870	2220	2667
Mo ft.-lbs.	246	508	992	1595	2403	3645	5000	6962	9005	11852
Mr ft.-lbs.	1474	1886	3005	4535	6364	8932	11678	15693	19425	25083
O.T.R.	5.982	3.712	3.028	2.842	2.648	2.451	2.336	2.254	2.157	2.116
$\overline{X}$ ft.	1.044	1.039	1.238	1.576	1.874	2.081	2.335	2.542	2.747	2.957
e ft.	-0.210	-0.122	-0.071	-0.076	-0.041	0.086	0.165	0.292	0.419	0.543
Me ft.-lbs.	-248	-162	-115	-142	-86	218	472	1002	1590	2431
S ft.^{3}	0.463	0.560	0.907	1.500	2.241	3.130	4.167	5.352	6.685	8.167
S.P.h p.s.f.	171	435	570	527	538	656	685	794	837	937
S.P.t p.s.f.	1241	1013	824	716	615	517	459	419	361	342
Friction	0.171	0.246	0.314	0.376	0.436	0.478	0.524	0.544	0.585	0.596
Inspct.	NO	NO	NO	NO	YES	YES	YES	YES	YES	YES

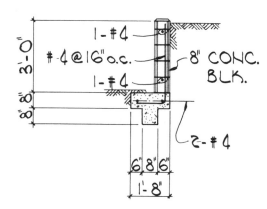

Axial load = 600 lbs./ft.
E.F.P. = 30 p.c.f.
Section 5.4.3

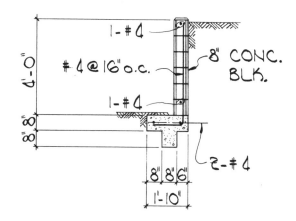

Axial load = 600 lbs./ft.
E.F.P. = 30 p.c.f.
Section 5.4.4

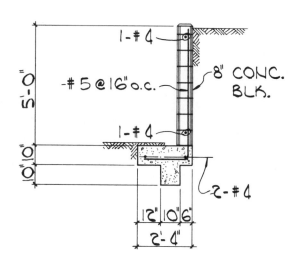

Axial load = 600 lbs./ft.
E.F.P. = 30 p.c.f.
Section 5.4.5

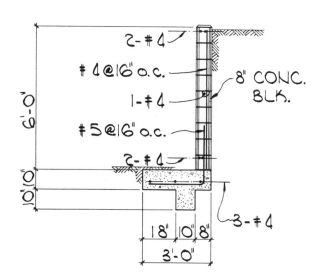

Axial load = 600 lbs./ft.
E.F.P. = 30 p.c.f.
Section 5.4.6

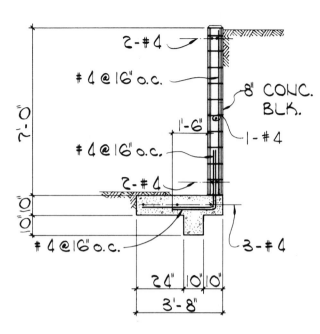

Axial load = 600 lbs./ft.
E.F.P. = 30 p.c.f.
Section 5.4.7

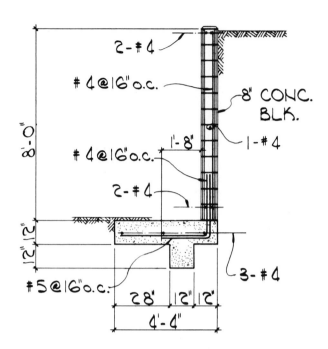

Axial load = 600 lbs./ft.
E.F.P. = 30 p.c.f.
Section 5.4.8

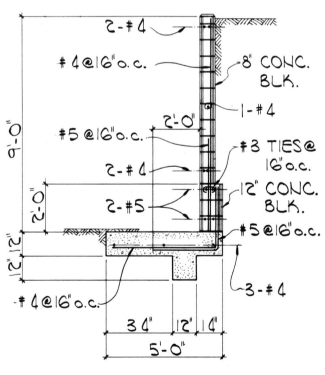

Axial load = 600 lbs./ft.
E.F.P. = 30 p.c.f.
Section 5.4.9

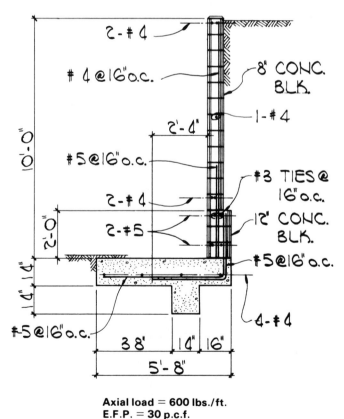

Axial load = 600 lbs./ft.
E.F.P. = 30 p.c.f.
Section 5.4.10

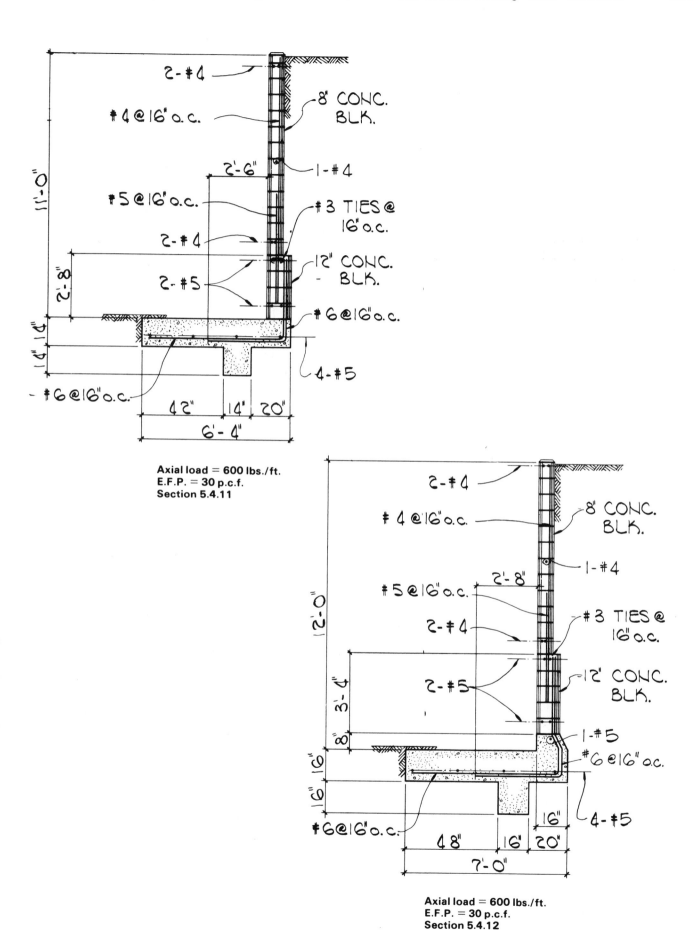

2-#4

8" CONC. BLK.

#4 @ 16" o.c.

2'-6"

1-#4

#5 @ 16" o.c.

#3 TIES @ 16" o.c.

2-#4

12" CONC. BLK.

2-#5

11'-0"

2'-8"

#6 @ 16" o.c.

1'-4"

4-#5

#6 @ 16" o.c.

42"   14"   20"

6'-4"

Axial load = 600 lbs./ft.
E.F.P. = 30 p.c.f.
Section 5.4.11

2-#4

8" CONC. BLK.

#4 @ 16" o.c.

2'-8"

1-#4

#5 @ 16" o.c.

#3 TIES @ 16" o.c.

2-#4

12" CONC. BLK.

2-#5

12'-0"

3'-4"

8"

1-#5

16"

#6 @ 16" o.c.

16"

16"   16"   20"

4-#5

#6 @ 16" o.c.

7'-0"

Axial load = 600 lbs./ft.
E.F.P. = 30 p.c.f.
Section 5.4.12

TABLE 5.5   PROPERTY LINE RETAINING WALLS – MASONRY STEM – SOIL NOT OVER FOOTING

SLOPE = 0 to 1   SURCHARGE = 0 lbs./sq. ft.   AXIAL = 800 lbs./ft.

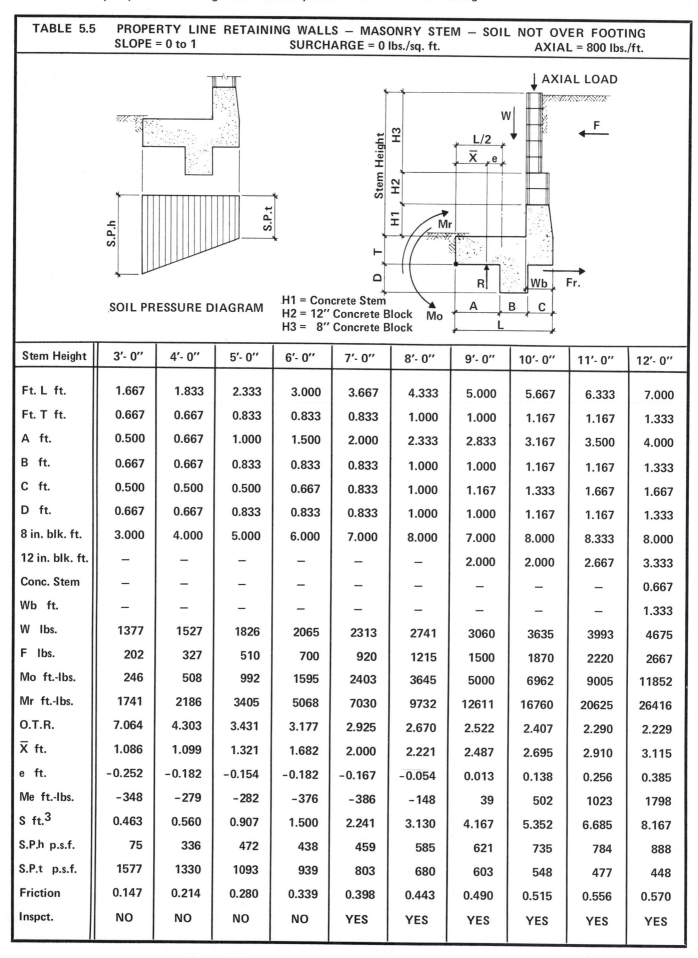

SOIL PRESSURE DIAGRAM

H1 = Concrete Stem
H2 = 12″ Concrete Block
H3 = 8″ Concrete Block

Stem Height	3′- 0″	4′- 0″	5′- 0″	6′- 0″	7′- 0″	8′- 0″	9′- 0″	10′- 0″	11′- 0″	12′- 0″
Ft. L ft.	1.667	1.833	2.333	3.000	3.667	4.333	5.000	5.667	6.333	7.000
Ft. T ft.	0.667	0.667	0.833	0.833	0.833	1.000	1.000	1.167	1.167	1.333
A ft.	0.500	0.667	1.000	1.500	2.000	2.333	2.833	3.167	3.500	4.000
B ft.	0.667	0.667	0.833	0.833	0.833	1.000	1.000	1.167	1.167	1.333
C ft.	0.500	0.500	0.500	0.667	0.833	1.000	1.167	1.333	1.667	1.667
D ft.	0.667	0.667	0.833	0.833	0.833	1.000	1.000	1.167	1.167	1.333
8 in. blk. ft.	3.000	4.000	5.000	6.000	7.000	8.000	7.000	8.000	8.333	8.000
12 in. blk. ft.	—	—	—	—	—	—	2.000	2.000	2.667	3.333
Conc. Stem	—	—	—	—	—	—	—	—	—	0.667
Wb ft.	—	—	—	—	—	—	—	—	—	1.333
W lbs.	1377	1527	1826	2065	2313	2741	3060	3635	3993	4675
F lbs.	202	327	510	700	920	1215	1500	1870	2220	2667
Mo ft.-lbs.	246	508	992	1595	2403	3645	5000	6962	9005	11852
Mr ft.-lbs.	1741	2186	3405	5068	7030	9732	12611	16760	20625	26416
O.T.R.	7.064	4.303	3.431	3.177	2.925	2.670	2.522	2.407	2.290	2.229
$\overline{X}$ ft.	1.086	1.099	1.321	1.682	2.000	2.221	2.487	2.695	2.910	3.115
e ft.	−0.252	−0.182	−0.154	−0.182	−0.167	−0.054	0.013	0.138	0.256	0.385
Me ft.-lbs.	−348	−279	−282	−376	−386	−148	39	502	1023	1798
S ft.^{3}	0.463	0.560	0.907	1.500	2.241	3.130	4.167	5.352	6.685	8.167
S.P.h p.s.f.	75	336	472	438	459	585	621	735	784	888
S.P.t p.s.f.	1577	1330	1093	939	803	680	603	548	477	448
Friction	0.147	0.214	0.280	0.339	0.398	0.443	0.490	0.515	0.556	0.570
Inspct.	NO	NO	NO	NO	YES	YES	YES	YES	YES	YES

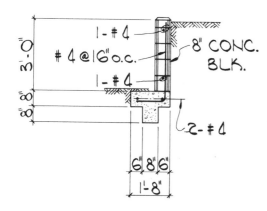

3'-0"

1-#4

#4@16"o.c.

1-#4

8" CONC. BLK.

8"

8"

2-#4

6" 8" 6"

1'-8"

Axial load = 800 lbs./ft.
E.F.P. = 30 p.c.f.
Section 5.5.3

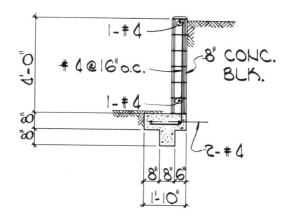

4'-0"

1-#4

#4@16"o.c.

1-#4

8" CONC. BLK.

8"

8"

2-#4

8" 8" 6"

1'-10"

Axial load = 800 lbs./ft.
E.F.P. = 30 p.c.f.
Section 5.5.4

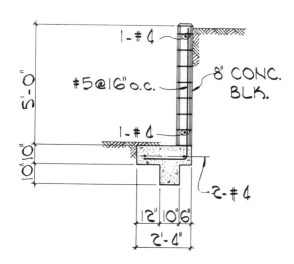

5'-0"

1-#4

#5@16"o.c.

1-#4

8" CONC. BLK.

10"

10"

2-#4

12" 10" 6"

2'-4"

Axial load = 800 lbs./ft.
E.F.P. = 30 p.c.f.
Section 5.5.5

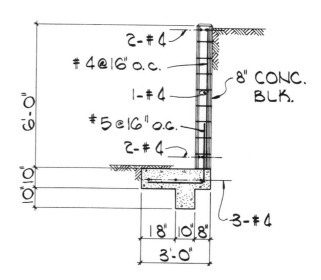

6'-0"

2-#4

#4@16"o.c.

1-#4

#5@16"o.c.

2-#4

8" CONC. BLK.

10"

10"

3-#4

18" 10" 8"

3'-0"

Axial load = 800 lbs./ft.
E.F.P. = 30 p.c.f.
Section 5.5.6

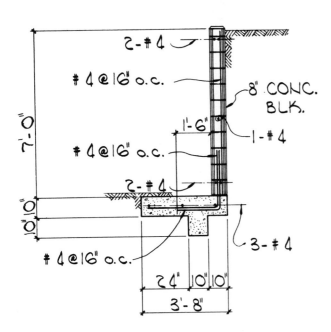

Axial load = 800 lbs./ft.
E.F.P. = 30 p.c.f.
Section 5.5.7

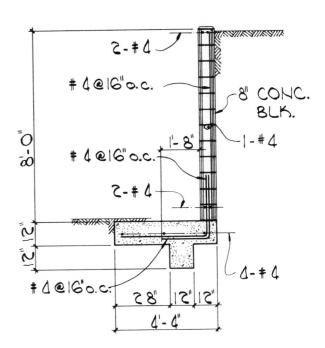

Axial load = 800 lbs./ft.
E.F.P. = 30 p.c.f.
Section 5.5.8

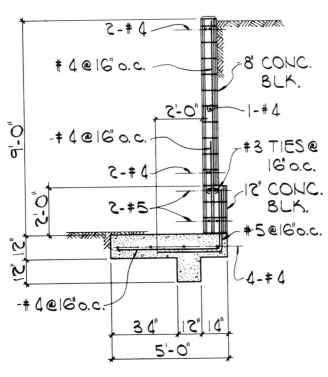

Axial load = 800 lbs./ft.
E.F.P. = 30 p.c.f.
Section 5.5.9

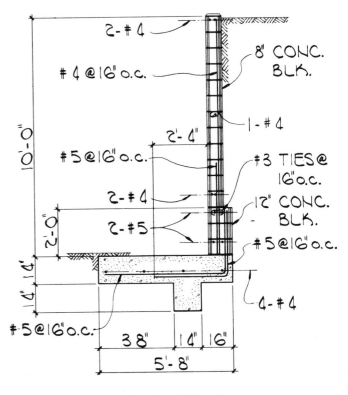

Axial load = 800 lbs./ft.
E.F.P. = 30 p.c.f.
Section 5.5.10

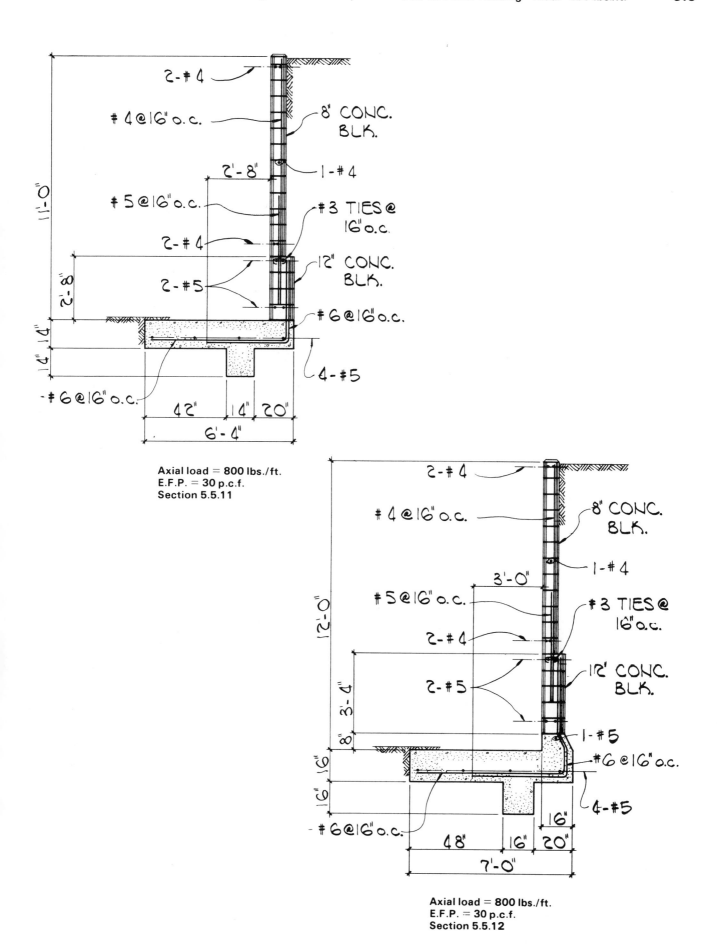

Axial load = 800 lbs./ft.
E.F.P. = 30 p.c.f.
Section 5.5.11

Axial load = 800 lbs./ft.
E.F.P. = 30 p.c.f.
Section 5.5.12

TABLE 5.6   PROPERTY LINE RETAINING WALLS – MASONRY STEM – SOIL NOT OVER FOOTING
SLOPE = 0 to 1   SURCHARGE = 50 lbs./sq. ft.   AXIAL = 0 lbs./ft.

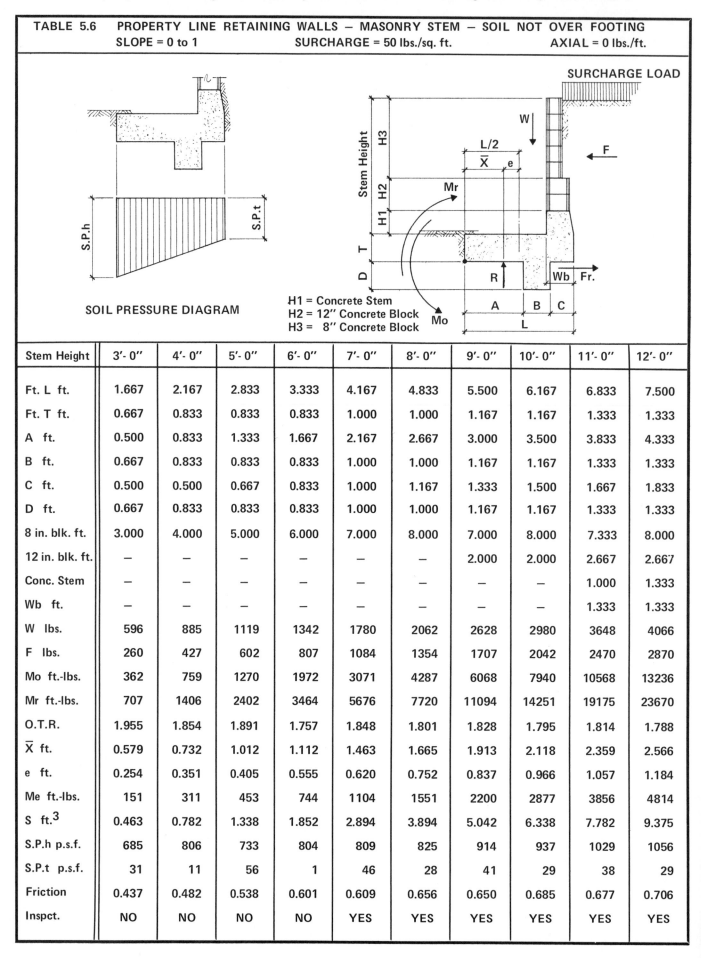

SOIL PRESSURE DIAGRAM

H1 = Concrete Stem
H2 = 12″ Concrete Block
H3 = 8″ Concrete Block

Stem Height	3′- 0″	4′- 0″	5′- 0″	6′- 0″	7′- 0″	8′- 0″	9′- 0″	10′- 0″	11′- 0″	12′- 0″
Ft. L ft.	1.667	2.167	2.833	3.333	4.167	4.833	5.500	6.167	6.833	7.500
Ft. T ft.	0.667	0.833	0.833	0.833	1.000	1.000	1.167	1.167	1.333	1.333
A ft.	0.500	0.833	1.333	1.667	2.167	2.667	3.000	3.500	3.833	4.333
B ft.	0.667	0.833	0.833	0.833	1.000	1.000	1.167	1.167	1.333	1.333
C ft.	0.500	0.500	0.667	0.833	1.000	1.167	1.333	1.500	1.667	1.833
D ft.	0.667	0.833	0.833	0.833	1.000	1.000	1.167	1.167	1.333	1.333
8 in. blk. ft.	3.000	4.000	5.000	6.000	7.000	8.000	7.000	8.000	7.333	8.000
12 in. blk. ft.	—	—	—	—	—	—	2.000	2.000	2.667	2.667
Conc. Stem	—	—	—	—	—	—	—	—	1.000	1.333
Wb ft.	—	—	—	—	—	—	—	—	1.333	1.333
W lbs.	596	885	1119	1342	1780	2062	2628	2980	3648	4066
F lbs.	260	427	602	807	1084	1354	1707	2042	2470	2870
Mo ft.-lbs.	362	759	1270	1972	3071	4287	6068	7940	10568	13236
Mr ft.-lbs.	707	1406	2402	3464	5676	7720	11094	14251	19175	23670
O.T.R.	1.955	1.854	1.891	1.757	1.848	1.801	1.828	1.795	1.814	1.788
$\overline{X}$ ft.	0.579	0.732	1.012	1.112	1.463	1.665	1.913	2.118	2.359	2.566
e ft.	0.254	0.351	0.405	0.555	0.620	0.752	0.837	0.966	1.057	1.184
Me ft.-lbs.	151	311	453	744	1104	1551	2200	2877	3856	4814
S ft.^{3}	0.463	0.782	1.338	1.852	2.894	3.894	5.042	6.338	7.782	9.375
S.P.h p.s.f.	685	806	733	804	809	825	914	937	1029	1056
S.P.t p.s.f.	31	11	56	1	46	28	41	29	38	29
Friction	0.437	0.482	0.538	0.601	0.609	0.656	0.650	0.685	0.677	0.706
Inspct.	NO	NO	NO	NO	YES	YES	YES	YES	YES	YES

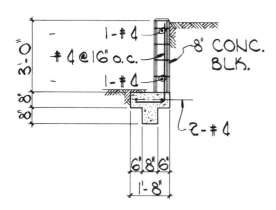

Surcharge load = 50 p.s.f.
E.F.P. = 30 p.c.f.
Section 5.6.3

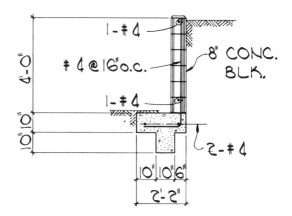

Surcharge load = 50 p.s.f.
E.F.P. = 30 p.c.f.
Section 5.6.4

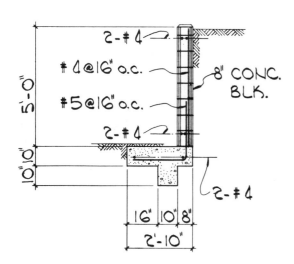

Surcharge load = 50 p.s.f.
E.F.P. = 30 p.c.f.
Section 5.6.5

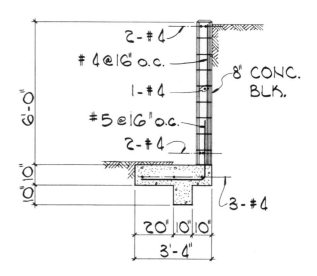

Surcharge load = 50 p.s.f.
E.F.P. = 30 p.c.f.
Section 5.6.6

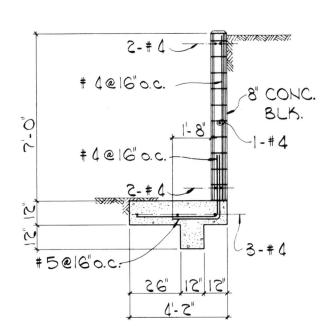

Surcharge load = 50 p.s.f.
E.F.P. = 30 p.c.f.
Section 5.6.7

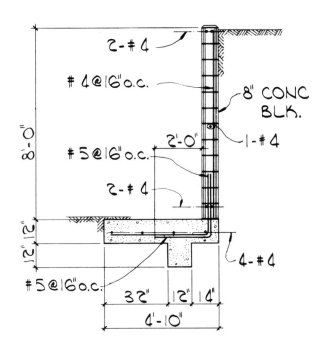

Surcharge load = 50 p.s.f.
E.F.P. = 30 p.c.f.
Section 5.6.8

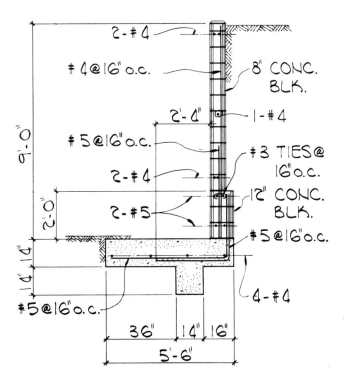

Surcharge load = 50 p.s.f.
E.F.P. = 30 p.c.f.
Section 5.6.9

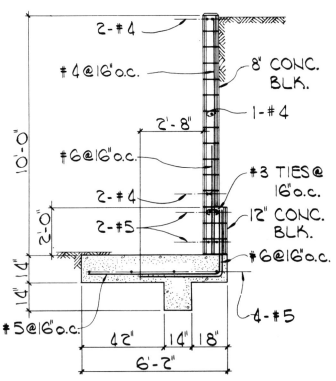

Surcharge load = 50 p.s.f.
E.F.P. = 30 p.c.f.
Section 5.6.10

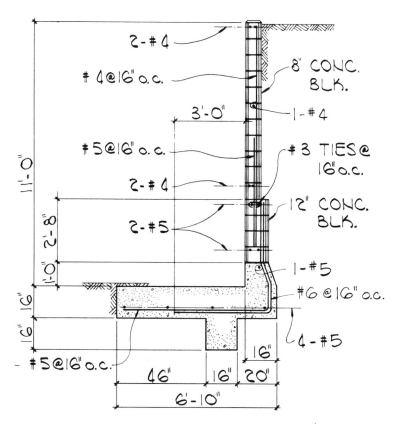

2-#4

#4@16" o.c.

8" CONC. BLK.

1-#4

3'-0"

#5@16" o.c.

#3 TIES @ 16" o.c.

2-#4

2-#5

12" CONC. BLK.

1-#5

#6 @16" o.c.

4-#5

11'-0"

2'-8"

1'-0"

6"

6"

#5@16" o.c.

46"   16"   20"

16"

6'-10"

Surcharge load = 50 p.s.f.
E.F.P. = 30 p.c.f.
Section 5.6.11

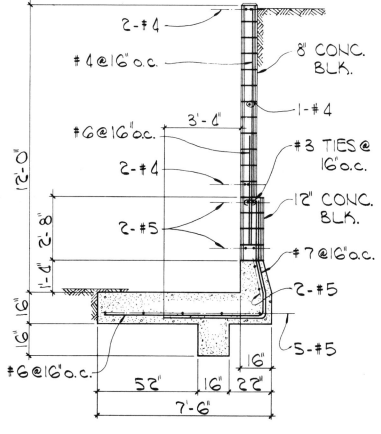

2-#4

#4@16" o.c.

8" CONC. BLK.

1-#4

3'-0"

#6@16" o.c.

#3 TIES @ 16" o.c.

2-#4

2-#5

12" CONC. BLK.

#7 @16" o.c.

2-#5

5-#5

12'-0"

2'-8"

1'-0"

6"

6"

#6@16" o.c.

52"   16"   22"

16"

7'-6"

Surcharge load = 50 p.s.f.
E.F.P. = 30 p.c.f.
Section 5.6.12

TABLE 5.7   PROPERTY LINE RETAINING WALLS — MASONRY STEM — SOIL NOT OVER FOOTING
SLOPE = 0 to 1   SURCHARGE = 100 lbs./sq. ft.   AXIAL = 0 lbs./ft.

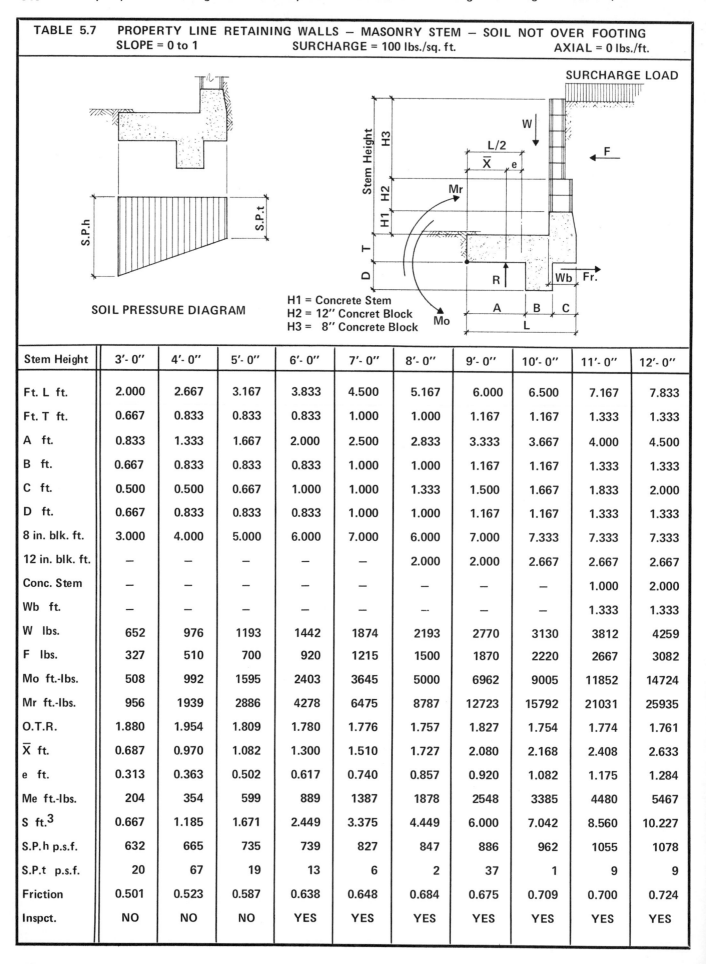

SOIL PRESSURE DIAGRAM

H1 = Concrete Stem
H2 = 12″ Concret Block
H3 = 8″ Concrete Block

Stem Height	3′- 0″	4′- 0″	5′- 0″	6′- 0″	7′- 0″	8′- 0″	9′- 0″	10′- 0″	11′- 0″	12′- 0″
Ft. L ft.	2.000	2.667	3.167	3.833	4.500	5.167	6.000	6.500	7.167	7.833
Ft. T ft.	0.667	0.833	0.833	0.833	1.000	1.000	1.167	1.167	1.333	1.333
A ft.	0.833	1.333	1.667	2.000	2.500	2.833	3.333	3.667	4.000	4.500
B ft.	0.667	0.833	0.833	0.833	1.000	1.000	1.167	1.167	1.333	1.333
C ft.	0.500	0.500	0.667	1.000	1.000	1.333	1.500	1.667	1.833	2.000
D ft.	0.667	0.833	0.833	0.833	1.000	1.000	1.167	1.167	1.333	1.333
8 in. blk. ft.	3.000	4.000	5.000	6.000	7.000	6.000	7.000	7.333	7.333	7.333
12 in. blk. ft.	—	—	—	—	—	2.000	2.000	2.667	2.667	2.667
Conc. Stem	—	—	—	—	—	—	—	—	1.000	2.000
Wb ft.	—	—	—	—	—	—	—	—	1.333	1.333
W lbs.	652	976	1193	1442	1874	2193	2770	3130	3812	4259
F lbs.	327	510	700	920	1215	1500	1870	2220	2667	3082
Mo ft.-lbs.	508	992	1595	2403	3645	5000	6962	9005	11852	14724
Mr ft.-lbs.	956	1939	2886	4278	6475	8787	12723	15792	21031	25935
O.T.R.	1.880	1.954	1.809	1.780	1.776	1.757	1.827	1.754	1.774	1.761
$\overline{X}$ ft.	0.687	0.970	1.082	1.300	1.510	1.727	2.080	2.168	2.408	2.633
e ft.	0.313	0.363	0.502	0.617	0.740	0.857	0.920	1.082	1.175	1.284
Me ft.-lbs.	204	354	599	889	1387	1878	2548	3385	4480	5467
S ft.^{3}	0.667	1.185	1.671	2.449	3.375	4.449	6.000	7.042	8.560	10.227
S.P.h p.s.f.	632	665	735	739	827	847	886	962	1055	1078
S.P.t p.s.f.	20	67	19	13	6	2	37	1	9	9
Friction	0.501	0.523	0.587	0.638	0.648	0.684	0.675	0.709	0.700	0.724
Inspct.	NO	NO	NO	YES	YES	YES	YES	YES	YES	YES

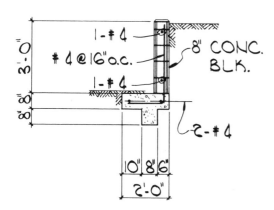

Surcharge load = 100 p.s.f.
E.F.P. = 30 p.c.f.
Section 5.7.3

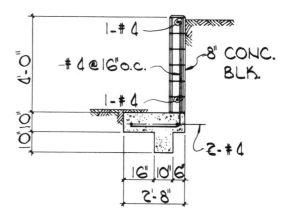

Surcharge load = 100 p.s.f.
E.F.P. = 30 p.c.f.
Section 5.7.4

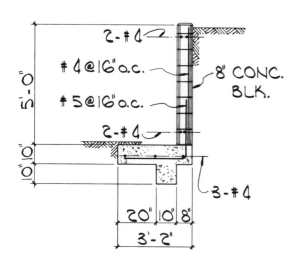

Surcharge load = 100 p.s.f.
E.F.P. = 30 p.c.f.
Section 5.7.5

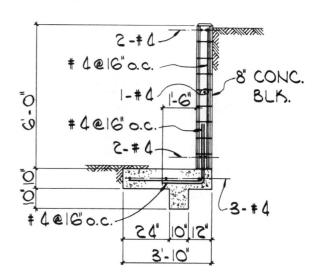

Surcharge load = 100 p.s.f.
E.F.P. = 30 p.c.f.
Section 5.7.6

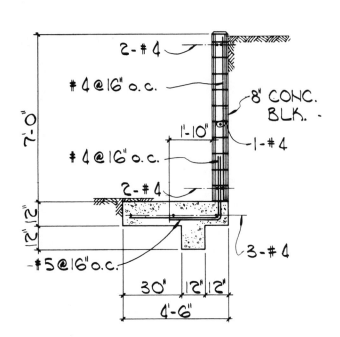

Surcharge load = 100 p.s.f.
E.F.P. = 30 p.c.f.
Section 5.7.7

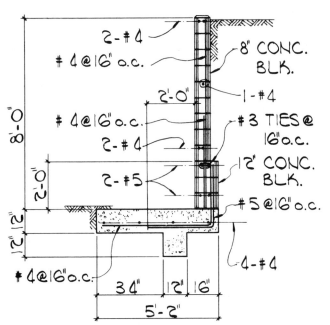

Surcharge load = 100 p.s.f.
E.F.P. = 30 p.c.f.
Section 5.7.8

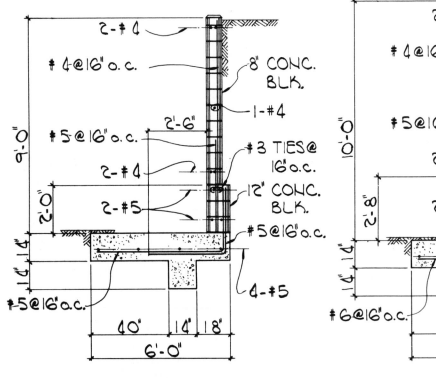

Surcharge load = 100 p.s.f.
E.F.P. = 30 p.c.f.
Section 5.7.9

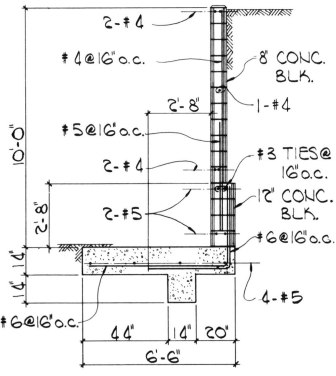

Surcharge load = 100 p.s.f.
E.F.P. = 30 p.c.f.
Section 5.7.10

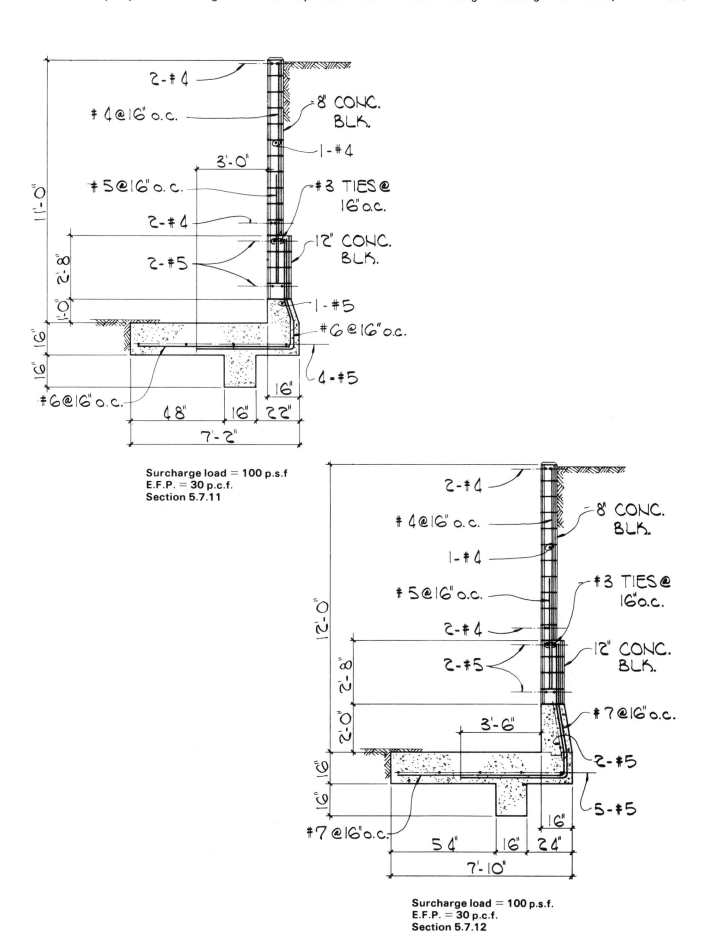

Top figure labels:

2-#4
#4@16" o.c.
3'-0"
#5@16" o.c.
2-#4
2-#5
1'-0"
2'-8"
1'-0"
6"
6"
6"
#6@16" o.c.
48"
16"
22"
7'-2"
8" CONC. BLK.
1-#4
#3 TIES@ 16" o.c.
12" CONC. BLK.
1-#5
#6@16" o.c.
4-#5
16"

Surcharge load = 100 p.s.f
E.F.P. = 30 p.c.f.
Section 5.7.11

Bottom figure labels:

2-#4
#4@16" o.c.
1-#4
#5@16" o.c.
2-#4
2-#5
12'-0"
2'-8"
2'-0"
6"
6"
3'-6"
#7@16" o.c.
54"
16"
24"
7'-10"
8" CONC. BLK.
#3 TIES@ 16" o.c.
12" CONC. BLK.
#7@16" o.c.
2-#5
5-#5
16"

Surcharge load = 100 p.s.f.
E.F.P. = 30 p.c.f.
Section 5.7.12

TABLE 5.8 PROPERTY LINE RETAINING WALLS – MASONRY STEM – SOIL NOT OVER FOOTING
SLOPE = 0 to 1   SURCHARGE = 200 lbs./sq. ft.   AXIAL = 0 lbs./ft.

SOIL PRESSURE DIAGRAM

SURCHARGE LOAD

H1 = Concrete Stem
H2 = 12″ Concrete Block
H3 = 8″ Concrete Block

Stem Height	3′- 0″	4′- 0″	5′- 0″	6′- 0″	7′- 0″	8′- 0″	9′- 0″	10′- 0″	11′- 0″	12′- 0″
Ft. L ft.	2.833	3.500	4.167	4.833	5.500	6.167	6.833	7.500	8.167	8.833
Ft. T ft.	0.833	0.833	0.833	1.000	1.000	1.167	1.167	1.333	1.333	1.333
A ft.	1.333	1.833	2.333	2.667	3.167	3.500	3.833	4.333	4.667	5.167
B ft.	0.833	0.833	0.833	1.000	1.000	1.167	1.167	1.333	1.333	1.333
C ft.	0.667	0.833	1.000	1.167	1.333	1.500	1.833	1.833	2.167	2.333
D ft.	0.833	0.833	0.833	1.000	1.000	1.167	1.167	1.333	1.333	1.333
8 in. blk. ft.	3.000	4.000	5.000	6.000	5.000	6.000	6.333	6.667	6.667	6.667
12 in. blk. ft.	–	–	–	–	2.000	2.000	2.667	2.667	2.667	2.667
Conc. Stem	–	–	–	–	–	–	–	0.667	1.667	2.667
Wb ft.	–	–	–	–	–	–	–	1.333	1.333	1.333
W lbs.	904	1143	1392	1832	2151	2707	3096	3759	4206	4662
F lbs.	510	700	920	1215	1500	1870	2220	2667	3082	3527
Mo ft.-lbs.	992	1595	2403	3645	5000	6962	9005	11852	14724	18025
Mr ft.-lbs.	1856	2982	4413	6669	9051	12650	16212	21465	26394	31961
O.T.R.	1.870	1.869	1.836	1.830	1.810	1.817	1.800	1.811	1.793	1.773
$\bar{X}$ ft.	0.955	1.213	1.444	1.650	1.883	2.101	2.328	2.557	2.775	2.989
e ft.	0.462	0.537	0.639	0.766	0.867	0.982	1.089	1.193	1.308	1.428
Me ft.-lbs.	418	613	890	1404	1865	2658	3372	4483	5502	6655
S ft.^{3}	1.338	2.042	2.894	3.894	5.042	6.338	7.782	9.375	11.116	13.005
S.P.h p.s.f.	631	627	642	740	761	858	886	979	1010	1040
S.P.t p.s.f.	7	26	27	18	21	20	20	23	20	16
Friction	0.564	0.613	0.661	0.663	0.697	0.691	0.717	0.709	0.733	0.756
Inspct.	NO	NO	YES	YES	YES	YES	YES	YES	YES	YES

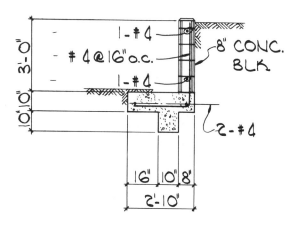

Surcharge load = 200 p.s.f.
E.F.P. = 30 p.c.f.
Section 5.8.3

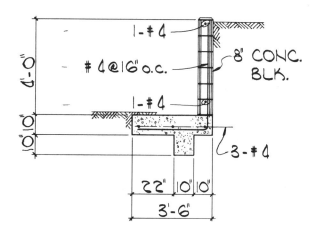

Surcharge load = 200 p.s.f.
E.F.P. = 30 p.c.f.
Section 5.8.4

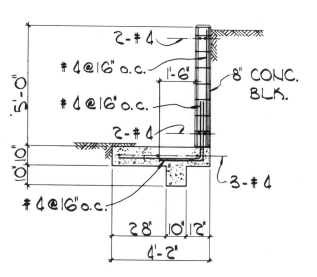

Surcharge load = 200 p.s.f.
E.F.P. = 30 p.c.f.
Section 5.8.5

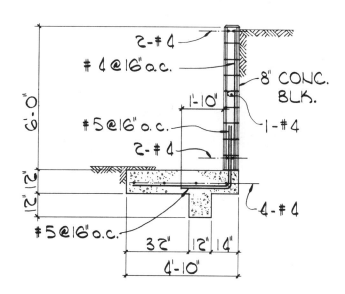

Surcharge load = 200 p.s.f.
E.F.P. = 30 p.c.f.
Section 5.8.6

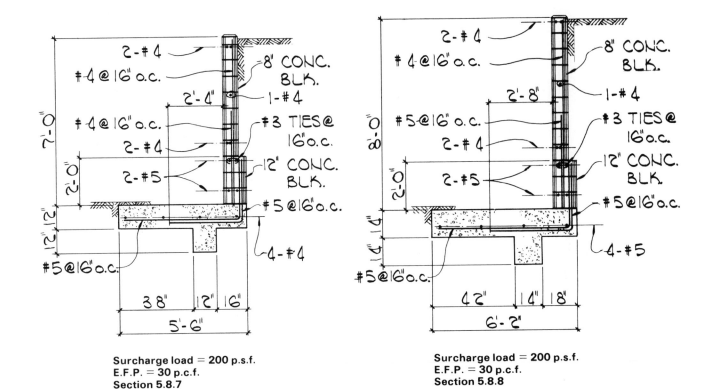

Surcharge load = 200 p.s.f.
E.F.P. = 30 p.c.f.
Section 5.8.7

Surcharge load = 200 p.s.f.
E.F.P. = 30 p.c.f.
Section 5.8.8

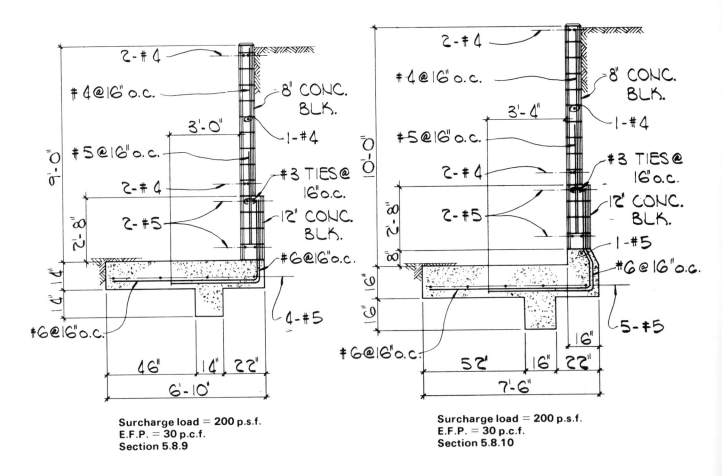

Surcharge load = 200 p.s.f.
E.F.P. = 30 p.c.f.
Section 5.8.9

Surcharge load = 200 p.s.f.
E.F.P. = 30 p.c.f.
Section 5.8.10

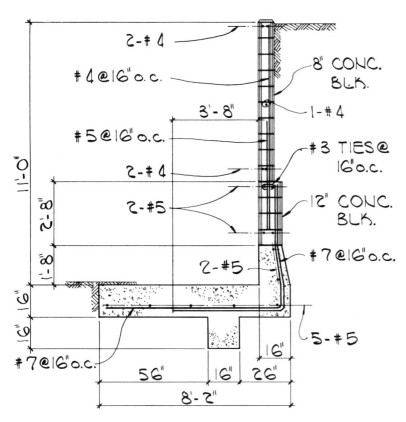

2-#4
#4@16"o.c.
8" CONC. BLK.
1-#4
3'-8"
#5@16"o.c.
#3 TIES@ 16"o.c.
2-#4
12" CONC. BLK.
2-#5
#7@16"o.c.
2-#5
11'-0"
2'-8"
1'-8"
6"
6"
#7@16"o.c.
5-#5
56"
16"
26"
16"
8'-2"

Surcharge load = 200 p.s.f.
E.F.P. = 30 p.c.f.
Section 5.8.11

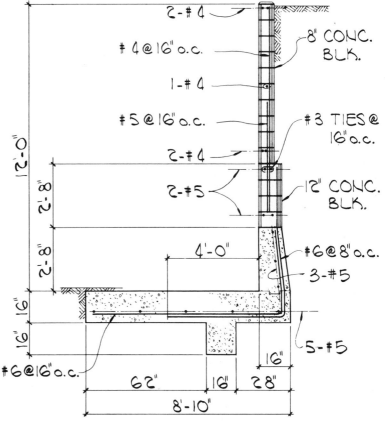

2-#4
#4@16"o.c.
8" CONC. BLK.
1-#4
#5@16"o.c.
#3 TIES@ 16"o.c.
2-#4
12" CONC. BLK.
2-#5
4'-0"
#6@8"o.c.
3-#5
12'-0"
2'-8"
2'-8"
6"
6"
#6@16"o.c.
5-#5
62"
16"
28"
16"
8'-10"

Surcharge load = 200 p.s.f.
E.F.P. = 30 p.c.f.
Section 5.8.12

TABLE 5.9   PROPERTY LINE RETAINING WALLS — MASONRY STEM — SOIL NOT OVER FOOTING
SLOPE = 5 to 1   SURCHARGE = 0 lbs./sq. ft.   AXIAL = 0 lbs./ft.

SOIL PRESSURE DIAGRAM

H1 = Concrete Stem
H2 = 12″ Concrete Block
H3 = 8″ Concrete Block

Stem Height	3′- 0″	4′- 0″	5′- 0″	6′- 0″	7′- 0″	8′- 0″	9′- 0″	10′- 0″	11′- 0″	12′- 0″
Ft. L ft.	1.667	2.000	2.500	3.167	3.833	4.500	5.167	5.833	6.500	7.167
Ft. T ft.	0.667	0.833	0.833	0.833	1.000	1.000	1.167	1.167	1.333	1.333
A ft.	0.500	0.667	1.167	1.667	2.000	2.500	2.833	3.167	3.667	4.000
B ft.	0.667	0.833	0.833	0.833	1.000	1.000	1.167	1.167	1.333	1.333
C ft.	0.500	0.500	0.500	0.667	0.833	1.000	1.167	1.500	1.500	1.833
D ft.	0.667	0.833	0.833	0.833	1.000	1.000	1.167	1.167	1.333	1.333
8 in. blk. ft.	3.000	4.000	5.000	6.000	7.000	8.000	7.000	8.000	8.333	8.000
12 in. blk. ft.	—	—	—	—	—	—	2.000	2.000	2.667	2.667
Conc. Stem	—	—	—	—	—	—	—	—	—	1.333
Wb ft.	—	—	—	—	—	—	—	—	—	1.333
W lbs.	581	847	1058	1301	1710	1993	2520	2906	3518	3991
F lbs.	215	374	544	747	1024	1296	1654	1995	2434	2844
Mo ft.-lbs.	263	602	1059	1702	2731	3888	5604	7426	10006	12642
Mr ft.-lbs.	682	1225	2006	3196	5040	6979	10028	13177	17638	22263
O.T.R.	2.594	2.035	1.895	1.878	1.846	1.795	1.789	1.774	1.763	1.761
$\overline{X}$ ft.	0.721	0.736	0.895	1.149	1.350	1.551	1.755	1.979	2.170	2.411
e ft.	0.112	0.264	0.355	0.435	0.567	0.699	0.828	0.938	1.080	1.172
Me ft.-lbs.	65	224	375	565	969	1393	2086	2726	3801	4679
S ft.^{3}	0.463	0.667	1.042	1.671	2.449	3.375	4.449	5.671	7.042	8.560
S.P.h p.s.f.	489	759	784	749	842	856	956	979	1081	1103
S.P.t p.s.f.	208	88	63	73	50	30	19	18	1	10
Friction	0.370	0.441	0.515	0.574	0.599	0.650	0.656	0.687	0.692	0.713
Inspct.	NO	NO	NO	NO	YES	YES	YES	YES	YES	YES

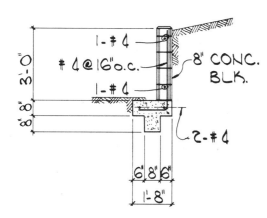

1 - #4
#4 @ 16" o.c.
1 - #4
8" CONC. BLK.
3'-0"
8"
8"
2 - #4
6" 8" 6"
1'-8"

Soil slope = 5 to 1
E.F.P. = 32 p.c.f.
Section 5.9.3

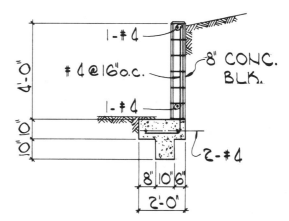

1 - #4
#4 @ 16" o.c.
1 - #4
8" CONC. BLK.
4'-0"
10"
2 - #4
8" 10" 6"
2'-0"

Soil slope = 5 to 1
E.F.P. = 32 p.c.f.
Section 5.9.4

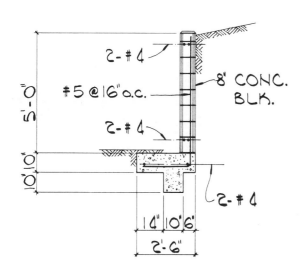

2 - #4
#5 @ 16" o.c.
2 - #4
8" CONC. BLK.
5'-0"
10"
2 - #4
14" 10" 6"
2'-6"

Soil slope = 5 to 1
E.F.P. = 32 p.c.f.
Section 5.9.5

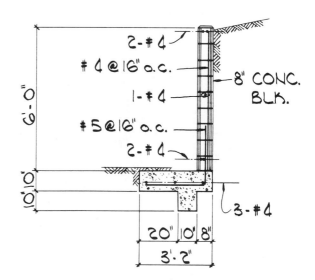

2 - #4
#4 @ 16" o.c.
1 - #4
#5 @ 16" o.c.
2 - #4
8" CONC. BLK.
6'-0"
10"
3 - #4
20" 10" 8"
3'-2"

Soil slope = 5 to 1
E.F.P. = 32 p.c.f.
Section 5.9.6

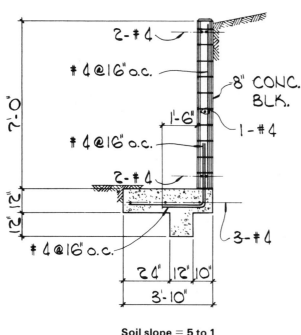

Soil slope = 5 to 1
E.F.P. = 32 p.c.f.
Section 5.9.7

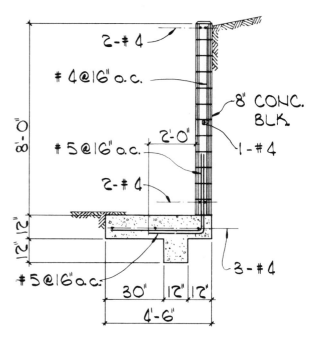

Soil slope = 5 to 1
E.F.P. = 32 p.c.f.
Section 5.9.8

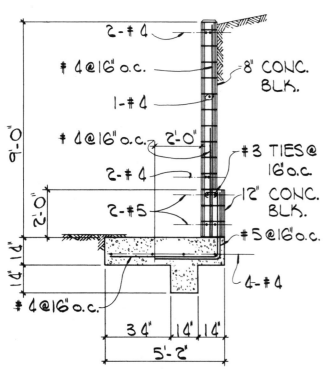

Soil slope = 5 to 1
E.F.P. = 32 p.c.f.
Section 5.9.9

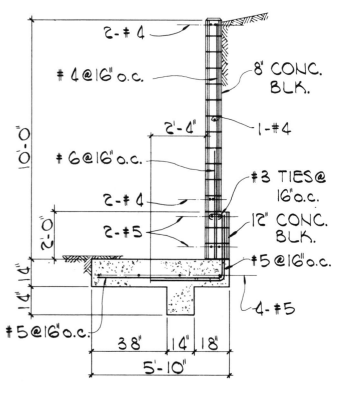

Soil slope = 5 to 1
E.F.P. = 32 p.c.f.
Section 5.9.10

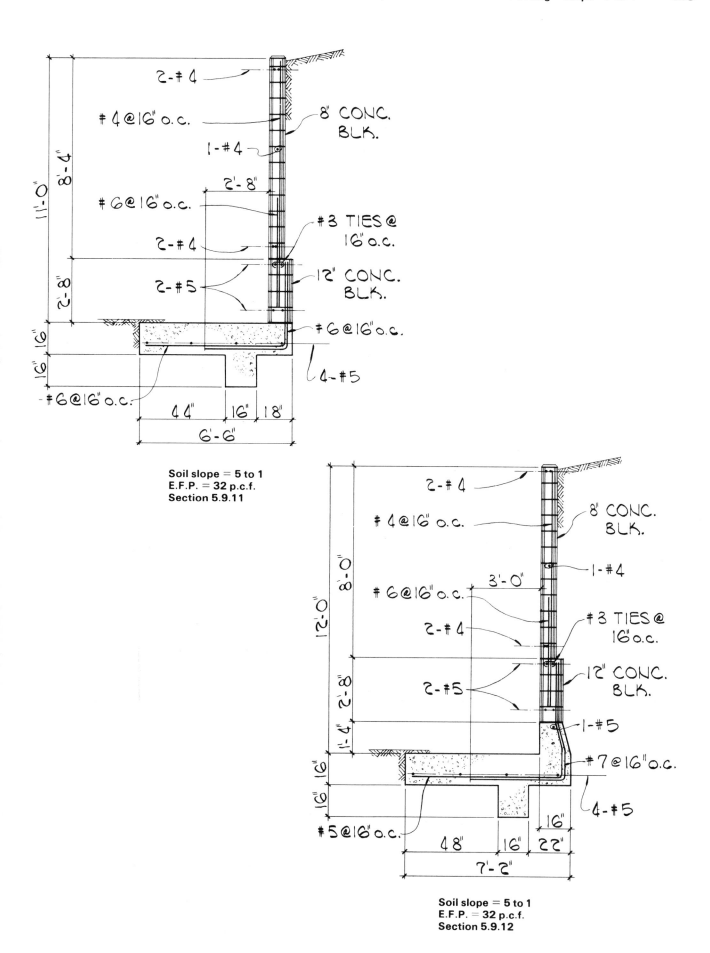

2-#4
#4@16" o.c.
1-#4
#6@16" o.c.
2-#4
2-#5

8" CONC. BLK.
#3 TIES @ 16" o.c.
12" CONC. BLK.
#6@16" o.c.
4-#5
#6@16" o.c.

8'-4"
2'-8"
11'-0"
6"
6"
2'-8"
44"   16"   18"
6'-6"

Soil slope = 5 to 1
E.F.P. = 32 p.c.f.
Section 5.9.11

2-#4
#4@16" o.c.
1-#4
#6@16" o.c.
2-#4
2-#5

8" CONC. BLK.
#3 TIES @ 16" o.c.
12" CONC. BLK.
1-#5
#7@16" o.c.
4-#5

8'-0"
12'-0"
2'-8"
1'-4"
6"
6"
3'-0"
#5@16" o.c.
48"   16"   22"
16"
7'-2"

Soil slope = 5 to 1
E.F.P. = 32 p.c.f.
Section 5.9.12

TABLE 5.10   PROPERTY LINE RETAINING WALLS — MASONRY STEM — SOIL NOT OVER FOOTING
SLOPE = 4 to 1   SURCHARGE = 0 lbs./sq. ft.   AXIAL = 0 lbs./ft.

SOIL PRESSURE DIAGRAM

H1 = Concrete Stem
H2 = 12" Concrete Block
H3 = 8" Concrete Block

Stem Height	3'-0"	4'-0"	5'-0"	6'-0"	7'-0"	8'-0"	9'-0"	10'-0"	11'-0"	12'-0"
Ft. L ft.	1.667	2.000	2.667	3.167	4.000	4.667	5.333	6.167	6.833	7.500
Ft. T ft.	0.667	0.833	0.833	0.833	1.000	1.000	1.167	1.167	1.333	1.333
A ft.	0.500	0.667	1.333	1.667	2.167	2.500	2.833	3.500	3.833	4.333
B ft.	0.667	0.833	0.833	0.833	1.000	1.000	1.167	1.167	1.333	1.333
C ft.	0.500	0.500	0.500	0.667	0.833	1.167	1.333	1.500	1.667	1.833
D ft.	0.667	0.833	0.833	0.833	1.000	1.000	1.167	1.167	1.333	1.333
8 in. blk. ft.	3.000	4.000	5.000	6.000	7.000	8.000	7.000	8.000	7.333	8.000
12 in. blk. ft.	—	—	—	—	—	—	2.000	2.000	2.667	2.667
Conc. Stem	—	—	—	—	—	—	—	—	1.000	1.333
Wb ft.	—	—	—	—	—	—	—	—	1.333	1.333
W lbs.	588	858	1096	1324	1767	2058	2632	3027	3712	4146
F lbs.	235	409	595	817	1120	1417	1809	2182	2662	3111
Mo ft.-lbs.	288	659	1158	1861	2987	4252	6130	8122	10944	13827
Mr ft.-lbs.	693	1249	2229	3270	5455	7478	10840	14539	19611	24272
O.T.R.	2.410	1.896	1.925	1.757	1.826	1.758	1.768	1.790	1.792	1.755
$\overline{X}$ ft.	0.690	0.687	0.978	1.064	1.396	1.567	1.789	2.120	2.335	2.519
e ft.	0.143	0.313	0.356	0.519	0.604	0.767	0.877	0.963	1.081	1.231
Me ft.-lbs.	84	268	390	688	1067	1578	2310	2915	4014	5105
S ft.^{3}	0.463	0.667	1.185	1.671	2.667	3.630	4.741	6.338	7.782	9.375
S.P.h p.s.f.	535	832	740	830	842	876	981	951	1059	1097
S.P.t p.s.f.	171	27	82	7	42	6	6	31	27	8
Friction	0.400	0.476	0.543	0.617	0.634	0.689	0.687	0.721	0.717	0.750
Inspct.	NO	NO	NO	NO	YES	YES	YES	YES	YES	YES

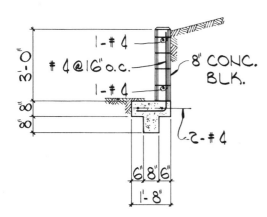

1-#4
#4@16" o.c.
1-#4
8" CONC. BLK.
3'-0"
8"
8"
2-#4
6" 8" 6"
1'-8"

Soil slope = 4 to 1
E.F.P. = 35 p.c.f.
Section 5.10.3

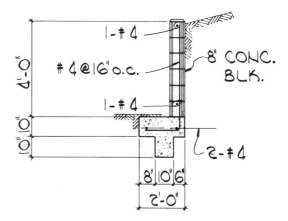

1-#4
#4@16" o.c.
1-#4
8" CONC. BLK.
4'-0"
1'-0"
2-#4
8" 10" 6"
2'-0"

Soil slope = 4 to 1
E.F.P. = 35 p.c.f.
Section 5.10.4

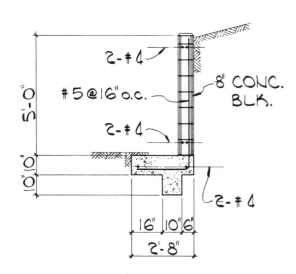

2-#4
#5@16" o.c.
2-#4
8" CONC. BLK.
5'-0"
1'-0"
2-#4
16" 10" 6"
2'-8"

Soil slope = 4 to 1
E.F.P. = 35 p.c.f.
Section 5.10.5

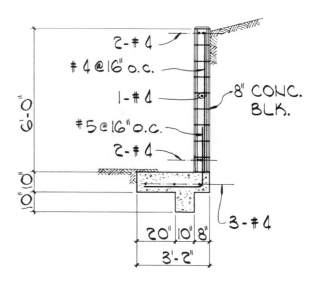

2-#4
#4@16" o.c.
1-#4
#5@16" o.c.
2-#4
8" CONC. BLK.
6'-0"
1'-0"
20" 10" 8"
3-#4
3'-2"

Soil slope = 4 to 1
E.F.P. = 35 p.c.f.
Section 5.10.6

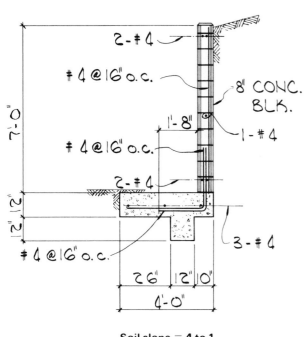

Soil slope = **4 to 1**
E.F.P. = **35** p.c.f.
Section 5.10.7

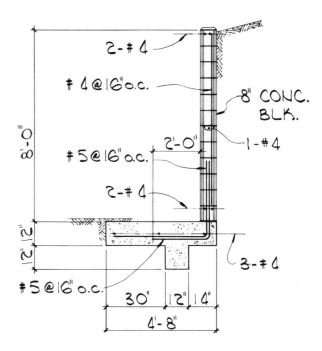

Soil slope = **4 to 1**
E.F.P. = **35** p.c.f.
Section 5.10.8

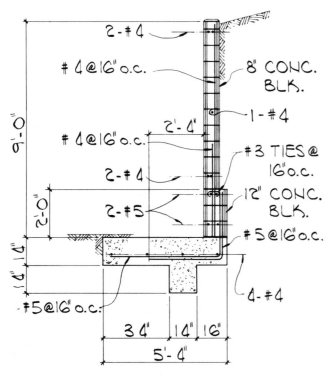

Soil slope = **4 to 1**
E.F.P. = **35** p.c.f.
Section 5.10.9

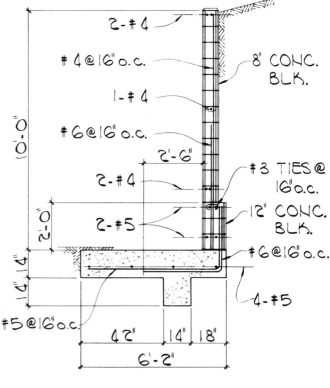

Soil slope = **4 to 1**
E.F.P. = **35** p.c.f.
Section 5.10.10

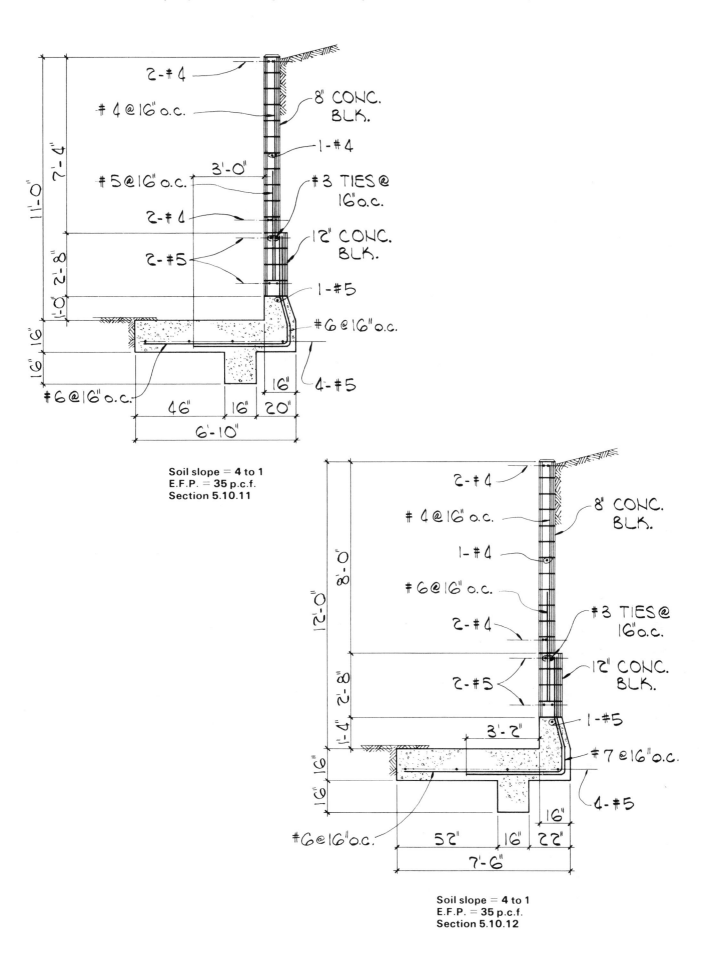

Soil slope = **4 to 1**
E.F.P. = **35** p.c.f.
Section 5.10.11

Soil slope = **4 to 1**
E.F.P. = **35** p.c.f.
Section 5.10.12

TABLE 5.11   PROPERTY LINE RETAINING WALLS — MASONRY STEM — SOIL NOT OVER FOOTING
SLOPE = 3 to 1   SURCHARGE = 0 lbs./sq. ft.   AXIAL = 0 lbs./ft.

SOIL PRESSURE DIAGRAM

H1 = Concrete Stem
H2 = 12″ Concrete Block
H3 = 8″ Concrete Block

Stem Height	3'- 0"	4'- 0"	5'- 0"	6'- 0"	7'- 0"	8'- 0"	9'- 0"	10'- 0"	11'- 0"	12'- 0"
Ft. L ft.	1.667	2.167	2.667	3.333	4.167	5.000	5.667	6.333	7.167	7.833
Ft. T ft.	0.667	0.833	0.833	0.833	1.000	1.000	1.167	1.167	1.333	1.333
A ft.	0.500	0.833	1.333	1.667	2.167	2.833	3.167	3.500	4.000	4.500
B ft.	0.667	0.833	0.833	0.833	1.000	1.000	1.167	1.167	1.333	1.333
C ft.	0.500	0.500	0.500	0.833	1.000	1.167	1.333	1.667	1.833	2.000
D ft.	0.667	0.833	0.833	0.833	1.000	1.000	1.167	1.167	1.333	1.333
8 in. blk. ft.	3.000	4.000	5.000	6.000	7.000	8.000	7.000	8.000	8.000	8.000
12 in. blk. ft.	—	—	—	—	—	—	2.000	2.000	2.000	2.000
Conc. Stem	—	—	—	—	—	—	—	—	1.000	2.000
Wb ft.	—	—	—	—	—	—	—	—	1.333	1.333
W lbs.	594	891	1113	1369	1824	2149	2742	3118	3854	4325
F lbs.	255	444	647	887	1216	1539	1964	2369	2890	3378
Mo ft.-lbs.	312	715	1257	2021	3243	4617	6655	8819	11882	15012
Mr ft.-lbs.	704	1419	2275	3553	5860	8375	12020	15407	21360	26484
O.T.R.	2.256	1.984	1.809	1.758	1.807	1.814	1.806	1.747	1.798	1.764
$\overline{X}$ ft.	0.660	0.790	0.914	1.120	1.434	1.749	1.956	2.113	2.459	2.652
e ft.	0.174	0.293	0.419	0.547	0.649	0.751	0.877	1.054	1.124	1.264
Me ft.-lbs.	103	261	466	749	1184	1615	2405	3286	4333	5469
S ft.^{3}	0.463	0.782	1.185	1.852	2.894	4.167	5.352	6.685	8.560	10.227
S.P.h p.s.f.	580	745	811	815	847	817	933	984	1044	1087
S.P.t p.s.f.	134	77	24	6	29	42	35	1	32	17
Friction	0.430	0.498	0.581	0.648	0.667	0.716	0.716	0.760	0.750	0.781
Inspct.	NO	NO	NO	NO	YES	YES	YES	YES	YES	YES

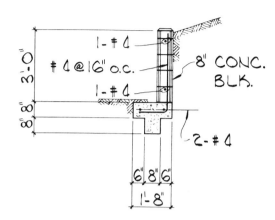

Soil slope = 3 to 1
E.F.P. = 38 p.c.f.
Section 5.11.3

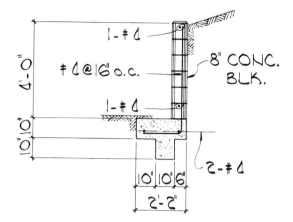

Soil slope = 3 to 1
E.F.P. = 38 p.c.f.
Section 5.11.4

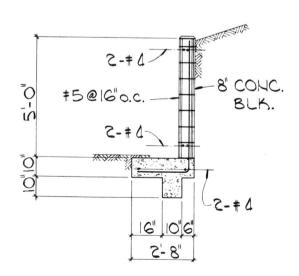

Soil slope = 3 to 1
E.F.P. = 38 p.c.f.
Section 5.11.5

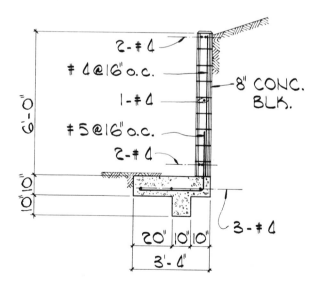

Soil slope = 3 to 1
E.F.P. = 38 p.c.f.
Section 5.11.6

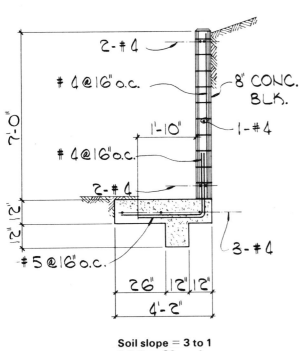

2-#4

#4@16"o.c.

8" CONC. BLK.

1'-10"

1-#4

7'-0"

#4@16"o.c.

2-#4

2'

12"

#5@16"o.c.

3-#4

26"   12"   12"

4'-2"

Soil slope = 3 to 1
E.F.P. = 38 p.c.f.
Section 5.11.7

2-#4

#4@16"o.c.

1-#4

8" CONC. BLK.

8'-0"

2'-2"

#5@16"o.c.

2-#4

2'

#5@16"o.c.

4-#4

30"   12"   14"

5'-0"

Soil slope = 3 to 1
E.F.P. = 38 p.c.f.
Section 5.11.8

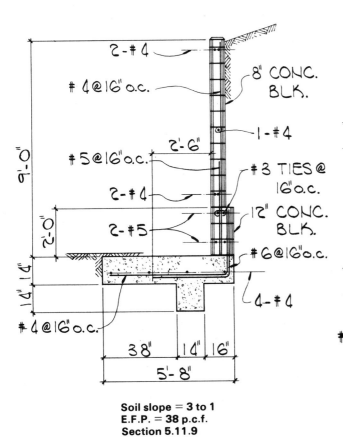

2-#4

#4@16"o.c.

8" CONC. BLK.

1-#4

2'-6"

9'-0"

#5@16"o.c.

#3 TIES @ 16"o.c.

2-#4

2-#5

12" CONC. BLK.

2'-0"

#6@16"o.c.

1'-4"

4-#4

#4@16"o.c.

38"   14"   16"

5'-8"

Soil slope = 3 to 1
E.F.P. = 38 p.c.f.
Section 5.11.9

2-#4

#4@16"o.c.

8" CONC. BLK.

1-#4

2'-8"

10'-0"

#6@16"o.c.

#3 TIES @ 16"o.c.

2-#4

2-#5

12" CONC. BLK.

2'-0"

#6@16"o.c.

1'-4"

4-#5

#6@16"o.c.

42"   14"   20"

6'-4"

Soil slope = 3 to 1
E.F.P. = 38 p.c.f.
Section 5.11.10

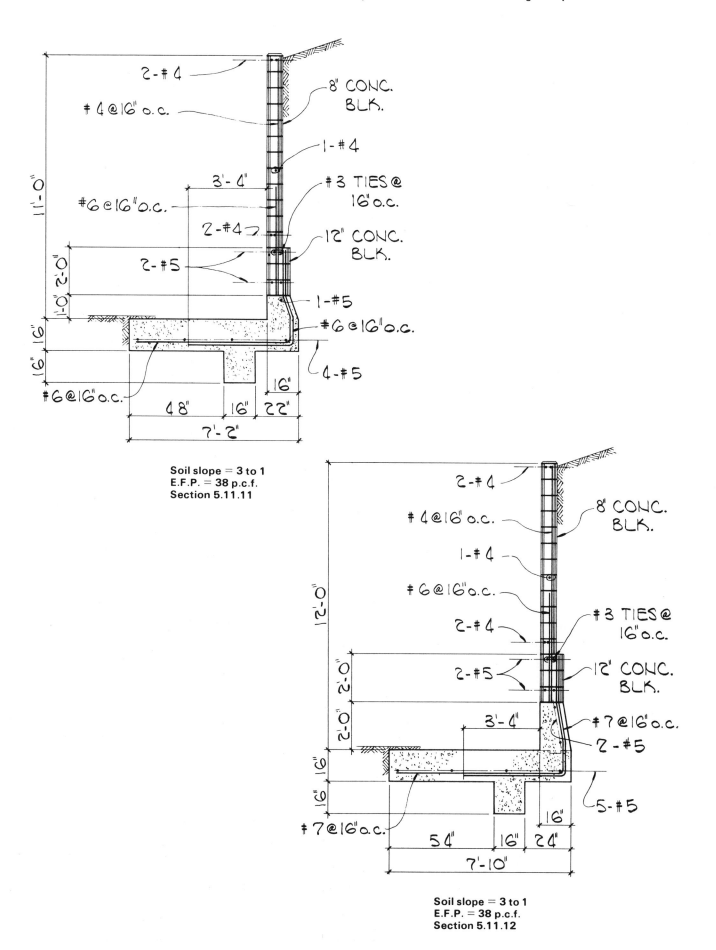

2-#4

#4 @ 16" o.c.

8" CONC. BLK.

1-#4

3'-4"

#3 TIES @ 16" o.c.

#6 @ 16" o.c.

2-#4

12" CONC. BLK.

2-#5

1-#5

#6 @ 16" o.c.

4-#5

#6 @ 16" o.c.

1'-0"   2'-0"   1'-0"   16"   16"

48"   16"   22"   16"

7'-2"

Soil slope = 3 to 1
E.F.P. = 38 p.c.f.
Section 5.11.11

2-#4

#4 @ 16" o.c.

8" CONC. BLK.

1-#4

#6 @ 16" o.c.

2-#4

#3 TIES @ 16" o.c.

2-#5

12" CONC. BLK.

3'-4"

#7 @ 16" o.c.

2-#5

5-#5

#7 @ 16" o.c.

12'-0"   2'-0"   2'-0"   16"   16"

54"   16"   24"   16"

7'-10"

Soil slope = 3 to 1
E.F.P. = 38 p.c.f.
Section 5.11.12

TABLE 5.12   PROPERTY LINE RETAINING WALLS – MASONRY STEM – SOIL NOT OVER FOOTING
SLOPE = 2 to 1   SURCHARGE = 0 lbs./sq. ft.   AXIAL = 0 lbs./ft.

SOIL PRESSURE DIAGRAM

H1 = Concrete Stem
H2 = 12″ Concrete Block
H3 = 8″ Concrete Block

Stem Height	3'- 0"	4'- 0"	5'- 0"	6'- 0"	7'- 0"	8'- 0"	9'- 0"	10'- 0"	11'- 0"	12'- 0"
Ft. L ft.	1.667	2.333	2.833	3.667	4.500	5.167	6.000	6.833	7.500	8.333
Ft. T ft.	0.667	0.833	0.833	0.833	1.000	1.167	1.167	1.333	1.333	1.333
A ft.	0.500	1.000	1.333	2.000	2.500	2.833	3.333	3.833	4.333	4.833
B ft.	0.667	0.833	0.833	0.833	1.000	1.167	1.167	1.333	1.333	1.333
C ft.	0.500	0.500	0.667	0.833	1.000	1.167	1.500	1.667	1.833	2.167
D ft.	0.667	0.833	0.833	0.833	1.000	1.167	1.167	1.333	1.333	1.333
8 in. blk. ft.	3.000	4.000	5.000	6.000	7.000	6.000	7.000	7.333	7.333	7.333
12 in. blk. ft.	—	—	—	—	—	2.000	2.000	2.667	2.667	2.667
Conc. Stem	—	—	—	—	—	—	—	—	1.000	2.000
Wb ft.	—	—	—	—	—	—	—	—	1.333	1.333
W lbs.	606	931	1162	1449	1928	2479	2887	3602	4080	4605
F lbs.	289	502	732	1004	1376	1807	2222	2762	3270	3822
Mo ft.-lbs.	353	809	1423	2287	3669	5520	7531	10433	13445	16988
Mr ft.-lbs.	723	1614	2525	4159	6716	9846	13427	18925	23822	30102
O.T.R.	2.047	1.995	1.775	1.819	1.830	1.784	1.783	1.814	1.772	1.772
$\overline{X}$ ft.	0.610	0.865	0.949	1.292	1.581	1.745	2.042	2.358	2.544	2.848
e ft.	0.223	0.302	0.468	0.541	0.669	0.838	0.958	1.059	1.206	1.319
Me ft.-lbs.	135	281	544	784	1290	2077	2765	3814	4922	6074
S ft.^{3}	0.463	0.907	1.338	2.241	3.375	4.449	6.000	7.782	9.375	11.574
S.P.h p.s.f.	655	709	817	745	811	947	942	1017	1069	1077
S.P.t p.s.f.	72	89	4	45	46	13	20	37	19	28
Friction	0.477	0.539	0.629	0.693	0.714	0.729	0.770	0.767	0.802	0.830
Inspct.	NO	NO	NO	NO	YES	YES	YES	YES	YES	YES

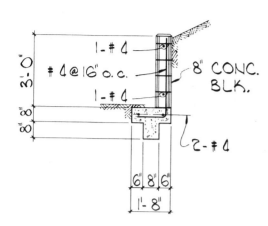

1-#4
#4@16" o.c.
1-#4
8" CONC. BLK.
2-#4
6" 8" 6"
1'-8"
3'-0"
8" 8"

Soil slope = 2 to 1
E.F.P. = 43 p.c.f.
Section 5.12.3

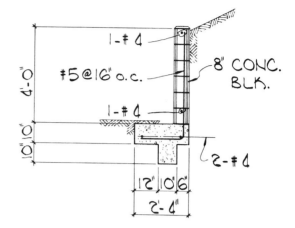

1-#4
#5@16" o.c.
1-#4
8" CONC. BLK.
2-#4
12" 10" 6"
2'-4"
4'-0"
1'-0"

Soil slope = 2 to 1
E.F.P. = 43 p.c.f.
Section 5.12.4

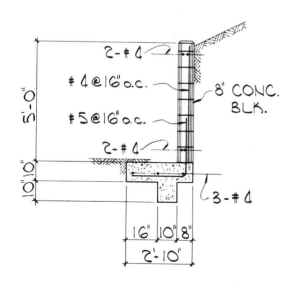

2-#4
#4@16" o.c.
#5@16" o.c.
2-#4
8" CONC. BLK.
3-#4
16" 10" 8"
2'-10"
5'-0"
1'-0"

Soil slope = 2 to 1
E.F.P. = 43 p.c.f.
Section 5.12.5

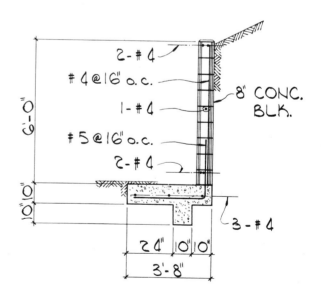

2-#4
#4@16" o.c.
1-#4
#5@16" o.c.
2-#4
8" CONC. BLK.
3-#4
24" 10" 10"
3'-8"
6'-0"
1'-0"

Soil slope = 2 to 1
E.F.P. = 43 p.c.f.
Section 5.12.6

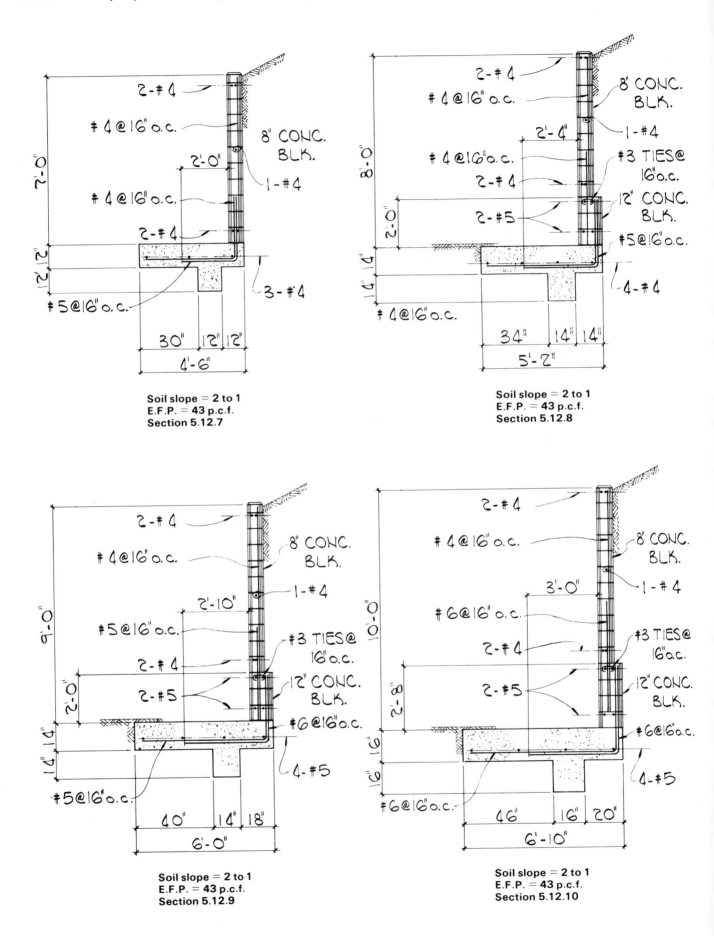

Soil slope = 2 to 1
E.F.P. = 43 p.c.f.
Section 5.12.7

Soil slope = 2 to 1
E.F.P. = 43 p.c.f.
Section 5.12.8

Soil slope = 2 to 1
E.F.P. = 43 p.c.f.
Section 5.12.9

Soil slope = 2 to 1
E.F.P. = 43 p.c.f.
Section 5.12.10

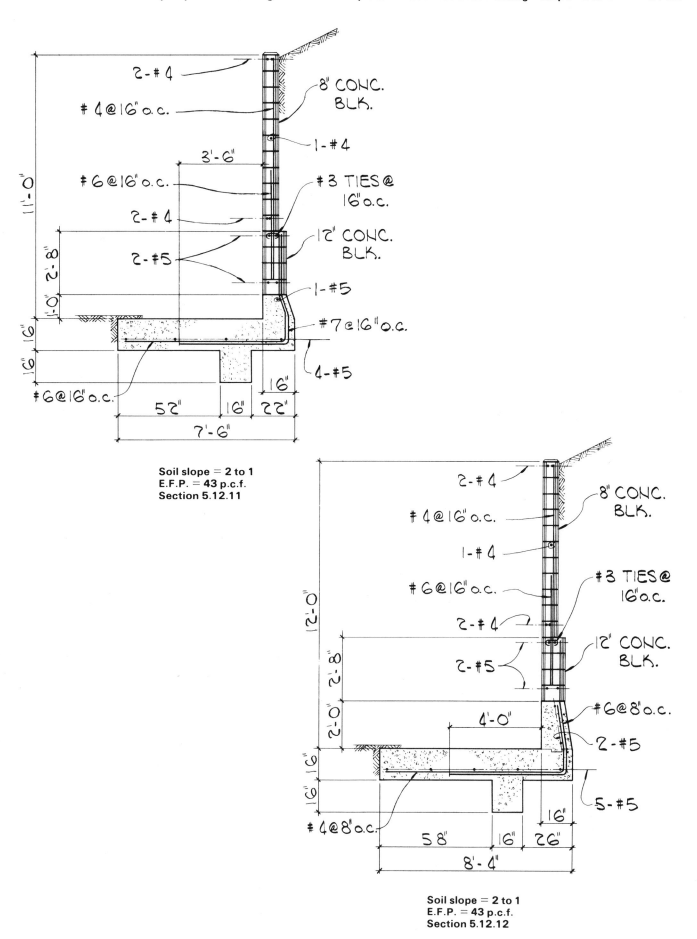

Soil slope = 2 to 1
E.F.P. = 43 p.c.f.
Section 5.12.11

Soil slope = 2 to 1
E.F.P. = 43 p.c.f.
Section 5.12.12

TABLE 5.13   PROPERTY LINE RETAINING WALLS – MASONRY STEM – SOIL NOT OVER FOOTING
SLOPE = 1.5 to 1   SURCHARGE = 0 lbs./sq. ft.   AXIAL = 0 lbs./ft.

SOIL PRESSURE DIAGRAM

H1 = Concrete Stem
H2 = 12″ Concrete Block
H3 = 8″ Concrete Block

Stem Height	3′- 0″	4′- 0″	5′- 0″	6′- 0″	7′- 0″	8′- 0″	9′- 0″	10′- 0″	11′- 0″	12′- 0″
Ft. L ft.	1.833	2.500	3.333	4.167	5.000	5.833	6.667	7.500	8.333	9.167
Ft. T ft.	0.667	0.833	0.833	1.000	1.000	1.167	1.333	1.333	1.333	1.500
A ft.	0.667	1.167	1.667	2.167	2.833	3.167	3.667	4.333	4.833	5.333
B ft.	0.667	0.833	0.833	1.000	1.000	1.167	1.333	1.333	1.333	1.500
C ft.	0.500	0.500	0.833	1.000	1.167	1.500	1.667	1.833	2.167	2.333
D ft.	0.667	0.833	0.833	1.000	1.000	1.167	1.333	1.333	1.333	1.500
8 in. blk. ft.	3.000	4.000	5.000	6.000	7.000	6.000	7.000	6.667	6.667	6.667
12 in. blk. ft.	–	–	–	–	–	2.000	2.000	2.667	2.667	2.667
Conc. Stem	–	–	–	–	–	–	–	0.667	1.667	2.667
Wb ft.	–	–	–	–	–	–	–	1.333	1.333	1.500
W lbs.	649	999	1293	1776	2131	2795	3503	4047	4606	5557
F lbs.	370	642	936	1347	1760	2311	2936	3532	4183	5012
Mo ft.-lbs.	452	1035	1820	3144	4693	7061	10114	13344	17197	22553
Mr ft.-lbs.	875	1888	3331	5690	8314	12614	17943	23629	30157	39801
O.T.R.	1.936	1.824	1.831	1.810	1.771	1.787	1.774	1.771	1.754	1.765
$\overline{X}$ ft.	0.651	0.854	1.169	1.433	1.699	1.987	2.235	2.541	2.814	3.104
e ft.	0.265	0.396	0.497	0.650	0.801	0.930	1.098	1.209	1.353	1.480
Me ft.-lbs.	172	395	643	1155	1706	2600	3847	4893	6231	8223
S ft.^{3}	0.560	1.042	1.852	2.894	4.167	5.671	7.407	9.375	11.574	14.005
S.P.h p.s.f.	662	779	735	825	836	938	1045	1062	1091	1193
S.P.t p.s.f.	46	20	41	27	17	21	6	18	14	19
Friction	0.569	0.643	0.724	0.759	0.826	0.827	0.838	0.873	0.908	0.902
Inspct.	NO	NO	NO	YES	YES	YES	YES	YES	YES	YES

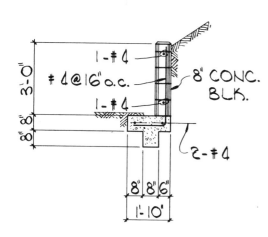

Soil slope = 1 ½ to 1
E.F.P. = 55 p.c.f.
Section 5.13.3

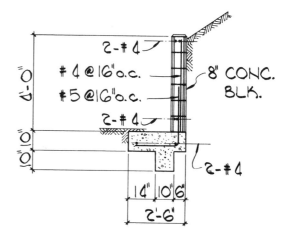

Soil slope = 1 ½ to 1
E.F.P. = 55 p.c.f.
Section 5.13.4

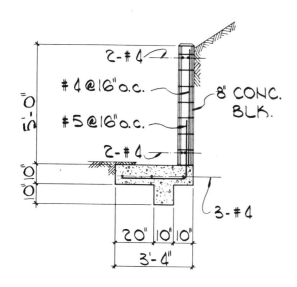

Soil slope = 1 ½ to 1
E.F.P. = 55 p.c.f.
Section 5.13.5

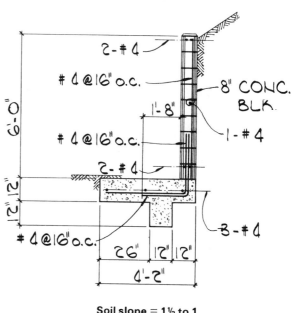

Soil slope = 1 ½ to 1
E.F.P. = 55 p.c.f.
Section 5.13.6

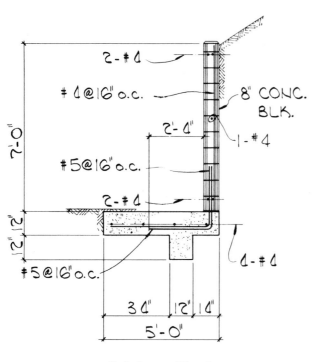

Soil slope = 1½ to 1
E.F.P. = 55 p.c.f.
Section 5.13.7

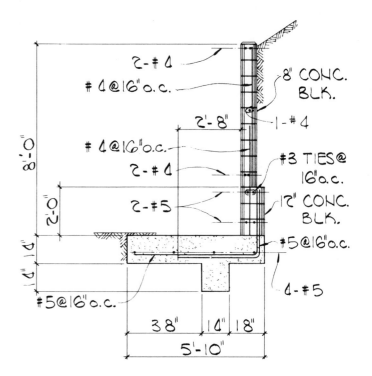

Soil slope = 1½ to 1
E.F.P. = 55 p.c.f.
Section 5.13.8

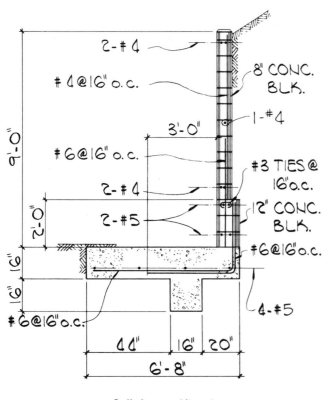

Soil slope = 1½ to 1
E.F.P. = 55 p.c.f.
Section 5.13.9

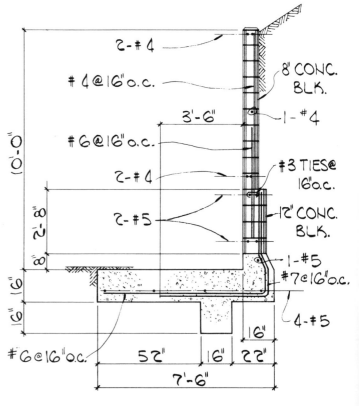

Soil slope = 1½ to 1
E.F.P. = 55 p.c.f.
Section 5.13.10

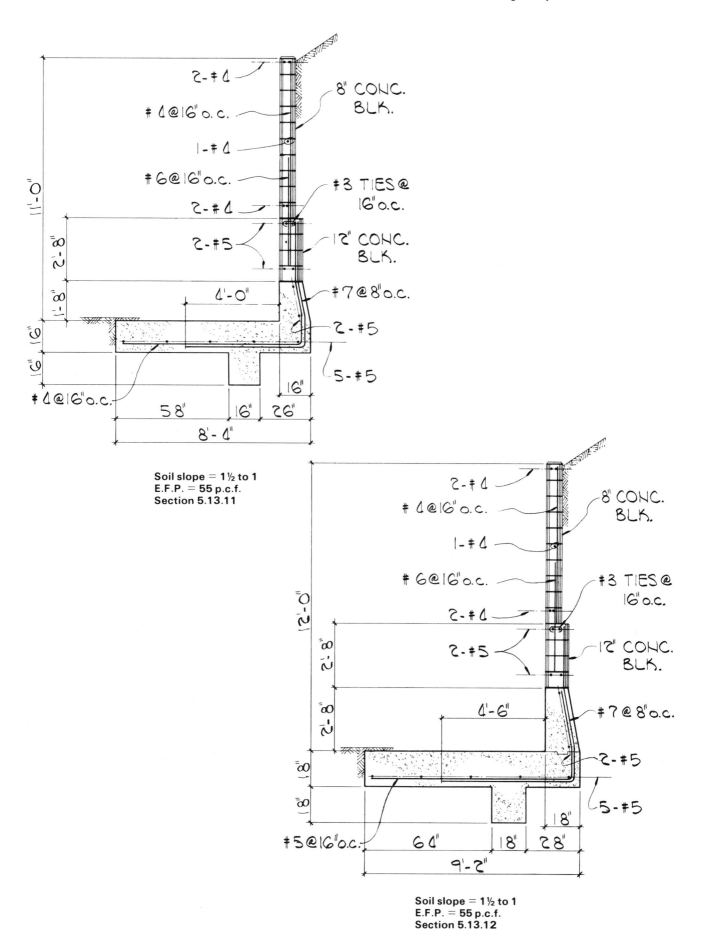

Soil slope = 1½ to 1
E.F.P. = 55 p.c.f.
Section 5.13.11

Soil slope = 1½ to 1
E.F.P. = 55 p.c.f.
Section 5.13.12

TABLE 5.14   PROPERTY LINE RETAINING WALLS – MASONRY STEM – SOIL NOT OVER FOOTING
SLOPE = 1 to 1   SURCHARGE = 0 lbs./sq. ft.   AXIAL = 0 lbs./ft.

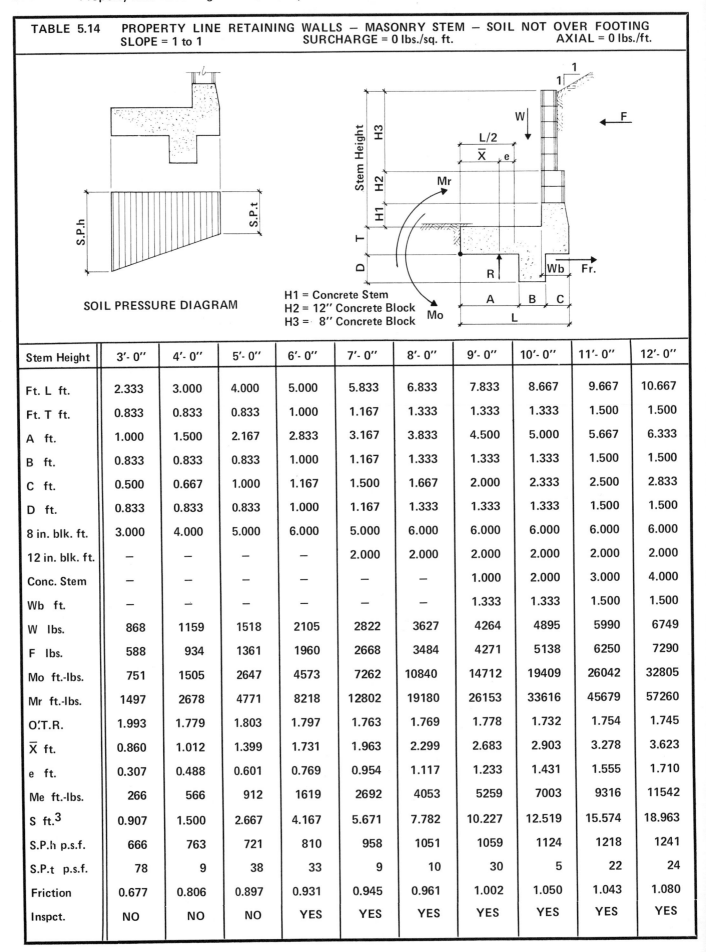

SOIL PRESSURE DIAGRAM

H1 = Concrete Stem
H2 = 12″ Concrete Block
H3 = 8″ Concrete Block

Stem Height	3'- 0"	4'- 0"	5'- 0"	6'- 0"	7'- 0"	8'- 0"	9'- 0"	10'- 0"	11'- 0"	12'- 0"
Ft. L ft.	2.333	3.000	4.000	5.000	5.833	6.833	7.833	8.667	9.667	10.667
Ft. T ft.	0.833	0.833	0.833	1.000	1.167	1.333	1.333	1.333	1.500	1.500
A ft.	1.000	1.500	2.167	2.833	3.167	3.833	4.500	5.000	5.667	6.333
B ft.	0.833	0.833	0.833	1.000	1.167	1.333	1.333	1.333	1.500	1.500
C ft.	0.500	0.667	1.000	1.167	1.500	1.667	2.000	2.333	2.500	2.833
D ft.	0.833	0.833	0.833	1.000	1.167	1.333	1.333	1.333	1.500	1.500
8 in. blk. ft.	3.000	4.000	5.000	6.000	5.000	6.000	6.000	6.000	6.000	6.000
12 in. blk. ft.	—	—	—	—	2.000	2.000	2.000	2.000	2.000	2.000
Conc. Stem	—	—	—	—	—	—	1.000	2.000	3.000	4.000
Wb ft.	—	—	—	—	—	—	1.333	1.333	1.500	1.500
W lbs.	868	1159	1518	2105	2822	3627	4264	4895	5990	6749
F lbs.	588	934	1361	1960	2668	3484	4271	5138	6250	7290
Mo ft.-lbs.	751	1505	2647	4573	7262	10840	14712	19409	26042	32805
Mr ft.-lbs.	1497	2678	4771	8218	12802	19180	26153	33616	45679	57260
O.T.R.	1.993	1.779	1.803	1.797	1.763	1.769	1.778	1.732	1.754	1.745
$\overline{X}$ ft.	0.860	1.012	1.399	1.731	1.963	2.299	2.683	2.903	3.278	3.623
e ft.	0.307	0.488	0.601	0.769	0.954	1.117	1.233	1.431	1.555	1.710
Me ft.-lbs.	266	566	912	1619	2692	4053	5259	7003	9316	11542
S ft.^{3}	0.907	1.500	2.667	4.167	5.671	7.782	10.227	12.519	15.574	18.963
S.P.h p.s.f.	666	763	721	810	958	1051	1059	1124	1218	1241
S.P.t p.s.f.	78	9	38	33	9	10	30	5	22	24
Friction	0.677	0.806	0.897	0.931	0.945	0.961	1.002	1.050	1.043	1.080
Inspct.	NO	NO	NO	YES	YES	YES	YES	YES	YES	YES

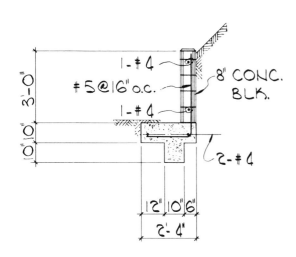

Soil slope = 1 to 1
E.F.P. = 80 p.c.f.
Section 5.14.3

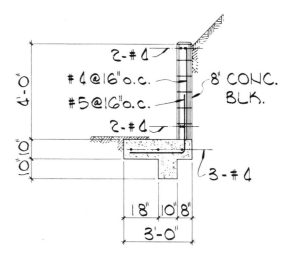

Soil slope = 1 to 1
E.F.P. = 80 p.c.f.
Section 5.14.4

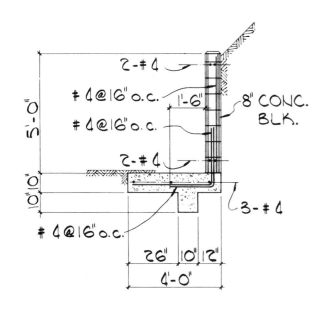

Soil slope = 1 to 1
E.F.P. = 80 p.c.f.
Section 5.14.5

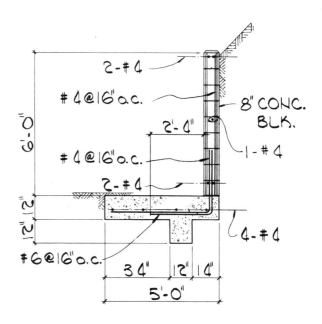

Soil slope = 1 to 1
E.F.P. = 80 p.c.f.
Section 5.14.6

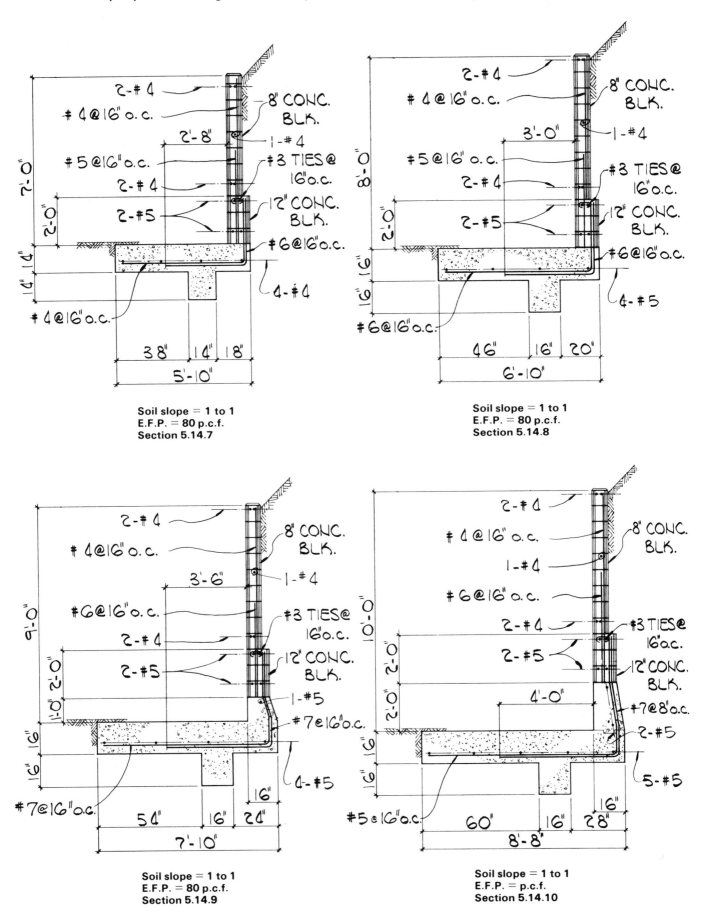

Soil slope = 1 to 1
E.F.P. = 80 p.c.f.
Section 5.14.7

Soil slope = 1 to 1
E.F.P. = 80 p.c.f.
Section 5.14.8

Soil slope = 1 to 1
E.F.P. = 80 p.c.f.
Section 5.14.9

Soil slope = 1 to 1
E.F.P. = p.c.f.
Section 5.14.10

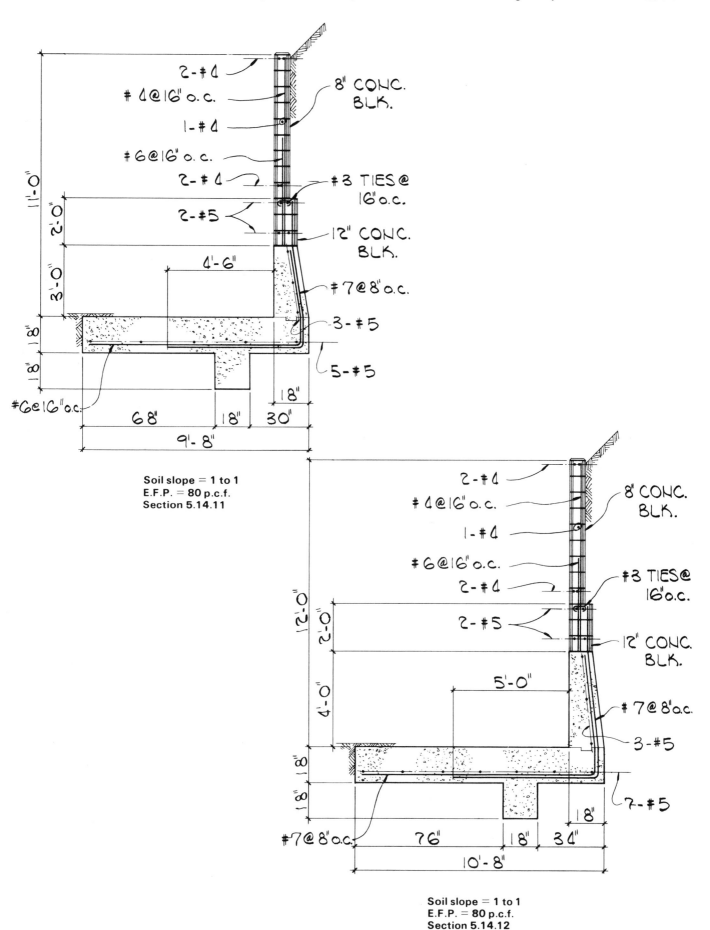

Section 5.14.11

2-#4
#4@16" o.c.
1-#4
#6@16" o.c.
2-#4
2-#5
8" CONC. BLK.
#3 TIES@ 16" o.c.
12" CONC. BLK.
#7@8" o.c.
3-#5
5-#5
#6@16" o.c.

11'-0"
2'-0"
3'-0"
8"
8"
4'-6"
68"   18"   30"
18"
9'-8"

Soil slope = 1 to 1
E.F.P. = 80 p.c.f.
Section 5.14.11

Section 5.14.12

2-#4
#4@16" o.c.
1-#4
#6@16" o.c.
2-#4
2-#5
8" CONC. BLK.
#3 TIES@ 16" o.c.
12" CONC. BLK.
#7@8" o.c.
3-#5
7-#5
#7@8" o.c.

12'-0"
2'-0"
4'-0"
8"
8"
5'-0"
76"   18"   34"
18"
10'-8"

Soil slope = 1 to 1
E.F.P. = 80 p.c.f.
Section 5.14.12

CHAPTER **6**

Undercut Footing Retaining Walls
Concrete Stem
Soil at the Toe Side of the Footing

The design data and drawings presented in this chapter are concerned with property line concrete retaining walls. The retained soil is not placed over the entire wall footing, but only at the toe side of the stem wall. The chapter consists of three design examples of retaining walls with various types of loading, a design data table for each of the fourteen loading conditions specified in Fig. 1.6, and a series of corresponding drawings following each design data table. The retaining wall drawings and the design data in this chapter are for a wall length of 1 foot.

The retaining walls in this chapter are designed using the following criteria:

Weight of soil = 100 p.c.f.   Weight of concrete = 150 p.c.f.
Concrete $f_c' = 2000$ p.s.i.   Reinforcing steel $f_s' = 20,000$ p.s.i.
Maximum allowable soil pressure = 4000 p.s.f.
Minimum allowable soil pressure = 0 p.s.f.
$\overline{X} \geqslant L/3$ Minimum O.T.R. = 1.50
Coefficient of soil friction = 0.40
Passive soil pressure = 300 p.c.f.

See Table C5 for reinforcement protection. Weep holes are not shown on the drawings.

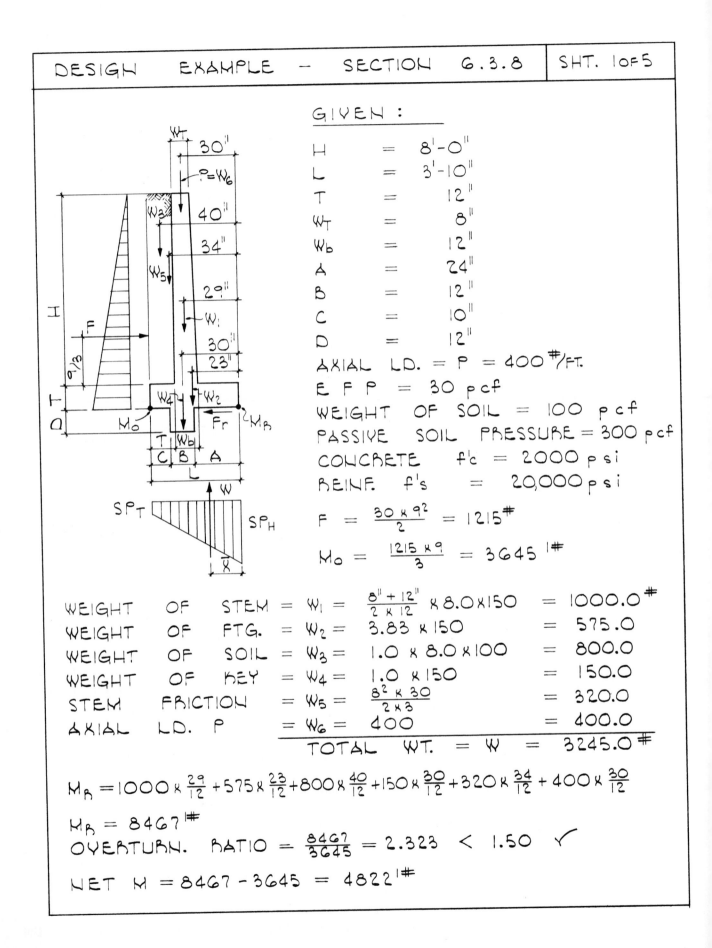

| DESIGN EXAMPLE — SECTION 6.3.8 | SHT. 1 of 5 |

GIVEN :

$H = 8'-0''$
$L = 3'-10''$
$T = 12''$
$W_T = 8''$
$W_b = 12''$
$A = 24''$
$B = 12''$
$C = 10''$
$D = 12''$

AXIAL LD. $= P = 400^\#/FT.$
E F P = 30 pcf
WEIGHT OF SOIL = 100 pcf
PASSIVE SOIL PRESSURE = 300 pcf
CONCRETE f'_c = 2000 psi
REINF. f'_s = 20,000 psi

$F = \frac{30 \times 9^2}{2} = 1215^\#$

$M_o = \frac{1215 \times 9}{3} = 3645^{'\#}$

WEIGHT OF STEM $= W_1 = \frac{8'' + 12''}{2 \times 12} \times 8.0 \times 150 = 1000.0^\#$
WEIGHT OF FTG. $= W_2 = 3.83 \times 150 = 575.0$
WEIGHT OF SOIL $= W_3 = 1.0 \times 8.0 \times 100 = 800.0$
WEIGHT OF KEY $= W_4 = 1.0 \times 150 = 150.0$
STEM FRICTION $= W_5 = \frac{8^2 \times 30}{2 \times 3} = 320.0$
AXIAL LD. P $= W_6 = 400 = 400.0$

TOTAL WT. $= W = 3245.0^\#$

$M_R = 1000 \times \frac{29}{12} + 575 \times \frac{23}{12} + 800 \times \frac{40}{12} + 150 \times \frac{30}{12} + 320 \times \frac{34}{12} + 400 \times \frac{30}{12}$

$M_R = 8467^{'\#}$
OVERTURN. RATIO $= \frac{8467}{3645} = 2.323 < 1.50$ ✓

NET M $= 8467 - 3645 = 4822^{'\#}$

316

DESIGN EXAMPLE - SECTION 6.3.8	SHT. 2 of 5

FOOTING   DESIGN :

$W = 3245^{\#}$;   NET $M = 4822^{'\#}$

$\bar{X} = \dfrac{4822}{3245} = 1.486'$

FTG. MIDDLE $\frac{1}{3} = \dfrac{3.83}{.3} = 1.277 < 1.486 \ \checkmark$

$e = \dfrac{3.83}{2} - 1.486 = 0.431'$

$M_e = 3245 \times .431 = 1398^{'\#}$

FTG. AREA $= 3.83$ SQ. FT.

FTG. SECT. MODULUS $= \dfrac{1 \times 3.83^2}{6} = 2.45^{'3}$

SOIL PRESSURE $= \dfrac{3245}{3.83} \pm \dfrac{1398}{2.45} = 846.5 \pm 570.5$

$\underline{SP_H = 1417 \ psf}$ $\qquad\qquad \underline{SP_T = 276 \ psf}$

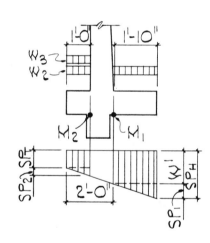

$W' = \dfrac{1417 - 276}{3.83} \times 2.0 + 276 = 871.4 \ psf$

$SP_1 = 1417 - 871.4 = 545.6 \ psf$

UNIF. LD. $= 871.4 - 150 = 721.4 \ ^{\#}/_{FT.}$

$+M_1 = \dfrac{545.6 \times 1.83^2 \times 2}{2 \times 3} + \dfrac{721.4 \times 1.83^2}{2}$

$+M_1 = 1823.6^{'\#}$

$V_1 = \dfrac{545.6 \times 1.83}{2} + 721.4 \times 1.83 = 1819.4^{\#}$

$SP_2 = \dfrac{1417 - 276}{3.83} \times 1.0 = 297.7 \ psf$

UNIF. LD. $= 276 - 150 - 800 = -674$

$-M_1 = -\dfrac{674 \times 1^2}{2} + \dfrac{297.7 \times 1^2}{2 \times 3} = -287^{'\#}$

d FOR $+A_s = 12 - (3 + .5 \times .62) = 8.70''$

d FOR $-A_s = 12 - (2 + .5 \times .5) = 9.70''$

$a = 1.13$; $j = 0.89$

DESIGN EXAMPLE — SECTION 6.3.8	SHT. 3 OF 5

FOOTING   DESIGN (CONT.) :

$+M = 1823.6\ ^{!\#}$; $-M = 287\ ^{!\#}$; $V_{MAX} = 1819.4$

$+A_s = \dfrac{1.819}{1.13 \times 8.70} = 0.185$ SQ. IN./FT.

$+A_s$ USE   #5 @ 16" O.C.

TENSION   IN   FTG.   AT $-M_2$

FTG. CROSS SECTION MODULUS $= \dfrac{12 \times 9.70^2}{6} = 188.2^{!!3}$

$f_T = \dfrac{12 \times 287}{188.2} = 18.3$ psi < 71 psi ✓

NO NEGATIVE REINF. REQ'D.

$\nu_{MAX} = \dfrac{1819.4}{12 \times .89 \times 8.7} = 19.6$ psi < 49 psi ✓

LONG. REINF. $= 12 \times 3.83 \times 12 \times .0015 = 0.83$ SQ. IN.

USE   3 - #4's

STEM   DESIGN   AT   TOP   OF   FTG.

$V = \dfrac{30 \times 8^2}{2} = 960\ ^{\#}$; $M = \dfrac{960 \times 8}{3} = 2560\ ^{!\#}$

$d = 12 - (2 + .5 \times .60) = 9.70"$; $a = 1.13$; $j = 0.89$

$A_s = \dfrac{2.56}{1.13 \times 9.70} = 0.233$ SQ. IN./FT.   USE   <u>#5 @ 16" O.C.</u>

$\nu = \dfrac{960}{12 \times .89 \times 9.70} = 9.3$ psi < 49 psi ✓

COMBINED STRESS AT BASE OF STEM :
AXIAL LD. $= 400 + 1000 + 320 = 1720\ ^{\#}$

$f_a = \dfrac{1720}{12 \times 12} = 11.9$ psi

$f_b = \dfrac{12 \times 2560 \times 6}{12 \times 9.7^2} = 163.2$ psi

COMBINED STRESS $= \dfrac{163.2}{900} + \dfrac{11.9}{750} = 0.197 < 1.0$ ✓

DESIGN EXAMPLE – SECTION 6.3.8	SHT. 4 oF 5

STEM DESIGN AT 4'-0" ABOVE TOP OF FTG.

$H = 8.0 - 4.0 = 4.0'$

$V = \dfrac{30 \times 4.0^2}{8} = 240^{\#}$; $M = \dfrac{240 \times 4}{3} = 320'^{\#}$

$d = 12 - \dfrac{12-8}{2} - (2 + .5 \times .62) = 7.70''$; $a = 1.13$; $j = 0.89$

$A_s = \dfrac{.32}{1.13 \times 7.70} = 0.036 \text{ sq.in./ft.}$ USE #4 @ 18" o.c.

$\nu = \dfrac{240}{12 \times .89 \times 7.70} = 2.9 \text{ psi} < 49 \text{ psi}$ ✓

LONG. REINF. = $0.0012 \times 8 \times 12 \times 10 = 1.15$ sq.in.

USE #5 @ 16" o.c.

SLIDING :

$Fr = \dfrac{1215}{3245} = 0.37 < 0.40$ ✓

CHECK SLIDING RESISTANCE WITHOUT AXIAL LD.

$W = 3245 - 400 = 2845^{\#}$

$Fr = \dfrac{1215}{2845} = 0.427 > 0.40$ ∴ KEY REQ'D.

PASSIVE SOIL PRESSURE = 300 pcf

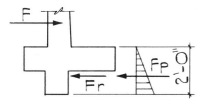

RESISTANCE TO SLIDING = $2845 \times .40 + \dfrac{300 \times 2^2}{2} = 1738^{\#} > 1215^{\#}$ ✓

DESIGN EXAMPLE — SECTION 6.3.8	SHT. 5 OF 5

CHECK WALL WITHOUT THE $400^{\#}$/FT. AXIAL LD.

TOTAL   WT. $= W = 3245 - 400 = 2845^{\#}$

$M_R = 8467 - 400 \times \dfrac{30}{12} = 7467^{'\#}$

OVERTURN.   RATIO $= \dfrac{7467}{3645} = 2.049 > 1.50$ ✓

NET M $= 7467 - 3645 = 3822^{'\#}$

$\bar{X} = \dfrac{3822}{2845} = 1.343'$;   FTG.   MIDDLE $\frac{1}{3} = 1.277$ ✓

$e = \dfrac{3.83}{2} - 1.343 = 0.573$

$M_e = 2845 \times .573 = 1631^{'\#}$

SOIL   PRESSURE $= \dfrac{2845}{3.83} \pm \dfrac{1631}{2.45} = 742 \pm 666$

$SP_H = 1408$ psf $SP_T = 76$ psf

CHECK FTG. MAX. M — (SEE CALC. SHT. No. 2)

$W' = \dfrac{1408 - 76}{3.83} \times 2.0 + 76 = 771$ psf

$SP_1 = 1408 - 771 = 637$ psf

UNIF. LD. $= 771 - 150 = 621^{\#}$/FT.

$+ M_1 = \dfrac{637 \times 1.83^2 \times 2}{2 \times 3} + \dfrac{621 \times 1.83^2}{2} = 1757.3^{'\#} < 1823.6^{'\#}$

STEM   WITHOUT AXIAL LD. IS O.K.

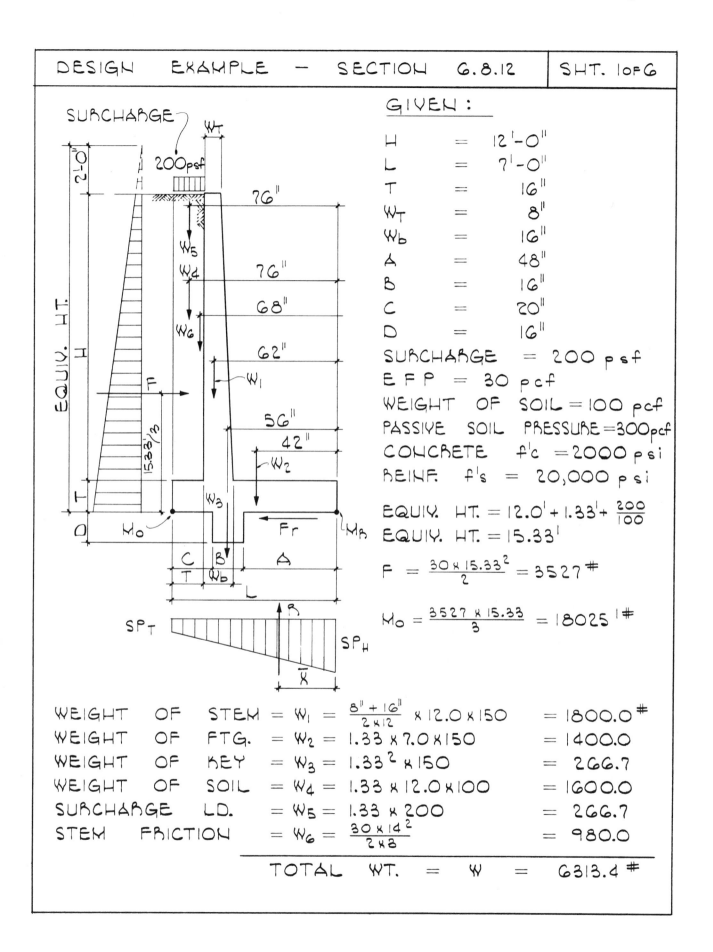

DESIGN EXAMPLE — SECTION 6.8.12 | SHT. 1 of 6

GIVEN :

H = 12'-0"
L = 7'-0"
T = 16"
W_T = 8"
W_b = 16"
A = 48"
B = 16"
C = 20"
D = 16"

SURCHARGE = 200 psf
E F P = 30 pcf
WEIGHT OF SOIL = 100 pcf
PASSIVE SOIL PRESSURE = 300 pcf
CONCRETE f'_c = 2000 psi
REINF. f'_s = 20,000 psi

EQUIV. HT. = $12.0' + 1.33' + \dfrac{200}{100}$
EQUIV. HT. = 15.33'

$F = \dfrac{30 \times 15.33^2}{2} = 3527^{\#}$

$M_o = \dfrac{3527 \times 15.33}{3} = 18025^{!\#}$

WEIGHT OF STEM = W_1 = $\dfrac{8'' + 16''}{2 \times 12} \times 12.0 \times 150$ = 1800.0 $^{\#}$
WEIGHT OF FTG. = W_2 = $1.33 \times 7.0 \times 150$ = 1400.0
WEIGHT OF KEY = W_3 = $1.33^2 \times 150$ = 266.7
WEIGHT OF SOIL = W_4 = $1.33 \times 12.0 \times 100$ = 1600.0
SURCHARGE LD. = W_5 = 1.33×200 = 266.7
STEM FRICTION = W_6 = $\dfrac{30 \times 14^2}{2 \times 8}$ = 980.0

TOTAL WT. = W = 6313.4 $^{\#}$

DESIGN EXAMPLE — SECTION 6.8.12 | SHT. 2 OF 6

FOOTING DESIGN:

$$M_R = 1800 \times \frac{62}{12} + 1400 \times \frac{42}{12} + 266.7 \times \frac{56}{12} + 1600 \times \frac{76}{12} + 2667 \times \frac{76}{12} + 980 \times \frac{68}{12}$$

$$M_R = 32820 \; '^\#$$

$$\text{OVERTURN. RATIO} = \frac{32820}{18025} = 1.821 \; < \; 1.50$$

$$\text{NET } M = 32820 - 18025 = 14795 \; '^\#$$

$$\bar{X} = \frac{14795}{6313.4} = 2.343'$$

$$\text{FTG. MIDDLE } \tfrac{1}{3} = \frac{7.0}{3} = 2.33' \approx 2.343' \; \checkmark$$

$$e = \frac{7.0}{2} - 2.343 = 1.157'$$

$$M_e = 6313.4 \times 1.157 = 7302 \; '^\#$$

FTG. AREA = 7.0 SQ. FT.

$$\text{FTG. SECT. MODULUS} = \frac{1 \times 7^2}{6} = 8.167 \; '^3$$

$$\text{SOIL PRESSURE} = \frac{6313.4}{7.0} \pm \frac{7302}{8.167} = 902 \pm 894$$

$$\underline{SP_H = 1796 \; psf} \qquad \underline{SP_T = 8.0 \; psf}$$

$$W' = \frac{1796 - 8.0}{7.0} \times 2.67 + 8.0 = 689 \; psf$$

$$SP_1 = 1796 - 689 = 1107 \; psf$$

$$\text{UNIF. LD.} = 689 - 200 = 489 \; ^\#/FT.$$

$$+M_1 = \frac{1107 \times 4.33^2 \times 2}{2 \times 3} + \frac{489 \times 4.33^2}{2}$$

$$+M_1 = 11520 \; '^\#$$

$$Y_1 = \frac{1107 \times 4.33}{2} + 489 \times 4.33$$

$$Y_1 = 4517.5 \; ^\#$$

$$SP_2 = \frac{1796 - 8.0}{7.0} \times 1.33 = 340.6 \; psf$$

$$\text{UNIF. LD.}_2 = 200 + 1200 + 200 - 8 = 1592 \; ^\#/FT.$$

$$-M_2 = \frac{1592 \times 1.33^2}{2} - \frac{340.6 \times 1.33^2}{2 \times 3} = 1314 \; '^\#$$

$$Y_2 = 1592 \times 1.33 - 340.6 \times 1.33 \times .5 = 1895.6 \; ^\#$$

DESIGN EXAMPLE — SECTION 6.8.12	SHT. 3 OF 6

FOOTING DESIGN (CONT.)

$+M_1 = 11520 \ '^{\#}$; $Y_1 = 4517.5^{\#}$

$-M_2 = 1314 \ '^{\#}$; $Y_2 = 1895.6^{\#}$

d_1 TO BOTT. REINF. $= 16 - (3 + .5 \times .62) = 12.7''$

d_2 TO TOP REINF. $= 16 - (2 + .5 \times .62) = 13.7''$

$a = 1.13$; $j = 0.89$

$+A_s = \dfrac{11.52}{1.13 \times 12.7} = 0.80$ SQ. IN./FT.   USE #6 @ 6" O.C.

$\nu_1 = \dfrac{4517.5}{12 \times .89 \times 12.7} = 33.3 \text{psi} < 49.0 \text{ psi} \ \checkmark$

CHECK $-M_2$ FOR TENSION IN CONCRETE

SECT. MODULUS $= \dfrac{12 \times 16^2}{6} = 512''^3$

$f_T = \dfrac{12 \times 1314}{512} = 30.8 \text{psi} < 71.0 \text{ psi} \ \checkmark$

$\therefore$ NO NEGATIVE REINF. REQ'D.

CHECK REINF. REQ'D. AT $+M_3$:

$W' = \dfrac{1796 - 8.0}{7.0} \times 5.0 + 8.0 = 1285 \text{ psf}$

$SP_1 = 1796 - 1285 = 511 \text{ psf}$

UNIF. LD. $= 1285 - 200 = 1085 \ ^{\#}/\text{FT.}$

$+M_3 = \dfrac{511 \times 2^2 \times 2}{2 \times 3} + \dfrac{1085 \times 2^2}{2} = 2851.3 \ '^{\#}$

$+A_s = \dfrac{2.85}{1.13 \times 12.7} = 0.198$ SQ. IN./FT.

FTG. LONG. REINF. :

$A_s = 0.0012 \times 16 \times 7.0 \times 12 = 1.61$ SQ. IN.

USE 5 - #5

DESIGN EXAMPLE - SECTION 6.8.12	SHT. 4 OF 6

STEM   DESIGN   AT   TOP   OF   FTG. :

EQUIV.   HT. $= 12.0' + \frac{200}{100} = 14.0'$;   E F P $= 30$ p.c.f.

$V = \frac{30 \times 14^2}{2} = 2940^{\#}$; $M = \frac{2940 \times 14}{3} = 13720'^{\#}$

$d = 16 - (2 + .5 \times .62) = 13.7''$; $a = 1.13$; $j = 0.89$

$A_s = \frac{13.72}{1.13 \times 13.7} = 0.886$ SQ. IN./FT.   USE #7 @ 8" o.c.

$\nu = \frac{2940}{12 \times .89 \times 13.7} = 20.1$ psi < 54.0 psi $\checkmark$

STEM   DESIGN   AT   4'-0"   ABOVE   TOP   OF   FTG.

EQUIVALENT   HT. $= 12.0' + \frac{200}{100} - 4.0' = 10.0'$

$V = \frac{30 \times 10^2}{2} = 1500^{\#}$; $M = \frac{1500 \times 10}{3} = 5000'^{\#}$

$d = 16 - \frac{8''}{3} - (2 + .50 \times .62) = 11.0''$; $a = 1.13$; $j = 0.89$

$A_s = \frac{5.00}{1.13 \times 11} = 0.40$ SQ. IN./FT.   USE #7 @ 16" o.c.

$\nu = \frac{1500}{12 \times .89 \times 11} = 12.8$ psi < 54 psi $\checkmark$

STEM   DESIGN   AT   8'-0"   ABOVE   TOP   OF   FTG.

EQUIVALENT   HT. $= 12.0' + \frac{200}{100} - 8.0' = 6.0'$

$V = \frac{30 \times 6^2}{2} = 540^{\#}$; $M = \frac{540 \times 6}{3} = 1080'^{\#}$

$d = 16'' - \frac{8 \times 2}{3} - (2 + .5 \times .62) = 8.3''$

$A_s = \frac{1.08}{1.13 \times 8.30} = 0.12$ SQ. IN./FT.   USE #5 @ 16" o.c.

STEM   LONG.   REINF. :

$A_s = 0.0012 \times 12'' \times 12.0' \times 12 = 2.07$ SQ. IN.

USE #4 @ 23" o.c.

DESIGN EXAMPLE — SECTION 6.8.12 | SHT. 5 OF 6

SLIDING :

$$Fr = \frac{3527}{6313.4} = 0.558 > 0.40$$

∴ KEY REQ'D.
PASSIVE SOIL PRESSURE = 300 pcf

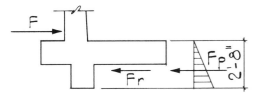

RESISTANCE TO SLIDING $= 6313.4 \times .40 + \frac{300 \times 2.67^2}{2} = 3592^\# > 3527^\# \checkmark$

CHECK WALL DESIGN WITHOUT SURCHARGE LD.

EQUIV. HT. $= 12.0 + 1.33 = 13.33'$

$F = \frac{13 \times 13.33^2}{2} = 2667^\#$; $M_o = \frac{2667 \times 13.33}{3} = 11852'^\#$

TOTAL WT. $= 6313.4 - 1.33 \times 200 - 980 + \frac{30 \times 12^2}{2 \times 3} = 5787^\#$

$M_R = 32820 - 266.7 \times \frac{76}{12} - 980 \times \frac{68}{12} + 720 \times \frac{68}{12} = 29658'^\#$

OVERTURN RATIO $= \frac{29658}{11852} = 2.502 > 1.5 \checkmark$

NET M $= 29658 - 11852 = 17806'^\#$

$\bar{X} = \frac{17806}{5787} = 3.077' > 2.33' \checkmark$

$e = \frac{7.0}{2} - 3.077 = 0.423'$

$M_e = 5787 \times .423 = 2448'^\#$

SOIL PRESSURE $= \frac{5787}{7.0} \pm \frac{2448}{8.167} = 826.7 \pm 299.7$

$\underline{SP_H = 1126.4 \ psf}$ $\qquad \underline{SP_T = 527 \ psf}$

DESIGN EXAMPLE — SECTION 6.8.12	SHT. 6 OF 6

CHECK WALL DESIGN WITHOUT SURCHARGE LD.

RE-CHECK FTG. M_1 (SEE SHT. No. 2)

$$W^1 = \frac{1126.4 - 527}{7.0} \times 2.67 + 527 = 755.3 \text{ psf}$$

$$SP_1 = 1126.4 - 755.3 = 371.1 \text{ psf}$$

UNIF. LD. $= 755.3 - 200 = 555.3 \text{ \#/FT.}$

$$+M_1 = \frac{371.1 \times 4.33^2 \times 2}{2 \times 3} + \frac{555.3 \times 4.33^2}{2} = 7536.3^{1\#} < 11520^{1\#} \checkmark$$

SLIDING :

$$F_r = \frac{2667}{5787} = 0.46 > 0.40$$

$\therefore$ KEY REQ'D.

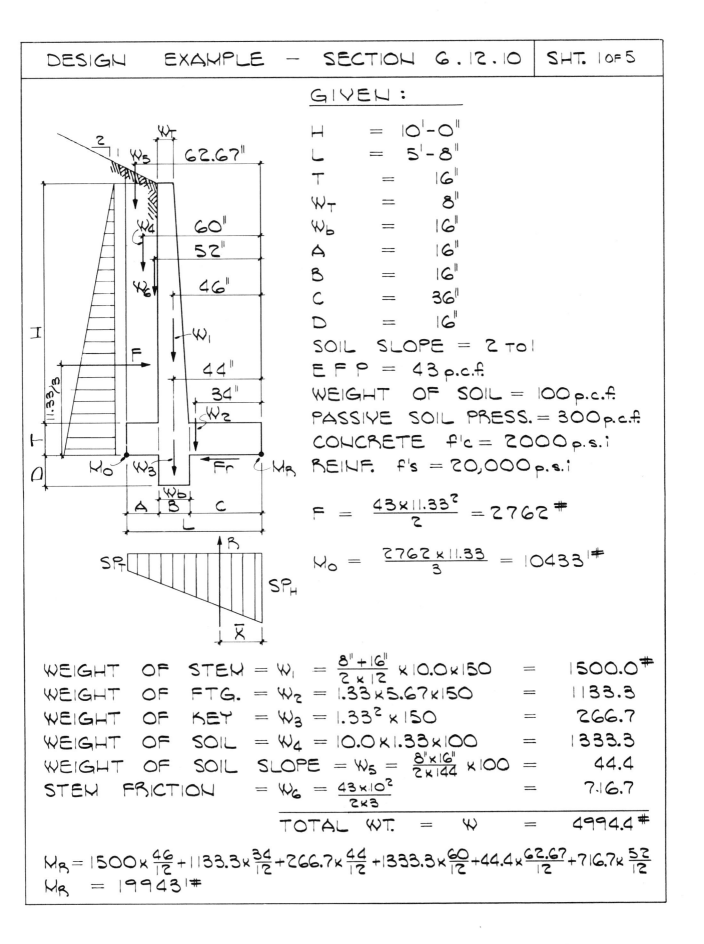

DESIGN EXAMPLE — SECTION 6.12.10 | SHT. 1 OF 5

GIVEN:

$$H = 10'-0''$$
$$L = 5'-8''$$
$$T = 16''$$
$$W_T = 8''$$
$$W_b = 16''$$
$$A = 16''$$
$$B = 16''$$
$$C = 36''$$
$$D = 16''$$

SOIL SLOPE = 2 TO 1
E F P = 43 p.c.f.
WEIGHT OF SOIL = 100 p.c.f.
PASSIVE SOIL PRESS. = 300 p.c.f.
CONCRETE $f'c$ = 2000 p.s.i
REINF. $f's$ = 20,000 p.s.i

$$F = \frac{43 \times 11.33^2}{2} = 2762^{\#}$$

$$M_0 = \frac{2762 \times 11.33}{3} = 10433'^{\#}$$

WEIGHT OF STEM = W_1 = $\frac{8''+16''}{2 \times 12} \times 10.0 \times 150$ = 1500.0$^{\#}$
WEIGHT OF FTG. = W_2 = $1.33 \times 5.67 \times 150$ = 1133.3
WEIGHT OF KEY = W_3 = $1.33^2 \times 150$ = 266.7
WEIGHT OF SOIL = W_4 = $10.0 \times 1.33 \times 100$ = 1333.3
WEIGHT OF SOIL SLOPE = W_5 = $\frac{8'' \times 16''}{2 \times 144} \times 100$ = 44.4
STEM FRICTION = W_6 = $\frac{43 \times 10^2}{2 \times 3}$ = 716.7

TOTAL WT. = W = 4994.4$^{\#}$

$M_R = 1500 \times \frac{46}{12} + 1133.3 \times \frac{34}{12} + 266.7 \times \frac{44}{12} + 1333.3 \times \frac{60}{12} + 44.4 \times \frac{62.67}{12} + 716.7 \times \frac{52}{12}$

$M_R = 19943'^{\#}$

DESIGN EXAMPLE — SECTION 6.12.10 | SHT. 2 OF 5

FOOTING DESIGN:

$W = 4994.4^{\#}$; $M_O = 10433^{\#}$; $M_R = 19943^{\#}$

OVERTURN. RATIO $= \dfrac{19943}{10433} = 1.912 > 1.50$ ✓

NET $M = 19943 - 10433 = 9510^{\#}$

$\bar{X} = \dfrac{9510}{4994.4} = 1.904'$

FTG. MIDDLE $\frac{1}{3} = \dfrac{5.67}{3} = 1.890' < 1.904'$ ✓

$e = \dfrac{5.67}{2} - 1.904 = 0.929'$

$M_e = 4994.4 \times .929 = 4639.8^{\#}$

FTG. AREA $= 5.67$ SQ.FT.

FTG. SECT. MODULUS $= \dfrac{1 \times 5.67^2}{6} = 5.352'^3$

SOIL PRESSURE $= \dfrac{4994.4}{5.67} \pm \dfrac{4639.8}{5.352} = 881 \pm 867$

$SP_H = 1748$ p.s.f. $SP_T = 14$ p.s.f.

$W' = \dfrac{1748-14}{5.67} \times 2.67 + 14.0 = 830$ p.s.f.

$SP' = 1748 - 830 = 918$ p.s.f.

UNIF. LD. $= 830 - 200 = 630^{\#}/$FT.

$+M_1 = \dfrac{918 \times 3^2 \times 2}{2 \times 3} + \dfrac{630 \times 3^2}{2} = 5589'^{\#}$

$V_1 = \dfrac{918 \times 3}{2} + 630 \times 3 = 3276'^{\#}$

$SP'' = \dfrac{1748-14}{5.67} \times 1.33 = 408$ p.s.f.

UNIF. LD. $= 14 - 1000 - 200 = -1186^{\#}/$FT.

$W_5 = -44.4$

$-M_2 = \dfrac{-1186 \times 1.33^2}{2} - \dfrac{44.4 \times 1.33^2 \times 2}{2 \times 3} + \dfrac{408 \times 1.33^2}{2 \times 3}$

$-M_2 = 960'^{\#}$

d TO TOP REINF. $= 16 - (2 + .5 \times .62) = 13.7''$; $a = 1.13$

d TO BOTT. REINF. $= 16 - (3 + .5 \times .62) = 12.7''$; $j = 0.89$

DESIGN   EXAMPLE - SECTION 6.12.10 | SHT. 3 OF 5

FOOTING   DESIGN   (CONT.) :

$+M_1 = 5589'^{\#}$; $V_1 = 3276^{\#}$; $d = 13.7''$

$A_s = \dfrac{5.589}{1.13 \times 13.7} = 0.36$ SQ. IN./FT.

$\nu = \dfrac{3276}{12 \times .89 \times 13.7} = 22.4$ p.s.i < 54 p.s.i ✓

CHECK M_2 FOR TENSION IN CONCRETE

$M_2 = 960'^{\#}$ SECT. MODULUS $= \dfrac{12 \times 12.7^2}{6} = 322.58''^3$

$f_T = \dfrac{12 \times 960}{322.58} = 35.7$ p.s.i < 71 p.s.i ✓

LONG.   REINF. :

USE   4 - #4

STEM   DESIGN   AT   TOP   OF   FTG.

$V = \dfrac{43 \times 10^2}{2} = 2150^{\#}$; $M = \dfrac{2150 \times 10}{3} = 7166.7'^{\#}$

$d = 16 - (2 + .5 \times .31) = 13.7''$; $a = 1.13$; $j = 0.89$

$A_s = \dfrac{7.16}{1.13 \times 13.7} = 0.463$ SQ. IN./FT.   <u>USE #5 @ 8'' o.c.</u>

$\nu = \dfrac{2150}{12 \times .89 \times 13.7} = 14.7$ p.s.i < 49 p.s.i ✓

STEM   DESIGN   AT   4'-0''   ABOVE   FTG. :

$H = 10.0' - 4.0' = 6.0'$; $d = 16'' - \dfrac{8''}{10} \times 4.0 - (2 + .5 \times .62) = 10.5''$

$V = \dfrac{43 \times 6^2}{2} = 774^{\#}$; $M = \dfrac{774 \times 6}{3} = 1548'^{\#}$

$A_s = \dfrac{1.55}{1.13 \times 10.5} = 0.13$ SQ. IN./FT.   <u>#5 @ 16'' o.c.</u>

$\nu = \dfrac{774}{12 \times .89 \times 10.5} = 6.9$ p.s.i < 49 p.s.i ✓

| DESIGN EXAMPLE – SECTION 6.12.10 | SHT. 4 OF 5 |

STEM DESIGN AT 6'-0" ABOVE FTG. :

$$H = 10.0 - 6.0 = 4.0 \quad ; \quad d = 16'' - \tfrac{8}{10} \times 6 - (2 + .5 \times .62) = 8.9''$$

$$V = \frac{43 \times 4^2}{2} = 344^{\#} \quad ; \quad M = \frac{344 \times 4}{3} = 458.7'^{\#}$$

$$A_s = \frac{0.458}{1.13 \times 8.9} = 0.045 \qquad \underline{\text{USE } \#4 @ 16'' \text{ o.c.}}$$

$$v = \frac{344}{12 \times .89 \times 8.9} = 3.6 \text{ p.s.i} \; < \; 54 \text{ p.s.i} \; \checkmark$$

LONG. REINF. :

$$A_s = 0.0015 \times 12 \times 10.0 \times 12 = 2.16 \text{ sq. in.}$$

$$\underline{\text{USE } \#4 @ 23'' \text{ o.c. HORIZ.}}$$

SLIDING :

$$Fr = \frac{2762}{4994.4} = 0.553 \; > \; 0.40$$

∴ KEY IS REQ'D.

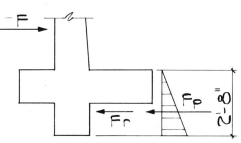

$$\text{RESISTANCE TO SLD'G} = 4994.4 \times .40 + \frac{300 \times 2.67^2}{2} = 3064.7^{\#} > 2762^{\#} \checkmark$$

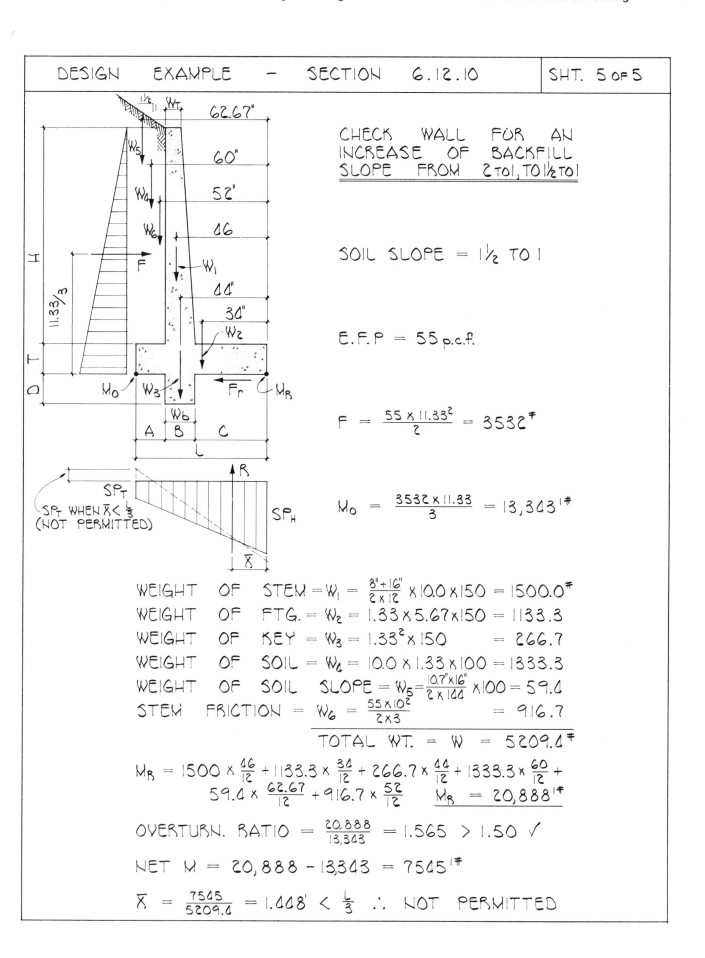

DESIGN EXAMPLE — SECTION 6.12.10 | SHT. 5 OF 5

CHECK WALL FOR AN INCREASE OF BACKFILL SLOPE FROM 2 TO 1, TO 1½ TO 1

SOIL SLOPE = 1½ TO 1

E.F.P = 55 p.c.f.

$$F = \frac{55 \times 11.33^2}{2} = 3532^{\#}$$

$$M_o = \frac{3532 \times 11.33}{3} = 13,343'^{\#}$$

WEIGHT OF STEM = $W_1 = \frac{8'' + 16''}{2 \times 12} \times 10.0 \times 150 = 1500.0^{\#}$

WEIGHT OF FTG. = $W_2 = 1.33 \times 5.67 \times 150 = 1133.3$

WEIGHT OF KEY = $W_3 = 1.33^2 \times 150 = 266.7$

WEIGHT OF SOIL = $W_4 = 10.0 \times 1.33 \times 100 = 1333.3$

WEIGHT OF SOIL SLOPE = $W_5 = \frac{10.7'' \times 16''}{2 \times 144} \times 100 = 59.4$

STEM FRICTION = $W_6 = \frac{55 \times 10^2}{2 \times 3} = 916.7$

TOTAL WT. = W = 5209.4$^{\#}$

$M_B = 1500 \times \frac{46}{12} + 1133.3 \times \frac{34}{12} + 266.7 \times \frac{44}{12} + 1333.3 \times \frac{60}{12} + 59.4 \times \frac{62.67}{12} + 916.7 \times \frac{52}{12}$ $M_B = 20,888'^{\#}$

OVERTURN. RATIO = $\frac{20,888}{13,343} = 1.565 > 1.50$ ✓

NET M = 20,888 − 13,343 = 7545$'^{\#}$

$\bar{X} = \frac{7545}{5209.4} = 1.448' < \frac{L}{3}$ ∴ NOT PERMITTED

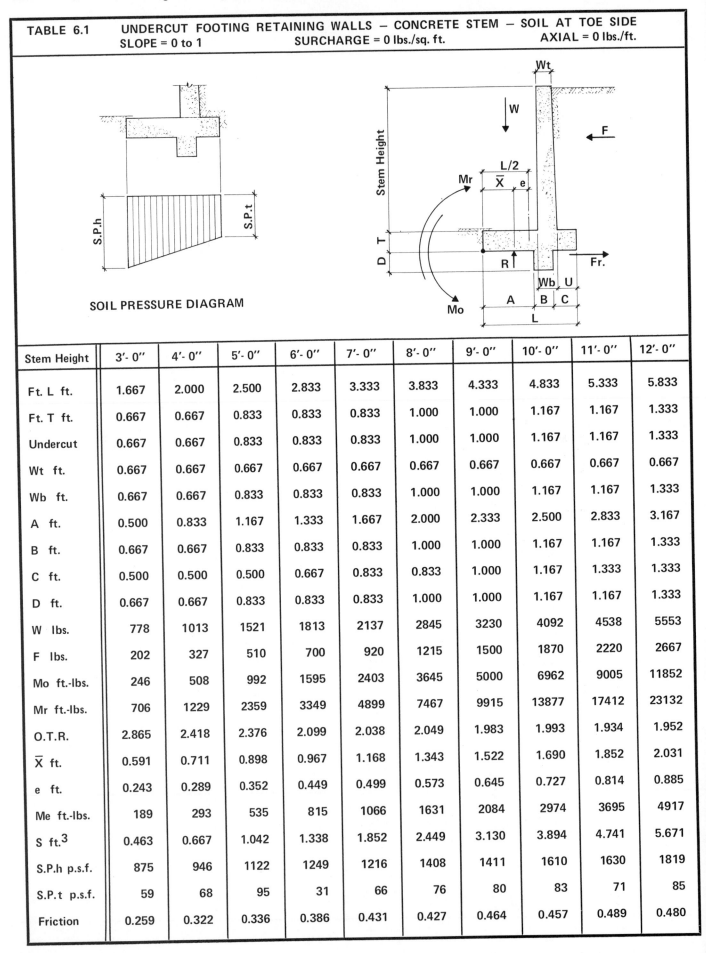

TABLE 6.1	UNDERCUT FOOTING RETAINING WALLS — CONCRETE STEM — SOIL AT TOE SIDE
	SLOPE = 0 to 1 SURCHARGE = 0 lbs./sq. ft. AXIAL = 0 lbs./ft.

SOIL PRESSURE DIAGRAM

Stem Height	3'- 0''	4'- 0''	5'- 0''	6'- 0''	7'- 0''	8'- 0''	9'- 0''	10'- 0''	11'- 0''	12'- 0''
Ft. L ft.	1.667	2.000	2.500	2.833	3.333	3.833	4.333	4.833	5.333	5.833
Ft. T ft.	0.667	0.667	0.833	0.833	0.833	1.000	1.000	1.167	1.167	1.333
Undercut	0.667	0.667	0.833	0.833	0.833	1.000	1.000	1.167	1.167	1.333
Wt ft.	0.667	0.667	0.667	0.667	0.667	0.667	0.667	0.667	0.667	0.667
Wb ft.	0.667	0.667	0.833	0.833	0.833	1.000	1.000	1.167	1.167	1.333
A ft.	0.500	0.833	1.167	1.333	1.667	2.000	2.333	2.500	2.833	3.167
B ft.	0.667	0.667	0.833	0.833	0.833	1.000	1.000	1.167	1.167	1.333
C ft.	0.500	0.500	0.500	0.667	0.833	0.833	1.000	1.167	1.333	1.333
D ft.	0.667	0.667	0.833	0.833	0.833	1.000	1.000	1.167	1.167	1.333
W lbs.	778	1013	1521	1813	2137	2845	3230	4092	4538	5553
F lbs.	202	327	510	700	920	1215	1500	1870	2220	2667
Mo ft.-lbs.	246	508	992	1595	2403	3645	5000	6962	9005	11852
Mr ft.-lbs.	706	1229	2359	3349	4899	7467	9915	13877	17412	23132
O.T.R.	2.865	2.418	2.376	2.099	2.038	2.049	1.983	1.993	1.934	1.952
$\overline{X}$ ft.	0.591	0.711	0.898	0.967	1.168	1.343	1.522	1.690	1.852	2.031
e ft.	0.243	0.289	0.352	0.449	0.499	0.573	0.645	0.727	0.814	0.885
Me ft.-lbs.	189	293	535	815	1066	1631	2084	2974	3695	4917
S ft.^{3}	0.463	0.667	1.042	1.338	1.852	2.449	3.130	3.894	4.741	5.671
S.P.h p.s.f.	875	946	1122	1249	1216	1408	1411	1610	1630	1819
S.P.t p.s.f.	59	68	95	31	66	76	80	83	71	85
Friction	0.259	0.322	0.336	0.386	0.431	0.427	0.464	0.457	0.489	0.480

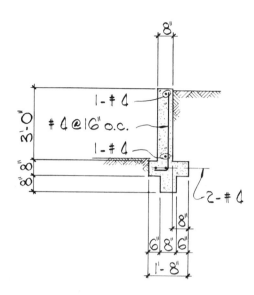

E.F.P. = 30 p.c.f.
Section 6.1.3

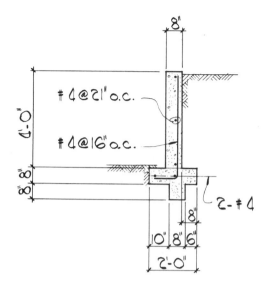

E.F.P. = 30 p.c.f.
Section 6.1.4

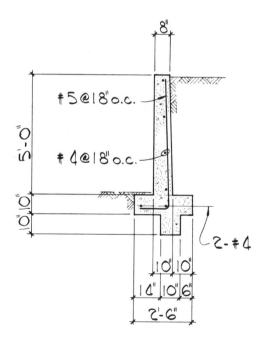

E.F.P. = 30 p.c.f.
Section 6.1.5

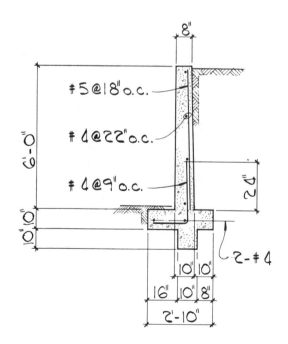

E.F.P. = 30 p.c.f.
Section 6.1.6

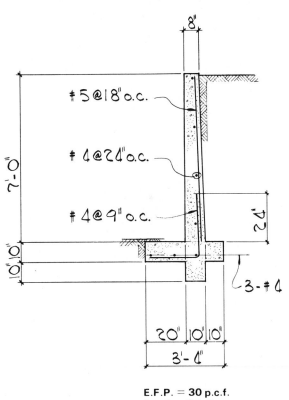

E.F.P. = 30 p.c.f.
Section 6.1.7

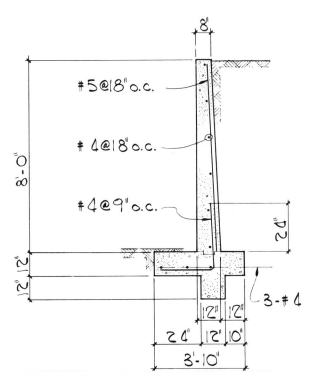

E.F.P. = 30 p.c.f.
Section 6.1.8

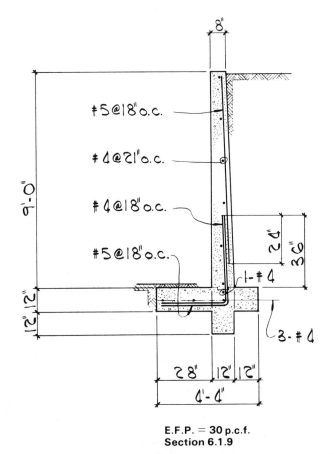

E.F.P. = 30 p.c.f.
Section 6.1.9

E.F.P. = 30 p.c.f.
Section 6.1.10

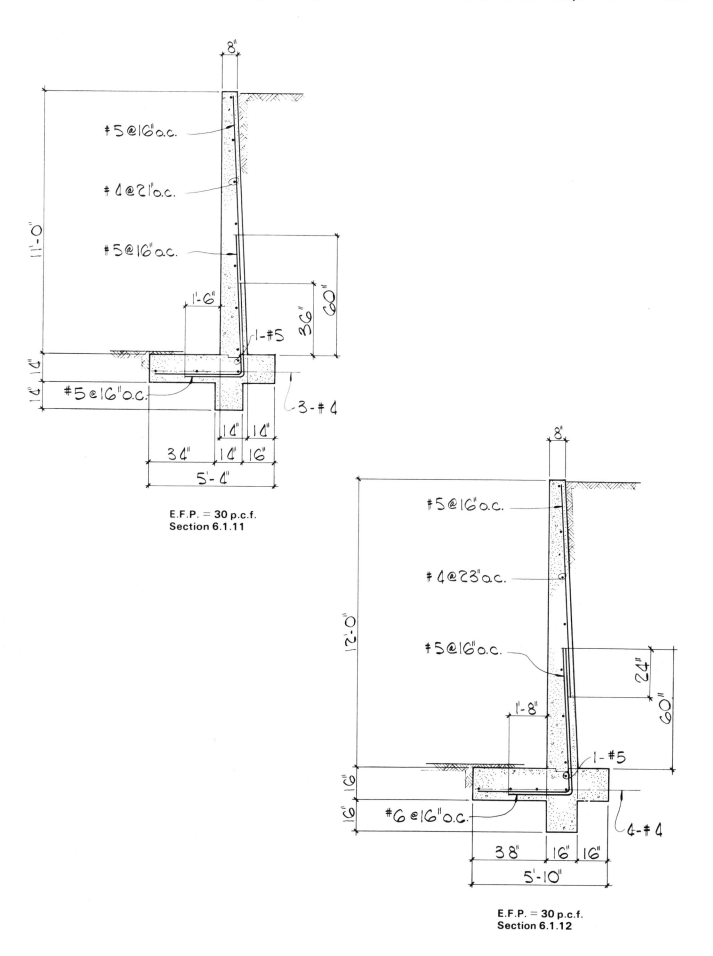

#5@16"o.c.

#4@21"o.c.

#5@16"o.c.

11'-0"

1'-6"

1-#5

36"

60"

4"

4"

#5@16"o.c.

3-#4

4" 4"

34"   4" 16"

5'-4"

E.F.P. = **30** p.c.f.
Section 6.1.11

8"

#5@16"o.c.

#4@23"o.c.

#5@16"o.c.

12'-0"

1'-8"

1-#5

24"

60"

6"

6"

#6@16"o.c.

4-#4

38"   16" 16"

5'-10"

E.F.P. = **30** p.c.f.
Section 6.1.12

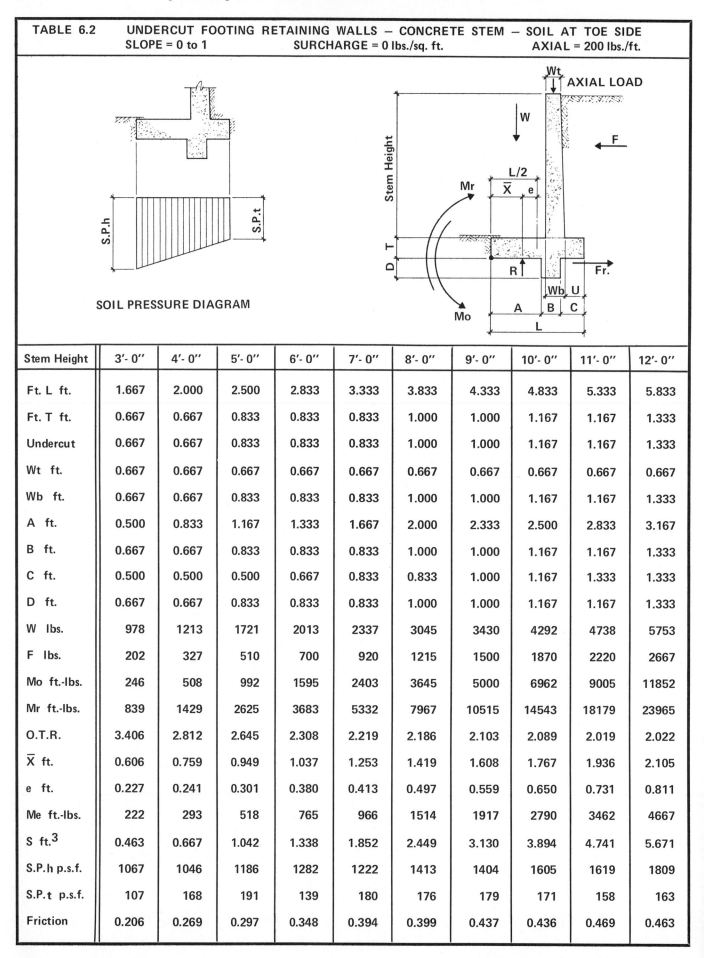

TABLE 6.2   UNDERCUT FOOTING RETAINING WALLS — CONCRETE STEM — SOIL AT TOE SIDE
SLOPE = 0 to 1   SURCHARGE = 0 lbs./sq. ft.   AXIAL = 200 lbs./ft.

SOIL PRESSURE DIAGRAM

Stem Height	3'- 0"	4'- 0"	5'- 0"	6'- 0"	7'- 0"	8'- 0"	9'- 0"	10'- 0"	11'- 0"	12'- 0"
Ft. L ft.	1.667	2.000	2.500	2.833	3.333	3.833	4.333	4.833	5.333	5.833
Ft. T ft.	0.667	0.667	0.833	0.833	0.833	1.000	1.000	1.167	1.167	1.333
Undercut	0.667	0.667	0.833	0.833	0.833	1.000	1.000	1.167	1.167	1.333
Wt ft.	0.667	0.667	0.667	0.667	0.667	0.667	0.667	0.667	0.667	0.667
Wb ft.	0.667	0.667	0.833	0.833	0.833	1.000	1.000	1.167	1.167	1.333
A ft.	0.500	0.833	1.167	1.333	1.667	2.000	2.333	2.500	2.833	3.167
B ft.	0.667	0.667	0.833	0.833	0.833	1.000	1.000	1.167	1.167	1.333
C ft.	0.500	0.500	0.500	0.667	0.833	0.833	1.000	1.167	1.333	1.333
D ft.	0.667	0.667	0.833	0.833	0.833	1.000	1.000	1.167	1.167	1.333
W lbs.	978	1213	1721	2013	2337	3045	3430	4292	4738	5753
F lbs.	202	327	510	700	920	1215	1500	1870	2220	2667
Mo ft.-lbs.	246	508	992	1595	2403	3645	5000	6962	9005	11852
Mr ft.-lbs.	839	1429	2625	3683	5332	7967	10515	14543	18179	23965
O.T.R.	3.406	2.812	2.645	2.308	2.219	2.186	2.103	2.089	2.019	2.022
$\overline{X}$ ft.	0.606	0.759	0.949	1.037	1.253	1.419	1.608	1.767	1.936	2.105
e ft.	0.227	0.241	0.301	0.380	0.413	0.497	0.559	0.650	0.731	0.811
Me ft.-lbs.	222	293	518	765	966	1514	1917	2790	3462	4667
S ft.^{3}	0.463	0.667	1.042	1.338	1.852	2.449	3.130	3.894	4.741	5.671
S.P.h p.s.f.	1067	1046	1186	1282	1222	1413	1404	1605	1619	1809
S.P.t p.s.f.	107	168	191	139	180	176	179	171	158	163
Friction	0.206	0.269	0.297	0.348	0.394	0.399	0.437	0.436	0.469	0.463

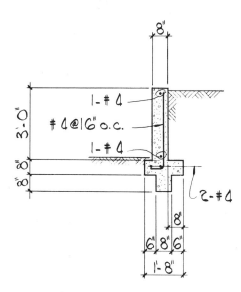

Axial load = 200 lbs./ft.
E.F.P. = 30 p.c.f.
Section 6.2.3

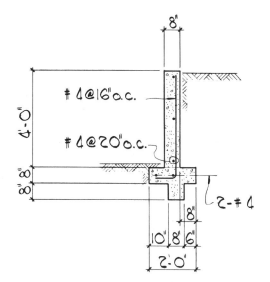

Axial load = 200 lbs./ft.
E.F.P. = 30 p.c.f.
Section 6.2.4

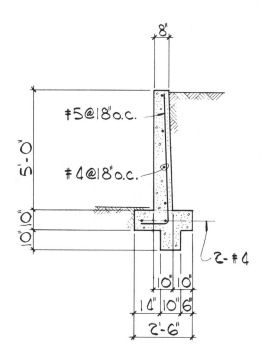

Axial load = 200 lbs./ft.
E.F.P. = 30 p.c.f.
Section 6.2.5

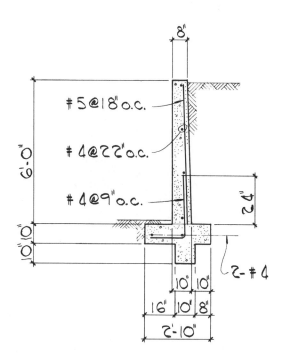

Axial load = 200 lbs./ft.
E.F.P. = 30 p.c.f.
Section 6.2.6

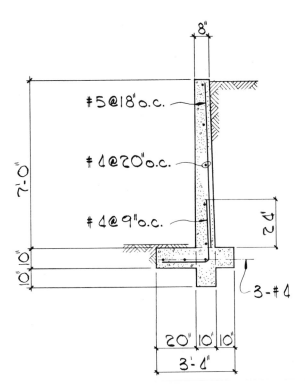

#5@18"o.c.

#4@20"o.c.

#4@9"o.c.

8"

7'-0"

24"

3-#4

20" 10" 10"

3'-4"

Axial load = 200 lbs./ft.
E.F.P. = 30 p.c.f.
Section 6.2.7

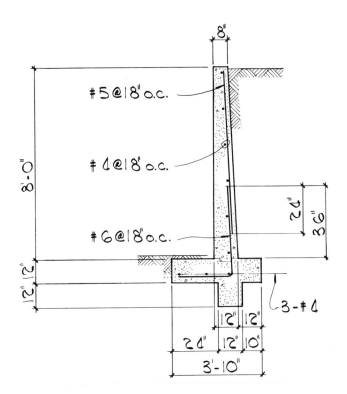

#5@18"o.c.

#4@18"o.c.

#6@18"o.c.

8"

8'-0"

24"

36"

3-#4

12" 12"

24" 12" 10"

3'-10"

Axial load = 200 lbs./ft.
E.F.P. = 30 p.c.f.
Section 6.2.8

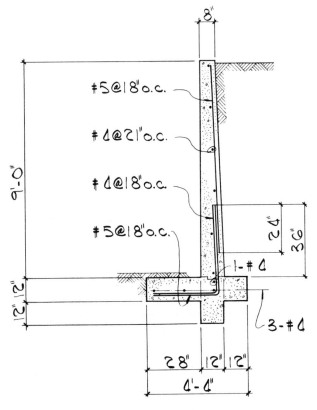

#5@18"o.c.

#4@21"o.c.

#4@18"o.c.

#5@18"o.c.

8"

9'-0"

24"

36"

1-#4

3-#4

28" 12" 12"

4'-4"

Axial load = 200 lbs./ft.
E.F.P. = 30 p.c.f.
Section 6.2.9

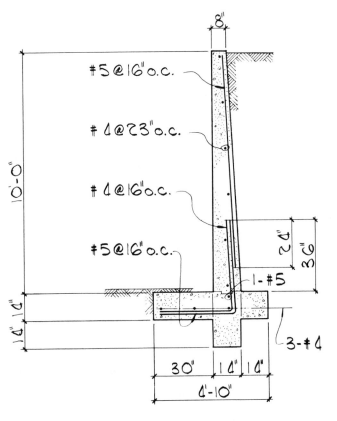

#5@16"o.c.

#4@23"o.c.

#4@16"o.c.

#5@16"o.c.

8"

10'-0"

24"

36"

1-#5

3-#4

30" 14" 14"

4'-10"

Axial load = 200 lbs./ft.
E.F.P. = 30 p.c.f.
Section 6.2.10

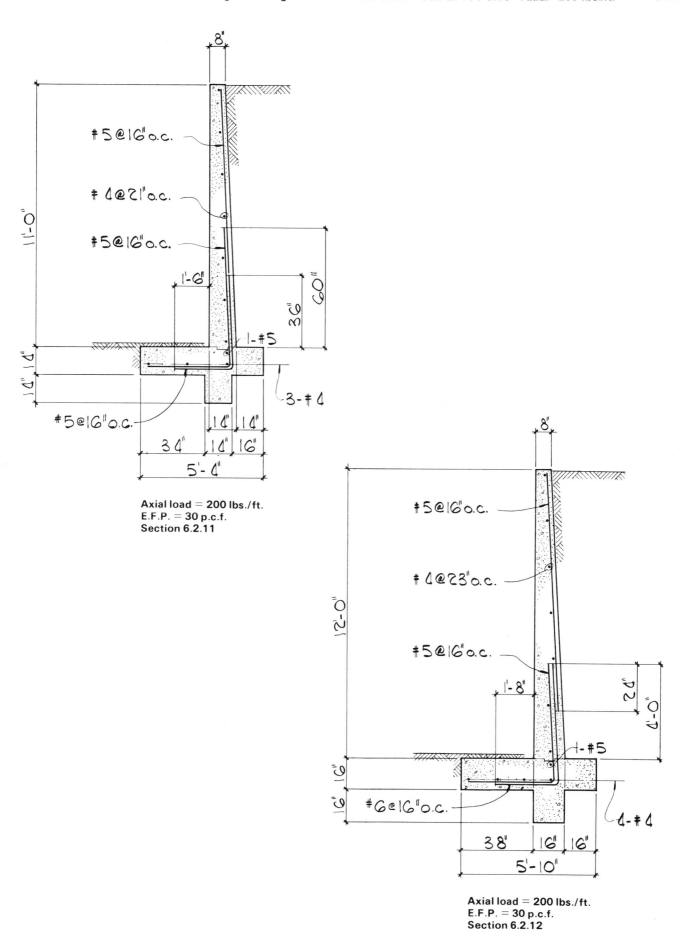

#5@16"o.c.

#4@21"o.c.

#5@16"o.c.

1-6"

36"

60"

1-#5

3-#4

#5@16"o.c.

1'-0"

4" 4"

34"   14"   16"

14" 14"

5'-4"

Axial load = 200 lbs./ft.
E.F.P. = 30 p.c.f.
Section 6.2.11

8"

#5@16"o.c.

#4@23"o.c.

#5@16"o.c.

1'-8"

24"

4'-0"

1-#5

#6@16"o.c.

4-#4

12'-0"

6"

6"

38"   16"   16"

5'-10"

Axial load = 200 lbs./ft.
E.F.P. = 30 p.c.f.
Section 6.2.12

TABLE 6.3	UNDERCUT FOOTING RETAINING WALLS — CONCRETE STEM — SOIL AT TOE SIDE
	SLOPE = 0 to 1 SURCHARGE = 0 lbs./sq. ft. AXIAL = 400 lbs./ft.

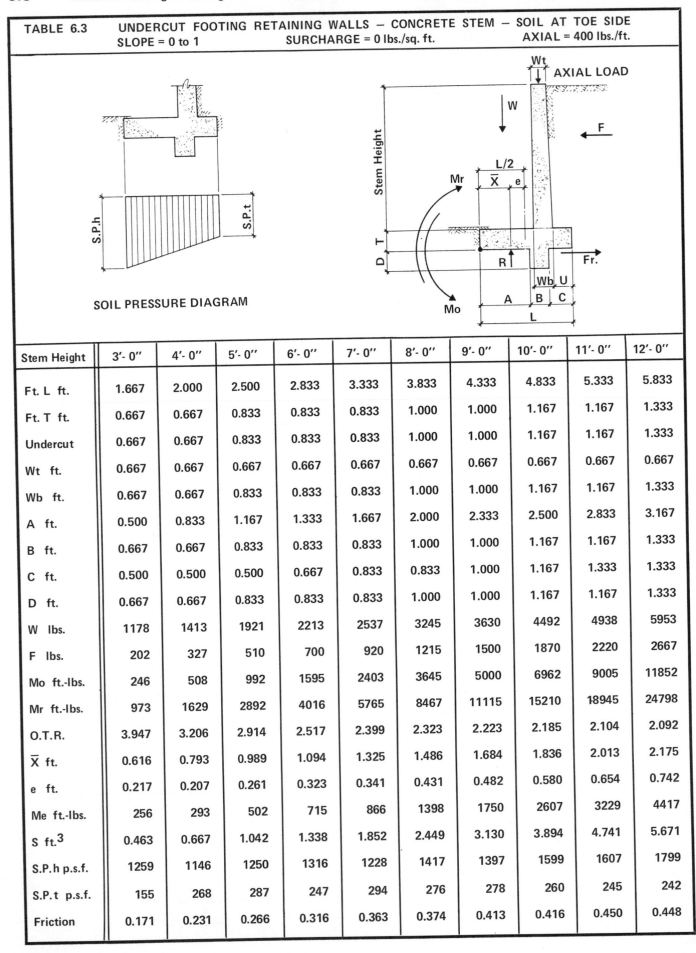

SOIL PRESSURE DIAGRAM

Stem Height	3'- 0''	4'- 0''	5'- 0''	6'- 0''	7'- 0''	8'- 0''	9'- 0''	10'- 0''	11'- 0''	12'- 0''
Ft. L ft.	1.667	2.000	2.500	2.833	3.333	3.833	4.333	4.833	5.333	5.833
Ft. T ft.	0.667	0.667	0.833	0.833	0.833	1.000	1.000	1.167	1.167	1.333
Undercut	0.667	0.667	0.833	0.833	0.833	1.000	1.000	1.167	1.167	1.333
Wt ft.	0.667	0.667	0.667	0.667	0.667	0.667	0.667	0.667	0.667	0.667
Wb ft.	0.667	0.667	0.833	0.833	0.833	1.000	1.000	1.167	1.167	1.333
A ft.	0.500	0.833	1.167	1.333	1.667	2.000	2.333	2.500	2.833	3.167
B ft.	0.667	0.667	0.833	0.833	0.833	1.000	1.000	1.167	1.167	1.333
C ft.	0.500	0.500	0.500	0.667	0.833	0.833	1.000	1.167	1.333	1.333
D ft.	0.667	0.667	0.833	0.833	0.833	1.000	1.000	1.167	1.167	1.333
W lbs.	1178	1413	1921	2213	2537	3245	3630	4492	4938	5953
F lbs.	202	327	510	700	920	1215	1500	1870	2220	2667
Mo ft.-lbs.	246	508	992	1595	2403	3645	5000	6962	9005	11852
Mr ft.-lbs.	973	1629	2892	4016	5765	8467	11115	15210	18945	24798
O.T.R.	3.947	3.206	2.914	2.517	2.399	2.323	2.223	2.185	2.104	2.092
$\overline{X}$ ft.	0.616	0.793	0.989	1.094	1.325	1.486	1.684	1.836	2.013	2.175
e ft.	0.217	0.207	0.261	0.323	0.341	0.431	0.482	0.580	0.654	0.742
Me ft.-lbs.	256	293	502	715	866	1398	1750	2607	3229	4417
S ft.^{3}	0.463	0.667	1.042	1.338	1.852	2.449	3.130	3.894	4.741	5.671
S.P.h p.s.f.	1259	1146	1250	1316	1228	1417	1397	1599	1607	1799
S.P.t p.s.f.	155	268	287	247	294	276	278	260	245	242
Friction	0.171	0.231	0.266	0.316	0.363	0.374	0.413	0.416	0.450	0.448

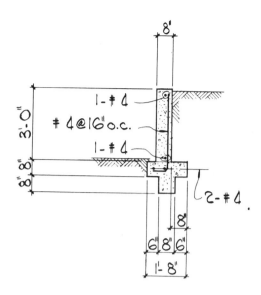

Axial load = 400 lbs./ft.
E.F.P. = 30 p.c.f.
Section 6.3.3

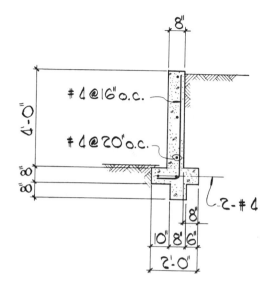

Axial load = 400 lbs./ft.
E.F.P. = 30 p.c.f.
Section 6.3.4

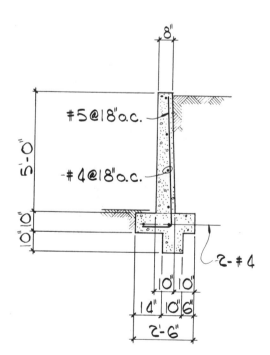

Axial load = 400 lbs./ft.
E.F.P. = 30 p.c.f.
Section 6.3.5

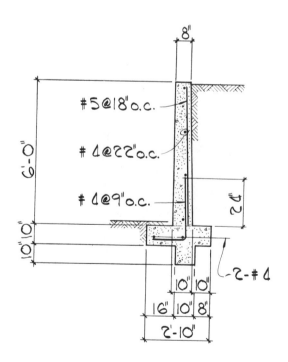

Axial load = 400 lbs./ft.
E.F.P. = 30 p.c.f.
Section 6.3.6

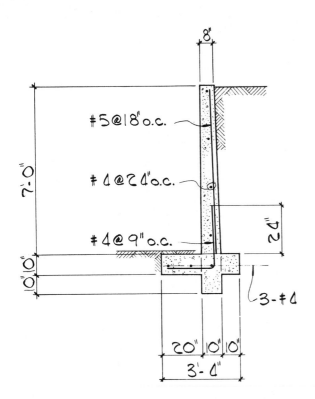

Axial load = 400 lbs./ft.
E.F.P. = 30 p.c.f.
Section 6.3.7

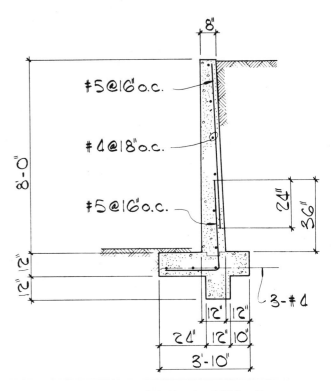

Axial load = 400 lbs./ft.
E.F.P. = 30 p.c.f.
Section 6.3.8

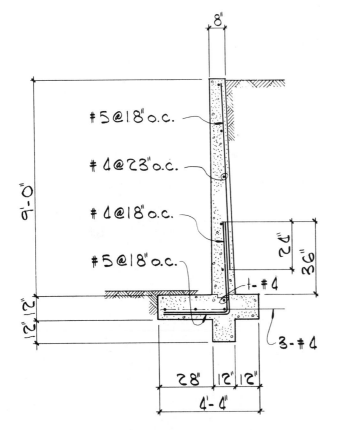

Axial load = 400 lbs./ft.
E.F.P. = 30 p.c.f.
Section 6.3.9

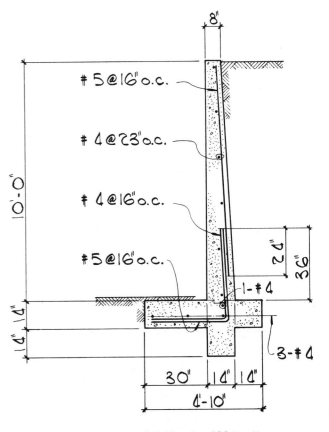

Axial load = 400 lbs./ft.
E.F.P. = 30 p.c.f.
Section 6.3.10

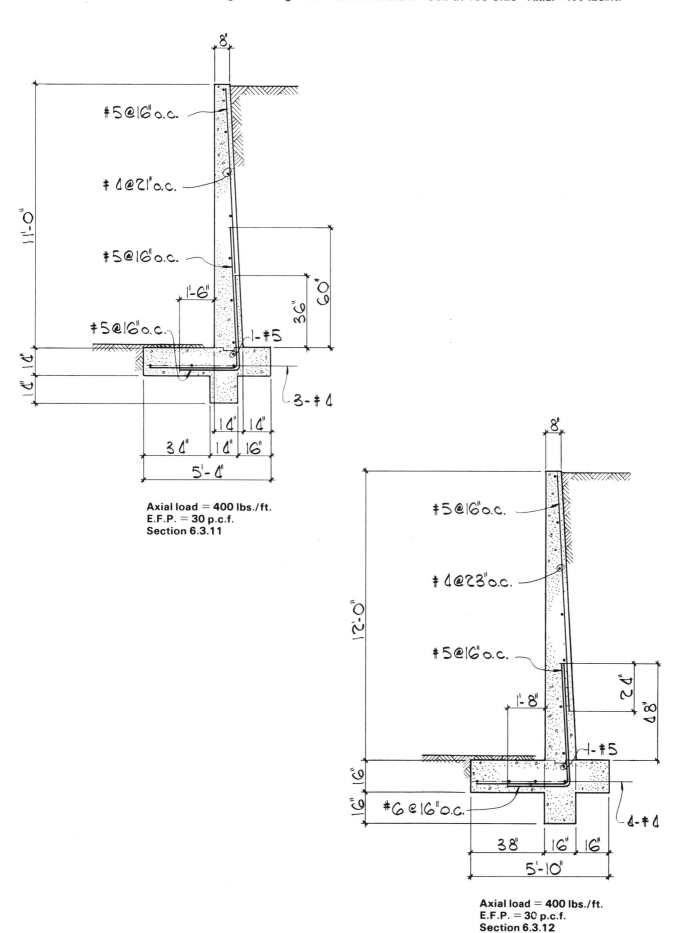

#5@16"o.c.

#4@21"o.c.

#5@16"o.c.

1'-6"

#5@16"o.c.

1-#5

3-#4

8'

11'-0"

60"

36"

14"

4"

4"

30"

14"

16"

14"

14"

5'-4"

Axial load = 400 lbs./ft.
E.F.P. = 30 p.c.f.
Section 6.3.11

#5@16"o.c.

#4@23"o.c.

#5@16"o.c.

1'-8"

1-#5

#6@16"o.c.

4-#4

8'

12'-0"

48"

24"

6"

6"

38"

16"

16"

5'-10"

Axial load = 400 lbs./ft.
E.F.P. = 30 p.c.f.
Section 6.3.12

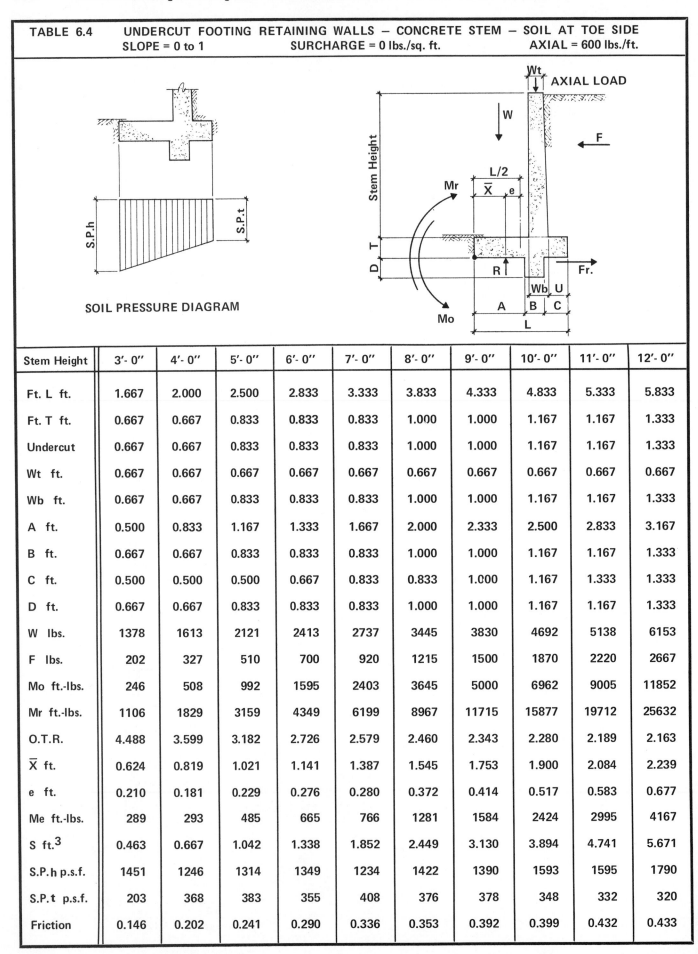

TABLE 6.4	UNDERCUT FOOTING RETAINING WALLS — CONCRETE STEM — SOIL AT TOE SIDE									
SLOPE = 0 to 1			SURCHARGE = 0 lbs./sq. ft.				AXIAL = 600 lbs./ft.			

SOIL PRESSURE DIAGRAM

Stem Height	3'- 0''	4'- 0''	5'- 0''	6'- 0''	7'- 0''	8'- 0''	9'- 0''	10'- 0''	11'- 0''	12'- 0''
Ft. L ft.	1.667	2.000	2.500	2.833	3.333	3.833	4.333	4.833	5.333	5.833
Ft. T ft.	0.667	0.667	0.833	0.833	0.833	1.000	1.000	1.167	1.167	1.333
Undercut	0.667	0.667	0.833	0.833	0.833	1.000	1.000	1.167	1.167	1.333
Wt ft.	0.667	0.667	0.667	0.667	0.667	0.667	0.667	0.667	0.667	0.667
Wb ft.	0.667	0.667	0.833	0.833	0.833	1.000	1.000	1.167	1.167	1.333
A ft.	0.500	0.833	1.167	1.333	1.667	2.000	2.333	2.500	2.833	3.167
B ft.	0.667	0.667	0.833	0.833	0.833	1.000	1.000	1.167	1.167	1.333
C ft.	0.500	0.500	0.500	0.667	0.833	0.833	1.000	1.167	1.333	1.333
D ft.	0.667	0.667	0.833	0.833	0.833	1.000	1.000	1.167	1.167	1.333
W lbs.	1378	1613	2121	2413	2737	3445	3830	4692	5138	6153
F lbs.	202	327	510	700	920	1215	1500	1870	2220	2667
Mo ft.-lbs.	246	508	992	1595	2403	3645	5000	6962	9005	11852
Mr ft.-lbs.	1106	1829	3159	4349	6199	8967	11715	15877	19712	25632
O.T.R.	4.488	3.599	3.182	2.726	2.579	2.460	2.343	2.280	2.189	2.163
$\overline{X}$ ft.	0.624	0.819	1.021	1.141	1.387	1.545	1.753	1.900	2.084	2.239
e ft.	0.210	0.181	0.229	0.276	0.280	0.372	0.414	0.517	0.583	0.677
Me ft.-lbs.	289	293	485	665	766	1281	1584	2424	2995	4167
S ft.^{3}	0.463	0.667	1.042	1.338	1.852	2.449	3.130	3.894	4.741	5.671
S.P.h p.s.f.	1451	1246	1314	1349	1234	1422	1390	1593	1595	1790
S.P.t p.s.f.	203	368	383	355	408	376	378	348	332	320
Friction	0.146	0.202	0.241	0.290	0.336	0.353	0.392	0.399	0.432	0.433

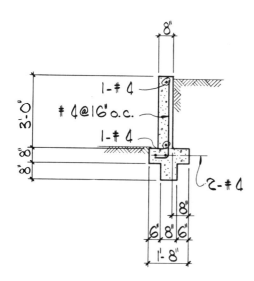

Axial load = 600 lbs./ft.
E.F.P. = 30 p.c.f.
Section 6.4.3

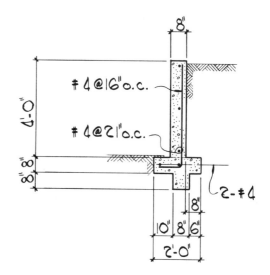

Axial load = 600 lbs./ft.
E.F.P. = 30 p.c.f.
Section 6.4.4

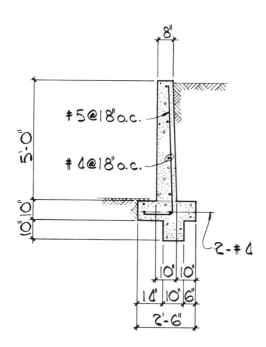

Axial load = 600 lbs./ft.
E.F.P. = 30 p.c.f.
Section 6.4.5

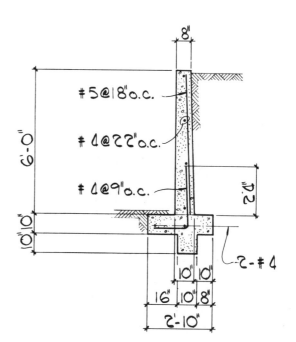

Axial load = 600 lbs./ft.
E.F.P. = 30 p.c.f.
Section 6.4.6

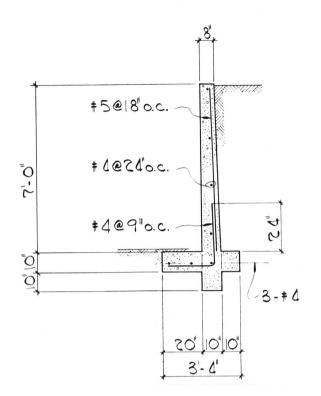

#5@18"o.c.

#4@24"o.c.

#4@9"o.c.

7'-0"

24"

3-#4

20" 10" 10"

3'-4"

Axial load = 600 lbs./ft.
E.F.P. = 30 p.c.f.
Section 6.4.7

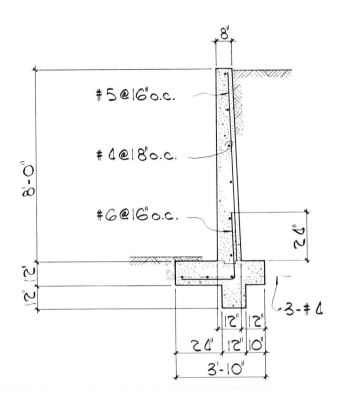

#5@16"o.c.

#4@18"o.c.

#6@16"o.c.

8'-0"

24"

3-#4

12" 12"
24" 12" 10"

3'-10"

Axial load = 600 lbs./ft.
E.F.P. = 30 p.c.f.
Section 6.4.8

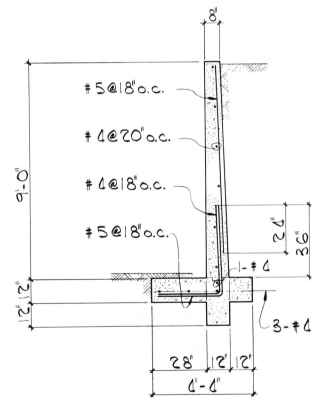

#5@18"o.c.

#4@20"o.c.

#4@18"o.c.

#5@18"o.c.

9'-0"

24"

36"

1-#4

3-#4

28" 12" 12"

4'-4"

Axial load = 600 lbs./ft.
E.F.P. = 30 p.c.f.
Section 6.4.9

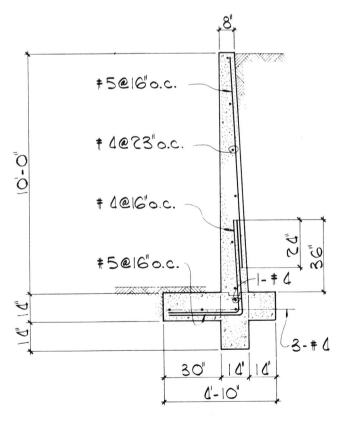

#5@16"o.c.

#4@23"o.c.

#4@16"o.c.

#5@16"o.c.

10'-0"

24"

36"

1-#4

3-#4

30" 14" 14"

4'-10"

Axial load = 600 lbs./ft.
E.F.P. = 30 p.c.f.
Section 6.4.10

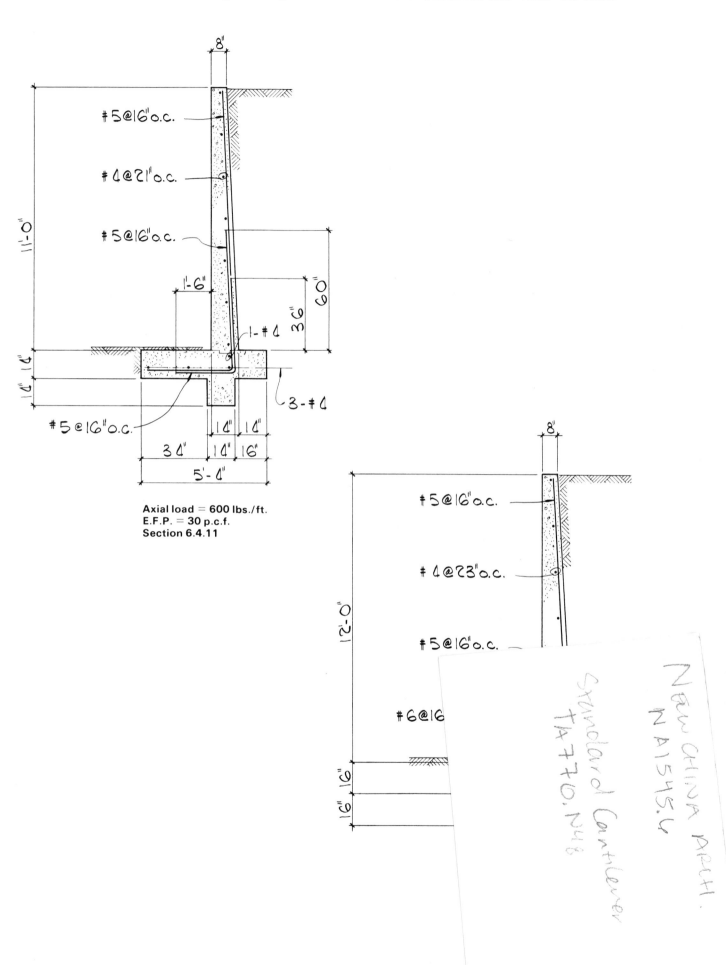

8"

#5@16"o.c.

#4@21"o.c.

#5@16"o.c.

11'-0"

1'-6"

1-#4

60"

36"

3-#4

#5@16"o.c.

1'4"

1'4"

1'4"

3'4"

1'4"

16"

5'-4"

Axial load = 600 lbs./ft.
E.F.P. = 30 p.c.f.
Section 6.4.11

8"

#5@16"o.c.

#4@23"o.c.

#5@16"o.c.

12'-0"

#6@16

16"

16"

New CHINA ARCH
NA1545.4
Standard Cantilever
TA770.N48

TABLE 6.5   UNDERCUT FOOTING RETAINING WALLS — CONCRETE STEM — SOIL AT TOE SIDE
SLOPE = 0 to 1   SURCHARGE = 0 lbs./sq. ft.   AXIAL = 800 lbs./ft.

SOIL PRESSURE DIAGRAM

Stem Height	3'- 0"	4'- 0"	5'- 0"	6'- 0"	7'- 0"	8'- 0"	9'- 0"	10'- 0"	11'- 0"	12'- 0"
Ft. L ft.	1.667	2.000	2.500	2.833	3.333	3.833	4.333	4.833	5.333	5.833
Ft. T ft.	0.667	0.667	0.833	0.833	0.833	1.000	1.000	1.167	1.167	1.333
Undercut	0.667	0.667	0.833	0.833	0.833	1.000	1.000	1.167	1.167	1.333
Wt ft.	0.667	0.667	0.667	0.667	0.667	0.667	0.667	0.667	0.667	0.667
Wb ft.	0.667	0.667	0.833	0.833	0.833	1.000	1.000	1.167	1.167	1.333
A ft.	0.500	0.833	1.167	1.333	1.667	2.000	2.333	2.500	2.833	3.167
B ft.	0.667	0.667	0.833	0.833	0.833	1.000	1.000	1.167	1.167	1.333
C ft.	0.500	0.500	0.500	0.667	0.833	0.833	1.000	1.167	1.333	1.333
D ft.	0.667	0.667	0.833	0.833	0.833	1.000	1.000	1.167	1.167	1.333
W lbs.	1578	1813	2321	2613	2937	3645	4030	4892	5338	6353
F lbs.	202	327	510	700	920	1215	1500	1870	2220	2667
Mo ft.-lbs.	246	508	992	1595	2403	3645	5000	6962	9005	11852
Mr ft.-lbs.	1239	2029	3425	4683	6632	9467	12315	16543	20479	26465
O.T.R.	5.029	3.993	3.451	2.935	2.760	2.597	2.463	2.376	2.274	2.233
$\overline{X}$ ft.	0.629	0.839	1.048	1.181	1.440	1.597	1.815	1.959	2.149	2.300
e ft.	0.204	0.161	0.202	0.235	0.227	0.319	0.352	0.458	0.517	0.617
Me ft.-lbs.	322	293	468	615	666	1164	1417	2240	2762	3917
S ft.^{3}	0.463	0.667	1.042	1.338	1.852	2.449	3.130	3.894	4.741	5.671
S.P.h p.s.f.	1643	1346	1378	1382	1240	1426	1383	1587	1584	1780
S.P.t p.s.f.	251	468	479	463	522	476	477	437	418	398
Friction	0.128	0.180	0.220	0.268	0.313	0.333	0.372	0.382	0.416	0.420

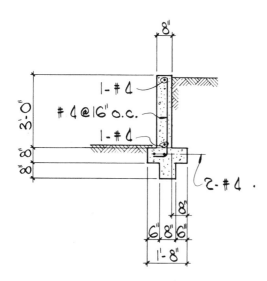

Axial load = 800 lbs./ft.
E.F.P. = 30 p.c.f.
Section 6.5.3

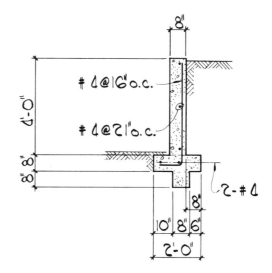

Axial load = 800 lbs./ft.
E.F.P. = 30 p.c.f.
Section 6.5.4

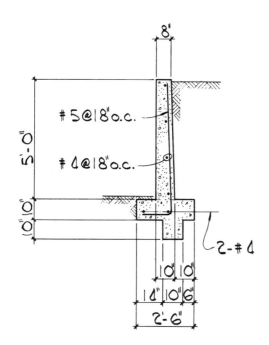

Axial load = 800 lbs./ft.
E.F.P. = 30 p.c.f.
Section 6.5.5

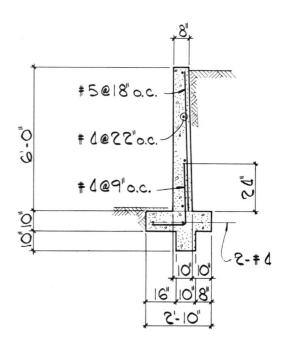

Axial load = 800 lbs./ft.
E.F.P. = 30 p.c.f.
Section 6.5.6

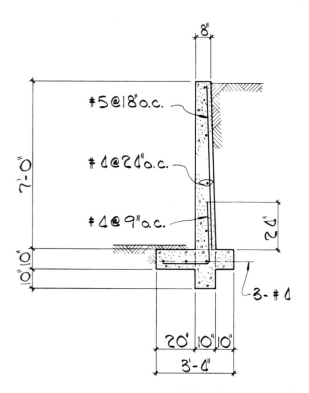

#5@18"o.c.

#4@24"o.c.

#4@9"o.c.

24"

3-#4

7'-0"

10"

20" 10" 10"

3'-4"

Axial load = 800 lbs./ft.
E.F.P. = 30 p.c.f.
Section 6.5.7

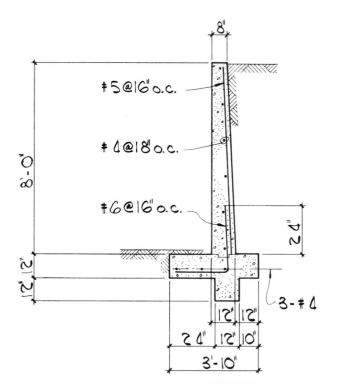

#5@16"o.c.

#4@18"o.c.

#6@16"o.c.

24"

3-#4

8'-0"

12" 12"

12" 12"

24" 12" 10"

3'-10"

Axial load = 800 lbs./ft.
E.F.P. = 30 p.c.f.
Section 6.5.8

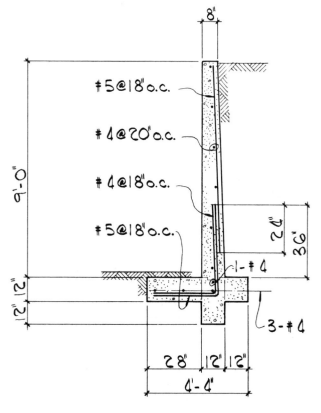

#5@18"o.c.

#4@20"o.c.

#4@18"o.c.

#5@18"o.c.

24"

36"

1-#4

3-#4

9'-0"

12" 12"

28" 12" 12"

4'-4"

Axial load = 800 lbs./ft.
E.F.P. = 30 p.c.f.
Section 6.5.9

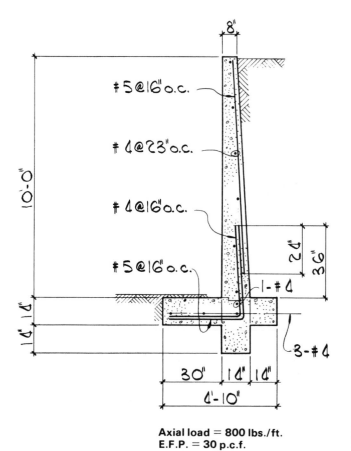

#5@16"o.c.

#4@23"o.c.

#4@16"o.c.

#5@16"o.c.

24"

36"

1-#4

3-#4

10'-0"

14" 14"

30" 14" 14"

4'-10"

Axial load = 800 lbs./ft.
E.F.P. = 30 p.c.f.
Section 6.5.10

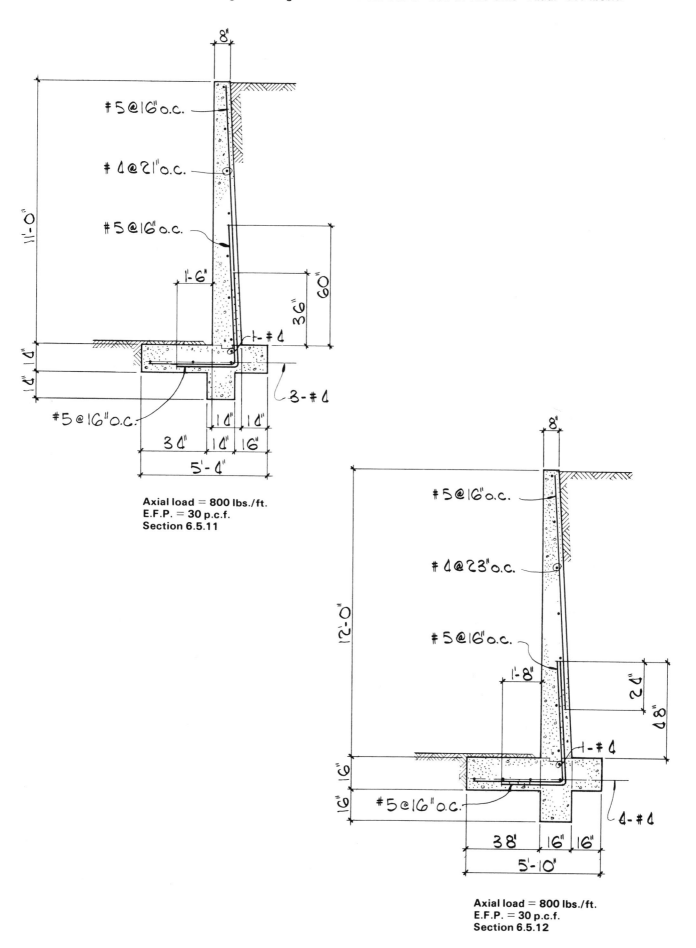

Axial load = 800 lbs./ft.
E.F.P. = 30 p.c.f.
Section 6.5.11

Axial load = 800 lbs./ft.
E.F.P. = 30 p.c.f.
Section 6.5.12

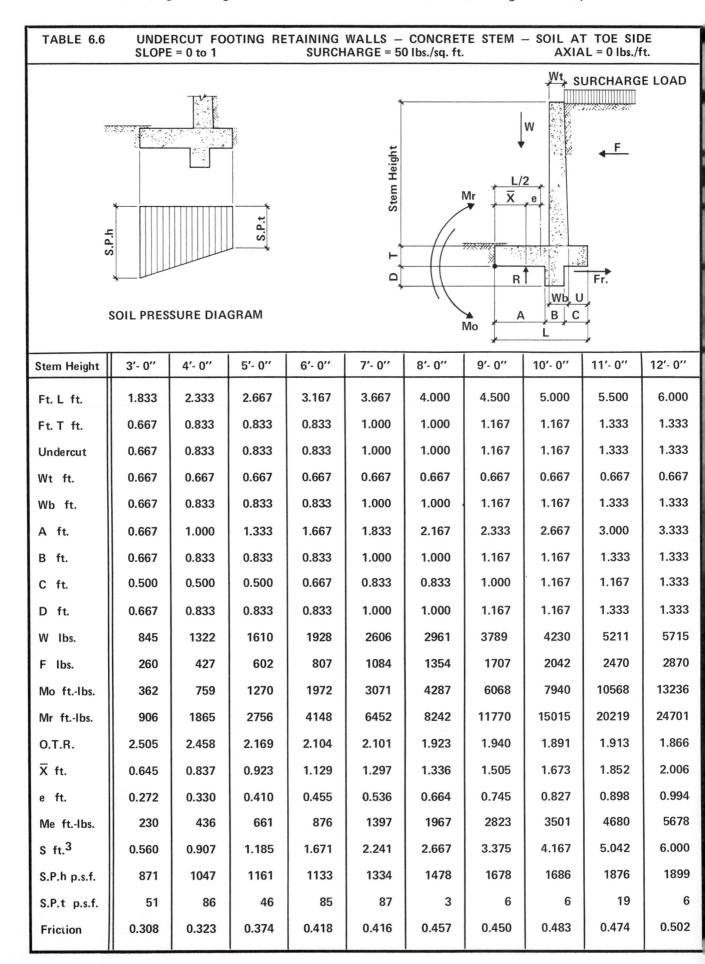

TABLE 6.6	UNDERCUT FOOTING RETAINING WALLS — CONCRETE STEM — SOIL AT TOE SIDE
SLOPE = 0 to 1	SURCHARGE = 50 lbs./sq. ft. AXIAL = 0 lbs./ft.

SOIL PRESSURE DIAGRAM

SURCHARGE LOAD

Stem Height	3'- 0''	4'- 0''	5'- 0''	6'- 0''	7'- 0''	8'- 0''	9'- 0''	10'- 0''	11'- 0''	12'- 0''
Ft. L ft.	1.833	2.333	2.667	3.167	3.667	4.000	4.500	5.000	5.500	6.000
Ft. T ft.	0.667	0.833	0.833	0.833	1.000	1.000	1.167	1.167	1.333	1.333
Undercut	0.667	0.833	0.833	0.833	1.000	1.000	1.167	1.167	1.333	1.333
Wt ft.	0.667	0.667	0.667	0.667	0.667	0.667	0.667	0.667	0.667	0.667
Wb ft.	0.667	0.833	0.833	0.833	1.000	1.000	1.167	1.167	1.333	1.333
A ft.	0.667	1.000	1.333	1.667	1.833	2.167	2.333	2.667	3.000	3.333
B ft.	0.667	0.833	0.833	0.833	1.000	1.000	1.167	1.167	1.333	1.333
C ft.	0.500	0.500	0.500	0.667	0.833	0.833	1.000	1.167	1.167	1.333
D ft.	0.667	0.833	0.833	0.833	1.000	1.000	1.167	1.167	1.333	1.333
W lbs.	845	1322	1610	1928	2606	2961	3789	4230	5211	5715
F lbs.	260	427	602	807	1084	1354	1707	2042	2470	2870
Mo ft.-lbs.	362	759	1270	1972	3071	4287	6068	7940	10568	13236
Mr ft.-lbs.	906	1865	2756	4148	6452	8242	11770	15015	20219	24701
O.T.R.	2.505	2.458	2.169	2.104	2.101	1.923	1.940	1.891	1.913	1.866
$\overline{X}$ ft.	0.645	0.837	0.923	1.129	1.297	1.336	1.505	1.673	1.852	2.006
e ft.	0.272	0.330	0.410	0.455	0.536	0.664	0.745	0.827	0.898	0.994
Me ft.-lbs.	230	436	661	876	1397	1967	2823	3501	4680	5678
S ft.^{3}	0.560	0.907	1.185	1.671	2.241	2.667	3.375	4.167	5.042	6.000
S.P.h p.s.f.	871	1047	1161	1133	1334	1478	1678	1686	1876	1899
S.P.t p.s.f.	51	86	46	85	87	3	6	6	19	6
Friction	0.308	0.323	0.374	0.418	0.416	0.457	0.450	0.483	0.474	0.502

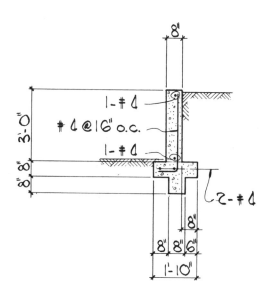

Surcharge load = **50** p.s.f.
E.F.P. = **30** p.c.f.
Section 6.6.3

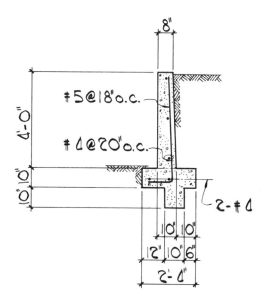

Surcharge load = **50** p.s.f.
E.F.P. = **30** p.c.f.
Section 6.6.4

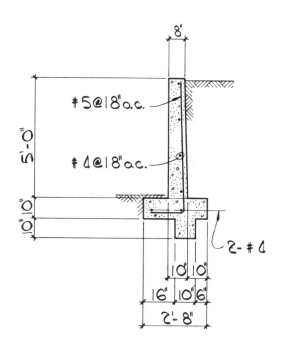

Surcharge load = **50** p.s.f.
E.F.P. = **30** p.c.f.
Section 6.6.5

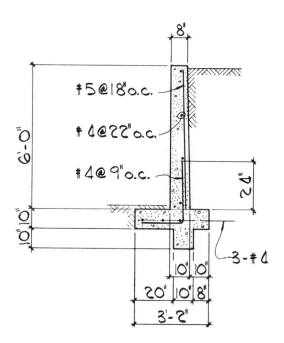

Surcharge load = **50** p.s.f.
E.F.P. = **30** p.c.f.
Section 6.6.6

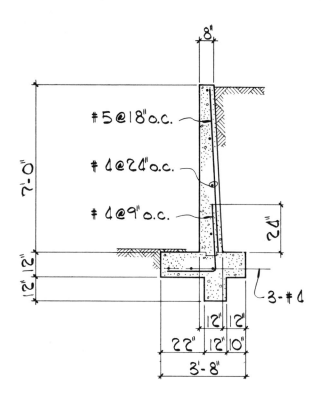

#5@18"o.c.

4@24"o.c.

4@9"o.c.

7'-0"

2" 12"

2d

3-#4

12" 12"

22"   12" 10"

3'-8"

Surcharge load = 50 p.s.f.
E.F.P. = 30 p.c.f.
Section 6.6.7

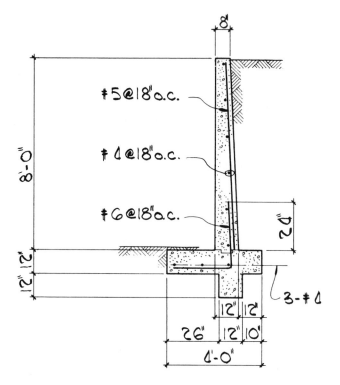

#5@18"o.c.

4@18"o.c.

#6@18"o.c.

8'-0"

12" 12"

2d

3-#4

12" 12"

26"   12" 10"

4'-0"

Surcharge load = 50 p.s.f.
E.F.P. = 30 p.c.f.
Section 6.6.8

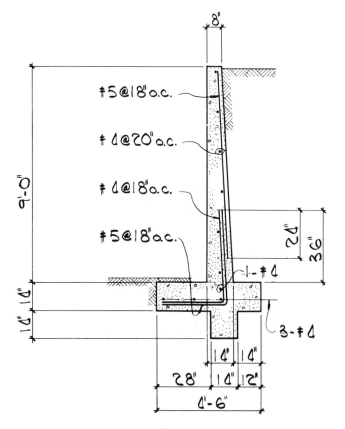

#5@18"o.c.

4@20"o.c.

4@18"o.c.

#5@18"o.c.

9'-0"

2d

36"

1-#4

14" 14"

3-#4

14" 14"

28"   14" 12"

4'-6"

Surcharge load = 50 p.s.f.
E.F.P. = 30 p.c.f.
Section 6.6.9

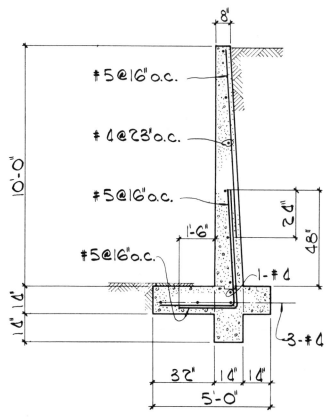

#5@16"o.c.

4@23"o.c.

#5@16"o.c.

#5@16"o.c.

10'-0"

1'-6"

2d

48"

1-#4

14" 14"

3-#4

14" 14"

32"   14" 14"

5'-0"

Surcharge load = 50 p.s.f.
E.F.P. = 30 p.c.f.
Section 6.6.10

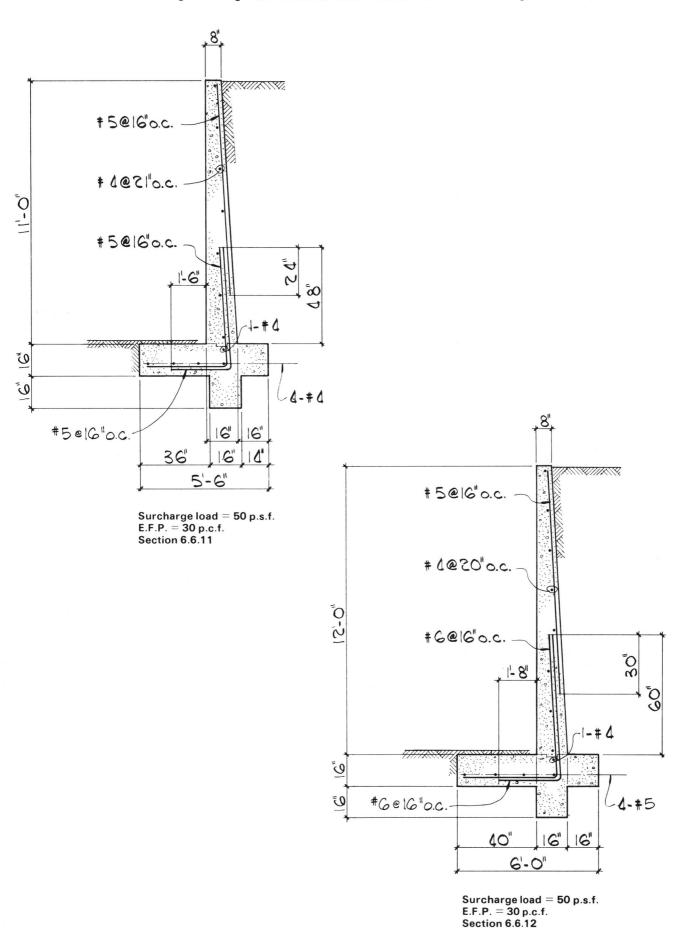

Surcharge load = 50 p.s.f.
E.F.P. = 30 p.c.f.
Section 6.6.11

Surcharge load = 50 p.s.f.
E.F.P. = 30 p.c.f.
Section 6.6.12

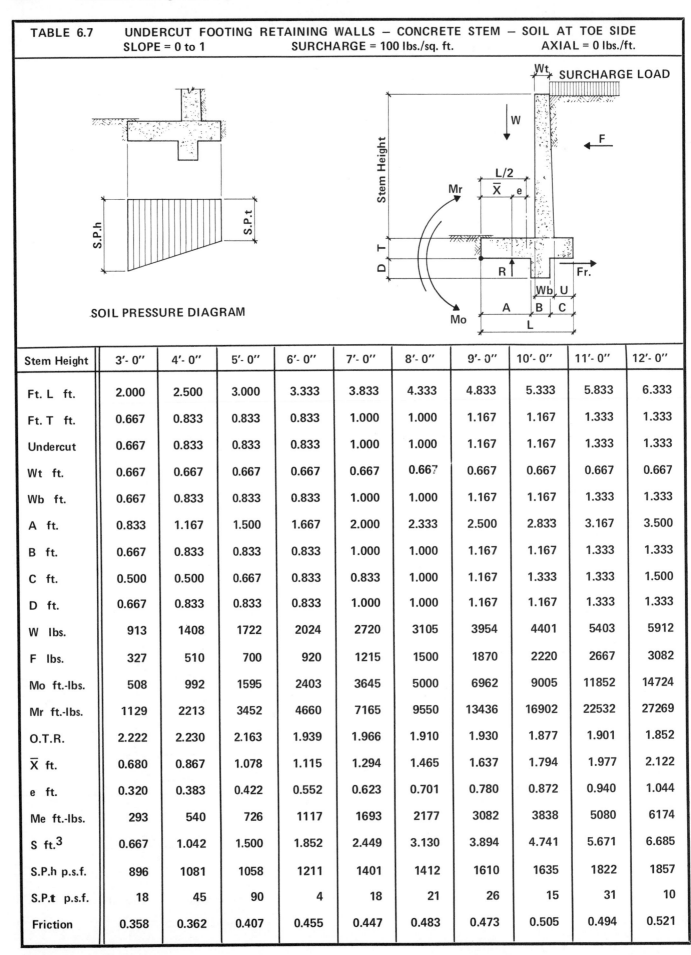

TABLE 6.7   UNDERCUT FOOTING RETAINING WALLS — CONCRETE STEM — SOIL AT TOE SIDE
SLOPE = 0 to 1   SURCHARGE = 100 lbs./sq. ft.   AXIAL = 0 lbs./ft.

SOIL PRESSURE DIAGRAM

Stem Height	3'- 0"	4'- 0"	5'- 0"	6'- 0"	7'- 0"	8'- 0"	9'- 0"	10'- 0"	11'- 0"	12'- 0"
Ft. L ft.	2.000	2.500	3.000	3.333	3.833	4.333	4.833	5.333	5.833	6.333
Ft. T ft.	0.667	0.833	0.833	0.833	1.000	1.000	1.167	1.167	1.333	1.333
Undercut	0.667	0.833	0.833	0.833	1.000	1.000	1.167	1.167	1.333	1.333
Wt ft.	0.667	0.667	0.667	0.667	0.667	0.667	0.667	0.667	0.667	0.667
Wb ft.	0.667	0.833	0.833	0.833	1.000	1.000	1.167	1.167	1.333	1.333
A ft.	0.833	1.167	1.500	1.667	2.000	2.333	2.500	2.833	3.167	3.500
B ft.	0.667	0.833	0.833	0.833	1.000	1.000	1.167	1.167	1.333	1.333
C ft.	0.500	0.500	0.667	0.833	0.833	1.000	1.167	1.333	1.333	1.500
D ft.	0.667	0.833	0.833	0.833	1.000	1.000	1.167	1.167	1.333	1.333
W lbs.	913	1408	1722	2024	2720	3105	3954	4401	5403	5912
F lbs.	327	510	700	920	1215	1500	1870	2220	2667	3082
Mo ft.-lbs.	508	992	1595	2403	3645	5000	6962	9005	11852	14724
Mr ft.-lbs.	1129	2213	3452	4660	7165	9550	13436	16902	22532	27269
O.T.R.	2.222	2.230	2.163	1.939	1.966	1.910	1.930	1.877	1.901	1.852
$\overline{X}$ ft.	0.680	0.867	1.078	1.115	1.294	1.465	1.637	1.794	1.977	2.122
e ft.	0.320	0.383	0.422	0.552	0.623	0.701	0.780	0.872	0.940	1.044
Me ft.-lbs.	293	540	726	1117	1693	2177	3082	3838	5080	6174
S ft.^{3}	0.667	1.042	1.500	1.852	2.449	3.130	3.894	4.741	5.671	6.685
S.P.h p.s.f.	896	1081	1058	1211	1401	1412	1610	1635	1822	1857
S.P.t p.s.f.	18	45	90	4	18	21	26	15	31	10
Friction	0.358	0.362	0.407	0.455	0.447	0.483	0.473	0.505	0.494	0.521

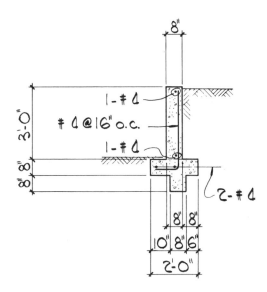

Surcharge load = **100** p.s.f.
E.F.P. = **30** p.c.f.
Section **6.7.3**

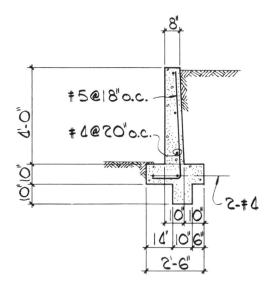

Surcharge load = **100** p.s.f.
E.F.P. = **30** p.c.f.
Section **6.7.4**

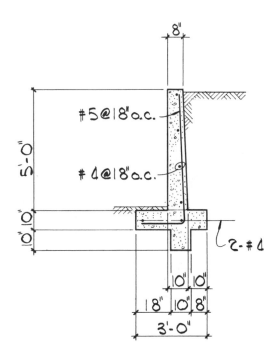

Surcharge load = **100** p.s.f.
E.F.P. = **30** p.c.f.
Section **6.7.5**

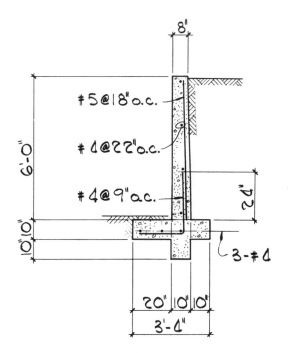

Surcharge load = **100** p.s.f.
E.F.P. = **30** p.c.f.
Section **6.7.6**

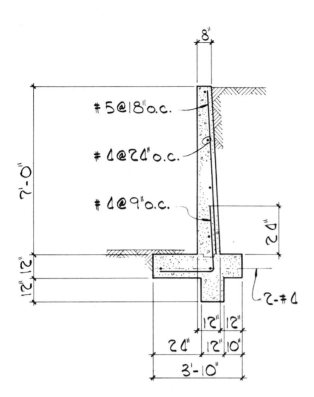

Surcharge load = 100 p.s.f.
E.F.P. = 30 p.c.f.
Section 6.7.8

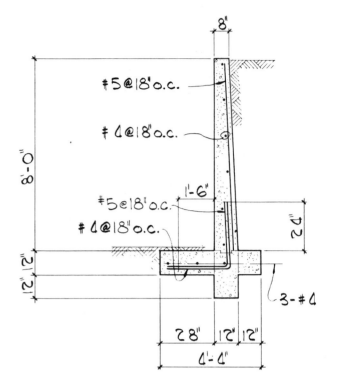

Surcharge load = 100 p.s.f.
E.F.P. = 30 p.c.f.
Section 6.7.7

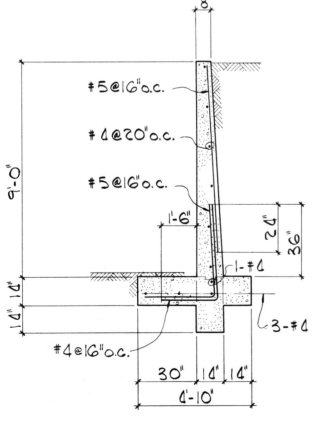

Surcharge load = 100 p.s.f.
E.F.P. = 30 p.c.f.
Section 6.7.9

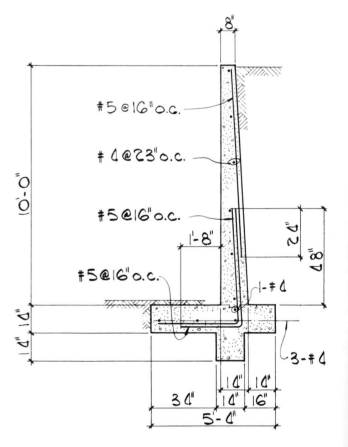

Surcharge load = 100 p.s.f.
E.F.P. = 30 p.c.f.
Section 6.7.10

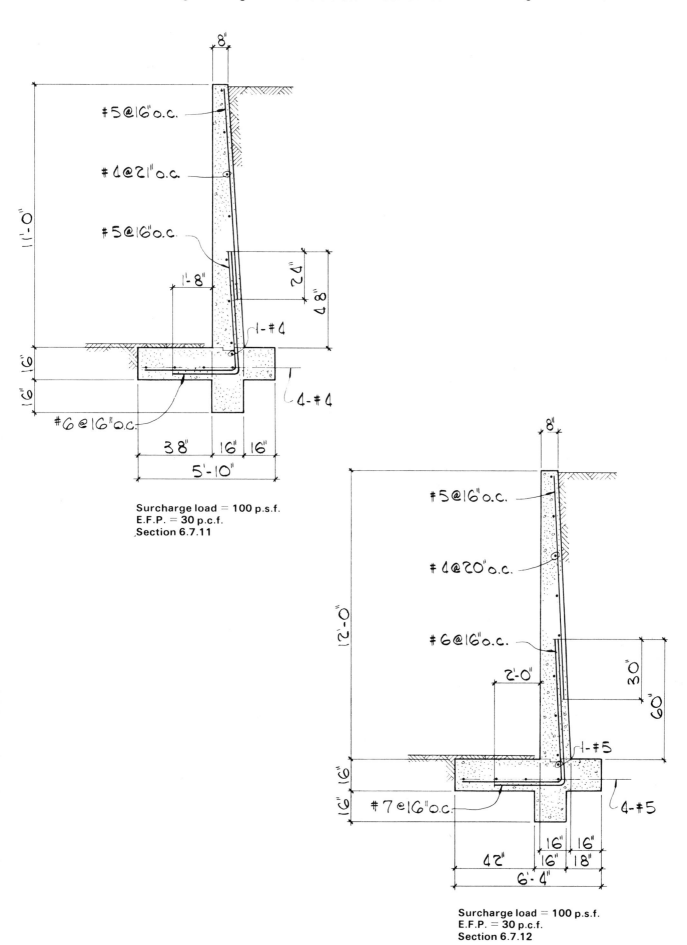

#5@16"o.c.

#4@21"o.c.

#5@16"o.c.

11'-0"

1'-8"

24"

48"

1-#4

6" 6" 6"

#6@16"o.c.

4-#4

38"   16" 16"

5'-10"

Surcharge load = 100 p.s.f.
E.F.P. = 30 p.c.f.
Section 6.7.11

8"

#5@16"o.c.

#4@20"o.c.

#6@16"o.c.

12'-0"

2'-0"

30"

60"

1-#5

6" 6"

#7@16"o.c.

4-#5

42"   16" 16"

16" 18"

6'-4"

Surcharge load = 100 p.s.f.
E.F.P. = 30 p.c.f.
Section 6.7.12

TABLE 6.8	UNDERCUT FOOTING RETAINING WALLS – CONCRETE STEM – SOIL AT TOE SIDE
	SLOPE = 0 to 1 SURCHARGE = 200 lbs./sq. ft. AXIAL = 0 lbs./ft.

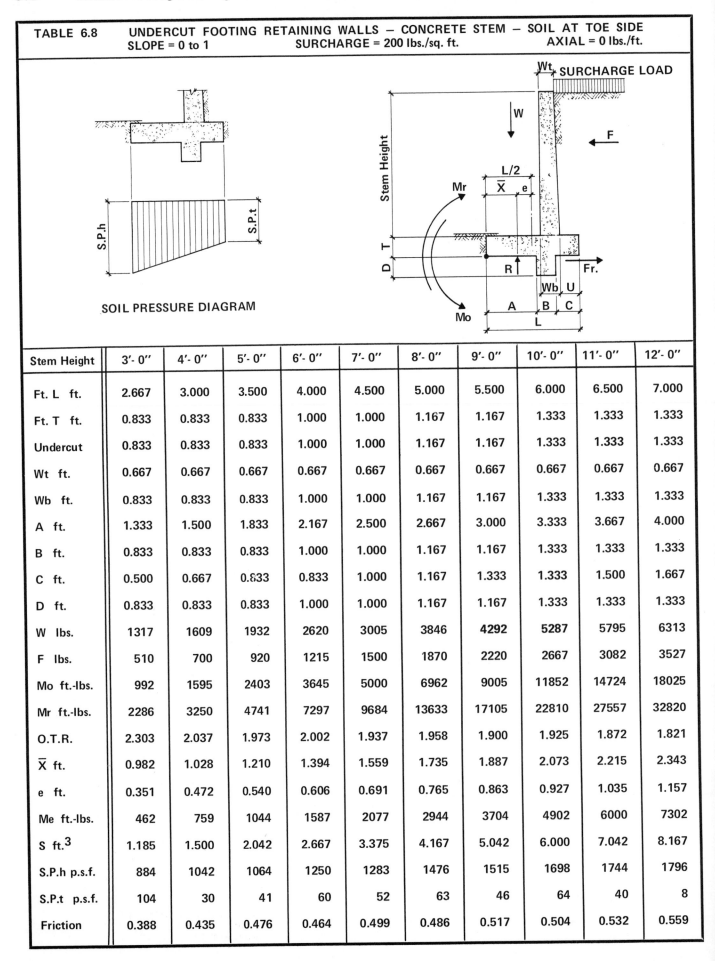

SOIL PRESSURE DIAGRAM

Stem Height	3'- 0''	4'- 0''	5'- 0''	6'- 0''	7'- 0''	8'- 0''	9'- 0''	10'- 0''	11'- 0''	12'- 0''
Ft. L ft.	2.667	3.000	3.500	4.000	4.500	5.000	5.500	6.000	6.500	7.000
Ft. T ft.	0.833	0.833	0.833	1.000	1.000	1.167	1.167	1.333	1.333	1.333
Undercut	0.833	0.833	0.833	1.000	1.000	1.167	1.167	1.333	1.333	1.333
Wt ft.	0.667	0.667	0.667	0.667	0.667	0.667	0.667	0.667	0.667	0.667
Wb ft.	0.833	0.833	0.833	1.000	1.000	1.167	1.167	1.333	1.333	1.333
A ft.	1.333	1.500	1.833	2.167	2.500	2.667	3.000	3.333	3.667	4.000
B ft.	0.833	0.833	0.833	1.000	1.000	1.167	1.167	1.333	1.333	1.333
C ft.	0.500	0.667	0.833	0.833	1.000	1.167	1.333	1.333	1.500	1.667
D ft.	0.833	0.833	0.833	1.000	1.000	1.167	1.167	1.333	1.333	1.333
W lbs.	1317	1609	1932	2620	3005	3846	4292	5287	5795	6313
F lbs.	510	700	920	1215	1500	1870	2220	2667	3082	3527
Mo ft.-lbs.	992	1595	2403	3645	5000	6962	9005	11852	14724	18025
Mr ft.-lbs.	2286	3250	4741	7297	9684	13633	17105	22810	27557	32820
O.T.R.	2.303	2.037	1.973	2.002	1.937	1.958	1.900	1.925	1.872	1.821
$\overline{X}$ ft.	0.982	1.028	1.210	1.394	1.559	1.735	1.887	2.073	2.215	2.343
e ft.	0.351	0.472	0.540	0.606	0.691	0.765	0.863	0.927	1.035	1.157
Me ft.-lbs.	462	759	1044	1587	2077	2944	3704	4902	6000	7302
S ft.^{3}	1.185	1.500	2.042	2.667	3.375	4.167	5.042	6.000	7.042	8.167
S.P.h p.s.f.	884	1042	1064	1250	1283	1476	1515	1698	1744	1796
S.P.t p.s.f.	104	30	41	60	52	63	46	64	40	8
Friction	0.388	0.435	0.476	0.464	0.499	0.486	0.517	0.504	0.532	0.559

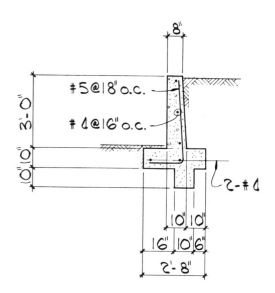

Surcharge load = 200 p.s.f.
E.F.P. = 30 p.c.f.
Section 6.8.3

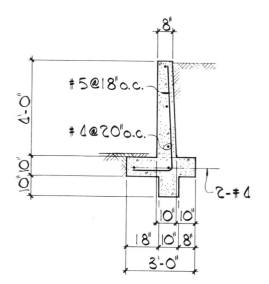

Surcharge load = 200 p.s.f.
E.F.P. = 30 p.c.f.
Section 6.8.4

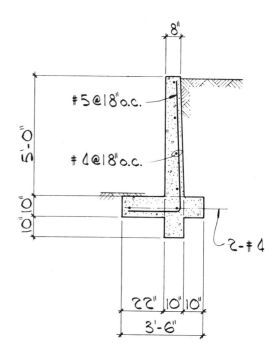

Surcharge load = 200 p.s.f.
E.F.P. = 30 p.c.f.
Section 6.8.5

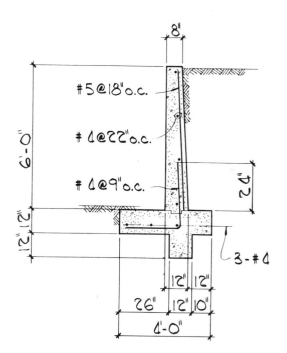

Surcharge load = 200 p.s.f.
E.F.P. = 30 p.c.f.
Section 6.8.6

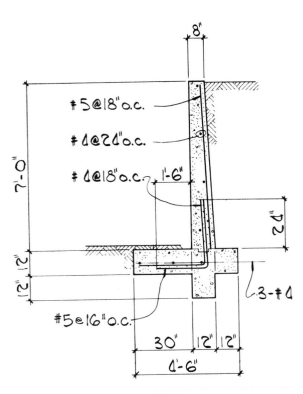

Surcharge load = 200 p.s.f.
E.F.P. = 30 p.c.f.
Section 6.8.7

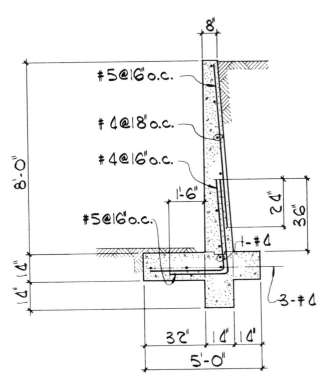

Surcharge load = 200 p.s.f.
E.F.P. = 30 p.c.f.
Section 6.8.8

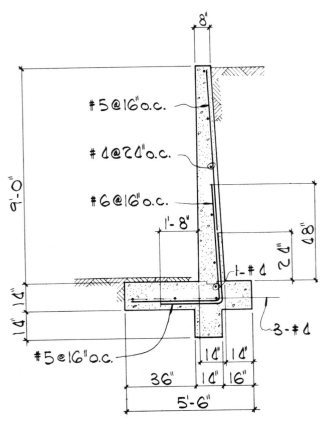

Surcharge load = 200 p.s.f.
E.F.P. = 30 p.c.f.
Section 6.8.9

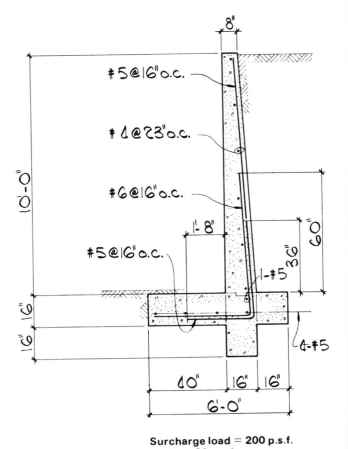

Surcharge load = 200 p.s.f.
E.F.P. = 30 p.c.f.
Section 6.8.10

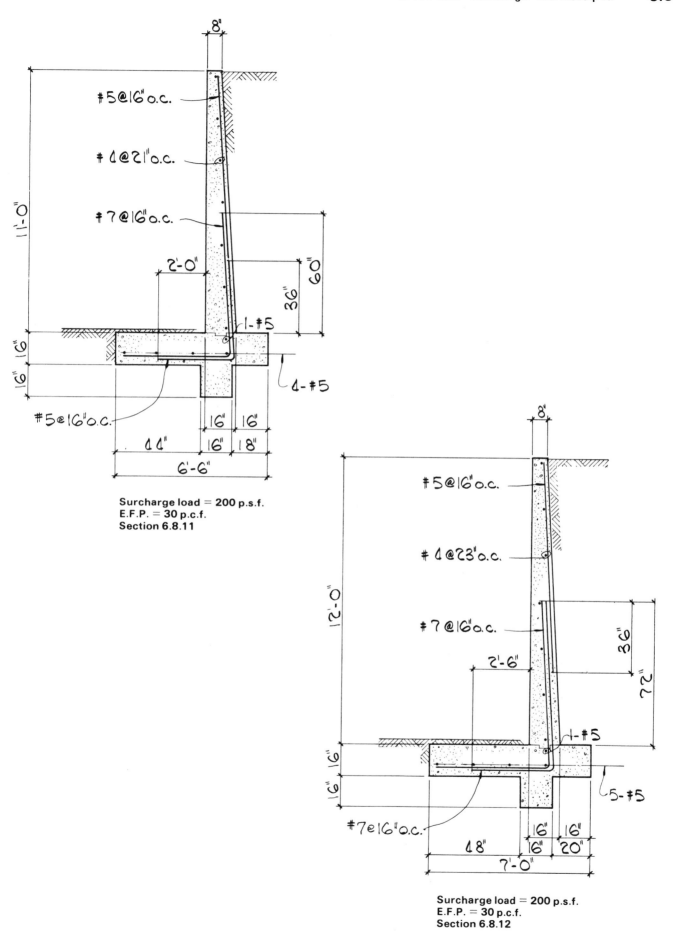

#5@16" o.c.

#4@21" o.c.

#7@16" o.c.

11'-0"

2'-0"

36"

60"

1-#5

16" 16"

4-#5

#5@16" o.c.

16" 18"

11"

6'-6"

Surcharge load = 200 p.s.f.
E.F.P. = 30 p.c.f.
Section 6.8.11

8"

#5@16" o.c.

#4@23" o.c.

#7@16" o.c.

12'-0"

2'-6"

36"

72"

1-#5

5-#5

#7@16" o.c.

16" 16"

18"

16" 20"

7'-0"

Surcharge load = 200 p.s.f.
E.F.P. = 30 p.c.f.
Section 6.8.12

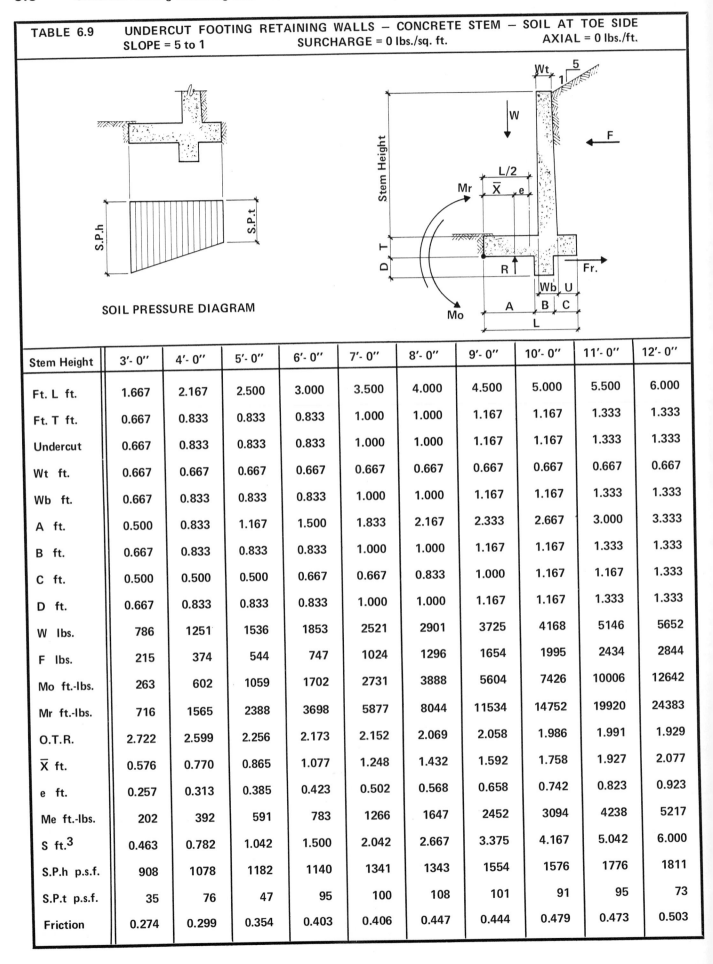

TABLE 6.9   UNDERCUT FOOTING RETAINING WALLS – CONCRETE STEM – SOIL AT TOE SIDE
SLOPE = 5 to 1   SURCHARGE = 0 lbs./sq. ft.   AXIAL = 0 lbs./ft.

SOIL PRESSURE DIAGRAM

Stem Height	3'- 0"	4'- 0"	5'- 0"	6'- 0"	7'- 0"	8'- 0"	9'- 0"	10'- 0"	11'- 0"	12'- 0"
Ft. L ft.	1.667	2.167	2.500	3.000	3.500	4.000	4.500	5.000	5.500	6.000
Ft. T ft.	0.667	0.833	0.833	0.833	1.000	1.000	1.167	1.167	1.333	1.333
Undercut	0.667	0.833	0.833	0.833	1.000	1.000	1.167	1.167	1.333	1.333
Wt ft.	0.667	0.667	0.667	0.667	0.667	0.667	0.667	0.667	0.667	0.667
Wb ft.	0.667	0.833	0.833	0.833	1.000	1.000	1.167	1.167	1.333	1.333
A ft.	0.500	0.833	1.167	1.500	1.833	2.167	2.333	2.667	3.000	3.333
B ft.	0.667	0.833	0.833	0.833	1.000	1.000	1.167	1.167	1.333	1.333
C ft.	0.500	0.500	0.500	0.667	0.667	0.833	1.000	1.167	1.167	1.333
D ft.	0.667	0.833	0.833	0.833	1.000	1.000	1.167	1.167	1.333	1.333
W lbs.	786	1251	1536	1853	2521	2901	3725	4168	5146	5652
F lbs.	215	374	544	747	1024	1296	1654	1995	2434	2844
Mo ft.-lbs.	263	602	1059	1702	2731	3888	5604	7426	10006	12642
Mr ft.-lbs.	716	1565	2388	3698	5877	8044	11534	14752	19920	24383
O.T.R.	2.722	2.599	2.256	2.173	2.152	2.069	2.058	1.986	1.991	1.929
$\overline{X}$ ft.	0.576	0.770	0.865	1.077	1.248	1.432	1.592	1.758	1.927	2.077
e ft.	0.257	0.313	0.385	0.423	0.502	0.568	0.658	0.742	0.823	0.923
Me ft.-lbs.	202	392	591	783	1266	1647	2452	3094	4238	5217
S ft.^{3}	0.463	0.782	1.042	1.500	2.042	2.667	3.375	4.167	5.042	6.000
S.P.h p.s.f.	908	1078	1182	1140	1341	1343	1554	1576	1776	1811
S.P.t p.s.f.	35	76	47	95	100	108	101	91	95	73
Friction	0.274	0.299	0.354	0.403	0.406	0.447	0.444	0.479	0.473	0.503

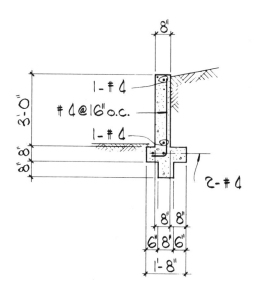

Soil slope = 5 to 1
E.F.P. = 32 p.c.f.
Section 6.9.3

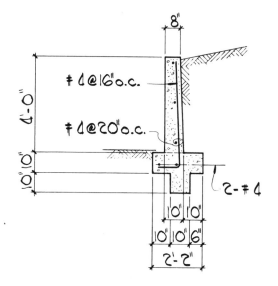

Soil slope = 5 to 1
E.F.P. = 32 p.c.f.
Section 6.9.4

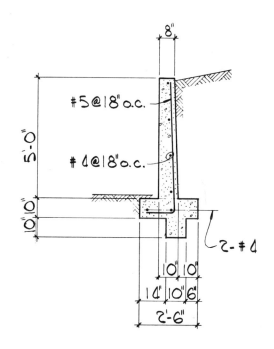

Soil slope = 5 to 1
E.F.P. = 32 p.c.f.
Section 6.9.5

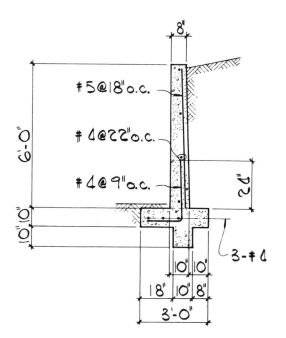

Soil slope = 5 to 1
E.F.P. = 32 p.c.f.
Section 6.9.6

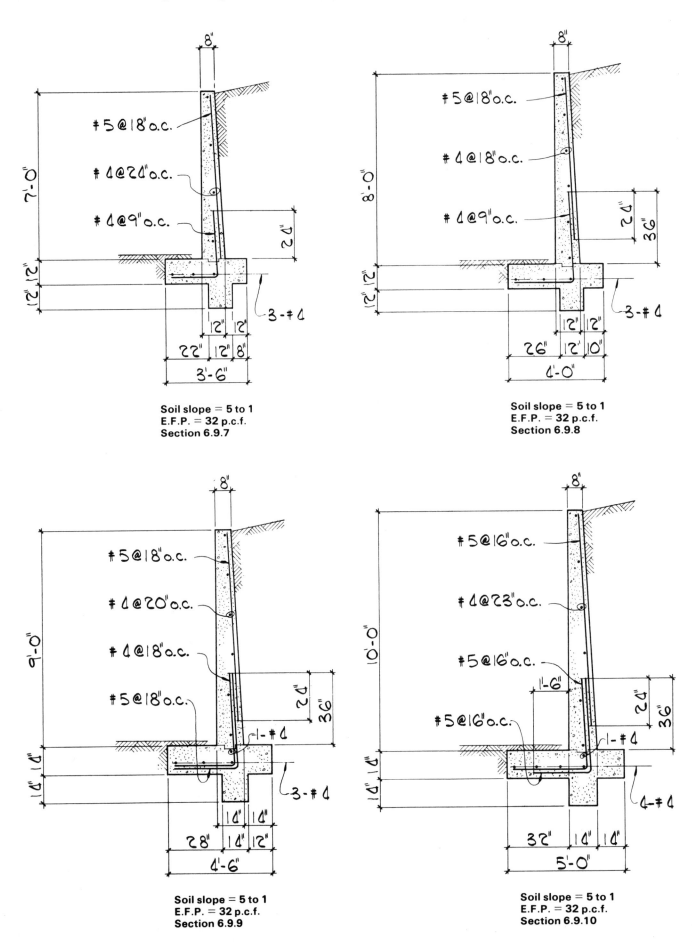

Soil slope = 5 to 1
E.F.P. = 32 p.c.f.
Section 6.9.7

Soil slope = 5 to 1
E.F.P. = 32 p.c.f.
Section 6.9.8

Soil slope = 5 to 1
E.F.P. = 32 p.c.f.
Section 6.9.9

Soil slope = 5 to 1
E.F.P. = 32 p.c.f.
Section 6.9.10

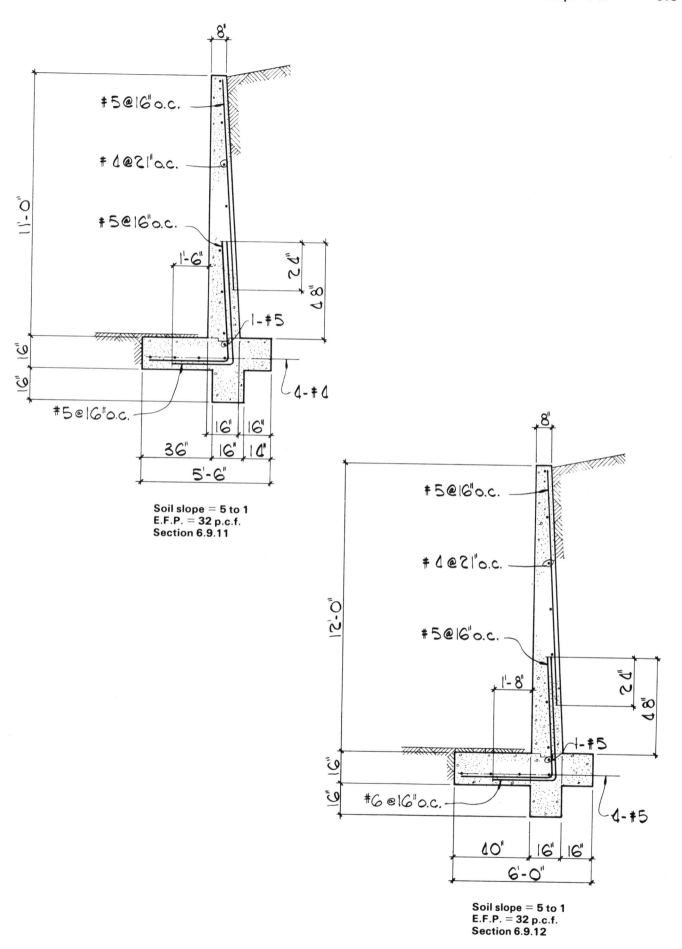

#5@16"o.c.

#4@21"o.c.

#5@16"o.c.

8'

1'-6"

2d

4 8"

1-#5

11'-0"

6"

6"

1-#4

#5@16"o.c.

16"   16"

36"   16" 10"

5'-6"

Soil slope = 5 to 1
E.F.P. = 32 p.c.f.
Section 6.9.11

#5@16"o.c.

#4@21"o.c.

#5@16"o.c.

8'

1'-8"

2d

4 8"

1-#5

12'-0"

6"

6"

#6@16"o.c.

1-#5

10"   16"   16"

6'-0"

Soil slope = 5 to 1
E.F.P. = 32 p.c.f.
Section 6.9.12

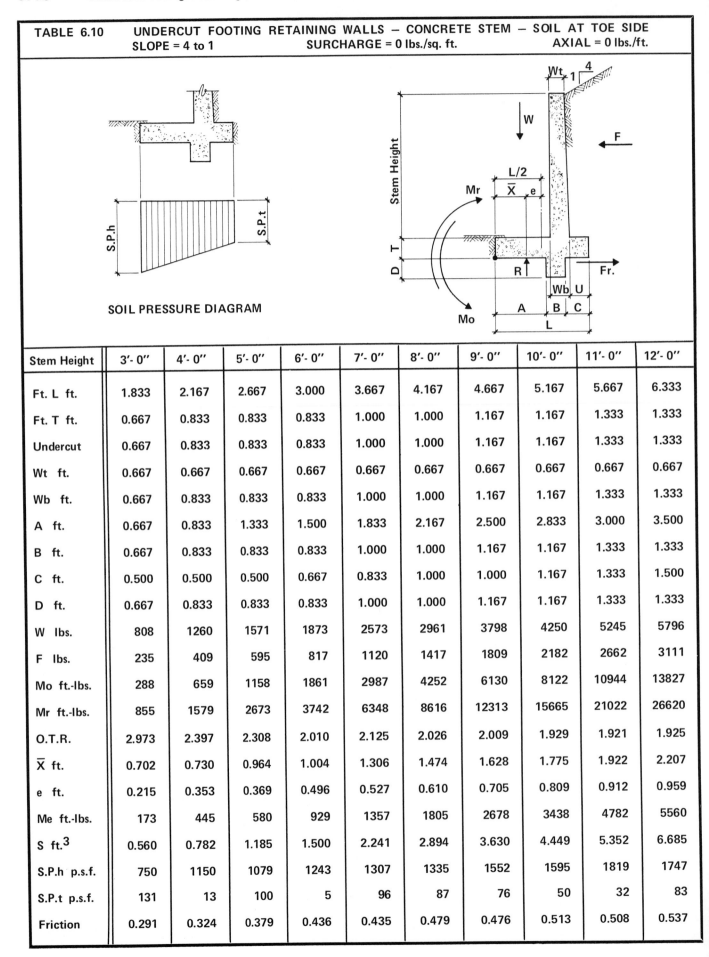

TABLE 6.10	UNDERCUT FOOTING RETAINING WALLS – CONCRETE STEM – SOIL AT TOE SIDE
SLOPE = 4 to 1	SURCHARGE = 0 lbs./sq. ft. AXIAL = 0 lbs./ft.

SOIL PRESSURE DIAGRAM

Stem Height	3'- 0''	4'- 0''	5'- 0''	6'- 0''	7'- 0''	8'- 0''	9'- 0''	10'- 0''	11'- 0''	12'- 0''
Ft. L ft.	1.833	2.167	2.667	3.000	3.667	4.167	4.667	5.167	5.667	6.333
Ft. T ft.	0.667	0.833	0.833	0.833	1.000	1.000	1.167	1.167	1.333	1.333
Undercut	0.667	0.833	0.833	0.833	1.000	1.000	1.167	1.167	1.333	1.333
Wt ft.	0.667	0.667	0.667	0.667	0.667	0.667	0.667	0.667	0.667	0.667
Wb ft.	0.667	0.833	0.833	0.833	1.000	1.000	1.167	1.167	1.333	1.333
A ft.	0.667	0.833	1.333	1.500	1.833	2.167	2.500	2.833	3.000	3.500
B ft.	0.667	0.833	0.833	0.833	1.000	1.000	1.167	1.167	1.333	1.333
C ft.	0.500	0.500	0.500	0.667	0.833	1.000	1.000	1.167	1.333	1.500
D ft.	0.667	0.833	0.833	0.833	1.000	1.000	1.167	1.167	1.333	1.333
W lbs.	808	1260	1571	1873	2573	2961	3798	4250	5245	5796
F lbs.	235	409	595	817	1120	1417	1809	2182	2662	3111
Mo ft.-lbs.	288	659	1158	1861	2987	4252	6130	8122	10944	13827
Mr ft.-lbs.	855	1579	2673	3742	6348	8616	12313	15665	21022	26620
O.T.R.	2.973	2.397	2.308	2.010	2.125	2.026	2.009	1.929	1.921	1.925
$\overline{X}$ ft.	0.702	0.730	0.964	1.004	1.306	1.474	1.628	1.775	1.922	2.207
e ft.	0.215	0.353	0.369	0.496	0.527	0.610	0.705	0.809	0.912	0.959
Me ft.-lbs.	173	445	580	929	1357	1805	2678	3438	4782	5560
S ft.^{3}	0.560	0.782	1.185	1.500	2.241	2.894	3.630	4.449	5.352	6.685
S.P.h p.s.f.	750	1150	1079	1243	1307	1335	1552	1595	1819	1747
S.P.t p.s.f.	131	13	100	5	96	87	76	50	32	83
Friction	0.291	0.324	0.379	0.436	0.435	0.479	0.476	0.513	0.508	0.537

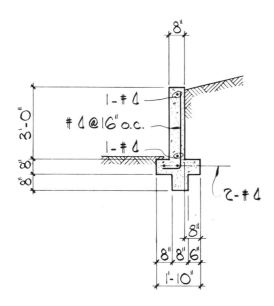

Soil slope = **4 to 1**
E.F.P. = **35** p.c.f.
Section 6.10.3

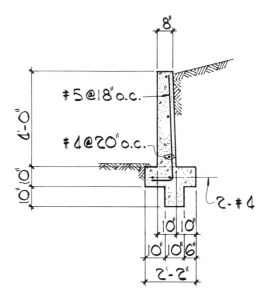

Soil slope = **4 to 1**
E.F.P. = **35** p.c.f.
Section 6.10.4

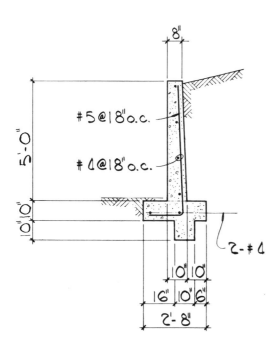

Soil slope = **4 to 1**
E.F.P. = **35** p.c.f.
Section 6.10.5

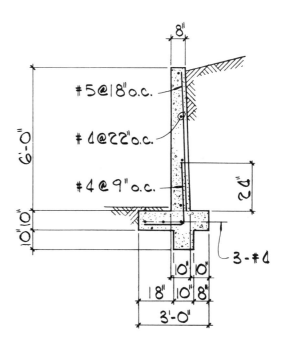

Soil slope = **4 to 1**
E.F.P. = **35** p.c.f.
Section 6.10.6

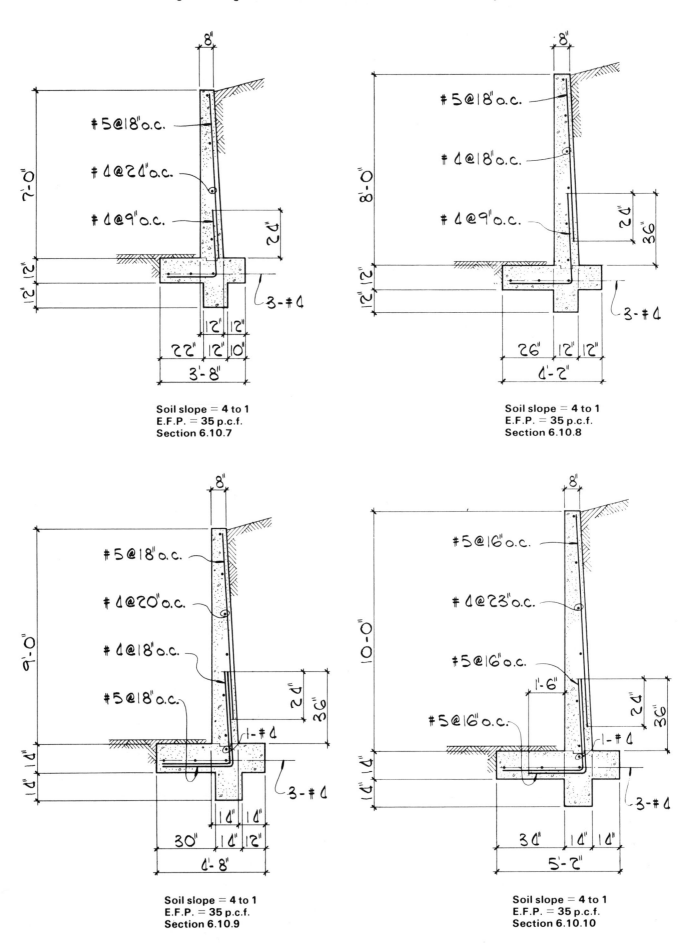

Soil slope = 4 to 1
E.F.P. = 35 p.c.f.
Section 6.10.7

Soil slope = 4 to 1
E.F.P. = 35 p.c.f.
Section 6.10.8

Soil slope = 4 to 1
E.F.P. = 35 p.c.f.
Section 6.10.9

Soil slope = 4 to 1
E.F.P. = 35 p.c.f.
Section 6.10.10

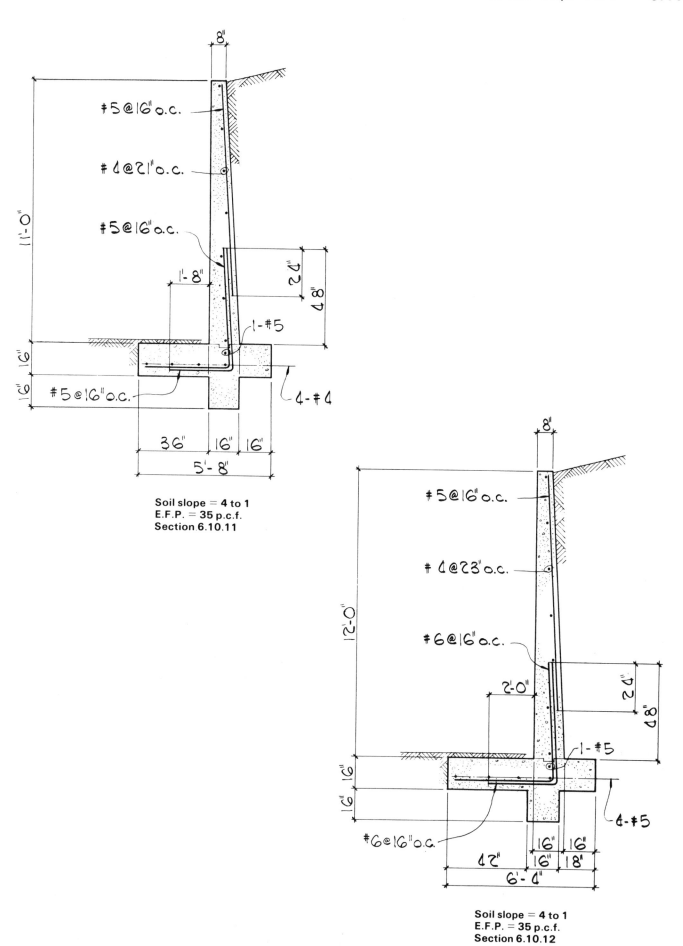

#5@16"o.c.

#4@21"o.c.

#5@16"o.c.

1'-0"

1'-8"

2'-4"

4'-8"

1-#5

6"

6"

#5@16"o.c.

4-#4

36"   16"   16"

5'-8"

Soil slope = **4 to 1**
E.F.P. = **35** p.c.f.
Section 6.10.11

8"

#5@16"o.c.

#4@23"o.c.

#6@16"o.c.

12'-0"

2'-0"

2'-4"

4'-8"

1-#5

6"   6"

6"

#6@16"o.c.

4-#5

24"   16"   16"

16"   18"

6'-4"

Soil slope = **4 to 1**
E.F.P. = **35** p.c.f.
Section 6.10.12

TABLE 6.11	UNDERCUT FOOTING RETAINING WALLS – CONCRETE STEM – SOIL AT TOE SIDE
	SLOPE = 3 to 1 SURCHARGE = 0 lbs./sq. ft. AXIAL = 0 lbs./ft.

SOIL PRESSURE DIAGRAM

Stem Height	3'- 0''	4'- 0''	5'- 0''	6'- 0''	7'- 0''	8'- 0''	9'- 0''	10'- 0''	11'- 0''	12'- 0''
Ft. L ft.	1.833	2.333	2.667	3.167	3.667	4.333	4.833	5.333	6.000	6.500
Ft. T ft.	0.667	0.833	0.833	0.833	1.000	1.000	1.167	1.167	1.333	1.333
Undercut	0.667	0.833	0.833	0.833	1.000	1.000	1.167	1.167	1.333	1.333
Wt ft.	0.667	0.667	0.667	0.667	0.667	0.667	0.667	0.667	0.667	0.667
Wb ft.	0.667	0.833	0.833	0.833	1.000	1.000	1.167	1.167	1.333	1.333
A ft.	0.667	1.000	1.333	1.667	1.833	2.333	2.500	2.833	3.333	3.667
B ft.	0.667	0.833	0.833	0.833	1.000	1.000	1.167	1.167	1.333	1.333
C ft.	0.500	0.500	0.500	0.667	0.833	1.000	1.167	1.333	1.333	1.500
D ft.	0.667	0.833	0.833	0.833	1.000	1.000	1.167	1.167	1.333	1.333
W lbs.	814	1292	1587	1915	2602	3022	3873	4335	5379	5908
F lbs.	255	444	647	887	1216	1539	1964	2369	2890	3378
Mo ft.-lbs.	312	715	1257	2021	3243	4617	6655	8819	11882	15012
Mr ft.-lbs.	863	1809	2702	4106	6427	9234	13088	16578	23105	28005
O.T.R.	2.765	2.529	2.150	2.032	1.982	2.000	1.967	1.880	1.945	1.865
$\overline{X}$ ft.	0.676	0.846	0.911	1.089	1.224	1.528	1.661	1.790	2.086	2.199
e ft.	0.240	0.320	0.422	0.494	0.610	0.639	0.756	0.877	0.914	1.051
Me ft.-lbs.	196	414	670	946	1586	1930	2927	3801	4915	6209
S ft.^{3}	0.560	0.907	1.185	1.671	2.241	3.130	3.894	4.741	6.000	7.042
S.P.h p.s.f.	793	1010	1160	1171	1418	1314	1553	1615	1716	1791
S.P.t p.s.f.	95	98	30	38	2	81	50	11	77	27
Friction	0.314	0.344	0.407	0.463	0.467	0.509	0.507	0.547	0.537	0.572

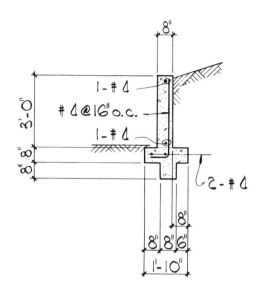

Soil slope = 3 to 1
E.F.P. = 38 p.c.f.
Section 6.11.3

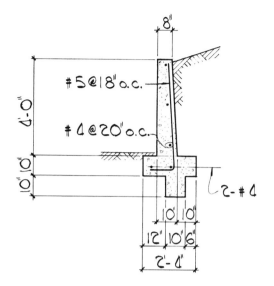

Soil slope = 3 to 1
E.F.P. = 38 p.c.f.
Section 6.11.4

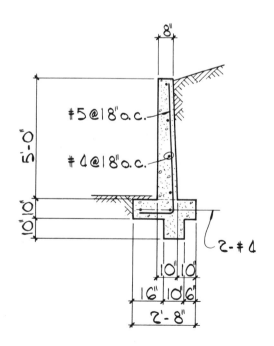

Soil slope = 3 to 1
E.F.P. = 38 p.c.f.
Section 6.11.5

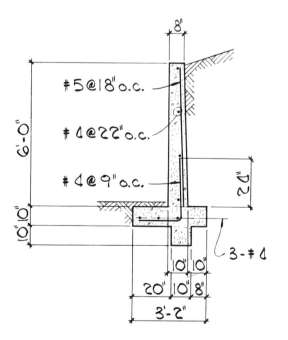

Soil slope = 3 to 1
E.F.P. = 38 p.c.f.
Section 6.11.6

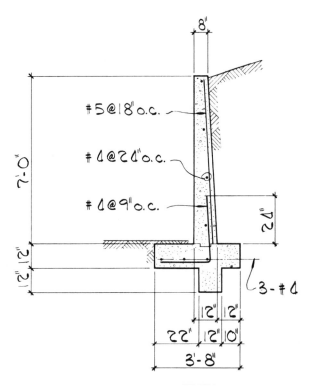

Soil slope = 3 to 1
E.F.P. = 38 p.c.f.
Section 6.11.7

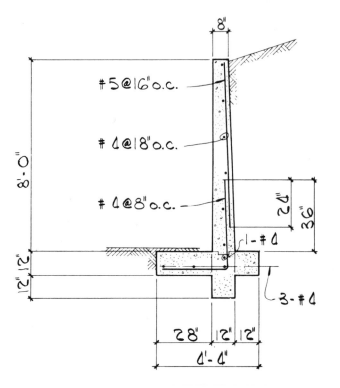

Soil slope = 3 to 1
E.F.P. = 38 p.c.f.
Section 6.11.8

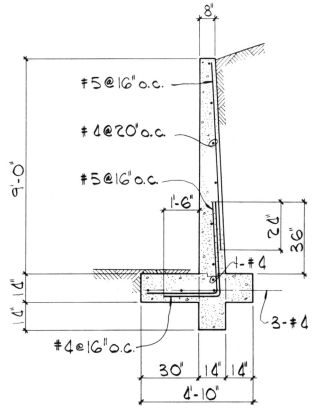

Soil slope = 3 to 1
E.F.P. = 38 p.c.f.
Section 6.11.9

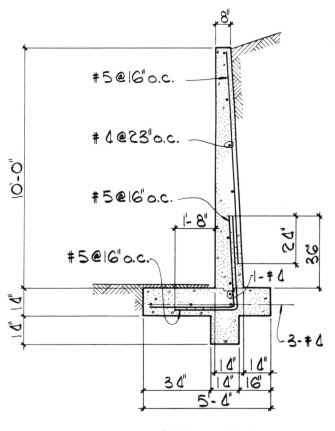

Soil slope = 3 to 1
E.F.P. = 38 p.c.f.
Section 6.11.10

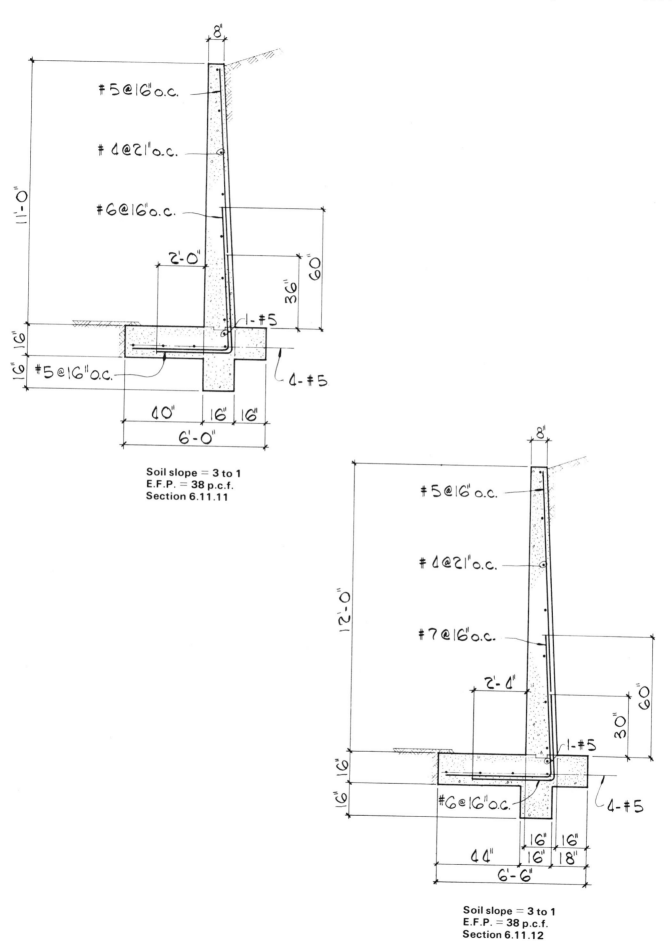

#5@16" o.c.

#4@21" o.c.

#6@16" o.c.

2'-0"

1-#5

4-#5

#5@16" o.c.

11'-0"

6"

6"

6"

36"

60"

10"   16"   16"

6'-0"

Soil slope = 3 to 1
E.F.P. = 38 p.c.f.
Section 6.11.11

#5@16" o.c.

#4@21" o.c.

#7@16" o.c.

2'-4"

1-#5

4-#5

#6@16" o.c.

12'-0"

6"

6"

30"

60"

44"   16"   16"   18"

6'-6"

Soil slope = 3 to 1
E.F.P. = 38 p.c.f.
Section 6.11.12

TABLE 6.12	UNDERCUT FOOTING RETAINING WALLS – CONCRETE STEM – SOIL AT TOE SIDE
SLOPE = 2 to 1	SURCHARGE = 0 lbs./sq. ft. AXIAL = 0 lbs./ft.

SOIL PRESSURE DIAGRAM

Stem Height	3'- 0''	4'- 0''	5'- 0''	6'- 0''	7'- 0''	8'- 0''	9'- 0''	10'- 0''	11'- 0''	12'- 0''
Ft. L ft.	1.833	2.333	2.833	3.333	4.000	4.500	5.167	5.667	6.333	7.000
Ft. T ft.	0.667	0.833	0.833	0.833	1.000	1.167	1.167	1.333	1.333	1.333
Undercut	0.667	0.833	0.833	0.833	1.000	1.167	1.167	1.333	1.333	1.333
Wt ft.	0.667	0.667	0.667	0.667	0.667	0.667	0.667	0.667	0.667	0.667
Wb ft.	0.667	0.833	0.833	0.833	1.000	1.167	1.167	1.333	1.333	1.333
A ft.	0.667	1.000	1.333	1.667	2.167	2.333	2.833	3.000	3.500	4.000
B ft.	0.667	0.833	0.833	0.833	1.000	1.167	1.167	1.333	1.333	1.333
C ft.	0.500	0.500	0.667	0.833	0.833	1.000	1.167	1.333	1.500	1.667
D ft.	0.667	0.833	0.833	0.833	1.000	1.167	1.167	1.333	1.333	1.333
W lbs.	826	1311	1634	1971	2701	3518	4010	4994	5562	6143
F lbs.	289	502	732	1004	1376	1807	2222	2762	3270	3822
Mo ft.-lbs.	353	809	1423	2287	3669	5520	7531	10433	13445	16988
Mr ft.-lbs.	878	1841	3008	4502	7456	10854	14713	19943	25456	31717
O.T.R.	2.485	2.275	2.114	1.969	2.032	1.966	1.954	1.912	1.893	1.867
$\overline{X}$ ft.	0.635	0.787	0.970	1.124	1.402	1.516	1.791	1.904	2.160	2.398
e ft.	0.281	0.380	0.447	0.543	0.598	0.734	0.792	0.929	1.007	1.102
Me ft.-lbs.	232	498	730	1070	1616	2581	3178	4640	5601	6771
S ft.^{3}	0.560	0.907	1.338	1.852	2.667	3.375	4.449	5.352	6.685	8.167
S.P.h p.s.f.	865	1111	1122	1169	1281	1546	1490	1748	1716	1707
S.P.t p.s.f.	36	13	31	14	69	17	62	14	40	48
Friction	0.350	0.383	0.448	0.509	0.509	0.514	0.554	0.553	0.588	0.622

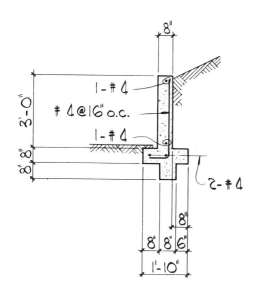

Soil slope = 2 to 1
E.F.P. = 43 p.c.f.
Section 6.12.3

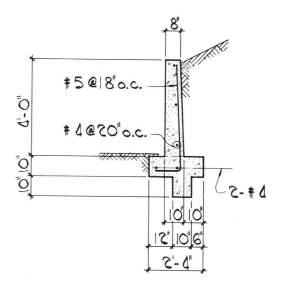

Soil slope = 2 to 1
E.F.P. = 43 p.c.f.
Section 6.12.4

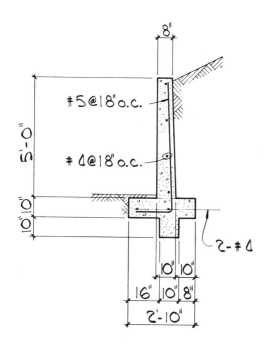

Soil slope = 2 to 1
E.F.P. = 43 p.c.f.
Section 6.12.5

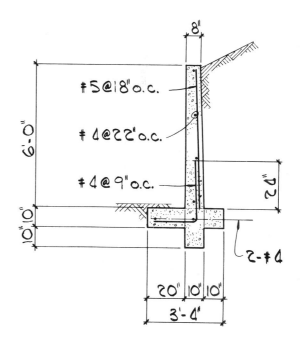

Soil slope = 2 to 1
E.F.P. = 43 p.c.f.
Section 6.12.6

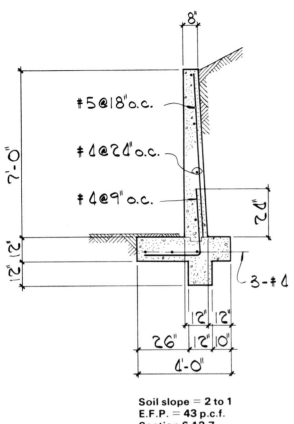

#5@18"o.c.

#4@24"o.c.

#4@9"o.c.

7'-0"

2"

12"

3-#4

12" 12"

26" 2" 10"

4'-0"

Soil slope = 2 to 1
E.F.P. = 43 p.c.f.
Section 6.12.7

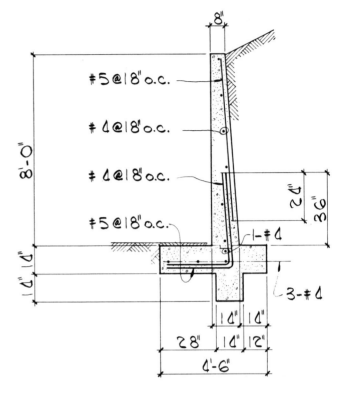

#5@18"o.c.

#4@18"o.c.

#4@18"o.c.

#5@18"o.c.

8'-0"

24"

36"

1-#4

3-#4

14"

28" 14" 12"

4'-6"

Soil slope = 2 to 1
E.F.P. = 43 p.c.f.
Section 6.12.8

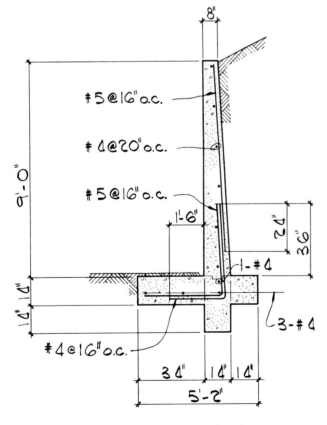

#5@16"o.c.

#4@20"o.c.

#5@16"o.c.

9'-0"

1'-6"

24"

36"

1-#4

3-#4

#4@16"o.c.

34" 14" 14"

5'-2"

Soil slope = 2 to 1
E.F.P. = 43 p.c.f.
Section 6.12.9

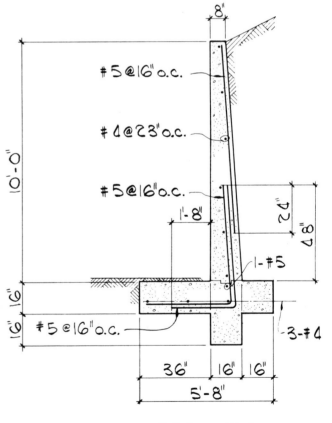

#5@16"o.c.

#4@23"o.c.

#5@16"o.c.

10'-0"

1'-8"

24"

48"

1-#5

3-#4

#5@16"o.c.

36" 16" 16"

5'-8"

Soil slope = 2 to 1
E.F.P. = 43 p.c.f.
Section 6.12.10

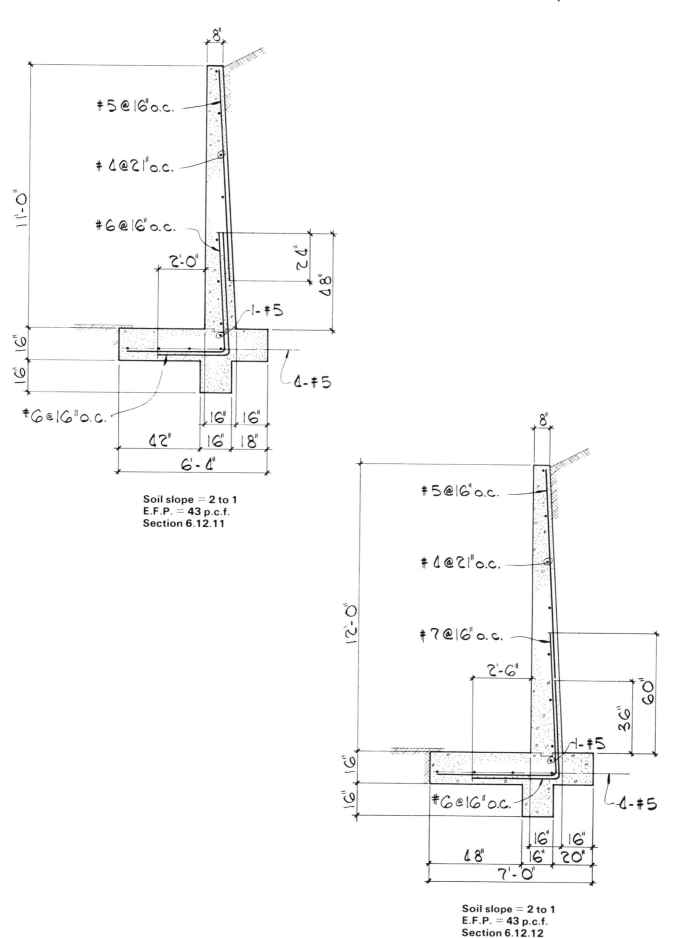

#5 @ 16" o.c.

#4 @ 21" o.c.

#6 @ 16" o.c.

2'-0"

24"

48"

1-#5

16"

16"

4-#5

#6 @ 16" o.c.

42"

16"

16"

16"

18"

6'-4"

8"

11'-0"

Soil slope = 2 to 1
E.F.P. = 43 p.c.f.
Section 6.12.11

#5 @ 16" o.c.

#4 @ 21" o.c.

#7 @ 16" o.c.

2'-6"

60"

36"

1-#5

16"

16"

4-#5

#6 @ 16" o.c.

48"

16"

16"

20"

7'-0"

8"

12'-0"

Soil slope = 2 to 1
E.F.P. = 43 p.c.f.
Section 6.12.12

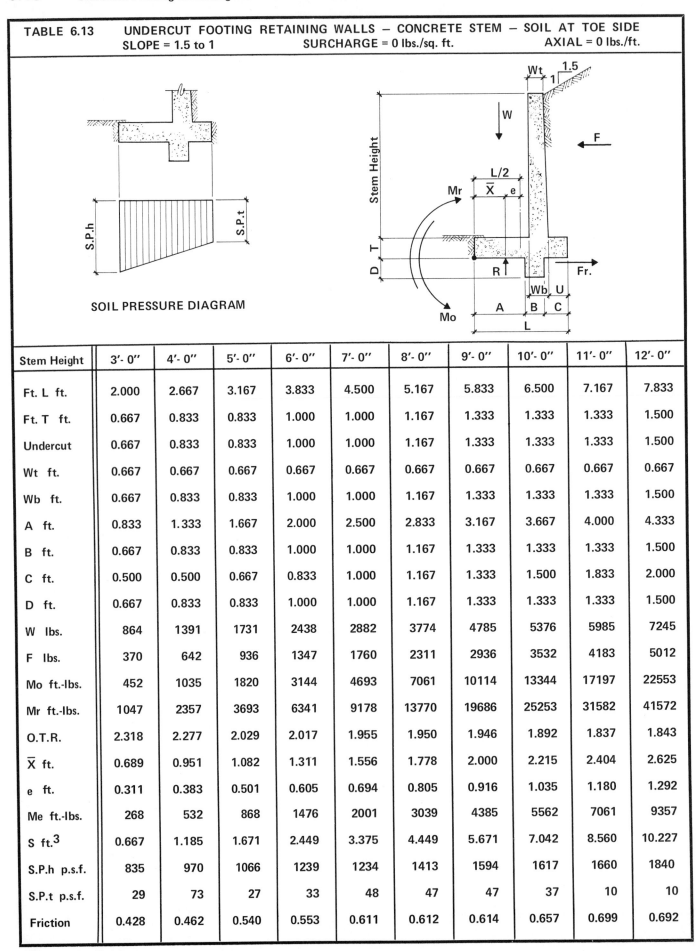

TABLE 6.13 UNDERCUT FOOTING RETAINING WALLS – CONCRETE STEM – SOIL AT TOE SIDE
SLOPE = 1.5 to 1   SURCHARGE = 0 lbs./sq. ft.   AXIAL = 0 lbs./ft.

SOIL PRESSURE DIAGRAM

Stem Height	3'-0"	4'-0"	5'-0"	6'-0"	7'-0"	8'-0"	9'-0"	10'-0"	11'-0"	12'-0"
Ft. L ft.	2.000	2.667	3.167	3.833	4.500	5.167	5.833	6.500	7.167	7.833
Ft. T ft.	0.667	0.833	0.833	1.000	1.000	1.167	1.333	1.333	1.333	1.500
Undercut	0.667	0.833	0.833	1.000	1.000	1.167	1.333	1.333	1.333	1.500
Wt ft.	0.667	0.667	0.667	0.667	0.667	0.667	0.667	0.667	0.667	0.667
Wb ft.	0.667	0.833	0.833	1.000	1.000	1.167	1.333	1.333	1.333	1.500
A ft.	0.833	1.333	1.667	2.000	2.500	2.833	3.167	3.667	4.000	4.333
B ft.	0.667	0.833	0.833	1.000	1.000	1.167	1.333	1.333	1.333	1.500
C ft.	0.500	0.500	0.667	0.833	1.000	1.167	1.333	1.500	1.833	2.000
D ft.	0.667	0.833	0.833	1.000	1.000	1.167	1.333	1.333	1.333	1.500
W lbs.	864	1391	1731	2438	2882	3774	4785	5376	5985	7245
F lbs.	370	642	936	1347	1760	2311	2936	3532	4183	5012
Mo ft.-lbs.	452	1035	1820	3144	4693	7061	10114	13344	17197	22553
Mr ft.-lbs.	1047	2357	3693	6341	9178	13770	19686	25253	31582	41572
O.T.R.	2.318	2.277	2.029	2.017	1.955	1.950	1.946	1.892	1.837	1.843
$\overline{X}$ ft.	0.689	0.951	1.082	1.311	1.556	1.778	2.000	2.215	2.404	2.625
e ft.	0.311	0.383	0.501	0.605	0.694	0.805	0.916	1.035	1.180	1.292
Me ft.-lbs.	268	532	868	1476	2001	3039	4385	5562	7061	9357
S ft.^{3}	0.667	1.185	1.671	2.449	3.375	4.449	5.671	7.042	8.560	10.227
S.P.h p.s.f.	835	970	1066	1239	1234	1413	1594	1617	1660	1840
S.P.t p.s.f.	29	73	27	33	48	47	47	37	10	10
Friction	0.428	0.462	0.540	0.553	0.611	0.612	0.614	0.657	0.699	0.692

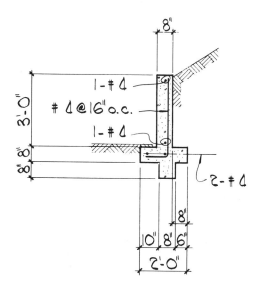

Soil slope = 1½ to 1
E.F.P. = 55 p.c.f.
Section 6.13.3

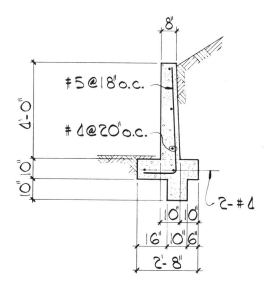

Soil slope = 1½ to 1
E.F.P. = 55 p.c.f.
Section 6.13.4

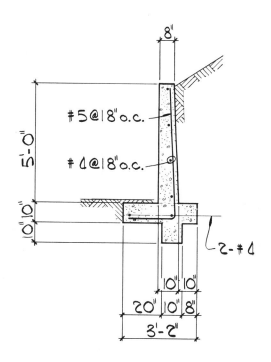

Soil slope = 1½ to 1
E.F.P. = 55 p.c.f.
Section 6.13.5

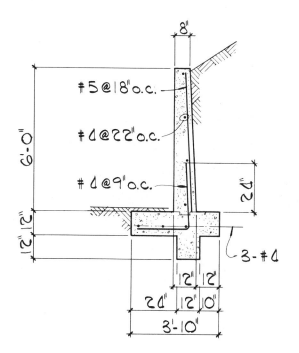

Soil slope = 1½ to 1
E.F.P. = 55 p.c.f.
Section 6.13.6

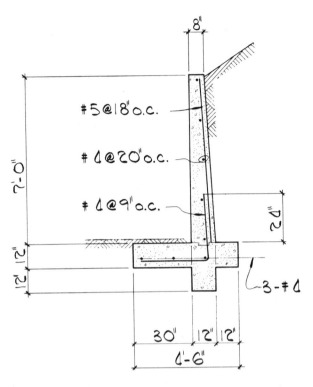

#5 @ 18" o.c.

#4 @ 20" o.c.

#4 @ 9" o.c.

7'-0"

2"

2"

2'-4"

3 - #4

30" 12" 12"

4'-6"

Soil slope = 1½ to 1
E.F.P. = 55 p.c.f.
Section 6.13.7

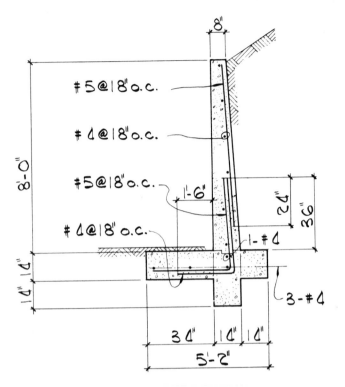

#5 @ 18" o.c.

#4 @ 18" o.c.

#5 @ 18" o.c.

#4 @ 18" o.c.

8'-0"

1'-4"

1'-4"

1'-6"

2'-4"

36"

1 - #4

3 - #4

34" 14" 14"

5'-2"

Soil slope = 1½ to 1
E.F.P. = 55 p.c.f.
Section 6.13.8

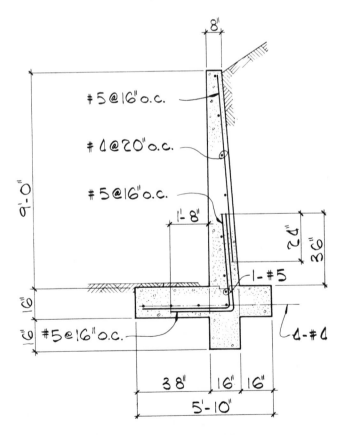

#5 @ 16" o.c.

#4 @ 20" o.c.

#5 @ 16" o.c.

#5 @ 16" o.c.

9'-0"

9"

9"

1'-8"

2'-4"

36"

1 - #5

4 - #4

38" 16" 16"

5'-10"

Soil slope = 1½ to 1
E.F.P. = 55 p.c.f.
Section 6.13.9

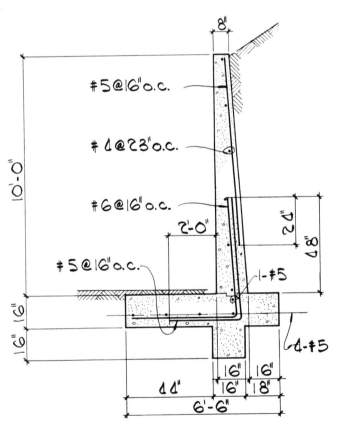

#5 @ 16" o.c.

#4 @ 23" o.c.

#6 @ 16" o.c.

#5 @ 16" o.c.

10'-0"

9"

9"

2'-0"

2'-4"

48"

1 - #5

4 - #5

44" 16" 18"
16" 16"

6'-6"

Soil slope = 1½ to 1
E.F.P. = 55 p.c.f.
Section 6.13.10

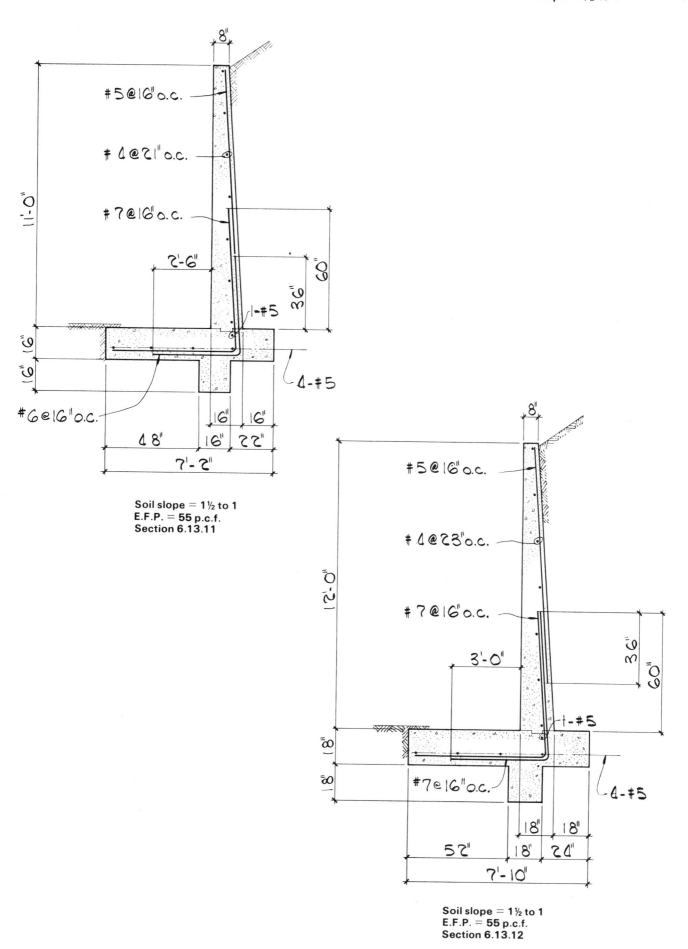

Soil slope = 1½ to 1
E.F.P. = 55 p.c.f.
Section 6.13.11

Soil slope = 1½ to 1
E.F.P. = 55 p.c.f.
Section 6.13.12

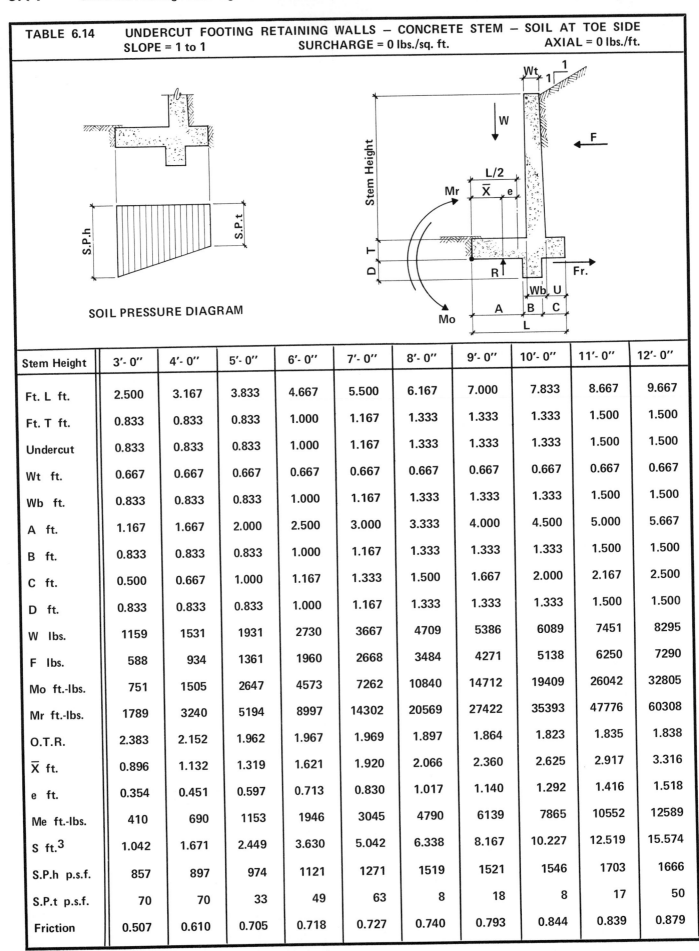

Stem Height	3'- 0"	4'- 0"	5'- 0"	6'- 0"	7'- 0"	8'- 0"	9'- 0"	10'- 0"	11'- 0"	12'- 0"
Ft. L ft.	2.500	3.167	3.833	4.667	5.500	6.167	7.000	7.833	8.667	9.667
Ft. T ft.	0.833	0.833	0.833	1.000	1.167	1.333	1.333	1.333	1.500	1.500
Undercut	0.833	0.833	0.833	1.000	1.167	1.333	1.333	1.333	1.500	1.500
Wt ft.	0.667	0.667	0.667	0.667	0.667	0.667	0.667	0.667	0.667	0.667
Wb ft.	0.833	0.833	0.833	1.000	1.167	1.333	1.333	1.333	1.500	1.500
A ft.	1.167	1.667	2.000	2.500	3.000	3.333	4.000	4.500	5.000	5.667
B ft.	0.833	0.833	0.833	1.000	1.167	1.333	1.333	1.333	1.500	1.500
C ft.	0.500	0.667	1.000	1.167	1.333	1.500	1.667	2.000	2.167	2.500
D ft.	0.833	0.833	0.833	1.000	1.167	1.333	1.333	1.333	1.500	1.500
W lbs.	1159	1531	1931	2730	3667	4709	5386	6089	7451	8295
F lbs.	588	934	1361	1960	2668	3484	4271	5138	6250	7290
Mo ft.-lbs.	751	1505	2647	4573	7262	10840	14712	19409	26042	32805
Mr ft.-lbs.	1789	3240	5194	8997	14302	20569	27422	35393	47776	60308
O.T.R.	2.383	2.152	1.962	1.967	1.969	1.897	1.864	1.823	1.835	1.838
$\bar{X}$ ft.	0.896	1.132	1.319	1.621	1.920	2.066	2.360	2.625	2.917	3.316
e ft.	0.354	0.451	0.597	0.713	0.830	1.017	1.140	1.292	1.416	1.518
Me ft.-lbs.	410	690	1153	1946	3045	4790	6139	7865	10552	12589
S ft.^{3}	1.042	1.671	2.449	3.630	5.042	6.338	8.167	10.227	12.519	15.574
S.P.h p.s.f.	857	897	974	1121	1271	1519	1521	1546	1703	1666
S.P.t p.s.f.	70	70	33	49	63	8	18	8	17	50
Friction	0.507	0.610	0.705	0.718	0.727	0.740	0.793	0.844	0.839	0.879

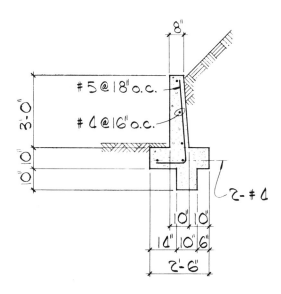

Soil slope = 1 to 1
E.F.P. = 80 p.c.f.
Section 6.14.3

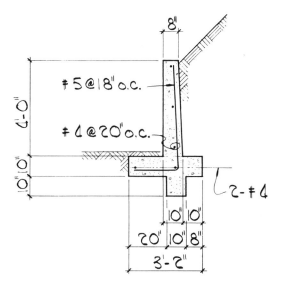

Soil slope = 1 to 1
E.F.P. = 80 p.c.f.
Section 6.14.4

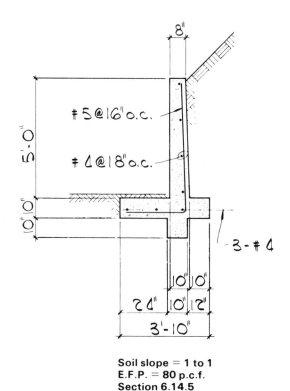

Soil slope = 1 to 1
E.F.P. = 80 p.c.f.
Section 6.14.5

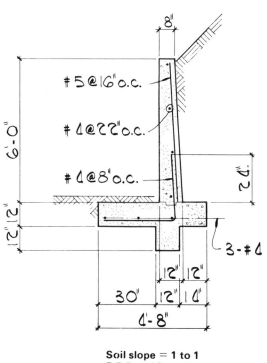

Soil slope = 1 to 1
E.F.P. = 80 p.c.f.
Section 6.14.6

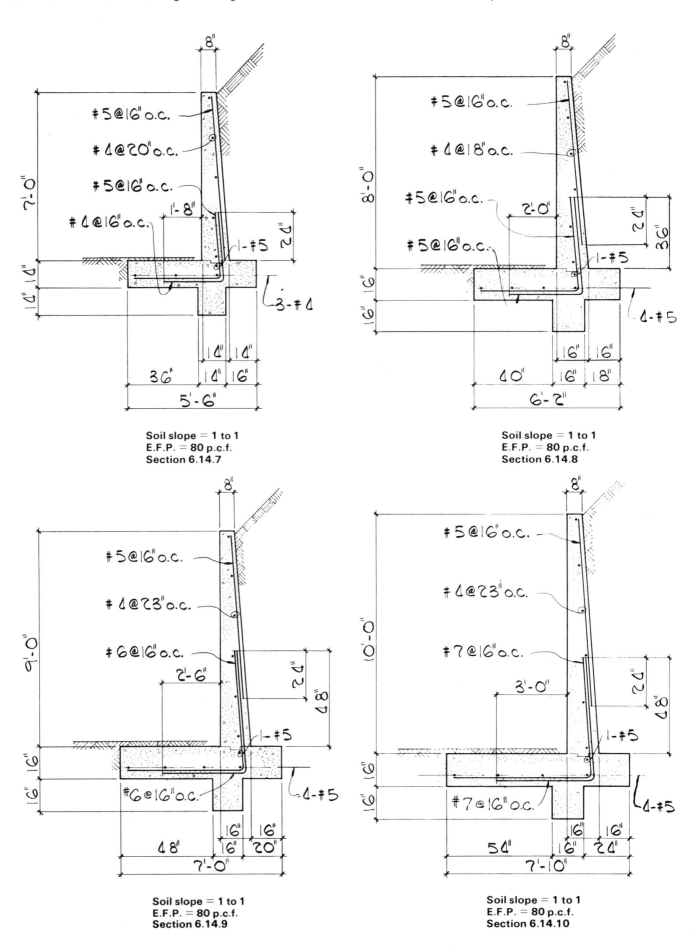

Soil slope = 1 to 1
E.F.P. = 80 p.c.f.
Section 6.14.7

Soil slope = 1 to 1
E.F.P. = 80 p.c.f.
Section 6.14.8

Soil slope = 1 to 1
E.F.P. = 80 p.c.f.
Section 6.14.9

Soil slope = 1 to 1
E.F.P. = 80 p.c.f.
Section 6.14.10

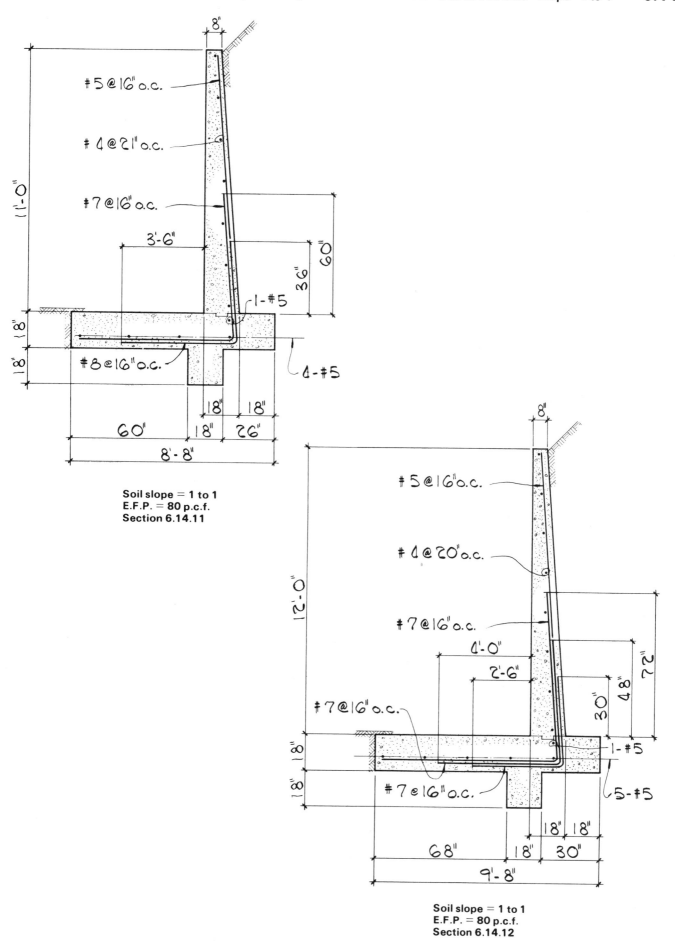

8"

#5@16" o.c.

#4@21" o.c.

#7@16" o.c.

1'-0"

3'-6"

60"

36"

1-#5

8"

18"

#8@16" o.c.

4-#5

18"

18"   18"   26"

60"

8'-8"

Soil slope = 1 to 1
E.F.P. = 80 p.c.f.
Section 6.14.11

8"

#5@16" o.c.

#4@20" o.c.

#7@16" o.c.

12'-0"

4'-0"

2'-6"

72"

48"

30"

#7@16" o.c.

8"

1-#5

8"

5-#5

#7@16" o.c.

18"   18"

68"

18"   30"

9'-8"

Soil slope = 1 to 1
E.F.P. = 80 p.c.f.
Section 6.14.12

CHAPTER 7

Undercut Footing Retaining Walls
Masonry Stem
Soil at the Toe Side of the Footing

The design data and drawings presented in this chapter are concerned with property line retaining walls constructed with concrete masonry stem walls. The retained soil is not placed over the entire wall footing, but only at the toe side of the stem wall. The chapter consists of three design examples of retaining walls with various types of loading, a design data table for each of the fourteen loading conditions specified in Fig. 1.6, and a series of corresponding drawings following each design data table. The retaining wall drawings and the design data in this chapter are for a wall length of 1 foot.

The retaining walls in this chapter are designed using the following criteria:

Weight of soil = 100 p.c.f.   Weight of concrete = 150 p.c.f.
Concrete $f_c' = 2000$ p.s.i.   Reinforcing steel $f_c' = 20,000$ p.s.i.
Hollow masonry units, grouted solid, grade N, $f_m' = 1500$ p.s.i.
Field inspection of masonry required (see the design data table)
Weight of 8-inch concrete block with all cells filled = 92 p.s.f.
Weight of 12-inch concrete block with all cells filled = 140 p.s.f.
Grout $f_c' = 2000$ p.s.i.   Mortar (type S) = 1800 p.s.i.
Maximum allowable soil pressure = 4000 p.s.f.
Minimum allowable soil pressure = 0 p.s.f.
$\overline{X} \geqslant L/3$ Minimum O.T.R. = 1.50
Coefficient of soil friction = 0.40
Passive soil pressure = 300 p.c.f.

See Tables C5 and C6 for reinforcement protection. Weep holes are not shown on the drawings.

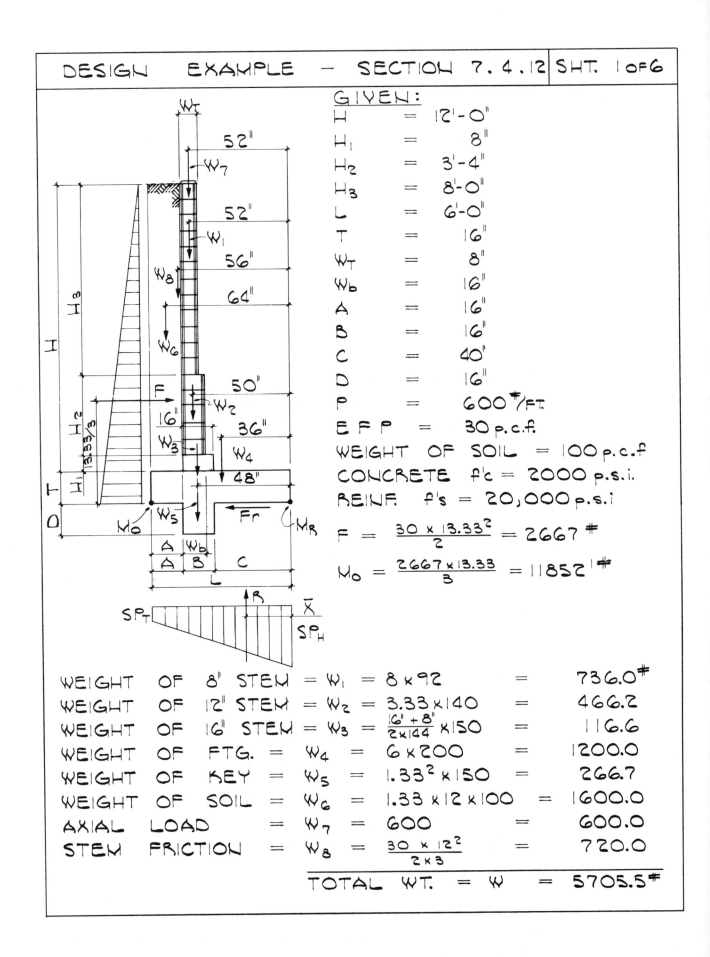

DESIGN EXAMPLE - SECTION 7.4.12 | SHT. 1 OF 6

GIVEN:

$$H = 12'\text{-}0''$$
$$H_1 = 8''$$
$$H_2 = 3'\text{-}4''$$
$$H_3 = 8'\text{-}0''$$
$$L = 6'\text{-}0''$$
$$T = 16''$$
$$W_T = 8''$$
$$W_b = 16''$$
$$A = 16''$$
$$B = 16''$$
$$C = 40''$$
$$D = 16''$$
$$P = 600\,{}^\#/\text{FT.}$$
$$EFP = 30 \text{ p.c.f.}$$

WEIGHT OF SOIL = 100 p.c.f

CONCRETE $f'c = 2000$ p.s.i.

REINF. $f's = 20,000$ p.s.i

$$F = \frac{30 \times 13.33^2}{2} = 2667\,{}^\#$$

$$M_O = \frac{2667 \times 13.33}{3} = 11852\,{}'^{\#}$$

WEIGHT OF 8" STEM	= W_1	= 8×92	=	736.0$^\#$
WEIGHT OF 12" STEM	= W_2	= 3.33×140	=	466.2
WEIGHT OF 16" STEM	= W_3	= $\frac{16'' + 8''}{2 \times 144} \times 150$	=	116.6
WEIGHT OF FTG.	= W_4	= 6×200	=	1200.0
WEIGHT OF KEY	= W_5	= $1.33^2 \times 150$	=	266.7
WEIGHT OF SOIL	= W_6	= $1.33 \times 12 \times 100$	=	1600.0
AXIAL LOAD	= W_7	= 600	=	600.0
STEM FRICTION	= W_8	= $\frac{30 \times 12^2}{2 \times 3}$	=	720.0

TOTAL WT. = W = 5705.5$^\#$

DESIGN EXAMPLE — SECTION 7.4.12 | SHT. 2 OF 6

$F = 2667^{\#}$; $M_O = 11852^{'\#}$; $W = 5705.5^{\#}$

$M_R = 736 \times \frac{52}{12} + 466.2 \times \frac{50}{12} + 116.6 \times \frac{48}{12} + 1200 \times \frac{36}{12} + 266.7 \times \frac{48}{12} + 1600 \times \frac{64}{12} +$

$\qquad 600 \times \frac{52}{12} + 720 \times \frac{56}{12} = 24758^{\#}$

OVERTURN. RATIO $= \frac{24758}{11852} = 2.089 > 1.50$ ✓

NET M $= 24758 - 11852 = 12906^{'\#}$

$\bar{X} = \frac{12906}{5705.5} = 2.262$

FTG. MIDDLE $\frac{1}{3} = \frac{6.0}{3} = 2.0' < 2.262'$

$e = \frac{6.0}{2} - 2.262 = 0.738'$

$M_e = 5705.5 \times .738 = 4210^{'\#}$

FOOTING DESIGN :

FTG. AREA $= 6.0$

FTG. SECT. MODULUS $= \frac{1 \times 6^2}{6} = 6.0$

SOIL PRESSURE $= \frac{5705.5}{6.0} + \frac{4210}{6} = 951 \pm 701.7$

$\underline{SP_H = 1652.6 \text{ p.s.f.}}$ $\qquad \underline{SP_T = 249.3 \text{ p.s.f.}}$

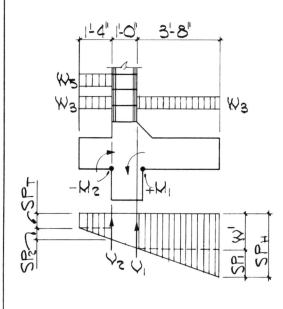

$W' = \frac{1652.6 - 249.3}{6.0} \times 2.33 + 249.3 = 794.2 \text{ p.s.f.}$

$SP_1 = 1652.6 - 794.2 = 858.4 \text{ p.s.f.}$

UNIF. LD. $= 794.2 - 200 = 594.2^{\#}/\text{FT.}$

$+M_1 = \frac{858.4 \times 3.67^2 \times 2}{3 \times 2} + \frac{594.2 \times 3.67^2}{2}$

$+M_1 = 7855.5^{'\#}$

$V_1 = \frac{858.4 \times 3.67}{2} + 594.2 \times 3.67 = 3755.9^{\#}$

$SP_2 = \frac{1652.6 - 249.3}{6.0} = 233.9 \text{ p.s.f.}$

UNIF. LD. $= 249.3 - 200 - 1200 = 1150.7^{\#}/\text{FT.}$

$M_2 = \frac{233.9 \times 1.33^2}{2 \times 3} - \frac{1150.7 \times 1.33^2}{2}$

$M_2 = -948.8^{'\#}$

391

DESIGN EXAMPLE - SECTION 7.4.12 | SHT. 3 OF 6

FOOTING DESIGN (CONT.):

$+M_1 = 7855.2^{'\#}$; $V_1 = 3755.9^{\#}$; $M_2 = -948.8^{'\#}$; $a = 1.13$; $j = 0.89$

d TO TOP REINF. $= 16 - (2 + .5 \times .62) = 13.7''$
d TO BOTT. REINF. $= 16 - (3 + .5 \times .62) = 12.7''$

$+A_s = \dfrac{7.85}{1.13 \times 13.7} = 0.505$ SQ.IN./FT.   __USE #7 @ 16''o.c.__

$v = \dfrac{3755.9}{12 \times .89 \times 13.7} = 25.6$ p.s.i < 54 p.s.i $\checkmark$

CHECK $-M_2$ FOR TENSION IN CONCRETE —
SECT. MODULUS $= \dfrac{12 \times 12.7^2}{6} = 322.58''^3$

$f_T = \dfrac{12 \times 958}{322.58} = 35.6$ p.s.i < 71 p.s.i $\checkmark$

LONG. REINF.:

$A_s = 0.0012 \times 12 \times 6.0 \times 16 = 1.38$ SQ.IN.

__USE 4 - #5's__

STEM DESIGN AT TOP OF FTG.:

$V = \dfrac{30 \times 12^2}{2} = 2160$; $M = \dfrac{2160 \times 12}{3} = 8640^{'\#}$
$d = 12 - (2 + .5 \times .62) = 9.70''$; $a = 1.13$; $j = 0.89$
$v = \dfrac{2160}{12 \times .89 \times 9.7} = 20.9$ p.s.i < 30 p.s.i (INSPECT. REQ'D.)
$A_s = \dfrac{8.64}{1.13 \times 9.70} = 0.79$ SQ.IN./FT.   TRY #7 @ 8''o.c.
$A_s = 0.90$ SQ.IN. ; $p = \dfrac{0.90}{12 \times 9.70} = 0.0077$; $np = 20 \times .0077 = 0.154$
$^2/kj = 5.52$; $f_m = \dfrac{12 \times 8640 \times 5.52}{12 \times 9.7^2} = 506.9$ p.s.i ≈ 500 p.s.i $\checkmark$ (+1.3%)

__USE #7 @ 8''o.c.__   (INSPECTION REQ'D.)

AXIAL LD. $= 736 + 560 + 600 + 720 = 2616^{\#}$

$f_a = \dfrac{2616}{144} = 18.2$ p.s.i

COMBINED STRESS RATIO $= \dfrac{506.9}{500} + \dfrac{18.2}{270} = 1.08 \approx 1.0$ $\checkmark$

DESIGN EXAMPLE – SECTION 7.4.12 | SHT. 4 OF 6

STEM DESIGN AT 2'-0" ABOVE FTG:

$H = 12.0' - 2.0' = 10.0'$

$V = \dfrac{30 \times 10^2}{2} = 1500^{\#}$; $M = \dfrac{1500 \times 10}{3} = 5000'^{\#}$

$As = \dfrac{5.0}{1.13 \times 9.7} = 0.45$ SQ. IN./FT. TRY #7 @ 16" o.c.

$As = 0.45$ SQ. IN. ; $p = \dfrac{0.45}{12 \times 9.7} = 0.00386$; $np = 20 \times 0.00386 = 0.077$

$^2\!/kj = 6.94$; $fm = \dfrac{12 \times 5000 \times 6.94}{12 \times 9.7^2} = 368.8$ p.s.i < 500 p.s.i $\checkmark$

USE #7 @ 16" o.c.

STEM DESIGN AT 4'-0" ABOVE FTG.:

$H = 12.0' - 4.0' = 8.0'$

$V = \dfrac{30 \times 8^2}{2} - 960^{\#}$; $M = \dfrac{960 \times 8}{3} = 2560'^{\#}$

$d = 8 - (2 + .5 \times .62) = 5.70''$; $a = 1.13$; $j = 0.89$

$v = \dfrac{960}{12 \times .89 \times 5.70} = 15.7$ p.s.i ≈ 15.0 p.s.i $\checkmark$

$As = \dfrac{2.56}{1.13 \times 5.70} = 0.397$ SQ. IN./FT. TRY #5 @ 8" o.c.

$As = 0.47$ SQ. IN./FT. ; $p = \dfrac{0.47}{12 \times 5.70} = 0.0069$; $np = 20 \times 0.0069 = 0.138$

$^2\!/kj = 5.71$; $fm = \dfrac{12 \times 2560 \times 5.71}{12 \times 5.7^2} = 450$ p.s.i < 500 p.s.i $\checkmark$

USE #5 @ 8" o.c.

AXIAL LD. $= 736 + 600 + 720 = 2056^{\#}$

$fa = \dfrac{2056}{12 \times 8} = 21.4$ p.s.i

COMBINED STRESS RATIO $= \dfrac{450}{500} + \dfrac{21.4}{270} = 0.98 < 1.0$

DESIGN   EXAMPLE - SECTION 7.4.12 | SHT. 5 OF 6

STEM DESIGN AT 7'-0" ABOVE FTG.:

$H = 12.0' - 7.0' = 5.0'$ $d = 5.70''$

$V = \dfrac{30 \times 5^2}{2} = 375^\#$; $M = \dfrac{375 \times 5}{3} = 625^{'\#}$

$A_s = \dfrac{0.625}{1.13 \times 5.70} = 0.097 \text{ sq.in./ft.}$ TRY #4 @ 16" o.c.

$A_s = 0.15 \text{ sq.in.}$; $p = \dfrac{0.15}{12 \times 5.7} = 0.0022$; $np = 20 \times 0.0022 = 0.044$

$2/kj = 8.54$; $f_m = \dfrac{12 \times 625 \times 8.54}{12 \times 5.7^2} = 164.3 \text{ p.s.i} < 500 \text{ p.s.i} \checkmark$

<u>USE #4 @ 16" o.c.</u>

<u>HORIZ. REINF.:</u>

12" STEM — $A_s = 0.0007 \times 4.0 \times 12 \times 12 = 0.40 \text{ sq.in.}$
<u>USE 2 - #4 TOP & BOTT.</u>

8" STEM — $A_s = 0.0007 \times 8.0 \times 12 \times 8 = 0.54 \text{ sq.in.}$
<u>USE 2 - #4 TOP & BOTT.</u>

<u>SLIDING:</u>

$F_r = \dfrac{2667}{5682.7} = 0.469 > 0.40$

∴ KEY IS REQ'D.

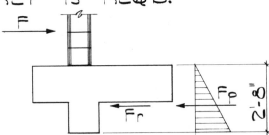

RESISTANCE TO SLIDING $= 5682.7 \times 0.40 + \dfrac{300 \times 2.67^2}{2} = 3342.4^\# > 2667^\#$ $\checkmark$

DESIGN EXAMPLE - SECTION 7.4.12 | SHT. 6 OF 6

CHECK WALL DESIGN WITHOUT AXIAL LOAD

TOTAL WT. $= 5705.5 - 600 = 5105.5^{\#}$

$M_R = 24758 - 600 \times \dfrac{52}{12} = 22158^{\prime\#}$

OVERTURN. RATIO $= \dfrac{22158}{11852} = 1.869 > 1.50$ ✓

NET $M = 22158 - 11852 = 10306^{\prime\#}$

$\bar{X} = \dfrac{10306}{5105.5} = 2.018^{\prime} > 2.0$ ✓

$e = \dfrac{6.0}{2} - 2.018 = 0.981^{\prime}$

$M_e = 5105.5 \times 0.981 = 5010.5^{\prime\#}$

SOIL PRESSURE $= \dfrac{5010.5}{6.0} \pm \dfrac{5105.5}{6.0} = 835 \pm 851$

$\underline{SP_H = 1686 \text{ p.s.f.}}$ $\qquad$ $\underline{SP_T = 16 \text{ p.s.f.}}$

CHECK FTG. M_1 (SEE CALC. SHT. No. 2)

$W^1 = \dfrac{1686 - 16}{6.0} \times 2.33 + 16 = 665.4 \text{ p.s.f.}$

$SP^1 = 1686 - 665.4 = 1020.6 \text{ p.s.f.}$

UNIF. LD. $= 665.4 - 200 = 465.4^{\#}/\text{FT.}$

$M_1 = \dfrac{1020.6 \times 3.67^2 \times 2}{3 \times 2} + \dfrac{465.4 \times 3.67^2}{2} = 7703.7 < 7820^{\prime\#}$ ✓

STEM DESIGN O.K.

CHECK SLIDING

$Fr = \dfrac{2667}{5105.5} = 0.522 > 0.40$

∴ KEY IS REQ'D.

RESISTANCE TO SLIDING $= 5105.5 \times .40 + \dfrac{300 \times 2.67^2}{2} = 3109.1^{\#} > 2667^{\#}$ ✓

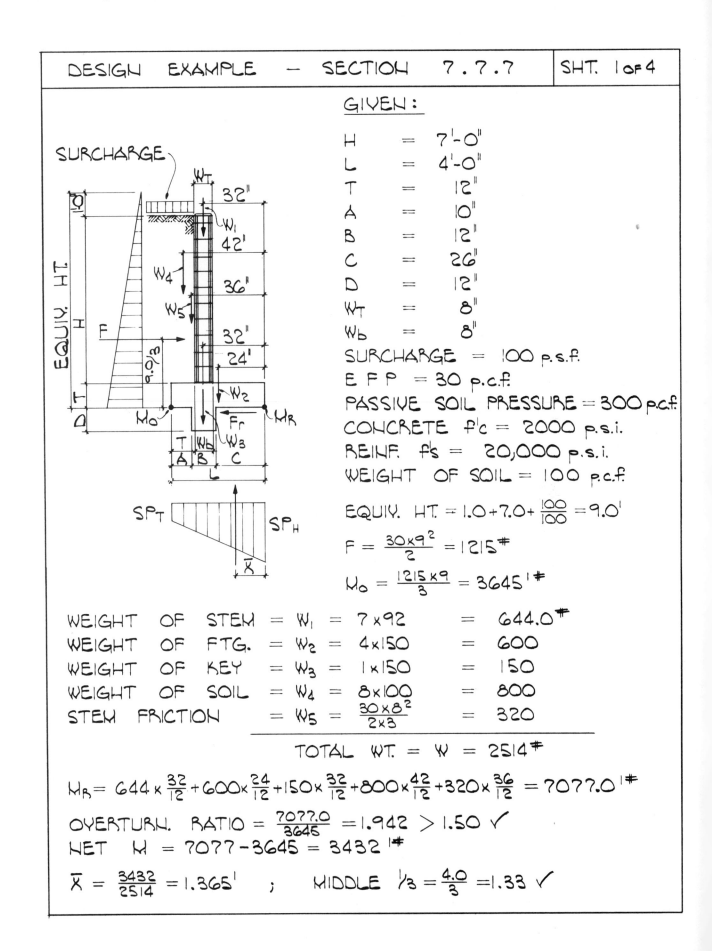

DESIGN   EXAMPLE   —   SECTION   7.7.7   SHT. 1 OF 4

GIVEN:

$H = 7'\text{-}0''$
$L = 4'\text{-}0''$
$T = 12''$
$A = 10''$
$B = 12''$
$C = 26''$
$D = 12''$
$W_T = 8''$
$W_b = 8''$

SURCHARGE = 100 p.s.f.
E F P = 30 p.c.f.
PASSIVE SOIL PRESSURE = 300 p.c.f.
CONCRETE f'_c = 2000 p.s.i.
REINF. f_s = 20,000 p.s.i.
WEIGHT OF SOIL = 100 p.c.f.

$$\text{EQUIV. HT.} = 1.0 + 7.0 + \frac{100}{100} = 9.0'$$

$$F = \frac{30 \times 9^2}{2} = 1215^{\#}$$

$$M_o = \frac{1215 \times 9}{3} = 3645'^{\#}$$

WEIGHT OF STEM = W_1 = 7 × 92 = 644.0$^{\#}$
WEIGHT OF FTG. = W_2 = 4 × 150 = 600
WEIGHT OF KEY = W_3 = 1 × 150 = 150
WEIGHT OF SOIL = W_4 = 8 × 100 = 800
STEM FRICTION = W_5 = $\frac{30 \times 8^2}{2 \times 3}$ = 320

TOTAL WT. = W = 2514$^{\#}$

$$M_R = 644 \times \frac{32}{12} + 600 \times \frac{24}{12} + 150 \times \frac{32}{12} + 800 \times \frac{42}{12} + 320 \times \frac{36}{12} = 7077.0'^{\#}$$

$$\text{OVERTURN. RATIO} = \frac{7077.0}{3645} = 1.942 > 1.50 \checkmark$$

NET M = 7077 − 3645 = 3432$'^{\#}$

$$\bar{X} = \frac{3432}{2514} = 1.365' \quad ; \quad \text{MIDDLE } \tfrac{1}{3} = \frac{4.0}{3} = 1.33 \checkmark$$

DESIGN EXAMPLE — SECTION 7.7.7	SHT. 2 of 4

FOOTING DESIGN:

$\bar{X} = 1.365'$; $e = \frac{4.0}{2} - 1.365 = 0.635'$; $W = 2514^{\#}$

$Me = 2514 \times .635 = 1596.4'^{\#}$

FTG. AREA $= 4.0$ SQ. FT.

SECT. MODULUS $= \frac{1 \times 4^2}{6} = 2.66'^3$

SOIL PRESSURE $= \frac{2514}{4.0} \pm \frac{1596.4}{2.66} = 628.5 \pm 598.8$

$\underline{SP_T = 29.7 \text{ p.s.f.}}$ $\qquad \underline{SP_H = 1227.3 \text{ p.s.f.}}$

$W' = \frac{1227.3 - 29.7}{4.0} \times 1.67 + 29.7 = 528.8 \text{ p.s.f.}$

$SP' = 1227.3 - 528.8 = 698.5 \text{ p.s.f.}$

UNIF. LD. $= 528.8 - 150 = 378.8 \text{ }^{\#}/\text{FT.}$

$M = \frac{378.8 \times 2.33^2}{2} - \frac{698.5 \times 2.33^2 \times 2}{2 \times 3}$

$M = 2298'^{\#}$

$V = 378.8 \times 2.33 - \frac{698.5 \times 2.33}{2}$

$V = 69.0^{\#}$

$d = 12'' - (3 + .5 \times .625) = 8.70''$; $a = 1.13$; $j = 0.89$

$\nu = \frac{69}{12 \times .89 \times 8.70} = 0.74 \text{ p.s.i.} < 54 \text{ p.s.i} \checkmark$

$As = \frac{2.298}{1.13 \times 8.70} = 0.233 \text{ SQ. IN./FT.}$

USE #5 @ 16" O.C.

LONG. REINF.

$As = 0.0012 \times 12 \times 4.0 \times 12 = 0.69 \text{ SQ. IN.}$

USE 3-#4

DESIGN EXAMPLE — SECTION 7.?.?	SHT. 3 OF 4

STEM DESIGN AT TOP OF FOOTING :

EQUIV. HT. $= 7.0 + \dfrac{100}{100} = 8.0'$

$V = \dfrac{30 \times 8^2}{2} = 960^{\#}$; $M = \dfrac{960 \times 8}{3} = 2560'^{\#}$

$d = 8 - (2 + .5 \times .62) = 5.70''$; $a = 1.13$; $j = 0.89$

$v = \dfrac{960}{12 \times .89 \times 5.70} = 15.8 \text{ p.s.i.} \approx 15.0 \text{ p.s.i } \checkmark$

$A_s = \dfrac{2.56}{1.13 \times 5.70} = 0.397 \text{ SQ. IN./FT}$ TRY #5 @ 8'' o.c.

$A_s = 0.47 \text{ SQ. IN.}$; $p = \dfrac{0.47}{12 \times 5.70} = 0.0068$; $np = 20 \times .0068 = 0.136$

$2/kj = 5.73$ $f_m = \dfrac{12 \times 2560 \times 5.73}{12 \times 5.70^2} = 451.5 \text{ p.s.i.} < 500 \text{ p.s.i.}$

USE #5 @ 8'' o.c. — CONT. INSPECTION REQ'D

STEM DESIGN AT 2'-0'' ABOVE FTG. :

EQUIV. HT. $= 7.0' - 2.0' + \dfrac{100}{100} = 6.0'$

$V = \dfrac{30 \times 6^2}{2} = 540^{\#}$; $M = \dfrac{540 \times 6}{3} = 1080'^{\#}$

$v = \dfrac{540}{12 \times .89 \times 5.70} = 8.9 \text{ p.s.i.} < 15.0 \text{ p.s.i}$

$A_s = \dfrac{1.08}{1.13 \times 5.70} = 0.167 \text{ SQ. IN./FT}$; TRY #5 @ 16'' o.c. ; $A_s = 0.23 \text{ SQ. IN.}$

$p = \dfrac{0.23}{12 \times 5.70} = 0.00336$; $np = 20 \times .00336 = 0.067$

$2/kj = 7.30$; $f_m = \dfrac{12 \times 1080 \times 7.30}{12 \times 5.70^2} = 242.7 \text{ p.s.i} < 500 \text{ p.s.i.} \checkmark$

USE #5 @ 16'' o.c.

SLIDING :

$F_r = \dfrac{1215}{2514} = 0.483 > 0.40$

A KEY IS REQ'D , MAKE KEY 12'' SQ.

KEY $v = \dfrac{1215}{12 \times 12} = 8.4 \text{ p.s.i.} < 54 \text{ p.s.i.} \checkmark$

RESISTANCE TO SLIDING $= 2514 \times .40 + \dfrac{300 \times 2^2}{2} = 1605^{\#} > 1215^{\#} \checkmark$

DESIGN EXAMPLE – SECTION 7.7.7	SHT. 4 OF 4

CHECK WALL WITHOUT SURCHARGE LOAD

EQUIV. HT. $= 1.0' + 7.0' = 8.0'$

$F = \dfrac{30 \times 8^2}{2} = 960^{\#}$; $M_0 = \dfrac{960 \times 8}{3} = 2560^{\#}$

$W = 2514 - 100 - 320 + \dfrac{30 \times 7^2}{2 \times 3} = 2339^{\#}$

$M_R = 644 \times \dfrac{32}{12} + 600 \times \dfrac{24}{12} + 150 \times \dfrac{32}{12} + 700 \times \dfrac{42}{12} + 245 \times \dfrac{36}{12} = 6502'^{\#}$

OVERTURN. RATIO $= \dfrac{6502}{2560} = 2.54 > 1.50$ ✓

NET M $= 6502 - 2560 = 3942'^{\#}$

$\bar{X} = \dfrac{3942}{2339} = 1.685' > 1.33'$ ✓

$e = 2.0 - 1.685 = 0.315'$

$M_e = 2339 \times .315 = 736.8'^{\#}$

SOIL PRESSURE $= \dfrac{2339}{4.0} \pm \dfrac{736.8}{2.66} = 584.7 \pm 276.4$

$\underline{SP_H = 861.1 \text{ p.s.f.}}$ $\underline{SP_T = 308.3 \text{ p.s.f.}}$

CHECK FTG. – (SEE CALC. SHT. No. 2)

$W = \dfrac{861.1 - 308.3}{4.0} \times 1.67 + 308.3 = 538.7 \text{ p.s.f.}$

$SP' = 861.1 - 538.7 = 322.4 \text{ p.s.f.}$

UNIF. LD. $= 538.7 - 150 = 388.7 ^{\#}/_{FT}$.

$M = \dfrac{388.7 \times 2.33^2}{2} - \dfrac{322.4 \times 2.33^2 \times 2}{2 \times 3} = 765.6'^{\#} < 2298'^{\#}$ ✓

CHECK SLIDING :

$F_r = \dfrac{960}{2339} = 0.41 > .40$

SLIDING RESISTANCE $= 2339 \times .40 + \dfrac{300 \times 2^2}{2} = 1535.6^{\#} > 960^{\#}$ ✓

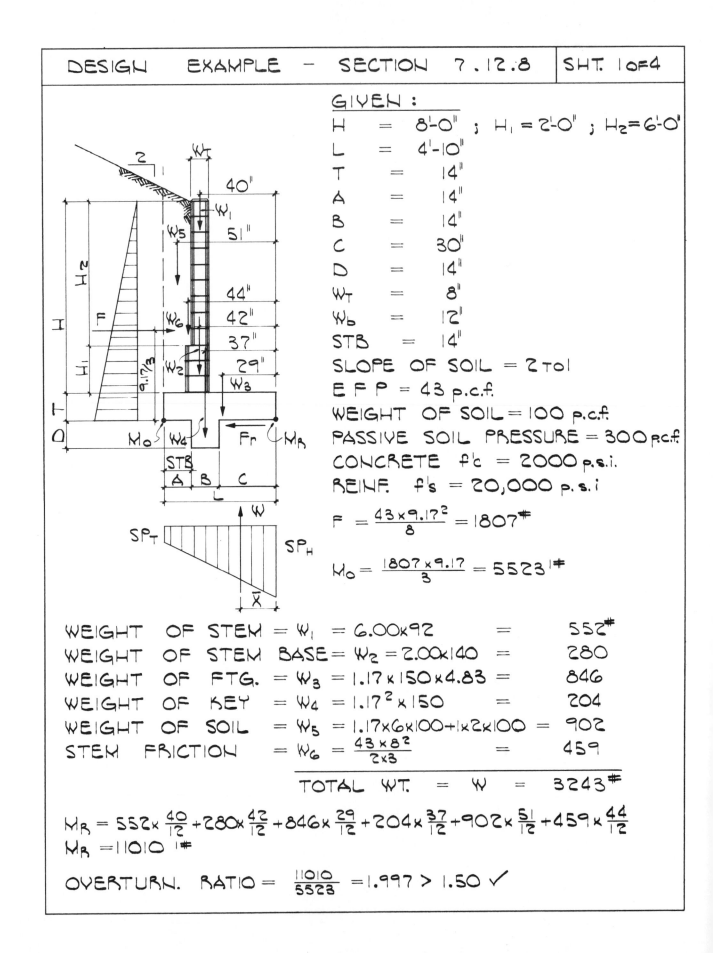

DESIGN EXAMPLE — SECTION 7.12.8 | SHT. 1 OF 4

GIVEN:

$H = 8'\text{-}0"$; $H_1 = 2'\text{-}0"$; $H_2 = 6'\text{-}0"$

$L = 4'\text{-}10"$

$T = 14"$

$A = 14"$

$B = 14"$

$C = 30"$

$D = 14"$

$W_T = 8"$

$W_b = 12"$

$STB = 14"$

SLOPE OF SOIL = 2 TO 1

E F P = 43 p.c.f.

WEIGHT OF SOIL = 100 p.c.f.

PASSIVE SOIL PRESSURE = 300 p.c.f.

CONCRETE f'_c = 2000 p.s.i.

REINF. f'_s = 20,000 p.s.i

$$F = \frac{43 \times 9.17^2}{8} = 1807^{\#}$$

$$M_O = \frac{1807 \times 9.17}{3} = 5523^{!\#}$$

WEIGHT OF STEM $= W_1 = 6.00 \times 92 \qquad = \qquad 552^{\#}$

WEIGHT OF STEM BASE $= W_2 = 2.00 \times 140 \qquad = \qquad 280$

WEIGHT OF FTG. $= W_3 = 1.17 \times 150 \times 4.83 = \qquad 846$

WEIGHT OF KEY $= W_4 = 1.17^2 \times 150 \qquad = \qquad 204$

WEIGHT OF SOIL $= W_5 = 1.17 \times 6 \times 100 + 1 \times 2 \times 100 = \quad 902$

STEM FRICTION $= W_6 = \dfrac{43 \times 8^2}{2 \times 3} \qquad = \qquad 459$

TOTAL WT. $= W = 3243^{\#}$

$$M_R = 552 \times \frac{40}{12} + 280 \times \frac{42}{12} + 846 \times \frac{29}{12} + 204 \times \frac{37}{12} + 902 \times \frac{51}{12} + 459 \times \frac{44}{12}$$

$$M_R = 11010^{!\#}$$

OVERTURN. RATIO $= \dfrac{11010}{5523} = 1.997 > 1.50$ ✓

DESIGN EXAMPLE – SECTION 7.12.8	SHT. 2 OF 4

FOOTING DESIGN:

$M_R = 11010\,'^{\#}$; $M_O = 5523\,'^{\#}$; $W = 3243\,^{\#}$

NET $M = 11010 - 5523 = 5487\,'^{\#}$

$\bar{X} = \dfrac{5487}{3243} = 1.692'$

FTG. MIDDLE $\frac{1}{3} = \dfrac{4.83}{3} = 1.611' < 1.692'$ ✓

$e = \dfrac{4.83}{2} - 1.692 = 0.724'$

$M_e = 3243 \times .724 = 2347\,'^{\#}$

FTG. AREA $= 4.83$ SQ. FT.

FTG. SECT. MODULUS $= \dfrac{1 \times 4.83^2}{6} = 3.89\,'^3$

SOIL PRESSURE $= \dfrac{3243}{4.83} \pm \dfrac{2347}{3.89} = 671.0 \pm 603.3$

$\underline{SP_T = 67.7\,p.s.f.}$ $\underline{SP_H = 1274.3\,p.s.f.}$

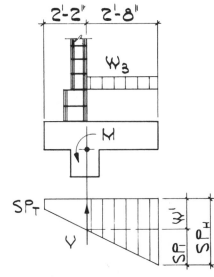

$W' = \dfrac{1274.3 - 67.7}{4.83} \times 2.17 + 67.7 = 609.1\ p.s.f.$

$SP_1 = 1274.3 - 609.1 = 665.2\ p.s.f.$

UNIF. LD. $= 609.1 - 175 = 434.1\ ^{\#}/FT.$

$M = \dfrac{434.1 \times 2.67^2}{2} + \dfrac{665.2 \times 2.67^2 \times 2}{2 \times 3}$

$M = 3124.6\,'^{\#}$

$V = 434.1 \times 2.67 + \dfrac{665.2 \times 2.67}{2}$

$V = 2047.1\,^{\#}$

$d = 14 - (3 + .5 \times .62) = 10.7''$; $a = 1.13$; $j = 0.89$

$\nu = \dfrac{2047.1}{12 \times .89 \times 10.7} = 17.9\ p.s.i < 54\ p.s.i$

$A_s = \dfrac{3.12}{1.13 \times 10.7} = 0.258\ sq.in./ft.$ USE #4 @ 16'' o.c. & #5 @ 16'' o.c.

LONG. REINF.:

$A_s = 0.0012 \times 12 \times 4.83 \times 14 = 0.97\ sq.in.$

USE 4 – #4's

DESIGN EXAMPLE – SECTION 7.12.8 | SHT. 3 OF 4

STEM DESIGN AT TOP OF FTG.

$H = 8.0'$ $EFP = 43$ p.c.f.

$V = \dfrac{43 \times 8^2}{2} = 1376^{\#}$; $M = \dfrac{1376 \times 8}{3} = 3669'^{\#}$

$d = 12 - (2 + .5 \times .62) = 9.70''$; $a = 1.13$; $j = 0.89$

$v = \dfrac{1376}{12 \times .89 \times 9.70} = 13.3$ p.s.i < 15 p.s.i ✓

$A_s = \dfrac{3.669}{1.13 \times 9.70} = 0.334$ sq.in. , TRY #5 @ 16''o.c. & #4 @ 16'' o.c.

$A_s = 0.38$ sq.in./ft. ; $p = \dfrac{0.38}{12 \times 9.7} = 0.0033$; $np = 40 \times .0031 = 0.131$

$^2/_{kj} = 5.80$; $fm = \dfrac{12 \times 3669 \times 5.80}{12 \times 9.70^2} = 226.2$ p.s.i < 250 p.s.i ✓

USE #5 @ 16''o.c. & #4 @ 16''o.c.

STEM DESIGN AT 1'-4'' ABOVE FTG.

$H = 8.0' - 2.00' = 6.00'$ $EFP = 43$ p.c.f.

$V = \dfrac{43 \times 6.0^2}{2} = 774'^{\#}$; $M = \dfrac{774 \times 6.0}{3} = 1548'^{\#}$

$d = 8'' - (2 + .5 \times .62) = 5.70''$; $a = 1.13$; $j = 0.89$

$v = \dfrac{774}{12 \times .89 \cdot 5.70} = 12.7$ p.s.i < 15.0 p.s.i ✓

$A_s = \dfrac{1.55}{1.13 \times 5.70} = 0.24$ sq. in./ft. , TRY #5 @ 16''o.c.

$A_s = 0.23$ sq. in./ft. ; $p = \dfrac{0.23}{12 \times 5.70} = 0.0033$; $np = 20 \times .0033 = 0.067$

$^2/_{kj} = 7.30$; $fm = \dfrac{12 \times 1548 \times 7.30}{12 \times 5.70^2} = 347.8$ p.s.i < 500 p.s.i

USE #5 @ 16''o.c. (CONT. INSPECT. REQ'D.)

DESIGN EXAMPLE SECTION 7.12.8 | SHT. 4 of 4

STEM DESIGN AT 4'-0" ABOVE FTG.

$H = 8.0' - 4.0' = 4.0'$; $EFP = 43$ p.c.f.

$V = \dfrac{43 \times 4^2}{2} = 344^{\#}$; $M = \dfrac{344 \times 4}{3} = 458.7'^{\#}$

$v = \dfrac{344}{12 \times .89 \times 5.70} = 5.6$ p.s.i < 15 p.s.i ✓

$A_s = \dfrac{0.458}{1.13 \times 5.70} = 0.071$ sq. in./ft.   TRY #4 @ 16" o.c.

$A_s = 0.15$ sq. in./ft. ; $p = \dfrac{0.15}{12 \times 5.70} = 0.0022$; $np = 20 \times .0022 = 0.044$

$2/kj = 8.54$; $fm = \dfrac{12 \times 458.7 \times 8.54}{12 \times 5.70^2} = 120.6$ p.s.i < 250 p.s.i

USE #4 @ 16" o.c.

LONG. REINF. :

$A_s = 0.0007 (6.67 \times 12 \times 8 + 1.33 \times 12 \times 12) = 0.582$ sq. in.

USE - 2 - #4's AT TOP & BOTTOM
2 - #5's AT TOP OF 12" BLK.

SLIDING :

$Fr = \dfrac{1807}{3244} = 0.557 > 0.40$ ∴   KEY IS REQ'D.
MAKE KEY 14" SQ.

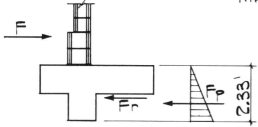

RESISTANCE TO SLIDING $= 3244 \times .40 + \dfrac{300 \times 2.33^2}{2} = 2112^{\#} > 1807^{\#}$

KEY $v = \dfrac{1807}{12 \times 14} = 10.7$ p.s.i < 54 p.s.i ✓

TABLE 7.1 UNDERCUT FOOTING RETAINING WALLS — MASONRY STEM — SOIL AT TOE SIDE
SLOPE = 0 to 1   SURCHARGE = 0 lbs./sq. ft.   AXIAL = 0 lbs./ft.

SOIL PRESSURE DIAGRAM

H1 = Concrete Stem
H2 = 12″ Concrete Block
H3 = 8″ Concrete Block

Stem Height	3′- 0″	4′- 0″	5′- 0″	6′- 0″	7′- 0″	8′- 0″	9′- 0″	10′- 0″	11′- 0″	12′- 0″
Ft. L ft.	1.667	2.000	2.500	3.000	3.333	4.000	4.500	5.000	5.500	6.000
Ft. T ft.	0.667	0.667	0.833	0.833	0.833	1.000	1.000	1.167	1.167	1.333
Undercut	0.667	0.667	0.833	0.833	0.833	1.000	1.000	1.167	1.167	1.333
A ft.	0.500	0.833	1.167	1.500	1.667	2.167	2.500	2.667	3.000	3.333
B ft.	0.667	0.667	0.833	0.833	0.833	1.000	1.000	1.167	1.167	1.333
C ft.	0.500	0.500	0.500	0.667	0.833	0.833	1.000	1.167	1.333	1.333
D ft.	0.667	0.667	0.833	0.833	0.833	1.000	1.000	1.167	1.167	1.333
8 in. blk. ft.	3.000	4.000	5.000	6.000	7.000	8.000	7.000	8.000	8.333	8.667
12 in. blk. ft.	—	—	—	—	—	—	2.000	2.000	2.667	2.667
Conc. Stem	—	—	—	—	—	—	—	—	—	0.667
Wb ft.	—	—	—	—	—	—	—	—	—	1.333
W lbs.	754	981	1418	1711	1993	2606	2990	3730	4195	5074
F lbs.	202	327	510	700	920	1215	1500	1870	2220	2667
Mo ft.-lbs.	246	508	992	1595	2403	3645	5000	6962	9005	11852
Mr ft.-lbs.	690	1197	2245	3456	4621	7323	9699	13343	16824	22073
O.T.R.	2.800	2.355	2.262	2.166	1.923	2.009	1.940	1.916	1.868	1.862
$\overline{X}$ ft.	0.588	0.702	0.883	1.087	1.112	1.411	1.572	1.711	1.864	2.014
e ft.	0.245	0.298	0.367	0.413	0.554	0.589	0.678	0.789	0.886	0.986
Me ft.-lbs.	185	293	520	706	1105	1534	2029	2944	3717	5001
S ft.^{3}	0.463	0.667	1.042	1.500	1.852	2.667	3.375	4.167	5.042	6.000
S.P.h p.s.f.	852	930	1067	1041	1194	1227	1265	1453	1500	1679
S.P.t p.s.f.	53	52	68	100	1	76	63	39	25	12
Friction	0.267	0.333	0.360	0.409	0.462	0.466	0.502	0.501	0.529	0.526
Inspct.	NO	NO	NO	NO	YES	YES	YES	YES	YES	YES

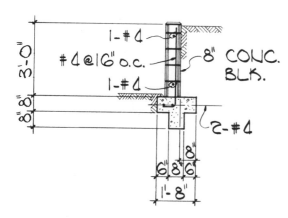

E.F.P. = 30 p.c.f.
Section 7.1.3

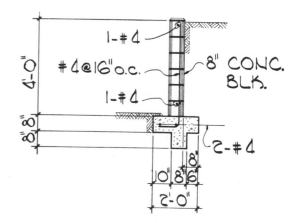

E.F.P. = 30 p.c.f.
Section 7.1.4

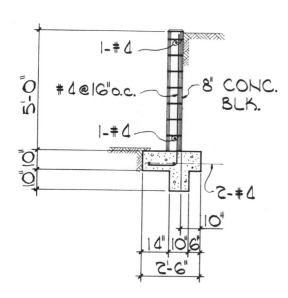

E.F.P. = 30 p.c.f.
Section 7.1.5

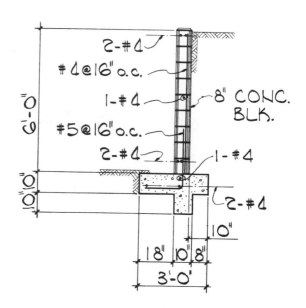

E.F.P. = 30 p.c.f.
Section 7.1.6

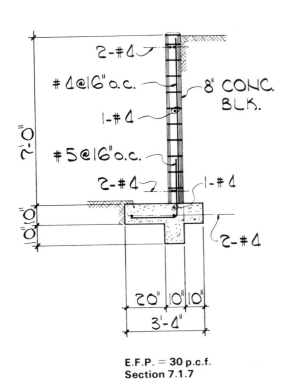

2-#4

#4@16" o.c.

1-#4

#5@16" o.c.

2-#4

1-#4

8" CONC. BLK.

2-#4

7'-0"

0'-0"

20" 10" 10"

3'-4"

E.F.P. = 30 p.c.f.
Section 7.1.7

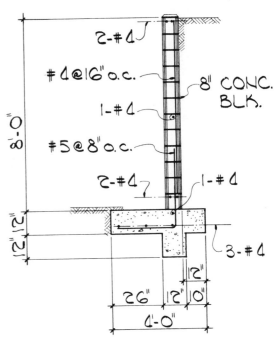

2-#4

#4@16" o.c.

1-#4

#5@8" o.c.

2-#4

1-#4

8" CONC. BLK.

3-#4

8'-0"

2'-2"

26" 12" 10"

12"

4'-0"

E.F.P. = 30 p.c.f.
Section 7.1.8

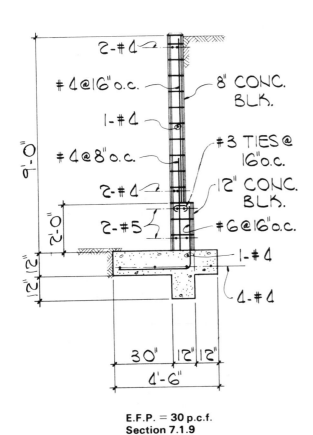

2-#4

#4@16" o.c.

1-#4

#4@8" o.c.

2-#4

2-#5

8" CONC. BLK.

#3 TIES @ 16" o.c.

12" CONC. BLK.

#6@16" o.c.

1-#4

4-#4

9'-0"

2'-0"

2'-2"

30" 12" 12"

4'-6"

E.F.P. = 30 p.c.f.
Section 7.1.9

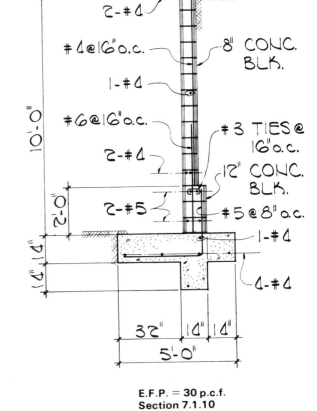

2-#4

#4@16" o.c.

1-#4

#6@16" o.c.

2-#4

2-#5

8" CONC. BLK.

#3 TIES @ 16" o.c.

12" CONC. BLK.

#5@8" o.c.

1-#4

4-#4

10'-0"

2'-0"

4'-4"

32" 14" 14"

5'-0"

E.F.P. = 30 p.c.f.
Section 7.1.10

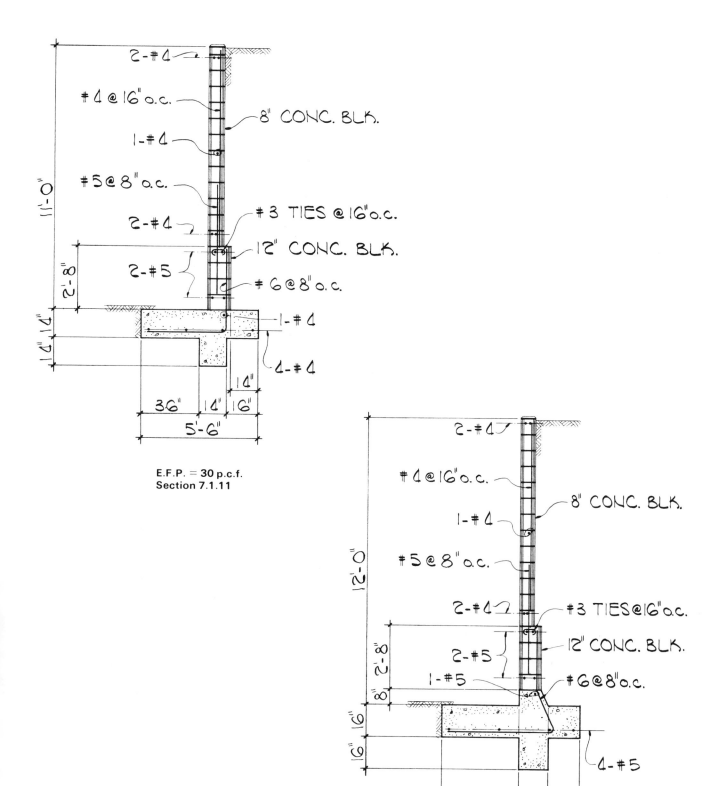

2-#4

#4 @ 16" o.c.

1-#4   8" CONC. BLK.

#5 @ 8" o.c.

2-#4   #3 TIES @ 16" o.c.

12" CONC. BLK.

2-#5   #6 @ 8" o.c.

11'-0"

2'-8"

1'-4"

1-#4

4-#4

36"   14"   16"

14"

5'-6"

E.F.P. = 30 p.c.f.
Section 7.1.11

2-#4

#4 @ 16" o.c.

1-#4   8" CONC. BLK.

#5 @ 8" o.c.

12'-0"

2-#4   #3 TIES @ 16" o.c.

12" CONC. BLK.

2-#5

1-#5   #6 @ 8" o.c.

2'-8"

8"

16"

16"

4-#5

40"   16"   16"

6'-0"

E.F.P. = 30 p.c.f.
Section 7.1.12

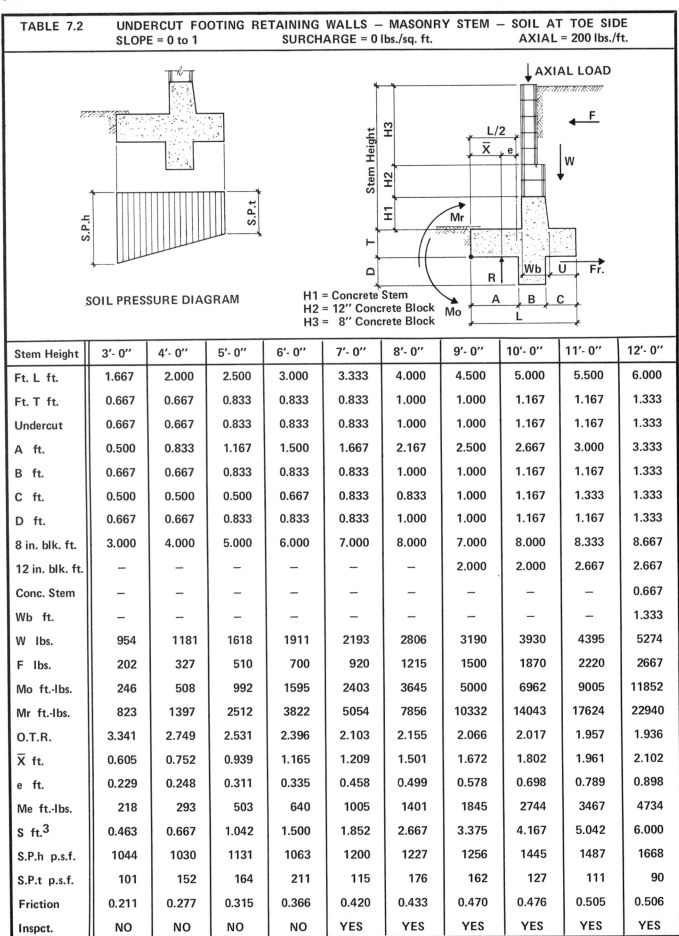

TABLE 7.2	UNDERCUT FOOTING RETAINING WALLS – MASONRY STEM – SOIL AT TOE SIDE
SLOPE = 0 to 1 SURCHARGE = 0 lbs./sq. ft. AXIAL = 200 lbs./ft.	

SOIL PRESSURE DIAGRAM

H1 = Concrete Stem
H2 = 12″ Concrete Block
H3 = 8″ Concrete Block

Stem Height	3'- 0"	4'- 0"	5'- 0"	6'- 0"	7'- 0"	8'- 0"	9'- 0"	10'- 0"	11'- 0"	12'- 0"
Ft. L ft.	1.667	2.000	2.500	3.000	3.333	4.000	4.500	5.000	5.500	6.000
Ft. T ft.	0.667	0.667	0.833	0.833	0.833	1.000	1.000	1.167	1.167	1.333
Undercut	0.667	0.667	0.833	0.833	0.833	1.000	1.000	1.167	1.167	1.333
A ft.	0.500	0.833	1.167	1.500	1.667	2.167	2.500	2.667	3.000	3.333
B ft.	0.667	0.667	0.833	0.833	0.833	1.000	1.000	1.167	1.167	1.333
C ft.	0.500	0.500	0.500	0.667	0.833	0.833	1.000	1.167	1.333	1.333
D ft.	0.667	0.667	0.833	0.833	0.833	1.000	1.000	1.167	1.167	1.333
8 in. blk. ft.	3.000	4.000	5.000	6.000	7.000	8.000	7.000	8.000	8.333	8.667
12 in. blk. ft.	—	—	—	—	—	—	2.000	2.000	2.667	2.667
Conc. Stem	—	—	—	—	—	—	—	—	—	0.667
Wb ft.	—	—	—	—	—	—	—	—	—	1.333
W lbs.	954	1181	1618	1911	2193	2806	3190	3930	4395	5274
F lbs.	202	327	510	700	920	1215	1500	1870	2220	2667
Mo ft.-lbs.	246	508	992	1595	2403	3645	5000	6962	9005	11852
Mr ft.-lbs.	823	1397	2512	3822	5054	7856	10332	14043	17624	22940
O.T.R.	3.341	2.749	2.531	2.396	2.103	2.155	2.066	2.017	1.957	1.936
$\overline{X}$ ft.	0.605	0.752	0.939	1.165	1.209	1.501	1.672	1.802	1.961	2.102
e ft.	0.229	0.248	0.311	0.335	0.458	0.499	0.578	0.698	0.789	0.898
Me ft.-lbs.	218	293	503	640	1005	1401	1845	2744	3467	4734
S ft.^{3}	0.463	0.667	1.042	1.500	1.852	2.667	3.375	4.167	5.042	6.000
S.P.h p.s.f.	1044	1030	1131	1063	1200	1227	1256	1445	1487	1668
S.P.t p.s.f.	101	152	164	211	115	176	162	127	111	90
Friction	0.211	0.277	0.315	0.366	0.420	0.433	0.470	0.476	0.505	0.506
Inspct.	NO	NO	NO	NO	YES	YES	YES	YES	YES	YES

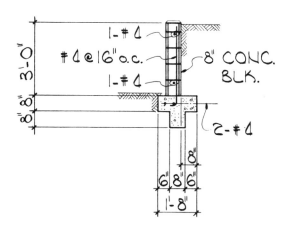

Axial load = 200 lbs./ft.
E.F.P. = 30 p.c.f.
Section 7.2.3

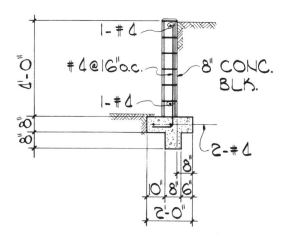

Axial load = 200 lbs./ft.
E.F.P. = 30 p.c.f.
Section 7.2.4

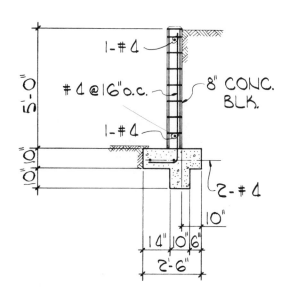

Axial load = 200 lbs./ft.
E.F.P. = 30 p.c.f.
Section 7.2.5

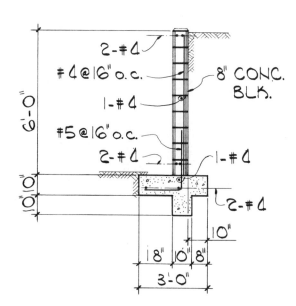

Axial load = 200 lbs./ft.
E.F.P. = 30 p.c.f.
Section 7.2.6

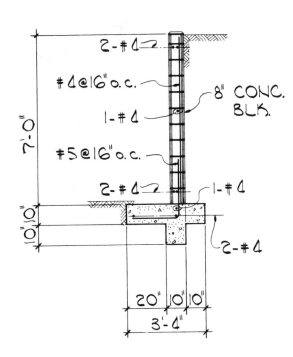

Axial load = 200 lbs./ft.
E.F.P. = 30 p.c.f.
Section 7.2.7

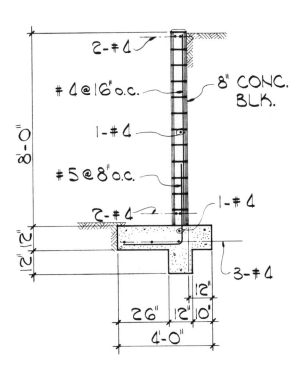

Axial load = 200 lbs./ft.
E.F.P. = 30 p.c.f.
Section 7.2.8

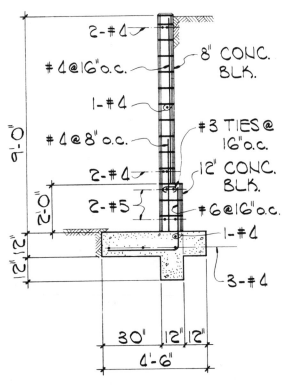

Axial load = 200 lbs./ft.
E.F.P. = 30 p.c.f.
Section 7.2.9

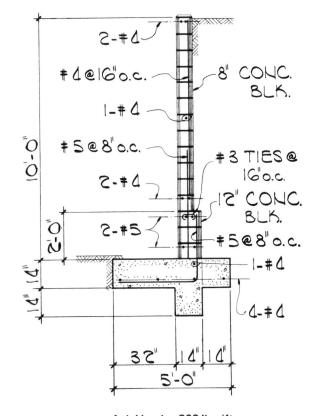

Axial load = 200 lbs./ft.
E.F.P. = 30 p.c.f.
Section 7.2.10

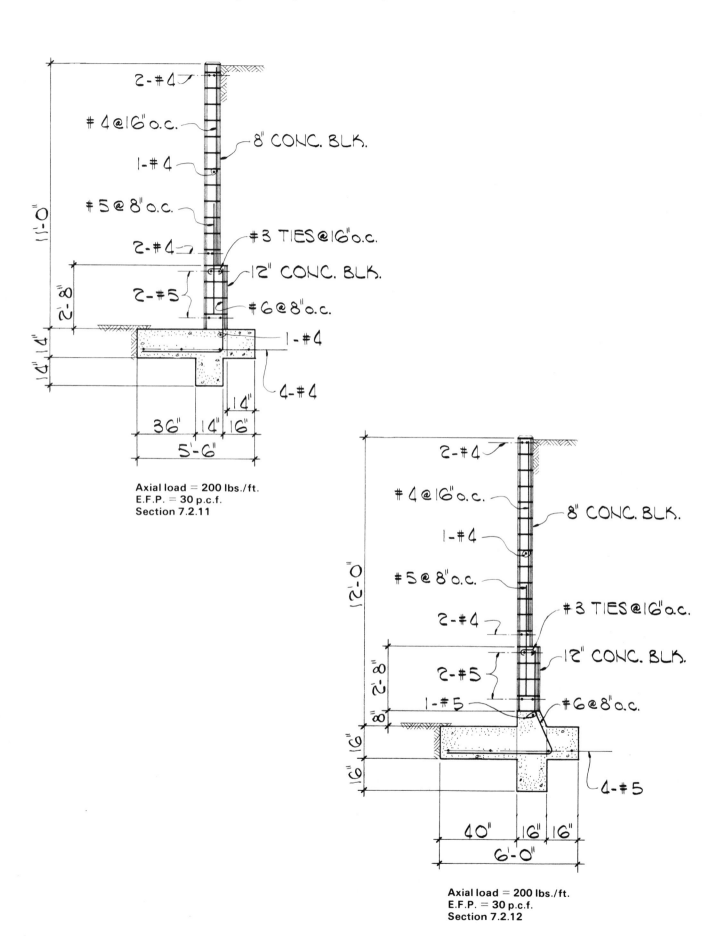

2-#4
#4@16"o.c.
1-#4
8" CONC. BLK.
#5@8"o.c.
2-#4
#3 TIES@16"o.c.
12" CONC. BLK.
2-#5
#6@8"o.c.
1-#4
4-#4

11'-0"
2'-8"
4"
4"
36" 14" 16"
14"
5'-6"

Axial load = 200 lbs./ft.
E.F.P. = 30 p.c.f.
Section 7.2.11

2-#4
#4@16"o.c.
1-#4
8" CONC. BLK.
#5@8"o.c.
2-#4
#3 TIES@16"o.c.
12" CONC. BLK.
2-#5
1-#5
#6@8"o.c.
4-#5

12'-0"
2'-8"
8"
16"
16"
40" 16" 16"
6'-0"

Axial load = 200 lbs./ft.
E.F.P. = 30 p.c.f.
Section 7.2.12

TABLE 7.3   'UNDERCUT FOOTING RETAINING WALLS — MASONRY STEM — SOIL AT TOE SIDE
SLOPE = 0 to 1   SURCHARGE = 0 lbs./sq. ft.   AXIAL = 400 lbs./ft.

SOIL PRESSURE DIAGRAM

H1 = Concrete Stem
H2 = 12" Concrete Block
H3 = 8" Concrete Block

Stem Height	3'- 0"	4'- 0"	5'- 0"	6'- 0"	7'- 0"	8'- 0"	9'- 0"	10'- 0"	11'- 0"	12'- 0"
Ft. L ft.	1.667	2.000	2.500	3.000	3.333	4.000	4.500	5.000	5.500	6.000
Ft. T ft.	0.667	0.667	0.833	0.833	0.833	1.000	1.000	1.167	1.167	1.333
Undercut	0.667	0.667	0.833	0.833	0.833	1.000	1.000	1.167	1.167	1.333
A ft.	0.500	0.833	1.167	1.500	1.667	2.167	2.500	2.667	3.000	3.333
B ft.	0.667	0.667	0.833	0.833	0.833	1.000	1.000	1.167	1.167	1.333
C ft.	0.500	0.500	0.500	0.667	0.833	0.833	1.000	1.167	1.333	1.333
D ft.	0.667	0.667	0.833	0.833	0.833	1.000	1.000	1.167	1.167	1.333
8 in. blk. ft.	3.000	4.000	5.000	6.000	7.000	8.000	7.000	8.000	8.333	8.667
12 in. blk. ft.	—	—	—	—	—	—	2.000	2.000	2.667	2.667
Conc. Stem	—	—	—	—	—	—	—	—	—	0.667
Wb ft.	—	—	—	—	—	—	—	—	—	1.333
W lbs.	1154	1381	1818	2111	2393	3006	3390	4130	4595	5474
F lbs.	202	327	510	700	920	1215	1500	1870	2220	2667
Mo ft.-lbs.	246	508	992	1595	2403	3645	5000	6962	9005	11852
Mr ft.-lbs.	957	1597	2779	4189	5487	8389	10966	14743	18424	23806
O.T.R.	3.882	3.143	2.800	2.626	2.283	2.302	2.193	2.118	2.046	2.009
$\overline{X}$ ft.	0.615	0.788	0.982	1.229	1.289	1.578	1.760	1.884	2.050	2.184
e ft.	0.218	0.212	0.268	0.271	0.378	0.422	0.490	0.616	0.700	0.816
Me ft.-lbs.	252	293	487	573	905	1268	1662	2544	3217	4468
S ft.^{3}	0.463	0.667	1.042	1.500	1.852	2.667	3.375	4.167	5.042	6.000
S.P.h p.s.f.	1236	1130	1195	1086	1206	1227	1246	1437	1474	1657
S.P.t p.s.f.	149	252	260	322	229	276	261	215	197	168
Friction	0.175	0.236	0.281	0.332	0.385	0.404	0.442	0.453	0.483	0.487
Inspct.	NO	NO	NO	NO	YES	YES	YES	YES	YES	YES

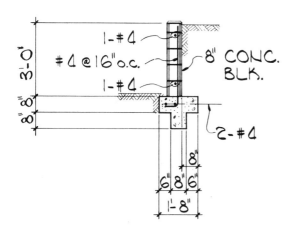

Axial load = 400 lbs./ft.
E.F.P. = 30 p.c.f.
Section 7.3.3

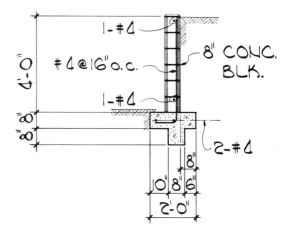

Axial load = 400 lbs./ft.
E.F.P. = 30 p.c.f.
Section 7.3.4

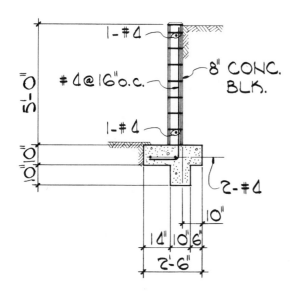

Axial load = 400 lbs./ft.
E.F.P. = 30 p.c.f.
Section 7.3.5

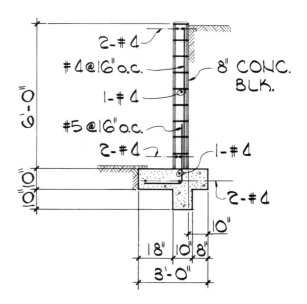

Axial load = 400 lbs./ft.
E.F.P. = 30 p.c.f.
Section 7.3.6

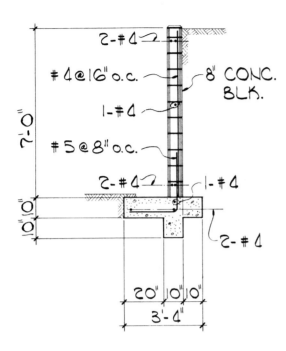

2-#4

#4@16" o.c.

1-#4

#5@8" o.c.

8" CONC. BLK.

2-#4   1-#4

2-#4

7'-0"

10" 10" 20"

3'-4"

Axial load = 400 lbs./ft.
E.F.P. = 30 p.c.f.
Section 7.3.7

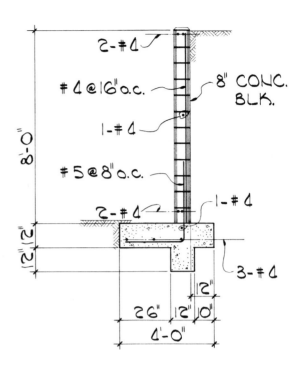

2-#4

#4@16" o.c.

1-#4

#5@8" o.c.

8" CONC. BLK.

2-#4   1-#4

3-#4

8'-0"

2'-2"

10" 12" 26"

12"

4'-0"

Axial load = 400 lbs./ft.
E.F.P. = 30 p.c.f.
Section 7.3.8

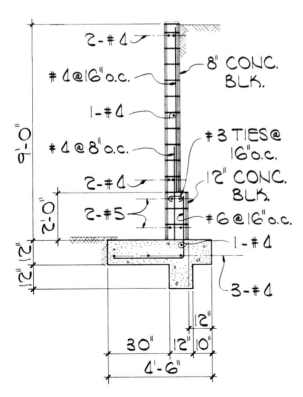

2-#4

#4@16" o.c.

1-#4

#4@8" o.c.

2-#4

2-#5

8" CONC. BLK.

#3 TIES@ 16" o.c.

12" CONC. BLK.

#6@16" o.c.

1-#4

3-#4

10'-0"

2'-0"

2'-2"

10" 12" 30"

12"

4'-6"

Axial load = 400 lbs./ft.
E.F.P. = 30 p.c.f.
Section 7.3.9

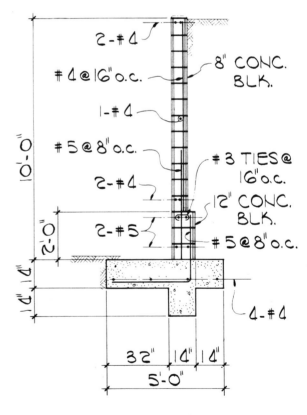

2-#4

#4@16" o.c.

1-#4

#5@8" o.c.

2-#4

2-#5

8" CONC. BLK.

#3 TIES@ 16" o.c.

12" CONC. BLK.

#5@8" o.c.

4-#4

10'-0"

2'-0"

4'-4"

14" 14" 32"

5'-0"

Axial load = 400 lbs./ft.
E.F.P. = 30 p.c.f.
Section 7.3.10

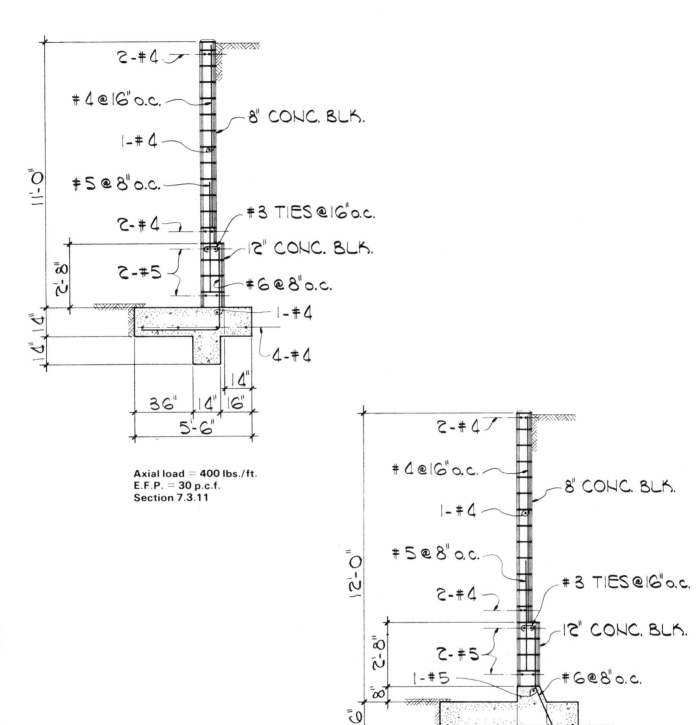

2-#4

#4 @ 16" o.c.

1-#4

#5 @ 8" o.c.

2-#4

2-#5

11'-0"

2'-8"

14"

8" CONC. BLK.

#3 TIES @ 16" o.c.

12" CONC. BLK.

#6 @ 8" o.c.

1-#4

4-#4

36"   14"   16"
14"
5'-6"

Axial load = 400 lbs./ft.
E.F.P. = 30 p.c.f.
Section 7.3.11

2-#4

#4 @ 16" o.c.

1-#4

#5 @ 8" o.c.

2-#4

2-#5

1-#5

12'-0"

2'-8"

8"

6"

6"

8" CONC. BLK.

#3 TIES @ 16" o.c.

12" CONC. BLK.

#6 @ 8" o.c.

4-#5

40"   16"   16"
6'-0"

Axial load = 400 lbs./ft.
E.F.P. = 30 p.c.f.
Section 7.3.12

TABLE 7.4   UNDERCUT FOOTING RETAINING WALLS — MASONRY STEM — SOIL AT TOE SIDE

SLOPE = 0 to 1   SURCHARGE = 0 lbs./sq. ft.   AXIAL = 600 lbs./ft.

SOIL PRESSURE DIAGRAM

H1 = Concrete Stem
H2 = 12″ Concrete Block
H3 = 8″ Concrete Block

Stem Height	3′- 0″	4′- 0″	5′- 0″	6′- 0″	7′- 0″	8′- 0″	9′- 0″	10′- 0″	11′- 0″	12′- 0″
Ft. L ft.	1.667	2.000	2.500	3.000	3.333	4.000	4.500	5.000	5.500	6.000
Ft. T ft.	0.667	0.667	0.833	0.833	0.833	1.000	1.000	1.167	1.167	1.333
Undercut	0.667	0.667	0.833	0.833	0.833	1.000	1.000	1.167	1.167	1.333
A ft.	0.500	0.833	1.167	1.500	1.667	2.167	2.500	2.667	3.000	3.333
B ft.	0.667	0.667	0.833	0.833	0.833	1.000	1.000	1.167	1.167	1.333
C ft.	0.500	0.500	0.500	0.667	0.833	0.833	1.000	1.167	1.333	1.333
D ft.	0.667	0.667	0.833	0.833	0.833	1.000	1.000	1.167	1.167	1.333
8 in. blk. ft.	3.000	4.000	5.000	6.000	7.000	8.000	7.000	8.000	8.333	8.000
12 in. blk.ft.	—	—	—	—	—	—	2.000	2.000	2.667	3.333
Conc. Stem	—	—	—	—	—	—	—	—	—	0.667
Wb ft.	—	—	—	—	—	—	—	—	—	1.333
W lbs.	1354	1581	2018	2311	2593	3206	3590	4362	4795	5706
F lbs.	202	327	510	700	920	1215	1500	1870	2220	2667
Mo ft.-lbs.	246	508	992	1595	2403	3645	5000	6962	9005	11852
Mr ft.-lbs.	1090	1797	3045	4556	5921	8923	11599	15543	19224	24798
O.T.R.	4.423	3.536	3.068	2.856	2.464	2.448	2.320	2.232	2.135	2.092
$\overline{X}$ ft.	0.623	0.815	1.017	1.281	1.356	1.646	1.838	1.967	2.131	2.269
e ft.	0.210	0.185	0.233	0.219	0.310	0.354	0.412	0.533	0.619	0.731
Me ft.-lbs.	285	293	470	506	805	1134	1479	2324	2967	4172
S ft.^{3}	0.463	0.667	1.042	1.500	1.852	2.667	3.375	4.167	5.042	6.000
S.P.h p.s.f.	1428	1230	1259	1108	1212	1227	1236	1430	1460	1646
S.P.t p.s.f.	197	352	356	433	343	376	360	315	283	256
Friction	0.149	0.207	0.253	0.303	0.355	0.379	0.418	0.429	0.463	0.467
Inspct.	NO	NO	NO	NO	YES	YES	YES	YES	YES	YES

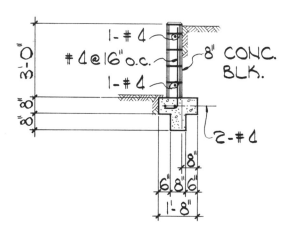

Axial load = 600 lbs./ft.
E.F.P. = 30 p.c.f.
Section 7.4.3

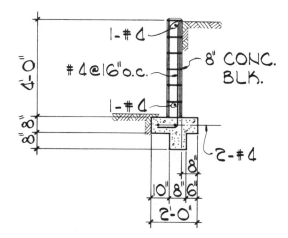

Axial load = 600 lbs./ft.
E.F.P. = 30 p.c.f.
Section 7.4.4

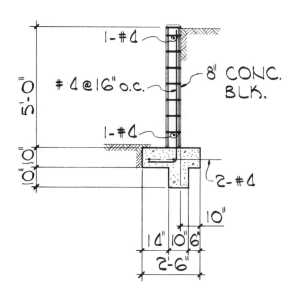

Axial load = 600 lbs./ft.
E.F.P. = 30 p.c.f.
Section 7.4.5

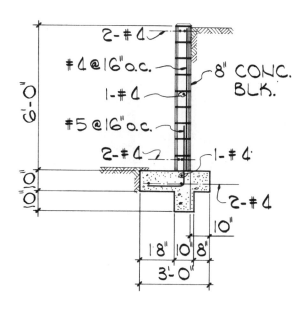

Axial load = 600 lbs./ft.
E.F.P. = 30 p.c.f.
Section 7.4.6

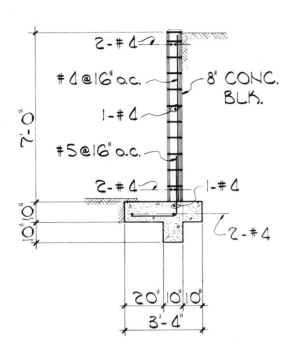

Axial load = 600 lbs./ft.
E.F.P. = 30 p.c.f.
Section 7.4.7

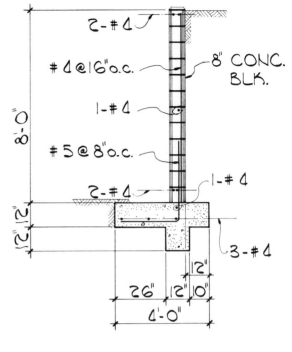

Axial load = 600 lbs./ft.
E.F.P. = 30 p.c.f.
Section 7.4.8

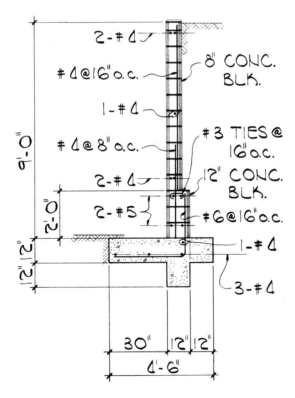

Axial load = 600 lbs./ft.
E.F.P. = 30 p.c.f.
Section 7.4.9

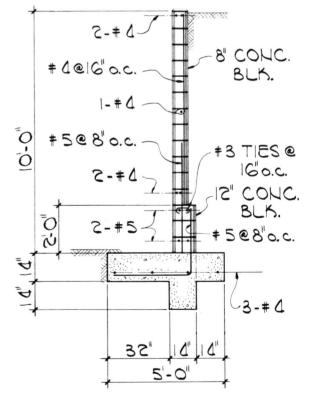

Axial load = 600 lbs./ft.
E.F.P. = 30 p.c.f.
Section 7.4.10

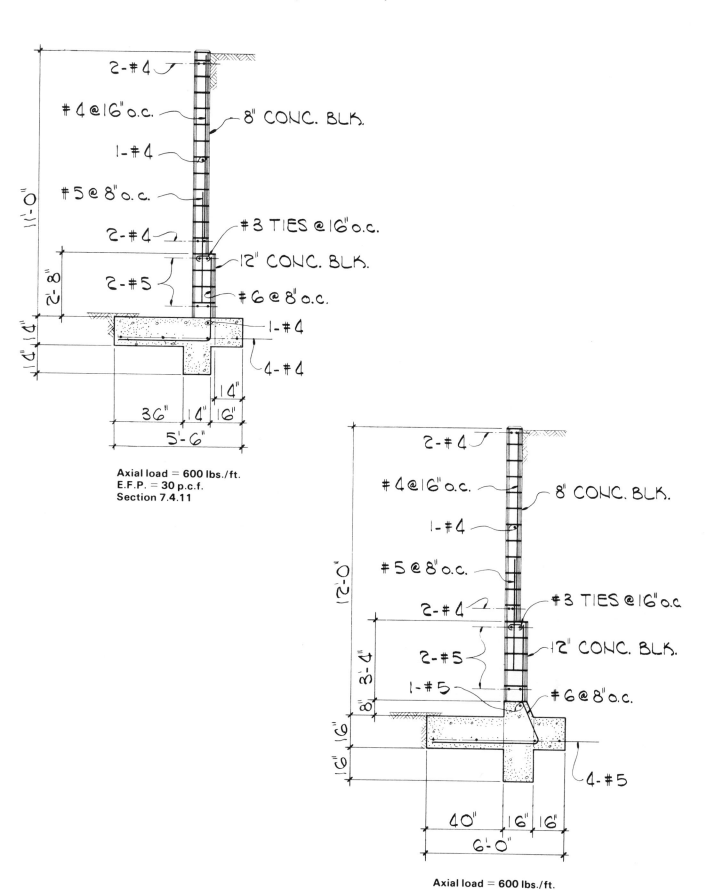

2-#4

#4 @ 16" o.c.

8" CONC. BLK.

1-#4

#5 @ 8" o.c.

2-#4

#3 TIES @ 16" o.c.

12" CONC. BLK.

2-#5

#6 @ 8" o.c.

1-#4

4-#4

11'-0"

2'-8"

1'-4"

36"   14"   16"

14"

5'-6"

Axial load = 600 lbs./ft.
E.F.P. = 30 p.c.f.
Section 7.4.11

2-#4

#4 @ 16" o.c.

8" CONC. BLK.

1-#4

#5 @ 8" o.c.

2-#4

#3 TIES @ 16" o.c.

12" CONC. BLK.

2-#5

1-#5

#6 @ 8" o.c.

4-#5

12'-0"

3'-4"

8"

6"

6"

40"   16"   16"

6'-0"

Axial load = 600 lbs./ft.
E.F.P. = 30 p.c.f.
Section 7.4.12

TABLE 7.5	UNDERCUT FOOTING RETAINING WALLS – MASONRY STEM – SOIL AT TOE SIDE
	SLOPE = 0 to 1 SURCHARGE = 0 lbs./sq. ft. AXIAL = 800 lbs./ft.

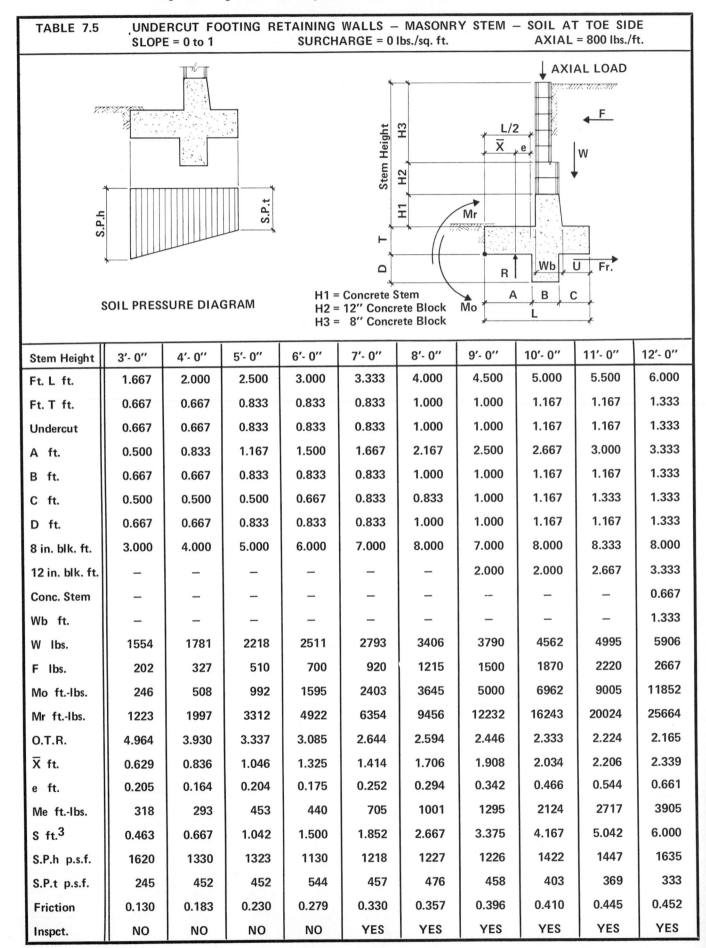

SOIL PRESSURE DIAGRAM

H1 = Concrete Stem
H2 = 12" Concrete Block
H3 = 8" Concrete Block

Stem Height	3'- 0"	4'- 0"	5'- 0"	6'- 0"	7'- 0"	8'- 0"	9'- 0"	10'- 0"	11'- 0"	12'- 0"
Ft. L ft.	1.667	2.000	2.500	3.000	3.333	4.000	4.500	5.000	5.500	6.000
Ft. T ft.	0.667	0.667	0.833	0.833	0.833	1.000	1.000	1.167	1.167	1.333
Undercut	0.667	0.667	0.833	0.833	0.833	1.000	1.000	1.167	1.167	1.333
A ft.	0.500	0.833	1.167	1.500	1.667	2.167	2.500	2.667	3.000	3.333
B ft.	0.667	0.667	0.833	0.833	0.833	1.000	1.000	1.167	1.167	1.333
C ft.	0.500	0.500	0.500	0.667	0.833	0.833	1.000	1.167	1.333	1.333
D ft.	0.667	0.667	0.833	0.833	0.833	1.000	1.000	1.167	1.167	1.333
8 in. blk. ft.	3.000	4.000	5.000	6.000	7.000	8.000	7.000	8.000	8.333	8.000
12 in. blk. ft.	–	–	–	–	–	–	2.000	2.000	2.667	3.333
Conc. Stem	–	–	–	–	–	–	–	–	–	0.667
Wb ft.	–	–	–	–	–	–	–	–	–	1.333
W lbs.	1554	1781	2218	2511	2793	3406	3790	4562	4995	5906
F lbs.	202	327	510	700	920	1215	1500	1870	2220	2667
Mo ft.-lbs.	246	508	992	1595	2403	3645	5000	6962	9005	11852
Mr ft.-lbs.	1223	1997	3312	4922	6354	9456	12232	16243	20024	25664
O.T.R.	4.964	3.930	3.337	3.085	2.644	2.594	2.446	2.333	2.224	2.165
$\overline{X}$ ft.	0.629	0.836	1.046	1.325	1.414	1.706	1.908	2.034	2.206	2.339
e ft.	0.205	0.164	0.204	0.175	0.252	0.294	0.342	0.466	0.544	0.661
Me ft.-lbs.	318	293	453	440	705	1001	1295	2124	2717	3905
S ft.^{3}	0.463	0.667	1.042	1.500	1.852	2.667	3.375	4.167	5.042	6.000
S.P.h p.s.f.	1620	1330	1323	1130	1218	1227	1226	1422	1447	1635
S.P.t p.s.f.	245	452	452	544	457	476	458	403	369	333
Friction	0.130	0.183	0.230	0.279	0.330	0.357	0.396	0.410	0.445	0.452
Inspct.	NO	NO	NO	NO	YES	YES	YES	YES	YES	YES

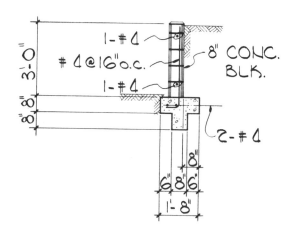

Axial load = 800 lbs./ft.
E.F.P. = 30 p.c.f.
Section 7.5.3

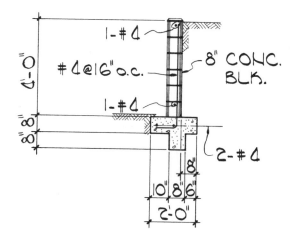

Axial load = 800 lbs./ft.
E.F.P. = 30 p.c.f.
Section 7.5.4

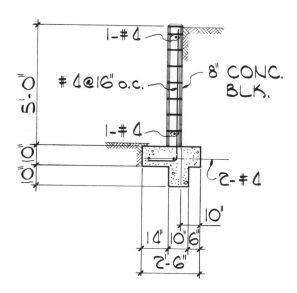

Axial load = 800 lbs./ft.
E.F.P. = 30 p.c.f.
Section 7.5.5

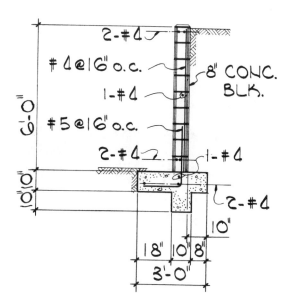

Axial load = 800 lbs./ft.
E.F.P. = 30 p.c.f.
Section 7.5.6

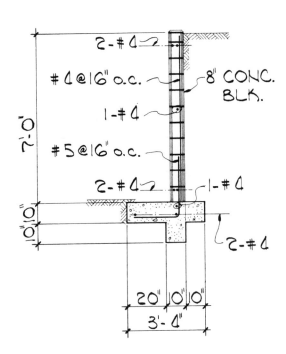

7'-0"

2-#4

#4 @ 16" o.c.

8" CONC. BLK.

1-#4

#5 @ 16" o.c.

2-#4

1-#4

2-#4

20" 10" 10"

3'-4"

Axial load = 800 lbs./ft.
E.F.P. = 30 p.c.f.
Section 7.5.7

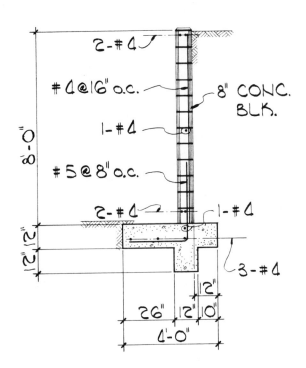

8'-0"

2-#4

#4 @ 16" o.c.

8" CONC. BLK.

1-#4

#5 @ 8" o.c.

2-#4

1-#4

3-#4

26" 12" 10"

4'-0"

Axial load = 800 lbs./ft.
E.F.P. = 30 p.c.f.
Section 7.5.8

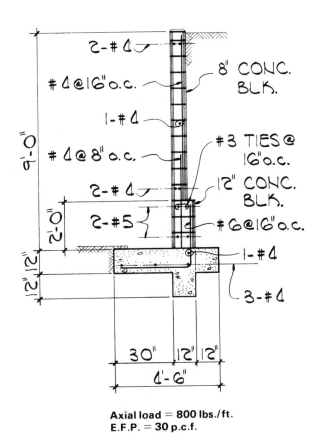

9'-0"

2-#4

#4 @ 16" o.c.

8" CONC. BLK.

1-#4

#4 @ 8" o.c.

#3 TIES @ 16" o.c.

2-#4

12" CONC. BLK.

2-#5

#6 @ 16" o.c.

1-#4

3-#4

30" 12" 12"

4'-6"

Axial load = 800 lbs./ft.
E.F.P. = 30 p.c.f.
Section 7.5.9

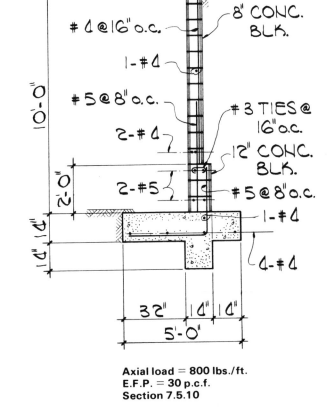

10'-0"

2-#4

#4 @ 16" o.c.

8" CONC. BLK.

1-#4

#5 @ 8" o.c.

#3 TIES @ 16" o.c.

2-#4

12" CONC. BLK.

2-#5

#5 @ 8" o.c.

1-#4

4-#4

32" 14" 14"

5'-0"

Axial load = 800 lbs./ft.
E.F.P. = 30 p.c.f.
Section 7.5.10

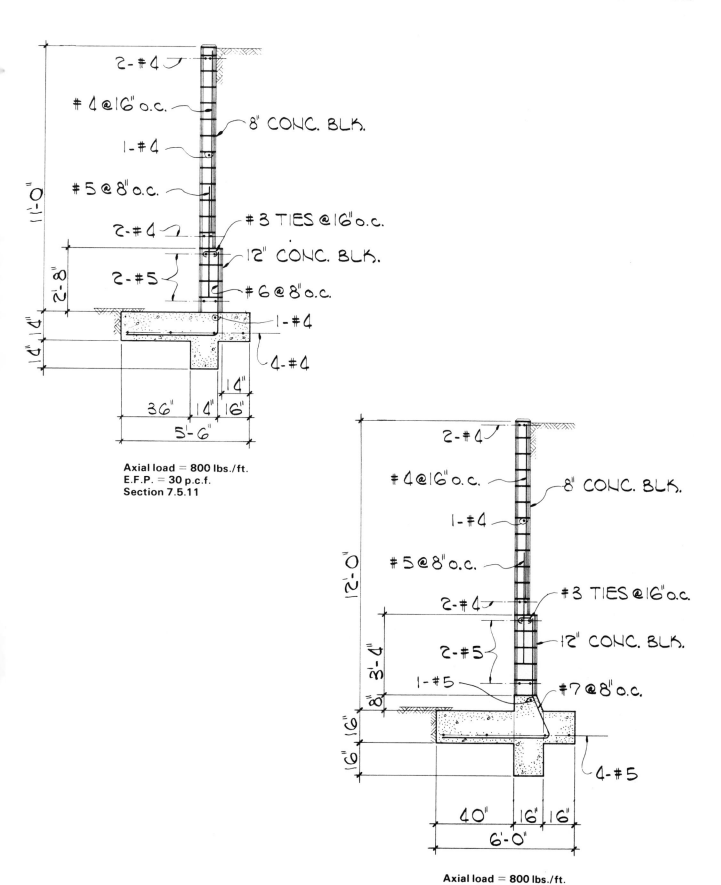

2-#4

#4 @ 16" o.c.

8" CONC. BLK.

1-#4

#5 @ 8" o.c.

2-#4

#3 TIES @ 16" o.c.

12" CONC. BLK.

2-#5

#6 @ 8" o.c.

1-#4

4-#4

11'-0"

2'-8"

4"

36" 14" 16"

14"

5'-6"

Axial load = 800 lbs./ft.
E.F.P. = 30 p.c.f.
Section 7.5.11

2-#4

#4 @ 16" o.c.

8" CONC. BLK.

1-#4

#5 @ 8" o.c.

2-#4

#3 TIES @ 16" o.c.

12" CONC. BLK.

2-#5

1-#5

#7 @ 8" o.c.

4-#5

12'-0"

3'-4"

8"

6"

6"

40" 16" 16"

6'-0"

Axial load = 800 lbs./ft.
E.F.P. = 30 p.c.f.
Section 7.5.12

TABLE 7.6   UNDERCUT FOOTING RETAINING WALLS – MASONRY STEM – SOIL AT TOE SIDE
SLOPE = 0 to 1   SURCHARGE = 50 lbs./sq. ft.   AXIAL = 0 lbs./ft.

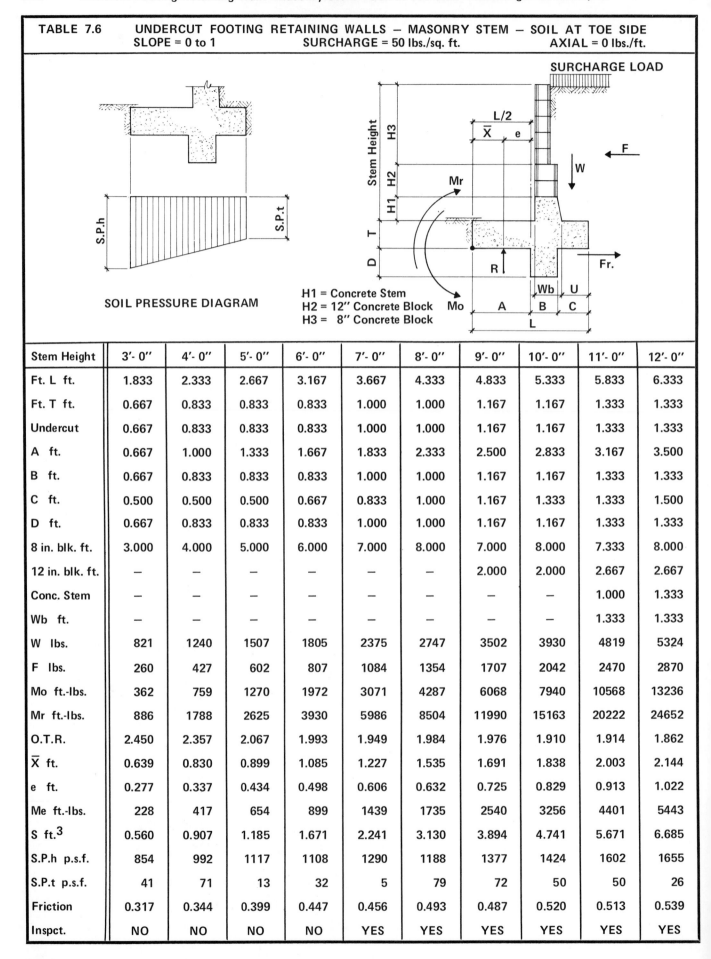

SOIL PRESSURE DIAGRAM

H1 = Concrete Stem
H2 = 12″ Concrete Block
H3 = 8″ Concrete Block

Stem Height	3'- 0″	4'- 0″	5'- 0″	6'- 0″	7'- 0″	8'- 0″	9'- 0″	10'- 0″	11'- 0″	12'- 0″
Ft. L ft.	1.833	2.333	2.667	3.167	3.667	4.333	4.833	5.333	5.833	6.333
Ft. T ft.	0.667	0.833	0.833	0.833	1.000	1.000	1.167	1.167	1.333	1.333
Undercut	0.667	0.833	0.833	0.833	1.000	1.000	1.167	1.167	1.333	1.333
A ft.	0.667	1.000	1.333	1.667	1.833	2.333	2.500	2.833	3.167	3.500
B ft.	0.667	0.833	0.833	0.833	1.000	1.000	1.167	1.167	1.333	1.333
C ft.	0.500	0.500	0.500	0.667	0.833	1.000	1.167	1.333	1.333	1.500
D ft.	0.667	0.833	0.833	0.833	1.000	1.000	1.167	1.167	1.333	1.333
8 in. blk. ft.	3.000	4.000	5.000	6.000	7.000	8.000	7.000	8.000	7.333	8.000
12 in. blk. ft.	—	—	—	—	—	—	2.000	2.000	2.667	2.667
Conc. Stem	—	—	—	—	—	—	—	—	1.000	1.333
Wb ft.	—	—	—	—	—	—	—	—	1.333	1.333
W lbs.	821	1240	1507	1805	2375	2747	3502	3930	4819	5324
F lbs.	260	427	602	807	1084	1354	1707	2042	2470	2870
Mo ft.-lbs.	362	759	1270	1972	3071	4287	6068	7940	10568	13236
Mr ft.-lbs.	886	1788	2625	3930	5986	8504	11990	15163	20222	24652
O.T.R.	2.450	2.357	2.067	1.993	1.949	1.984	1.976	1.910	1.914	1.862
$\overline{X}$ ft.	0.639	0.830	0.899	1.085	1.227	1.535	1.691	1.838	2.003	2.144
e ft.	0.277	0.337	0.434	0.498	0.606	0.632	0.725	0.829	0.913	1.022
Me ft.-lbs.	228	417	654	899	1439	1735	2540	3256	4401	5443
S ft.^{3}	0.560	0.907	1.185	1.671	2.241	3.130	3.894	4.741	5.671	6.685
S.P.h p.s.f.	854	992	1117	1108	1290	1188	1377	1424	1602	1655
S.P.t p.s.f.	41	71	13	32	5	79	72	50	50	26
Friction	0.317	0.344	0.399	0.447	0.456	0.493	0.487	0.520	0.513	0.539
Inspct.	NO	NO	NO	NO	YES	YES	YES	YES	YES	YES

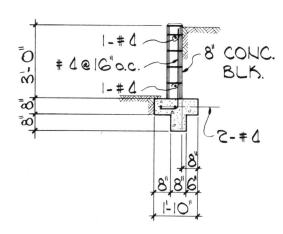

Surcharge load = **50** p.s.f.
E.F.P. = **30** p.c.f.
Section 7.6.3

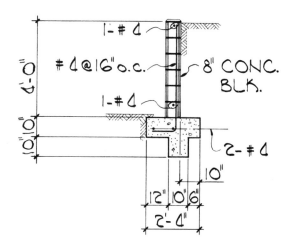

Surcharge load = **50** p.s.f.
E.F.P. = **30** p.c.f.
Section 7.6.4

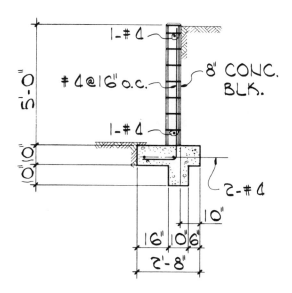

Surcharge load = **50** p.s.f.
E.F.P. = **30** p.c.f.
Section 7.6.5

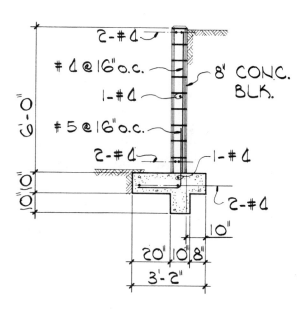

Surcharge load = **50** p.s.f.
E.F.P. = **30** p.c.f.
Section 7.6.6

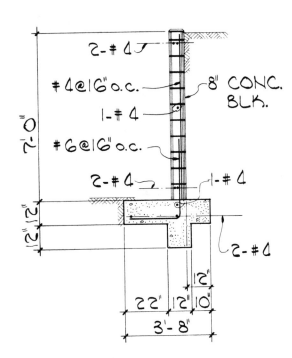

Surcharge load = 50 p.s.f.
E.F.P. = 30 p.c.f.
Section 7.6.7

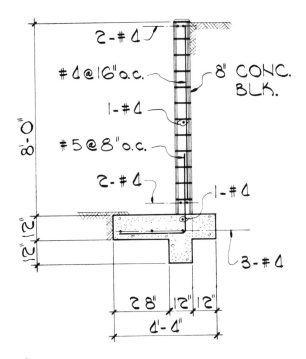

Surcharge load = 50 p.s.f.
E.F.P. = 30 p.c.f.
Section 7.6.8

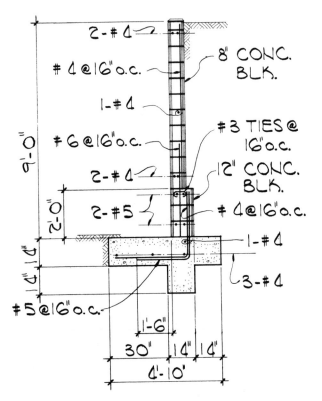

Surcharge load = 50 p.s.f.
E.F.P. = 30 p.c.f.
Section 7.6.9

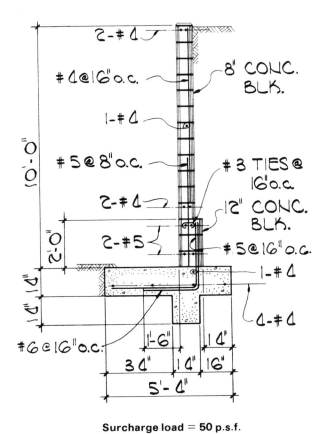

Surcharge load = 50 p.s.f.
E.F.P. = 30 p.c.f.
Section 7.6.10

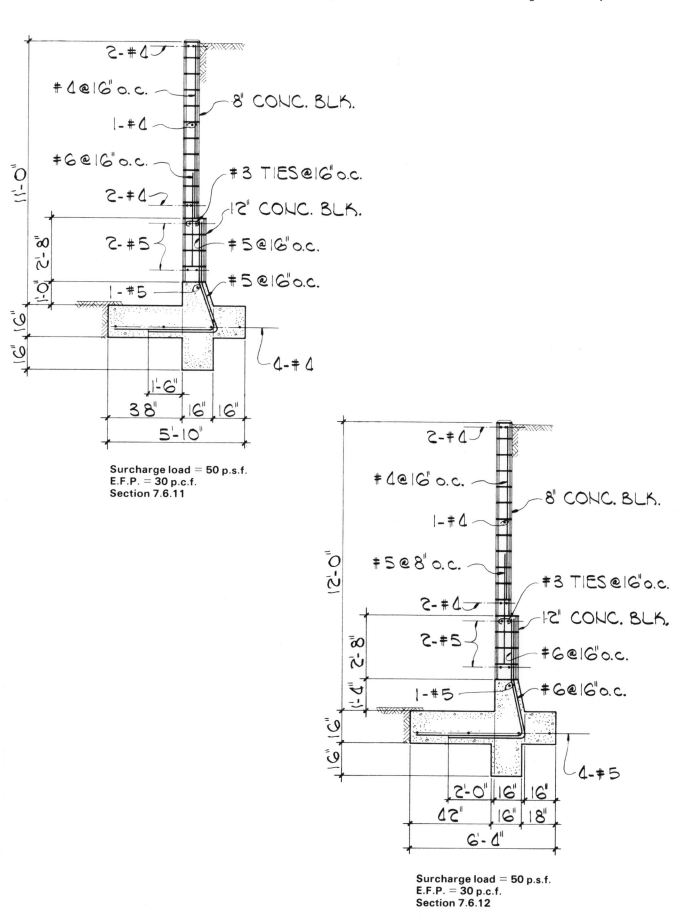

Top diagram:

2-#4

#4@16" o.c.

8" CONC. BLK.

1-#4

#6@16" o.c.

2-#4

#3 TIES@16" o.c.

12" CONC. BLK.

2-#5

#5@16" o.c.

1-#5

#5@16" o.c.

4-#4

11'-0"

2'-8"

1'-0"

6"

6"

1'-6"

38"   16"   16"

5'-10"

Surcharge load = 50 p.s.f.
E.F.P. = 30 p.c.f.
Section 7.6.11

Bottom diagram:

2-#4

#4@16" o.c.

8" CONC. BLK.

1-#4

#5@8" o.c.

#3 TIES@16" o.c.

2-#4

12" CONC. BLK.

2-#5

#6@16" o.c.

1-#5

#6@16" o.c.

4-#5

12'-0"

2'-8"

1'-4"

6"

6"

2'-0"   16"   16"

42"   16"   18"

6'-4"

Surcharge load = 50 p.s.f.
E.F.P. = 30 p.c.f.
Section 7.6.12

TABLE 7.7		**UNDERCUT FOOTING RETAINING WALLS – MASONRY STEM – SOIL AT TOE SIDE**								
		SLOPE = 0 to 1		**SURCHARGE = 100 lbs./sq. ft.**				**AXIAL = 0 lbs./ft.**		

SOIL PRESSURE DIAGRAM

H1 = Concrete Stem
H2 = 12" Concrete Block
H3 = 8" Concrete Block

Stem Height	3'- 0"	4'- 0"	5'- 0"	6'- 0"	7'- 0"	8'- 0"	9'- 0"	10'- 0"	11'- 0"	12'- 0"
Ft. T ft.	2.000	2.500	3.000	3.500	4.000	4.500	5.167	5.667	6.167	6.667
Ft. L ft.	0.667	0.833	0.833	0.833	1.000	1.000	1.167	1.167	1.333	1.333
Undercut	0.667	0.833	0.833	0.833	1.000	1.000	1.167	1.167	1.333	1.333
A ft.	0.833	1.167	1.500	1.833	2.167	2.500	2.833	3.167	3.333	3.667
B ft.	0.667	0.833	0.833	0.833	1.000	1.000	1.167	1.167	1.333	1.333
C ft.	0.500	0.500	0.667	0.833	0.833	1.000	1.167	1.333	1.500	1.667
D ft.	0.667	0.833	0.833	0.833	1.000	1.000	1.167	1.167	1.333	1.333
8 in. blk. ft.	3.000	4.000	5.000	6.000	7.000	6.000	7.000	7.333	7.333	7.333
12 in. blk. ft.	—	—	—	—	—	2.000	2.000	2.667	2.667	2.667
Conc. Stem	—	—	—	—	—	—	—	—	1.000	2.000
Wb ft.	—	—	—	—	—	—	—	—	1.333	1.333
W lbs.	889	1326	1619	1922	2514	2898	3667	4132	5043	5576
F lbs.	327	510	700	920	1215	1500	1870	2220	2667	3082
Mo ft.-lbs.	508	992	1595	2403	3645	5000	6962	9005	11852	14724
Mr ft.-lbs.	1105	2123	3287	4740	7077	9408	13629	17142	22576	27387
O.T.R.	2.174	2.139	2.060	1.972	1.942	1.882	1.958	1.904	1.905	1.860
$\overline{X}$ ft.	0.671	0.852	1.042	1.216	1.365	1.521	1.818	1.969	2.126	2.271
e ft.	0.329	0.398	0.455	0.534	0.635	0.729	0.765	0.864	0.957	1.062
Me ft.-lbs.	293	528	737	1027	1596	2113	2806	3571	4825	5925
S ft.^{3}	0.667	1.042	1.500	2.042	2.667	3.375	4.449	5.352	6.338	7.407
S.P.h p.s.f.	884	1037	1031	1052	1227	1270	1340	1397	1579	1636
S.P.t p.s.f.	6	24	48	46	30	18	79	62	56	37
Friction	0.367	0.385	0.433	0.479	0.483	0.518	0.510	0.537	0.529	0.553
Inspct.	NO	NO	NO	YES	YES	YES	YES	YES	YES	YES

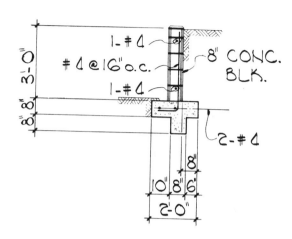

Surcharge load = 100 p.s.f.
E.F.P. = 30 p.c.f.
Section 7.7.3

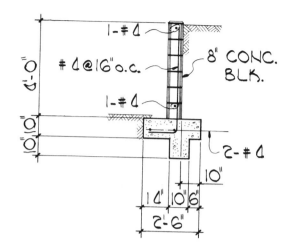

Surcharge load = 100 p.s.f.
E.F.P. = 30 p.c.f.
Section 7.7.4

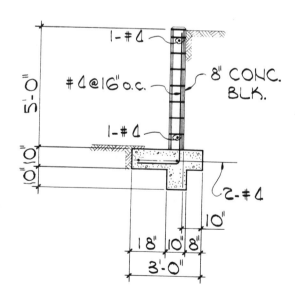

Surcharge load = 100 p.s.f.
E.F.P. = 30 p.c.f.
Section 7.7.5

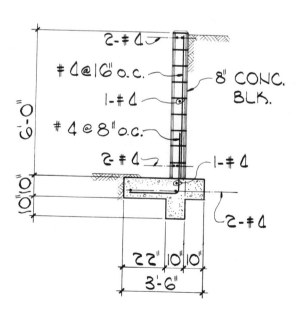

Surcharge load = 100 p.s.f.
E.F.P. = 30 p.c.f.
Section 7.7.6

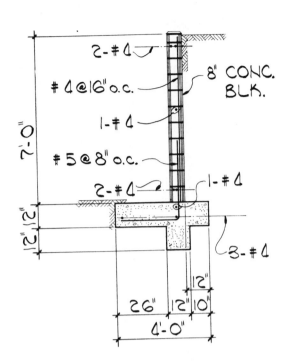

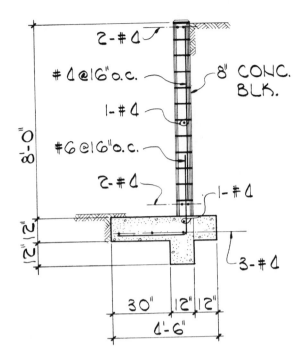

Surcharge load = 100 p.s.f.
E.F.P. = 30 p.c.f.
Section 7.7.7

Surcharge load = 100 p.s.f.
E.F.P. = 30 p.c.f.
Section 7.7.8

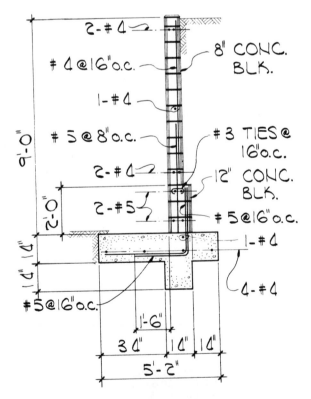

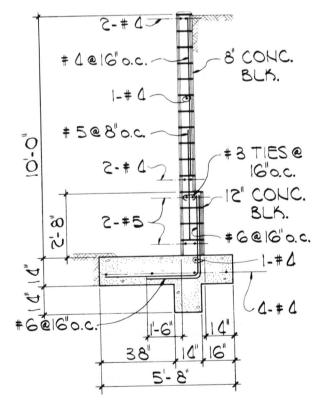

Surcharge load = 100 p.s.f.
E.F.P. = 30 p.c.f.
Section 7.7.9

Surcharge load = 100 p.s.f.
E.F.P. = 30 p.c.f.
Section 7.7.10

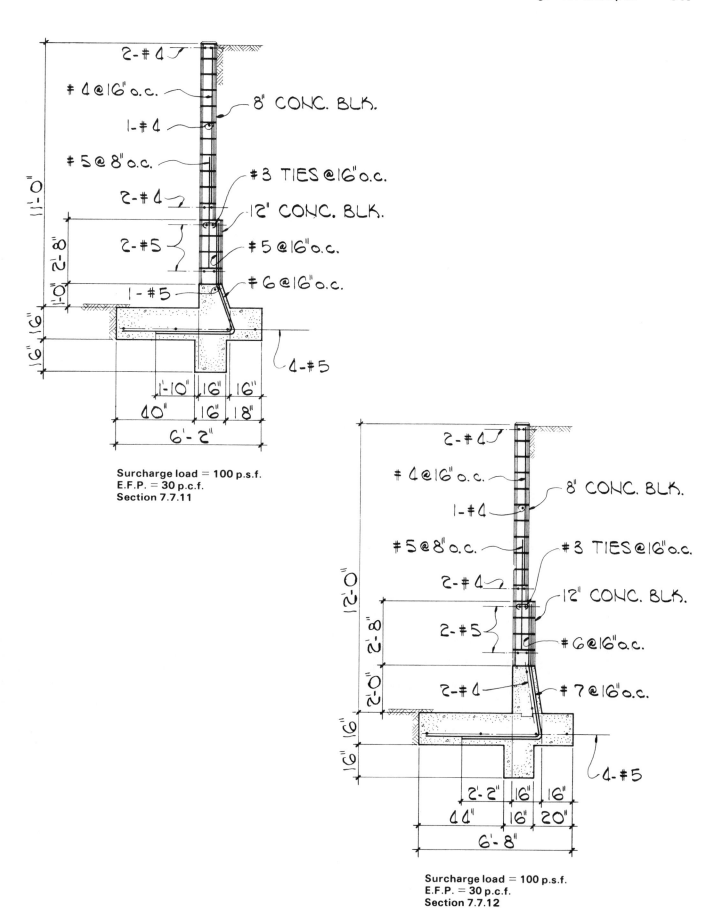

2-#4

#4@16" o.c.

1-#4

#5@8" o.c.

2-#4

2-#5

1-#5

8" CONC. BLK.

#3 TIES @16" o.c.

12" CONC. BLK.

#5 @16" o.c.

#6 @16" o.c.

4-#5

11'-0"

2'-8"

1'-0"

6"

6"

1'-10" 16" 16"

10" 16" 18"

6'-2"

Surcharge load = 100 p.s.f.
E.F.P. = 30 p.c.f.
Section 7.7.11

2-#4

#4@16" o.c.

1-#4

#5@8" o.c.

2-#4

2-#5

2-#4

8" CONC. BLK.

#3 TIES @16" o.c.

12" CONC. BLK.

#6@16" o.c.

#7@16" o.c.

4-#5

12'-0"

2'-8"

2'-0"

6"

6"

2'-2" 16" 16"

14" 16" 20"

6'-8"

Surcharge load = 100 p.s.f.
E.F.P. = 30 p.c.f.
Section 7.7.12

TABLE 7.8   UNDERCUT FOOTING RETAINING WALLS – MASONRY STEM – SOIL AT TOE SIDE
SLOPE = 0 to 1   SURCHARGE = 200 lbs./sq. ft.   AXIAL = 0 lbs./ft.

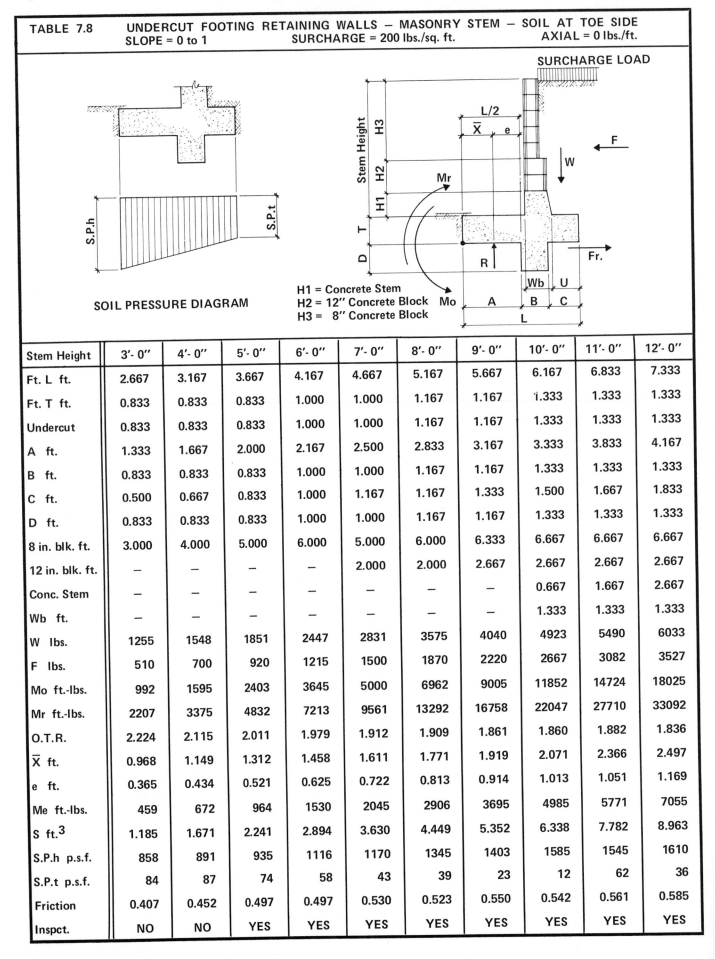

SOIL PRESSURE DIAGRAM

H1 = Concrete Stem
H2 = 12″ Concrete Block
H3 = 8″ Concrete Block

Stem Height	3′- 0″	4′- 0″	5′- 0″	6′- 0″	7′- 0″	8′- 0″	9′- 0″	10′- 0″	11′- 0″	12′- 0″
Ft. L ft.	2.667	3.167	3.667	4.167	4.667	5.167	5.667	6.167	6.833	7.333
Ft. T ft.	0.833	0.833	0.833	1.000	1.000	1.167	1.167	1.333	1.333	1.333
Undercut	0.833	0.833	0.833	1.000	1.000	1.167	1.167	1.333	1.333	1.333
A ft.	1.333	1.667	2.000	2.167	2.500	2.833	3.167	3.333	3.833	4.167
B ft.	0.833	0.833	0.833	1.000	1.000	1.167	1.167	1.333	1.333	1.333
C ft.	0.500	0.667	0.833	1.000	1.167	1.167	1.333	1.500	1.667	1.833
D ft.	0.833	0.833	0.833	1.000	1.000	1.167	1.167	1.333	1.333	1.333
8 in. blk. ft.	3.000	4.000	5.000	6.000	5.000	6.000	6.333	6.667	6.667	6.667
12 in. blk. ft.	—	—	—	—	2.000	2.000	2.667	2.667	2.667	2.667
Conc. Stem	—	—	—	—	—	—	—	0.667	1.667	2.667
Wb ft.	—	—	—	—	—	—	—	1.333	1.333	1.333
W lbs.	1255	1548	1851	2447	2831	3575	4040	4923	5490	6033
F lbs.	510	700	920	1215	1500	1870	2220	2667	3082	3527
Mo ft.-lbs.	992	1595	2403	3645	5000	6962	9005	11852	14724	18025
Mr ft.-lbs.	2207	3375	4832	7213	9561	13292	16758	22047	27710	33092
O.T.R.	2.224	2.115	2.011	1.979	1.912	1.909	1.861	1.860	1.882	1.836
$\overline{X}$ ft.	0.968	1.149	1.312	1.458	1.611	1.771	1.919	2.071	2.366	2.497
e ft.	0.365	0.434	0.521	0.625	0.722	0.813	0.914	1.013	1.051	1.169
Me ft.-lbs.	459	672	964	1530	2045	2906	3695	4985	5771	7055
S ft.^{3}	1.185	1.671	2.241	2.894	3.630	4.449	5.352	6.338	7.782	8.963
S.P.h p.s.f.	858	891	935	1116	1170	1345	1403	1585	1545	1610
S.P.t p.s.f.	84	87	74	58	43	39	23	12	62	36
Friction	0.407	0.452	0.497	0.497	0.530	0.523	0.550	0.542	0.561	0.585
Inspct.	NO	NO	YES	YES	YES	YES	YES	YES	YES	YES

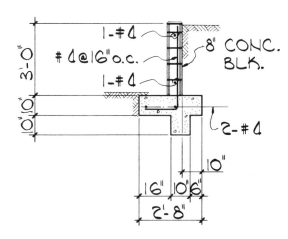

1-#4
#4@16"o.c.
1-#4
8" CONC. BLK.
2-#4
3'-0"
0'-0"
10"
16" 10" 6"
2'-8"

Surcharge load = 200 p.s.f.
E.F.P. = 30 p.c.f.
Section 7.8.3

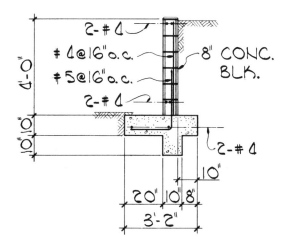

2-#4
#4@16"o.c.
#5@16"o.c.
2-#4
8" CONC. BLK.
2-#4
4'-0"
0'-0"
10"
20" 10" 8"
3'-2"

Surcharge load = 200 p.s.f.
E.F.P. = 30 p.c.f.
Section 7.8.4

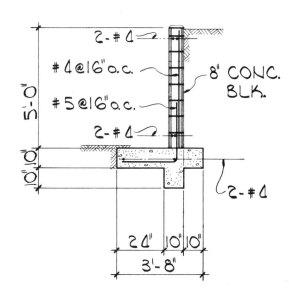

2-#4
#4@16"o.c.
#5@16"o.c.
2-#4
8" CONC. BLK.
2-#4
5'-0"
0'-0"
24" 10" 10"
3'-8"

Surcharge load = 200 p.s.f.
E.F.P. = 30 p.c.f.
Section 7.8.5

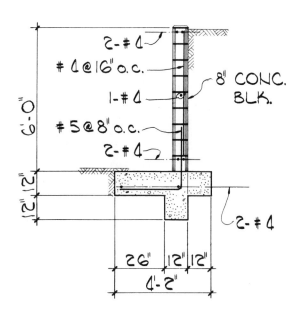

2-#4
#4@16"o.c.
1-#4
#5@8"o.c.
2-#4
8" CONC. BLK.
2-#4
6'-0"
1'-2"
26" 12" 12"
4'-2"

Surcharge load = 200 p.s.f.
E.F.P. = 30 p.c.f.
Section 7.8.6

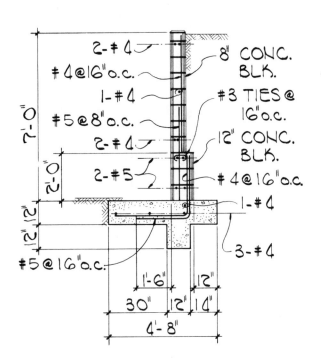

2-#4
#4@16"o.c.
1-#4
#5@8"o.c.
2-#4
2-#5
#5@16"o.c.

8" CONC. BLK.
#3 TIES @ 16"o.c.
12" CONC. BLK.
#4@16"o.c.
1-#4
3-#4

2'-0"
2'-0"
2'-2"

1'-6"   12"
30"   12"   14"
4'-8"

Surcharge load = 200 p.s.f.
E.F.P. = 30 p.c.f.
Section 7.8.7

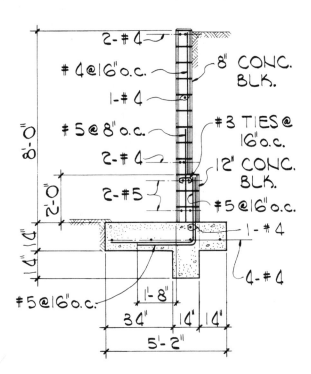

2-#4
#4@16"o.c.
1-#4
#5@8"o.c.
2-#4
2-#5
#5@16"o.c.

8" CONC. BLK.
#3 TIES @ 16"o.c.
12" CONC. BLK.
#5@16"o.c.
1-#4
4-#4

8'-0"
2'-0"
1'-4"

1'-8"
34"   14"   14"
5'-2"

Surcharge load = 200 p.s.f.
E.F.P. = 30 p.c.f.
Section 7.8.8

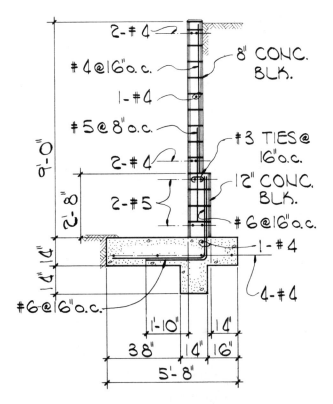

2-#4
#4@16"o.c.
1-#4
#5@8"o.c.
2-#4
2-#5
#6@16"o.c.

8" CONC. BLK.
#3 TIES @ 16"o.c.
12" CONC. BLK.
1-#4
4-#4

9'-0"
2'-8"
1'-4"

1'-10"   14"
38"   14"   16"
5'-8"

Surcharge load = 200 p.s.f.
E.F.P. = 30 p.c.f.
Section 7.8.9

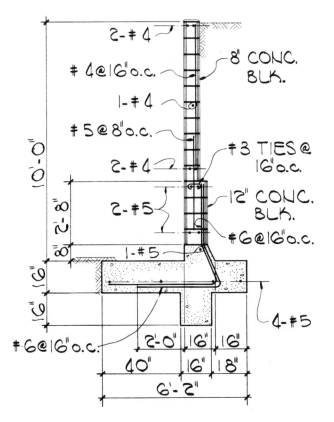

2-#4
#4@16"o.c.
1-#4
#5@8"o.c.
2-#4
2-#5
#6@16"o.c.

8" CONC. BLK.
#3 TIES @ 16"o.c.
12" CONC. BLK.
#6@16"o.c.
1-#5
4-#5

10'-0"
2'-8"
8"
6"
6"

2'-0"   16"   16"
40"   16"   18"
6'-2"

Surcharge load = 200 p.s.f.
E.F.P. = 30 p.c.f.
Section 7.8.10

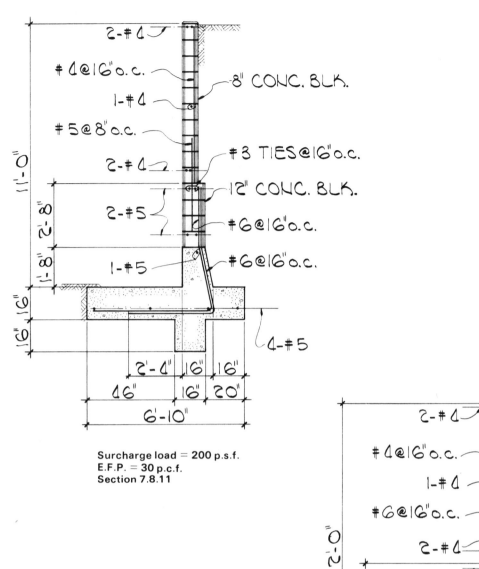

2-#4
#4@16"o.c.
1-#4
#5@8"o.c.
2-#4
2-#5
1-#5

8" CONC. BLK.
#3 TIES@16"o.c.
12" CONC. BLK.
#6@16"o.c.
#6@16"o.c.
4-#5

1'-0"
2'-8"
1'-8"
6"
6"

2'-4" 16" 16"
16" 16" 20"
6'-10"

Surcharge load = 200 p.s.f.
E.F.P. = 30 p.c.f.
Section 7.8.11

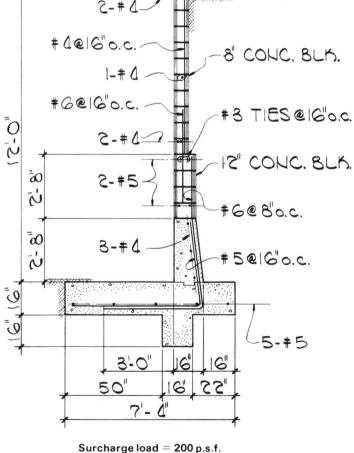

2-#4
#4@16"o.c.
1-#4
#6@16"o.c.
2-#4
2-#5
3-#4

8" CONC. BLK.
#3 TIES@16"o.c.
12" CONC. BLK.
#6@8"o.c.
#5@16"o.c.
5-#5

12'-0"
2'-8"
2'-8"
6"
6"

3'-0" 16" 16"
50" 16" 22"
7'-4"

Surcharge load = 200 p.s.f.
E.F.P. = 30 p.c.f.
Section 7.8.12

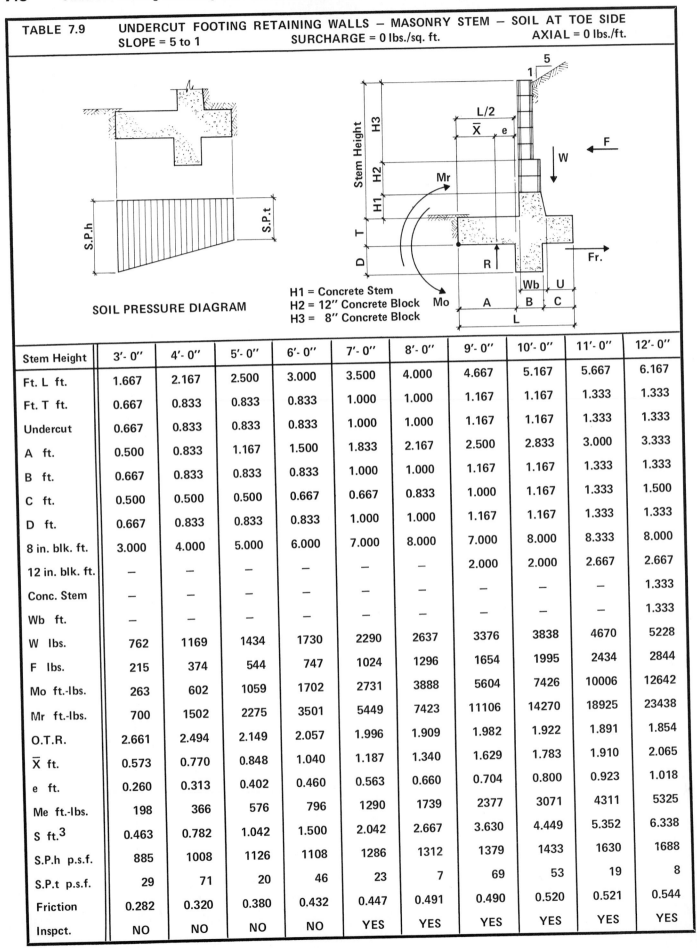

TABLE 7.9		UNDERCUT FOOTING RETAINING WALLS – MASONRY STEM – SOIL AT TOE SIDE								
	SLOPE = 5 to 1			SURCHARGE = 0 lbs./sq. ft.				AXIAL = 0 lbs./ft.		

SOIL PRESSURE DIAGRAM

H1 = Concrete Stem
H2 = 12" Concrete Block
H3 = 8" Concrete Block

Stem Height	3'- 0"	4'- 0"	5'- 0"	6'- 0"	7'- 0"	8'- 0"	9'- 0"	10'- 0"	11'- 0"	12'- 0"
Ft. L ft.	1.667	2.167	2.500	3.000	3.500	4.000	4.667	5.167	5.667	6.167
Ft. T ft.	0.667	0.833	0.833	0.833	1.000	1.000	1.167	1.167	1.333	1.333
Undercut	0.667	0.833	0.833	0.833	1.000	1.000	1.167	1.167	1.333	1.333
A ft.	0.500	0.833	1.167	1.500	1.833	2.167	2.500	2.833	3.000	3.333
B ft.	0.667	0.833	0.833	0.833	1.000	1.000	1.167	1.167	1.333	1.333
C ft.	0.500	0.500	0.500	0.667	0.667	0.833	1.000	1.167	1.333	1.500
D ft.	0.667	0.833	0.833	0.833	1.000	1.000	1.167	1.167	1.333	1.333
8 in. blk. ft.	3.000	4.000	5.000	6.000	7.000	8.000	7.000	8.000	8.333	8.000
12 in. blk. ft.	—	—	—	—	—	—	2.000	2.000	2.667	2.667
Conc. Stem	—	—	—	—	—	—	—	—	—	1.333
Wb ft.	—	—	—	—	—	—	—	—	—	1.333
W lbs.	762	1169	1434	1730	2290	2637	3376	3838	4670	5228
F lbs.	215	374	544	747	1024	1296	1654	1995	2434	2844
Mo ft.-lbs.	263	602	1059	1702	2731	3888	5604	7426	10006	12642
Mr ft.-lbs.	700	1502	2275	3501	5449	7423	11106	14270	18925	23438
O.T.R.	2.661	2.494	2.149	2.057	1.996	1.909	1.982	1.922	1.891	1.854
$\overline{X}$ ft.	0.573	0.770	0.848	1.040	1.187	1.340	1.629	1.783	1.910	2.065
e ft.	0.260	0.313	0.402	0.460	0.563	0.660	0.704	0.800	0.923	1.018
Me ft.-lbs.	198	366	576	796	1290	1739	2377	3071	4311	5325
S ft.^{3}	0.463	0.782	1.042	1.500	2.042	2.667	3.630	4.449	5.352	6.338
S.P.h p.s.f.	885	1008	1126	1108	1286	1312	1379	1433	1630	1688
S.P.t p.s.f.	29	71	20	46	23	7	69	53	19	8
Friction	0.282	0.320	0.380	0.432	0.447	0.491	0.490	0.520	0.521	0.544
Inspct.	NO	NO	NO	NO	YES	YES	YES	YES	YES	YES

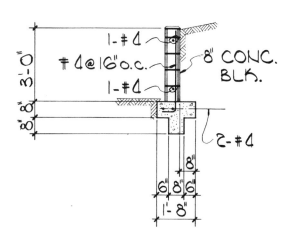

Soil slope = 5 to 1
E.F.P. = 32 p.c.f.
Section 7.9.3

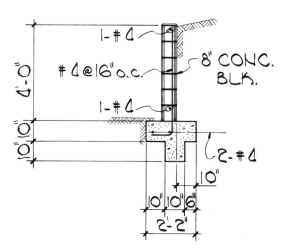

Soil slope = 5 to 1
E.F.P. = 32 p.c.f.
Section 7.9.4

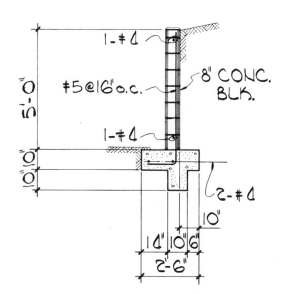

Soil slope = 5 to 1
E.F.P. = 32 p.c.f.
Section 7.9.5

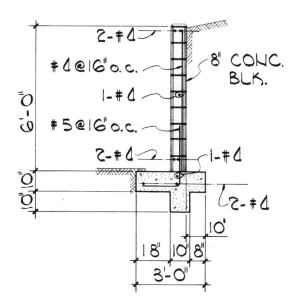

Soil slope = 5 to 1
E.F.P. = 32 p.c.f.
Section 7.9.6

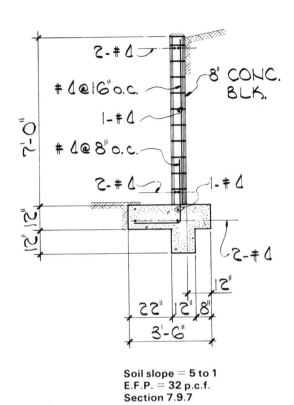

Soil slope = 5 to 1
E.F.P. = 32 p.c.f.
Section 7.9.7

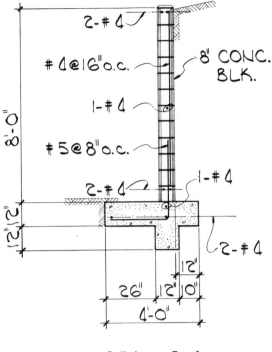

Soil slope = 5 to 1
E.F.P. = 32 p.c.f.
Section 7.9.8

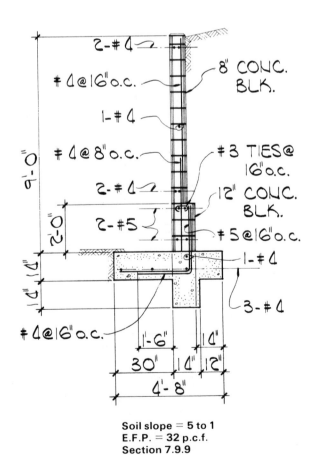

Soil slope = 5 to 1
E.F.P. = 32 p.c.f.
Section 7.9.9

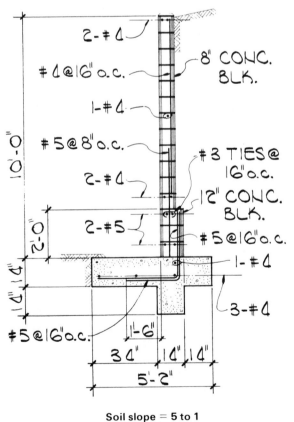

Soil slope = 5 to 1
E.F.P. = 32 p.c.f.
Section 7.9.10

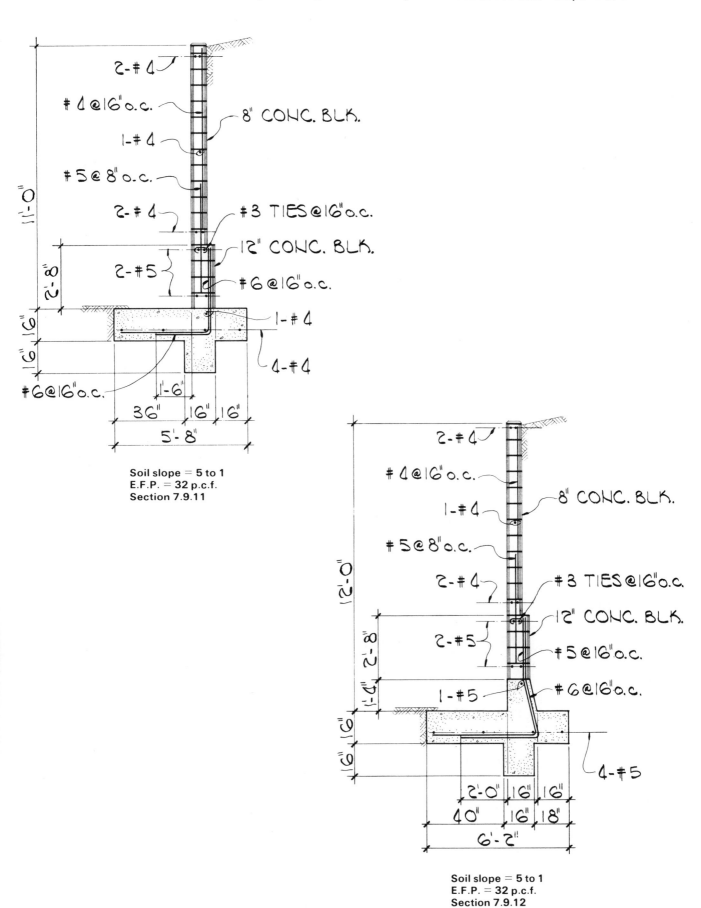

Top diagram labels:

2-#4

#4@16"o.c.

1-#4

#5@8"o.c.

2-#4

2-#5

8" CONC. BLK.

#3 TIES@16"o.c.

12" CONC. BLK.

#6@16"o.c.

1-#4

4-#4

#6@16"o.c.

11'-0"

2'-8"

6"

6"

1'-6"

36" | 16" | 16"

5'-8"

Soil slope = 5 to 1
E.F.P. = 32 p.c.f.
Section 7.9.11

Bottom diagram labels:

2-#4

#4@16"o.c.

1-#4

#5@8"o.c.

2-#4

2-#5

1-#5

8" CONC. BLK.

#3 TIES@16"o.c.

12" CONC. BLK.

#5@16"o.c.

#6@16"o.c.

4-#5

12'-0"

2'-8"

1'-4"

6"

6"

2'-0" | 16" | 16"

40" | 16" | 18"

6'-2"

Soil slope = 5 to 1
E.F.P. = 32 p.c.f.
Section 7.9.12

TABLE 7.10   UNDERCUT FOOTING RETAINING WALLS — MASONRY STEM — SOIL AT TOE SIDE
SLOPE = 4 to 1   SURCHARGE = 0 lbs./sq. ft.   AXIAL = 0 lbs./ft.

SOIL PRESSURE DIAGRAM

H1 = Concrete Stem
H2 = 12" Concrete Block
H3 = 8" Concrete Block

Stem Height	3'- 0''	4'- 0''	5'- 0''	6'- 0''	7'- 0''	8'- 0''	9'- 0''	10'- 0''	11'- 0''	12'- 0''
Ft. L ft.	1.833	2.167	2.667	3.167	3.667	4.333	4.833	5.333	6.000	6.500
Ft. T ft.	0.667	0.833	0.833	0.833	1.000	1.000	1.167	1.167	1.333	1.333
Undercut	0.667	0.833	0.833	0.833	1.000	1.000	1.167	1.167	1.333	1.333
A ft.	0.667	0.833	1.333	1.667	1.833	2.333	2.500	2.833	3.333	3.667
B ft.	0.667	0.833	0.833	0.833	1.000	1.000	1.167	1.167	1.333	1.333
C ft.	0.500	0.500	0.500	0.667	0.833	1.000	1.167	1.333	1.333	1.500
D ft.	0.667	0.833	0.833	0.833	1.000	1.000	1.167	1.167	1.333	1.333
8 in. blk. ft.	3.000	4.000	5.000	6.000	7.000	8.000	7.000	8.000	7.333	8.000
12 in. blk. ft.	—	—	—	—	—	—	2.000	2.000	2.667	2.667
Conc. Stem	—	—	—	—	—	—	—	—	1.000	1.333
Wb ft.	—	—	—	—	—	—	—	—	1.333	1.333
W lbs.	784	1178	1469	1771	2342	2722	3482	3921	4852	5372
F lbs.	235	409	595	817	1120	1417	1809	2182	2662	3111
Mo ft.-lbs.	288	659	1158	1861	2987	4252	6130	8122	10944	13827
Mr ft.-lbs.	835	1516	2542	3838	5882	8402	11896	15104	21004	25591
O.T.R.	2.903	2.301	2.196	2.062	1.969	1.976	1.941	1.859	1.919	1.851
$\overline{X}$ ft.	0.698	0.727	0.943	1.116	1.236	1.525	1.656	1.781	2.073	2.190
e ft.	0.219	0.356	0.391	0.467	0.597	0.642	0.761	0.886	0.927	1.060
Me ft.-lbs.	171	419	574	827	1399	1747	2648	3474	4497	5694
S ft.^{3}	0.560	0.782	1.185	1.671	2.241	3.130	3.894	4.741	6.000	7.042
S.P.h p.s.f.	734	1080	1035	1054	1263	1186	1400	1468	1558	1635
S.P.t p.s.f.	122	8	67	64	14	70	40	2	59	18
Friction	0.300	0.347	0.405	0.461	0.478	0.521	0.520	0.557	0.549	0.579
Inspct.	NO	NO	NO	NO	YES	YES	YES	YES	YES	YES

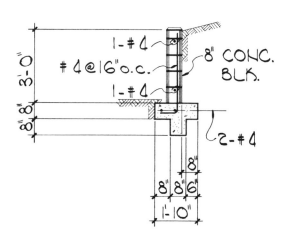

Soil slope = 4 to 1
E.F.P. = 35 p.c.f.
Section 7.10.3

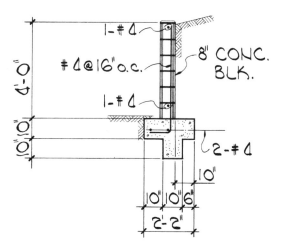

Soil slope = 4 to 1
E.F.P. = 35 p.c.f.
Section 7.10.4

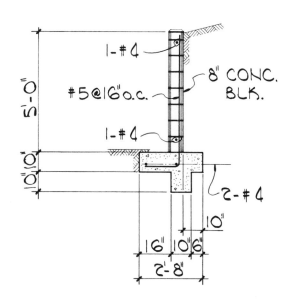

Soil slope = 4 to 1
E.F.P. = 35 p.c.f.
Section 7.10.5

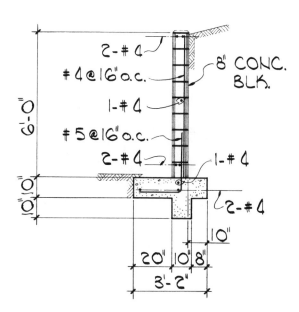

Soil slope = 4 to 1
E.F.P. = 35 p.c.f.
Section 7.10.6

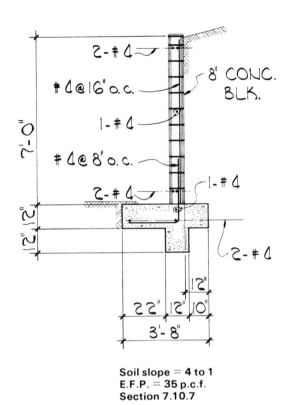

Soil slope = 4 to 1
E.F.P. = 35 p.c.f.
Section 7.10.7

8' CONC. BLK.
2-#4
#4@16" o.c.
1-#4
#4@8" o.c.
2-#5
1-#4
2-#4
3'-8"
22" 12" 10"
12"

Soil slope = 4 to 1
E.F.P. = 35 p.c.f.
Section 7.10.8

8' CONC. BLK.
2-#4
#4@16" o.c.
1-#4
#5@16" o.c.
2-#4
1-#4
3-#4
#5@16" o.c.
4'-4"
28" 12" 12"
1'-6"

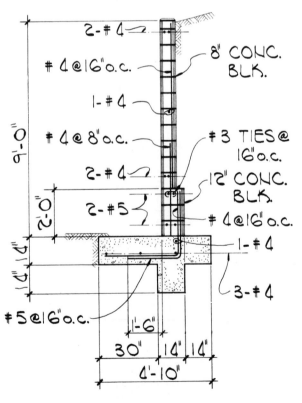

Soil slope = 4 to 1
E.F.P. = 35 p.c.f.
Section 7.10.9

8' CONC. BLK.
2-#4
#4@16" o.c.
1-#4
#4@8" o.c.
2-#4
2-#5
#3 TIES@ 16" o.c.
12" CONC. BLK.
#4@16" o.c.
1-#4
3-#4
#5@16" o.c.
4'-10"
30" 14" 14"
1'-6"

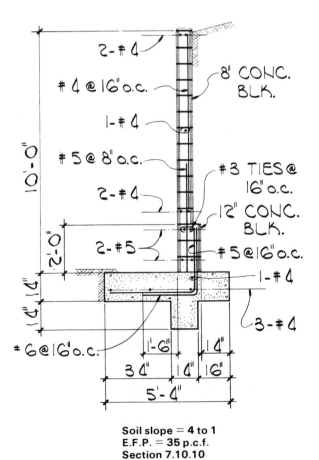

Soil slope = 4 to 1
E.F.P. = 35 p.c.f.
Section 7.10.10

8' CONC. BLK.
2-#4
#4@16" o.c.
1-#4
#5@8" o.c.
2-#4
2-#5
#3 TIES@ 16" o.c.
12" CONC. BLK.
#5@16" o.c.
1-#4
3-#4
#6@16" o.c.
5'-4"
34" 14" 16"
1'-6"

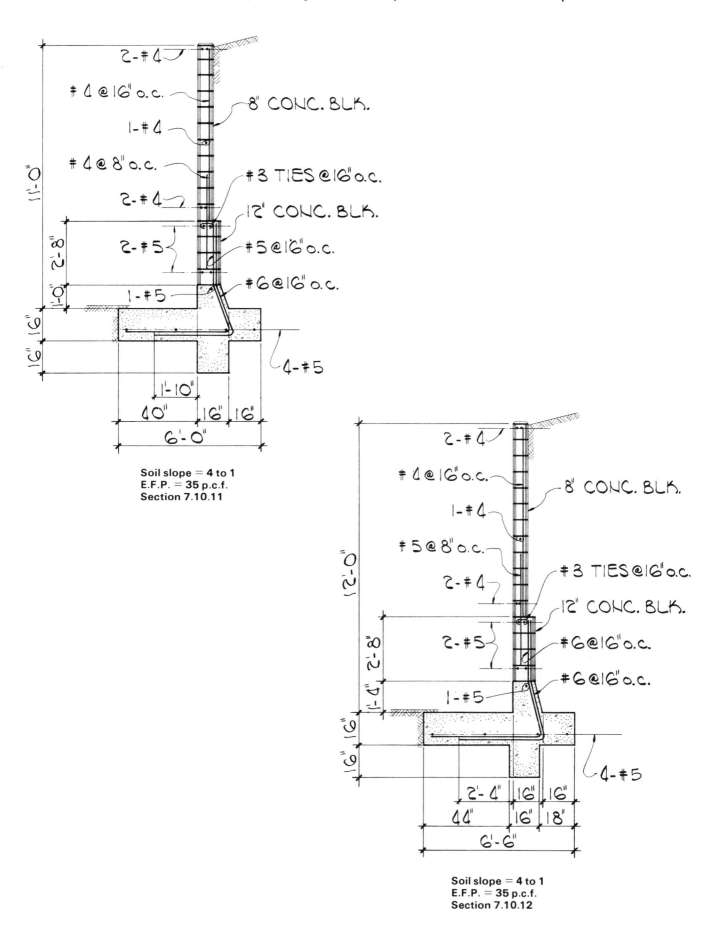

2-#4
#4 @ 16" o.c.
1-#4
#4 @ 8" o.c.
2-#4
#3 TIES @ 16" o.c.
12" CONC. BLK.
2-#5
#5 @ 16" o.c.
1-#5
#6 @ 16" o.c.
8" CONC. BLK.
4-#5

1'-0"
2'-8"
1'-0"
6"
6"

1'-10"
10"   16"   16"
6'-0"

Soil slope = 4 to 1
E.F.P. = 35 p.c.f.
Section 7.10.11

2-#4
#4 @ 16" o.c.
1-#4
#5 @ 8" o.c.
2-#4
#3 TIES @ 16" o.c.
12" CONC. BLK.
2-#5
#6 @ 16" o.c.
1-#5
#6 @ 16" o.c.
8" CONC. BLK.
4-#5

12'-0"
2'-8"
1'-4"
6"
6"

2'-4"   16"   16"
44"   16"   18"
6'-6"

Soil slope = 4 to 1
E.F.P. = 35 p.c.f.
Section 7.10.12

TABLE 7.11	UNDERCUT FOOTING RETAINING WALLS — MASONRY STEM — SOIL AT TOE SIDE
	SLOPE = 3 to 1 SURCHARGE = 0 lbs./sq. ft. AXIAL = 0 lbs./ft.

SOIL PRESSURE DIAGRAM

H1 = Concrete Stem
H2 = 12″ Concrete Block
H3 = 8″ Concrete Block

Stem Height	3'- 0''	4'- 0''	5'- 0''	6'- 0''	7'- 0''	8'- 0''	9'- 0''	10'- 0''	11'- 0''	12'- 0''
Ft. L ft.	1.833	2.333	2.833	3.333	3.833	4.500	5.000	5.667	6.167	6.833
Ft. T ft.	0.667	0.833	0.833	0.833	1.000	1.000	1.167	1.167	1.333	1.333
Undercut	0.667	0.833	0.833	0.833	1.000	1.000	1.167	1.167	1.333	1.333
A ft.	0.667	1.000	1.333	1.667	2.000	2.500	2.667	3.167	3.333	3.833
B ft.	0.667	0.833	0.833	0.833	1.000	1.000	1.167	1.167	1.333	1.333
C ft.	0.500	0.500	0.667	0.833	0.833	1.000	1.167	1.333	1.500	1.667
D ft.	0.667	0.833	0.833	0.833	1.000	1.000	1.167	1.167	1.333	1.333
8 in. blk. ft.	3.000	4.000	5.000	6.000	7.000	8.000	7.000	8.000	8.000	8.000
12 in. blk. ft.	—	—	—	—	—	—	2.000	2.000	2.000	2.000
Conc. Stem	—	—	—	—	—	—	—	—	1.000	2.000
Wb ft.	—	—	—	—	—	—	—	—	1.333	1.333
W lbs.	790	1210	1505	1812	2396	2783	3557	4035	4954	5541
F lbs.	255	444	647	887	1216	1539	1964	2369	2890	3378
Mo ft.-lbs.	312	715	1257	2021	3243	4617	6655	8819	11882	15012
Mr ft.-lbs.	843	1732	2804	4171	6358	8988	12660	16675	22106	27904
O.T.R.	2.701	2.422	2.230	2.064	1.961	1.947	1.902	1.891	1.861	1.859
$\overline{X}$ ft.	0.672	0.840	1.028	1.186	1.300	1.570	1.688	1.947	2.064	2.327
e ft.	0.245	0.326	0.389	0.480	0.616	0.680	0.812	0.886	1.019	1.090
Me ft.-lbs.	194	395	585	870	1477	1891	2888	3575	5049	6040
S ft.^{3}	0.560	0.907	1.338	1.852	2.449	3.375	4.167	5.352	6.338	7.782
S.P.h p.s.f.	777	954	969	1014	1228	1179	1404	1380	1600	1587
S.P.t p.s.f.	86	83	94	74	22	58	18	44	7	35
Friction	0.323	0.367	0.430	0.490	0.508	0.553	0.552	0.587	0.583	0.610
Inspct.	NO	NO	NO	NO	YES	YES	YES	YES	YES	YES

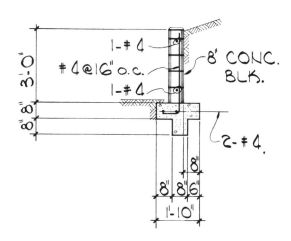

1-#4
#4@16" o.c.
1-#4
8" CONC. BLK.
3'-0"
8"
8"
2-#4
8"
8" 8" 6"
1'-10"

Soil slope = 3 to 1
E.F.P. = 38 p.c.f.
Section 7.11.3

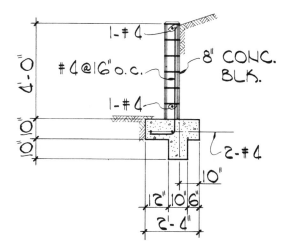

1-#4
#4@16" o.c.
1-#4
8" CONC. BLK.
4'-0"
10"
10"
2-#4
10"
12" 10" 6"
2'-4"

Soil slope = 3 to 1
E.F.P. = 38 p.c.f.
Section 7.11.4

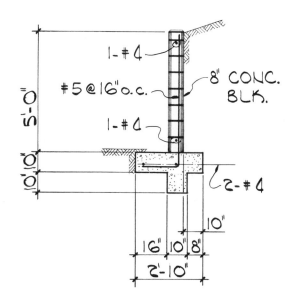

1-#4
#5@16" o.c.
1-#4
8" CONC. BLK.
5'-0"
10"
10"
2-#4
10"
16" 10" 8"
2'-10"

Soil slope = 3 to 1
E.F.P. = 38 p.c.f.
Section 7.11.5

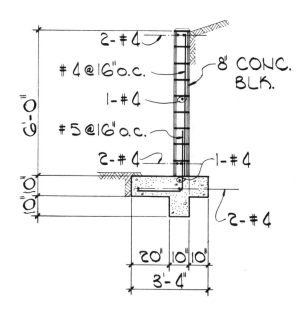

2-#4
#4@16" o.c.
1-#4
#5@16" o.c.
2-#4
8" CONC. BLK.
1-#4
6'-0"
10"
10"
2-#4
20" 10" 10"
3'-4"

Soil slope = 3 to 1
E.F.P. = 38 p.c.f.
Section 7.11.6

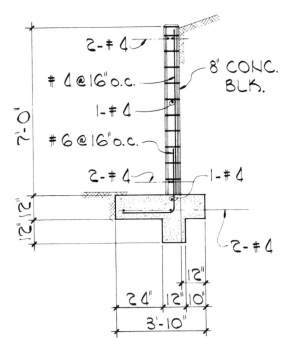

2-#4
#4 @ 16" o.c.
8" CONC. BLK.
1-#4
#6 @ 16" o.c.
2-#4
1-#4
7'-0"
2"
12"
2-#4
2"
12"
24"   12" 10"
3'-10"

Soil slope = 3 to 1
E.F.P. = 38 p.c.f.
Section 7.11.7

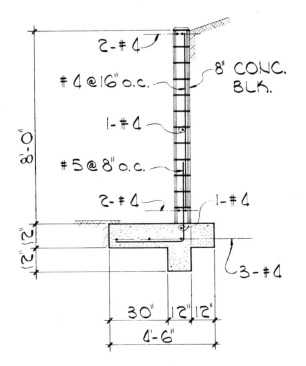

2-#4
#4 @ 16" o.c.
8" CONC. BLK.
1-#4
#5 @ 8" o.c.
2-#4
1-#4
8'-0"
2"
3-#4
30"   12" 12"
4'-6"

Soil slope = 3 to 1
E.F.P. = 38 p.c.f.
Section 7.11.8

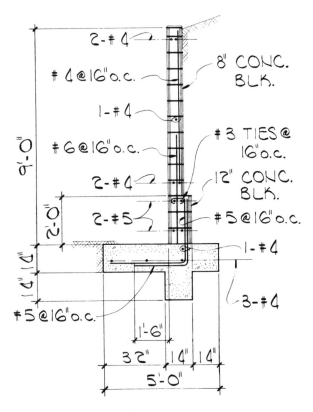

2-#4
#4 @ 16" o.c.
8" CONC. BLK.
1-#4
#6 @ 16" o.c.
#3 TIES @ 16" o.c.
2-#4
12" CONC. BLK.
2-#5
#5 @ 16" o.c.
1-#4
9'-0"
2'-0"
1'-4"
#5 @ 16" o.c.
3-#4
1'-6"
32"   14" 14"
5'-0"

Soil slope = 3 to 1
E.F.P. = 38 p.c.f.
Section 7.11.9

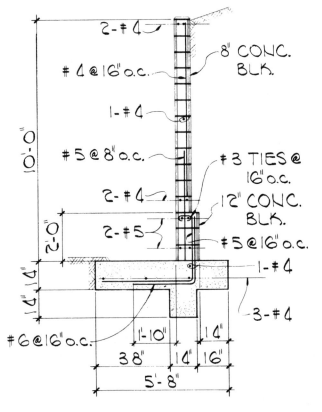

2-#4
#4 @ 16" o.c.
8" CONC. BLK.
1-#4
#5 @ 8" o.c.
#3 TIES @ 16" o.c.
2-#4
12" CONC. BLK.
2-#5
#5 @ 16" o.c.
1-#4
10'-0"
2'-0"
1'-4"
#6 @ 16" o.c.
3-#4
1'-10"
38"   14" 16"
5'-8"

Soil slope = 3 to 1
E.F.P. = 38 p.c.f.
Section 7.11.10

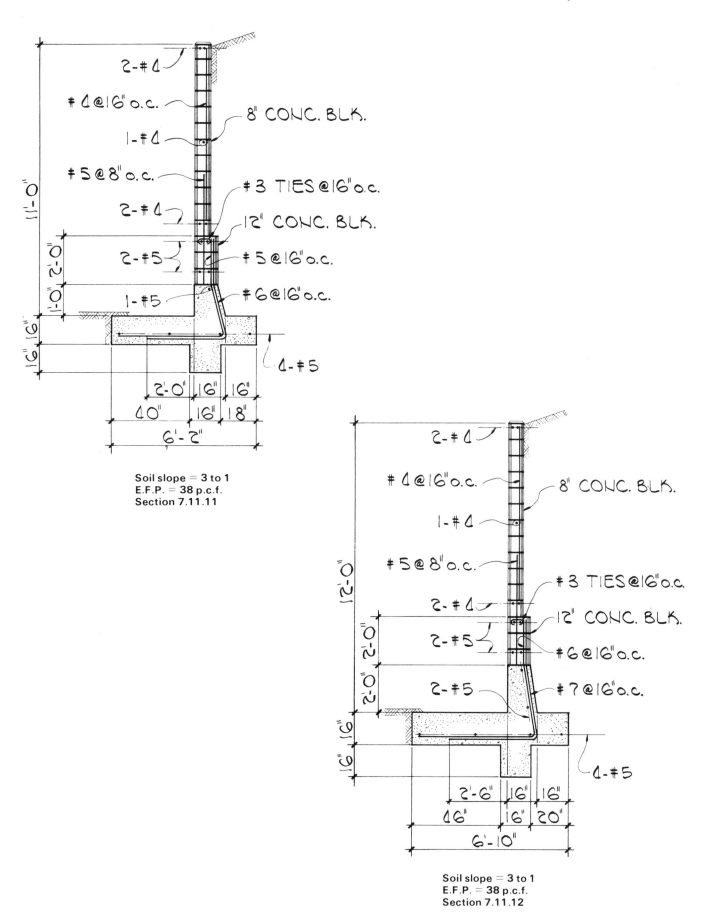

2-#4

#4 @ 16" o.c.

1-#4

#5 @ 8" o.c.

2-#4

2-#5

1-#5

8" CONC. BLK.

#3 TIES @ 16" o.c.

12" CONC. BLK.

#5 @ 16" o.c.

#6 @ 16" o.c.

4-#5

11'-0"

2'-0"

1'-0"

16"

16"

2'-0"

16"

16"

40"

16"

18"

6'-2"

Soil slope = 3 to 1
E.F.P. = 38 p.c.f.
Section 7.11.11

2-#4

#4 @ 16" o.c.

1-#4

#5 @ 8" o.c.

2-#4

2-#5

2-#5

8" CONC. BLK.

#3 TIES @ 16" o.c.

12" CONC. BLK.

#6 @ 16" o.c.

#7 @ 16" o.c.

4-#5

12'-0"

2'-0"

2'-0"

16"

16"

2'-6"

16"

16"

16"

16"

20"

6'-10"

Soil slope = 3 to 1
E.F.P. = 38 p.c.f.
Section 7.11.12

TABLE 7.12	UNDERCUT FOOTING RETAINING WALLS — MASONRY STEM — SOIL AT TOE SIDE
	SLOPE = 2 to 1 SURCHARGE = 0 lbs./sq. ft. AXIAL = 0 lbs./ft.

SOIL PRESSURE DIAGRAM

H1 = Concrete Stem
H2 = 12" Concrete Block
H3 = 8" Concrete Block

Stem Height	3'- 0"	4'- 0"	5'- 0"	6'- 0"	7'- 0"	8'- 0"	9'- 0"	10'- 0"	11'- 0"	12'- 0"
Ft. L ft.	1.833	2.500	3.000	3.500	4.167	4.833	5.333	6.000	6.667	7.333
Ft. T ft.	0.667	0.833	0.833	0.833	1.000	1.167	1.167	1.333	1.333	1.333
Undercut	0.667	0.833	0.833	0.833	1.000	1.167	1.167	1.333	1.333	1.333
A ft.	0.667	1.167	1.500	1.833	2.167	2.500	2.833	3.333	3.667	4.167
B ft.	0.667	0.833	0.833	0.833	1.000	1.167	1.167	1.333	1.333	1.333
C ft.	0.500	0.500	0.667	0.833	1.000	1.167	1.333	1.333	1.667	1.833
D ft.	0.667	0.833	0.833	0.833	1.000	1.167	1.167	1.333	1.333	1.333
8 in. blk. ft.	3.000	4.000	5.000	6.000	7.000	6.000	7.000	7.333	7.333	7.333
12 in. blk. ft.	—	—	—	—	—	2.000	2.000	2.667	2.667	2.667
Conc. Stem	—	—	—	—	—	—	—	—	1.000	2.000
Wb ft.	—	—	—	—	—	—	—	—	1.333	1.333
W lbs.	802	1250	1552	1869	2495	3244	3694	4609	5201	5808
F lbs.	289	502	732	1004	1376	1807	2222	2762	3270	3822
Mo ft.-lbs.	353	809	1423	2287	3669	5520	7531	10433	13445	16988
Mr ft.-lbs.	858	1970	3117	4574	7301	11023	14158	19864	25339	31643
O.T.R.	2.428	2.435	2.191	2.000	1.990	1.997	1.880	1.904	1.885	1.863
$\overline{X}$ ft.	0.629	0.929	1.092	1.224	1.456	1.696	1.794	2.046	2.287	2.523
e ft.	0.287	0.321	0.408	0.526	0.628	0.720	0.873	0.954	1.047	1.143
Me ft.-lbs.	230	401	634	984	1566	2337	3223	4396	5443	6639
S ft.^{3}	0.560	1.042	1.500	2.042	2.894	3.894	4.741	6.000	7.407	8.963
S.P.h p.s.f.	848	885	940	1016	1140	1271	1373	1501	1515	1533
S.P.t p.s.f.	26	115	95	52	58	71	13	36	45	51
Friction	0.361	0.402	0.471	0.537	0.551	0.557	0.602	0.599	0.629	0.658
Inspct.	NO	NO	NO	NO	YES	YES	YES	YES	YES	YES

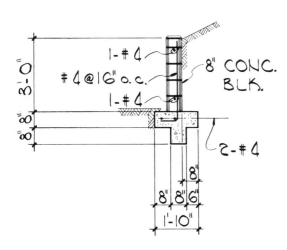

Soil slope = 2 to 1
E.F.P. = 43 p.c.f.
Section 7.12.3

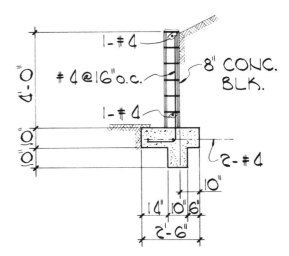

Soil slope = 2 to 1
E.F.P. = 43 p.c.f.
Section 7.12.4

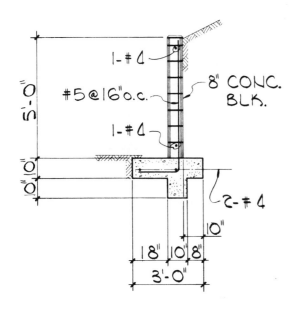

Soil slope = 2 to 1
E.F.P. = 43 p.c.f.
Section 7.12.5

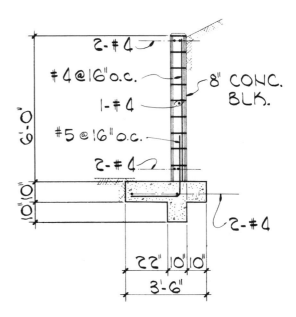

Soil slope = 2 to 1
E.F.P. = 43 p.c.f.
Section 7.12.6

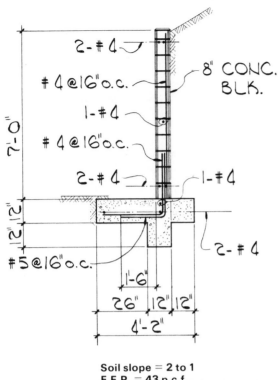

Soil slope = 2 to 1
E.F.P. = **43** p.c.f.
Section 7.12.7

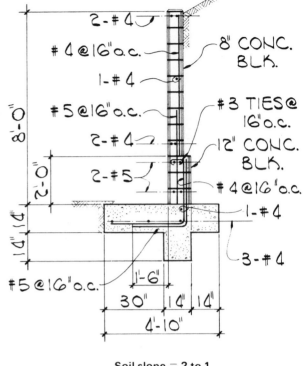

Soil slope = 2 to 1
E.F.P. = **43** p.c.f.
Section 7.12.8

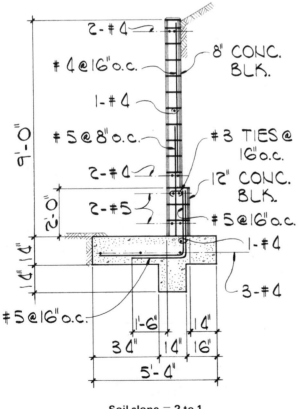

Soil slope = 2 to 1
E.F.P. = **43** p.c.f.
Section 7.12.9

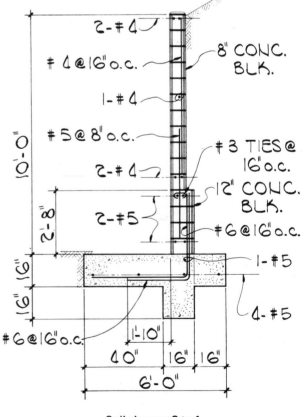

Soil slope = 2 to 1
E.F.P. = **43** p.c.f.
Section 7.12.10

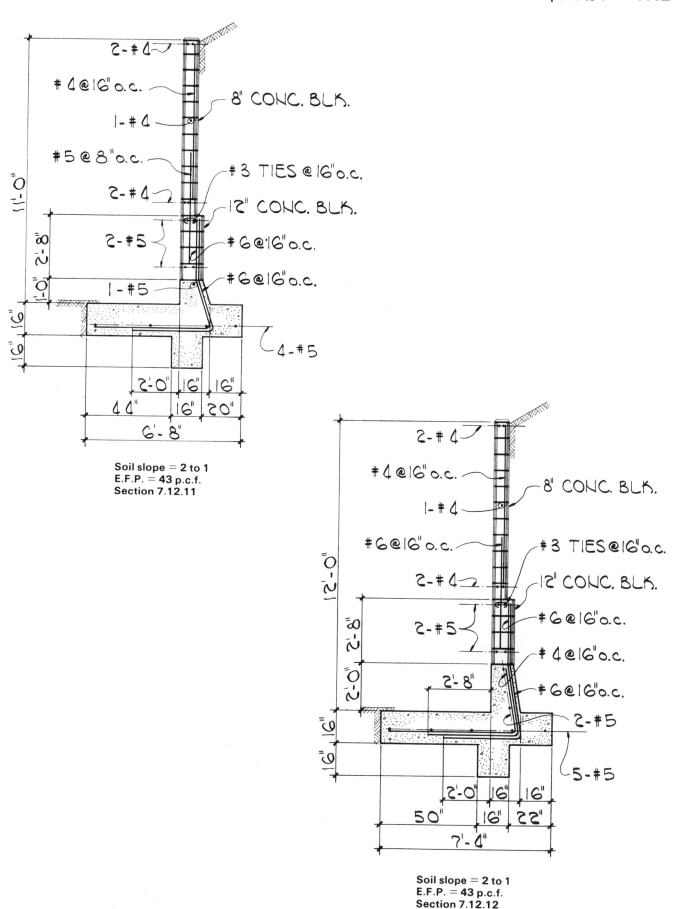

2-#4
#4 @16" o.c.
1-#4
#5 @ 8" o.c.
2-#4
2-#5
1-#5

8" CONC. BLK.
#3 TIES @16" o.c.
12" CONC. BLK.
#6 @16" o.c.
#6 @16" o.c.
4-#5

11'-0"
2'-8"
1'-0"
16"
16"

2'-0"   16"   16"
44"   16"   20"
6'-8"

Soil slope = 2 to 1
E.F.P. = 43 p.c.f.
Section 7.12.11

2-#4
#4 @16" o.c.
1-#4
#6 @16" o.c.
2-#4
2-#5

8" CONC. BLK.
#3 TIES @16" o.c.
12" CONC. BLK.
#6 @16" o.c.
#4 @16" o.c.
#6 @16" o.c.
2-#5
5-#5

12'-0"
2'-8"
2'-0"
16"
16"

2'-8"
2'-0"   16"   16"
50"   16"   22"
7'-4"

Soil slope = 2 to 1
E.F.P. = 43 p.c.f.
Section 7.12.12

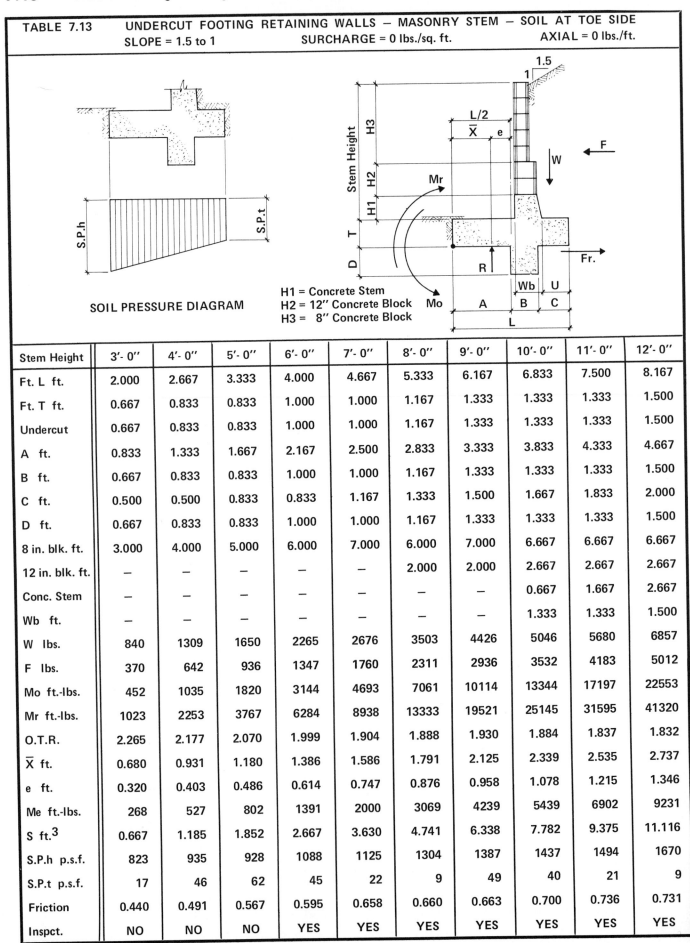

TABLE 7.13   UNDERCUT FOOTING RETAINING WALLS — MASONRY STEM — SOIL AT TOE SIDE
SLOPE = 1.5 to 1   SURCHARGE = 0 lbs./sq. ft.   AXIAL = 0 lbs./ft.

H1 = Concrete Stem
H2 = 12″ Concrete Block
H3 = 8″ Concrete Block

SOIL PRESSURE DIAGRAM

Stem Height	3'- 0"	4'- 0"	5'- 0"	6'- 0"	7'- 0"	8'- 0"	9'- 0"	10'- 0"	11'- 0"	12'- 0"
Ft. L ft.	2.000	2.667	3.333	4.000	4.667	5.333	6.167	6.833	7.500	8.167
Ft. T ft.	0.667	0.833	0.833	1.000	1.000	1.167	1.333	1.333	1.333	1.500
Undercut	0.667	0.833	0.833	1.000	1.000	1.167	1.333	1.333	1.333	1.500
A ft.	0.833	1.333	1.667	2.167	2.500	2.833	3.333	3.833	4.333	4.667
B ft.	0.667	0.833	0.833	1.000	1.000	1.167	1.333	1.333	1.333	1.500
C ft.	0.500	0.500	0.833	0.833	1.167	1.333	1.500	1.667	1.833	2.000
D ft.	0.667	0.833	0.833	1.000	1.000	1.167	1.333	1.333	1.333	1.500
8 in. blk. ft.	3.000	4.000	5.000	6.000	7.000	6.000	7.000	6.667	6.667	6.667
12 in. blk. ft.	–	–	–	–	–	2.000	2.000	2.667	2.667	2.667
Conc. Stem	–	–	–	–	–	–	–	0.667	1.667	2.667
Wb ft.	–	–	–	–	–	–	–	1.333	1.333	1.500
W lbs.	840	1309	1650	2265	2676	3503	4426	5046	5680	6857
F lbs.	370	642	936	1347	1760	2311	2936	3532	4183	5012
Mo ft.-lbs.	452	1035	1820	3144	4693	7061	10114	13344	17197	22553
Mr ft.-lbs.	1023	2253	3767	6284	8938	13333	19521	25145	31595	41320
O.T.R.	2.265	2.177	2.070	1.999	1.904	1.888	1.930	1.884	1.837	1.832
$\overline{X}$ ft.	0.680	0.931	1.180	1.386	1.586	1.791	2.125	2.339	2.535	2.737
e ft.	0.320	0.403	0.486	0.614	0.747	0.876	0.958	1.078	1.215	1.346
Me ft.-lbs.	268	527	802	1391	2000	3069	4239	5439	6902	9231
S ft.^{3}	0.667	1.185	1.852	2.667	3.630	4.741	6.338	7.782	9.375	11.116
S.P.h p.s.f.	823	935	928	1088	1125	1304	1387	1437	1494	1670
S.P.t p.s.f.	17	46	62	45	22	9	49	40	21	9
Friction	0.440	0.491	0.567	0.595	0.658	0.660	0.663	0.700	0.736	0.731
Inspct.	NO	NO	NO	YES	YES	YES	YES	YES	YES	YES

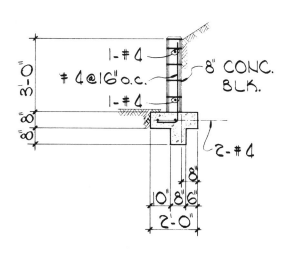

Soil slope = 1½ to 1
E.F.P. = 55 p.c.f.
Section 7.13.3

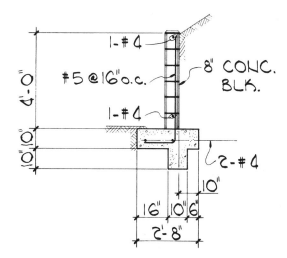

Soil slope = 1½ to 1
E.F.P. = 55 p.c.f.
Section 7.13.4

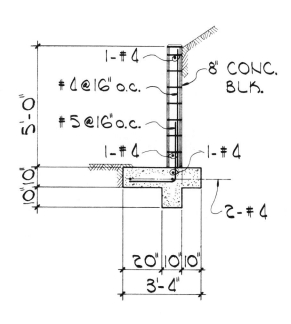

Soil slope = 1½ to 1
E.F.P. = 55 p.c.f.
Section 7.13.5

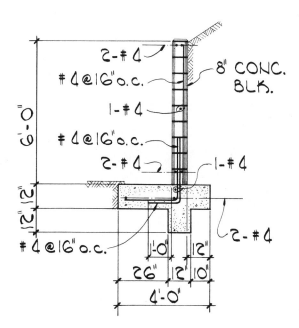

Soil slope = 1½ to 1
E.F.P. = 55 p.c.f.
Section 7.13.6

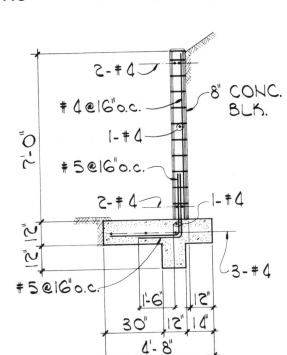

2-#4
#4@16" o.c.
1-#4
#5@16" o.c.
2-#4
1-#4
#5@16" o.c.
8" CONC. BLK.
3-#4

2'-0"
2"

30"   12"   14"
1'-6"   12"
4'-8"

Soil slope = 1½ to 1
E.F.P. = 55 p.c.f.
Section 7.13.7

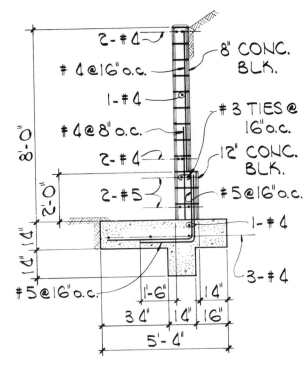

2-#4
#4@16" o.c.
1-#4
#4@8" o.c.
2-#4
2-#5
8" CONC. BLK.
#3 TIES @ 16" o.c.
12" CONC. BLK.
#5@16" o.c.
1-#4
3-#4
#5@16" o.c.

8'-0"
2'-0"
1'-4"

34"   14"   16"
1'-6"   14"
5'-4"

Soil slope = 1½ to 1
E.F.P. = 55 p.c.f.
Section 7.13.8

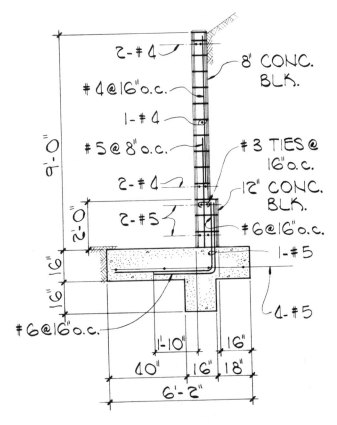

2-#4
#4@16" o.c.
1-#4
#5@8" o.c.
2-#4
2-#5
8" CONC. BLK.
#3 TIES @ 16" o.c.
12" CONC. BLK.
#6@16" o.c.
1-#5
4-#5
#6@16" o.c.

9'-0"
2'-0"
6"
6"

40"   16"   18"
1'-10"   16"
6'-2"

Soil slope = 1½ to 1
E.F.P. = 55 p.c.f.
Section 7.13.9

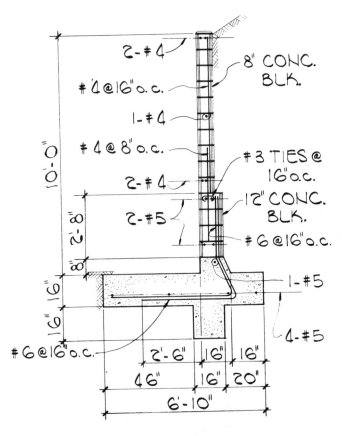

2-#4
#4@16" o.c.
1-#4
#4@8" o.c.
2-#4
2-#5
8" CONC. BLK.
#3 TIES @ 16" o.c.
12" CONC. BLK.
#6@16" o.c.
1-#5
4-#5
#6@16" o.c.

10'-0"
2'-8"
8"
16"
16"

46"   16"   20"
2'-6"   16"   16"
6'-10"

Soil slope = 1½ to 1
E.F.P. = 55 p.c.f.
Section 7.13.10

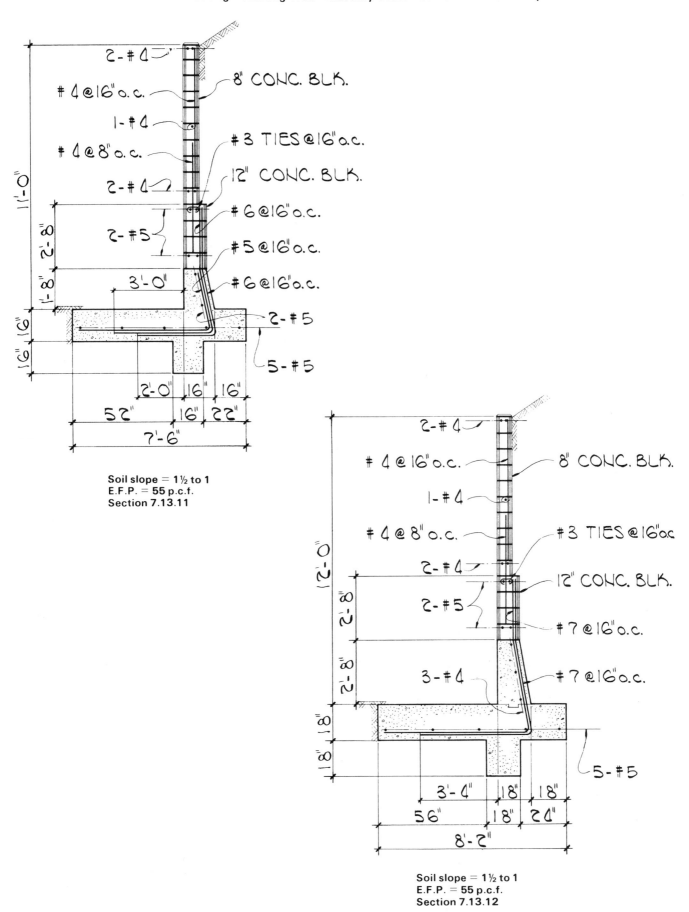

Soil slope = 1 ½ to 1
E.F.P. = 55 p.c.f.
Section 7.13.11

Soil slope = 1 ½ to 1
E.F.P. = 55 p.c.f.
Section 7.13.12

TABLE 7.14   UNDERCUT FOOTING RETAINING WALLS – MASONRY STEM – SOIL AT TOE SIDE
SLOPE = 1 to 1   SURCHARGE = 0 lbs./sq. ft.   AXIAL = 0 lbs. /ft.

SOIL PRESSURE DIAGRAM

H1 = Concrete Stem
H2 = 12″ Concrete Block
H3 = 8″ Concrete Block

Stem Height	3′- 0″	4′- 0″	5′- 0″	6′- 0″	7′- 0″	8′- 0″	9′- 0″	10′- 0″	11′- 0″	12′- 0″
Ft. L ft.	2.500	3.167	4.000	4.833	5.667	6.500	7.333	8.167	9.000	10.000
Ft. T ft.	0.833	0.833	0.833	1.000	1.167	1.333	1.333	1.333	1.500	1.500
Undercut	0.833	0.833	0.833	1.000	1.167	1.333	1.333	1.333	1.500	1.500
A ft.	1.167	1.667	2.167	2.667	3.167	3.667	4.167	4.667	5.167	5.833
B ft.	0.833	0.833	0.833	1.000	1.167	1.333	1.333	1.333	1.500	1.500
C ft.	0.500	0.667	1.000	1.167	1.333	1.500	1.833	2.167	2.333	2.667
D ft.	0.833	0.833	0.833	1.000	1.167	1.333	1.333	1.333	1.500	1.500
8 in. blk. ft.	3.000	4.000	5.000	6.000	5.000	6.000	6.000	6.000	6.000	6.000
12 in. blk. ft.	–	–	–	–	2.000	2.000	2.000	2.000	2.000	2.000
Conc. Stem	–	–	–	–	–	–	1.000	2.000	3.000	4.000
Wb ft.	–	–	–	–	–	–	1.333	1.333	1.500	1.500
W lbs.	1097	1449	1849	2557	3442	4408	5109	5838	7133	8002
F lbs.	588	934	1361	1960	2668	3484	4271	5138	6250	7290
Mo ft.-lbs.	751	1505	2647	4573	7262	10840	14712	19409	26042	32805
Mr ft.-lbs.	1722	3095	5250	8824	13953	20537	27400	35474	47619	60233
O.T.R.	2.292	2.055	1.984	1.929	1.921	1.894	1.862	1.828	1.829	1.836
$\overline{X}$ ft.	0.884	1.096	1.408	1.662	1.944	2.200	2.483	2.752	3.025	3.428
e ft.	0.366	0.487	0.592	0.754	0.889	1.050	1.183	1.331	1.475	1.572
Me ft.-lbs.	401	706	1094	1929	3061	4628	6045	7772	10521	12582
S ft.^{3}	1.042	1.671	2.667	3.894	5.352	7.042	8.963	11.116	13.500	16.667
S.P.h p.s.f.	824	880	873	1024	1179	1335	1371	1414	1572	1555
S.P.t p.s.f.	54	35	52	34	35	21	22	16	13	45
Friction	0.536	0.645	0.736	0.767	0.775	0.791	0.836	0.880	0.876	0.911
Inspct.	NO	NO	NO	YES	YES	YES	YES	YES	YES	YES

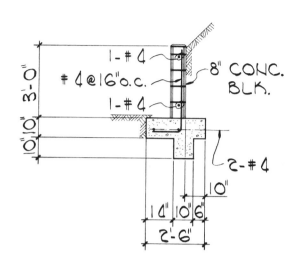

Soil slope = 1 to 1
E.F.P. = 80 p.c.f.
Section 7.14.3

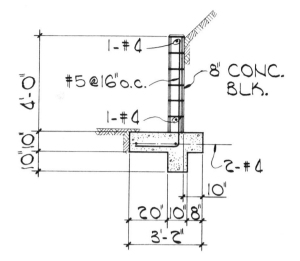

Soil slope = 1 to 1
E.F.P. = 80 p.c.f.
Section 7.14.4

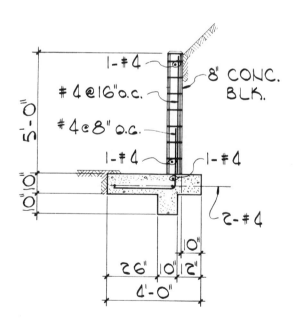

Soil slope = 1 to 1
E.F.P. = 80 p.c.f.
Section 7.14.5

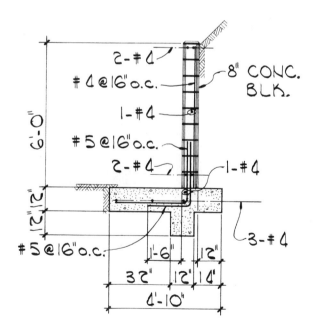

Soil slope = 1 to 1
E.F.P. = 80 p.c.f.
Section 7.14.6

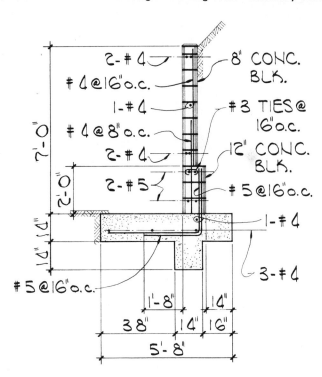

Soil slope = 1 to 1
E.F.P. = 80 p.c.f.
Section 7.14.7

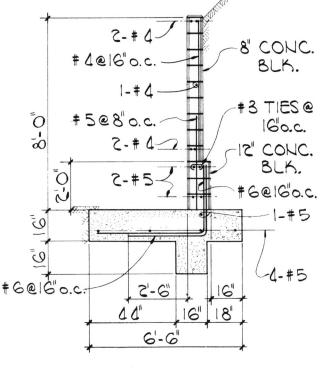

Soil slope = 1 to 1
E.F.P. = 80 p.c.f.
Section 7.14.8

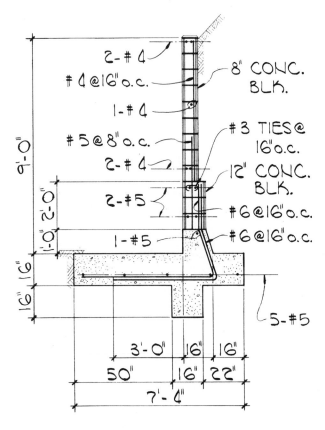

Soil slope = 1 to 1
E.F.P. = 80 p.c.f.
Section 7.14.9

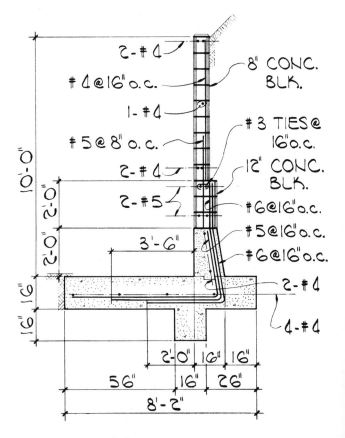

Soil slope = 1 to 1
E.F.P. = 80 p.c.f.
Section 7.14.10

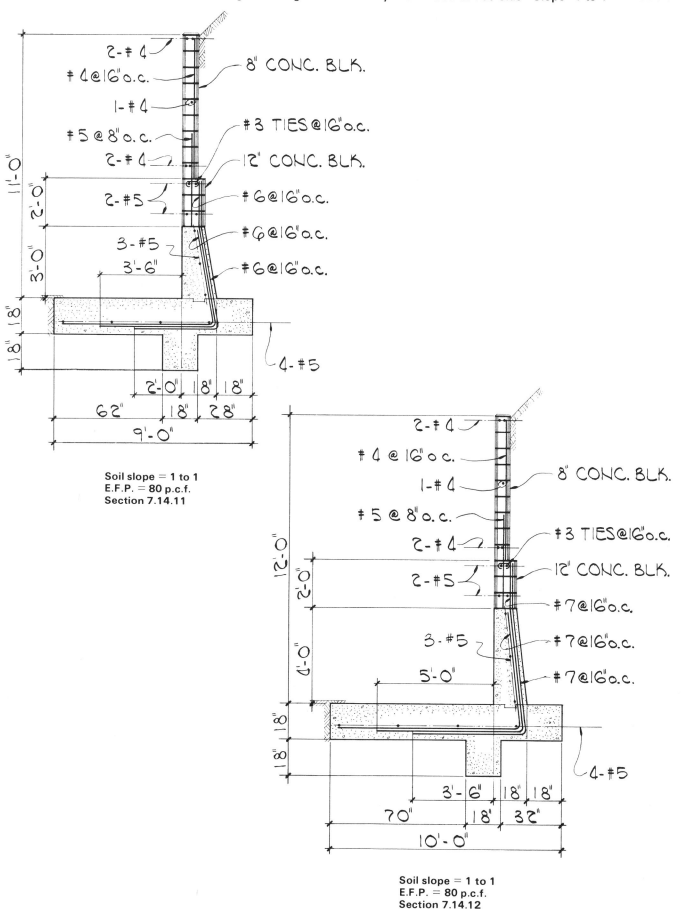

2-#4
#4@16"o.c.
1-#4
#5@8"o.c.
2-#4
2-#5
3-#5
3'-6"

8" CONC. BLK.
#3 TIES@16"o.c.
12" CONC. BLK.
#6@16"o.c.
#6@16"o.c.
#6@16"o.c.
4-#5

11'-0"
2'-0"
3'-0"
8"
8"

2'-0" 18" 18"
18"
62"   18"   28"
9'-0"

Soil slope = 1 to 1
E.F.P. = 80 p.c.f.
Section 7.14.11

2-#4
#4@16"o.c.
1-#4
#5@8"o.c.
2-#4
2-#5
3-#5
5'-0"

8" CONC. BLK.
#3 TIES@16"o.c.
12" CONC. BLK.
#7@16"o.c.
#7@16"o.c.
#7@16"o.c.
4-#5

12'-0"
2'-0"
4'-0"
18"
18"

3'-6" 18" 18"
18"
70"   18"   32"
10'-0"

Soil slope = 1 to 1
E.F.P. = 80 p.c.f.
Section 7.14.12

CHAPTER 8

General Retaining Walls
Concrete Stem
Soil at the Heel Side of the Footing

The design data and drawings presented in this chapter are concerned with property line concrete retaining walls. The retained soil is not placed over the entire wall footing, but only at the heel side of the stem wall. The chapter consists of three design examples of retaining walls with various types of loading, a design data table for each of the fourteen loading conditions specified in Fig. 1.6, and a series of corresponding drawings following each design data table. The retaining wall drawings and the design data in this chapter are for a wall length of 1 foot.

The retaining walls in this chapter are designed using the following criteria:

Weight of soil = 100 p.c.f.   Weight of concrete = 150 p.c.f.
Concrete $f_c' = 2000$ p.s.i.   Reinforcing steel $f_s' = 20,000$ p.s.i.
Maximum allowable soil pressure = 4000 p.s.f.
Minimum allowable soil pressure = 0 p.s.f.
$\overline{X} \geqslant L/3$ Minimum O.T.R. = 1.50
Coefficient of soil friction = 0.40
Passive soil pressure = 300 p.c.f.

See Table C5 for reinforcement protection. Weep holes are not shown on the drawings.

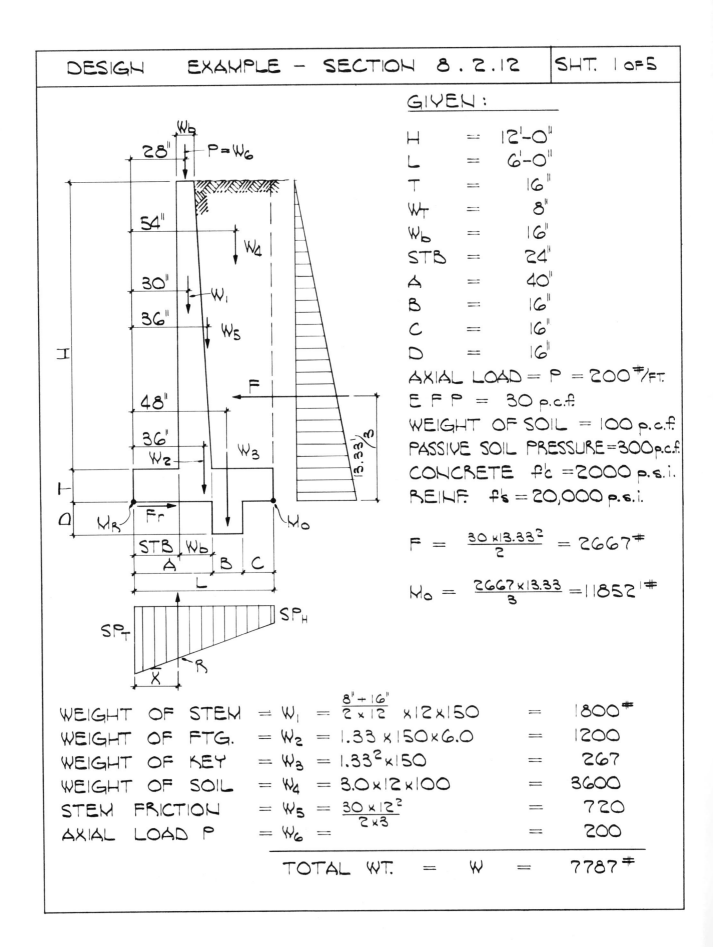

DESIGN EXAMPLE – SECTION 8.2.12 | SHT. 1 of 5

GIVEN:

$$H = 12'-0''$$
$$L = 6'-0''$$
$$T = 16''$$
$$W_T = 8''$$
$$W_b = 16''$$
$$STB = 24''$$
$$A = 40''$$
$$B = 16''$$
$$C = 16''$$
$$D = 16''$$

AXIAL LOAD = P = 200 #/FT.
E F P = 30 p.c.f.
WEIGHT OF SOIL = 100 p.c.f.
PASSIVE SOIL PRESSURE = 300 p.c.f.
CONCRETE f'_c = 2000 p.s.i.
REINF. f'_s = 20,000 p.s.i.

$$F = \frac{30 \times 13.33^2}{2} = 2667^\#$$

$$M_o = \frac{2667 \times 13.33}{3} = 11852'^\#$$

WEIGHT OF STEM	$= W_1 =$	$\frac{8'' + 16''}{2 \times 12} \times 12 \times 150$	$=$	$1800^\#$
WEIGHT OF FTG.	$= W_2 =$	$1.33 \times 150 \times 6.0$	$=$	1200
WEIGHT OF KEY	$= W_3 =$	$1.33^2 \times 150$	$=$	267
WEIGHT OF SOIL	$= W_4 =$	$3.0 \times 12 \times 100$	$=$	3600
STEM FRICTION	$= W_5 =$	$\frac{30 \times 12^2}{2 \times 3}$	$=$	720
AXIAL LOAD P	$= W_6 =$		$=$	200
	TOTAL WT.	$= W =$		$7787^\#$

| DESIGN EXAMPLE — SECTION 8.2.12 | SHT. 2 OF 5 |

FOOTING DESIGN:

$$M_R = 1800 \times \frac{30}{12} + 1200 \times \frac{36}{12} + 267 \times \frac{48}{12} + 3600 \times \frac{54}{12} + 720 \times \frac{36}{12} + 200 \times \frac{28}{12}$$

$$M_R = 27993'^{\#}$$

$$\text{OVERTURN. RATIO} = \frac{27993}{11852} = 2.362 > 1.50 \checkmark$$

$$\text{NET } M = 27993 - 11852 = 16141'^{\#}$$

$$\bar{X} = \frac{16141}{7787} = 2.073'$$

$$\text{FTG. MIDDLE } \frac{1}{3} = \frac{6.0}{3} = 2.0' < 2.073' \checkmark$$

$$e = \frac{6.0}{2} - 2.073 = 0.927'$$

$$M_e = 7787 \times .927 = 7219'^{\#}$$

FTG. AREA = 6.0 SQ. FT.

$$\text{FTG. SECT. MODULUS} = \frac{1 \times 6^2}{6} = 6.0'^3$$

$$\text{SOIL PRESSURE} = \frac{7787}{6.0} \pm \frac{7219}{6.0} = 1298 \pm 1203$$

$$SP_T = 2501 \text{ p.s.f.} \qquad SP_H = 95.0 \text{ p.s.f.}$$

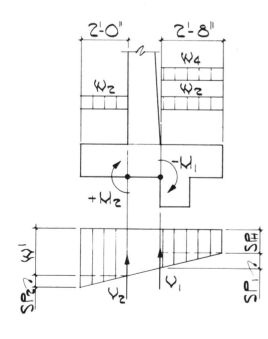

$-M_1:$

$$SP_1 = \frac{2501 - 92}{6.0} \times 2.67 = 1072 \text{ p.s.f.}$$

$$\text{UNIF. LD.} = 1200 + 200 - 95 = 1305^{\#}/\text{FT.}$$

$$-M_1 = \frac{1305 \times 2.67^2}{2} - \frac{1072 \times 2.67^2}{2 \times 3} = 3378'^{\#}$$

$$V_1 = 1305 \times 2.67 - \frac{1072 \times 2.67}{2} = 2053^{\#}$$

$+M_2:$

$$W' = \frac{2501 - 95}{6.0} \times 4.0 + 95 = 1699 \text{ p.s.f.}$$

$$SP_2 = 2501 - 1699 = 802 \text{ p.s.f.}$$

$$\text{UNIF. LD.} = 1699 - 200 = 1499^{\#}/\text{FT.}$$

$$+M_2 = \frac{1499 \times 2^2}{2} + \frac{802 \times 2^2 \times 2}{2 \times 3} = 4067'^{\#}$$

$$V_2 = 1499 \times 2 + \frac{802 \times 2}{2} = 3800^{\#}$$

DESIGN EXAMPLE – SECTION 8.2.12	SHT. 3 OF 5

FOOTING DESIGN – (CONT.)

$-M_1 = 3378'^{\#}$; $V_1 = 2053^{\#}$

$+M_2 = 4067'^{\#}$; $V_2 = 3800^{\#}$

d FOR $+A_s = 16 - (3 + .5 \times .62) = 12.70''$

d FOR $-A_s = 16 - (2 + .5 \times .62) = 13.70''$

$a = 1.13$; $j = 0.89$

$+A_s = \dfrac{4.067}{1.13 \times 12.70} = 0.285$ SQ. IN./FT.

$v_2 = \dfrac{3800}{12 \times .89 \times 12.70} = 28.0$ p.s.i. < 54 p.s.i $\checkmark$

$-A_s = \dfrac{3.378}{1.13 \times 13.70} = 0.218$ SQ. IN./FT.

$v_1 = \dfrac{2053}{12 \times .89 \times 13.70} = 14.0$ p.s.i < 54 p.s.i $\checkmark$

USE #6 @ 8" o.c. AT BOTTOM

#5 @ 16" o.c. & #4 @ 16" o.c. AT TOP

LONG. REINF.:

$A_s = 0.0015 \times 16 \times 72 = 1.73$ SQ. IN.

USE 3-#5 & 2-#4

STEM DESIGN AT TOP OF FOOTING:

$H = 12'-0''$ E F P = 30 p.c.f.

$V = \dfrac{30 \times 12^2}{2} = 2160^{\#}$; $M = \dfrac{2160 \times 12}{3} = 8640'^{\#}$

$d = 16 - (2 + .5 \times .62) = 13.70''$; $a = 1.13$; $j = 0.89$

$v = \dfrac{2160}{12 \times .89 \times 13.7} = 14.8$ p.s.i < 54 p.s.i $\checkmark$

$A_s = \dfrac{8.64}{1.13 \times 13.70} = 0.558$ SQ. IN./FT.

USE #6 @ 8" o.c.

DESIGN EXAMPLE — SECTION 8.2.12	SHT. 4 of 5

STEM DESIGN AT 3'-0" ABOVE FTG.

$H = 12.0' - 3.0' = 9.0'$ $EFP = 30.0$ p.c.f.

$V = \dfrac{30 \times 9^2}{2} = 1215^{\#}$; $M = \dfrac{1215 \times 9}{3} = 3645'^{\#}$

$d = 16 - \dfrac{8''}{12} \times 3 - (2 + .5 \times .62) = 11.0''$

$v = \dfrac{1215}{12 \times .89 \times 11} = 10.3$ p.s.i < 54 p.s.i. ✓

$A_s = \dfrac{3.645}{1.13 \times 11} = 0.293$ SQ. IN./FT.

USE #6 @ 8" o.c.

STEM DESIGN AT 6'-0" ABOVE FTG.

$H = 12.0' - 6.0' = 6.0'$

$V = \dfrac{30 \times 6^2}{2} = 540^{\#}$; $M = \dfrac{540 \times 6.0}{3} = 1080'^{\#}$

$d = 16 - 4.0 - (2 + .5 \times .62) = 9.70''$

$v = \dfrac{540}{12 \times .89 \times 9.70} = 5.20$ p.s.i. < 54 p.s.i ✓

$A_s = \dfrac{1.08}{1.13 \times 9.70} = 0.098$ SQ. IN./FT.

USE #5 @ 16" o.c.

LONG. REINF.:

$A_s = 0.0015 \times 12 \times 12 \times \dfrac{8'' + 16''}{2} = 2.59$ SQ. IN.   **USE #5 @ 18" o.c. HORIZ.**

SLIDING:

$F_r = \dfrac{2667}{7787} = .342 < 0.40$

CHECK WITHOUT AXIAL LD. — $W = 7587^{\#}$

$F_r = \dfrac{2667}{7587} = 0.35 < 0.40$

FTG. KEY IS OPTIONAL. HOWEVER IT IS USED FOR STABILITY CALCS.

KEY IS 16" SQ.

$v = \dfrac{2667}{16^2} = 10.4$ p.s.i < 54.0 p.s.i ✓

DESIGN EXAMPLE — SECTION 8.2.12 | SHT. 5 of 5

CHECK WALL WITHOUT 200 #/FT. AXIAL LD.

$W = 7787^{\#} - 200^{\#} = 7587^{\#}$

$F = 2667^{\#}$; $M_0 = 11852^{|\#}$

$M_R = 27993 - 200 \times \dfrac{28}{12} = 27526^{|\#}$

OVERTURN. RATIO $= \dfrac{27526}{11852} = 2.322 > 1.50 \checkmark$

NET $M = 27526 - 11852 = 15674^{|\#}$

$\bar{X} = \dfrac{15674}{7587} = 2.066^{|} > 2.0^{|} \checkmark$

$e = \dfrac{6.0}{2} - 2.066 = 0.934^{|}$

$M_e = 7587 \times .934 = 7087^{|\#}$

SOIL PRESSURE $= \dfrac{7587}{6.0} \pm \dfrac{7087}{6.0} = 1264.5 \pm 1181.2$

$\underline{SP_T = 2445.7 \text{ p.s.f.}}$ $\underline{SP_H = 83.3 \text{ p.s.f.}}$

FTG. BENDING & SHEAR -(SEE CALC. SHT. No. 2)

$-M_1$: $SP_1 = \dfrac{2445.7 - 83.3}{6.0} \times 2.67 = 1051.3 \text{ p.s.f.}$

UNIF. LD. $= 1200 + 200 - 83.3 = 1316.7 \ ^{\#}/\text{FT.}$

$-M_1 = \dfrac{1316.7 \times 2.67^2}{2} - \dfrac{1051.3 \times 2.67^2}{2 \times 3} = 3435^{|\#} \approx 3413^{|\#}$

$V_1 = 1316.7 \times 2.67 - \dfrac{1051.3 \times 2.67}{2} = 2109^{\#} \approx 2076^{\#}$

$+M_2$: $W_1 = \dfrac{2445.7 - 83.3}{6.0} \times 4.0 + 83.3 = 1658.2 \text{ p.s.f.}$

$SP_2 = 2445.7 - 1658.2 = 787.5 \text{ p.s.f.}$

UNIF. LD. $= 1658.2 - 200 = 1458.2 \ ^{\#}/\text{FT.}$

$+M_2 = \dfrac{1458.2 \times 2^2}{2} + \dfrac{787.5 \times 2^2 \times 2}{2 \times 3} = 3966^{|\#} \checkmark$

$V_2 = 1458.2 \times 2 + 787.5 \times \dfrac{2}{2} = 3704^{\#} \checkmark$

NO CHANGE IN FTG. DESIGN
ALSO NO CHANGE IN STEM DESIGN.

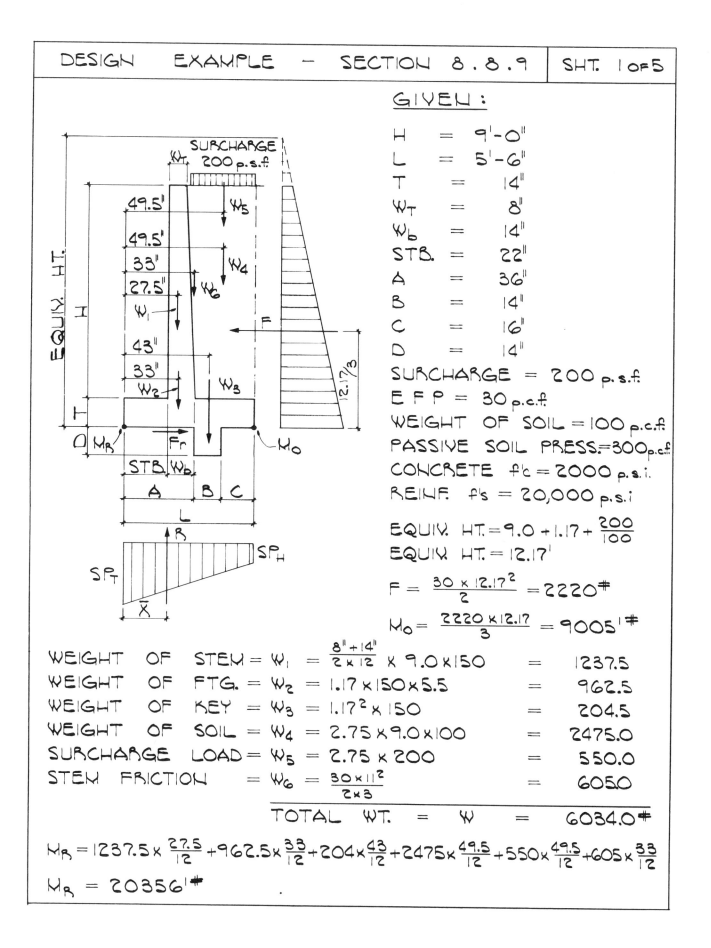

DESIGN   EXAMPLE — SECTION 8.8.9 | SHT. 1 OF 5

GIVEN:

$H = 9'-0''$
$L = 5'-6''$
$T = 14''$
$W_T = 8''$
$W_b = 14''$
$STB. = 22''$
$A = 36''$
$B = 14''$
$C = 16''$
$D = 14''$

SURCHARGE $= 200$ p.s.f.
EFP $= 30$ p.c.f.
WEIGHT OF SOIL $= 100$ p.c.f.
PASSIVE SOIL PRESS. $= 300$ p.c.f.
CONCRETE $f'c = 2000$ p.s.i.
REINF. $f's = 20,000$ p.s.i

EQUIV. HT. $= 9.0 + 1.17 + \dfrac{200}{100}$
EQUIV. HT. $= 12.17'$

$F = \dfrac{30 \times 12.17^2}{2} = 2220^{\#}$

$M_0 = \dfrac{2220 \times 12.17}{3} = 9005'^{\#}$

WEIGHT OF STEM $= W_1 = \dfrac{8'' + 14''}{2 \times 12} \times 9.0 \times 150 = 1237.5$
WEIGHT OF FTG. $= W_2 = 1.17 \times 150 \times 5.5 = 962.5$
WEIGHT OF KEY $= W_3 = 1.17^2 \times 150 = 204.5$
WEIGHT OF SOIL $= W_4 = 2.75 \times 9.0 \times 100 = 2475.0$
SURCHARGE LOAD $= W_5 = 2.75 \times 200 = 550.0$
STEM FRICTION $= W_6 = \dfrac{30 \times 11^2}{2 \times 3} = 605.0$

TOTAL WT. $= W = 6034.0^{\#}$

$M_R = 1237.5 \times \dfrac{27.5}{12} + 962.5 \times \dfrac{33}{12} + 204 \times \dfrac{43}{12} + 2475 \times \dfrac{49.5}{12} + 550 \times \dfrac{49.5}{12} + 605 \times \dfrac{33}{12}$

$M_R = 20356'^{\#}$

DESIGN EXAMPLE — SECTION 8.8.9 |

FOOTING DESIGN:

$M_O = 9005^{'\#}$; $M_R = 20356^{'\#}$; $W = 6034^{\#}$

OVERTURN. RATIO $= \dfrac{20356}{9005} = 2.26 > 1.50$ ✓

NET $M = 20356 - 9005 = 11351^{'\#}$

$\bar{X} = \dfrac{11351}{6034} = 1.881'$

FTG. MIDDLE $\frac{1}{3} = \dfrac{5.50}{3} = 1.83' < 1.881'$ ✓

$e = \dfrac{5.50}{2} - 1.881 = 0.869'$

$M_e = 6034 \times .869 = 5243.5^{'\#}$

FTG. AREA $= 5.50^{'2}$

FTG. SECT. MODULUS $= \dfrac{1 \times 5.50^2}{6} = 5.042^{'3}$

SOIL PRESSURE $= \dfrac{6034}{5.50} \pm \dfrac{5243.5}{5.042} = 1097 \pm 1040$

$\underline{SP_T = 2137 \text{ p.s.f.}}$ $\underline{SP_H = 57.0 \text{ p.s.f.}}$

$-M_1:$

$SP_1 = \dfrac{2137-57}{5.50} \times 2.50 = 945.5 \text{ p.s.f.}$

UNIF. LD. $= 200 + 900 + 175 - 57 = 1218^{\#}/\text{FT.}$

$-M_1 = \dfrac{1218 \times 2.5^2}{2} - \dfrac{945.5 \times 2.5^2}{2 \times 3} = 2821^{'\#}$

$V_1 = 1218 \times 2.5 - \dfrac{945.5 \times 2.5}{2} = 1863^{\#}$

$+M_2:$

$W' = \dfrac{2137-57}{5.50} \times 3.67 + 57 = 1444 \text{ p.s.f.}$

$SP_2 = 2137 - 1444 = 693 \text{ p.s.f.}$

UNIF. LD. $= 1444 - 175 = 1269^{\#}/\text{FT.}$

$+M_2 = \dfrac{1269 \times 1.83^2}{2} + \dfrac{693 \times 1.83^2 \times 2}{2 \times 3} = 2900^{'\#}$

$V_2 = 1269 \times 1.83 + \dfrac{693 \times 1.83}{2} = 2956^{\#}$

DESIGN   EXAMPLE — SECTION 8.8.9 | SHT. 3 of 5

FOOTING DESIGN : (CONT.)

$-M_1 = 2821^{'\#}$; $V_1 = 1863^{\#}$

$+M_2 = 2900^{'\#}$; $V_2 = 2956^{\#}$

d FOR $+A_s$ = 14−(3+.5×.62) = 10.70″
d FOR $-A_s$ = 14−(2+.5×.62) = 11.70″

$a = 1.13$; $j = 0.89$

$+A_s = \dfrac{2.90}{1.13 \times 10.70} = 0.24$ SQ. IN./FT.

$v_2 = \dfrac{2956}{12 \times .89 \times 10.7} = 2.15$ p.s.i $<$ 54 p.s.i ✓

$-A_s = \dfrac{2.82}{1.13 \times 11.70} = 0.213$ SQ. IN./FT.

$v_1 = \dfrac{1863}{12 \times .89 \times 11.70} = 14.9$ p.s.i $<$ 54 p.s.i ✓

USE   #4 @ 16″ o.c. & #6 @ 16″ o.c. AT BOTT.
      #6 @ 16″ o.c. AT TOP

LONG. REINF. :

$A_s = 0.0015 \times 5.50 \times 12 \times \dfrac{8+14}{2} = 1.09$ SQ. IN.

USE   5 - #4's   HORIZ.

STEM DESIGN AT TOP OF FTG. :

$H = 9'-0''$; EQUIV. HT. $= 9.0 + \dfrac{200}{100} = 11.0'$; EFP = 30 p.c.f.

$V = \dfrac{30 \times 11^2}{2} = 1815^{\#}$; $M = \dfrac{1815 \times 11}{3} = 6655^{'\#}$

$d = 14-(2+.5 \times .62) = 11.70''$; $a = 1.13$; $j = 0.89$

$v = \dfrac{1815}{12 \times .89 \times 11.70} = 14.5$ p.s.i $<$ 54 p.s.i

$A_s = \dfrac{6.655}{1.13 \times 11.70} = 0.50$ SQ. IN./FT.

USE #4 @ 16″ o.c. & #6 @ 16″ o.c.

DESIGN EXAMPLE — SECTION 8.8.9	SHT. 4 OF 5

STEM DESIGN AT 3'-0" ABOVE FTG.

EQUIV. HT. $= 9.0' - 3.0' + \frac{200}{100} = 8.0'$; $EFP = 30$ p.c.f.

$V = \frac{30 \times 8^2}{2} = 960^\#$; $M = \frac{960 \times 8}{3} = 2560'^\#$

$d = 14 - \frac{6"}{9} \times 3 - (2 + .5 \times .62) = 9.70"$

$v = \frac{960}{12 \times .89 \times 9.7} = 9.3$ p.s.i < 54 p.s.i. $\checkmark$

$A_s = \frac{2.56}{1.13 \times 9.70} = 0.233$ SQ. IN./FT.

USE #4 @ 16" o.c. & #6 @ 16" o.c.

STEM DESIGN AT 6'-0" ABOVE FTG.

EQUIV. HT. $= 9.0' - 6.0' + \frac{200}{100} = 5.0'$; $EFP = 30$ p.c.f.

$V = \frac{30 \times 5^2}{2} = 375^\#$; $M = \frac{375 \times 5}{3} = 625'^\#$

$d = 14 - \frac{6"}{9} \times 6 - (2 + .5 \times .62) = 7.70"$

$v = \frac{375}{12 \times .89 \times 7.70} = 4.6$ p.s.i. < 54 p.s.i $\checkmark$

$A_s = \frac{0.625}{1.13 \times 7.70} = 0.072$ SQ. IN./FT.

USE #5 @ 16" o.c.

LONG. REINF. :

$A_s = 0.0015 \times 9 \times 12 \times \frac{8" + 14"}{2} = 1.782$ SQ. IN.

USE #4 @ 22" o.c. HORIZ.

SLIDING :

$F_r = \frac{2220}{6034} = 0.368 < 0.40 \checkmark$

CHECK WITHOUT SURCHARGE LD.

$W = 6034 - 605 - 550 + \frac{30 \times 9^2}{2 \times 3} = 5284^\#$; $F = \frac{30 \times 10.17^2}{2} = 1551^\#$

$F_r = \frac{1551}{5284} = 0.293 < 0.40 \checkmark$

KEY IS USED FOR STABILITY CALC'S.

KEY IS 14" SQ.

$v = \frac{2220}{14^2} = 11.3$ p.s.i < 54 p.s.i $\checkmark$

DESIGN EXAMPLE - SECTION 8.8.9	SHT. 5 of 5

CHECK   WALL   WITHOUT   200 p.s.f.   SURCHARGE   LOAD

$W = 6034 - 605 - 550 + \dfrac{30 \times 9^2}{2 \times 3} = 5284^{\#}$

$F = \dfrac{30 \times 10.17^2}{2} = 1550^{\#} \quad ; \quad M_0 = \dfrac{1550 \times 10.17}{3} = 5254^{'\#}$

$M_R = 20356 - 550 \times \dfrac{49.5}{12} - 605 \times \dfrac{33}{12} + \dfrac{30 \times 9^2}{2 \times 3} \times \dfrac{33}{12} = 17538^{'\#}$

OVERTURN. RATIO $= \dfrac{17538}{5254} = 3.338 > 1.50$ ✓

NET M $= 17538 - 5254 = 12284^{'\#}$

$\overline{X} = \dfrac{12284}{5284} = 2.325' > 1.83'$ ✓

$e = \dfrac{5.50}{2} - 2.325 = 0.425'$

$M_e = 5284 \times .425 = 2245.7^{'\#}$

SOIL PRESSURE $= \dfrac{5284}{5.50} \pm \dfrac{2245.7}{5.04^2} = 960.7 \pm 445.4$

$\underline{SP_T = 1406.1 \, p.s.f.} \qquad \underline{SP_H = 515.3 \, p.s.f.}$

FTG. BENDING & SHEAR - (SEE CALC. SHT. No.2)

$-M_1 :\quad SP_1 = \dfrac{1406.1 - 515.3}{5.50} \times 2.50 = 405 \, p.s.f.$

$\qquad$ UNIF. LD. $= 900 + 175 - 515.3 = 559.7 \, ^{\#}/_{FT.}$

$\qquad -M_1 = \dfrac{559.7 \times 2.5^2}{2} - \dfrac{405 \times 2.5^2}{2 \times 3} = 1327^{'\#} < 2821^{'\#}$ ✓

$\qquad V_1 = 559.7 \times 2.5 - \dfrac{405 \times 2.5}{2} = 893^{\#} < 1863^{\#}$ ✓

$+M_2 :\quad W' = \dfrac{1406.1 - 553}{5.50} \times 3.67 + 553 = 1121.8 \, p.s.f.$

$\qquad SP_2 = 1406.1 - 1121.8 = 284.3 \, p.s.f.$

$\qquad$ UNIF. LD. $= 1121.8 - 175 = 946.8 \, ^{\#}/_{FT.}$

$\qquad +M_2 = \dfrac{946.8 \times 1.83^2}{2} + \dfrac{284.3 \times 1.83^2 \times 2}{2 \times 3} = 1909.5^{'\#} < 2900^{\#}$ ✓

$\qquad V_2 = 946.8 \times 1.83 + \dfrac{284.3 \times 1.83}{2} = 1992.8^{\#} < 2956^{\#}$ ✓

NO CHANGE REQ'D IN FTG. DESIGN,
ALSO NO CHANGE REQ'D FOR STEM DESIGN
SINCE SURCHARGE IS MAX. LOAD TO STEM.

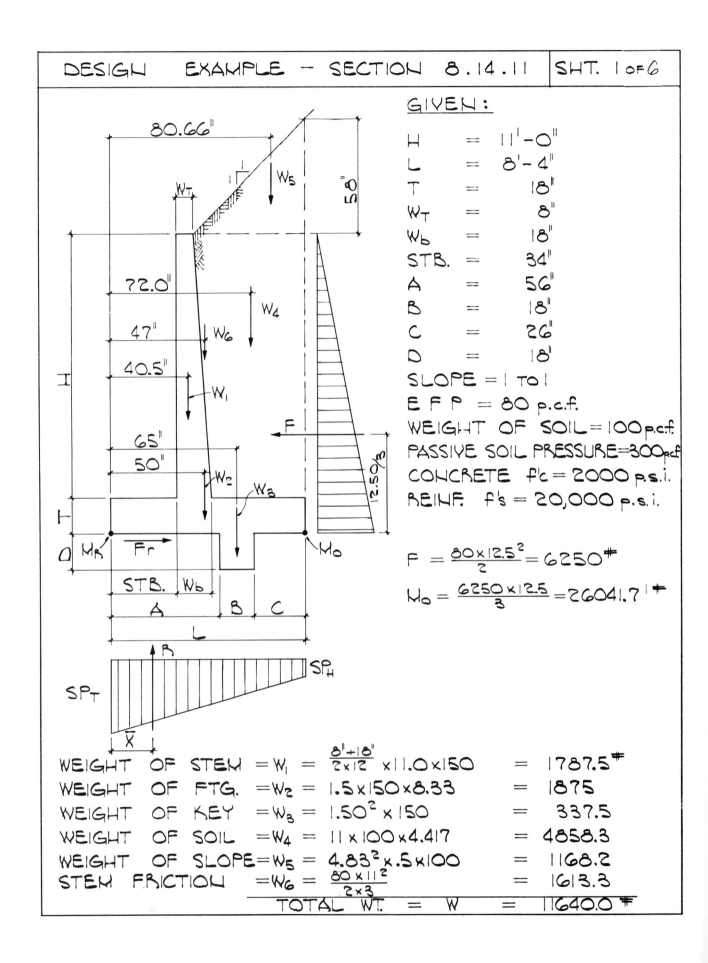

DESIGN   EXAMPLE — SECTION 8.14.11 │ SHT. 1 OF 6

GIVEN :

$$H = 11'-0''$$
$$L = 8'-4''$$
$$T = 18''$$
$$W_T = 8''$$
$$W_b = 18''$$
$$STB. = 34''$$
$$A = 56''$$
$$B = 18''$$
$$C = 26''$$
$$D = 18''$$

SLOPE = 1 TO 1
E F P = 80 p.c.f.
WEIGHT OF SOIL = 100 p.c.f.
PASSIVE SOIL PRESSURE = 300 pcf
CONCRETE f'c = 2000 p.s.i.
REINF. f's = 20,000 p.s.i.

$$F = \frac{80 \times 12.5^2}{2} = 6250^{\#}$$

$$M_0 = \frac{6250 \times 12.5}{3} = 26041.7'^{\#}$$

WEIGHT OF STEM = $W_1 = \frac{8''+18''}{2 \times 12} \times 11.0 \times 150 = 1787.5^{\#}$

WEIGHT OF FTG. = $W_2 = 1.5 \times 150 \times 8.33 = 1875$

WEIGHT OF KEY = $W_3 = 1.50^2 \times 150 = 337.5$

WEIGHT OF SOIL = $W_4 = 11 \times 100 \times 4.417 = 4858.3$

WEIGHT OF SLOPE = $W_5 = 4.83^2 \times .5 \times 100 = 1168.2$

STEM FRICTION = $W_6 = \frac{80 \times 11^2}{2 \times 3} = 1613.3$

TOTAL WT. = W = 11640.0$^{\#}$

| DESIGN EXAMPLE – SECTION 8.14.11 | SHT. 2 OF 6 |

FOOTING DESIGN :

$$M_0 = 26041.7^{'\#} \quad ; \quad W = 11640^{\#}$$

$$M_R = 1787.5 \times \frac{40.5}{12} + 1875 \times \frac{50}{12} + 337.5 \times \frac{65}{12} + 4858.3 \times \frac{72.0}{12} + $$

$$1168.2 \times \frac{80.66}{12} + 1613.3 \times \frac{47}{12} = 58994^{'\#}$$

$$\text{OVERTURN. RATIO} = \frac{58994}{26041.7} = 2.265 > 1.50$$

$$\text{NET } M = 58994 - 26041.7 = 32952.3^{'\#}$$

$$\bar{X} = \frac{32952.3}{11640.0} = 2.831'$$

$$\text{FTG. MIDDLE } \frac{1}{3} = \frac{8.33}{3} = 2.777' < 2.831'$$

$$e = \frac{8.33}{2} - 2.831 = 1.334'$$

$$M_e = 11640 \times 1.334 = 15527^{'\#}$$

$$\text{FTG. AREA} = 8.33^{'2}$$

$$\text{FTG. SECT. MODULUS} = \frac{1 \times 8.33^2}{6} = 11.573^{'3}$$

$$\text{SOIL PRESSURE} = \frac{11640}{8.33} \pm \frac{15527}{11.573} = 1397 \pm 1342$$

$$\underline{SP_T = 2739.0 \text{ p.s.f}} \qquad \underline{SP_H = 55.0 \text{ p.s.f}}$$

$-M_1$:

$$SP_1 = \frac{2739.4 - 55}{8.33} \times 4.0 = 1288.3 \text{ p.s.f}$$

$$\text{UNIF. LD.} = 1100 + 225 - 55 = 1270 \; \#/\text{FT.}$$

$$W_5 = \frac{4.0^2 \times 100}{2} = 800^{\#}$$

$$-M_1 = \frac{1270 \times 4^2}{2} + \frac{800 \times 4.0 \times 2}{3} - $$

$$\frac{1288.3 \times 4.0^2}{2 \times 3} = 8857.8^{'\#}$$

$$V_1 = 1270 \times 4 + 800 - \frac{1288.3 \times 4}{2}$$

$$V_1 = 3303^{\#}$$

DESIGN EXAMPLE — SECTION 8.14.11 | SHT. 3 OF 6

FOOTING DESIGN: (CONT.)

$+M_2:$ $\quad W' = \dfrac{2739.4 - 55}{8.33} \times 5.50 + 55 = 1826.4 \text{ p.s.f.}$

$\quad SP_2 = 2739.4 - 1826.4 = 912.6 \text{ p.s.f.}$

$\quad \text{UNIF. LD} = 1826.4 - 225 = 1601.4 \text{ \#/FT.}$

$\quad +M_2 = \dfrac{1601.4 \times 2.83^2}{2} + \dfrac{912.6 \times 2.83^2 \times 2}{2 \times 3} = 8854.2^{'\#}$

$\quad V_2 = 1601.4 \times 2.83 + \dfrac{912.6 \times 2.83}{2} = 5830^{\#}$

$-M_1 = 8857.3^{'\#} \quad ; \quad V_1 = 3303^{\#}$

$+M_2 = 8854.2^{'\#} \quad ; \quad V_2 = 5830^{\#}$

$d \text{ FOR } +A_s = 18 - (3 + .5 \times .62) = 14.70''$

$d \text{ FOR } -A_s = 18 - (2 + .5 \times .62) = 15.70''$

$a = 1.13 \quad ; \quad j = 0.89$

$+A_s = \dfrac{8.85}{1.13 \times 15.70} = 0.498 \text{ sq. in./ft.}$

$\quad v_2 = \dfrac{5830}{12 \times .89 \times 15.7} = 34.7 \text{ p.s.i} < 54 \text{ p.s.i} \checkmark$

$-A_s = \dfrac{8.86}{1.13 \times 14.70} = 0.533 \text{ sq. in./ft.}$

$\quad v_1 = \dfrac{3303}{12 \times .85 \times 14.70} = 21.0 \text{ p.s.i} < 54 \text{ p.s.i} \checkmark$

USE #6 @ 9" o.c. AT BOTTOM

$\quad$ #5 @ 18" o.c. & #6 @ 18" o.c. AT TOP

LONG. REINF.

$A_s = 0.0015 \times 11 \times 12 \times \dfrac{8'' + 18'}{2} = 2.57 \text{ sq. in.}$

USE 4-#5 & 3-#4

DESIGN EXAMPLE – SECTION 8.14.11	SHT. 4 OF 6

STEM DESIGN AT TOP OF FOOTING:

$H = 11.0'$; $EFP = 80$ p.c.f.

$V = \dfrac{80 \times 11^2}{2} = 4840^{\#}$; $M = \dfrac{4840 \times 11}{3} = 17,746.7'^{\#}$

$d = 18 - (2 + .5 \times .75) = 15.60''$; $a = 1.13$; $j = 0.89$

$v = \dfrac{4840}{12 \times .89 \times 15.6} = 29.0$ p.s.i < 54 p.s.i $\checkmark$

$A_s = \dfrac{17.746}{1.13 \times 15.60} = 1.00$ SQ. IN./FT.

USE #5 & #6 @ 9″ o.c.

STEM DESIGN AT 3'-6″ ABOVE FOOTING:

$H = 11.0 - 3.50 = 7.50'$; $EFP = 80$ p.c.f.

$V = \dfrac{80 \times 7.5^2}{2} = 2250^{\#}$; $M = \dfrac{2250 \times 7.5}{3} = 5625'^{\#}$

$d = 18.0 - \dfrac{10}{11} \times 3.5 - (2 + .5 \times .62) = 12.5''$

$v = \dfrac{2250}{12 \times .89 \times 12.5} = 16.9$ p.s.i. < 54 p.s.i $\checkmark$

$A_s = \dfrac{5.625}{1.13 \times 12.5} = 0.398$ SQ. IN./FT.

USE #5 @ 9″ o.c.

STEM DESIGN AT 6'-0″ ABOVE FOOTING:

$H = 11.0' - 6.0' = 5.0'$; $EFP = 80$ p.c.f.

$V = \dfrac{80 \times 5^2}{2} = 1000^{\#}$; $M = \dfrac{1000 \times 5}{3} = 1666'^{\#}$

$d = 18.0 - \dfrac{10}{11} \times 6.0 - (2 + .5 \times .62) = 10.2''$

$v = \dfrac{1000}{12 \times .89 \times 10.2} = 9.2$ p.s.i < 54 p.s.i $\checkmark$

$A_s = \dfrac{1.666}{1.13 \times 10.2} = 0.145$ SQ. IN./FT.

USE #5 @ 18″ o.c.

DESIGN EXAMPLE - SECTION 8.14.11	SHT. 5 of 6

STEM   DESIGN - (CONT.)

$$\text{LONG. REINF.} = 0.0015 \times 11.0 \times 12 \times \frac{8+18}{2} = 2.574 \text{ SQ. IN.}$$

USE #5 @ 16" o.c. HORIZ.

SLIDING :

$$F_r = \frac{6250}{11640} = 0.537 > 0.40$$

$$\therefore \text{ A KEY IS REQ'D}$$

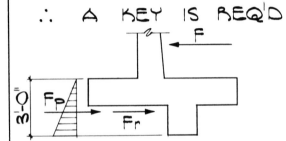

$$\text{RESISTANCE TO SLIDING} = 11640 \times .40 + \frac{300 \times 3^2}{2} = 6006^{\#} \approx 6250^{\#}$$
$$(+4\%)$$

$$\nu = \frac{6250}{18^2} = 19.3 \text{ p.s.i} < 54 \text{ p.s.i} \checkmark$$

DESIGN EXAMPLE — SECTION 8.2.12	SHT. 6 OF 6

CHECK WALL WITH 200 #/FT. AXIAL LOAD

$W = 7787^{\#} + 200^{\#} = 7987^{\#}$

$F = 2667^{\#}$; $M_O = 11852^{\prime\#}$

$M_R = 27993 + 200 \times \frac{28}{12} = 28460^{\prime\#}$

OVERTURN. RATIO $= \frac{28460}{11852} = 2.401 > 1.50 \checkmark$

NET $M = 28460 - 11852 = 16608^{\prime\#}$

$\bar{X} = \frac{16608}{7987} = 2.08^{\prime} > 2.0^{\prime} \checkmark$

$e = \frac{6.0}{2} - 2.08 = 0.920^{\prime}$

$M_e = 7987 \times .92 = 7348^{\prime\#}$

SOIL PRESSURE $= \frac{7987}{6.0} \pm \frac{7348}{6.0} = 1331.2 \pm 1224.7$

$\underline{SP_T = 2555.9 \text{ p.s.f.}}$ $\qquad$ $\underline{SP_H = 106.5 \text{ p.s.f.}}$

FTG. BENDING & SHEAR - (SEE CALC. SHT. No. 2)

$-M_1$: $SP_1 = \frac{2555.9 - 106.5}{6.0} \times 2.67 = 1090 \text{ p.s.f.}$

$\qquad$ UNIF. LD. $= 1200 + 200 - 106.5 = 1293.5 \ ^{\#}/_{FT.}$

$\qquad -M_1 = \frac{1293.5 \times 2.67^2}{2} - \frac{1090 \times 2.67^2}{2 \times 3} = 3315.5^{\prime\#} \approx 3413^{\prime\#}$

$\qquad V_1 = 1293.5 \times 2.67 - \frac{1090 \times 2.67}{2} = 1998.5^{\#} \approx 2076^{\#}$

$+M_2$: $W_1 = \frac{2555.9 - 106.5}{6.0} \times 4.0 + 106.5 = 1739.4 \text{ p.s.f.}$

$\qquad SP_2 = 2555.9 - 1739.4 = 816.5 \text{ p.s.f.}$

$\qquad$ UNIF. LD. $= 1739.4 - 200 = 1539.4 \ ^{\#}/_{FT.}$

$\qquad +M_2 = \frac{1539.4 \times 2^2}{2} + \frac{816.5 \times 2^2 \times 2}{2 \times 3} = 4167^{\prime\#} \approx 4067^{\prime\#} \checkmark$

$\qquad V_2 = 1539.4 \times 2 + 816.5 \times \frac{2}{2} = 3895.3^{\#} \approx 3800^{\#} \checkmark$

$\qquad$ NO CHANGE IN FTG. DESIGN

$\qquad$ ALSO NO CHANGE IN STEM DESIGN.

TABLE 8.1	GENERAL RETAINING WALLS – CONCRETE STEM – SOIL AT HEEL SIDE
	SLOPE = 0 to 1 SURCHARGE = 0 lbs./sq. ft. AXIAL = 0 lbs./ft.

SOIL PRESSURE DIAGRAM

Stem Height	3'- 0"	4'- 0"	5'- 0"	6'- 0"	7'- 0"	8'- 0"	9'- 0"	10'- 0"	11'- 0"	12'- 0"
Ft. L ft.	1.667	2.000	2.500	3.000	3.500	4.000	4.500	5.000	5.500	6.000
Ft. T ft.	0.667	0.667	0.833	0.833	0.833	1.000	1.000	1.167	1.167	1.333
Setback	0.500	0.667	0.833	1.000	1.167	1.333	1.500	1.667	1.833	2.000
Wt ft.	0.667	0.667	0.667	0.667	0.667	0.667	0.667	0.667	0.667	0.667
Wb ft.	0.667	0.667	0.833	0.833	0.833	1.000	1.000	1.167	1.167	1.333
A ft.	0.500	0.833	1.167	1.500	1.833	2.167	2.500	2.667	3.000	3.333
B ft.	0.667	0.667	0.833	0.833	0.833	1.000	1.000	1.167	1.167	1.333
C ft.	0.500	0.500	0.500	0.667	0.833	0.833	1.000	1.167	1.333	1.333
D ft.	0.667	0.667	0.833	0.833	0.833	1.000	1.000	1.167	1.167	1.333
W lbs.	728	1013	1562	2084	2682	3537	4305	5371	6309	7587
F lbs.	202	327	510	700	920	1215	1500	1870	2220	2667
Mo ft.-lbs.	246	508	992	1595	2403	3645	5000	6962	9005	11852
Mr ft.-lbs.	709	1229	2369	3787	5685	8566	11732	16228	20986	27527
O.T.R.	2.878	2.418	2.387	2.373	2.366	2.350	2.347	2.331	2.331	2.323
$\overline{X}$ ft.	0.636	0.711	0.881	1.051	1.224	1.391	1.564	1.725	1.899	2.066
e ft.	0.198	0.289	0.369	0.449	0.526	0.609	0.686	0.775	0.851	0.934
Me ft.-lbs.	144	293	577	935	1412	2153	2954	4161	5369	7085
S ft.^{3}	0.463	0.667	1.042	1.500	2.042	2.667	3.375	4.167	5.042	6.000
S.P.t p.s.f.	748	946	1179	1318	1458	1691	1832	2073	2212	2445
S.P.h p.s.f.	126	68	71	71	75	77	81	75	82	84
Friction	0.277	0.322	0.327	0.336	0.343	0.344	0.348	0.348	0.352	0.351

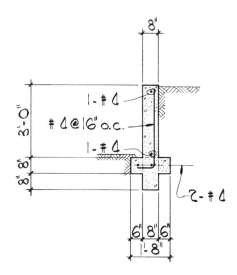

E.F.P. = 30 p.c.f.
Section 8.1.3

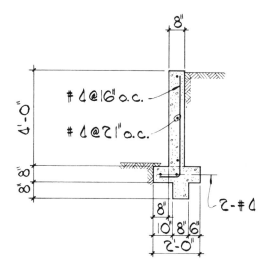

E.F.P. = 30 p.c.f.
Section 8.1.4

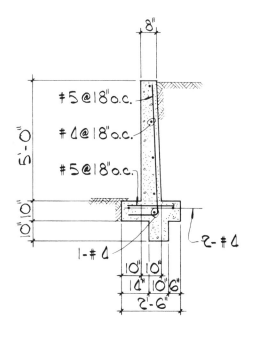

E.F.P. = 30 p.c.f.
Section 8.1.5

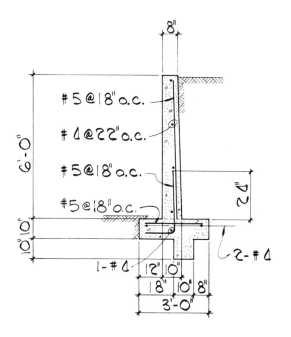

E.F.P. = 30 p.c.f.
Section 8.1.6

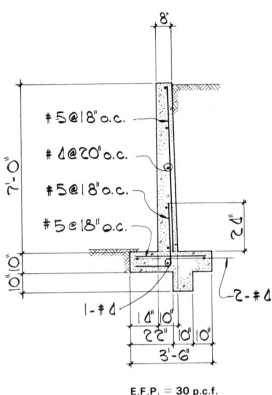

E.F.P. = 30 p.c.f.
Section 8.1.7

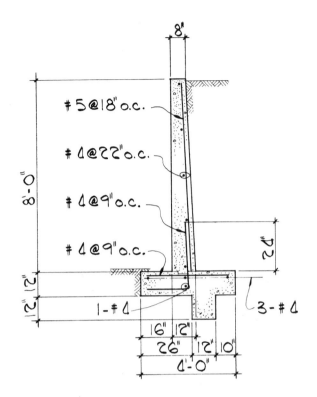

E.F.P. = 30 p.c.f.
Section 8.1.8

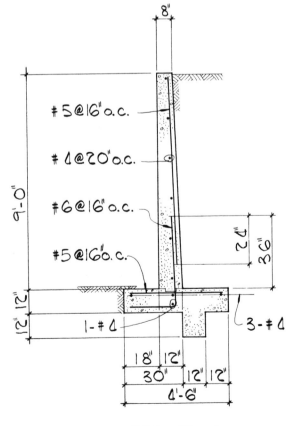

E.F.P. = 30 p.c.f.
Section 8.1.9

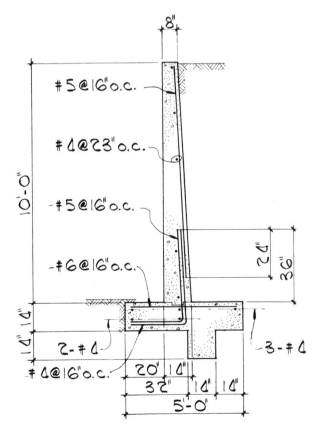

E.F.P. = 30 p.c.f.
Section 8.1.10

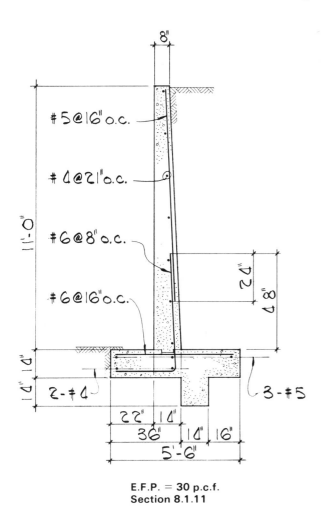

#5@16" o.c.

#4@21" o.c.

#6@8" o.c.

#6@16" o.c.

11'-0"

2-#4

3-#5

24"

8"

4" 4"

22" 14"

36" 14" 16"

5'-6"

E.F.P. = 30 p.c.f.
Section 8.1.11

#5@16" o.c.

#4@23" o.c.

#6@8" o.c.

#4@16" o.c.

12'-0"

2-#4

3-#5

#5@16" o.c.

24"

8"

6"

6"

24"

24" 16"

40" 16" 16"

6'-0"

E.F.P. = 30 p.c.f.
Section 8.1.12

481

TABLE 8.2	GENERAL RETAINING WALLS – CONCRETE STEM – SOIL AT HEEL SIDE
	SLOPE = 0 to 1 SURCHARGE = 0 lbs./sq. ft. AXIAL = 200 lbs./ft.

SOIL PRESSURE DIAGRAM

Stem Height	3'- 0''	4'- 0''	5'- 0''	6'- 0''	7'- 0''	8'- 0''	9'- 0''	10'- 0''	11'- 0''	12'- 0''
Ft. L ft.	1.667	2.000	2.500	3.000	3.500	4.000	4.500	5.000	5.500	6.000
Ft. T ft.	0.667	0.667	0.833	0.833	0.833	1.000	1.000	1.167	1.167	1.333
Setback	0.500	0.667	0.833	1.000	1.167	1.333	1.500	1.667	1.833	2.000
Wt ft.	0.667	0.667	0.667	0.667	0.667	0.667	0.667	0.667	0.667	0.667
Wb ft.	0.667	0.667	0.833	0.833	0.833	1.000	1.000	1.167	1.167	1.333
A ft.	0.500	0.833	1.167	1.500	1.833	2.167	2.500	2.667	3.000	3.333
B ft.	0.667	0.667	0.833	0.833	0.833	1.000	1.000	1.167	1.167	1.333
C ft.	0.500	0.500	0.500	0.667	0.833	0.833	1.000	1.167	1.333	1.333
D ft.	0.667	0.667	0.833	0.833	0.833	1.000	1.000	1.167	1.167	1.333
W lbs.	928	1213	1762	2284	2882	3737	4505	5571	6509	7787
F lbs.	202	327	510	700	920	1215	1500	1870	2220	2667
Mo ft.-lbs.	246	508	992	1595	2403	3645	5000	6962	9005	11852
Mr ft.-lbs.	876	1429	2602	4053	5985	8899	12099	16628	21420	27993
O.T.R.	3.554	2.812	2.622	2.541	2.490	2.441	2.420	2.388	2.379	2.362
$\overline{X}$ ft.	0.678	0.759	0.913	1.076	1.243	1.406	1.576	1.735	1.907	2.073
e ft.	0.155	0.241	0.337	0.424	0.507	0.594	0.674	0.765	0.843	0.927
Me ft.-lbs.	144	293	593	968	1462	2219	3037	4261	5485	7219
S ft.^{3}	0.463	0.667	1.042	1.500	2.042	2.667	3.375	4.167	5.042	6.000
S.P.t p.s.f.	868	1046	1275	1407	1540	1766	1901	2137	2271	2501
S.P.h p.s.f.	246	168	135	116	107	102	101	91	95	95
Friction	0.217	0.269	0.290	0.307	0.319	0.325	0.333	0.336	0.341	0.342

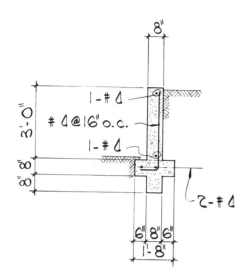

Axial load = 200 lbs./ft.
E.F.P. = 30 p.c.f.
Section 8.2.3

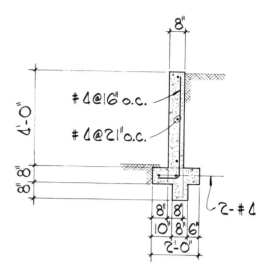

Axial load = 200 lbs./ft.
E.F.P. = 30 p.c.f.
Section 8.2.4

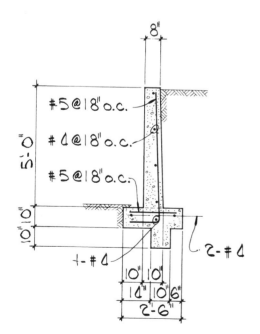

Axial load = 200 lbs./ft.
E.F.P. = 30 p.c.f.
Section 8.2.5

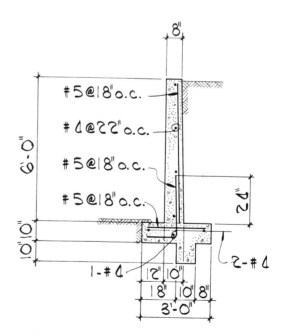

Axial load = 200 lbs./ft.
E.F.P. = 30 p.c.f.
Section 8.2.6

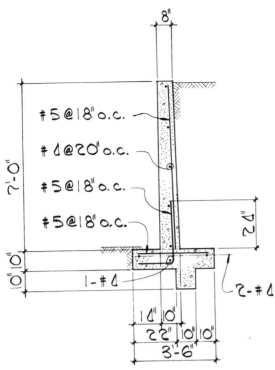

Axial load = 200 lbs./ft.
E.F.P. = 30 p.c.f.
Section 8.2.7

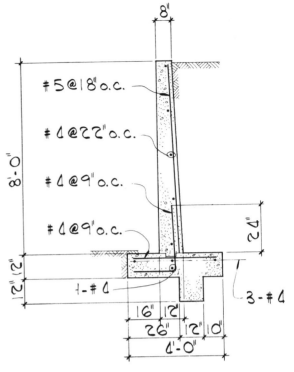

Axial load = 200 lbs./ft.
E.F.P. = 30 p.c.f.
Section 8.2.8

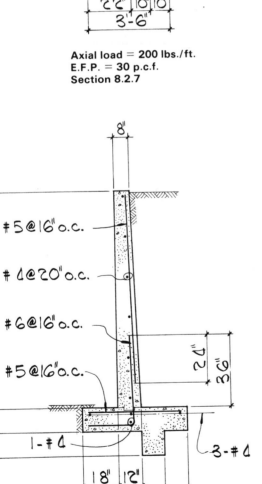

Axial load = 200 lbs./ft.
E.F.P. = 30 p.c.f.
Section 8.2.9

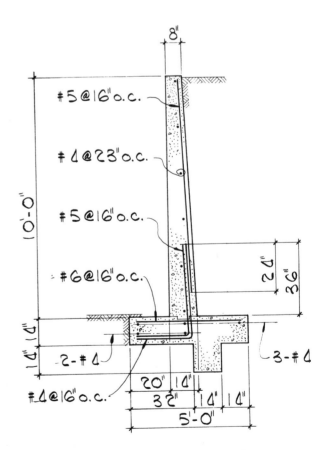

Axial load = 200 lbs./ft.
E.F.P. = 30 p.c.f.
Section 8.2.10

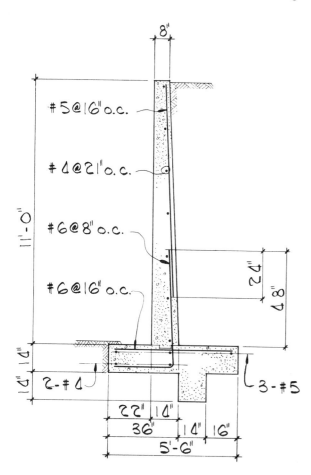

#5@16"o.c.

#4@21"o.c.

#6@8"o.c.

#6@16"o.c.

24"

48"

11'-0"

1'-4"

2-#4

3-#5

22" | 14"
36" | 14" | 16"
5'-6"

Axial load = 200 lbs./ft.
E.F.P. = 30 p.c.f.
Section 8.2.11

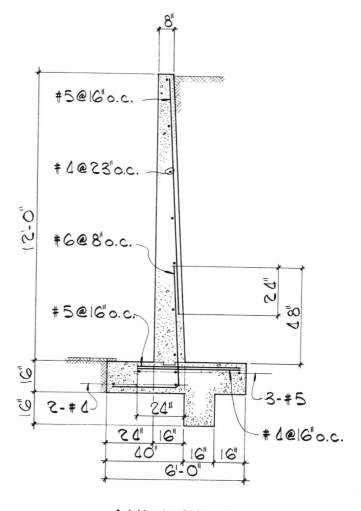

8"

#5@16"o.c.

#4@23"o.c.

#6@8"o.c.

#5@16"o.c.

24"

48"

12'-0"

1'-6"

1'-6"

2-#4

24"

3-#5

#4@16"o.c.

24" | 16"
40" | 16" | 16"
6'-0"

Axial load = 200 lbs./ft.
E.F.P. = 30 p.c.f.
Section 8.2.12

485

TABLE 8.3	GENERAL RETAINING WALLS – CONCRETE STEM – SOIL AT HEEL SIDE
	SLOPE = 0 to 1 SURCHARGE = 0 lbs./sq. ft. AXIAL = 400 lbs./ft.

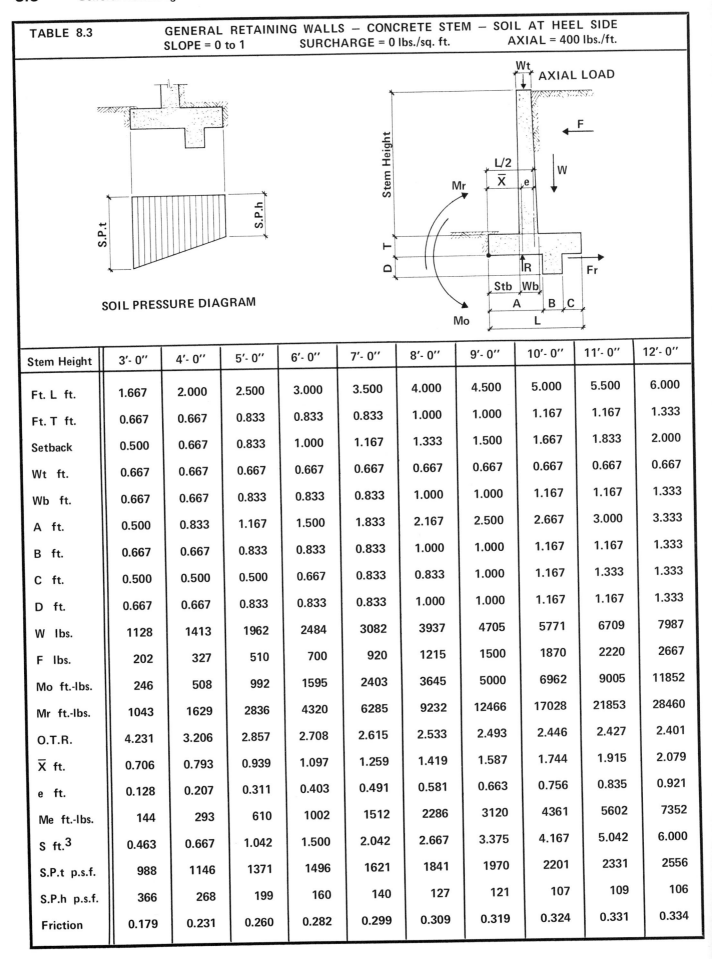

SOIL PRESSURE DIAGRAM

Stem Height	3'- 0''	4'- 0''	5'- 0''	6'- 0''	7'- 0''	8'- 0''	9'- 0''	10'- 0''	11'- 0''	12'- 0''
Ft. L ft.	1.667	2.000	2.500	3.000	3.500	4.000	4.500	5.000	5.500	6.000
Ft. T ft.	0.667	0.667	0.833	0.833	0.833	1.000	1.000	1.167	1.167	1.333
Setback	0.500	0.667	0.833	1.000	1.167	1.333	1.500	1.667	1.833	2.000
Wt ft.	0.667	0.667	0.667	0.667	0.667	0.667	0.667	0.667	0.667	0.667
Wb ft.	0.667	0.667	0.833	0.833	0.833	1.000	1.000	1.167	1.167	1.333
A ft.	0.500	0.833	1.167	1.500	1.833	2.167	2.500	2.667	3.000	3.333
B ft.	0.667	0.667	0.833	0.833	0.833	1.000	1.000	1.167	1.167	1.333
C ft.	0.500	0.500	0.500	0.667	0.833	0.833	1.000	1.167	1.333	1.333
D ft.	0.667	0.667	0.833	0.833	0.833	1.000	1.000	1.167	1.167	1.333
W lbs.	1128	1413	1962	2484	3082	3937	4705	5771	6709	7987
F lbs.	202	327	510	700	920	1215	1500	1870	2220	2667
Mo ft.-lbs.	246	508	992	1595	2403	3645	5000	6962	9005	11852
Mr ft.-lbs.	1043	1629	2836	4320	6285	9232	12466	17028	21853	28460
O.T.R.	4.231	3.206	2.857	2.708	2.615	2.533	2.493	2.446	2.427	2.401
$\overline{X}$ ft.	0.706	0.793	0.939	1.097	1.259	1.419	1.587	1.744	1.915	2.079
e ft.	0.128	0.207	0.311	0.403	0.491	0.581	0.663	0.756	0.835	0.921
Me ft.-lbs.	144	293	610	1002	1512	2286	3120	4361	5602	7352
S ft.^{3}	0.463	0.667	1.042	1.500	2.042	2.667	3.375	4.167	5.042	6.000
S.P.t p.s.f.	988	1146	1371	1496	1621	1841	1970	2201	2331	2556
S.P.h p.s.f.	366	268	199	160	140	127	121	107	109	106
Friction	0.179	0.231	0.260	0.282	0.299	0.309	0.319	0.324	0.331	0.334

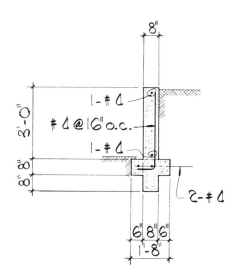

Axial load = 400 lbs./ft.
E.F.P. = 30 p.c.f.
Section 8.3.3

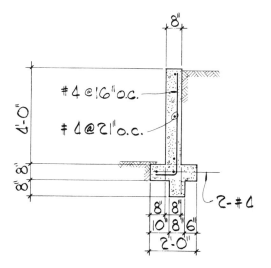

Axial load = 400 lbs./ft.
E.F.P. = 30 p.c.f.
Section 8.3.4

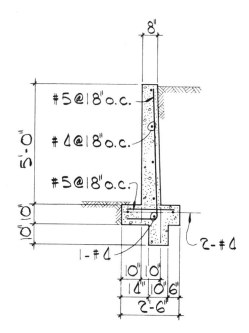

Axial load = 400 lbs./ft.
E.F.P. = 30 p.c.f.
Section 8.3.5

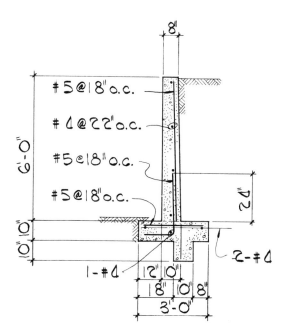

Axial load = 400 lbs./ft.
E.F.P. = 30 p.c.f.
Section 8.3.6

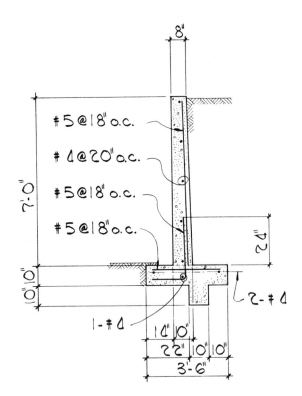

#5@18" o.c.

#4@20" o.c.

#5@18" o.c.

#5@18" o.c.

7'-0"

24"

1-#4

2-#4

10" 10"

22" 10" 10"

3'-6"

Axial load = 400 lbs./ft.
E.F.P. = 30 p.c.f.
Section 8.3.7

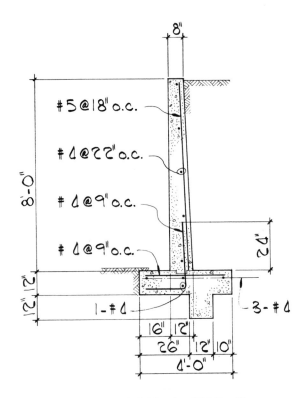

#5@18" o.c.

#4@22" o.c.

#4@9" o.c.

#4@9" o.c.

8'-0"

24"

12"

1-#4

3-#4

16" 12"

26" 12" 10"

4'-0"

Axial load = 400 lbs./ft.
E.F.P. = 30 p.c.f.
Section 8.3.8

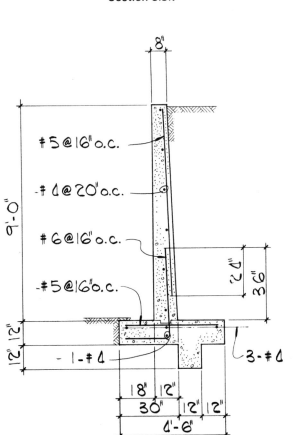

#5@16" o.c.

#4@20" o.c.

#6@16" o.c.

#5@16" o.c.

9'-0"

24"

36"

12"

1-#4

3-#4

18" 12"

30" 12" 12"

4'-6"

Axial load = 400 lbs./ft.
E.F.P. = 30 p.c.f.
Section 8.3.9

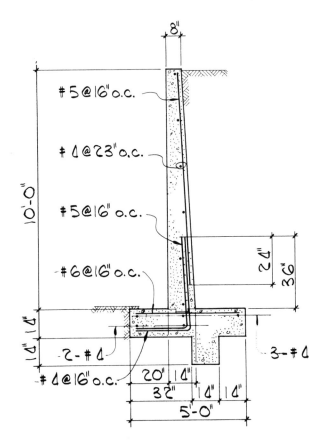

#5@16" o.c.

#4@23" o.c.

#5@16" o.c.

#6@16" o.c.

10'-0"

24"

36"

14"

2-#4

#4@16" o.c.

3-#4

20" 14"

32" 14" 14"

5'-0"

Axial load = 400 lbs./ft.
E.F.P. = 30 p.c.f.
Section 8.3.10

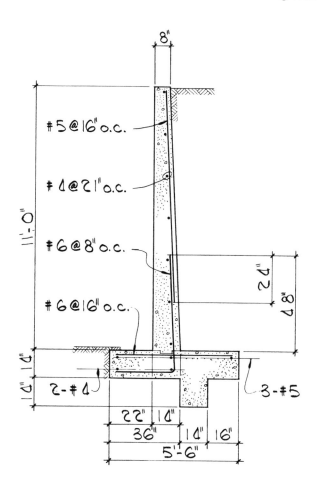

Axial load = 400 lbs./ft.
E.F.P. = 30 p.c.f.
Section 8.3.11

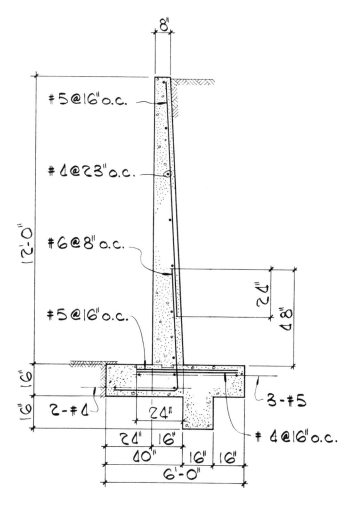

Axial load = 400 lbs./ft.
E.F.P. = 30 p.c.f.
Section 8.3.12

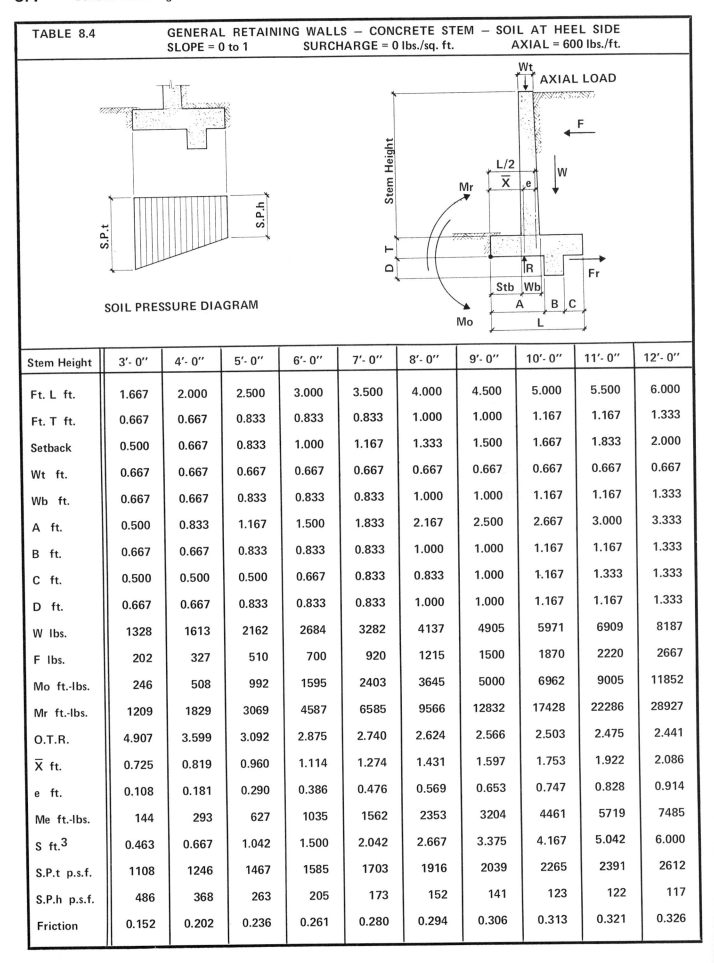

TABLE 8.4   GENERAL RETAINING WALLS — CONCRETE STEM — SOIL AT HEEL SIDE
SLOPE = 0 to 1   SURCHARGE = 0 lbs./sq. ft.   AXIAL = 600 lbs./ft.

SOIL PRESSURE DIAGRAM

Stem Height	3'- 0''	4'- 0''	5'- 0''	6'- 0''	7'- 0''	8'- 0''	9'- 0''	10'- 0''	11'- 0''	12'- 0''
Ft. L ft.	1.667	2.000	2.500	3.000	3.500	4.000	4.500	5.000	5.500	6.000
Ft. T ft.	0.667	0.667	0.833	0.833	0.833	1.000	1.000	1.167	1.167	1.333
Setback	0.500	0.667	0.833	1.000	1.167	1.333	1.500	1.667	1.833	2.000
Wt ft.	0.667	0.667	0.667	0.667	0.667	0.667	0.667	0.667	0.667	0.667
Wb ft.	0.667	0.667	0.833	0.833	0.833	1.000	1.000	1.167	1.167	1.333
A ft.	0.500	0.833	1.167	1.500	1.833	2.167	2.500	2.667	3.000	3.333
B ft.	0.667	0.667	0.833	0.833	0.833	1.000	1.000	1.167	1.167	1.333
C ft.	0.500	0.500	0.500	0.667	0.833	0.833	1.000	1.167	1.333	1.333
D ft.	0.667	0.667	0.833	0.833	0.833	1.000	1.000	1.167	1.167	1.333
W lbs.	1328	1613	2162	2684	3282	4137	4905	5971	6909	8187
F lbs.	202	327	510	700	920	1215	1500	1870	2220	2667
Mo ft.-lbs.	246	508	992	1595	2403	3645	5000	6962	9005	11852
Mr ft.-lbs.	1209	1829	3069	4587	6585	9566	12832	17428	22286	28927
O.T.R.	4.907	3.599	3.092	2.875	2.740	2.624	2.566	2.503	2.475	2.441
$\overline{X}$ ft.	0.725	0.819	0.960	1.114	1.274	1.431	1.597	1.753	1.922	2.086
e ft.	0.108	0.181	0.290	0.386	0.476	0.569	0.653	0.747	0.828	0.914
Me ft.-lbs.	144	293	627	1035	1562	2353	3204	4461	5719	7485
S ft.^{3}	0.463	0.667	1.042	1.500	2.042	2.667	3.375	4.167	5.042	6.000
S.P.t p.s.f.	1108	1246	1467	1585	1703	1916	2039	2265	2391	2612
S.P.h p.s.f.	486	368	263	205	173	152	141	123	122	117
Friction	0.152	0.202	0.236	0.261	0.280	0.294	0.306	0.313	0.321	0.326

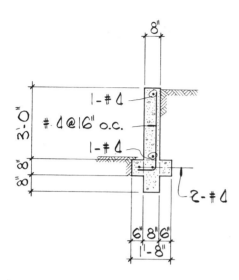

Axial load = 600 lbs./ft.
E.F.P. = 30 p.c.f.
Section 8.4.3

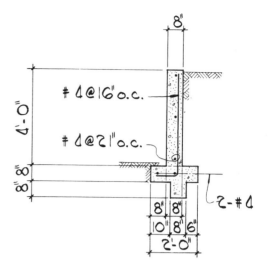

Axial load = 600 lbs./ft.
E.F.P. = 30 p.c.f.
Section 8.4.4

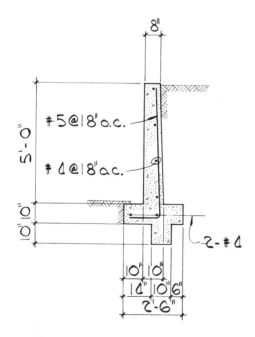

Axial load = 600 lbs./ft.
E.F.P. = 30 p.c.f.
Section 8.4.5

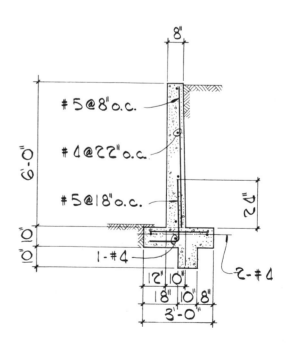

Axial load = 600 lbs./ft.
E.F.P. = 30 p.c.f.
Section 8.4.6

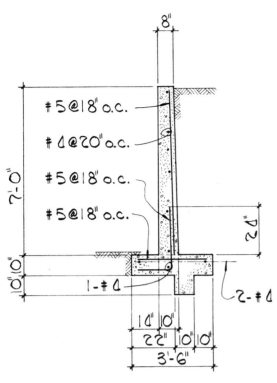

Axial load = 600 lbs./ft.
E.F.P. = 30 p.c.f.
Section 8.4.7

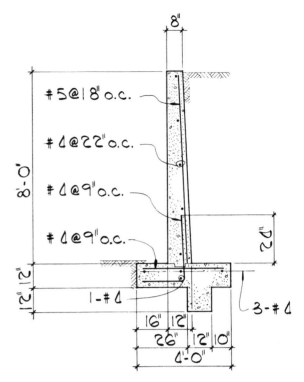

Axial load = 600 lbs./ft.
E.F.P. = 30 p.c.f.
Section 8.4.8

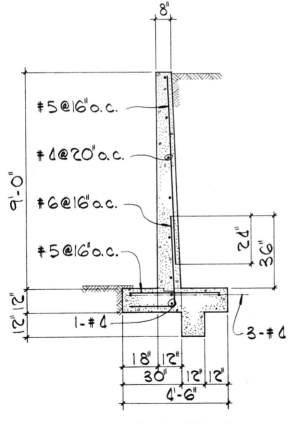

Axial load = 600 lbs./ft.
E.F.P. = 30 p.c.f.
Section 8.4.9

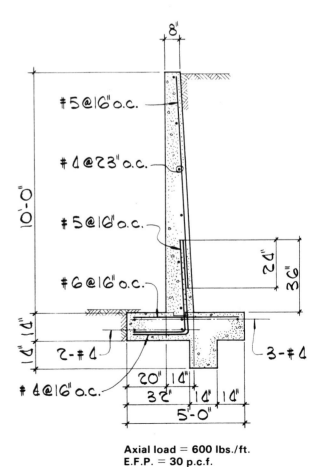

Axial load = 600 lbs./ft.
E.F.P. = 30 p.c.f.
Section 8.4.10

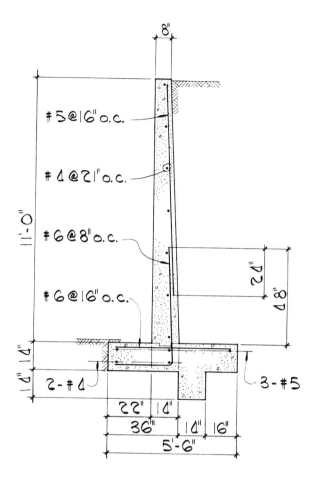

#5@16" o.c.

#4@21" o.c.

#6@8" o.c.

#6@16" o.c.

11'-0"

24"

48"

4"4"

4"

2-#4

3-#5

22"   14"

36"   14"   16"

5'-6"

Axial load = 600 lbs./ft.
E.F.P. = 30 p.c.f.
Section 8.4.11

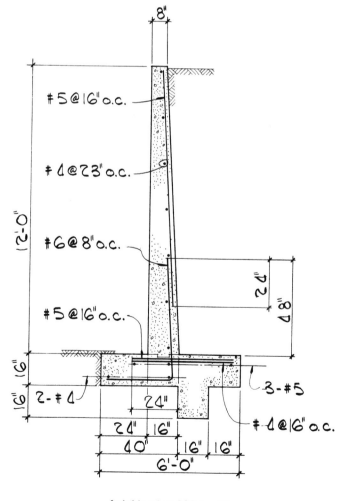

#5@16" o.c.

#4@23" o.c.

#6@8" o.c.

#5@16" o.c.

12'-0"

24"

48"

6"   6"

6"

2-#4

24"

3-#5

#4@16" o.c.

24"   16"

40"   16"   16"

6'-0"

Axial load = 600 lbs./ft.
E.F.P. = 30 p.c.f.
Section 8.4.12

TABLE 8.5	GENERAL RETAINING WALLS — CONCRETE STEM — SOIL AT HEEL SIDE
	SLOPE = 0 to 1 SURCHARGE = 0 lbs./sq. ft. AXIAL = 800 lbs./ft.

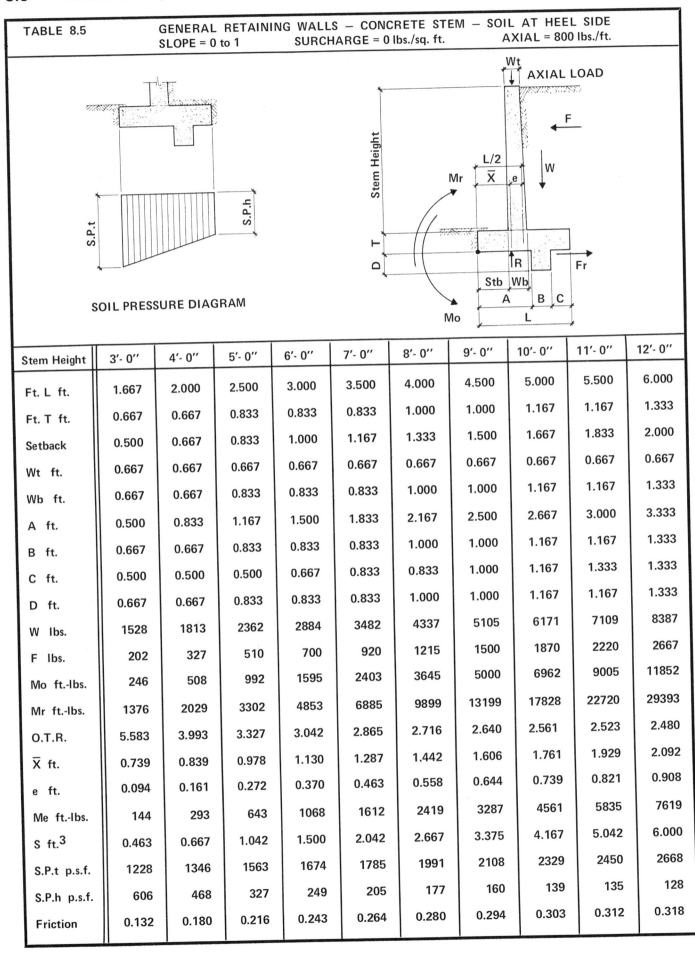

SOIL PRESSURE DIAGRAM

Stem Height	3'- 0''	4'- 0''	5'- 0''	6'- 0''	7'- 0''	8'- 0''	9'- 0''	10'- 0''	11'- 0''	12'- 0''
Ft. L ft.	1.667	2.000	2.500	3.000	3.500	4.000	4.500	5.000	5.500	6.000
Ft. T ft.	0.667	0.667	0.833	0.833	0.833	1.000	1.000	1.167	1.167	1.333
Setback	0.500	0.667	0.833	1.000	1.167	1.333	1.500	1.667	1.833	2.000
Wt ft.	0.667	0.667	0.667	0.667	0.667	0.667	0.667	0.667	0.667	0.667
Wb ft.	0.667	0.667	0.833	0.833	0.833	1.000	1.000	1.167	1.167	1.333
A ft.	0.500	0.833	1.167	1.500	1.833	2.167	2.500	2.667	3.000	3.333
B ft.	0.667	0.667	0.833	0.833	0.833	1.000	1.000	1.167	1.167	1.333
C ft.	0.500	0.500	0.500	0.667	0.833	0.833	1.000	1.167	1.333	1.333
D ft.	0.667	0.667	0.833	0.833	0.833	1.000	1.000	1.167	1.167	1.333
W lbs.	1528	1813	2362	2884	3482	4337	5105	6171	7109	8387
F lbs.	202	327	510	700	920	1215	1500	1870	2220	2667
Mo ft.-lbs.	246	508	992	1595	2403	3645	5000	6962	9005	11852
Mr ft.-lbs.	1376	2029	3302	4853	6885	9899	13199	17828	22720	29393
O.T.R.	5.583	3.993	3.327	3.042	2.865	2.716	2.640	2.561	2.523	2.480
$\overline{X}$ ft.	0.739	0.839	0.978	1.130	1.287	1.442	1.606	1.761	1.929	2.092
e ft.	0.094	0.161	0.272	0.370	0.463	0.558	0.644	0.739	0.821	0.908
Me ft.-lbs.	144	293	643	1068	1612	2419	3287	4561	5835	7619
S ft.^{3}	0.463	0.667	1.042	1.500	2.042	2.667	3.375	4.167	5.042	6.000
S.P.t p.s.f.	1228	1346	1563	1674	1785	1991	2108	2329	2450	2668
S.P.h p.s.f.	606	468	327	249	205	177	160	139	135	128
Friction	0.132	0.180	0.216	0.243	0.264	0.280	0.294	0.303	0.312	0.318

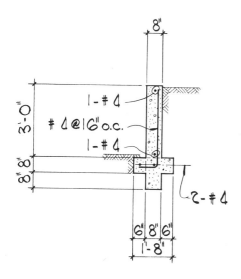

Axial load = 800 lbs./ft.
E.F.P. = 30 p.c.f.
Section 8.5.3

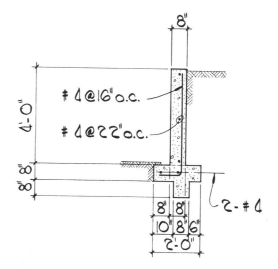

Axial load = 800 lbs./ft.
E.F.P. = 30 p.c.f.
Section 8.5.4

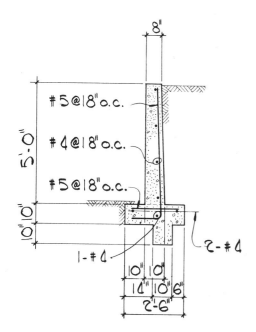

Axial load = 800 lbs./ft.
E.F.P. = 30 p.c.f.
Section 8.5.5

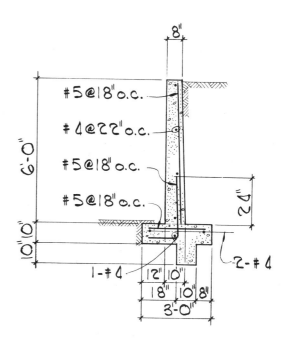

Axial load = 800 lbs./ft.
E.F.P. = 30 p.c.f.
Section 8.5.6

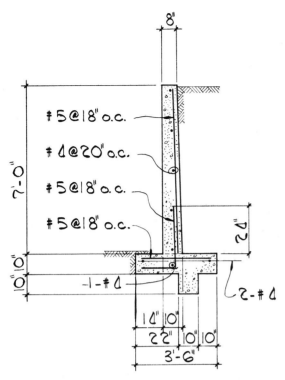

#5@18" o.c.
#4@20" o.c.
#5@18" o.c.
#5@18" o.c.
7'-0"
1-#4
2-#4
24"
1'0"
22"
0"10"
3'-6"

Axial load = 800 lbs./ft.
E.F.P. = 30 p.c.f.
Section 8.5.7

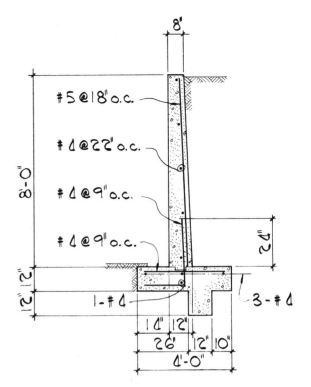

#5@18" o.c.
#4@22" o.c.
#4@9" o.c.
#4@9" o.c.
8'-0"
1-#4
3-#4
24"
12"
1'2"
26"
12"10"
4'-0"

Axial load = 800 lbs./ft.
E.F.P. = 30 p.c.f.
Section 8.5.8

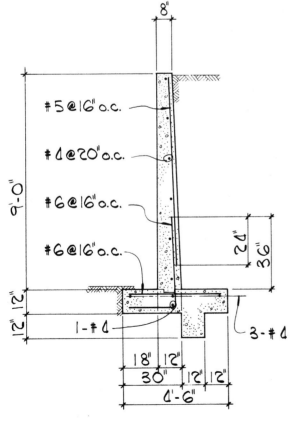

#5@16" o.c.
#4@20" o.c.
#6@16" o.c.
#6@16" o.c.
9'-0"
1-#4
3-#4
24"
36"
12"
18"
12"
30"
12"12"
4'-6"

Axial load = 800 lbs./ft.
E.F.P. = 30 p.c.f.
Section 8.5.9

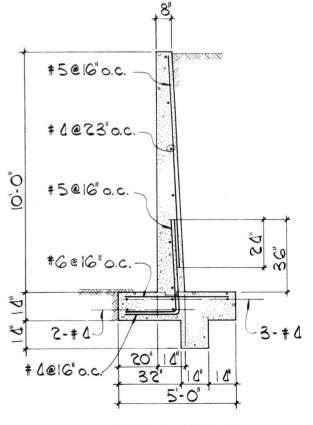

#5@16" o.c.
#4@23" o.c.
#5@16" o.c.
#6@16" o.c.
#4@16" o.c.
10'-0"
2-#4
3-#4
24"
36"
14"
20"
14"
32"
14"14"
5'-0"

Axial load = 800 lbs./ft.
E.F.P. = 30 p.c.f.
Section 8.5.10

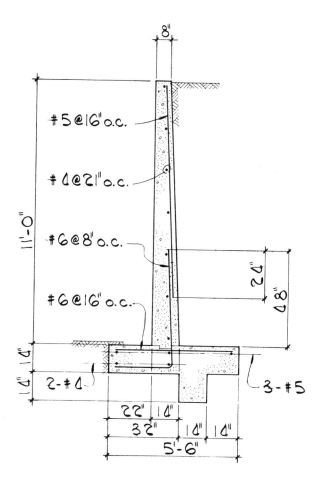

#5@16" o.c.

#4@21" o.c.

#6@8" o.c.

#6@16" o.c.

11'-0"

24"

48"

8"

4"

4"

2-#4

3-#5

22" 14"

32" 14" 14"

5'-6"

Axial load = 800 lbs./ft.
E.F.P. = 30 p.c.f.
Section 8.5.11

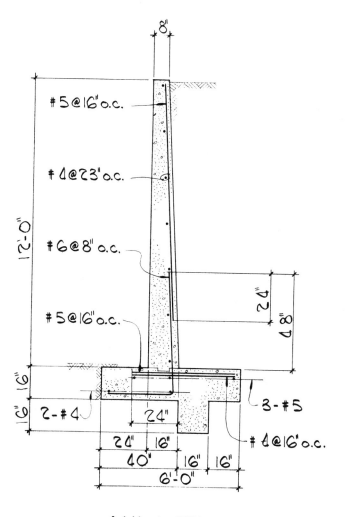

#5@16" o.c.

#4@23" o.c.

#6@8" o.c.

#5@16" o.c.

12'-0"

24"

48"

8"

6"

6"

2-#4

24"

3-#5

#4@16" o.c.

24" 16"

40" 16" 16"

6'-0"

Axial load = 800 lbs./ft.
E.F.P. = 30 p.c.f.
Section 8.5.12

TABLE 8.6	GENERAL RETAINING WALLS – CONCRETE STEM – SOIL AT HEEL SIDE
	SLOPE = 0 to 1 SURCHARGE = 50 lbs./sq. ft. AXIAL = 0 lbs./ft.

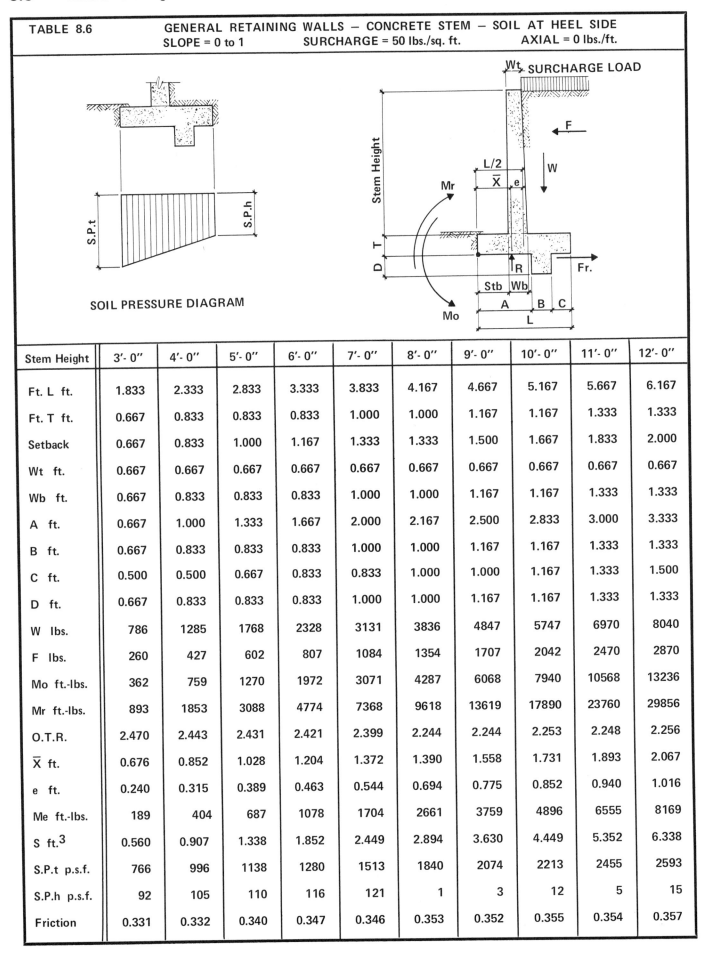

SOIL PRESSURE DIAGRAM

Stem Height	3'- 0"	4'- 0"	5'- 0"	6'- 0"	7'- 0"	8'- 0"	9'- 0"	10'- 0"	11'- 0"	12'- 0"
Ft. L ft.	1.833	2.333	2.833	3.333	3.833	4.167	4.667	5.167	5.667	6.167
Ft. T ft.	0.667	0.833	0.833	0.833	1.000	1.000	1.167	1.167	1.333	1.333
Setback	0.667	0.833	1.000	1.167	1.333	1.333	1.500	1.667	1.833	2.000
Wt ft.	0.667	0.667	0.667	0.667	0.667	0.667	0.667	0.667	0.667	0.667
Wb ft.	0.667	0.833	0.833	0.833	1.000	1.000	1.167	1.167	1.333	1.333
A ft.	0.667	1.000	1.333	1.667	2.000	2.167	2.500	2.833	3.000	3.333
B ft.	0.667	0.833	0.833	0.833	1.000	1.000	1.167	1.167	1.333	1.333
C ft.	0.500	0.500	0.667	0.833	0.833	1.000	1.000	1.167	1.333	1.500
D ft.	0.667	0.833	0.833	0.833	1.000	1.000	1.167	1.167	1.333	1.333
W lbs.	786	1285	1768	2328	3131	3836	4847	5747	6970	8040
F lbs.	260	427	602	807	1084	1354	1707	2042	2470	2870
Mo ft.-lbs.	362	759	1270	1972	3071	4287	6068	7940	10568	13236
Mr ft.-lbs.	893	1853	3088	4774	7368	9618	13619	17890	23760	29856
O.T.R.	2.470	2.443	2.431	2.421	2.399	2.244	2.244	2.253	2.248	2.256
$\overline{X}$ ft.	0.676	0.852	1.028	1.204	1.372	1.390	1.558	1.731	1.893	2.067
e ft.	0.240	0.315	0.389	0.463	0.544	0.694	0.775	0.852	0.940	1.016
Me ft.-lbs.	189	404	687	1078	1704	2661	3759	4896	6555	8169
S ft.^{3}	0.560	0.907	1.338	1.852	2.449	2.894	3.630	4.449	5.352	6.338
S.P.t p.s.f.	766	996	1138	1280	1513	1840	2074	2213	2455	2593
S.P.h p.s.f.	92	105	110	116	121	1	3	12	5	15
Friction	0.331	0.332	0.340	0.347	0.346	0.353	0.352	0.355	0.354	0.357

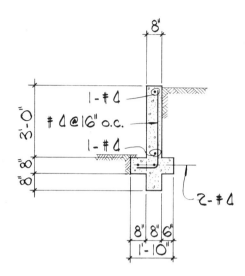

Surcharge load = 50 p.s.f.
E.F.P. = 30 p.c.f.
Section 8.6.3

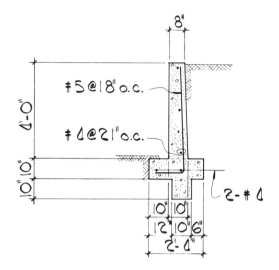

Surcharge load = 50 p.s.f.
E.F.P. = 30 p.c.f.
Section 8.6.4

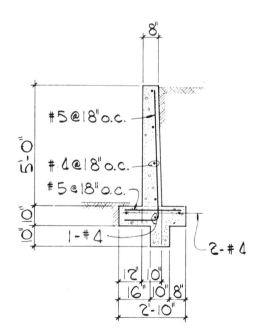

Surcharge load = 50 p.s.f.
E.F.P. = 30 p.c.f.
Section 8.6.5

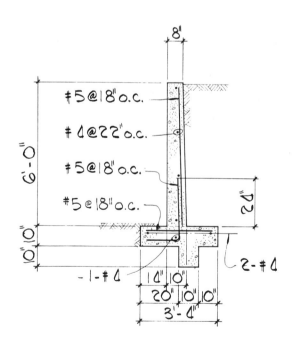

Surcharge load = 50 p.s.f.
E.F.P. = 30 p.c.f.
Section 8.6.6

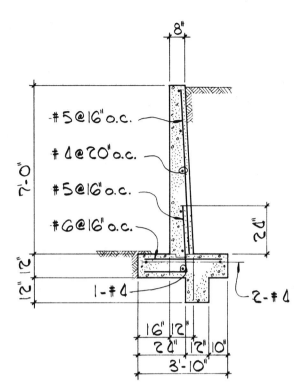

Surcharge load = 50 p.s.f.
E.F.P. = 30 p.c.f.
Section 8.6.7

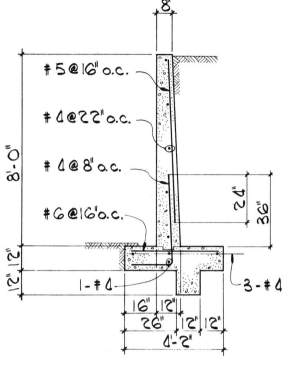

Surcharge load = 50 p.s.f.
E.F.P. = 30 p.c.f.
Section 8.6.8

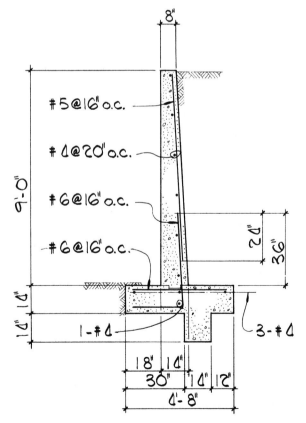

Surcharge load = 50 p.s.f.
E.F.P. = 30 p.c.f.
Section 8.6.9

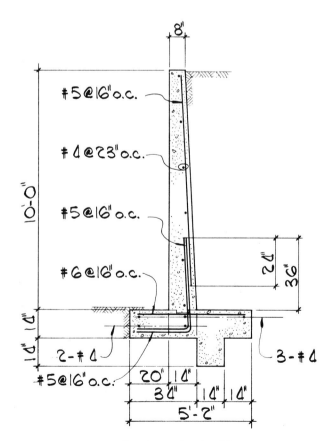

Surcharge load = 50 p.s.f.
E.F.P. = 30 p.c.f.
Section 8.6.10

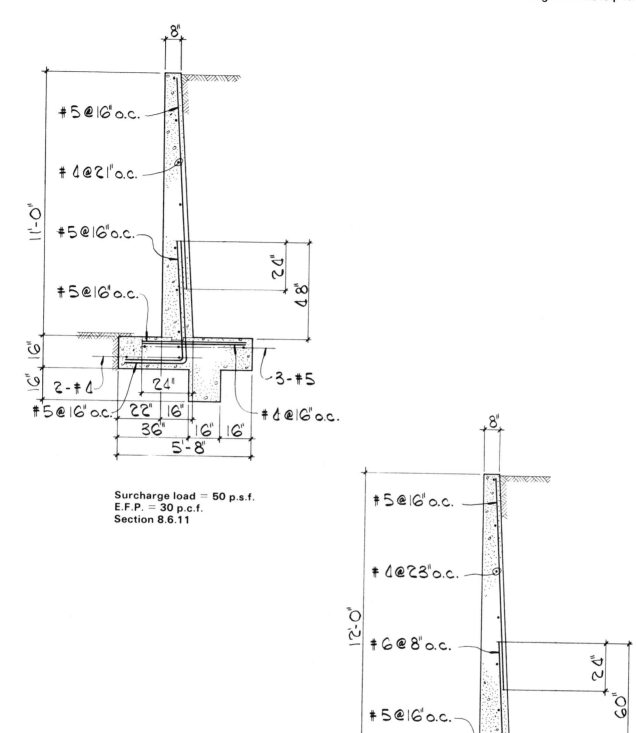

Surcharge load = 50 p.s.f.
E.F.P. = 30 p.c.f.
Section 8.6.11

Surcharge load = 50 p.s.f.
E.F.P. = 30 p.c.f.
Section 8.6.12

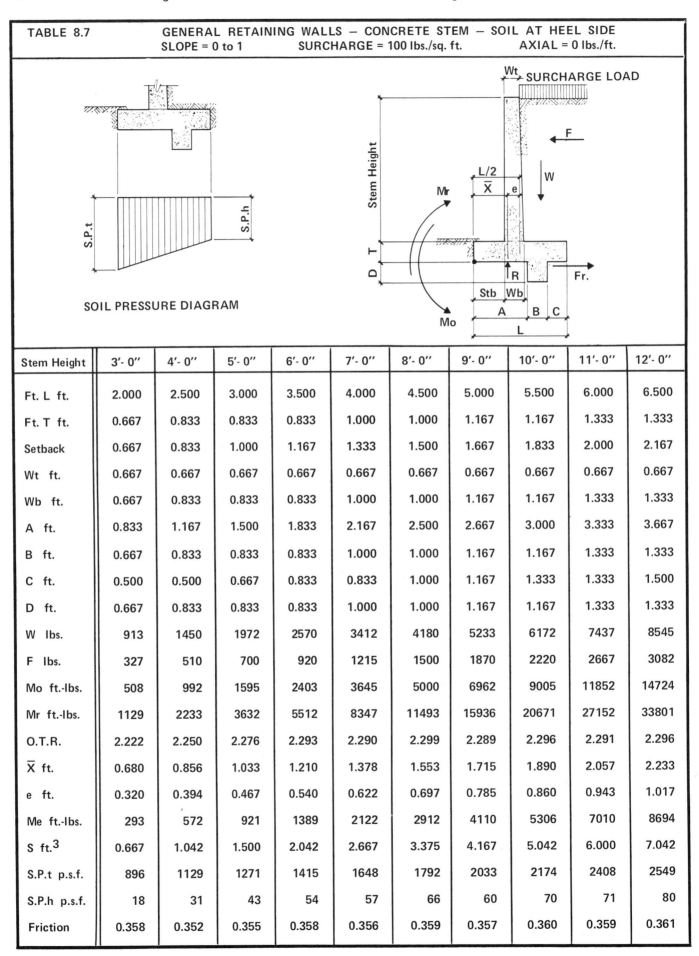

SOIL PRESSURE DIAGRAM

TABLE 8.7	GENERAL RETAINING WALLS – CONCRETE STEM – SOIL AT HEEL SIDE
	SLOPE = 0 to 1 SURCHARGE = 100 lbs./sq. ft. AXIAL = 0 lbs./ft.

Stem Height	3'- 0"	4'- 0"	5'- 0"	6'- 0"	7'- 0"	8'- 0"	9'- 0"	10'- 0"	11'- 0"	12'- 0"
Ft. L ft.	2.000	2.500	3.000	3.500	4.000	4.500	5.000	5.500	6.000	6.500
Ft. T ft.	0.667	0.833	0.833	0.833	1.000	1.000	1.167	1.167	1.333	1.333
Setback	0.667	0.833	1.000	1.167	1.333	1.500	1.667	1.833	2.000	2.167
Wt ft.	0.667	0.667	0.667	0.667	0.667	0.667	0.667	0.667	0.667	0.667
Wb ft.	0.667	0.833	0.833	0.833	1.000	1.000	1.167	1.167	1.333	1.333
A ft.	0.833	1.167	1.500	1.833	2.167	2.500	2.667	3.000	3.333	3.667
B ft.	0.667	0.833	0.833	0.833	1.000	1.000	1.167	1.167	1.333	1.333
C ft.	0.500	0.500	0.667	0.833	0.833	1.000	1.167	1.333	1.333	1.500
D ft.	0.667	0.833	0.833	0.833	1.000	1.000	1.167	1.167	1.333	1.333
W lbs.	913	1450	1972	2570	3412	4180	5233	6172	7437	8545
F lbs.	327	510	700	920	1215	1500	1870	2220	2667	3082
Mo ft.-lbs.	508	992	1595	2403	3645	5000	6962	9005	11852	14724
Mr ft.-lbs.	1129	2233	3632	5512	8347	11493	15936	20671	27152	33801
O.T.R.	2.222	2.250	2.276	2.293	2.290	2.299	2.289	2.296	2.291	2.296
$\overline{X}$ ft.	0.680	0.856	1.033	1.210	1.378	1.553	1.715	1.890	2.057	2.233
e ft.	0.320	0.394	0.467	0.540	0.622	0.697	0.785	0.860	0.943	1.017
Me ft.-lbs.	293	572	921	1389	2122	2912	4110	5306	7010	8694
S ft.^{3}	0.667	1.042	1.500	2.042	2.667	3.375	4.167	5.042	6.000	7.042
S.P.t p.s.f.	896	1129	1271	1415	1648	1792	2033	2174	2408	2549
S.P.h p.s.f.	18	31	43	54	57	66	60	70	71	80
Friction	0.358	0.352	0.355	0.358	0.356	0.359	0.357	0.360	0.359	0.361

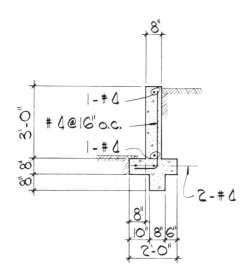

Surcharge load = 100 p.s.f.
E.F.P. = 30 p.c.f.
Section 8.7.3

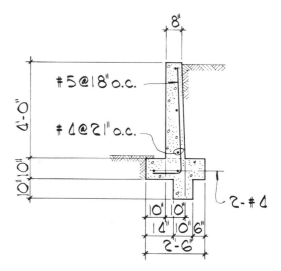

Surcharge load = 100 p.s.f.
E.F.P. = 30 p.c.f.
Section 8.7.4

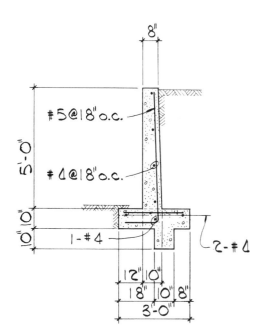

Surcharge load = 100 p.s.f.
E.F.P. = 30 p.c.f.
Section 8.7.5

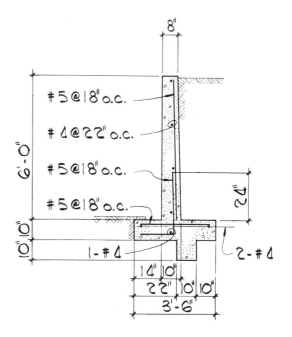

Surcharge load = 100 p.s.f.
E.F.P. = 30 p.c.f.
Section 8.7.6

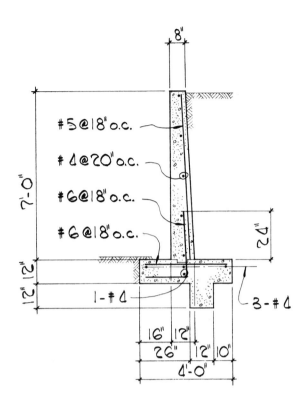

8"

#5@18" o.c.

#4@20" o.c.

#6@18" o.c.

#6@18" o.c.

7'-0"

24"

12" 12"

1-#4

3-#4

16" 12"

26" 12" 10"

4'-0"

Surcharge load = 100 p.s.f.
E.F.P. = 30 p.c.f.
Section 8.7.7

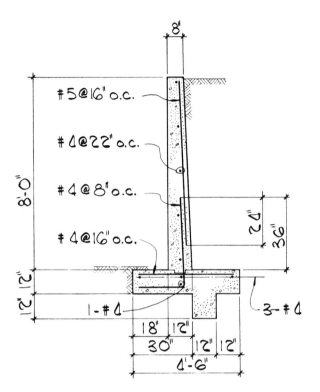

8"

#5@16" o.c.

#4@22" o.c.

#4@8" o.c.

#4@16" o.c.

8'-0"

24"

36"

12" 12"

1-#4

3-#4

18" 12"

30" 12" 12"

4'-6"

Surcharge load = 100 p.s.f.
E.F.P. = 30 p.c.f.
Section 8.7.8

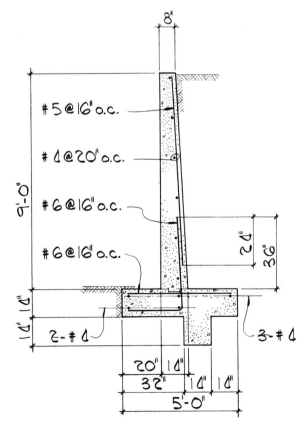

8"

#5@16" o.c.

#4@20" o.c.

#6@16" o.c.

#6@16" o.c.

9'-0"

24"

36"

14" 14"

2-#4

3-#4

20" 14"

32" 14" 14"

5'-0"

Surcharge load = 100 p.s.f.
E.F.P. = 30 p.c.f.
Section 8.7.9

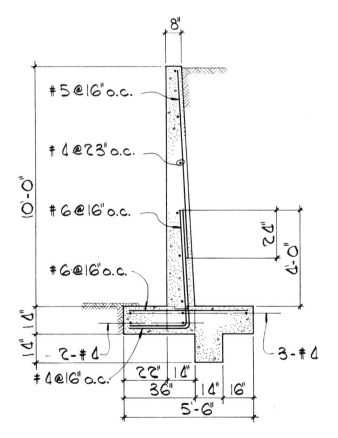

8"

#5@16" o.c.

#4@23" o.c.

#6@16" o.c.

#6@16" o.c.

10'-0"

24"

4'-0"

14" 14"

2-#4

3-#4

#4@16" o.c.

22" 14"

36" 14" 16"

5'-6"

Surcharge load = 100 p.s.f.
E.F.P. = 30 p.c.f.
Section 8.7.10

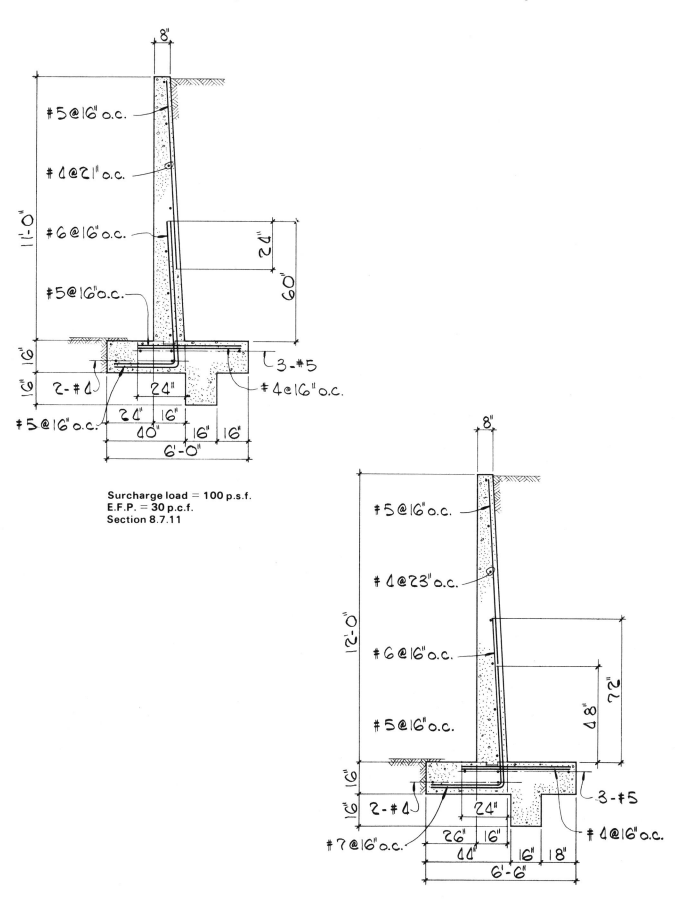

8"

#5@16" o.c.

#4@21" o.c.

#6@16" o.c.

24"

60"

#5@16" o.c.

1'-0"

6"

3-#5

6"

#4@16" o.c.

2-#4

24"

24"   16"

#5@16" o.c.

40"

16"   16"

6'-0"

Surcharge load = 100 p.s.f.
E.F.P. = 30 p.c.f.
Section 8.7.11

8"

#5@16" o.c.

#4@23" o.c.

#6@16" o.c.

72"

48"

#5@16" o.c.

12'-0"

6"

6"

2-#4

24"

3-#5

#4@16" o.c.

#7@16" o.c.

26"   16"

44"

16"   18"

6'-6"

Surcharge load = 100 p.s.f.
E.F.P. = 30 p.c.f.
Section 8.7.12

TABLE 8.8	GENERAL RETAINING WALLS – CONCRETE STEM – SOIL AT HEEL SIDE
	SLOPE = 0 to 1 SURCHARGE = 200 lbs./sq. ft. AXIAL = 0 lbs./ft.

SOIL PRESSURE DIAGRAM

Stem Height	3'- 0"	4'- 0"	5'- 0"	6'- 0"	7'- 0"	8'- 0"	9'- 0"	10'- 0"	11'- 0"	12'- 0"
Ft. L ft.	2.667	3.000	3.500	4.000	4.500	5.000	5.500	6.000	6.500	7.000
Ft. T ft.	0.833	0.833	0.833	1.000	1.000	1.167	1.167	1.333	1.333	1.333
Setback	0.833	1.000	1.167	1.333	1.500	1.667	1.833	2.000	2.167	2.333
Wt ft.	0.667	0.667	0.667	0.667	0.667	0.667	0.667	0.667	0.667	0.667
Wb ft.	0.833	0.833	0.833	1.000	1.000	1.167	1.167	1.333	1.333	1.333
A ft.	1.333	1.500	1.833	2.167	2.500	2.667	3.000	3.333	3.667	4.000
B ft.	0.833	0.833	0.833	1.000	1.000	1.167	1.167	1.333	1.333	1.333
C ft.	0.500	0.667	0.833	0.833	1.000	1.167	1.333	1.333	1.500	1.667
D ft.	0.833	0.833	0.833	1.000	1.000	1.167	1.167	1.333	1.333	1.333
W lbs.	1442	1859	2457	3287	4055	5096	6034	7287	8395	9580
F lbs.	510	700	920	1215	1500	1870	2220	2667	3082	3527
Mo ft.-lbs.	992	1595	2403	3645	5000	6962	9005	11852	14724	18025
Mr ft.-lbs.	2384	3477	5339	8128	11253	15643	20356	26777	33401	41033
O.T.R.	2.402	2.179	2.221	2.230	2.251	2.247	2.261	2.259	2.269	2.276
$\overline{X}$ ft.	0.965	1.012	1.194	1.364	1.542	1.704	1.881	2.048	2.225	2.402
e ft.	0.368	0.488	0.556	0.636	0.708	0.796	0.869	0.952	1.025	1.098
Me ft.-lbs.	531	907	1365	2090	2870	4058	5243	6935	8606	10522
S ft.^{3}	1.185	1.500	2.042	2.667	3.375	4.167	5.042	6.000	7.042	8.167
S.P.t p.s.f.	989	1224	1371	1606	1752	1993	2137	2370	2514	2657
S.P.h p.s.f.	92	15	33	38	51	45	57	59	69	80
Friction	0.354	0.377	0.375	0.370	0.370	0.367	0.368	0.366	0.367	0.368

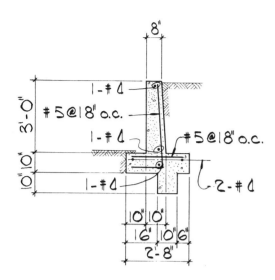

Surcharge load = 200 p.s.f.
E.F.P. = 30 p.c.f.
Section 8.8.3

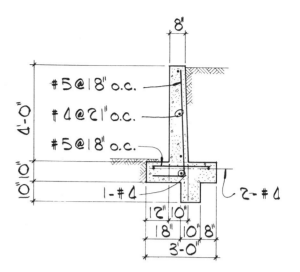

Surcharge load = 200 p.s.f.
E.F.P. = 30 p.c.f.
Section 8.8.4

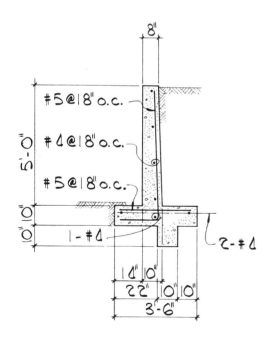

Surcharge load = 200 p.s.f.
E.F.P. = 30 p.c.f.
Section 8.8.5

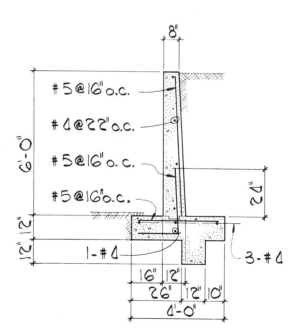

Surcharge load = 200 p.s.f.
E.F.P. = 30 p.c.f.
Section 8.8.6

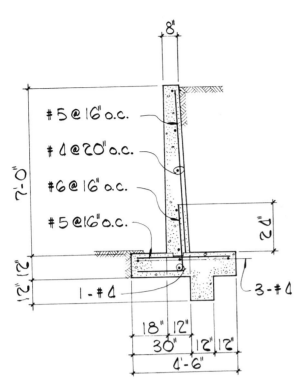

Surcharge load = 200 p.s.f.
E.F.P. = 30 p.c.f.
Section 8.8.7

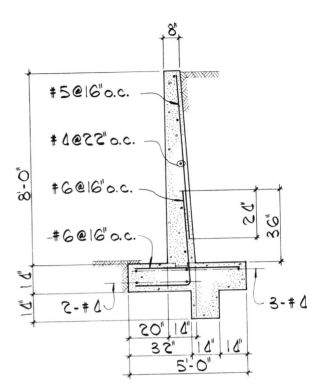

Surcharge load = 200 p.s.f.
E.F.P. = 30 p.c.f.
Section 8.8.8

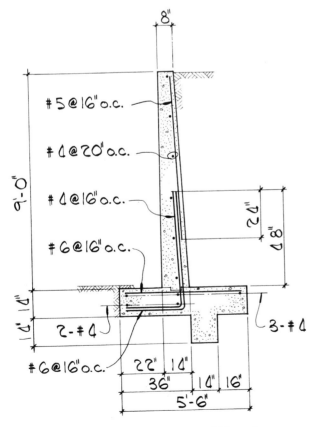

Surcharge load = 200 p.s.f.
E.F.P. = 30 p.c.f.
Section 8.8.9

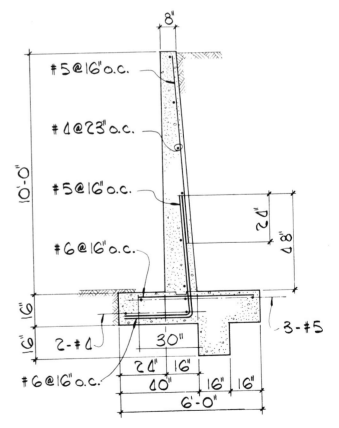

Surcharge load = 200 p.s.f.
E.F.P. = 30 p.c.f.
Section 8.8.10

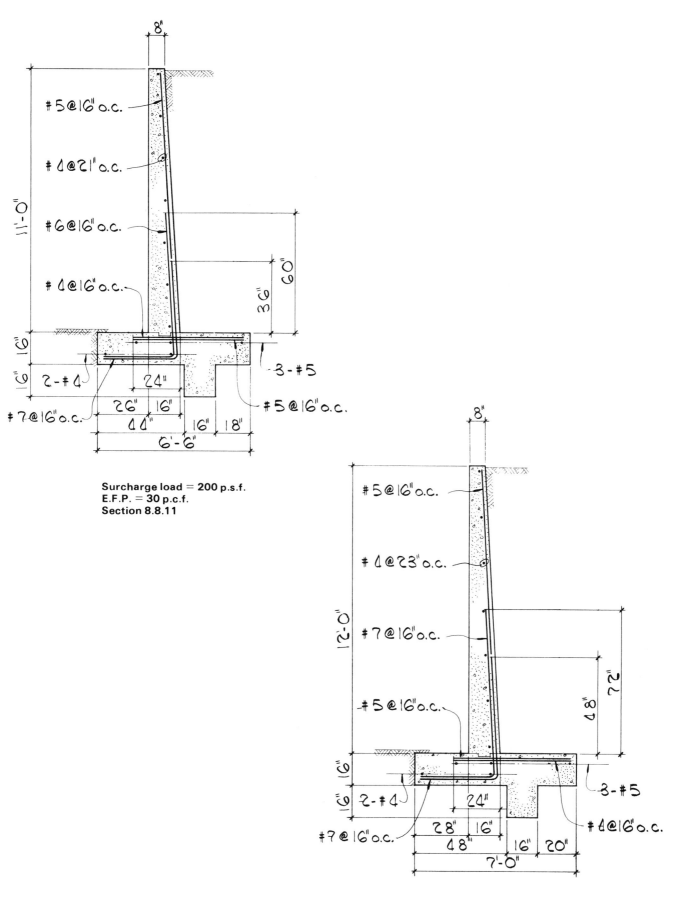

Surcharge load = 200 p.s.f.
E.F.P. = 30 p.c.f.
Section 8.8.11

Surcharge load = 200 p.s.f.
E.F.P. = 30 p.c.f.
Section 8.8.12

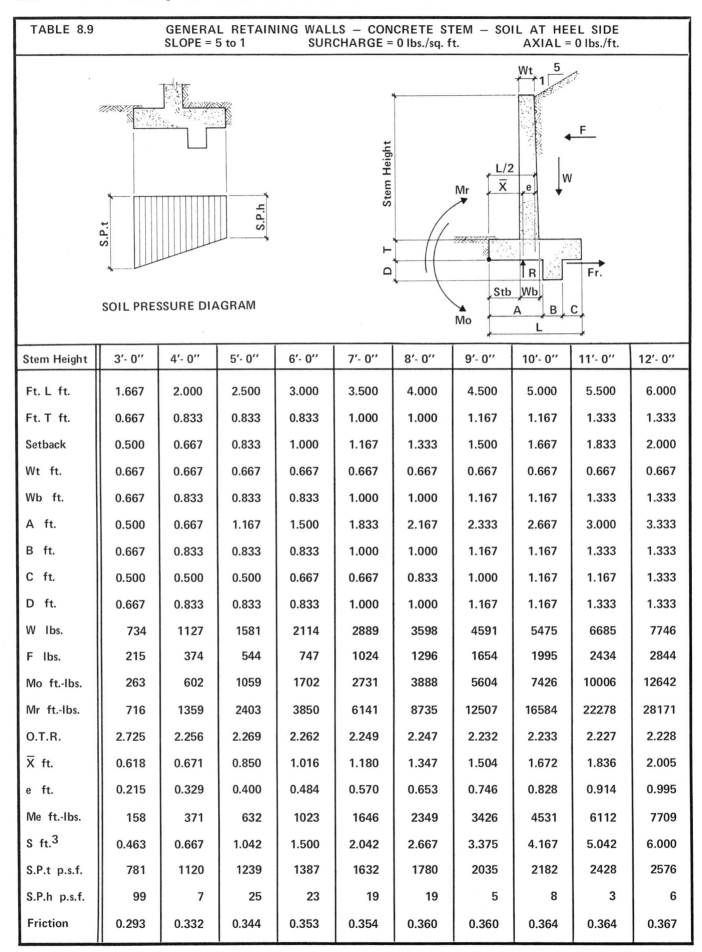

SOIL PRESSURE DIAGRAM

Stem Height	3'- 0"	4'- 0"	5'- 0"	6'- 0"	7'- 0"	8'- 0"	9'- 0"	10'- 0"	11'- 0"	12'- 0"
Ft. L ft.	1.667	2.000	2.500	3.000	3.500	4.000	4.500	5.000	5.500	6.000
Ft. T ft.	0.667	0.833	0.833	0.833	1.000	1.000	1.167	1.167	1.333	1.333
Setback	0.500	0.667	0.833	1.000	1.167	1.333	1.500	1.667	1.833	2.000
Wt ft.	0.667	0.667	0.667	0.667	0.667	0.667	0.667	0.667	0.667	0.667
Wb ft.	0.667	0.833	0.833	0.833	1.000	1.000	1.167	1.167	1.333	1.333
A ft.	0.500	0.667	1.167	1.500	1.833	2.167	2.333	2.667	3.000	3.333
B ft.	0.667	0.833	0.833	0.833	1.000	1.000	1.167	1.167	1.333	1.333
C ft.	0.500	0.500	0.500	0.667	0.667	0.833	1.000	1.167	1.167	1.333
D ft.	0.667	0.833	0.833	0.833	1.000	1.000	1.167	1.167	1.333	1.333
W lbs.	734	1127	1581	2114	2889	3598	4591	5475	6685	7746
F lbs.	215	374	544	747	1024	1296	1654	1995	2434	2844
Mo ft.-lbs.	263	602	1059	1702	2731	3888	5604	7426	10006	12642
Mr ft.-lbs.	716	1359	2403	3850	6141	8735	12507	16584	22278	28171
O.T.R.	2.725	2.256	2.269	2.262	2.249	2.247	2.232	2.233	2.227	2.228
$\overline{X}$ ft.	0.618	0.671	0.850	1.016	1.180	1.347	1.504	1.672	1.836	2.005
e ft.	0.215	0.329	0.400	0.484	0.570	0.653	0.746	0.828	0.914	0.995
Me ft.-lbs.	158	371	632	1023	1646	2349	3426	4531	6112	7709
S ft.^{3}	0.463	0.667	1.042	1.500	2.042	2.667	3.375	4.167	5.042	6.000
S.P.t p.s.f.	781	1120	1239	1387	1632	1780	2035	2182	2428	2576
S.P.h p.s.f.	99	7	25	23	19	19	5	8	3	6
Friction	0.293	0.332	0.344	0.353	0.354	0.360	0.360	0.364	0.364	0.367

TABLE 8.9 — GENERAL RETAINING WALLS – CONCRETE STEM – SOIL AT HEEL SIDE
SLOPE = 5 to 1   SURCHARGE = 0 lbs./sq. ft.   AXIAL = 0 lbs./ft.

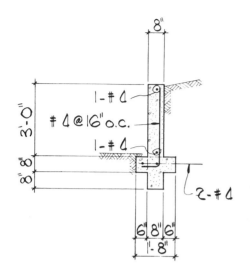

Soil slope = 5 to 1
E.F.P. = 32 p.c.f.
Section 8.9.3

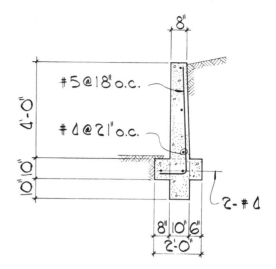

Soil slope = 5 to 1
E.F.P. = 32 p.c.f.
Section 8.9.4

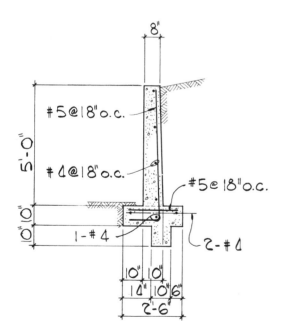

Soil slope = 5 to 1
E.F.P. = 32 p.c.f.
Section 8.9.5

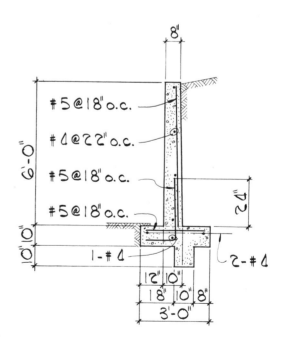

Soil slope = 5 to 1
E.F.P. = 32 p.c.f.
Section 8.9.6

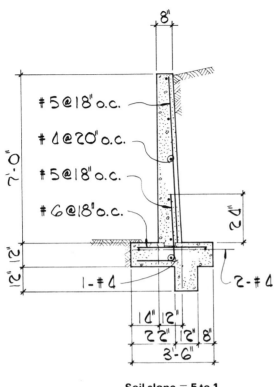

8"

#5@18" o.c.

#4@20" o.c.

#5@18" o.c.

#6@18" o.c.

7'-0"

24"

2"

1-#4

2-#4

2" 2"

10" 12"

22" 12" 2" 8"

3'-6"

Soil slope = 5 to 1
E.F.P. = 32 p.c.f.
Section 8.9.7

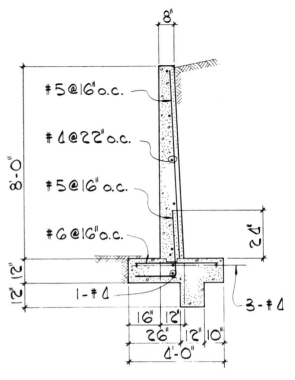

8"

#5@16" o.c.

#4@22" o.c.

#5@16" o.c.

#6@16" o.c.

8'-0"

24"

2" 2"

1-#4

3-#4

16" 12"

26" 12" 10"

4'-0"

Soil slope = 5 to 1
E.F.P. = 32 p.c.f.
Section 8.9.8

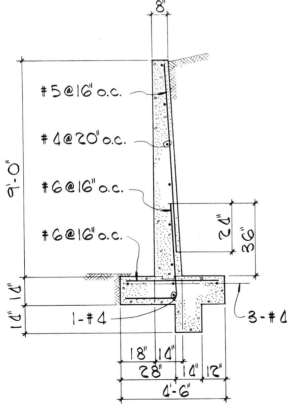

8"

#5@16" o.c.

#4@20" o.c.

#6@16" o.c.

#6@16" o.c.

9'-0"

24"

36"

1" 1"

1-#4

3-#4

18" 1'-0"

28" 1'-0" 12"

4'-6"

Soil slope = 5 to 1
E.F.P. = 32 p.c.f.
Section 8.9.9

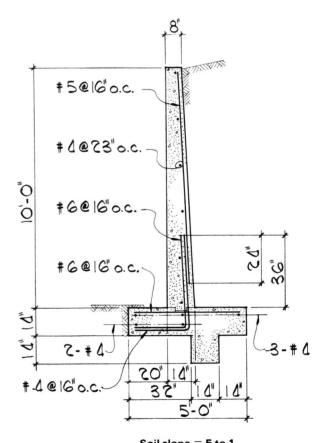

8"

#5@16" o.c.

#4@23" o.c.

#6@16" o.c.

#6@16" o.c.

10'-0"

24"

36"

1" 1"

2-#4

3-#4

#4@16" o.c.

20" 1'-0"

32" 1'-0" 1'-0"

5'-0"

Soil slope = 5 to 1
E.F.P. = 32 p.c.f.
Section 8.9.10

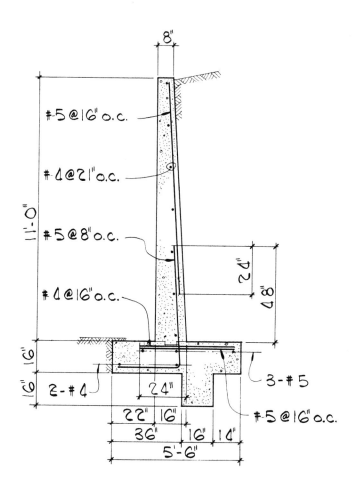

Soil slope = **5 to 1**
E.F.P. = **32 p.c.f.**
Section **8.9.11**

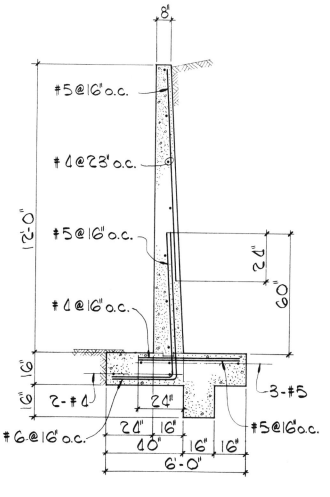

Soil slope = **5 to 1**
E.F.P. = **32 p.c.f.**
Section **8.9.12**

TABLE 8.10	GENERAL RETAINING WALLS — CONCRETE STEM — SOIL AT HEEL SIDE
	SLOPE = 4 to 1 SURCHARGE = 0 lbs./sq. ft. AXIAL = 0 lbs./ft.

SOIL PRESSURE DIAGRAM

Stem Height	3'- 0''	4'- 0''	5'- 0''	6'- 0''	7'- 0''	8'- 0''	9'- 0''	10'- 0''	11'- 0''	12'- 0''
Ft. L ft.	1.667	2.167	2.667	3.167	3.667	4.333	4.833	5.333	5.833	6.333
Ft. T ft.	0.667	0.833	0.833	0.833	1.000	1.000	1.167	1.167	1.333	1.333
Setback	0.500	0.667	0.833	1.000	1.167	1.500	1.667	1.833	2.000	2.167
Wt ft.	0.667	0.667	0.667	0.667	0.667	0.667	0.667	0.667	0.667	0.667
Wb ft.	0.667	0.833	0.833	0.833	1.000	1.000	1.167	1.167	1.333	1.333
A ft.	0.500	0.833	1.333	1.667	1.833	2.333	2.500	2.833	3.167	3.500
B ft.	0.667	0.833	0.833	0.833	1.000	1.000	1.167	1.167	1.333	1.333
C ft.	0.500	0.500	0.500	0.667	0.833	1.000	1.167	1.333	1.333	1.500
D ft.	0.667	0.833	0.833	0.833	1.000	1.000	1.167	1.167	1.333	1.333
W lbs.	739	1227	1705	2263	3070	3832	4863	5780	7031	8126
F lbs.	235	409	595	817	1120	1417	1809	2182	2662	3111
Mo ft.-lbs.	288	659	1158	1861	2987	4252	6130	8122	10944	13827
Mr ft.-lbs.	723	1578	2725	4298	6740	10150	14323	18788	24986	31360
O.T.R.	2.513	2.395	2.353	2.309	2.257	2.387	2.337	2.313	2.283	2.268
$\overline{X}$ ft.	0.589	0.749	0.919	1.077	1.223	1.539	1.685	1.845	1.997	2.157
e ft.	0.245	0.334	0.414	0.507	0.611	0.628	0.732	0.821	0.919	1.009
Me ft.-lbs.	181	410	706	1146	1874	2405	3559	4746	6465	8201
S ft.[3]	0.463	0.782	1.185	1.671	2.241	3.130	3.894	4.741	5.671	6.685
S.P.t p.s.f.	834	1091	1235	1401	1674	1653	1920	2085	2345	2510
S.P.h p.s.f.	53	42	43	29	1	116	92	82	65	56
Friction	0.318	0.333	0.349	0.361	0.365	0.370	0.372	0.378	0.379	0.383

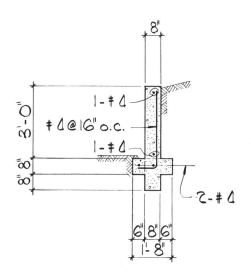

Soil slope = **4 to 1**
E.F.P. = **35 p.c.f.**
Section **8.10.3**

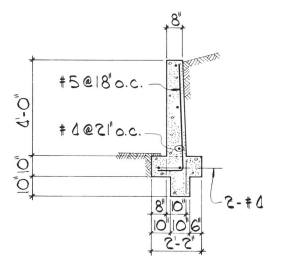

Soil slope = **4 to 1**
E.F.P. = **35 p.c.f.**
Section **8.10.4**

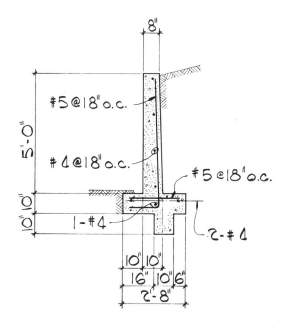

Soil slope = **4 to 1**
E.F.P. = **35 p.c.f.**
Section **8.10.5**

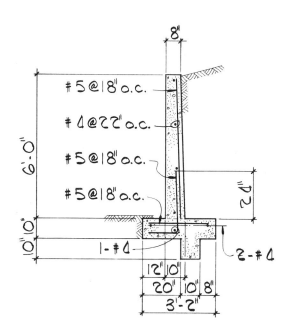

Soil slope = **4 to 1**
E.F.P. = **35 p.c.f.**
Section **8.10.6**

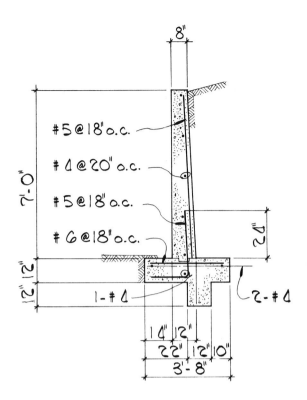

Soil slope = 4 to 1
E.F.P. = 35 p.c.f.
Section 8.10.7

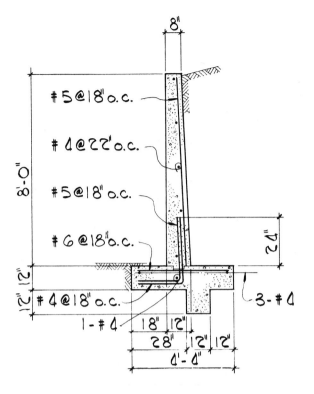

Soil slope = 4 to 1
E.F.P. = 35 p.c.f.
Section 8.10.8

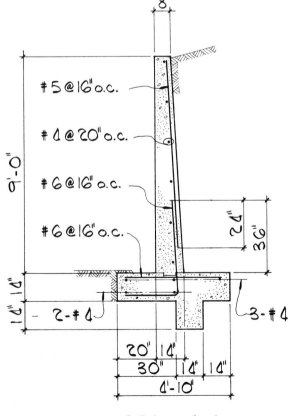

Soil slope = 4 to 1
E.F.P. = 35 p.c.f.
Section 8.10.9

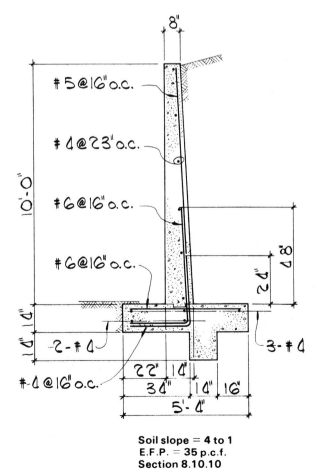

Soil slope = 4 to 1
E.F.P. = 35 p.c.f.
Section 8.10.10

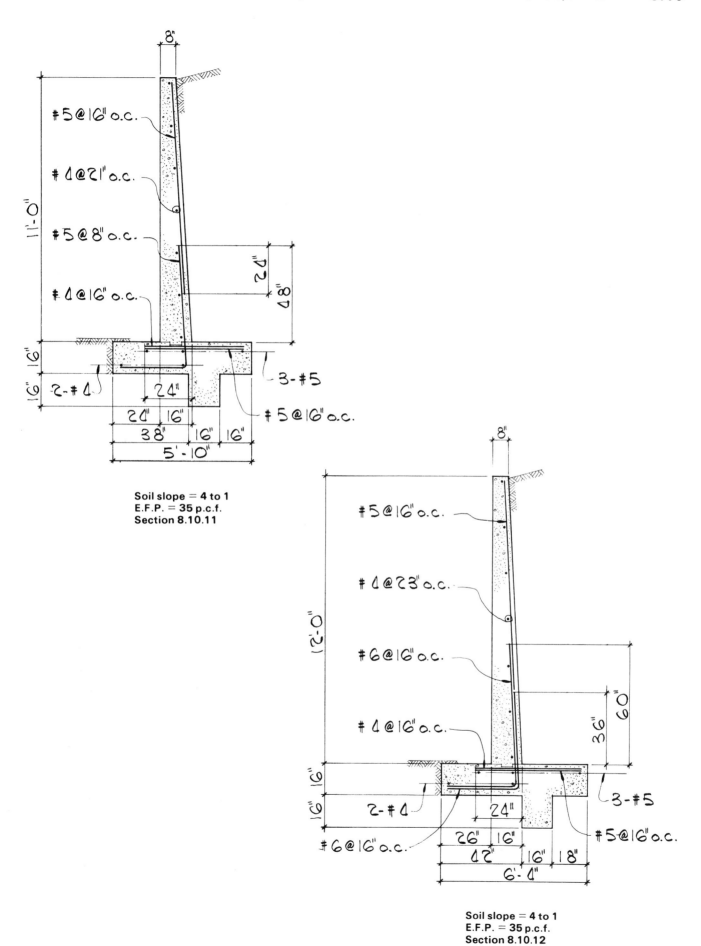

Soil slope = 4 to 1
E.F.P. = 35 p.c.f.
Section 8.10.11

Soil slope = 4 to 1
E.F.P. = 35 p.c.f.
Section 8.10.12

TABLE 8.11	GENERAL RETAINING WALLS − CONCRETE STEM − SOIL AT HEEL SIDE
	SLOPE = 3 to 1 SURCHARGE = 0 lbs./sq. ft. AXIAL = 0 lbs./ft.

SOIL PRESSURE DIAGRAM

Stem Height	3'- 0"	4'- 0"	5'- 0"	6'- 0"	7'- 0"	8'- 0"	9'- 0"	10'- 0"	11'- 0"	12'- 0"
Ft. L ft.	1.667	2.333	2.833	3.333	3.833	4.333	5.000	5.500	6.000	6.500
Ft. T ft.	0.667	0.833	0.833	0.833	1.000	1.000	1.167	1.167	1.333	1.333
Setback	0.500	0.833	1.000	1.167	1.333	1.500	1.667	1.833	2.000	2.167
Wt ft.	0.667	0.667	0.667	0.667	0.667	0.667	0.667	0.667	0.667	0.667
Wb ft.	0.667	0.833	0.833	0.833	1.000	1.000	1.167	1.167	1.333	1.333
A ft.	0.500	1.000	1.333	1.667	2.000	2.333	2.667	3.000	3.333	3.667
B ft.	0.667	0.833	0.833	0.833	1.000	1.000	1.167	1.167	1.333	1.333
C ft.	0.500	0.500	0.667	0.833	0.833	1.000	1.167	1.333	1.333	1.500
D ft.	0.667	0.833	0.833	0.833	1.000	1.000	1.167	1.167	1.333	1.333
W lbs.	744	1259	1744	2311	3133	3884	5123	6075	7368	8503
F lbs.	255	444	647	887	1216	1539	1964	2369	2890	3378
Mo ft.-lbs.	312	715	1257	2021	3243	4617	6655	8819	11882	15012
Mr ft.-lbs.	729	1802	3028	4719	7349	10290	15502	20234	26774	33485
O.T.R.	2.336	2.520	2.409	2.335	2.266	2.229	2.329	2.294	2.253	2.230
$\overline{X}$ ft.	0.560	0.864	1.016	1.167	1.311	1.461	1.727	1.879	2.021	2.173
e ft.	0.273	0.303	0.401	0.499	0.606	0.706	0.773	0.871	0.979	1.077
Me ft.-lbs.	203	381	699	1154	1899	2741	3961	5291	7212	9161
S ft.^{3}	0.463	0.907	1.338	1.852	2.449	3.130	4.167	5.042	6.000	7.042
S.P.t p.s.f.	886	960	1138	1317	1593	1772	1975	2154	2430	2609
S.P.h p.s.f.	8	119	93	70	42	20	74	55	26	7
Friction	0.343	0.353	0.371	0.384	0.388	0.396	0.383	0.390	0.392	0.397

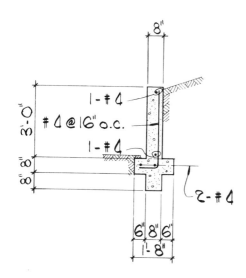

Soil slope = 3 to 1
E.F.P. = 38 p.c.f.
Section 8.11.3

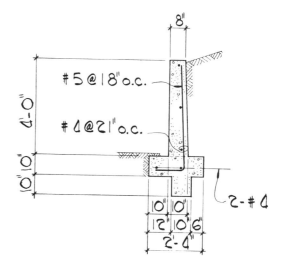

Soil slope = 3 to 1
E.F.P. = 38 p.c.f.
Section 8.11.4

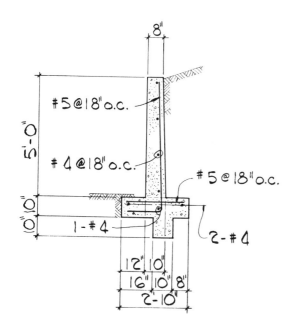

Soil slope = 3 to 1
E.F.P. = 38 p.c.f.
Section 8.11.5

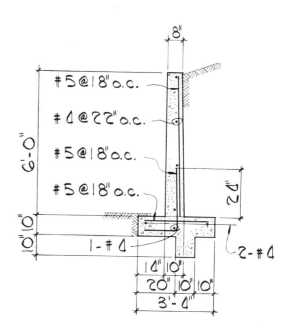

Soil slope = 3 to 1
E.F.P. = 38 p.c.f.
Section 8.11.6

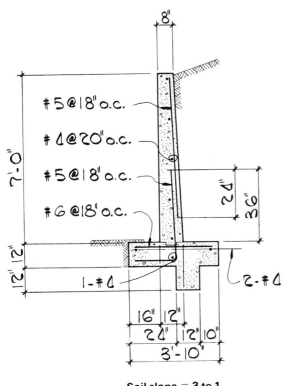

#5 @ 18" o.c.

#4 @ 20" o.c.

#5 @ 18" o.c.

#6 @ 18" o.c.

7'-0"

24"

36"

1 - #4

2 - #4

12" 12"

16" 12"

24"

12" 10"

3'-10"

Soil slope = 3 to 1
E.F.P. = 38 p.c.f.
Section 8.11.7

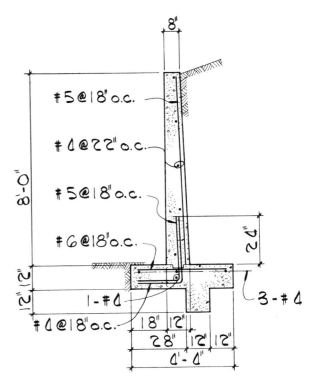

#5 @ 18" o.c.

#4 @ 22" o.c.

#5 @ 18" o.c.

#6 @ 18" o.c.

8'-0"

24"

1 - #4

3 - #4

#4 @ 18" o.c.

12" 12"

18" 12"

28"

12" 12"

4'-4"

Soil slope = 3 to 1
E.F.P. = 38 p.c.f.
Section 8.11.8

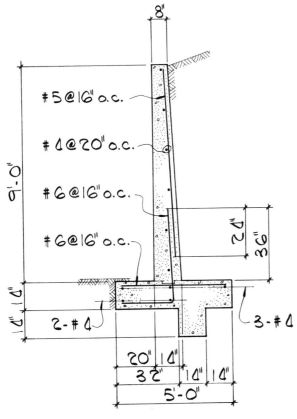

#5 @ 16" o.c.

#4 @ 20" o.c.

#6 @ 16" o.c.

#6 @ 16" o.c.

9'-0"

24"

36"

2 - #4

3 - #4

12" 12"

20" 12"

32"

12" 12"

5'-0"

Soil slope = 3 to 1
E.F.P. = 38 p.c.f.
Section 8.11.9

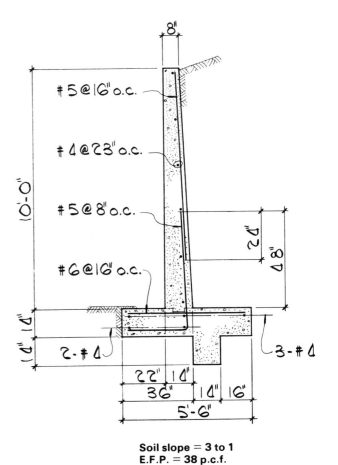

#5 @ 16" o.c.

#4 @ 23" o.c.

#5 @ 8" o.c.

#6 @ 16" o.c.

10'-0"

24"

48"

2 - #4

3 - #4

12" 12"

22" 12"

36"

12" 16"

5'-6"

Soil slope = 3 to 1
E.F.P. = 38 p.c.f.
Section 8.11.10

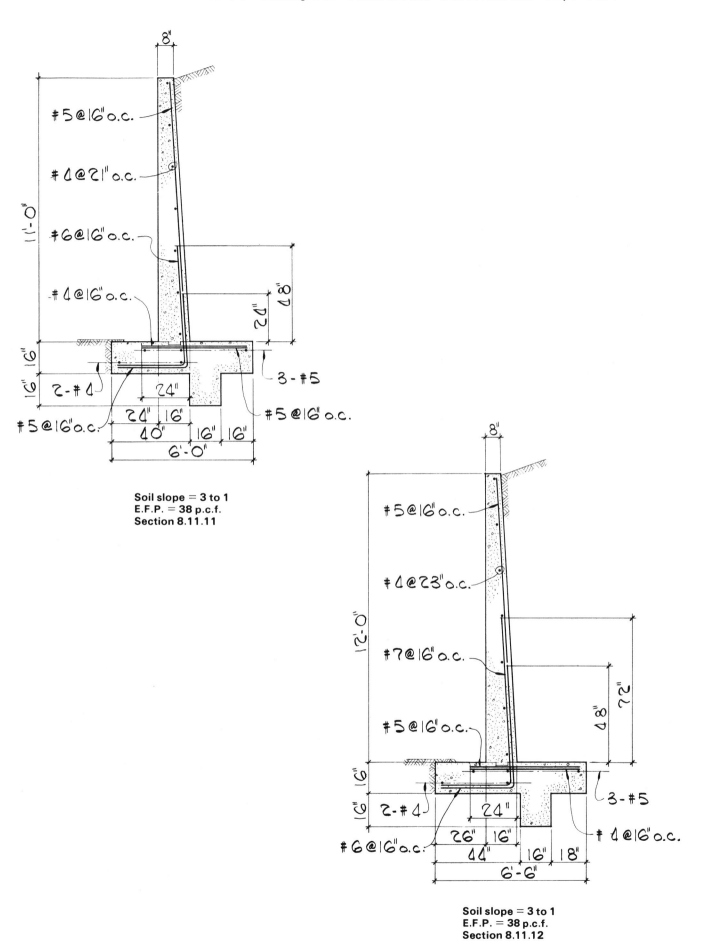

#5 @ 16" o.c.

#4 @ 21" o.c.

#6 @ 16" o.c.

#4 @ 16" o.c.

11'-0"

6"

6"

2 - #4

#5 @ 16" o.c.

24"

3 - #5

#5 @ 16" o.c.

8"

22"

8"

24"

40"   16"   16"

6'-0"

Soil slope = 3 to 1
E.F.P. = 38 p.c.f.
Section 8.11.11

8"

#5 @ 16" o.c.

#4 @ 23" o.c.

#7 @ 16" o.c.

#5 @ 16" o.c.

12'-0"

6"

6"

2 - #4

#6 @ 16" o.c.

24"

3 - #5

#4 @ 16" o.c.

72"

48"

26"   16"

44"   16"   18"

6'-6"

Soil slope = 3 to 1
E.F.P. = 38 p.c.f.
Section 8.11.12

TABLE 8.12	GENERAL RETAINING WALLS — CONCRETE STEM — SOIL AT HEEL SIDE
	SLOPE = 2 to 1 SURCHARGE = 0 lbs./sq. ft. AXIAL = 0 lbs./ft.

SOIL PRESSURE DIAGRAM

Stem Height	3'- 0"	4'- 0"	5'- 0"	6'- 0"	7'- 0"	8'- 0"	9'- 0"	10'- 0"	11'- 0"	12'- 0"
Ft. L ft.	1.833	2.333	3.000	3.500	4.000	4.667	5.167	5.833	6.333	6.833
Ft. T ft.	0.667	0.833	0.833	0.833	1.000	1.167	1.167	1.333	1.333	1.333
Setback	0.667	0.833	1.000	1.167	1.333	1.500	1.667	2.000	2.167	2.333
Wt ft.	0.667	0.667	0.667	0.667	0.667	0.667	0.667	0.667	0.667	0.667
Wb ft.	0.667	0.833	0.833	0.833	1.000	1.167	1.167	1.333	1.333	1.333
A ft.	0.667	1.000	1.500	1.833	2.167	2.500	2.833	3.167	3.500	3.833
B ft.	0.667	0.833	0.833	0.833	1.000	1.167	1.167	1.333	1.333	1.333
C ft.	0.500	0.500	0.667	0.833	0.833	1.000	1.167	1.333	1.500	1.667
D ft.	0.667	0.833	0.833	0.833	1.000	1.167	1.167	1.333	1.333	1.333
W lbs.	771	1278	1890	2494	3359	4536	5452	6734	7840	9033
F lbs.	289	502	732	1004	1376	1870	2222	2762	3270	3822
Mo ft.-lbs.	353	809	1423	2287	3669	5520	7531	10433	13445	16988
Mr ft.-lbs.	868	1835	3439	5296	8157	12726	16950	23946	30269	37627
O.T.R.	2.457	2.267	2.418	2.316	2.223	2.305	2.251	2.295	2.251	2.215
$\overline{X}$ ft.	0.668	0.802	1.067	1.207	1.336	1.589	1.728	2.007	2.146	2.285
e ft.	0.249	0.364	0.433	0.543	0.664	0.745	0.856	0.910	1.021	1.132
Me ft.-lbs.	192	465	819	1355	2231	3378	4666	6127	8003	10222
S ft.^{3}	0.560	0.907	1.500	2.042	2.667	3.630	4.449	5.671	6.685	7.782
S.P.t p.s.f.	763	1061	1176	1376	1676	1902	2104	2235	2435	2635
S.P.h p.s.f.	78	35	84	49	3	41	7	74	41	8
Friction	0.375	0.393	0.387	0.403	0.410	0.398	0.408	0.410	0.417	0.423

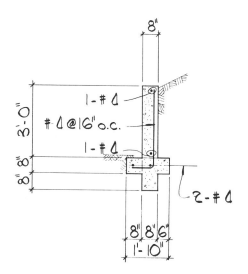

Soil slope = **2 to 1**
E.F.P. = **43** p.c.f.
Section **8.12.3**

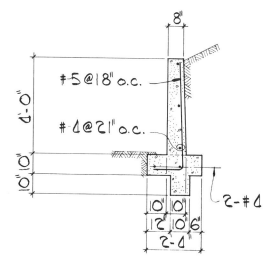

Soil slope = **2 to 1**
E.F.P. = **43** p.c.f.
Section **8.12.4**

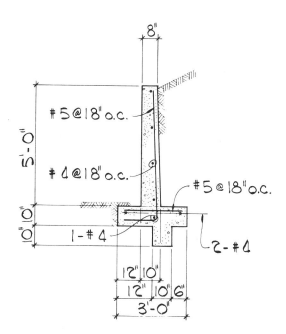

Soil slope = **2 to 1**
E.F.P. = **43** p.c.f.
Section **8.12.5**

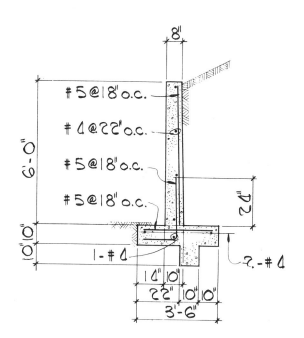

Soil slope = **2 to 1**
E.F.P. = **43** p.c.f.
Section **8.12.6**

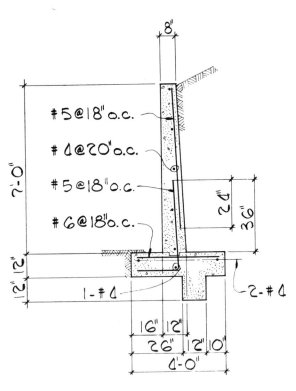

Soil slope = 2 to 1
E.F.P. = 43 p.c.f.
Section 8.12.7

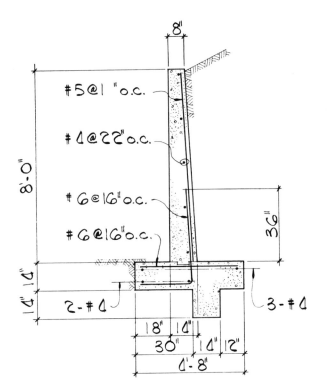

Soil slope = 2 to 1
E.F.P. = 43 p.c.f.
Section 8.12.8

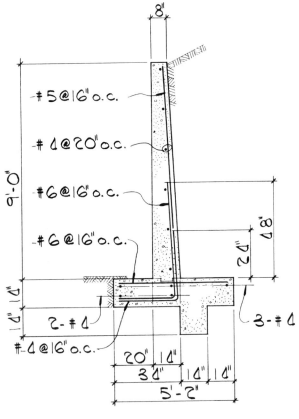

Soil slope = 2 to 1
E.F.P. = 43 p.c.f.
Section 8.12.9

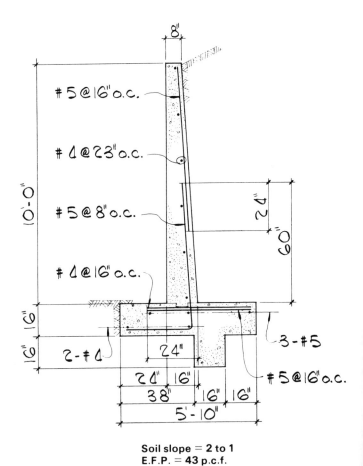

Soil slope = 2 to 1
E.F.P. = 43 p.c.f.
Section 8.12.10

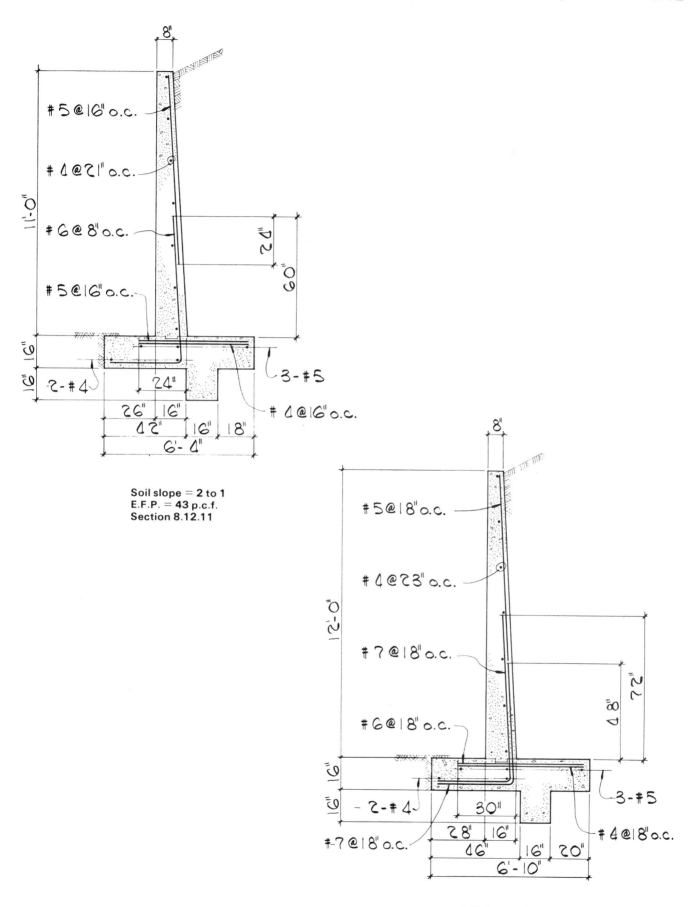

Soil slope = **2 to 1**
E.F.P. = **43** p.c.f.
Section 8.12.11

Soil slope = **2 to 1**
E.F.P. = **43** p.c.f.
Section 8.12.12

TABLE 8.13	GENERAL RETAINING WALLS — CONCRETE STEM — SOIL AT HEEL SIDE
	SLOPE = 1.5 to 1 SURCHARGE = 0 lbs./sq. ft. AXIAL = 0 lbs./ft.

SOIL PRESSURE DIAGRAM

Stem Height	3'- 0"	4'- 0"	5'- 0"	6'- 0"	7'- 0"	8'- 0"	9'- 0"	10'- 0"	11'- 0"	12'- 0"
Ft. L ft.	2.000	2.667	3.333	4.000	4.500	5.167	5.833	6.500	7.167	7.833
Ft. T ft.	0.667	0.833	0.833	1.000	1.000	1.167	1.333	1.333	1.333	1.500
Setback	0.667	0.833	1.167	1.333	1.500	1.667	2.000	2.167	2.333	2.667
Wt ft.	0.667	0.667	0.667	0.667	0.667	0.667	0.667	0.667	0.667	0.667
Wb ft.	0.667	0.833	0.833	1.000	1.000	1.167	1.333	1.333	1.333	1.500
A ft.	0.833	1.333	1.667	2.167	2.500	2.833	3.167	3.667	4.000	4.333
B ft.	0.667	0.833	0.833	1.000	1.000	1.167	1.333	1.333	1.333	1.500
C ft.	0.500	0.500	0.833	0.833	1.000	1.167	1.333	1.500	1.833	2.000
D ft.	0.667	0.833	0.833	1.000	1.000	1.167	1.333	1.333	1.333	1.500
W lbs.	864	1513	2096	3063	3847	5129	6410	7765	9255	10945
F lbs.	370	642	936	1347	1760	2311	2936	3532	4183	5012
Mo ft.-lbs.	452	1035	1820	3144	4693	7061	10114	13344	17197	22553
Mr ft.-lbs.	1046	2420	4274	7430	10496	15932	22766	30560	39928	52114
O.T.R.	2.314	2.338	2.349	2.363	2.236	2.256	2.251	2.290	2.322	2.311
$\overline{X}$ ft.	0.687	0.915	1.171	1.399	1.508	1.729	1.974	2.217	2.456	2.701
e ft.	0.313	0.418	0.495	0.601	0.742	0.854	0.943	1.033	1.127	1.216
Me ft.-lbs.	270	632	1038	1841	2854	4380	6044	8019	10431	13308
S ft.^{3}	0.667	1.185	1.852	2.667	3.375	4.449	5.671	7.042	8.560	10.227
S.P.t p.s.f.	837	1101	1190	1456	1701	1977	2165	2333	2510	2698
S.P.h p.s.f.	27	34	68	76	9	8	33	56	73	96
Friction	0.428	0.425	0.446	0.440	0.457	0.451	0.458	0.455	0.452	0.458

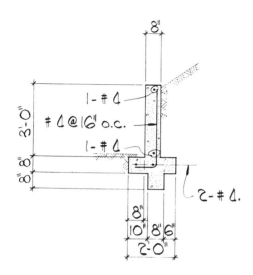

Soil slope = 1½ to 1
E.F.P. = 55 p.c.f.
Section 8.13.3

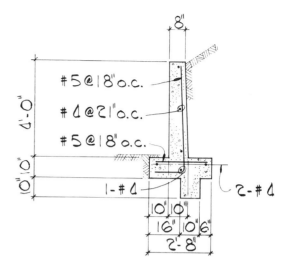

Soil slope = 1½ to 1
E.F.P. = 55 p.c.f.
Section 8.13.4

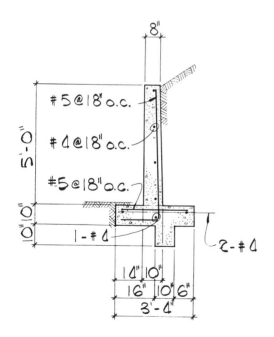

Soil slope = 1½ to 1
E.F.P. = 55 p.c.f.
Section 8.13.5

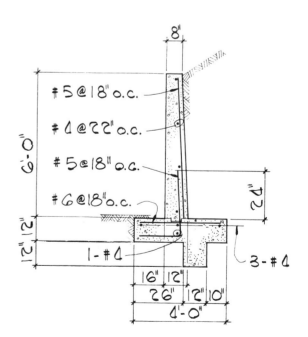

Soil slope = 1½ to 1
E.F.P. = 55 p.c.f.
Section 8.13.6

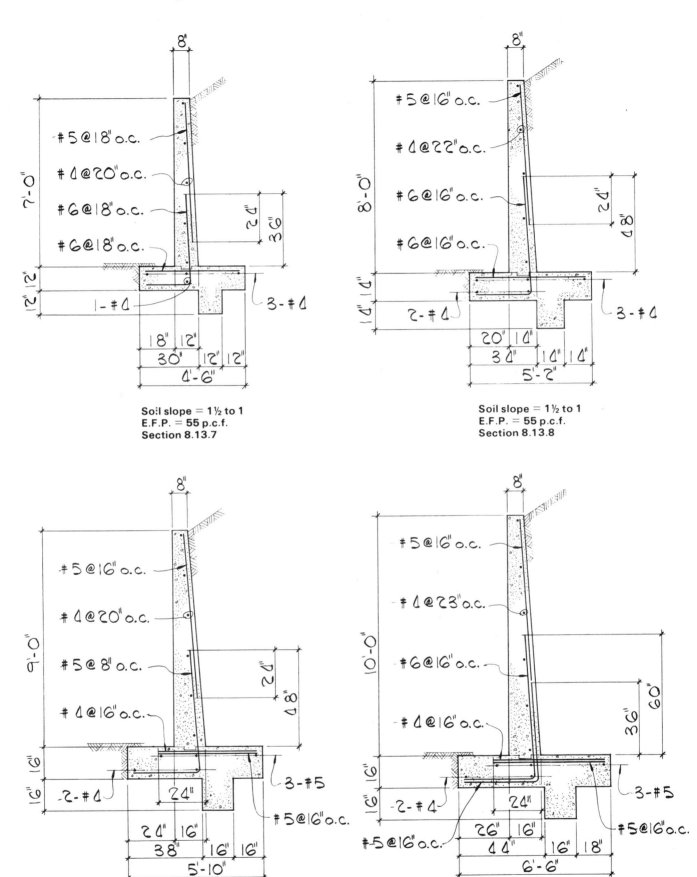

Soil slope = 1½ to 1
E.F.P. = 55 p.c.f.
Section 8.13.7

Soil slope = 1½ to 1
E.F.P. = 55 p.c.f.
Section 8.13.8

Soil slope = 1½ to 1
E.F.P. = 55 p.c.f.
Section 8.13.9

Soil slope = 1½ to 1
E.F.P. = 55 p.c.f.
Section 8.13.10

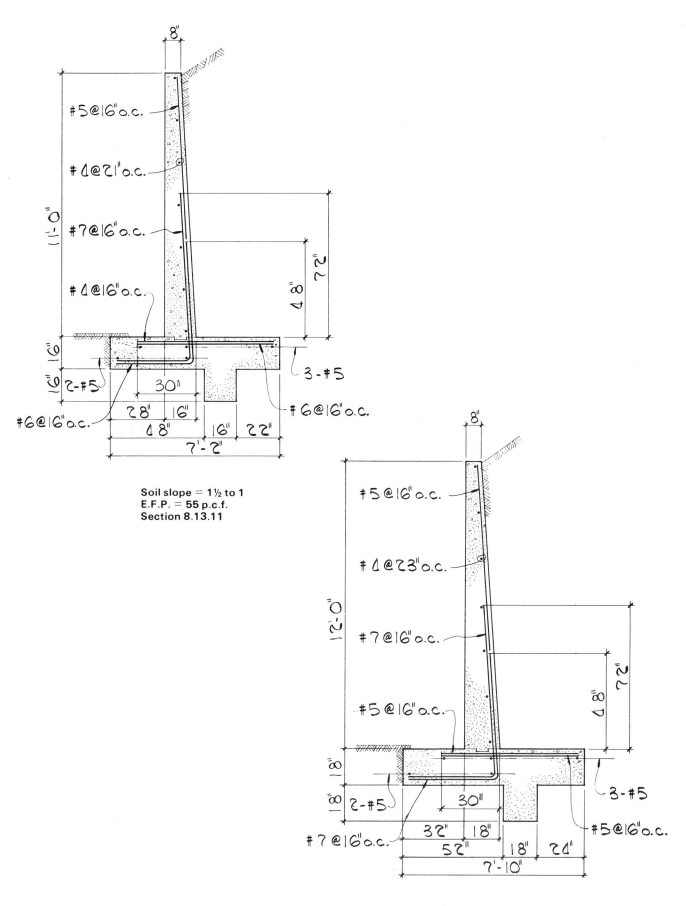

Soil slope = 1½ to 1
E.F.P. = 55 p.c.f.
Section 8.13.11

Soil slope = 1½ to 1
E.F.P. = 55 p.c.f.
Section 8.13.12

TABLE 8.14	GENERAL RETAINING WALLS – CONCRETE STEM – SOIL AT HEEL SIDE
	SLOPE = 1 to 1 SURCHARGE = 0 lbs./sq. ft. AXIAL = 0 lbs./ft.

SOIL PRESSURE DIAGRAM

Stem Height	3'- 0''	4'- 0''	5'- 0''	6'- 0''	7'- 0''	8'- 0''	9'- 0''	10'- 0''	11'- 0''	12'- 0''
Ft. L ft.	2.500	3.167	3.833	4.667	5.333	6.167	6.833	7.500	8.333	9.000
Ft. T ft.	0.833	0.833	0.833	1.000	1.167	1.333	1.333	1.333	1.500	1.500
Setback	0.833	1.000	1.333	1.500	1.833	2.000	2.333	2.500	2.833	3.000
Wt ft.	0.667	0.667	0.667	0.667	0.667	0.667	0.667	0.667	0.667	0.667
Wb ft.	0.833	0.833	0.833	1.000	1.167	1.333	1.333	1.333	1.500	1.500
A ft.	1.167	1.667	2.000	2.500	2.833	3.333	3.833	4.333	4.667	5.167
B ft.	0.833	0.833	0.833	1.000	1.167	1.333	1.333	1.333	1.500	1.500
C ft.	0.500	0.667	1.000	1.167	1.333	1.500	1.667	1.833	2.167	2.333
D ft.	0.833	0.833	0.833	1.000	1.167	1.333	1.333	1.333	1.500	1.500
W lbs.	1199	1842	2522	3792	4963	6699	7948	9539	11640	13555
F lbs.	588	934	1361	1960	2668	3484	4271	5138	6250	7290
Mo ft.-lbs.	751	1505	2647	4573	7262	10840	14712	19409	26042	32805
Mr ft.-lbs.	1817	3505	5911	10635	16120	24848	33042	43289	58904	73755
O.T.R.	2.419	2.328	2.234	2.325	2.220	2.292	2.246	2.230	2.262	2.248
$\overline{X}$ ft.	0.889	1.085	1.294	1.598	1.785	2.091	2.306	2.503	2.823	3.021
e ft.	0.361	0.498	0.622	0.735	0.882	0.992	1.110	1.247	1.343	1.479
Me ft.-lbs.	433	917	1570	2788	4377	6648	8826	11891	15637	20046
S ft.^{3}	1.042	1.671	2.449	3.630	4.741	6.338	7.782	9.375	11.574	13.500
S.P.t p.s.f.	895	1131	1299	1581	1854	2135	2297	2540	2748	2991
S.P.h p.s.f.	64	33	17	45	7	37	29	3	46	21
Friction	0.490	0.507	0.540	0.517	0.538	0.520	0.537	0.539	0.537	0.538

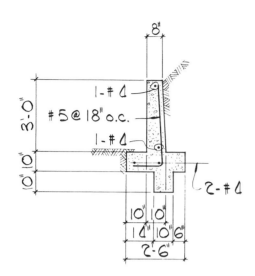

Soil slope = 1 to 1
E.F.P. = 80 p.c.f.
Section 8.14.3

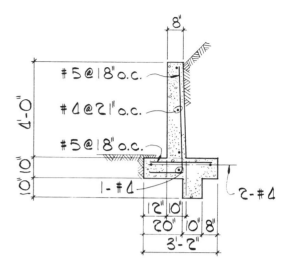

Soil slope = 1 to 1
E.F.P. = 80 p.c.f.
Section 8.14.4

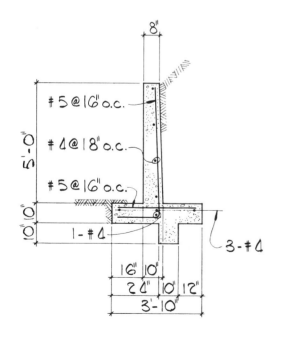

Soil slope = 1 to 1
E.F.P. = 80 p.c.f.
Section 8.14.5

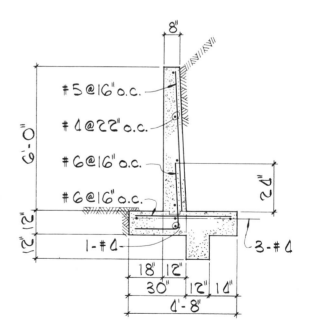

Soil slope = 1 to 1
E.F.P. = 80 p.c.f.
Section 8.14.6

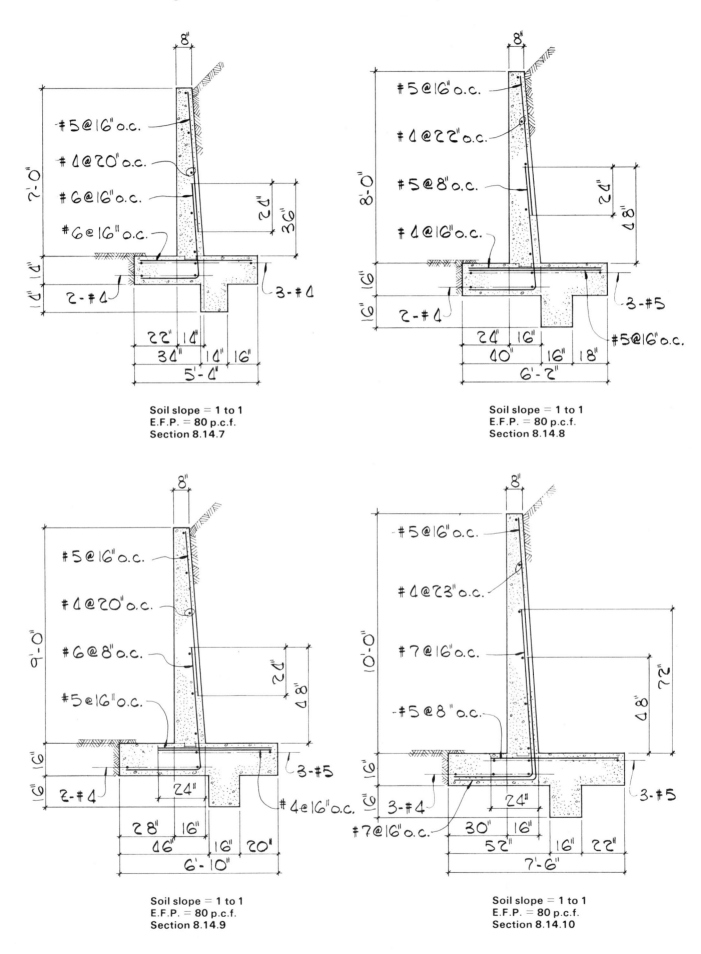

Soil slope = 1 to 1
E.F.P. = 80 p.c.f.
Section 8.14.7

Soil slope = 1 to 1
E.F.P. = 80 p.c.f.
Section 8.14.8

Soil slope = 1 to 1
E.F.P. = 80 p.c.f.
Section 8.14.9

Soil slope = 1 to 1
E.F.P. = 80 p.c.f.
Section 8.14.10

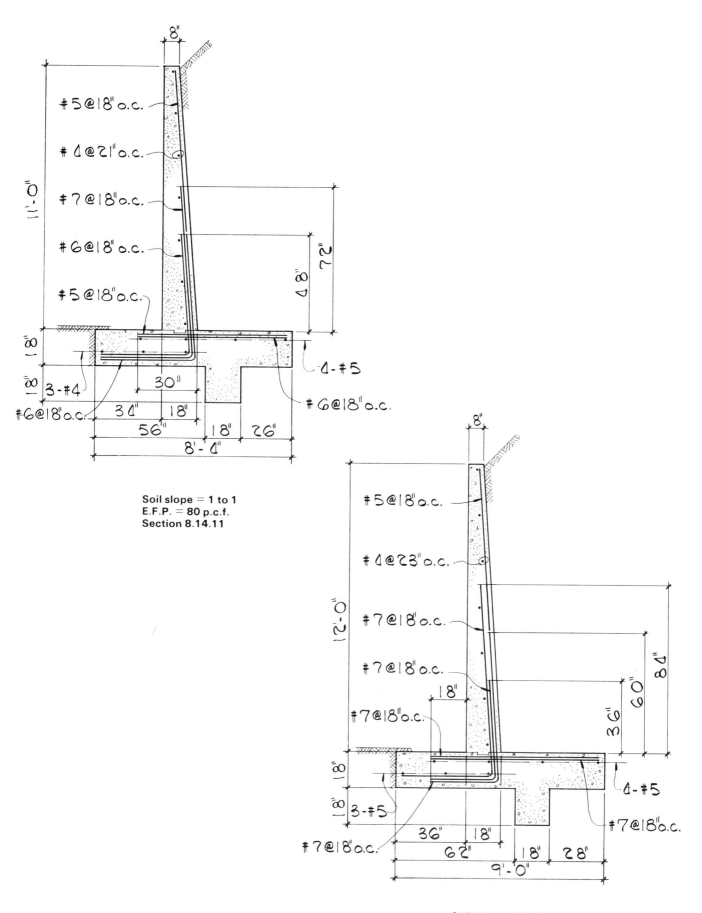

Soil slope = 1 to 1
E.F.P. = 80 p.c.f.
Section 8.14.11

Soil slope = 1 to 1
E.F.P. = 80 p.c.f.
Section 8.14.12

CHAPTER 9

General Retaining Walls
Masonry Stem
Soil at the Heel Side of the Footing

The design data and drawings presented in this chapter are concerned with property line retaining walls constructed with concrete masonry stem walls. The retained soil is not placed over the entire wall footing, but only at the heel side of the stem wall. The chapter consists of three design examples of retaining walls with various types of loading, a design data table for each of the fourteen loading conditions specified in Fig. 1.6, and a series of corresponding drawings following each design data table. The retaining wall drawings and the design data in this chapter are for a wall length of 1 foot.

The retaining walls in this chapter are designed using the following criteria:

Weight of soil = 100 p.c.f.   Weight of concrete = 150 p.c.f.
Concrete $f_c' = 2000$ p.s.i.   Reinforcing steel $f_c' = 20{,}000$ p.s.i.
Hollow masonry units, grouted solid, grade N, $f_m' = 1500$ p.s.i.
Field inspection of masonry required (see the design data table)
Weight of 8-inch concrete block with all cells filled = 92 p.s.f.
Weight of 12-inch concrete block with all cells filled = 140 p.s.f.
Grout $f_c' = 2000$ p.s.i.   Mortar (type S) = 1800 p.s.i.
Maximum allowable soil pressure = 4000 p.s.f.
Minimum allowable soil pressure = 0 p.s.f.
$\overline{X} \geqslant L/3$ Minimum O.T.R. = 1.50
Coefficient of soil friction = 0.40
Passive soil pressure = 300 p.c.f.

See Tables C5 and C6 for reinforcement protection. Weep holes are not shown on the drawings.

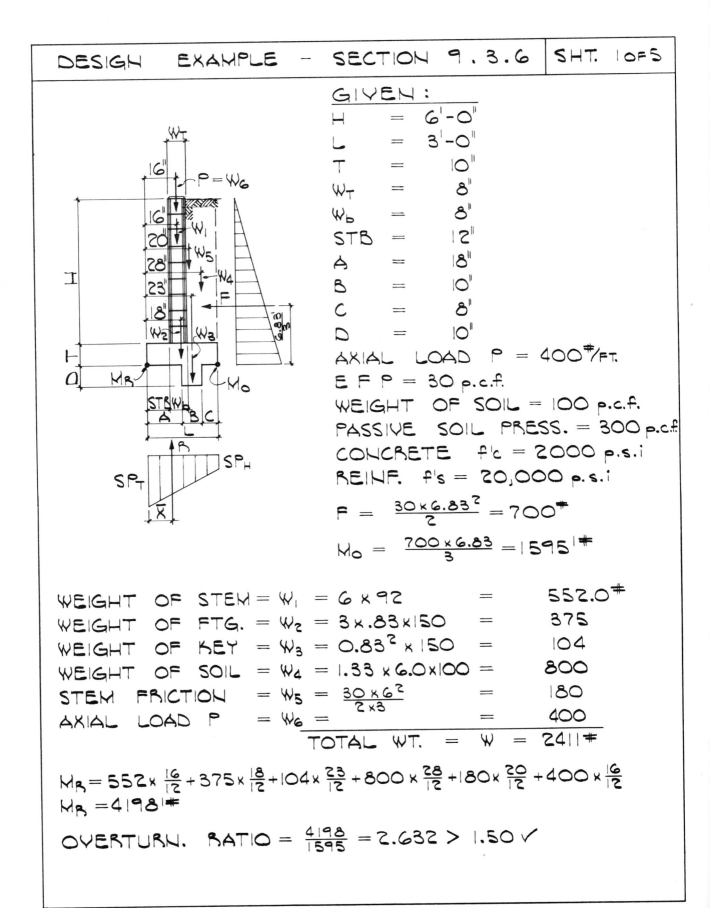

DESIGN EXAMPLE — SECTION 9.3.6 | SHT. 1 OF 5

GIVEN:

$H = 6'-0''$

$L = 3'-0''$

$T = 10''$

$W_T = 8''$

$W_b = 8''$

$STB = 12''$

$A = 18''$

$B = 10''$

$C = 8''$

$D = 10''$

AXIAL LOAD $P = 400^\#/FT.$

$EFP = 30$ p.c.f.

WEIGHT OF SOIL = 100 p.c.f.

PASSIVE SOIL PRESS. = 300 p.c.f.

CONCRETE $f'_c = 2000$ p.s.i

REINF. $f's = 20,000$ p.s.i

$$F = \frac{30 \times 6.83^2}{2} = 700^\#$$

$$M_o = \frac{700 \times 6.83}{3} = 1595'^\#$$

WEIGHT OF STEM = W_1 = 6 × 92	=	552.0$^\#$
WEIGHT OF FTG. = W_2 = 3 × .83 × 150	=	375
WEIGHT OF KEY = W_3 = 0.83^2 × 150	=	104
WEIGHT OF SOIL = W_4 = 1.33 × 6.0 × 100	=	800
STEM FRICTION = W_5 = $\dfrac{30 \times 6^2}{2 \times 3}$	=	180
AXIAL LOAD P = W_6 =	=	400
		TOTAL WT. = W = 2411$^\#$

$$M_R = 552 \times \frac{16}{12} + 375 \times \frac{18}{12} + 104 \times \frac{23}{12} + 800 \times \frac{28}{12} + 180 \times \frac{20}{12} + 400 \times \frac{16}{12}$$

$$M_R = 4198'^\#$$

$$\text{OVERTURN. RATIO} = \frac{4198}{1595} = 2.632 > 1.50 \checkmark$$

DESIGN EXAMPLE - SECTION 9.3.6 | SHT. 2 OF 5

FOOTING DESIGN:

$W = 2411^{\#}$; $M_O = 1595^{'\#}$; $M_R = 4198^{'\#}$

NET $M = 4198 - 1595 = 2603^{'\#}$

$\bar{X} = \dfrac{2603}{2411} = 1.079'$

FTG. MIDDLE $\frac{1}{3} = \dfrac{3.0}{3} = 1.00' < 1.079'$

$e = \dfrac{3.0}{2} - 1.079 = 0.421'$

$M_e = 2411 \times .421 = 1015^{'\#}$

FTG. AREA $= 3.0$ SQ. FT.

FTG. SECT. MODULUS $= \dfrac{1 \times 3^2}{6} = 1.50^{'3}$

SOIL PRESSURE $= \dfrac{2411}{3.0} \pm \dfrac{1015}{1.50} = 803.7 \pm 676.7$

$\underline{SP_T = 1480.7 \text{ p.s.f.}}$ $\qquad$ $\underline{SP_H = 127.0 \text{ p.s.f.}}$

$-M_1:$
$SP_1 = \dfrac{1480.7 - 127.0}{3} \times 1.33 = 601.5 \text{ p.s.f.}$

UNIF. LD. $= 600 + 125 - 127 = 598 \, ^{\#}/\text{FT.}$

$-M_1 = \dfrac{598 \times 1.33^2}{2} - \dfrac{601.5 \times 1.33^2}{2 \times 3}$

$-M_1 = 351^{'\#}$

$V_1 = 598 \times 1.33 - \dfrac{601.5 \times 1.33}{2}$

$V_1 = 395.3^{\#}$

$+M_2:$
$W' = \dfrac{1480.7 - 127.0}{3.0} = \times 2.0 + 127 = 1029.5 \text{ p.s.f.}$

$SP_2 = 1480.7 - 1029.5 = 451.2 \text{ p.s.f.}$

UNIF. LD. $= 1029.5 - 125.0 = 904.5 \, ^{\#}/\text{FT.}$

$M_2 = \dfrac{904.5 \times 1.0^2}{2} + \dfrac{451.2 \times 1.0^2 \times 2}{2 \times 3} = 602.6^{'\#}$

$V_2 = 904.5 \times 1.0 + \dfrac{451.2 \times 1.0}{2} = 1130.1^{\#}$

DESIGN   EXAMPLE — SECTION 9.3.6 | SHT. 3 OF 5

FOOTING   DESIGN :

$-M_1 = 351^{'\#}$; $V_1 = 395.3^{\#}$
$+M_2 = 602.6^{'\#}$; $V_2 = 1130.1^{\#}$

MOMENTS ARE LOW, CHECK
FOR TENSION IN CONCRETE

CROSS SECTION SECT. MODULUS $= \dfrac{12 \times 10^2}{6} = 200^{''3}$

MAX. $M = 602.6^{'\#}$

f TENSION $= \dfrac{12 \times 602.6}{200} = 36.1$ p.s.i < 71 p.s.i ✓

MIN. As REQ'D. EA. WAY $= 0.012 \times 10 \times 36 = 0.43$ SQ. IN.

USE   # 4 @ 16" o.c. TOP & BOTTOM & LONG. DIRECTION

CHECK FTG. SHEAR ; $v = \dfrac{1130.1}{12 \times .89 \times 6.75} = 15.7$ p.s.i < 54 p.s.i ✓

STEM DESIGN AT TOP OF FTG.

$H = 6.0'$;   E F P $= 30$ p.c.f.

$V = \dfrac{30 \times 6^2}{2} = 540^{\#}$; $M = \dfrac{540 \times 6}{3} = 1080^{'\#}$

$d = 8'' - (2 + .5 \times .5) = 5.75''$; $a = 1.13$; $j = 0.89$

$v = \dfrac{540}{12 \times .89 \times 5.75} = 8.8$ p.s.i < 15 p.s.i ✓

$As = \dfrac{1.08}{1.13 \times 5.75} = 0.166$ SQ. IN./FT.   TRY # 5 @ 16" o.c.

$As = 0.23$ SQ. IN. ; $p = \dfrac{0.23}{12 \times 5.75} = 0.0033$; $np = 40 \times .0033 = 0.132$

$^2/_{kj} = 5.79$; $fm = \dfrac{12 \times 1080 \times 5.79}{12 \times 5.75^2} = 189.1$ p.s.i < 250 p.s.i

USE #5 @ 16" o.c. (NO CONT. INSPECTION REQ'D.)

AXIAL $fa = \dfrac{400 + 552}{12 \times 8} = 9.92$ p.s.i < 135 p.s.i

COMBINED STRESS $= \dfrac{189.1}{250.0} + \dfrac{9.92}{135.0} = 0.83 < 1.0$ ✓

DESIGN EXAMPLE — SECTION 9.3.6	SHT. 4 of 5

STEM DESIGN AT 2'-0" ABOVE FTG.

$H = 6.0' - 2.0' \quad 4.0' \qquad E F P = 30 \text{ p.c.f.}$

$V = \dfrac{30 \times 4^2}{2} = 240^{\#} \quad ; \quad M = \dfrac{240 \times 4}{3} = 320'^{\#}$

$\nu = \dfrac{240}{12 \times .89 \times 5.75} = 3.90 \text{ p.s.i} < 15 \text{ p.s.i} \checkmark$

$A_s = \dfrac{.32}{1.13 \times 5.75} = 0.05 \text{ sq.in./ft} \qquad \text{TRY } \#4 @ 16" \text{ o.c.}$

$A_s = 0.15 \text{ sq.in.} \quad ; \quad p = \dfrac{0.15}{12 \times 5.75} = 0.00217 \quad ; \quad np = 40 \times 0.00217 = 0.087$

$^2/_{kj} = 6.68 \quad ; \quad f_m = \dfrac{12 \times 320 \times 6.68}{12 \times 5.75^2} = 64.6 \text{ p.s.i} < 250 \text{ p.s.i}$

USE #4 @ 16" o.c.

SLIDING:

$F_r = \dfrac{700}{2411} = 0.290 < 0.40 \checkmark$

CHECK F_r WITHOUT AXIAL LD.

$W = 2411 - 400 = 2011^{\#}$

$F_r = \dfrac{700}{2011} = 0.348 < 0.40 \checkmark$

FTG. KEY IS OPTIONAL. HOWEVER IT IS USED FOR STABILITY CALC'S.

KEY IS 10" SQ.

KEY $\nu = \dfrac{700}{10^2} = 7.0 \text{ p.s.i} < 54 \text{ p.s.i} \checkmark$

DESIGN EXAMPLE – SECTION 9.3.6	SHT. 5 OF 5

CHECK WALL WITHOUT 400 #/FT. AXIAL LD.

$F = 700^{\#}$; $M_o = 1595^{'\#}$

$W = 2411 - 400 = 2011^{\#}$

$M_R = 4198 - 400 \times \frac{16}{12} = 3665^{'\#}$

OVERTURN. RATIO $= \frac{3665}{1595} = 2.297 > 1.50$ ✓

NET M $= 3665 - 1595 = 2070^{'\#}$

$\bar{X} = \frac{2070}{2011} = 1.029' > 1.00$ ✓

$e = \frac{3.0}{2} - 1.029 = 0.471'$

$M_e = 2011 \times .471 = 947^{'\#}$

SOIL PRESSURE $= \frac{2011}{3.0} \pm \frac{947}{1.50} = 670.3 \pm 631.3$

$\underline{SP_T = 1301.6 \text{ p.s.f.}}$ $\qquad \underline{SP_H = 39 \text{ p.s.f.}}$

SOIL PRESSURES ARE LESS THAN ORIGINAL DESIGN ∴ AXIAL LOAD IS MAX. CASE FOR DESIGN.

STEM DESIGN UNAFFECTED IN BENDING.

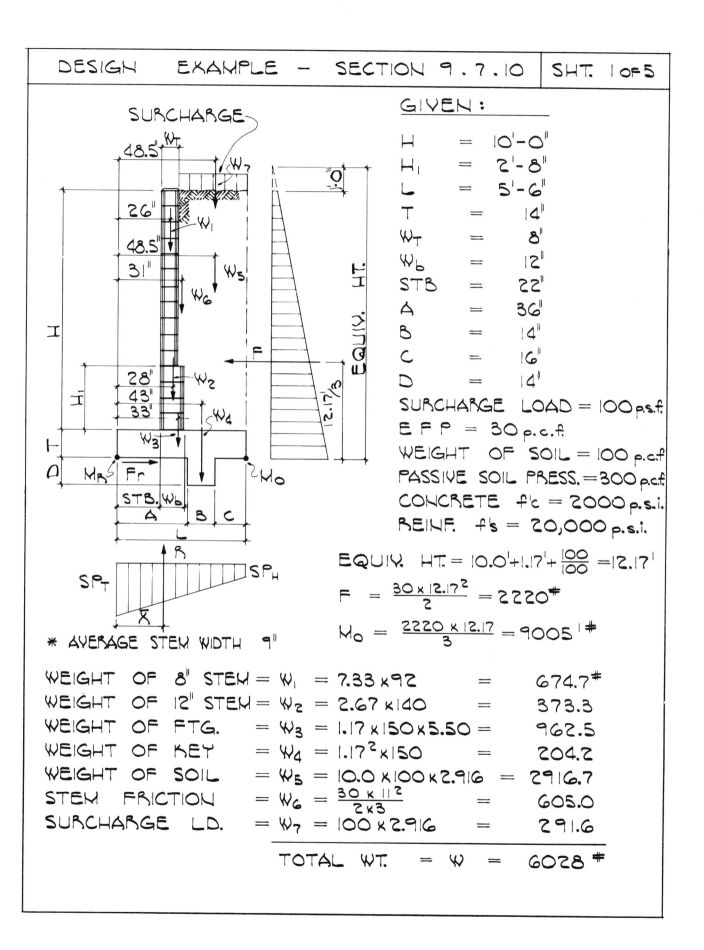

DESIGN EXAMPLE - SECTION 9.7.10 | SHT. 1 OF 5

SURCHARGE

GIVEN:

$$H = 10'-0''$$
$$H_1 = 2'-8''$$
$$L = 5'-6''$$
$$T = 14''$$
$$W_T = 8''$$
$$W_b = 12''$$
$$STB = 22''$$
$$A = 36''$$
$$B = 14''$$
$$C = 16''$$
$$D = 14''$$

SURCHARGE LOAD = 100 p.s.f.
E F P = 30 p.c.f.
WEIGHT OF SOIL = 100 p.c.f.
PASSIVE SOIL PRESS. = 300 p.c.f.
CONCRETE f'_c = 2000 p.s.i.
REINF. f'_s = 20,000 p.s.i.

$$\text{EQUIV. HT.} = 10.0' + 1.17' + \frac{100}{100} = 12.17'$$

$$F = \frac{30 \times 12.17^2}{2} = 2220^\#$$

$$M_0 = \frac{2220 \times 12.17}{3} = 9005'^\#$$

* AVERAGE STEM WIDTH 9"

WEIGHT OF 8" STEM	$= W_1 = 7.33 \times 92$	$=$	$674.7^\#$
WEIGHT OF 12" STEM	$= W_2 = 2.67 \times 140$	$=$	373.3
WEIGHT OF FTG.	$= W_3 = 1.17 \times 150 \times 5.50$	$=$	962.5
WEIGHT OF KEY	$= W_4 = 1.17^2 \times 150$	$=$	204.2
WEIGHT OF SOIL	$= W_5 = 10.0 \times 100 \times 2.916$	$=$	2916.7
STEM FRICTION	$= W_6 = \dfrac{30 \times 11^2}{2 \times 3}$	$=$	605.0
SURCHARGE LD.	$= W_7 = 100 \times 2.916$	$=$	291.6

TOTAL WT. = W = 6028 $^\#$

DESIGN   EXAMPLE   —   SECTION   9.7.10 | SHT. 2 OF 5

FOOTING DESIGN:

$$W = 6028^{\#} \quad ; \quad M_0 = 9005^{'\#} \quad ; \quad F = 2220^{\#}$$

$$M_R = 674.7 \times \frac{26}{12} + 373.3 \times \frac{28}{12} + 962.5 \times \frac{33}{12} + 204.2 \times \frac{43}{12} + 2916.7 \times \frac{48.5}{12}$$
$$+ 605 \times \frac{31}{12} + 291.6 \times \frac{48.5}{12}$$

$$M_R = 20,182^{'\#}$$

$$\text{OVERTURN. RATIO} = \frac{20182}{9005} = 2.241 > 1.50 \checkmark$$

$$\text{NET } M = 20182 - 9005 = 11177^{'\#}$$

$$\bar{X} = \frac{11177}{6028} = 1.854'$$

$$\text{FTG. MIDDLE } \tfrac{1}{3} = \frac{5.50}{3} = 1.833' < 1.854' \checkmark$$

$$e = \frac{5.50}{2} - 1.854 = 0.896'$$

$$M_e = 6028 \times .896 = 5401^{'\#}$$

FTG. AREA = 5.50 SQ. FT.

$$\text{FTG. SECT. MODULUS} = \frac{1 \times 5.50^2}{6} = 5.04^{'3}$$

$$\text{SOIL PRESSURE} = \frac{6028}{5.50} \pm \frac{5401}{5.04} = 1096 \pm 1070.7$$

$$\underline{SP_T = 2166.7 \text{ p.s.f.}} \qquad \underline{SP_H = 25.3 \text{ p.s.f.}}$$

$-M_1:$

$$SP_1 = \frac{2166.7 - 25.3}{5.50} \times 2.67 = 1038.4 \text{ p.s.f.}$$

$$\text{UNIF. LD.} = 100 + 1000 + 125 - 25.3 = 1199.7 \,{}^{\#}\!/_{FT.}$$

$$-M_1 = \frac{1199.7 \times 2.67^2}{2} - \frac{1038.4 \times 2.67^2}{2 \times 3} = 3035^{'\#}$$

$$V_1 = 1199.7 \times 2.67 - \frac{1038.4 \times 2.67}{2} = 1818.5^{\#}$$

$+M_2:$

$$W' = \frac{2166.7 - 25.3}{5.50} \times 3.67 + 25.3 = 1453 \text{ p.s.f.}$$

$$SP_2 = 2166.7 - 1453 = 713.7 \text{ p.s.f.}$$

$$\text{UNIF. LD.} = 1453 - 125 = 1328 \,{}^{\#}\!/_{FT.}$$

$$+M_2 = \frac{1328 \times 1.83^2}{2} + \frac{713.7 \times 1.83^2 \times 2}{2 \times 3} = 3031^{'\#}$$

$$V_2 = 1328 \times 1.83 + \frac{713.7 \times 1.83}{2} = 3083.3^{\#}$$

DESIGN EXAMPLE — SECTION 9.7.10 | SHT. 3 of 5

FOOTING DESIGN : (CONT.)

$-M_1 = 3035^{'\#}$; $V_1 = 1818.5^{\#}$

$+M_2 = 3031^{'\#}$; $V_2 = 3083.3^{\#}$

d FOR $+A_S$ = $14 - (3 + .5 \times .5) = 10.7''$

d FOR $-A_S$ = $14 - (2 + .5 \times .5) = 11.7''$

$a = 1.13$; $j = 0.89$

$+A_S = \dfrac{3.03}{1.13 \times 10.7} = 0.25$ SQ. IN./FT.

$v_2 = \dfrac{3083.3}{12 \times .89 \times 10.7} = 27$ p.s.i < 54 p.s.i ✓

$-A_S = \dfrac{3.03}{1.13 \times 11.7} = 0.23$ SQ. IN./FT.

USE #6 @ 16" o.c. TOP & BOTTOM

LONG. REINF. :

$A_S = 0.0012 \times 12 \times 5.5 \times 14 = 1.11$ SQ. IN.

USE 5 - #4's

STEM DESIGN AT TOP OF FTG.

$H = 10.0' + 1.0' = 11.0'$ $EFP = 30$ p.c.f.

$V = \dfrac{30 \times 11^2}{2} = 1815^{\#}$ $M = \dfrac{1815 \times 11}{3} = 6655^{'\#}$

$d = 12 - (2 + .5 \times .75) = 9.60''$; $a = 1.13$; $j = 0.89$

$v = \dfrac{1815}{12 \times .89 \times 9.60} = 17.7$ p.s.i > 15.0 p.s.i ∴ INSPEC. REQ'D

$A_S = \dfrac{6.655}{1.13 \times 9.70} = 0.607$ SQ. IN./FT.   TRY #6 @ 8" o.c.

$A_S = 0.66$ SQ. IN. ; $p = \dfrac{0.66}{12 \times 9.6} = 0.00573$; $np = 20 \times .00573 = 0.115$

$2/kj = 6.04$; $f_m = \dfrac{12 \times 6655 \times 6.04}{12 \times 9.6^2} = 436.1$ p.s.i > 500 p.s.i ✓

USE #6 @ 8" o.c. CONT. INSPECTION REQ'D

DESIGN EXAMPLE — SECTION 9.7.10 | SHT. 4 OF 5

STEM DESIGN AT 2'-8" ABOVE FTG.

$H = 10.0' + 1.0 - 2.67 = 8.33'$; $EFP = 30$ p.c.f.

$V = \dfrac{30 \times 8.33^2}{2} = 1041^\#$; $M = \dfrac{1041 \times 8.33}{3} = 2890'^\#$

$d = 8 - (2 + .5 \times .62) = 5.7"$

$v = \dfrac{1041}{12 \times .89 \times 5.7} = 17.1$ p.s.i > 15 p.s.i ∴ INSPECT. REQ'D.✓

$A_s = \dfrac{2.89}{1.13 \times 5.7} = 0.448$ SQ.IN./FT.   TRY #5 @ 8" o.c.

$A_s = 0.47$ SQ.IN. ; $p = \dfrac{0.47}{12 \times 5.7} = 0.0069$; $np = 20 \times .0069 = 0.138$

$2/kj = 5.71$; $fm = \dfrac{12 \times 2890 \times 5.71}{12 \times 5.7^2} = 508$ p.s.i ≈ 500 p.s.i ✓ +1.5%

USE #5 @ 16" o.c. CONT. INSPECTION REQ'D.

STEM DESIGN AT 6'-0" ABOVE FTG.

$H = 10.0' + 1.0' - 6.0' = 5.0'$; $EFP = 30$ p.c.f.

$V = \dfrac{30 \times 5^2}{2} = 375^\#$; $M = \dfrac{375 \times 5.0}{3} = 625'^\#$

$d = 5.7"$

$v = \dfrac{375}{12 \times .89 \times 5.7} = 6.2$ p.s.i < 15 p.s.i ✓

$A_s = \dfrac{.625}{1.13 \times 5.7} = 0.097$ SQ.IN./FT.   TRY #4 @ 16" o.c.

$A_s = 0.15$ SQ.IN. ; $p = \dfrac{0.15}{12 \times 5.7} = 0.0022$; $np = 20 \times .0022 = 0.044$

$2/kj = 8.54$; $fm = \dfrac{12 \times 625 \times 8.54}{12 \times 5.7^2} = 164.3$ p.s.i > 250 p.s.i ✓

USE #4 @ 16" o.c.

LONG. REINF.:

$A_s = 0.0007(12 \times 32 + 8 \times 88) = 0.76$ SQ.IN.

USE 2-#4's AT TOP & BOTT. OF 8" WALL

   2-#5's AT TOP & BOTT. OF 12" WALL

SLIDING :

$F_r = \dfrac{2220}{6028} = 0.368 \approx 0.40$ USE 14" SQ. KEY

KEY $v = \dfrac{2220}{14^2} = 11.3$ p.s.i < 54 p.s.i ✓

DESIGN EXAMPLE — SECTION 9.7.10	SHT. 5 of 5

CHECK WALL WITHOUT 100 p.s.f. SURCHARGE LD.

EQUIV. HT. = $10.0' + 1.17' = 11.17'$; E F P = 30 p.c.f.

$F = \frac{30 \times 11.17^2}{2} = 1870^\#$; $M_0 = \frac{1870 \times 11.17}{3} = 6962'^\#$

$W = 6028 - 605 + \frac{30 \times 10^2}{2 \times 3} - 291.6 = 5631^\#$

$M_R = 20182 - 605 \times \frac{31}{12} + \frac{30 \times 10^2}{2 \times 3} \times \frac{31}{12} - 291.6 \times \frac{48.5}{12} = 18,732'^\#$

OVERTURN. RATIO = $\frac{18732}{6962} = 2.691 > 1.50$ ✓

NET M = $18,732 - 6962 = 11770'^\#$

$\bar{X} = \frac{11770}{5631} = 2.09' > 1.833'$ ✓

$e = \frac{5.50}{2} - 2.09 = 0.66'$

$M_e = 5631 \times .66 = 3716.5'^\#$

SOIL PRESSURE = $\frac{5631}{5.50} \pm \frac{3716.4}{5.04} = 1023.8 \pm 737.4$

$\underline{SP_T = 1761.2 \text{ p.s.f.}}$ $\underline{SP_H = 286.4 \text{ p.s.f.}}$

SOIL PRESSURES ARE LESS THAN WALL WITH SURCHARGE LD. ∴ FTG. DESIGN IS O.K.

MAX. CONDITION OF STEM DESIGN IS WITH THE SURCHARGE LD.

CHECK SLIDING :

$W = 5631^\#$; $F = 1870^\#$

$F_r = \frac{1870}{5631} = 0.332 < 0.40$ ✓

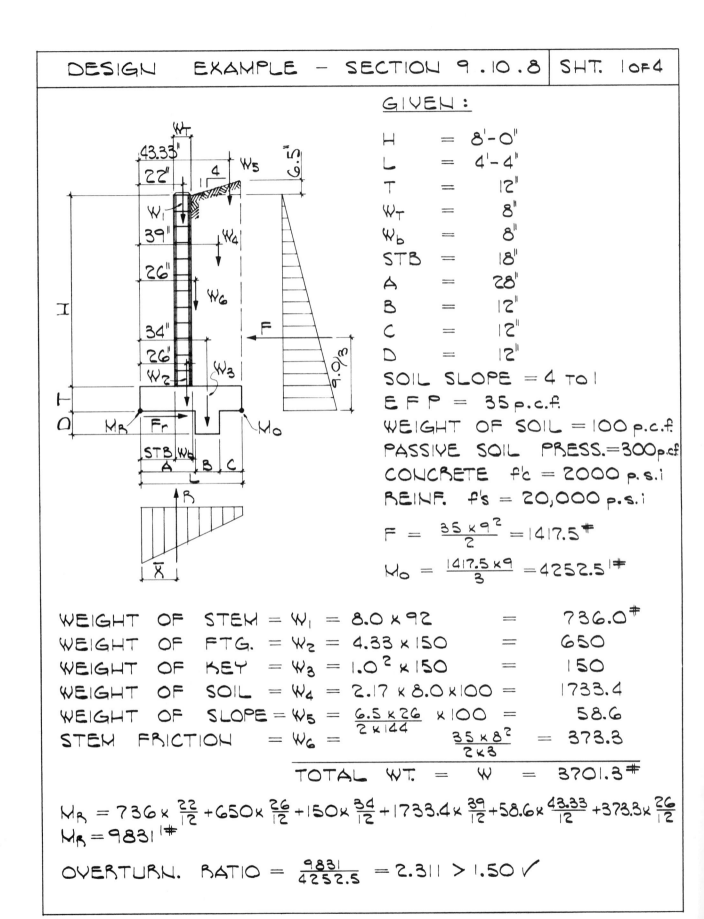

DESIGN EXAMPLE - SECTION 9.10.8 | SHT. 1 OF 4

GIVEN :

H = 8'-0"
L = 4'-4"
T = 12"
W_T = 8"
W_b = 8"
STB = 18"
A = 28"
B = 12"
C = 12"
D = 12"

SOIL SLOPE = 4 TO 1
E F P = 35 p.c.f.
WEIGHT OF SOIL = 100 p.c.f.
PASSIVE SOIL PRESS. = 300 p.c.f.
CONCRETE f'_c = 2000 p.s.i
REINF. f'_s = 20,000 p.s.i

$F = \dfrac{35 \times 9^2}{2} = 1417.5^{\#}$

$M_O = \dfrac{1417.5 \times 9}{3} = 4252.5^{\, l\#}$

WEIGHT OF STEM = W_1 = 8.0 × 92 = 736.0 #
WEIGHT OF FTG. = W_2 = 4.33 × 150 = 650
WEIGHT OF KEY = W_3 = 1.0² × 150 = 150
WEIGHT OF SOIL = W_4 = 2.17 × 8.0 × 100 = 1733.4
WEIGHT OF SLOPE = W_5 = $\dfrac{6.5 \times 26}{2 \times 144}$ × 100 = 58.6
STEM FRICTION = W_6 = $\dfrac{35 \times 8^2}{2 \times 3}$ = 373.3
_____
TOTAL WT. = W = 3701.3 #

$M_R = 736 \times \dfrac{22}{12} + 650 \times \dfrac{26}{12} + 150 \times \dfrac{34}{12} + 1733.4 \times \dfrac{39}{12} + 58.6 \times \dfrac{43.33}{12} + 373.3 \times \dfrac{26}{12}$

$M_R = 9831^{\, l\#}$

OVERTURN. RATIO = $\dfrac{9831}{4252.5}$ = 2.311 > 1.50 ✓

DESIGN EXAMPLE – SECTION 9.10.8 | SHT. 2 OF 4

FOOTING DESIGN:

$W = 3701.3^{\#}$; $M_O = 4252.5^{'\#}$; $M_R = 9831^{'\#}$

NET $M = 9831 - 4252.5 = 5578.5^{'\#}$

$\bar{X} = \dfrac{5578.5}{3701.3} = 1.507'$

FTG. MIDDLE $\frac{1}{3} = \dfrac{4.33}{3} = 1.444' < 1.507'$ ✓

$e = \dfrac{4.33}{2} - 1.507 = 0.658'$

$M_e = 3701.3 \times .658 = 2435.5^{'\#}$

FTG. AREA $= 4.33$ SQ. FT.

FTG. SECT. MODULUS $= \dfrac{1 \times 4.33^2}{6} = 3.13^{'3}$

SOIL PRESSURE $= \dfrac{3701.3}{4.33} \pm \dfrac{2435.5}{3.13} = 854.8 \pm 774.9$

$\underline{SP_T = 1629.7 \text{ p.s.f.}}$ $\qquad$ $\underline{SP_H = 79.9 \text{ p.s.f.}}$

$-M:$

$SP_1 = \dfrac{1629.7 - 79.9}{4.33} \times 2.17 = 775 \text{ p.s.f.}$

UNIF. LD. $= 800 + 150 - 79.9 = 870.1^{\#}/\text{FT.}$

$-M_1 = \dfrac{870.1 \times 2.17^2}{2} + 58.6 \times 2.17 \times .67 - \dfrac{775 \times 2.17^2}{2 \times 3}$

$-M_1 = 1521^{'\#}$

$V_1 = 870.1 \times 2.17 + 58.6 - \dfrac{775 \times 2.17}{2}$

$V_1 = 1104.4^{\#}$

$+M_2:$

$W' = \dfrac{1629.7 - 79.9}{4.33} \times 2.83 + 79.9 = 1093.3 \text{ p.s.f.}$

$SP_2 = 1629.7 - 1093.3 = 536.4 \text{ p.s.f.}$

UNIF. LD. $= 1093.3 - 150 = 943.3^{\#}/\text{FT.}$

$+M_2 = \dfrac{943.3 \times 1.5^2}{2} + \dfrac{536.4 \times 1.5^2 \times 2}{2 \times 3}$

$+M_2 = 1463.5^{'\#}$

$V_2 = 943.3 \times 1.5 + \dfrac{536.4 \times 1.5}{2}$

$V_2 = 1817.3^{\#}$

DESIGN EXAMPLE – SECTION 9.10.8	SHT. 3 OF 4

FOOTING DESIGN: (CONT.)

$-M_1 = 1521^{'\#}$; $V_1 = 1104.4^{\#}$

$+M_2 = 1463.5^{'\#}$; $V_2 = 1817.3^{\#}$

d FOR $+A_s = 12 - (3 + .5 \times .5) = 8.75''$

d FOR $-A_s = 12 - (2 + .5 \times .5) = 9.75''$

$a = 1.13$; $j = 0.89$

$+A_s = \dfrac{1.463}{1.13 \times 8.75} = 0.148$ SQ. IN./FT.

<u>USE #6 @ 16" o.c. TOP & BOTTOM</u>

LONG. REINF.:

$A_s = 0.0012 \times 12 \times 52 = 0.748$ SQ. IN.

<u>USE 3 - #4's</u>

STEM DESIGN AT TOP OF FTG.

$H = 8.0'$; $EFP = 35$ p.c.f.

$V = \dfrac{35 \times 8.0^2}{2} = 1120^{\#}$; $M = \dfrac{1120 \times 8.0}{3} = 2986.7^{'\#}$

$d = 8 - (2 + .5 \times .62) = 5.70''$; $a = 1.13$; $j = 0.89$

$v = \dfrac{1120}{12 \times .89 \times 5.7} = 18.4$ p.s.i < 30 p.s.i ✓ CONT. INSPECT. REQ'D.

$A_s = \dfrac{2.986}{1.13 \times 5.7} = 0.463$ SQ. IN./FT. TRY #5 @ 8" o.c.

$A_s = 0.47$ SQ. IN. ; $p = \dfrac{0.47}{12 \times 5.70} = 0.0069$; $np = 20 \times .0069 = 0.138$

$\frac{2}{kj} = 5.71$; $f_m = \dfrac{12 \times 2986.7 \times 5.71}{12 \times 5.70^2} = 525$ p.s.i ≈ 500 p.s.i ✓ +5%

<u>USE #5 @ 8" o.c. CONT. INSPECTION REQ'D.</u>

DESIGN   EXAMPLE - SECTION 9.10.8 | SHT. 4 OF 4

STEM AT 3'-0" ABOVE FTG.

$H = 8.0' - 3.0' = 5.0$ $EFP = 35 \text{ p.c.f}$

$V = \dfrac{35 \times 5^2}{2} = 437.5^{\#}$; $M = \dfrac{437.5 \times 5}{3} = 729.2'^{\#}$

$v = \dfrac{437.5}{12 \times .89 \times 5.70} = 7.2 \text{ p.s.i} < 30 \text{ p.s.i} \checkmark$

$A_s = \dfrac{0.729}{1.13 \times 5.70} = 0.113 \text{ sq. in./ft.}$ TRY #4 @ 16" o.c.

$A_s = 0.15 \text{ sq. in.}$; $p = \dfrac{0.15}{12 \times 5.70} = 0.0022$; $np = 20 \times .0022 = 0.044$

$2/kj = 8.54$; $f_m = \dfrac{12 \times 729.2 \times 8.54}{12 \times 5.7^2} = 191.7 \text{ p.s.i} < 500 \text{ p.s.i} \checkmark$

USE #4 @ 16" o.c. CONT. INSPECTION REQ'D.

LONG. REINF. $= 0.0007 \times 8 \times 12 \times 8 = 0.54 \text{ sq. in.}$

USE 5 - #4's

SLIDING :

$F_r = \dfrac{1417.5}{3701.3} = 0.383 \approx 0.40$

USE 12" SQ. KEY

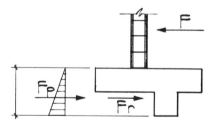

RESISTANCE TO SLIDING $= 3701.3 \times .40 + \dfrac{300 \times 2^2}{2} = 1780.5^{\#} \checkmark$

KEY $v = \dfrac{1417.5}{12^2} = 9.8 \text{ p.s.i} < 54 \text{ p.s.i} \checkmark$

TABLE 9.1   GENERAL RETAINING WALLS – MASONRY STEM – SOIL AT HEEL SIDE
SLOPE = 0 to 1   SURCHARGE = 0 lbs./sq. ft.   AXIAL = 0 lbs./ft.

SOIL PRESSURE DIAGRAM

H1 = Concrete Stem
H2 = 12″ Concrete Block
H3 = 8″ Concrete Block

Stem Height	3′- 0″	4′- 0″	5′- 0″	6′- 0″	7′- 0″	8′- 0″	9′- 0″	10′- 0″	11′- 0″.	12′- 0″
Ft. L ft.	1.667	2.000	2.500	3.000	3.500	4.000	4.500	5.000	5.500	6.000
Ft. T ft.	0.667	0.667	0.833	0.833	0.833	1.000	1.000	1.167	1.167	1.333
Setback	0.500	0.667	0.833	1.000	1.167	1.333	1.500	1.667	1.833	2.000
A ft.	0.500	0.833	1.167	1.500	1.833	2.167	2.500	2.667	3.000	3.333
B ft.	0.667	0.667	0.833	0.833	0.833	1.000	1.000	1.167	1.167	1.333
C ft.	0.500	0.500	0.500	0.667	0.833	0.833	1.000	1.167	1.333	1.333
D ft.	0.667	0.667	0.833	0.833	0.833	1.000	1.000	1.167	1.167	1.333
8 in. blk. ft.	3.000	4.000	5.000	6.000	7.000	8.000	7.000	8.000	8.333	8.667
12 in. blk. ft.	—	—	—	—	—	—	2.000	2.000	2.667	2.667
Conc. Stem	—	—	—	—	—	—	—	—	—	0.667
Wb ft.	—	—	—	—	—	—	—	—	—	1.333
W lbs.	704	981	1502	2011	2597	3406	4168	5185	6123	7352
F lbs.	202	327	510	700	920	1215	1500	1870	2220	2667
Mo ft.-lbs.	246	508	992	1595	2403	3645	5000	6962	9005	11852
Mr ft.-lbs.	689	1197	2280	3665	5526	8267	11395	15703	20430	26731
O.T.R.	2.797	2.355	2.297	2.297	2.299	2.268	2.279	2.255	2.269	2.255
$\overline{X}$ ft.	0.629	0.702	0.857	1.029	1.202	1.357	1.534	1.686	1.866	2.024
e ft.	0.204	0.298	0.393	0.471	0.548	0.643	0.716	0.814	0.884	0.976
Me ft.-lbs.	144	293	590	947	1422	2190	2982	4223	5413	7177
S ft.^{3}	0.463	0.667	1.042	1.500	2.042	2.667	3.375	4.167	5.042	6.000
S.P.t p.s.f.	734	930	1167	1302	1439	1673	1810	2051	2187	2421
S.P.h p.s.f.	112	52	34	39	45	30	43	24	40	29
Friction	0.286	0.333	0.340	0.348	0.354	0.357	0.360	0.361	0.363	0.363
Inspct.	NO	NO	NO	NO	YES	YES	YES	YES	YES	YES

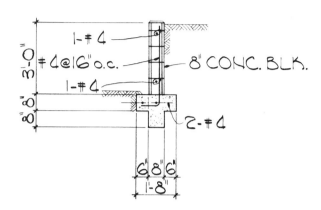

E.F.P. = 30 p.c.f.
Section 9.1.3

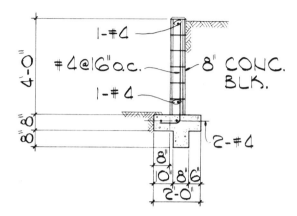

E.F.P. = 30 p.c.f.
Section 9.1.4

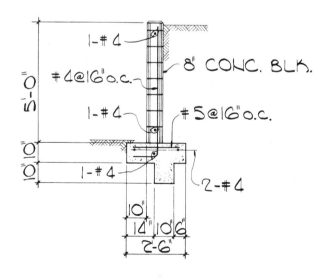

E.F.P. = 30 p.c.f.
Section 9.1.5

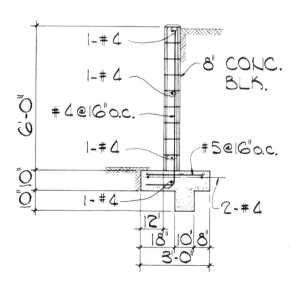

E.F.P. = 30 p.c.f.
Section 9.1.6

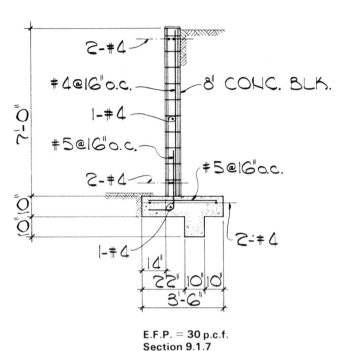

E.F.P. = 30 p.c.f.
Section 9.1.7

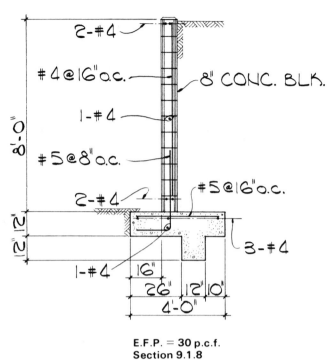

E.F.P. = 30 p.c.f.
Section 9.1.8

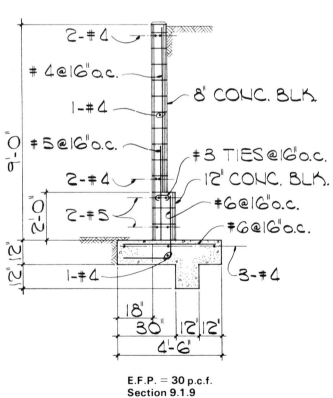

E.F.P. = 30 p.c.f.
Section 9.1.9

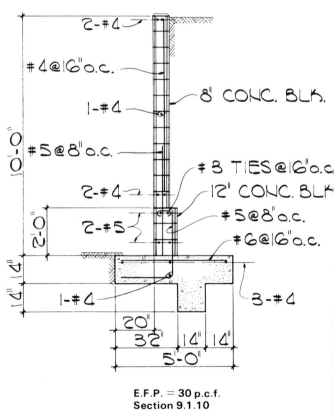

E.F.P. = 30 p.c.f.
Section 9.1.10

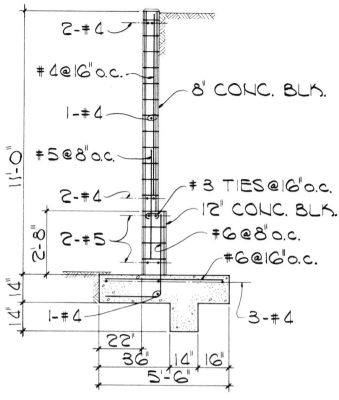

2-#4

#4@16"o.c.

1-#4

#5@8"o.c.

8" CONC. BLK.

2-#4

#3 TIES@16"o.c.
12" CONC. BLK.
#6@8"o.c.
#6@16"o.c.

2-#5

1-#4

3-#4

11'-0"

2'-8"

1'-4"

22"

36"   14"   16"

5'-6"

E.F.P. = 30 p.c.f.
Section 9.1.11

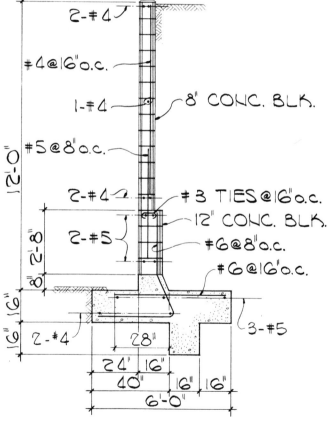

2-#4

#4@16"o.c.

1-#4

#5@8"o.c.

8" CONC. BLK.

2-#4

#3 TIES@16"o.c.
12" CONC. BLK.
#6@8"o.c.
#6@16"o.c.

2-#5

2-#4

3-#5

12'-0"

2'-8"

8"

16"

16"

16"

24"   16"

40"   16"   16"

28"

6'-0"

E.F.P. = 30 p.c.f.
Section 9.1.12

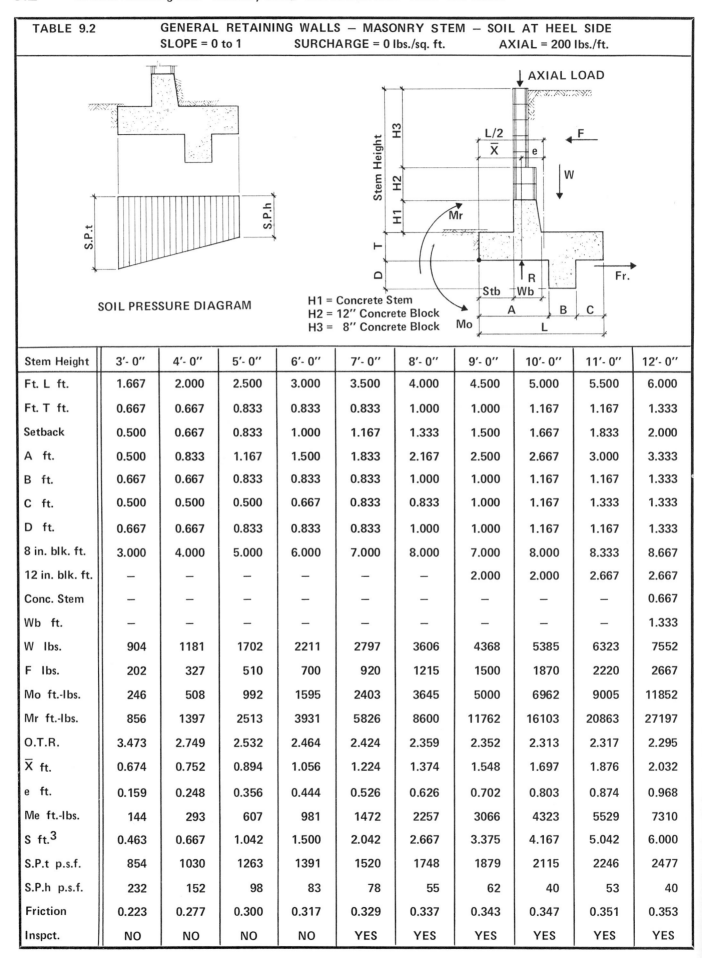

TABLE 9.2	GENERAL RETAINING WALLS – MASONRY STEM – SOIL AT HEEL SIDE SLOPE = 0 to 1 SURCHARGE = 0 lbs./sq. ft. AXIAL = 200 lbs./ft.

SOIL PRESSURE DIAGRAM

H1 = Concrete Stem
H2 = 12" Concrete Block
H3 = 8" Concrete Block

Stem Height	3'-0"	4'-0"	5'-0"	6'-0"	7'-0"	8'-0"	9'-0"	10'-0"	11'-0"	12'-0"
Ft. L ft.	1.667	2.000	2.500	3.000	3.500	4.000	4.500	5.000	5.500	6.000
Ft. T ft.	0.667	0.667	0.833	0.833	0.833	1.000	1.000	1.167	1.167	1.333
Setback	0.500	0.667	0.833	1.000	1.167	1.333	1.500	1.667	1.833	2.000
A ft.	0.500	0.833	1.167	1.500	1.833	2.167	2.500	2.667	3.000	3.333
B ft.	0.667	0.667	0.833	0.833	0.833	1.000	1.000	1.167	1.167	1.333
C ft.	0.500	0.500	0.500	0.667	0.833	0.833	1.000	1.167	1.333	1.333
D ft.	0.667	0.667	0.833	0.833	0.833	1.000	1.000	1.167	1.167	1.333
8 in. blk. ft.	3.000	4.000	5.000	6.000	7.000	8.000	7.000	8.000	8.333	8.667
12 in. blk. ft.	—	—	—	—	—	—	2.000	2.000	2.667	2.667
Conc. Stem	—	—	—	—	—	—	—	—	—	0.667
Wb ft.	—	—	—	—	—	—	—	—	—	1.333
W lbs.	904	1181	1702	2211	2797	3606	4368	5385	6323	7552
F lbs.	202	327	510	700	920	1215	1500	1870	2220	2667
Mo ft.-lbs.	246	508	992	1595	2403	3645	5000	6962	9005	11852
Mr ft.-lbs.	856	1397	2513	3931	5826	8600	11762	16103	20863	27197
O.T.R.	3.473	2.749	2.532	2.464	2.424	2.359	2.352	2.313	2.317	2.295
$\overline{X}$ ft.	0.674	0.752	0.894	1.056	1.224	1.374	1.548	1.697	1.876	2.032
e ft.	0.159	0.248	0.356	0.444	0.526	0.626	0.702	0.803	0.874	0.968
Me ft.-lbs.	144	293	607	981	1472	2257	3066	4323	5529	7310
S ft.^{3}	0.463	0.667	1.042	1.500	2.042	2.667	3.375	4.167	5.042	6.000
S.P.t p.s.f.	854	1030	1263	1391	1520	1748	1879	2115	2246	2477
S.P.h p.s.f.	232	152	98	83	78	55	62	40	53	40
Friction	0.223	0.277	0.300	0.317	0.329	0.337	0.343	0.347	0.351	0.353
Inspct.	NO	NO	NO	NO	YES	YES	YES	YES	YES	YES

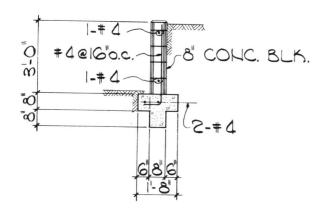

Axial load = 200 lbs./ft.
E.F.P. = 30 p.c.f.
Section 9.2.3

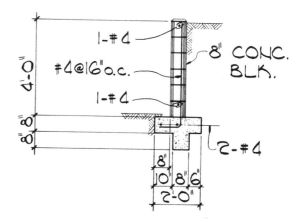

Axial load = 200 lbs./ft.
E.F.P. = 30 p.c.f.
Section 9.2.4

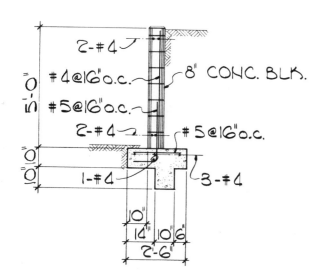

Axial load = 200 lbs./ft.
E.F.P. = 30 p.c.f.
Section 9.2.5

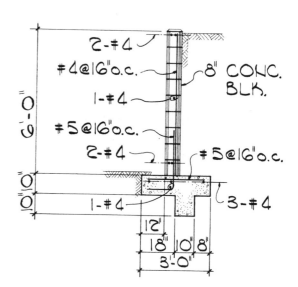

Axial load = 200 lbs./ft.
E.F.P. = 30 p.c.f.
Section 9.2.6

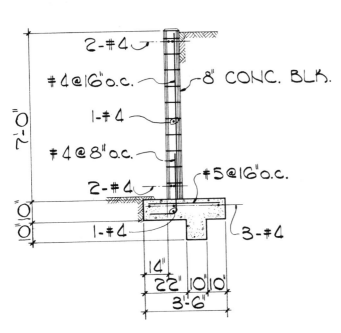

Axial load = 200 lbs./ft.
E.F.P. = 30 p.c.f.
Section 9.2.7

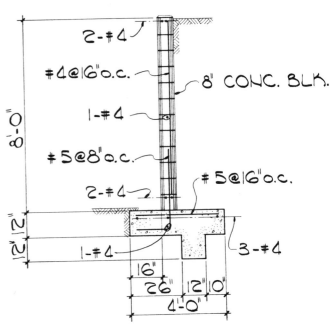

Axial load = 200 lbs./ft.
E.F.P. = 30 p.c.f.
Section 9.2.8

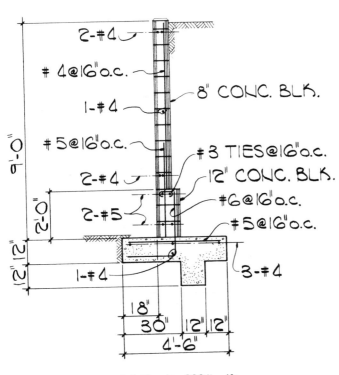

Axial load = 200 lbs./ft.
E.F.P. = 30 p.c.f.
Section 9.2.9

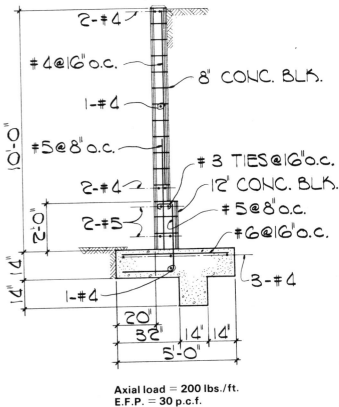

Axial load = 200 lbs./ft.
E.F.P. = 30 p.c.f.
Section 9.2.10

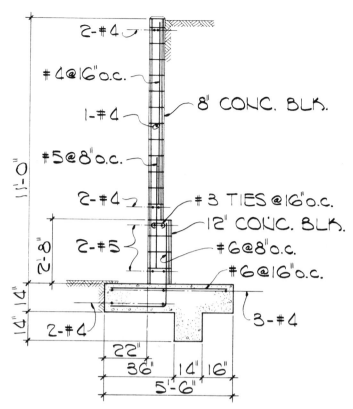

Axial load = 200 lbs./ft.
E.F.P. = 30 p.c.f.
Section 9.2.11

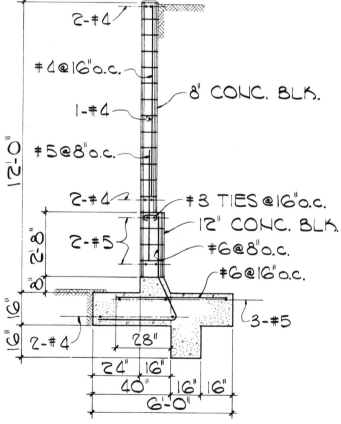

Axial load = 200 lbs./ft.
E.F.P. = 30 p.c.f.
Section 9.2.12

TABLE 9.3	GENERAL RETAINING WALLS — MASONRY STEM — SOIL AT HEEL SIDE
	SLOPE = 0 to 1 SURCHARGE = 0 lbs./sq. ft. AXIAL = 400 lbs./ft.

SOIL PRESSURE DIAGRAM

H1 = Concrete Stem
H2 = 12″ Concrete Block
H3 = 8″ Concrete Block

Stem Height	3′- 0″	4′- 0″	5′- 0″	6′- 0″	7′- 0″	8′- 0″	9′- 0″	10′- 0″	11′- 0″	12′- 0″
Ft. L ft.	1.667	2.000	2.500	3.000	3.500	4.000	4.500	5.000	5.500	6.000
Ft. T ft.	0.667	0.667	0.833	0.833	0.833	1.000	1.000	1.167	1.167	1.333
Setback	0.500	0.667	0.833	1.000	1.167	1.333	1.500	1.667	1.833	2.000
A ft.	0.500	0.833	1.167	1.500	1.833	2.167	2.500	2.667	3.000	3.333
B ft.	0.667	0.667	0.833	0.833	0.833	1.000	1.000	1.167	1.167	1.333
C ft.	0.500	0.500	0.500	0.667	0.833	0.833	1.000	1.167	1.333	1.333
D ft.	0.667	0.667	0.833	0.833	0.833	1.000	1.000	1.167	1.167	1.333
8 in. blk. ft.	3.000	4.000	5.000	6.000	7.000	8.000	7.000	8.000	8.333	8.667
12 in. blk. ft.	—	—	—	—	—	—	2.000	2.000	2.667	2.667
Conc. Stem	—	—	—	—	—	—	—	—	—	0.667
Wb ft.	—	—	—	—	—	—	—	—	—	1.333
W lbs.	1104	1381	1902	2411	2997	3806	4568	5585	6523	7752
F lbs.	202	327	510	700	920	1215	1500	1870	2220	2667
Mo ft.-lbs.	246	508	992	1595	2403	3645	5000	6962	9005	11852
Mr ft.-lbs.	1023	1597	2746	4198	6126	8933	12128	16503	21297	27664
O.T.R.	4.150	3.143	2.767	2.631	2.549	2.451	2.426	2.370	2.365	2.334
$\overline{X}$ ft.	0.703	0.788	0.922	1.079	1.242	1.389	1.561	1.708	1.884	2.040
e ft.	0.130	0.212	0.328	0.421	0.508	0.611	0.689	0.792	0.866	0.960
Me ft.-lbs.	144	293	623	1014	1522	2324	3149	4423	5646	7443
S ft.^{3}	0.463	0.667	1.042	1.500	2.042	2.667	3.375	4.167	5.042	6.000
S.P.t p.s.f.	974	1130	1359	1480	1602	1823	1948	2179	2306	2533
S.P.h p.s.f.	352	252	162	128	111	80	82	56	66	51
Friction	0.183	0.236	0.268	0.290	0.307	0.319	0.328	0.335	0.340	0.344
Inspct.	NO	NO	NO	NO	YES	YES	YES	YES	YES	YES

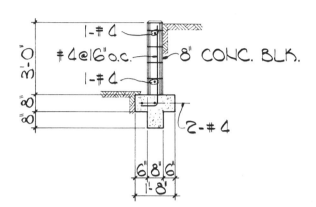

Axial load = 400 lbs./ft.
E.F.P. = 30 p.c.f.
Section 9.3.3

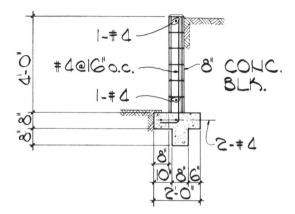

Axial load = 400 lbs./ft.
E.F.P. = 30 p.c.f.
Section 9.3.4

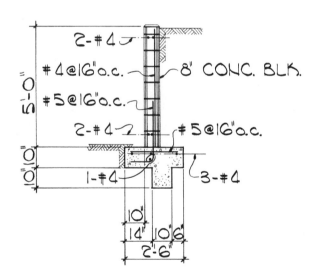

Axial load = 400 lbs./ft.
E.F.P. = 30 p.c.f.
Section 9.3.5

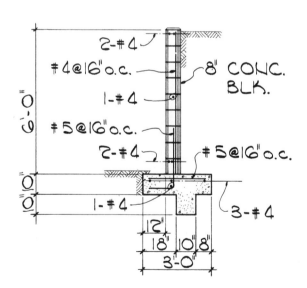

Axial load = 400 lbs./ft.
E.F.P. = 30 p.c.f.
Section 9.3.6

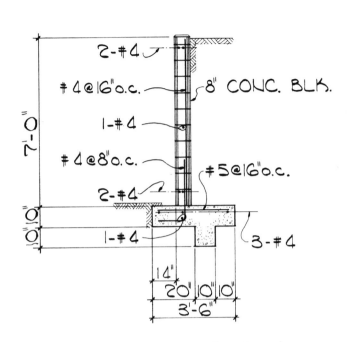

Axial load = 400 lbs./ft.
E.F.P. = 30 p.c.f.
Section 9.3.7

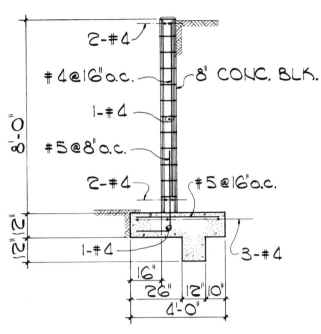

Axial load = 400 lbs./ft.
E.F.P. = 30 p.c.f.
Section 9.3.8

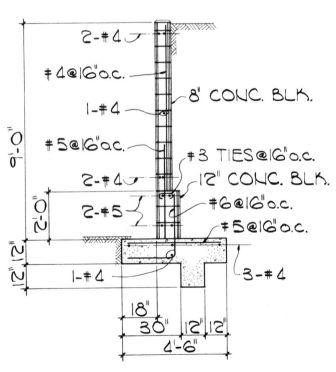

Axial load = 400 lbs./ft.
E.F.P. = 30 p.c.f.
Section 9.3.9

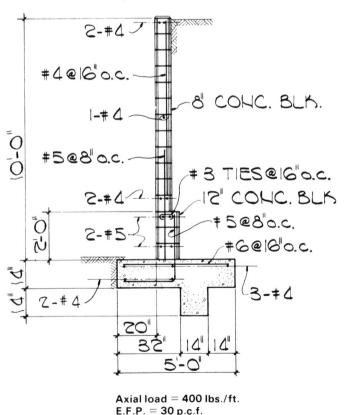

Axial load = 400 lbs./ft.
E.F.P. = 30 p.c.f.
Section 9.3.10

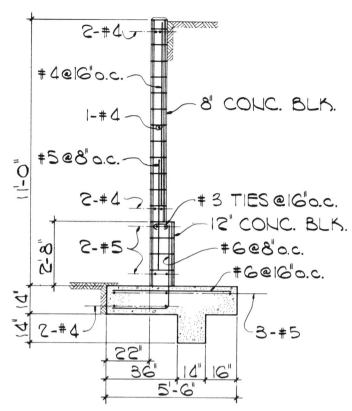

2-#4

#4@16"o.c.

1-#4

#5@8"o.c.

8" CONC. BLK.

2-#4

#5@8"o.c.

2-#4

#3 TIES @16"o.c.

12" CONC. BLK.

2-#5

#6@8"o.c.

#6@16"o.c.

3-#5

2-#4

1'-0"

2'-8"

4"

4"

22"

36" 14" 16"

5'-6"

Axial load = 400 lbs./ft.
E.F.P. = 30 p.c.f.
Section 9.3.11

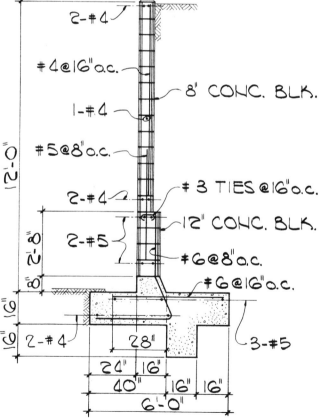

2-#4

#4@16"o.c.

1-#4

#5@8"o.c.

8" CONC. BLK.

#3 TIES @16"o.c.

2-#4

12" CONC. BLK.

2-#5

#6@8"o.c.

#6@16"o.c.

3-#5

2-#4

2'-0"

2'-8"

8"

6"

6"

28"

24" 16"

40" 16" 16"

6'-0"

Axial load = 400 lbs./ft.
E.F.P. = 30 p.c.f.
Section 9.3.12

TABLE 9.4	GENERAL RETAINING WALLS – MASONRY STEM – SOIL AT HEEL SIDE
	SLOPE = 0 to 1 SURCHARGE = 0 lbs./sq. ft. AXIAL = 600 lbs./ft.

SOIL PRESSURE DIAGRAM

AXIAL LOAD

H1 = Concrete Stem
H2 = 12″ Concrete Block
H3 = 8″ Concrete Block

Stem Height	3′- 0″	4′- 0″	5′- 0″	6′- 0″	7′- 0″	8′- 0″	9′- 0″	10′- 0″	11′- 0″	12′- 0″
Ft. L ft.	1.667	2.000	2.500	3.000	3.500	4.000	4.500	5.000	5.500	6.000
Ft. T ft.	0.667	0.667	0.833	0.833	0.833	1.000	1.000	1.167	1.167	1.333
Setback	0.500	0.667	0.833	1.000	1.167	1.333	1.500	1.667	1.833	2.000
A ft.	0.500	0.833	1.167	1.500	1.833	2.167	2.500	2.667	3.000	3.333
B ft.	0.667	0.667	0.833	0.833	0.833	1.000	1.000	1.167	1.167	1.333
C ft.	0.500	0.500	0.500	0.667	0.833	0.833	1.000	1.167	1.333	1.333
D ft.	0.667	0.667	0.833	0.833	0.833	1.000	1.000	1.167	1.167	1.333
8 in. blk. ft.	3.000	4.000	5.000	6.000	7.000	8.000	7.000	8.000	8.333	8.000
12 in. blk. ft.	—	—	—	—	—	—	2.000	2.000	2.667	3.333
Conc. Stem	—	—	—	—	—	—	—	—	—	0.667
Wb ft.	—	—	—	—	—	—	—	—	—	1.333
W lbs.	1304	1581	2102	2611	3197	4006	4768	5795	6723	7962
F lbs.	202	327	510	700	920	1215	1500	1870	2220	2667
Mo ft.-lbs.	246	508	992	1595	2403	3645	5000	6962	9005	11852
Mr ft.-lbs.	1189	1797	2980	4465	6426	9267	12495	16937	21730	28171
O.T.R.	4.826	3.536	3.002	2.799	2.674	2.542	2.499	2.433	2.413	2.377
$\overline{X}$ ft.	0.723	0.815	0.946	1.099	1.258	1.403	1.572	1.721	1.893	2.050
e ft.	0.110	0.185	0.304	0.401	0.492	0.597	0.678	0.779	0.857	0.950
Me ft.-lbs.	144	293	640	1047	1572	2390	3232	4513	5763	7566
S ft.^{3}	0.463	0.667	1.042	1.500	2.042	2.667	3.375	4.167	5.042	6.000
S.P.t p.s.f.	1094	1230	1455	1569	1684	1898	2017	2242	2365	2588
S.P.h p.s.f.	472	352	226	172	143	105	102	76	79	66
Friction	0.155	0.207	0.243	0.268	0.288	0.303	0.315	0.323	0.330	0.335
Inspct.	NO	NO	NO	NO	YES	YES	YES	YES	YES	YES

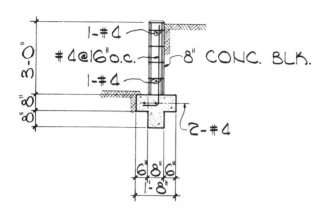

Axial load = 600 lbs./ft.
E.F.P. = 30 p.c.f.
Section 9.4.3

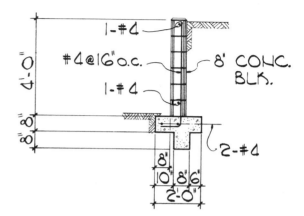

Axial load = 600 lbs./ft.
E.F.P. = 30 p.c.f.
Section 9.4.4

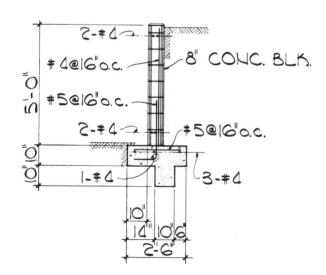

Axial load = 600 lbs./ft.
E.F.P. = 30 p.c.f.
Section 9.4.5

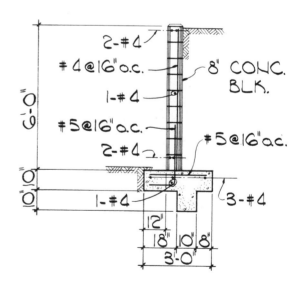

Axial load = 600 lbs./ft.
E.F.P. = 30 p.c.f.
Section 9.4.6

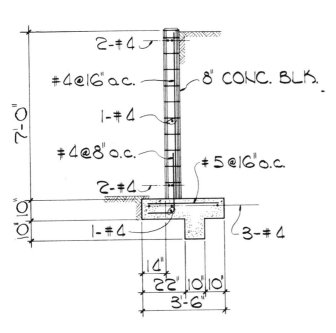

Axial load = 600 lbs./ft.
E.F.P. = 30 p.c.f.
Section 9.4.7

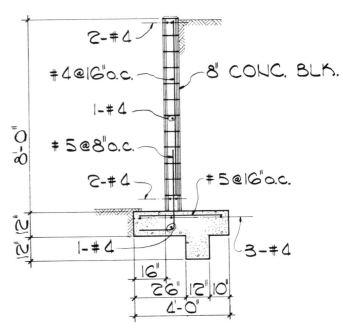

Axial load = 600 lbs./ft.
E.F.P. = 30 p.c.f.
Section 9.4.8

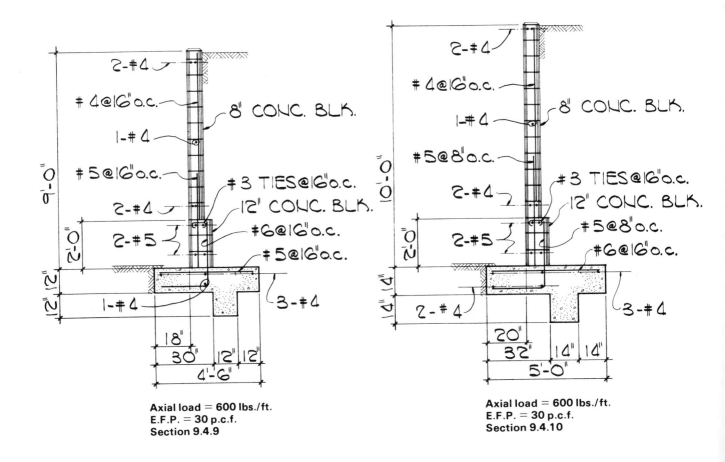

Axial load = 600 lbs./ft.
E.F.P. = 30 p.c.f.
Section 9.4.9

Axial load = 600 lbs./ft.
E.F.P. = 30 p.c.f.
Section 9.4.10

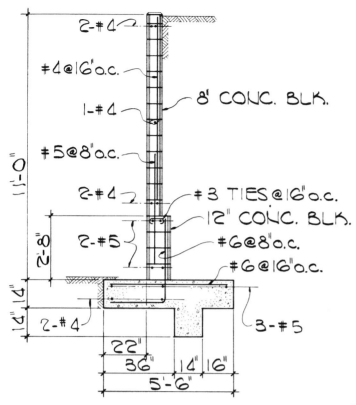

2-#4

#4@16"o.c.

1-#4

8" CONC. BLK.

#5@8"o.c.

2-#4

#3 TIES@16"o.c.

12" CONC. BLK.

2-#5

#6@8"o.c.

#6@16"o.c.

13'-0"

2'-8"

14"

2-#4

3-#5

22"

36"

14" 16"

5'-6"

Axial load = 600 lbs./ft.
E.F.P. = 30 p.c.f.
Section 9.4.11

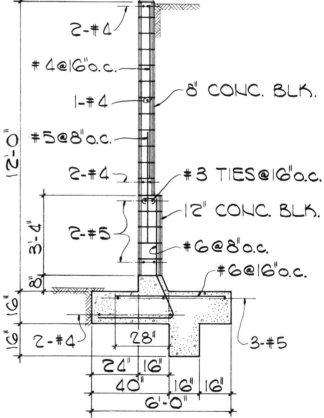

2-#4

#4@16"o.c.

1-#4

8" CONC. BLK.

#5@8"o.c.

2-#4

#3 TIES@16"o.c.

12" CONC. BLK.

2-#5

#6@8"o.c.

#6@16"o.c.

12'-0"

3'-4"

8"

16"

16"

2-#4

3-#5

28"

24" 16"

40"

16" 16"

6'-0"

Axial load = 600 lbs./ft.
E.F.P. = 30 p.c.f.
Section 9.4.12

TABLE 9.5	GENERAL RETAINING WALLS – MASONRY STEM – SOIL AT HEEL SIDE
	SLOPE = 0 to 1 SURCHARGE = 0 lbs./sq. ft. AXIAL = 800 lbs./ft.

SOIL PRESSURE DIAGRAM

H1 = Concrete Stem
H2 = 12″ Concrete Block
H3 = 8″ Concrete Block

Stem Height	3'- 0"	4'- 0"	5'- 0"	6'- 0"	7'- 0"	8'- 0"	9'- 0"	10'- 0"	11'- 0"	12'- 0"
Ft L ft.	1.667	2.000	2.500	3.000	3.500	4.000	4.500	5.000	5.500	6.000
Ft T ft.	0.667	0.667	0.833	0.833	0.833	1.000	1.000	1.167	1.167	1.333
Setback ft.	0.500	0.667	0.833	1.000	1.167	1.333	1.500	1.667	1.833	2.000
A ft.	0.500	0.833	1.167	1.500	1.833	2.167	2.500	2.667	3.000	3.333
B ft.	0.667	0.667	0.833	0.833	0.833	1.000	1.000	1.167	1.167	1.333
C ft.	0.500	0.500	0.500	0.667	0.833	0.833	1.000	1.167	1.333	1.333
D ft.	0.667	0.667	0.833	0.833	0.833	1.000	1.000	1.167	1.167	1.333
8 in. blk. ft.	3.000	4.000	5.000	6.000	7.000	8.000	7.000	8.000	8.333	8.000
12 in. blk. ft.	–	–	–	–	–	–	2.000	2.000	2.667	3.333
Conc. Stem	–	–	–	–	–	–	–	–	–	0.667
Wb ft.	–	–	–	–	–	–	–	–	–	1.333
W lbs.	1504	1781	2302	2811	3397	4206	4968	5995	6923	8162
F lbs.	202	327	510	700	920	1215	1500	1870	2220	2667
Mo ft.-lbs.	246	508	992	1595	2403	3645	5000	6962	9005	11852
Mr ft.-lbs.	1356	1997	3213	4731	6726	9600	12862	17337	22163	28637
O.T.R;	5.502	3.930	3.237	2.966	2.799	2.634	2.572	2.490	2.461	2.416
$\overline{X}$ ft.	0.738	0.836	0.965	1.116	1.272	1.416	1.583	1.731	1.901	2.057
e ft.	0.096	0.164	0.285	0.384	0.478	0.584	0.667	0.769	0.849	0.943
Me ft.-lbs.	144	293	657	1081	1622	2457	3316	4613	5879	7699
S ft.[3]	0.463	0.667	1.042	1.500	2.042	2.667	3.375	4.167	5.042	6.000
S.P.t p.s.f.	1214	1330	1551	1657	1765	1973	2086	2306	2425	2643
S.P.h p.s.f.	592	452	290	217	176	130	122	92	93	77
Friction	0.134	0.183	0.222	0.249	0.271	0.289	0.302	0.312	0.321	0.327
Inspct.	NO	NO	NO	NO	YES	YES	YES	YES	YES	YES

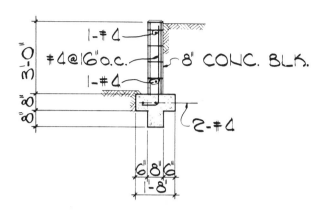

Axial load = 800 lbs./ft.
E.F.P. = 30 p.c.f.
Section 9.5.3

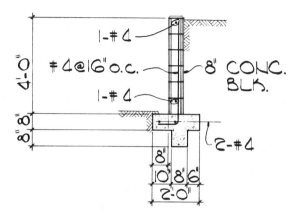

Axial load = 800 lbs./ft.
E.F.P. = 30 p.c.f.
Section 9.5.4

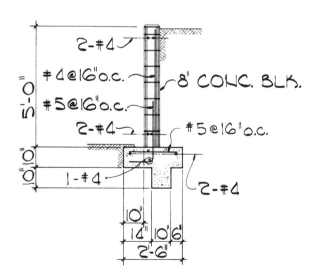

Axial load = 800 lbs./ft.
E.F.P. = 30 p.c.f.
Section 9.5.5

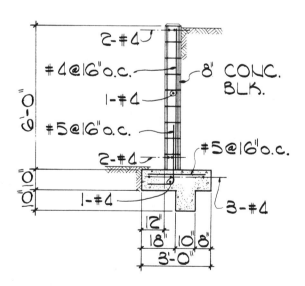

Axial load = 800 lbs./ft.
E.F.P. = 30 p.c.f.
Section 9.5.6

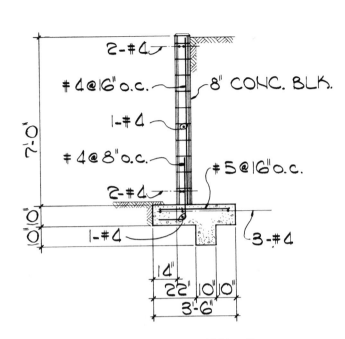

Axial load = 800 lbs./ft.
E.F.P. = 30 p.c.f.
Section 9.5.7

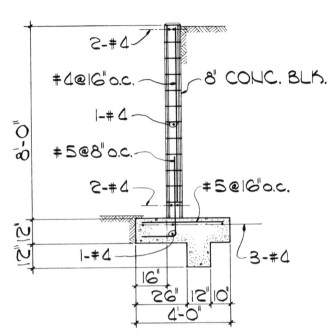

Axial load = 800 lbs./ft.
E.F.P. = 30 p.c.f.
Section 9.5.8

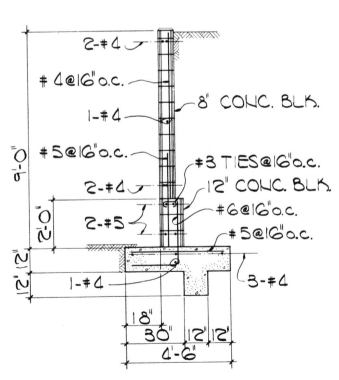

Axial load = 800 lbs./ft.
E.F.P. = 30 p.c.f.
Section 9.5.9

Axial load = 800 lbs./ft.
E.F.P. = 30 p.c.f.
Section 9.5.10

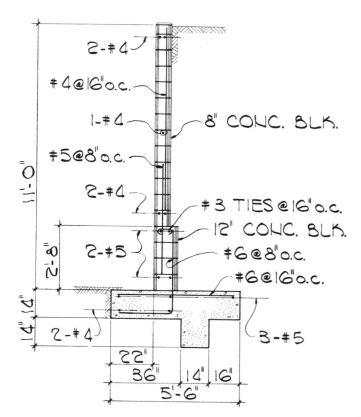

2-#4

#4@16"o.c.

1-#4   8" CONC. BLK.

#5@8"o.c.

2-#4

#3 TIES @16"o.c.

12" CONC. BLK.

2-#5   #6@8"o.c.

#6@16"o.c.

2-#4   3-#5

22"

36" 14" 16"

5'-6"

1'-0"

2'-8"

1'-4"

Axial load = 800 lbs./ft.
E.F.P. = 30 p.c.f.
Section 9.5.11

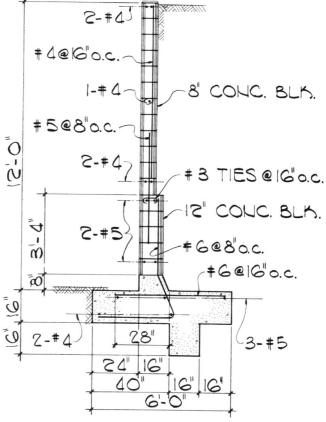

2-#4

#4@16"o.c.

1-#4   8" CONC. BLK.

#5@8"o.c.

2-#4   #3 TIES @16"o.c.

12" CONC. BLK.

2-#5   #6@8"o.c.

#6@16"o.c.

2-#4   28"   3-#5

24" 16"

40" 16" 16"

6'-0"

2'-0"

3'-4"

8"

6" 6"

Axial load = 800 lbs./ft.
E.F.P. = 30 p.c.f.
Section 9.5.12

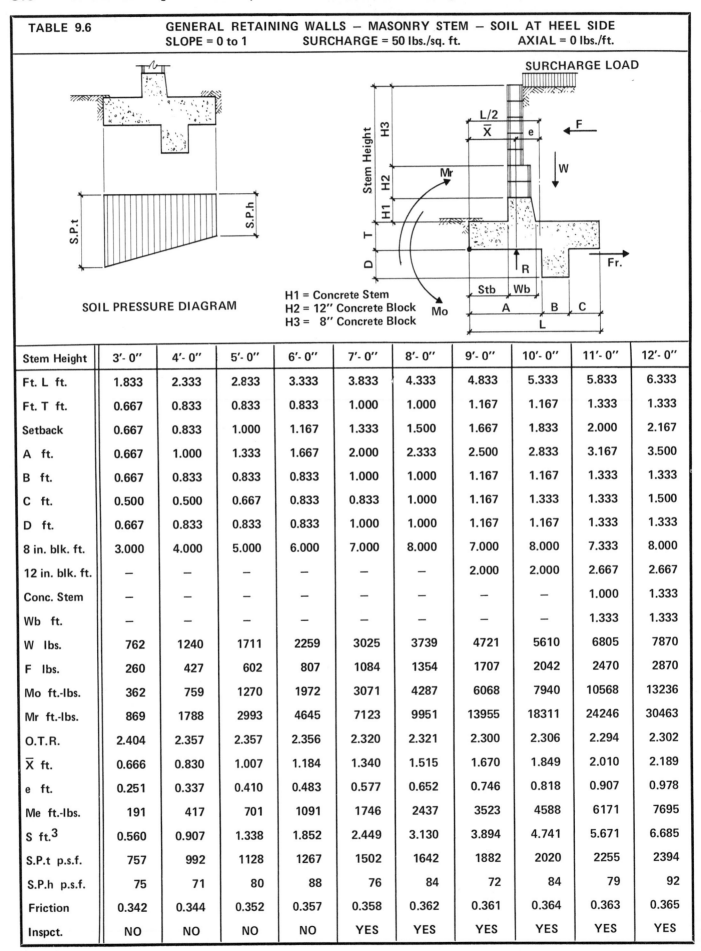

TABLE 9.6	GENERAL RETAINING WALLS — MASONRY STEM — SOIL AT HEEL SIDE SLOPE = 0 to 1 SURCHARGE = 50 lbs./sq. ft. AXIAL = 0 lbs./ft.

SOIL PRESSURE DIAGRAM

H1 = Concrete Stem
H2 = 12″ Concrete Block
H3 = 8″ Concrete Block

Stem Height	3′- 0″	4′- 0″	5′- 0″	6′- 0″	7′- 0″	8′- 0″	9′- 0″	10′- 0″	11′- 0″	12′- 0″
Ft. L ft.	1.833	2.333	2.833	3.333	3.833	4.333	4.833	5.333	5.833	6.333
Ft. T ft.	0.667	0.833	0.833	0.833	1.000	1.000	1.167	1.167	1.333	1.333
Setback	0.667	0.833	1.000	1.167	1.333	1.500	1.667	1.833	2.000	2.167
A ft.	0.667	1.000	1.333	1.667	2.000	2.333	2.500	2.833	3.167	3.500
B ft.	0.667	0.833	0.833	0.833	1.000	1.000	1.167	1.167	1.333	1.333
C ft.	0.500	0.500	0.667	0.833	0.833	1.000	1.167	1.333	1.333	1.500
D ft.	0.667	0.833	0.833	0.833	1.000	1.000	1.167	1.167	1.333	1.333
8 in. blk. ft.	3.000	4.000	5.000	6.000	7.000	8.000	7.000	8.000	7.333	8.000
12 in. blk. ft.	—	—	—	—	—	—	2.000	2.000	2.667	2.667
Conc. Stem	—	—	—	—	—	—	—	—	1.000	1.333
Wb ft.	—	—	—	—	—	—	—	—	1.333	1.333
W lbs.	762	1240	1711	2259	3025	3739	4721	5610	6805	7870
F lbs.	260	427	602	807	1084	1354	1707	2042	2470	2870
Mo ft.-lbs.	362	759	1270	1972	3071	4287	6068	7940	10568	13236
Mr ft.-lbs.	869	1788	2993	4645	7123	9951	13955	18311	24246	30463
O.T.R.	2.404	2.357	2.357	2.356	2.320	2.321	2.300	2.306	2.294	2.302
$\overline{X}$ ft.	0.666	0.830	1.007	1.184	1.340	1.515	1.670	1.849	2.010	2.189
e ft.	0.251	0.337	0.410	0.483	0.577	0.652	0.746	0.818	0.907	0.978
Me ft.-lbs.	191	417	701	1091	1746	2437	3523	4588	6171	7695
S ft.^{3}	0.560	0.907	1.338	1.852	2.449	3.130	3.894	4.741	5.671	6.685
S.P.t p.s.f.	757	992	1128	1267	1502	1642	1882	2020	2255	2394
S.P.h p.s.f.	75	71	80	88	76	84	72	84	79	92
Friction	0.342	0.344	0.352	0.357	0.358	0.362	0.361	0.364	0.363	0.365
Inspct.	NO	NO	NO	NO	YES	YES	YES	YES	YES	YES

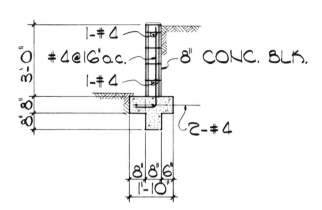

Surcharge load = 50 p.s.f.
E.F.P. = 30 p.c.f.
Section 9.6.3

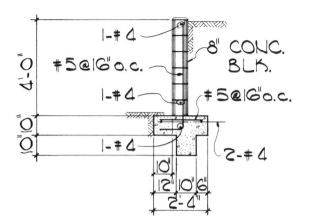

Surcharge load = 50 p.s.f.
E.F.P. = 30 p.c.f.
Section 9.6.4

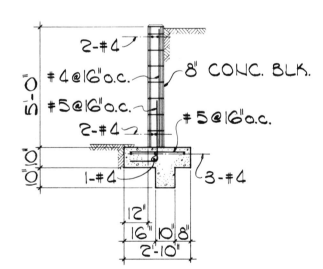

Surcharge load = 50 p.s.f.
E.F.P. = 30 p.c.f.
Section 9.6.5

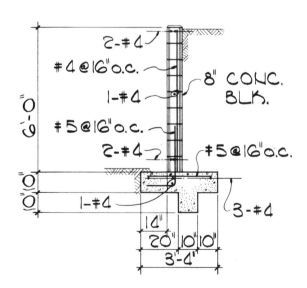

Surcharge load = 50 p.s.f.
E.F.P. = 30 p.c.f.
Section 9.6.6

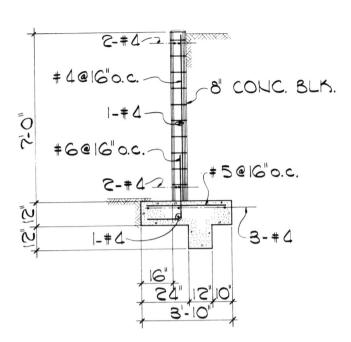

Surcharge load = 50 p.s.f.
E.F.P. = 30 p.c.f.
Section 9.6.7

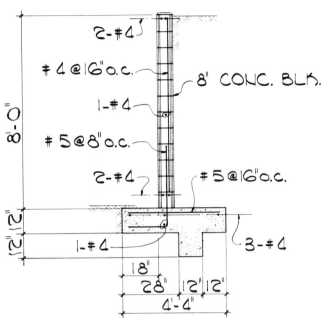

Surcharge load = 50 p.s.f.
E.F.P. = 30 p.c.f.
Section 9.6.8

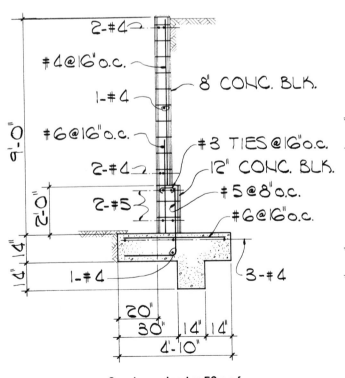

Surcharge load = 50 p.s.f.
E.F.P. = 30 p.c.f.
Section 9.6.9

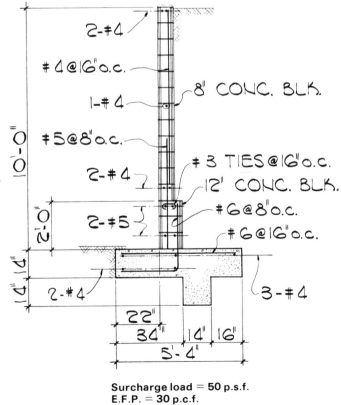

Surcharge load = 50 p.s.f.
E.F.P. = 30 p.c.f.
Section 9.6.10

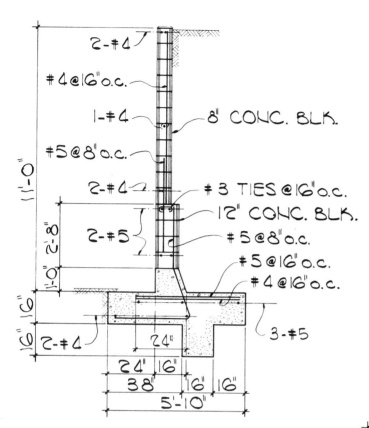

2-#4

#4@16"o.c.

1-#4 — 8" CONC. BLK.

#5@8"o.c.

2-#4 — #3 TIES @16"o.c.
12" CONC. BLK.

2-#5 — #5@8"o.c.
#5@16"o.c.
#4@16"o.c.

11'-0"

2'-8"

1'-0"

6"

6"

2-#4

3-#5

24"

24" 16"

38"   16" 16"

5'-0"

Surcharge load = 50 p.s.f.
E.F.P. = 30 p.c.f.
Section 9.6.11

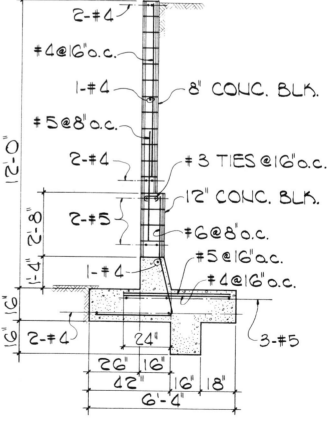

2-#4

#4@16"o.c.

1-#4 — 8" CONC. BLK.

#5@8"o.c.

2-#4 — #3 TIES @16"o.c.

12" CONC. BLK.

2-#5 — #6@8"o.c.
#5@16"o.c.

1-#4 — #4@16"o.c.

12'-0"

2'-8"

1'-4"

6"

6"

2-#4

3-#5

24"

26" 16"

42"   16" 18"

6'-4"

Surcharge load = 50 p.s.f.
E.F.P. = 30 p.c.f.
Section 9.6.12

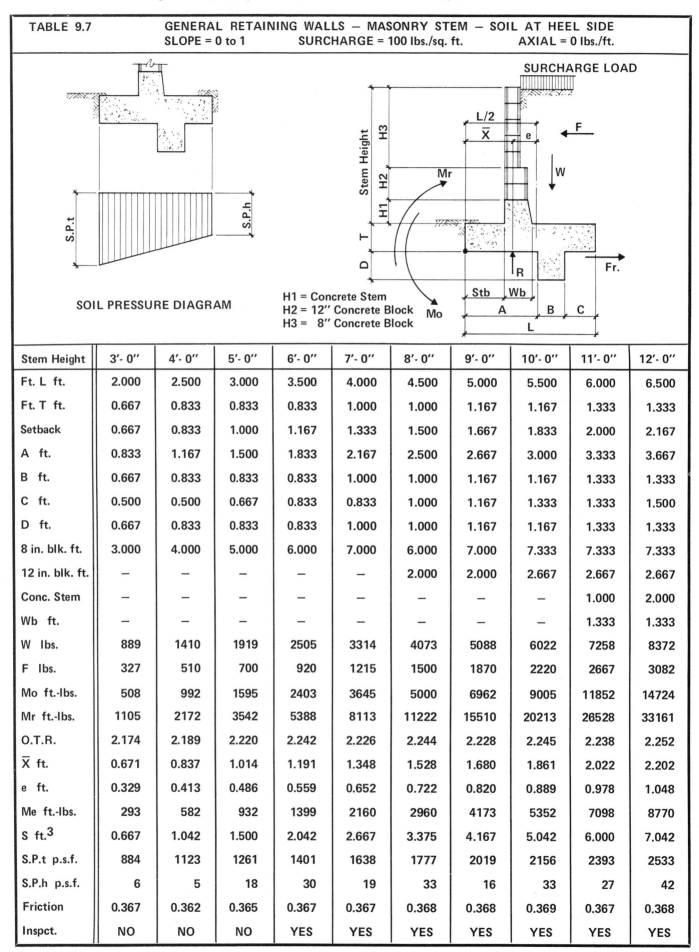

TABLE 9.7 — GENERAL RETAINING WALLS — MASONRY STEM — SOIL AT HEEL SIDE
SLOPE = 0 to 1   SURCHARGE = 100 lbs./sq. ft.   AXIAL = 0 lbs./ft.

SOIL PRESSURE DIAGRAM

H1 = Concrete Stem
H2 = 12″ Concrete Block
H3 = 8″ Concrete Block

Stem Height	3′- 0″	4′- 0″	5′- 0″	6′- 0″	7′- 0″	8′- 0″	9′- 0″	10′- 0″	11′- 0″	12′- 0″
Ft. L ft.	2.000	2.500	3.000	3.500	4.000	4.500	5.000	5.500	6.000	6.500
Ft. T ft.	0.667	0.833	0.833	0.833	1.000	1.000	1.167	1.167	1.333	1.333
Setback	0.667	0.833	1.000	1.167	1.333	1.500	1.667	1.833	2.000	2.167
A ft.	0.833	1.167	1.500	1.833	2.167	2.500	2.667	3.000	3.333	3.667
B ft.	0.667	0.833	0.833	0.833	1.000	1.000	1.167	1.167	1.333	1.333
C ft.	0.500	0.500	0.667	0.833	0.833	1.000	1.167	1.333	1.333	1.500
D ft.	0.667	0.833	0.833	0.833	1.000	1.000	1.167	1.167	1.333	1.333
8 in. blk. ft.	3.000	4.000	5.000	6.000	7.000	6.000	7.000	7.333	7.333	7.333
12 in. blk. ft.	—	—	—	—	—	2.000	2.000	2.667	2.667	2.667
Conc. Stem	—	—	—	—	—	—	—	—	1.000	2.000
Wb ft.	—	—	—	—	—	—	—	—	1.333	1.333
W lbs.	889	1410	1919	2505	3314	4073	5088	6022	7258	8372
F lbs.	327	510	700	920	1215	1500	1870	2220	2667	3082
Mo ft.-lbs.	508	992	1595	2403	3645	5000	6962	9005	11852	14724
Mr ft.-lbs.	1105	2172	3542	5388	8113	11222	15510	20213	26528	33161
O.T.R.	2.174	2.189	2.220	2.242	2.226	2.244	2.228	2.245	2.238	2.252
$\overline{X}$ ft.	0.671	0.837	1.014	1.191	1.348	1.528	1.680	1.861	2.022	2.202
e ft.	0.329	0.413	0.486	0.559	0.652	0.722	0.820	0.889	0.978	1.048
Me ft.-lbs.	293	582	932	1399	2160	2960	4173	5352	7098	8770
S ft.^{3}	0.667	1.042	1.500	2.042	2.667	3.375	4.167	5.042	6.000	7.042
S.P.t p.s.f.	884	1123	1261	1401	1638	1777	2019	2156	2393	2533
S.P.h p.s.f.	6	5	18	30	19	33	16	33	27	42
Friction	0.367	0.362	0.365	0.367	0.367	0.368	0.368	0.369	0.367	0.368
Inspct.	NO	NO	NO	YES	YES	YES	YES	YES	YES	YES

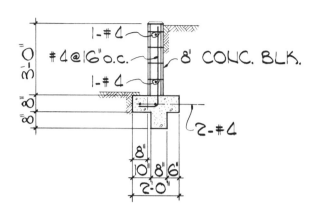

Surcharge load = 100 p.s.f.
E.F.P. = 30 p.c.f.
Section 9.7.3

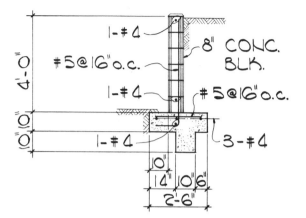

Surcharge load = 100 p.s.f.
E.F.P. = 30 p.c.f.
Section 9.7.4

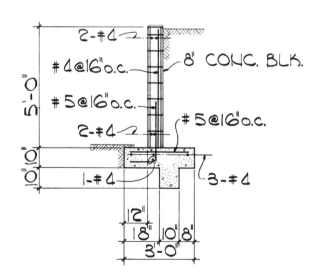

Surcharge load = 100 p.s.f.
E.F.P. = 30 p.c.f.
Section 9.7.5

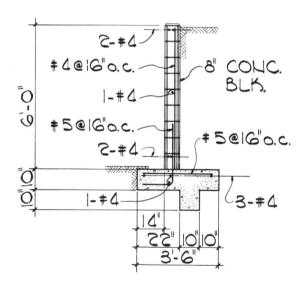

Surcharge load = 100 p.s.f.
E.F.P. = 30 p.c.f.
Section 9.7.6

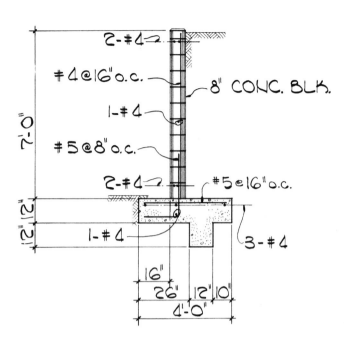

Surcharge load = 100 p.s.f.
E.F.P. = 30 p.c.f.
Section 9.7.7

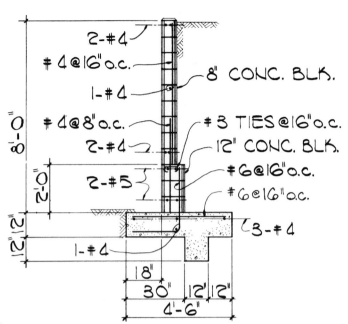

Surcharge load = 100 p.s.f.
E.F.P. = 30 p.c.f.
Section 9.7.8

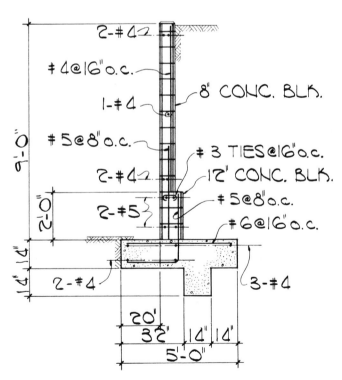

Surcharge load = 100 p.s.f.
E.F.P. = 30 p.c.f.
Section 9.7.9

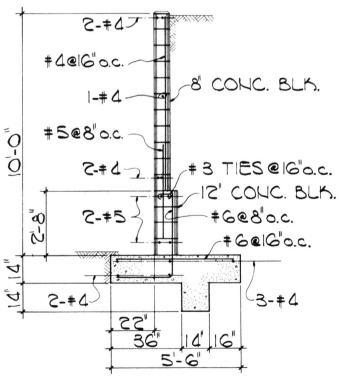

Surcharge load = 100 p.s.f.
E.F.P. = 30 p.c.f.
Section 9.7.10

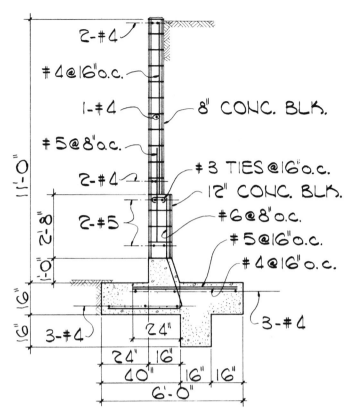

2-#4
#4@16"o.c.
1-#4
#5@8"o.c.
2-#4
2-#5
8" CONC. BLK.
#3 TIES@16"o.c.
12" CONC. BLK.
#6@8"o.c.
#5@16"o.c.
#4@16"o.c.
3-#4
3-#4
24"
24" 16"
40" 16" 16"
6'-0"
11'-0"
2'-8"
1'-0"
6"
6"

Surcharge load = 100 p.s.f.
E.F.P. = 30 p.c.f.
Section 9.7.11

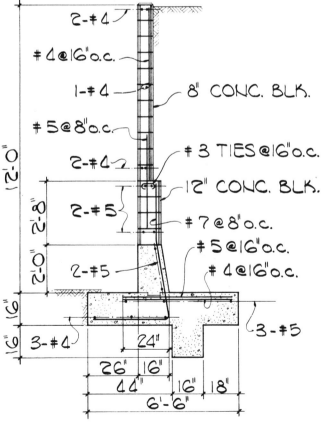

2-#4
#4@16"o.c.
1-#4
#5@8"o.c.
2-#4
2-#5
2-#5
8" CONC. BLK.
#3 TIES@16"o.c.
12" CONC. BLK.
#7@8"o.c.
#5@16"o.c.
#4@16"o.c.
3-#4
3-#5
24"
26" 16"
44" 16" 18"
6'-6"
12'-0"
2'-8"
2'-0"
6"
6"

Surcharge load = 100 p.s.f.
E.F.P. = 30 p.c.f.
Section 9.7.12

TABLE 9.8	GENERAL RETAINING WALLS — MASONRY STEM — SOIL AT HEEL SIDE
	SLOPE = 0 to 1 SURCHARGE = 200 lbs./sq. ft. AXIAL = 0 lbs./ft.

SOIL PRESSURE DIAGRAM

H1 = Concrete Stem
H2 = 12″ Concrete Block
H3 = 8″ Concrete Block

Stem Height	3'- 0"	4'- 0"	5'- 0"	6'- 0"	7'- 0"	8'- 0"	9'- 0"	10'- 0"	11'- 0"	12'- 0"
Ft. L ft.	2.667	3.167	3.500	4.000	4.500	5.000	5.500	6.000	6.500	7.000
Ft. T ft.	0.833	0.833	0.833	1.000	1.000	1.167	1.167	1.333	1.333	1.333
Setback	0.833	1.000	1.167	1.333	1.500	1.667	1.833	2.000	2.167	2.333
A ft.	1.333	1.667	1.833	2.167	2.500	2.667	3.000	3.333	3.667	4.000
B ft.	0.833	0.833	0.833	1.000	1.000	1.167	1.167	1.333	1.333	1.333
C ft.	0.500	0.667	0.833	0.833	1.000	1.167	1.333	1.333	1.500	1.667
D ft.	0.833	0.833	0.833	1.000	1.000	1.167	1.167	1.333	1.333	1.333
8 in. blk. ft.	3.000	4.000	5.000	6.000	5.000	6.000	6.333	6.667	6.667	6.667
12 in. blk. ft.	—	—	—	—	2.000	2.000	2.667	2.667	2.667	2.667
Conc. Stem	—	—	—	—	—	—	—	0.667	1.667	2.667
Wb ft.	—	—	—	—	—	—	—	1.333	1.333	1.333
W lbs.	1422	1948	2413	3222	3977	4990	5919	7143	8253	9441
F lbs.	510	700	920	1215	1500	1870	2220	2667	3082	3527
Mo ft.-lbs.	992	1595	2403	3645	5000	6962	9005	11852	14724	18025
Mr ft.-lbs.	2352	3809	5250	7960	11047	15314	19992	26250	32855	40474
O.T.R.	2.369	2.388	2.185	2.184	2.209	2.200	2.220	2.215	2.231	2.245
$\overline{X}$ ft.	0.956	1.137	1.180	1.339	1.520	1.674	1.856	2.016	2.197	2.378
e ft.	0.377	0.447	0.570	0.661	0.730	0.826	0.894	0.984	1.053	1.122
Me ft.-lbs.	537	870	1376	2129	2902	4123	5290	7032	8691	10594
S ft.^{3}	1.185	1.671	2.042	2.667	3.375	4.167	5.042	6.000	7.042	8.167
S.P.t p.s.f.	986	1136	1364	1604	1744	1988	2125	2363	2504	2646
S.P.h p.s.f.	80	94	15	7	24	8	27	19	35	52
Friction	0.359	0.360	0.381	0.377	0.377	0.375	0.375	0.373	0.373	0.374
Inspct.	NO	NO	YES	YES	YES	YES	YES	YES	YES	YES

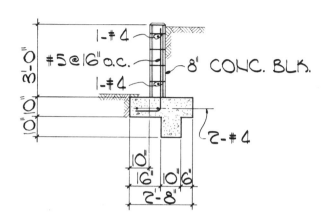

Surcharge load = 200 p.s.f.
E.F.P. = 30 p.c.f.
Section 9.8.3

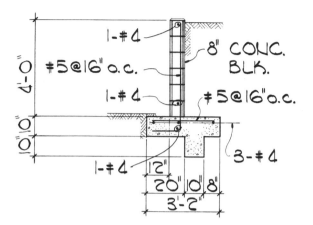

Surcharge load = 200 p.s.f.
E.F.P. = 30 p.c.f.
Section 9.8.4

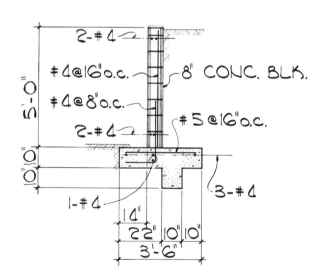

Surcharge load = 200 p.s.f.
E.F.P. = 30 p.c.f.
Section 9.8.5

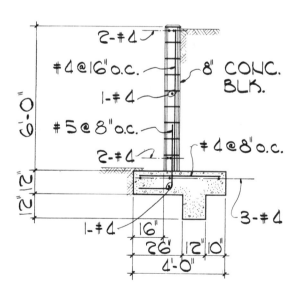

Surcharge load = 200 p.s.f.
E.F.P. = 30 p.c.f.
Section 9.8.6

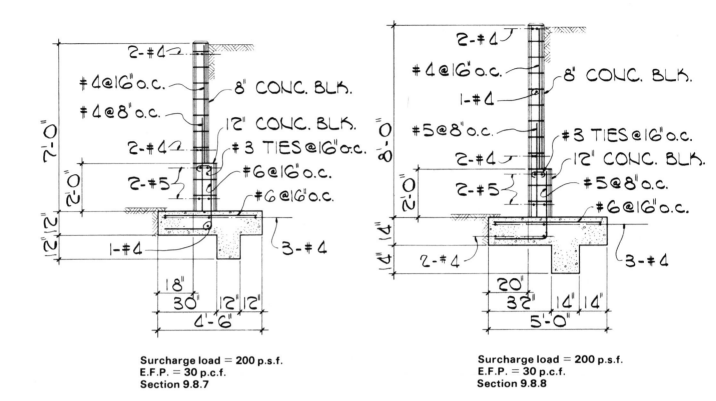

Surcharge load = 200 p.s.f.
E.F.P. = 30 p.c.f.
Section 9.8.7

Surcharge load = 200 p.s.f.
E.F.P. = 30 p.c.f.
Section 9.8.8

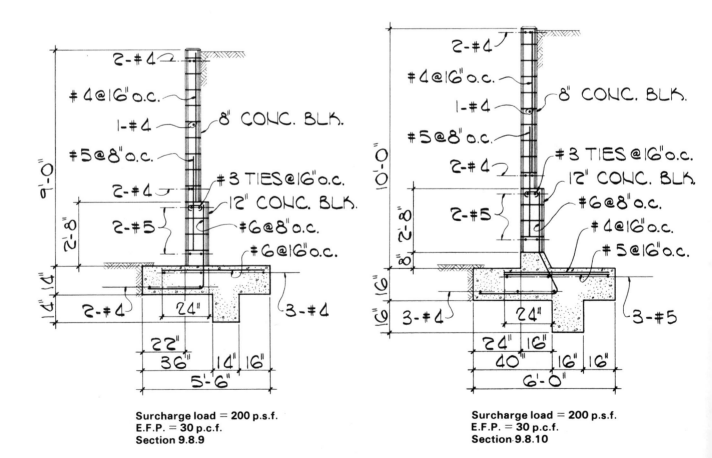

Surcharge load = 200 p.s.f.
E.F.P. = 30 p.c.f.
Section 9.8.9

Surcharge load = 200 p.s.f.
E.F.P. = 30 p.c.f.
Section 9.8.10

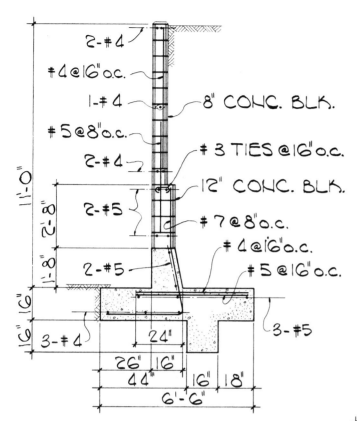

2-#4

#4 @ 16" o.c.

1-#4   8" CONC. BLK.

#5 @ 8" o.c.

2-#4   #3 TIES @ 16" o.c.

12" CONC. BLK.

2-#5

#7 @ 8" o.c.

#4 @ 16" o.c.

2-#5   #5 @ 16" o.c.

11'-0"

2'-8"

1'-8"

6"

6"

3-#4

24"

3-#5

26" | 16"

44"

16" | 18"

6'-6"

Surcharge load = 200 p.s.f.
E.F.P. = 30 p.c.f.
Section 9.8.11

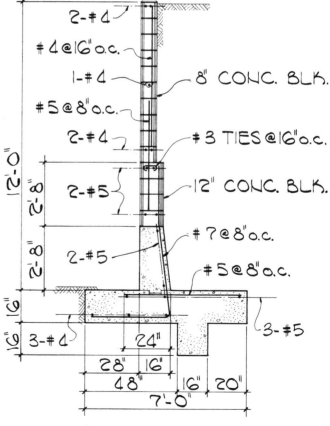

2-#4

#4 @ 16" o.c.

1-#4   8" CONC. BLK.

#5 @ 8" o.c.

2-#4   #3 TIES @ 16" o.c.

12" CONC. BLK.

2-#5

#7 @ 8" o.c.

2-#5

#5 @ 8" o.c.

12'-0"

2'-8"

2'-8"

6"

6"

3-#4

24"

3-#5

28" | 16"

48"

16" | 20"

7'-0"

Surcharge load = 200 p.s.f.
E.F.P. = 30 p.c.f.
Section 9.8.12

TABLE 9.9	GENERAL RETAINING WALLS — MASONRY STEM — SOIL AT HEEL SIDE
	SLOPE = 5 to 1 SURCHARGE = 0 lbs./sq. ft. AXIAL = 0 lbs./ft.

SOIL PRESSURE DIAGRAM

H1 = Concrete Stem
H2 = 12" Concrete Block
H3 = 8" Concrete Block

Stem Height	3'- 0"	4'- 0"	5'- 0"	6'- 0"	7'- 0"	8'- 0"	9'- 0"	10'- 0"	11'- 0"	12'- 0"
Ft. L ft.	1.667	2.167	2.667	3.167	3.667	4.167	4.667	5.167	5.667	6.167
Ft. T ft.	0.667	0.833	0.833	0.833	1.000	1.000	1.167	1.167	1.333	1.333
Setback	0.500	0.667	0.833	1.000	1.167	1.333	1.500	1.667	1.833	2.000
A ft.	0.500	0.833	1.333	1.667	1.833	2.167	2.500	2.833	3.000	3.333
B ft.	0.667	0.833	0.833	0.833	1.000	1.000	1.167	1.167	1.333	1.333
C ft.	0.500	0.500	0.500	0.667	0.833	1.000	1.000	1.167	1.333	1.500
D ft.	0.667	0.833	0.833	0.833	1.000	1.000	1.167	1.167	1.333	1.333
8 in. blk. ft.	3.000	4.000	5.000	6.000	7.000	8.000	7.000	8.000	8.333	8.000
12 in. blk. ft.	—	—	—	—	—	—	2.000	2.000	2.667	2.667
Conc. Stem	—	—	—	—	—	—	—	—	—	1.333
Wb ft.	—	—	—	—	—	—	—	—	—	1.333
W lbs.	710	1169	1628	2166	2922	3633	4603	5505	6680	7778
F lbs.	215	374	544	747	1024	1296	1654	1995	2434	2844
Mo ft.-lbs.	263	602	1059	1702	2731	3888	5604	7426	10006	12642
Mr ft.-lbs.	696	1501	2607	4129	6425	9101	12922	17150	22782	28911
O.T.R.	2.649	2.492	2.463	2.426	2.353	2.341	2.306	2.309	2.277	2.287
$\overline{X}$ ft.	0.611	0.769	0.951	1.120	1.264	1.435	1.590	1.766	1.913	2.092
e ft.	0.223	0.314	0.382	0.463	0.569	0.648	0.744	0.817	0.921	0.992
Me ft.-lbs.	158	367	622	1003	1663	2355	3423	4496	6151	7712
S ft.^{3}	0.463	0.782	1.185	1.671	2.241	2.894	3.630	4.449	5.352	6.338
S.P.t p.s.f.	767	1009	1135	1284	1539	1686	1929	2076	2328	2478
S.P.h p.s.f.	85	70	86	84	55	58	43	55	30	44
Friction	0.303	0.320	0.334	0.345	0.350	0.357	0.359	0.362	0.364	0.366
Inspct.	NO	NO	NO	NO	YES	YES	YES	YES	YES	YES

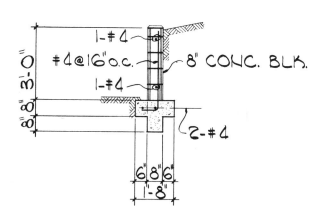

Soil slope = 5 to 1
E.F.P. = 32 p.c.f.
Section 9.9.3

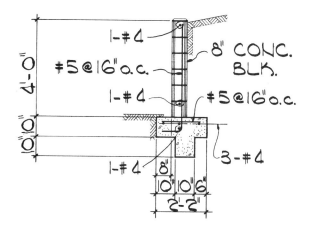

Soil slope = 5 to 1
E.F.P. = 32 p.c.f.
Section 9.9.4

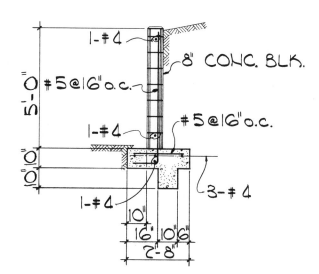

Soil slope = 5 to 1
E.F.P. = 32 p.c.f.
Section 9.9.5

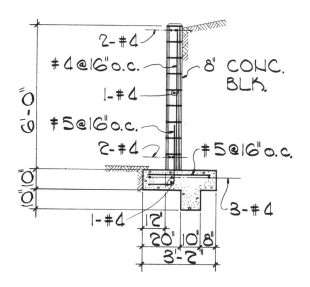

Soil slope = 5 to 1
E.F.P. = 32 p.c.f.
Section 9.9.6

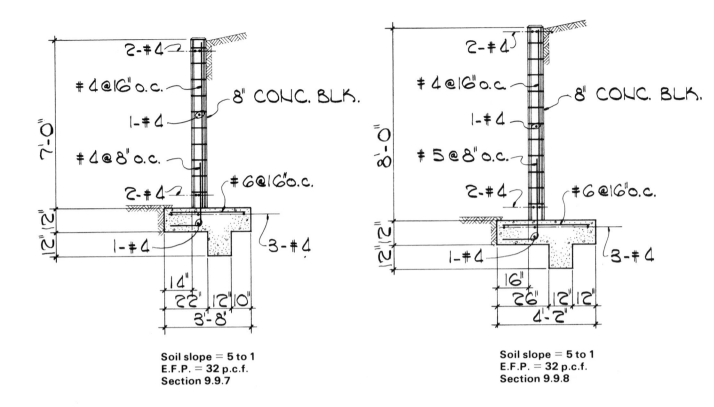

Soil slope = 5 to 1
E.F.P. = 32 p.c.f.
Section 9.9.7

Soil slope = 5 to 1
E.F.P. = 32 p.c.f.
Section 9.9.8

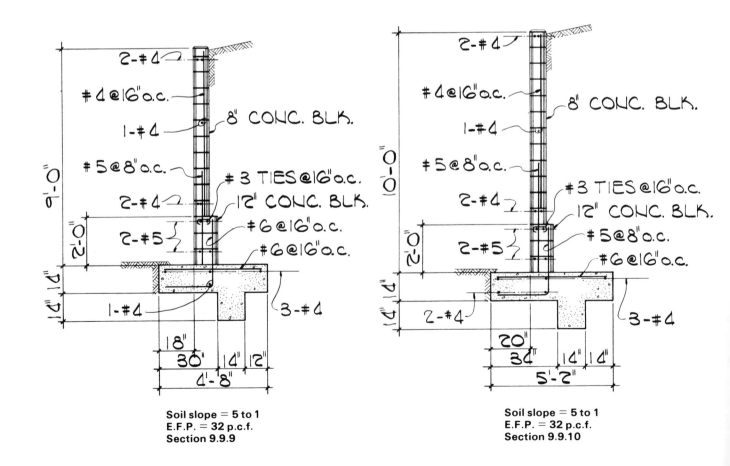

Soil slope = 5 to 1
E.F.P. = 32 p.c.f.
Section 9.9.9

Soil slope = 5 to 1
E.F.P. = 32 p.c.f.
Section 9.9.10

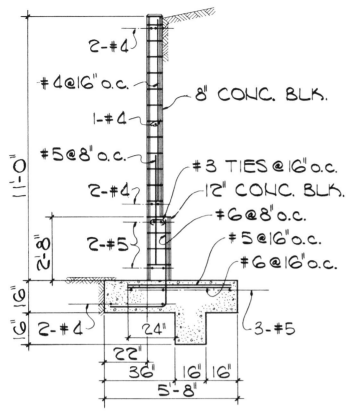

2-#4

#4@16" o.c.

1-#4

#5@8" o.c.

8" CONC. BLK.

2-#4

#3 TIES @ 16" o.c.

12" CONC. BLK.

#6@8" o.c.

2-#5

#5@16" o.c.

#6@16" o.c.

11'-0"

2'-8"

6"

2-#4

6"

24"

3-#5

22"

36"   16"   16"

5'-8"

Soil slope = 5 to 1
E.F.P. = 32 p.c.f.
Section 9.9.11

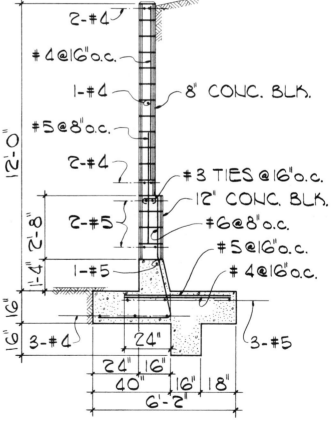

2-#4

#4@16" o.c.

1-#4

8" CONC. BLK.

#5@8" o.c.

2-#4

#3 TIES @ 16" o.c.

12" CONC. BLK.

2-#5

#6@8" o.c.

#5@16" o.c.

1-#5

#4@16" o.c.

12'-0"

2'-8"

1'-4"

6"

3-#4

6"

24"

3-#5

24"   16"

40"   16"   18"

6'-2"

Soil slope = 5 to 1
E.F.P. = 32 p.c.f.
Section 9.9.12

TABLE 9.10	GENERAL RETAINING WALLS — MASONRY STEM — SOIL AT HEEL SIDE
	SLOPE = 4 to 1 SURCHARGE = 0 lbs./sq. ft. AXIAL = 0 lbs./ft.

SOIL PRESSURE DIAGRAM

H1 = Concrete Stem
H2 = 12" Concrete Block
H3 = 8" Concrete Block

Stem Height	3'- 0"	4'- 0"	5'- 0"	6'- 0"	7'- 0"	8'- 0"	9'- 0"	10'- 0"	11'- 0"	12'- 0"
Ft. L ft.	1.667	2.167	2.667	3.333	3.833	4.333	4.833	5.333	5.833	6.333
Ft. T ft.	0.667	0.833	0.833	0.833	1.000	1.000	1.167	1.167	1.333	1.333
Setback	0.500	0.667	0.833	1.167	1.333	1.500	1.667	1.833	2.000	2.167
A ft.	0.500	0.833	1.333	1.667	2.000	2.333	2.500	2.833	3.167	3.500
B ft.	0.667	0.833	0.833	0.833	1.000	1.000	1.167	1.167	1.333	1.333
C ft.	0.500	0.500	0.500	0.833	0.833	1.000	1.167	1.333	1.333	1.500
D ft.	0.667	0.833	0.833	0.833	1.000	1.000	1.167	1.167	1.333	1.333
8 in. blk. ft.	3.000	4.000	5.000	6.000	7.000	8.000	7.000	8.000	7.333	8.000
12 in. blk. ft.	—	—	—	—	—	—	2.000	2.000	2.667	2.667
Conc. Stem	—	—	—	—	—	—	—	—	1.000	1.333
Wb ft.	—	—	—	—	—	—	—	—	1.333	1.333
W lbs.	715	1178	1644	2211	2980	3701	4698	5604	6822	7914
F lbs.	235	409	595	817	1120	1417	1809	2182	2662	3111
Mo ft.-lbs.	288	659	1158	1861	2987	4252	6130	8122	10944	13827
Mr ft.-lbs.	703	1515	2633	4522	6988	9816	13845	18244	24250	30566
O.T.R.	2.443	2.300	2.274	2.430	2.340	2.308	2.259	2.246	2.216	2.211
$\overline{X}$ ft.	0.581	0.726	0.897	1.203	1.343	1.503	1.642	1.806	1.951	2.115
e ft.	0.253	0.357	0.436	0.463	0.574	0.664	0.775	0.861	0.966	1.051
Me ft.-lbs.	181	421	717	1024	1711	2456	3639	4822	6591	8321
S ft.^{3}	0.463	0.782	1.185	1.852	2.449	3.130	3.894	4.741	5.671	6.685
S.P.t p.s.f.	819	1081	1221	1216	1476	1639	1907	2068	2332	2494
S.P.h p.s.f.	39	6	12	110	79	69	37	33	7	5
Friction	0.329	0.347	0.362	0.370	0.376	0.383	0.385	0.389	0.390	0.393
Inspct.	NO	NO	NO	NO	YES	YES	YES	YES	YES	YES

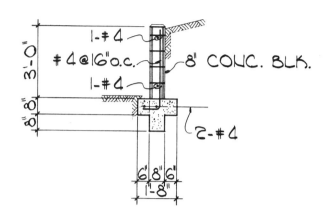

Soil slope = 4 to 1
E.F.P. = 35 p.c.f.
Section 9.10.3

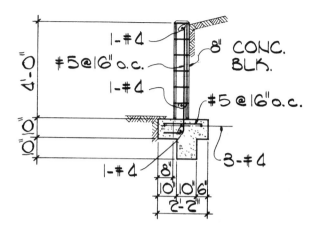

Soil slope = 4 to 1
E.F.P. = 35 p.c.f.
Section 9.10.4

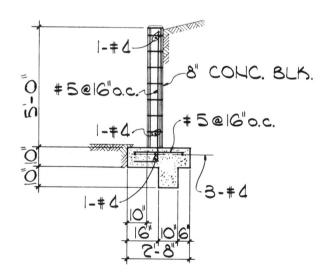

Soil slope = 4 to 1
E.F.P. = 35 p.c.f.
Section 9.10.5

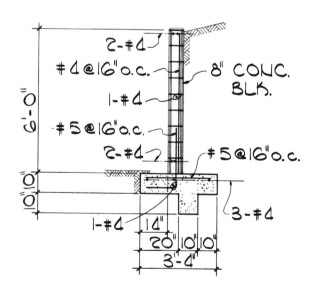

Soil slope = 4 to 1
E.F.P. = 35 p.c.f.
Section 9.10.6

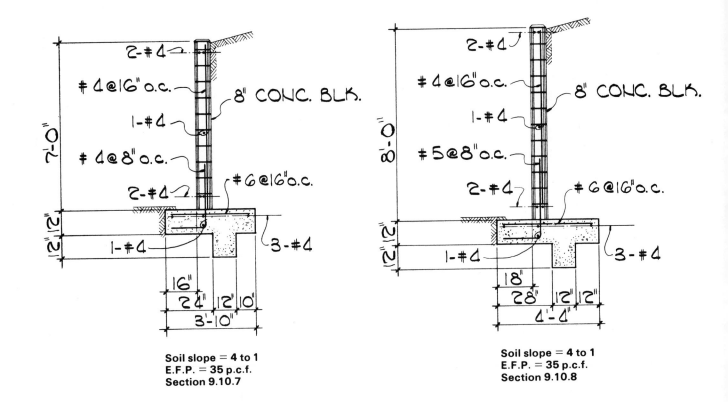

Soil slope = 4 to 1
E.F.P. = 35 p.c.f.
Section 9.10.7

Soil slope = 4 to 1
E.F.P. = 35 p.c.f.
Section 9.10.8

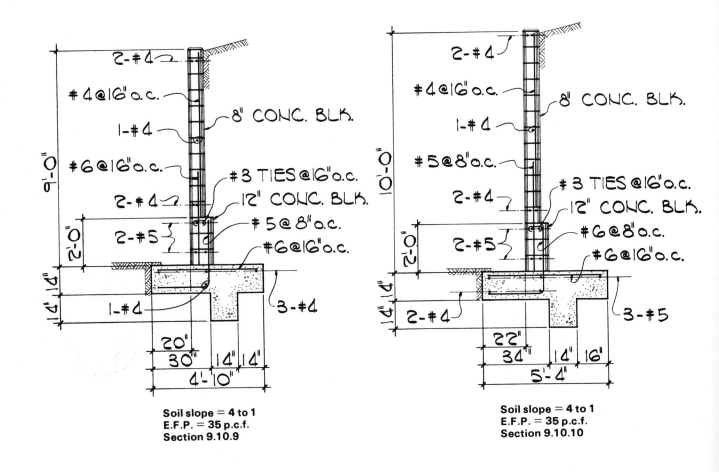

Soil slope = 4 to 1
E.F.P. = 35 p.c.f.
Section 9.10.9

Soil slope = 4 to 1
E.F.P. = 35 p.c.f.
Section 9.10.10

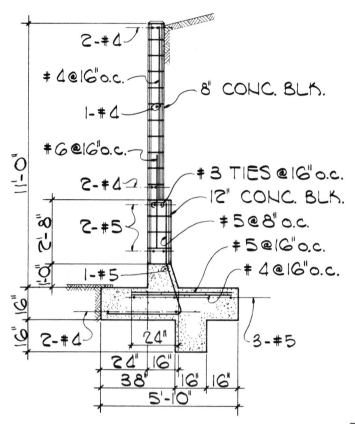

2-#4

#4 @ 16" o.c.

1-#4

8" CONC. BLK.

#6 @ 16" o.c.

2-#4

#3 TIES @ 16" o.c.

2-#5

12" CONC. BLK.

#5 @ 8" o.c.

#5 @ 16" o.c.

1-#5

#4 @ 16" o.c.

2-#4

3-#5

11'-0"

2'-8"

1'-0"

6"

6"

24"

24"   16"

38"   16"   16"

5'-0"

Soil slope = 4 to 1
E.F.P. = 35 p.c.f.
Section 9.10.11

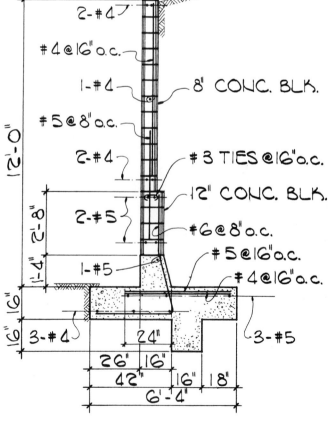

2-#4

#4 @ 16" o.c.

1-#4

8" CONC. BLK.

#5 @ 8" o.c.

2-#4

#3 TIES @ 16" o.c.

2-#5

12" CONC. BLK.

#6 @ 8" o.c.

1-#5

#5 @ 16" o.c.

#4 @ 16" o.c.

3-#4

24"

3-#5

12'-0"

2'-8"

1'-4"

6"

6"

6"

26"   16"

42"   16"   18"

6'-4"

Soil slope = 4 to 1
E.F.P. = 35 p.c.f.
Section 9.10.12

TABLE 9.11   GENERAL RETAINING WALLS — MASONRY STEM — SOIL AT HEEL SIDE
SLOPE = 3 to 1   SURCHARGE = 0 lbs./sq. ft.   AXIAL = 0 lbs./ft.

SOIL PRESSURE DIAGRAM

H1 = Concrete Stem
H2 = 12″ Concrete Block
H3 = 8″ Concrete Block

Stem Height	3′- 0″	4′- 0″	5′- 0″	6′- 0″	7′- 0″	8′- 0″	9′- 0″	10′- 0″	11′- 0″	12′- 0″
Ft. L ft.	1.833	2.333	2.833	3.333	4.000	4.500	5.000	5.500	6.167	6.667
Ft. T ft.	0.667	0.833	0.833	0.833	1.000	1.000	1.167	1.167	1.333	1.333
Setback	0.667	0.833	1.000	1.167	1.333	1.500	1.667	1.833	2.000	2.167
A ft.	0.667	1.000	1.333	1.667	2.167	2.500	2.667	3.000	3.333	3.667
B ft.	0.667	0.833	0.833	0.833	1.000	1.000	1.167	1.167	1.333	1.333
C ft.	0.500	0.500	0.667	0.833	0.833	1.000	1.167	1.333	1.500	1.667
D ft.	0.667	0.833	0.833	0.833	1.000	1.000	1.167	1.167	1.333	1.333
8 in. blk. ft.	3.000	4.000	5.000	6.000	7.000	8.000	7.000	8.000	8.000	8.000
12 in. blk. ft.	–	–	–	–	–	–	2.000	2.000	2.000	2.000
Conc. Stem	–	–	–	–	–	–	–	–	1.000	2.000
Wb ft.	–	–	–	–	–	–	–	–	1.333	1.333
W lbs.	737	1210	1683	2238	3171	3924	4958	5899	7395	8556
F lbs.	255	444	647	887	1216	1539	1964	2369	2890	3378
Mo ft.-lbs.	312	715	1257	2021	3243	4617	6655	8819	11882	15012
Mr ft.-lbs.	831	1730	2925	4579	7694	10721	15012	19675	27436	34367
O.T.R.	2.661	2.420	2.327	2.266	2.373	2.322	2.256	2.231	2.309	2.289
$\overline{X}$ ft.	0.704	0.839	0.991	1.143	1.404	1.556	1.685	1.840	2.103	2.262
e ft.	0.213	0.328	0.426	0.524	0.596	0.694	0.815	0.910	0.980	1.071
Me ft.-lbs.	157	397	716	1172	1891	2724	4039	5366	7246	9166
S ft.^{3}	0.560	0.907	1.338	1.852	2.667	3.375	4.167	5.042	6.338	7.407
S.P.t p.s.f.	683	956	1129	1304	1502	1679	1961	2137	2343	2521
S.P.h p.s.f.	122	82	59	39	84	65	22	8	56	46
Friction	0.347	0.367	0.384	0.396	0.383	0.392	0.396	0.402	0.391	0.395
Inspct.	NO	NO	NO	NO	YES	YES	YES	YES	YES	YES

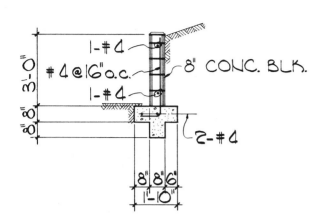

Soil slope = 3 to 1
E.F.P. = 38 p.c.f.
Section 9.11.3

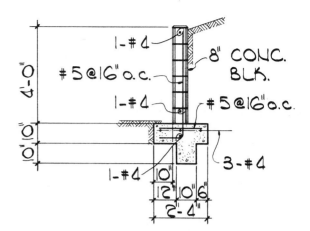

Soil slope = 3 to 1
E.F.P. = 38 p.c.f.
Section 9.11.4

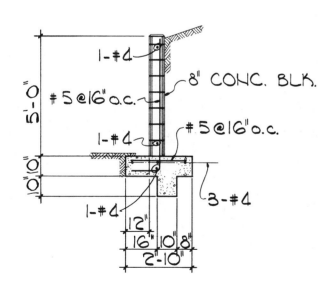

Soil slope = 3 to 1
E.F.P. = 38 p.c.f.
Section 9.11.5

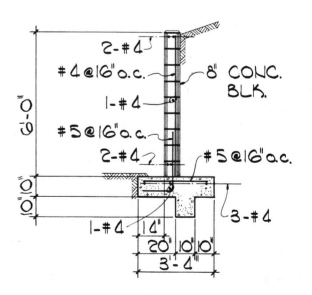

Soil slope = 3 to 1
E.F.P. = 38 p.c.f.
Section 9.11.6

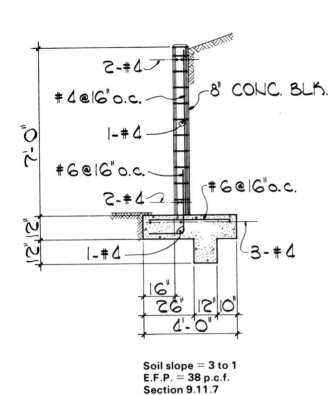

2-#4
#4 @ 16" o.c.
1-#4
#6 @ 16" o.c.
2-#4
8" CONC. BLK.
#6 @ 16" o.c.
7'-0"
2"
2"
1-#4
3-#4
16"
26"
12" 10"
4'-0"

Soil slope = 3 to 1
E.F.P. = 38 p.c.f.
Section 9.11.7

2-#4
#4 @ 16" o.c.
1-#4
#5 @ 8" o.c.
2-#4
8" CONC. BLK.
#6 @ 16" o.c
8'-0"
2"
2"
1-#4
3-#4
18"
30"
12" 12"
4'-6"

Soil slope = 3 to 1
E.F.P. = 38 p.c.f.
Section 9.11.8

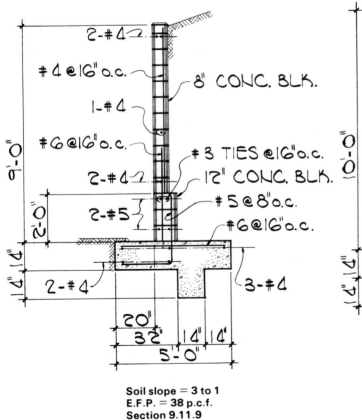

2-#4
#4 @ 16" o.c.
1-#4
#6 @ 16" o.c.
2-#4
2-#5
2-#4
8" CONC. BLK.
#3 TIES @ 16" o.c.
12" CONC. BLK.
#5 @ 8" o.c.
#6 @ 16" o.c.
3-#4
9'-0"
2'-0"
4"
4"
20"
32"
14" 14"
5'-0"

Soil slope = 3 to 1
E.F.P. = 38 p.c.f.
Section 9.11.9

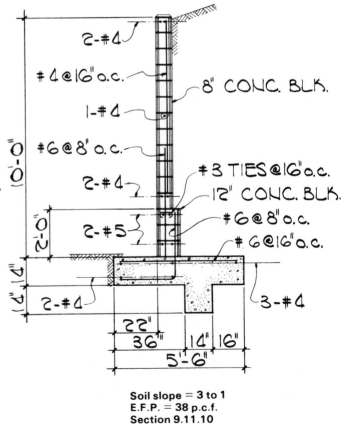

2-#4
#4 @ 16" o.c.
1-#4
#6 @ 8" o.c.
2-#4
2-#5
2-#4
8" CONC. BLK.
#3 TIES @ 16" o.c.
12" CONC. BLK.
#6 @ 8" o.c.
#6 @ 16" o.c.
3-#4
10'-0"
2'-0"
4"
4"
22"
36"
14" 16"
5'-6"

Soil slope = 3 to 1
E.F.P. = 38 p.c.f.
Section 9.11.10

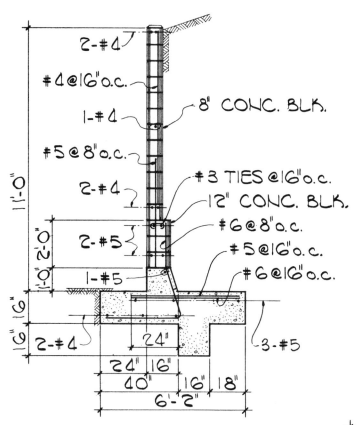

2-#4

#4@16"o.c.

1-#4

#5@8"o.c.

2-#4

2-#5

1-#5

8" CONC. BLK.

#3 TIES @16"o.c.

12" CONC. BLK.

#6@8"o.c.

#5@16"o.c.

#6@16"o.c.

2-#4

3-#5

1'-0"

2'-0"

6"

6"

24"

24"   16"

40"   16"   18"

6'-2"

Soil slope = 3 to 1
E.F.P. = 38 p.c.f.
Section 9.11.11

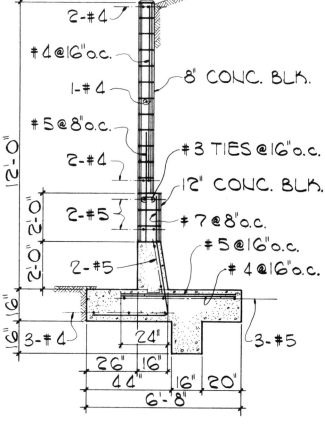

2-#4

#4@16"o.c.

1-#4

#5@8"o.c.

2-#4

2-#5

2-#5

8" CONC. BLK.

#3 TIES @16"o.c.

12" CONC. BLK.

#7@8"o.c.

#5@16"o.c.

#4@16"o.c.

3-#4

3-#5

12'-0"

2'-0"

2'-0"

6"

6"

24"

26"   16"

44"   16"   20"

6'-8"

Soil slope = 3 to 1
E.F.P. = 38 p.c.f.
Section 9.11.12

TABLE 9.12	GENERAL RETAINING WALLS – MASONRY STEM – SOIL AT HEEL SIDE
	SLOPE = 2 to 1 SURCHARGE = 0 lbs./sq. ft. AXIAL = 0 lbs./ft.

SOIL PRESSURE DIAGRAM

H1 = Concrete Stem
H2 = 12" Concrete Block
H3 = 8" Concrete Block

Stem Height	3'- 0"	4'- 0"	5'- 0"	6'- 0"	7'- 0"	8'- 0"	9'- 0"	10'- 0"	11'- 0"	12'- 0"
Ft. L ft.	1.833	2.500	3.000	3.500	4.167	4.833	5.333	5.833	6.500	7.000
Ft. T ft.	0.667	0.833	0.833	0.833	1.000	1.167	1.167	1.333	1.333	1.333
Setback	0.667	0.833	1.000	1.167	1.333	1.667	1.833	2.000	2.167	2.333
A ft.	0.667	1.167	1.500	1.833	2.167	2.500	2.833	3.167	3.667	4.000
B ft.	0.667	0.833	0.833	0.833	1.000	1.167	1.167	1.333	1.333	1.333
C ft.	0.500	0.500	0.667	0.833	1.000	1.167	1.333	1.333	1.500	1.667
D ft.	0.667	0.833	0.833	0.833	1.000	1.167	1.167	1.333	1.333	1.333
8 in. blk. ft.	3.000	4.000	5.000	6.000	7.000	6.000	7.000	7.333	7.333	7.333
12 in. blk. ft.	—	—	—	—	—	2.000	2.000	2.667	2.667	2.667
Conc. Stem	—	—	—	—	—	—	—	—	1.000	2.000
Wb ft.	—	—	—	—	—	—	—	—	1.333	1.333
W lbs.	747	1324	1829	2421	3404	4411	5316	6526	7887	9108
F lbs.	289	502	732	1004	1376	1807	2222	2762	3270	3822
Mo ft.-lbs.	353	809	1423	2287	3669	5520	7531	10433	13445	16988
Mr ft.-lbs.	844	2007	3333	5153	8516	12980	17287	23186	31091	38691
O.T.R.	2.389	2.480	2.343	2.253	2.321	2.351	2.295	2.222	2.312	2.278
$\overline{X}$ ft.	0.657	0.904	1.044	1.184	1.424	1.691	1.835	1.954	2.237	2.383
e ft.	0.259	0.346	0.456	0.566	0.660	0.725	0.831	0.963	1.013	1.117
Me ft.-lbs.	194	458	833	1371	2245	3200	4420	6282	7988	10174
S ft.^{3}	0.560	1.042	1.500	2.042	2.894	3.894	4.741	5.671	7.042	8.167
S.P.t p.s.f.	753	969	1165	1363	1593	1734	1929	2227	2348	2547
S.P.h p.s.f.	62	90	54	20	41	91	64	11	79	55
Friction	0.387	0.379	0.400	0.415	0.404	0.410	0.418	0.423	0.415	0.420
Inspct.	NO	NO	NO	NO	YES	YES	YES	YES	YES	YES

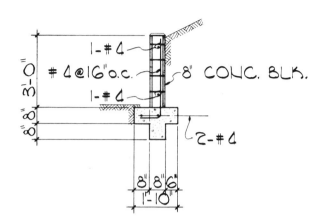

Soil slope = **2 to 1**
E.F.P. = **43** p.c.f.
Section **9.12.3**

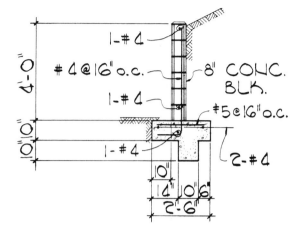

Soil slope = **2 to 1**
E.F.P. = **43** p.c.f.
Section **9.12.4**

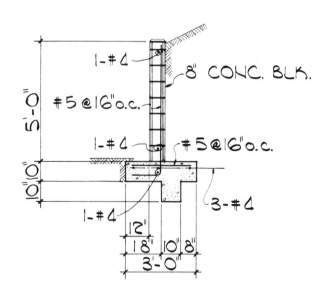

Soil slope = **2 to 1**
E.F.P. = **43** p.c.f.
Section **9.12.5**

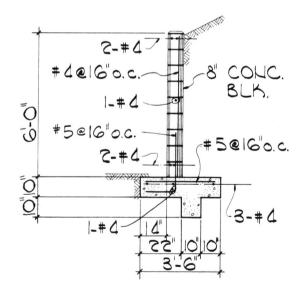

Soil slope = **2 to 1**
E.F.P. = **43** p.c.f.
Section **9.12.6**

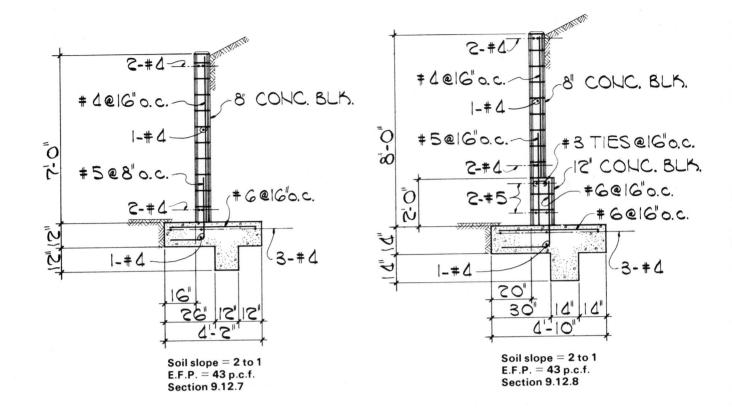

Soil slope = 2 to 1
E.F.P. = 43 p.c.f.
Section 9.12.7

Soil slope = 2 to 1
E.F.P. = 43 p.c.f.
Section 9.12.8

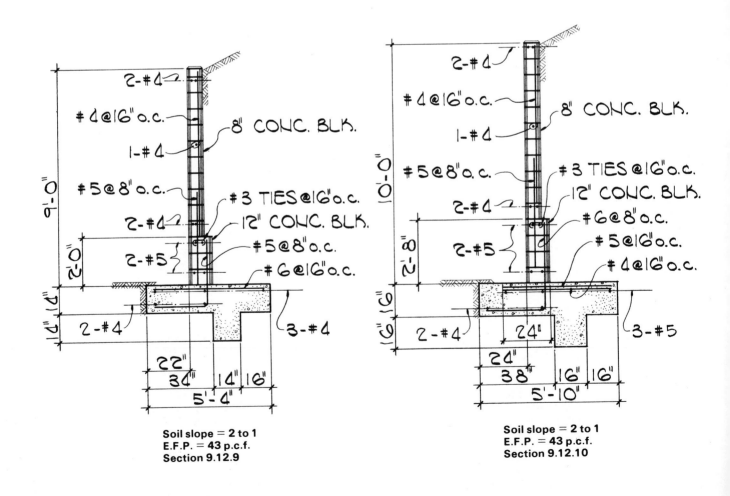

Soil slope = 2 to 1
E.F.P. = 43 p.c.f.
Section 9.12.9

Soil slope = 2 to 1
E.F.P. = 43 p.c.f.
Section 9.12.10

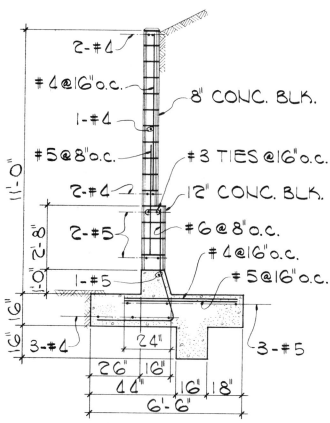

Soil slope = 2 to 1
E.F.P. = 43 p.c.f.
Section 9.12.11

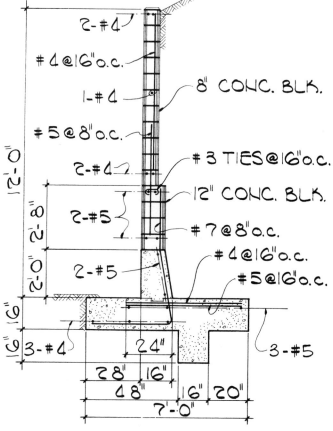

Soil slope = 2 to 1
E.F.P. = 43 p.c.f.
Section 9.12.12

9.13 General Retaining Wall—Masonry Stem—Soil at Heel Side—Slope = 1½ to 1

TABLE 9.13 — GENERAL RETAINING WALLS — MASONRY STEM — SOIL AT HEEL SIDE
SLOPE = 1.5 to 1   SURCHARGE = 0 lbs./sq. ft.   AXIAL = 0 lbs./ft.

SOIL PRESSURE DIAGRAM

H1 = Concrete Stem
H2 = 12″ Concrete Block
H3 = 8″ Concrete Block

Stem Height	3'-0"	4'-0"	5'-0"	6'-0"	7'-0"	8'-0"	9'-0"	10'-0"	11'-0"	12'-0"
Ft. L ft.	2.000	2.667	3.333	4.000	4.667	5.333	6.000	6.500	7.167	7.833
Ft. T ft.	0.667	0.833	0.833	1.000	1.000	1.167	1.333	1.333	1.333	1.500
Setback	0.667	0.833	1.167	1.333	1.500	1.833	2.000	2.167	2.333	2.667
A ft.	0.833	1.333	1.667	2.167	2.500	2.833	3.333	3.667	4.000	4.333
B ft.	0.667	0.833	0.833	1.000	1.000	1.167	1.333	1.333	1.333	1.500
C ft.	0.500	0.500	0.833	0.833	1.167	1.333	1.333	1.500	1.833	2.000
D ft.	0.667	0.833	0.833	1.000	1.000	1.167	1.333	1.333	1.333	1.500
8 in. blk. ft.	3.000	4.000	5.000	6.000	7.000	6.000	7.000	6.667	6.667	6.667
12 in. blk. ft.	–	–	–	–	–	2.000	2.000	2.667	2.667	2.667
Conc. Stem	–	–	–	–	–	–	–	0.667	1.667	2.667
Wb ft.	–	–	–	–	–	–	–	1.333	1.333	1.500
W lbs.	840	1464	2035	2965	3901	5014	6437	7579	9077	10737
F lbs.	370	642	936	1347	1760	2311	2936	3532	4183	5012
Mo ft.-lbs.	452	1035	1820	3144	4693	7061	10114	13344	17197	22553
Mr ft.-lbs.	1022	2343	4153	7180	10928	16262	23330	29792	39146	51047
O.T.R.	2.261	2.264	2.282	2.284	2.328	2.303	2.307	2.233	2.276	2.263
$\overline{X}$ ft.	0.678	0.893	1.147	1.361	1.598	1.835	2.053	2.170	2.418	2.654
e ft.	0.322	0.440	0.520	0.639	0.735	0.831	0.947	1.080	1.165	1.263
Me ft.-lbs.	270	644	1058	1895	2869	4169	6095	8185	10578	13561
S ft.^{3}	0.667	1.185	1.852	2.667	3.630	4.741	6.000	7.042	8.560	10.227
S.P.t p.s.f.	825	1093	1182	1452	1626	1820	2089	2328	2502	2697
S.P.h p.s.f.	15	5	39	31	46	61	57	4	31	45
Friction	0.440	0.439	0.460	0.454	0.451	0.461	0.456	0.466	0.461	0.467
Inspct.	NO	NO	NO	YES	YES	YES	YES	YES	YES	YES

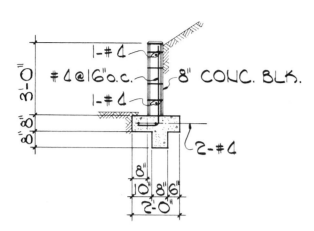

Soil slope = 1½ to 1
E.F.P. = 55 p.c.f.
Section 9.13.3

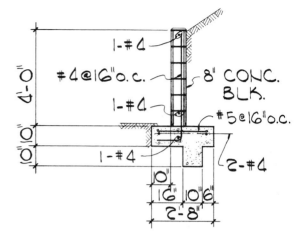

Soil slope = 1½ to 1
E.F.P. = 55 p.c.f.
Section 9.13.4

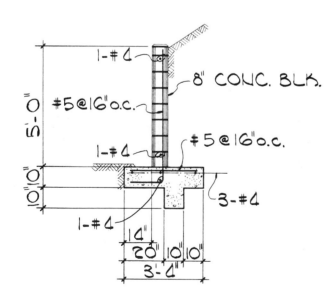

Soil slope = 1½ to 1
E.F.P. = 55 p.c.f.
Section 9.13.5

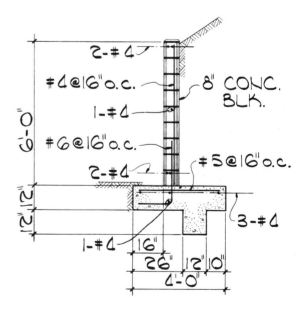

Soil slope = 1½ to 1
E.F.P. = 55 p.c.f.
Section 9.13.6

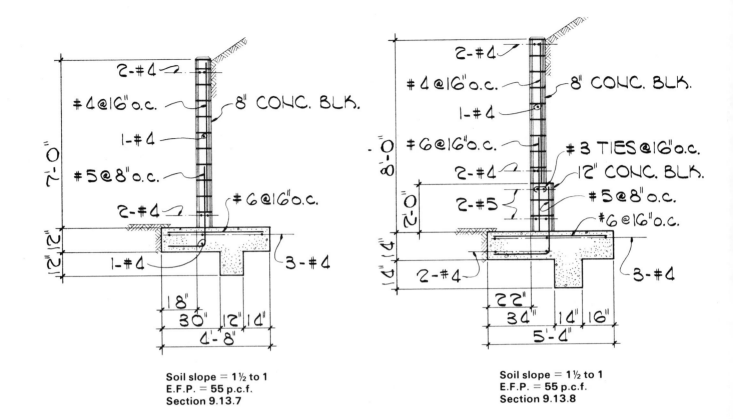

Soil slope = 1½ to 1
E.F.P. = 55 p.c.f.
Section 9.13.7

Soil slope = 1½ to 1
E.F.P. = 55 p.c.f.
Section 9.13.8

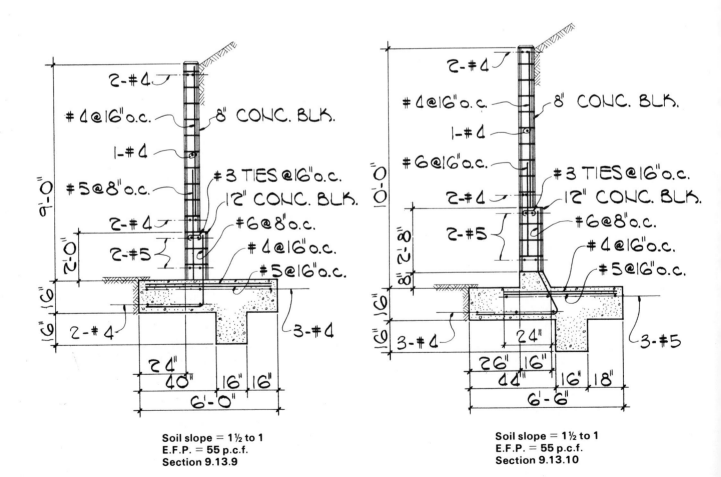

Soil slope = 1½ to 1
E.F.P. = 55 p.c.f.
Section 9.13.9

Soil slope = 1½ to 1
E.F.P. = 55 p.c.f.
Section 9.13.10

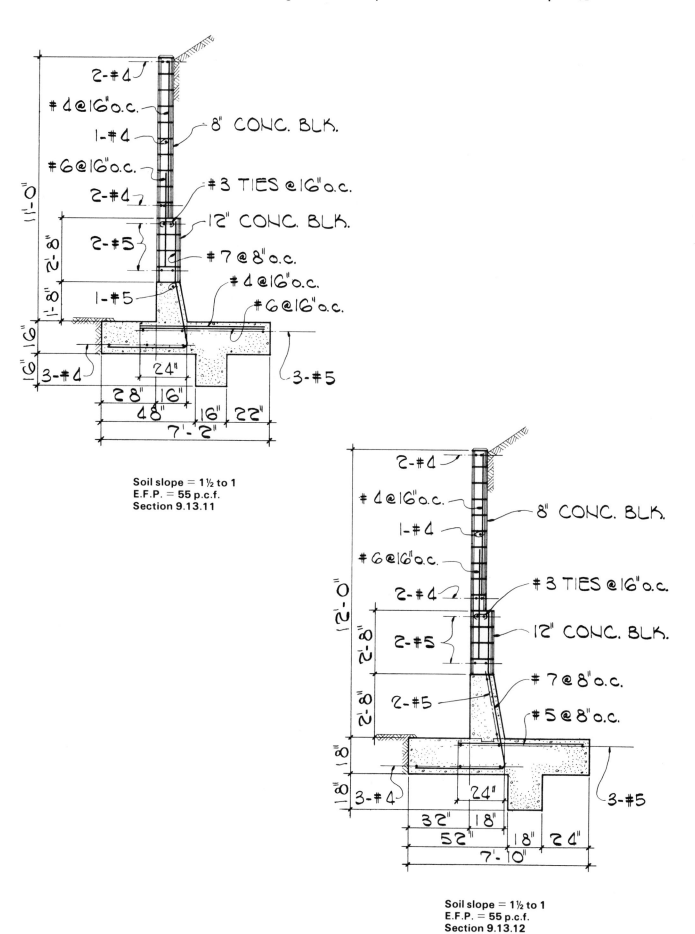

2-#4

#4@16"o.c.

8" CONC. BLK.

1-#4

#6@16"o.c.

2-#4

#3 TIES @16"o.c.

12" CONC. BLK.

2-#5

#7 @ 8"o.c.

1-#5

#4@16"o.c.

#6@16"o.c.

3-#4

3-#5

24"

28"   16"

48"   16"   22"

7'-2"

11'-0"   2'-8"   1'-8"   6"   6"

Soil slope = 1½ to 1
E.F.P. = 55 p.c.f.
Section 9.13.11

2-#4

#4@16"o.c.

8" CONC. BLK.

1-#4

#6@16"o.c.

2-#4

#3 TIES @16"o.c.

12" CONC. BLK.

2-#5

2-#5

#7 @ 8"o.c.

#5 @ 8"o.c.

3-#4

24"

3-#5

32"   18"

52"   18"   24"

7'-10"

12'-0"   2'-8"   2'-8"   8"   8"

Soil slope = 1½ to 1
E.F.P. = 55 p.c.f.
Section 9.13.12

TABLE 9.14

GENERAL RETAINING WALLS — MASONRY STEM — SOIL AT HEEL SIDE
SLOPE = 1 to 1   SURCHARGE = 0 lbs./sq. ft.   AXIAL = 0 lbs./ft.

SOIL PRESSURE DIAGRAM

H1 = Concrete Stem
H2 = 12″ Concrete Block
H3 = 8″ Concrete Block

Stem Height	3′- 0″	4′- 0″	5′- 0″	6′- 0″	7′- 0″	8′- 0″	9′- 0″	10′- 0″	11′- 0″	12′- 0″
Ft. L ft.	2.500	3.167	4.000	4.667	5.500	6.333	7.000	7.667	8.500	9.167
Ft. T ft.	0.833	0.833	0.833	1.000	1.167	1.333	1.333	1.333	1.500	1.500
Setback	0.833	1.000	1.333	1.500	1.833	2.167	2.333	2.500	2.833	3.000
A ft.	1.167	1.667	2.167	2.500	3.000	3.500	4.000	4.333	4.833	5.333
B ft.	0.833	0.833	0.833	1.000	1.167	1.333	1.333	1.333	1.500	1.500
C ft.	0.500	0.667	1.000	1.167	1.333	1.500	1.667	2.000	2.167	2.333
D ft.	0.833	0.833	0.833	1.000	1.167	1.333	1.333	1.333	1.500	1.500
8 in. blk. ft.	3.000	4.000	5.000	6.000	5.000	6.000	6.000	6.000	6.000	6.000
12 in. blk. ft.	—	—	—	—	2.000	2.000	2.000	2.000	2.000	2.000
Conc. Stem	—	—	—	—	—	—	1.000	2.000	3.000	4.000
Wb ft.	—	—	—	—	—	—	1.333	1.333	1.500	1.500
W lbs.	1163	1794	2597	3694	5034	6564	8037	9661	11766	13714
F lbs.	588	934	1361	1960	2668	3484	4271	5138	6250	7290
Mo ft.-lbs.	751	1505	2647	4573	7262	10840	14712	19409	26042	32805
Mr ft.-lbs.	1758	3412	6302	10328	16712	25200	33969	44505	60348	75604
O.T.R.	2.340	2.266	2.381	2.258	2.301	2.325	2.309	2.293	2.317	2.305
$\overline{X}$ ft.	0.866	1.063	1.407	1.558	1.877	2.187	2.396	2.598	2.916	3.121
e ft.	0.384	0.521	0.593	0.776	0.873	0.979	1.104	1.236	1.334	1.463
Me ft.-lbs.	447	934	1539	2866	4393	6428	8872	11939	15700	20059
S ft.^{3}	1.042	1.671	2.667	3.630	5.042	6.685	8.167	9.796	12.042	14.005
S.P.t p.s.f.	894	1125	1227	1581	1787	1998	2234	2479	2688	2928
S.P.h p.s.f.	36	8	72	2	44	75	62	41	80	64
Friction	0.506	0.521	0.524	0.531	0.530	0.531	0.531	0.532	0.531	0.532
Inspct.	NO	NO	NO	YES	YES	YES	YES	YES	YES	YES

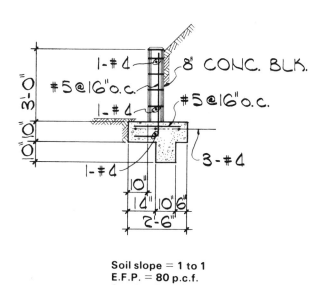

Soil slope = 1 to 1
E.F.P. = 80 p.c.f.
Section 9.14.3

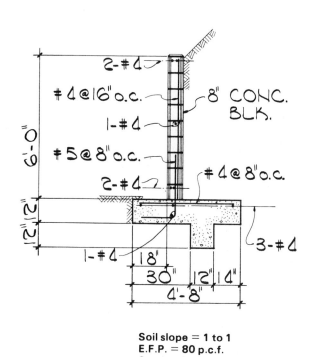

Soil slope = 1 to 1
E.F.P. = 80 p.c.f.
Section 9.14.4

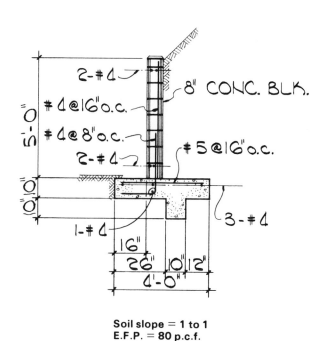

Soil slope = 1 to 1
E.F.P. = 80 p.c.f.
Section 9.14.5

Soil slope = 1 to 1
E.F.P. = 80 p.c.f.
Section 9.14.6

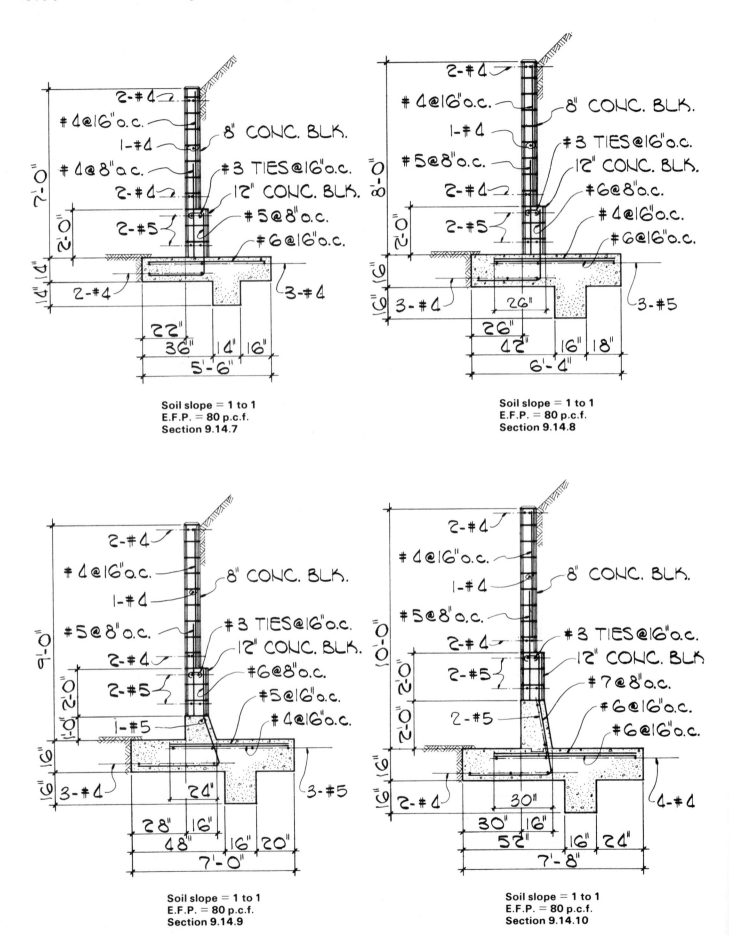

Soil slope = 1 to 1
E.F.P. = 80 p.c.f.
Section 9.14.7

Soil slope = 1 to 1
E.F.P. = 80 p.c.f.
Section 9.14.8

Soil slope = 1 to 1
E.F.P. = 80 p.c.f.
Section 9.14.9

Soil slope = 1 to 1
E.F.P. = 80 p.c.f.
Section 9.14.10

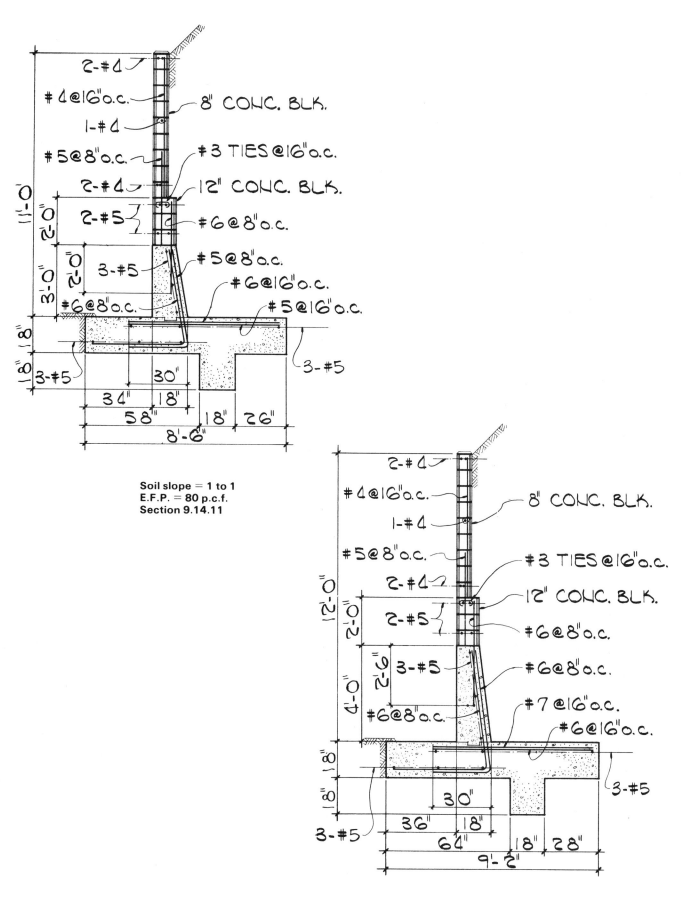

Soil slope = 1 to 1
E.F.P. = 80 p.c.f.
Section 9.14.11

Soil slope = 1 to 1
E.F.P. = 80 p.c.f.
Section 9.14.12

APPENDIX A

Retaining Wall Design Data Tables for Retaining Walls Designed for Loading Conditions Nos. 2 to 8 Inclusive without the Externally Applied Load

The table numbers in this appendix correspond to the table numbers in each chapter.

TABLE 2.2a	PROPERTY LINE RETAINING WALLS — CONCRETE STEM — SOIL OVER FOOTING NO AXIAL LOAD									
Stem Height	3'- 0''	4'- 0''	5'- 0''	6'- 0''	7'- 0''	8'- 0''	9'- 0''	10'- 0''	11'- 0''	12'- 0''
W lbs.	1145	1797	2604	3547	4495	5878	7255	8987	10459	12520
F lbs.	202	327	510	700	920	1215	1500	1870	2220	2667
Mo ft.-lbs.	246	508	992	1595	2403	3645	5000	6962	9005	11852
Mr ft.-lbs.	1313	2605	4365	7026	9919	14774	20458	28101	35069	45820
O.T.R.	5.328	5.126	4.399	4.404	4.127	4.053	4.092	4.036	3.894	3.866
$\overline{X}$ ft.	0.932	1.167	1.295	1.531	1.672	1.893	2.131	2.352	2.492	2.713
e ft.	0.318	0.416	0.538	0.635	0.745	0.857	0.953	1.065	1.175	1.287
Me ft.-lbs.	364	748	1401	2254	3348	5036	6911	9568	12286	16112
S ft.^{3}	1.042	1.671	2.241	3.130	3.894	5.042	6.338	7.782	8.963	10.667
S.P.t p.s.f.	808	1015	1336	1539	1790	2068	2267	2545	2797	3075
S.P.h p.s.f.	108	120	85	98	70	70	86	86	55	55
Friction	0.176	0.182	0.196	0.197	0.205	0.207	0.207	0.208	0.212	0.213

TABLE 2.3a	PROPERTY LINE RETAINING WALLS — CONCRETE STEM — SOIL OVER FOOTING NO AXIAL LOAD									
Stem Height	3'- 0''	4'- 0''	5'- 0''	6'- 0''	7'- 0''	8'- 0''	9'- 0''	10'- 0''	11'- 0''	12'- 0''
W lbs.	1345	1963	2812	3788	4770	6195	7430	9183	10672	12753
F lbs.	202	327	510	700	920	1215	1500	1870	2220	2667
Mo ft.-lbs.	246	508	992	1595	2403	3645	5000	6962	9005	11852
Mr ft.-lbs.	1863	3160	5164	8114	11294	16569	21552	29456	36645	47706
O.T.R.	7.560	6.219	5.203	5.086	4.699	4.546	4.310	4.231	4.069	4.025
$\overline{X}$ ft.	1.202	1.351	1.483	1.721	1.864	2.086	2.228	2.449	2.590	2.811
e ft.	0.298	0.399	0.517	0.613	0.720	0.831	0.939	1.051	1.160	1.272
Me ft.-lbs.	401	784	1453	2321	3432	5145	6976	9648	12379	16222
S ft.^{3}	1.500	2.042	2.667	3.630	4.449	5.671	6.685	8.167	9.375	11.116
S.P.t p.s.f.	715	945	1248	1451	1695	1969	2217	2493	2743	3021
S.P.h p.s.f.	181	177	158	172	152	155	130	131	102	102
Friction	0.150	0.166	0.181	0.185	0.193	0.196	0.202	0.204	0.208	0.209

TABLE 2.4a	PROPERTY LINE RETAINING WALLS — CONCRETE STEM — SOIL OVER FOOTING NO AXIAL LOAD									
Stem Height	3'- 0''	4'- 0''	5'- 0''	6'- 0''	7'- 0''	8'- 0''	9'- 0''	10'- 0''	11'- 0''	12'- 0''
W lbs.	1612	2213	3021	3909	4907	6353	7605	9379	10884	12987
F lbs.	202	327	510	700	920	1215	1500	1870	2220	2667
Mo ft.-lbs.	246	508	992	1595	2403	3645	5000	6962	9005	11852
Mr ft.-lbs.	2752	4098	6032	8688	12016	17506	22675	30843	38256	49631
O.T.R.	11.166	8.064	6.078	5.446	5.000	4.803	4.535	4.430	4.248	4.188
$\overline{X}$ ft.	1.555	1.622	1.668	1.814	1.959	2.182	2.324	2.546	2.688	2.909
e ft.	0.279	0.378	0.498	0.602	0.708	0.818	0.926	1.037	1.146	1.258
Me ft.-lbs.	449	837	1505	2355	3474	5199	7041	9728	12471	16332
S ft.^{3}	2.241	2.667	3.130	3.894	4.741	6.000	7.042	8.560	9.796	11.574
S.P.t p.s.f.	640	867	1178	1414	1653	1925	2170	2445	2693	2969
S.P.h p.s.f.	239	239	216	204	187	192	170	172	147	147
Friction	0.125	0.148	0.169	0.179	0.188	0.191	0.197	0.199	0.204	0.205

TABLE 2.5a	PROPERTY LINE RETAINING WALLS — CONCRETE STEM — SOIL OVER FOOTING NO SURCHARGE LOAD									
Stem Height	3'- 0''	4'- 0''	5'- 0''	6'- 0''	7'- 0''	8'- 0''	9'- 0''	10'- 0''	11'- 0''	12'- 0''
W lbs.	1945	2463	3333	4151	5182	6512	7780	9575	11097	13220
F lbs.	202	327	510	700	920	1215	1500	1870	2220	2667
Mo ft.-lbs.	246	508	992	1595	2403	3645	5000	6962	9005	11852
Mr ft.-lbs.	4113	5160	7464	9896	13528	18469	23827	32263	39903	51595
O.T.R.	16.688	10.155	7.521	6.203	5.629	5.067	4.765	4.634	4.431	4.353
$\overline{X}$ ft.	1.988	1.889	1.942	2.000	2.147	2.276	2.420	2.642	2.784	3.006
e ft.	0.262	0.361	0.475	0.584	0.687	0.807	0.913	1.024	1.132	1.244
Me ft.-lbs.	509	890	1584	2422	3559	5254	7106	9808	12564	16442
S ft.^{3}	3.375	3.375	3.894	4.449	5.352	6.338	7.407	8.963	10.227	12.042
S.P.t p.s.f.	583	811	1096	1348	1580	1885	2126	2400	2645	2921
S.P.h p.s.f.	281	284	283	259	250	227	208	211	188	190
Friction	0.104	0.133	0.153	0.169	0.178	0.187	0.193	0.195	0.200	0.202

TABLE 2.6a	PROPERTY LINE RETAINING WALLS — CONCRETE STEM — SOIL OVER FOOTING NO SURCHARGE LOAD									
Stem Height	3'- 0"	4'- 0"	5'- 0"	6'- 0"	7'- 0"	8'- 0"	9'- 0"	10'- 0"	11'- 0"	12'- 0"
W lbs.	1078	1805	2604	3547	4787	6037	7626	8987	10905	12753
F lbs.	202	350	510	700	960	1215	1550	1870	2282	2667
Mo ft.-lbs.	246	565	992	1595	2560	3645	5254	6962	9380	11852
Mr ft.-lbs.	1152	2479	4365	7026	10951	15658	22120	28101	37442	47706
O.T.R.	4.675	4.391	4.399	4.404	4.278	4.296	4.210	4.036	3.992	4.025
$\overline{X}$ ft.	0.840	1.060	1.295	1.531	1.753	1.990	2.212	2.352	2.573	2.811
e ft.	0.327	0.440	0.538	0.635	0.747	0.843	0.955	1.065	1.177	1.272
Me ft.-lbs.	352	793	1401	2254	3576	5091	7283	9568	12831	16222
S ft.^{3}	0.907	1.500	2.241	3.130	4.167	5.352	6.685	7.782	9.375	11.116
S.P.t p.s.f.	850	1131	1336	1539	1816	2016	2293	2545	2823	3021
S.P.h p.s.f.	74	73	85	98	99	114	115	86	85	102
Friction	0.187	0.194	0.196	0.197	0.201	0.201	0.203	0.208	0.209	0.209

TABLE 2.7a	PROPERTY LINE RETAINING WALLS — CONCRETE STEM — SOIL OVER FOOTING NO SURCHARGE LOAD									
Stem Height	3'- 0"	4'- 0"	5'- 0"	6'- 0"	7'- 0"	8'- 0"	9'- 0"	10'- 0"	11'- 0"	12'- 0"
W lbs.	1212	1980	2812	3667	4928	6195	7805	9379	11338	13220
F lbs.	202	350	510	700	960	1215	1550	1870	2282	2667
Mo ft.-lbs.	246	565	992	1595	2560	3645	5254	6962	9380	11852
Mr ft.-lbs.	1486	3033	5164	7560	11671	16569	23270	30843	40765	51595
O.T.R.	6.027	5.372	5.203	4.739	4.559	4.546	4.429	4.430	4.346	4.353
$\overline{X}$ ft.	1.023	1.247	1.483	1.626	1.849	2.086	2.308	2.546	2.768	3.006
e ft.	0.311	0.420	0.517	0.624	0.735	0.831	0.942	1.037	1.149	1.244
Me ft.-lbs.	376	832	1453	2287	3621	5145	7351	9728	13024	16442
S ft.^{3}	1.185	1.852	2.667	3.375	4.449	5.671	7.042	8.560	10.227	12.042
S.P.t p.s.f.	772	1043	1248	1493	1768	1969	2245	2445	2721	2921
S.P.h p.s.f.	137	145	158	137	140	155	157	172	174	190
Friction	0.166	0.177	0.181	0.191	0.195	0.196	0.199	0.199	0.201	0.202

TABLE 2.8a	PROPERTY LINE RETAINING WALLS — CONCRETE STEM — SOIL OVER FOOTING NO SURCHARGE LOAD									
Stem Height	3'- 0''	4'- 0''	5'- 0''	6'- 0''	7'- 0''	8'- 0''	9'- 0''	10'- 0''	11'- 0''	12'- 0''
W lbs.	1503	2242	3125	4305	5353	7024	8342	10200	11988	13920
F lbs.	220	350	510	735	960	1260	1550	1927	2282	2667
Mo ft.-lbs.	282	565	992	1715	2560	3851	5254	7279	9380	11852
Mr ft.-lbs.	2207	3974	6492	10265	13973	21058	26898	36017	46019	57720
O.T.R.	7.836	7.038	6.541	5.985	5.458	5.468	5.119	4.948	4.906	4.870
$\overline{X}$ ft.	1.281	1.520	1.760	1.986	2.132	2.450	2.594	2.817	3.056	3.295
e ft.	0.303	0.396	0.490	0.597	0.701	0.800	0.906	1.016	1.110	1.205
Me ft.-lbs.	455	889	1532	2572	3755	5622	7555	10362	13313	16772
S ft.^{3}	1.671	2.449	3.375	4.449	5.352	7.042	8.167	9.796	11.574	13.500
S.P.t p.s.f.	747	948	1148	1411	1646	1879	2117	2388	2589	2789
S.P.h p.s.f.	202	222	241	255	243	282	267	273	288	304
Friction	0.147	0.156	0.163	0.171	0.179	0.179	0.186	0.189	0.190	0.192

TABLE 3.2a	PROPERTY LINE RETAINING WALLS — CONCRETE STEM — SOIL OVER FOOTING NO AXIAL LOAD									
Stem Height	3'- 0''	4'- 0''	5'- 0''	6'- 0''	7'- 0''	8'- 0''	9'- 0''	10'- 0''	11'- 0''	12'- 0''
W lbs.	1121	1681	2543	3353	4410	5748	6943	8606	10273	12285
F lbs.	202	327	510	700	920	1215	1500	1870	2220	2667
Mo ft.-lbs.	246	508	992	1595	2403	3645	5000	6962	9005	11852
Mr ft.-lbs.	1305	2337	4327	6464	9859	14650	19262	26564	34854	45494
O.T.R.	5.296	4.599	4.360	4.052	4.102	4.019	3.852	3.815	3.871	3.839
$\overline{X}$ ft.	0.945	1.088	1.311	1.452	1.691	1.915	2.054	2.278	2.516	2.738
e ft.	0.305	0.412	0.522	0.631	0.726	0.835	0.946	1.056	1.150	1.262
Me ft.-lbs.	342	693	1328	2116	3202	4801	6566	9086	11818	15499
S ft.^{3}	1.042	1.500	2.241	2.894	3.894	5.042	6.000	7.407	8.963	10.667
S.P.t p.s.f.	777	1022	1286	1536	1735	1997	2251	2518	2719	2989
S.P.h p.s.f.	120	98	101	73	90	93	63	64	82	83
Friction	0.180	0.194	0.201	0.209	0.209	0.211	0.216	0.217	0.216	0.217

TABLE 3.3a	PROPERTY LINE RETAINING WALLS — CONCRETE STEM — SOIL OVER FOOTING NO AXIAL LOAD									
Stem Height	3'- 0''	4'- 0''	5'- 0''	6'- 0''	7'- 0''	8'- 0''	9'- 0''	10'- 0''	11'- 0''	12'- 0''
W lbs.	1321	1931	2752	3594	4547	5906	7118	8802	10485	12518
F lbs.	202	327	510	700	920	1215	1500	1870	2220	2667
Mo ft.-lbs.	246	508	992	1595	2403	3645	5000	6962	9005	11852
Mr ft.-lbs.	1855	3150	5126	7511	10535	15534	20327	27886	36430	47380
O.T.R.	7.527	6.198	5.164	4.708	4.384	4.262	4.065	4.005	4.046	3.998
$\overline{X}$ ft.	1.218	1.368	1.502	1.646	1.788	2.013	2.153	2.377	2.616	2.838
e ft.	0.282	0.382	0.498	0.604	0.712	0.820	0.930	1.040	1.134	1.245
Me ft.-lbs.	373	738	1370	2172	3237	4845	6620	9150	11894	15589
S ft.^{3}	1.500	2.042	2.667	3.375	4.167	5.352	6.338	7.782	9.375	11.116
S.P.t p.s.f.	689	913	1202	1442	1686	1948	2199	2464	2667	2935
S.P.h p.s.f.	192	190	174	155	133	137	110	112	129	130
Friction	0.153	0.169	0.185	0.195	0.202	0.206	0.211	0.212	0.212	0.213

TABLE 3.4a	PROPERTY LINE RETAINING WALLS — CONCRETE STEM — SOIL OVER FOOTING NO AXIAL LOAD									
Stem Height	3'- 0''	4'- 0''	5'- 0''	6'- 0''	7'- 0''	8'- 0''	9'- 0''	10'- 0''	11'- 0''	12'- 0''
W lbs.	1588	2098	2960	3836	4822	6064	7468	9008	10698	12528
F lbs.	202	327	510	700	920	1215	1500	1870	2220	2667
Mo ft.-lbs.	246	508	992	1595	2403	3645	5000	6962	9005	11852
Mr ft.-lbs.	2744	3761	5994	8639	11956	16444	22543	29258	38042	47400
O.T.R.	11.134	7.401	6.039	5.415	4.975	4.511	4.509	4.202	4.225	3.999
$\overline{X}$ ft.	1.573	1.550	1.690	1.836	1.981	2.111	2.349	2.475	2.714	2.837
e ft.	0.260	0.366	0.477	0.581	0.686	0.806	0.901	1.025	1.119	1.246
Me ft.-lbs.	413	769	1412	2227	3307	4889	6727	9231	11971	15608
S ft.^{3}	2.241	2.449	3.130	3.894	4.741	5.671	7.042	8.167	9.796	11.116
S.P.t p.s.f.	617	861	1134	1366	1602	1902	2104	2417	2617	2938
S.P.h p.s.f.	249	233	232	222	207	178	194	156	173	130
Friction	0.127	0.156	0.172	0.183	0.191	0.200	0.201	0.208	0.208	0.213

TABLE 3.5a	PROPERTY LINE RETAINING WALLS — CONCRETE STEM — SOIL OVER FOOTING NO AXIAL LOAD									
Stem Height	3'- 0''	4'- 0''	5'- 0''	6'- 0''	7'- 0''	8'- 0''	9'- 0''	10'- 0''	11'- 0''	12'- 0''
W lbs.	1921	2431	3168	4078	4960	6381	7643	9203	10910	12762
F lbs.	202	327	510	700	920	1215	1500	1870	2220	2667
Mo ft.-lbs.	246	508	992	1595	2403	3645	5000	6962	9005	11852
Mr ft.-lbs.	4105	5150	6931	9847	12701	18344	23696	30645	39689	49325
O.T.R.	16.656	10.134	6.984	6.172	5.285	5.033	4.739	4.402	4.407	4.162
$\overline{X}$ ft.	2.009	1.909	1.874	2.024	2.076	2.304	2.446	2.573	2.812	2.936
e ft.	0.241	0.341	0.459	0.560	0.674	0.780	0.887	1.010	1.104	1.230
Me ft.-lbs.	463	829	1454	2282	3342	4976	6780	9296	12048	15700
S ft.^{3}	3.375	3.375	3.630	4.449	5.042	6.338	7.407	8.560	10.227	11.574
S.P.t p.s.f.	564	786	1080	1302	1565	1820	2062	2370	2571	2888
S.P.h p.s.f.	290	295	278	276	239	250	231	198	215	175
Friction	0.105	0.134	0.161	0.172	0.186	0.190	0.196	0.203	0.204	0.209

TABLE 3.6a	PROPERTY LINE RETAINING WALLS — CONCRETE STEM — SOIL OVER FOOTING NO SURCHARGE LOAD									
Stem Height	3'- 0''	4'- 0''	5'- 0''	6'- 0''	7'- 0''	8'- 0''	9'- 0''	10'- 0''	11'- 0''	12'- 0''
W lbs.	1054	1756	2543	3474	4672	5748	7282	8812	10696	12540
F lbs.	202	350	510	700	960	1215	1550	1870	2282	2667
Mo ft.-lbs.	246	565	992	1595	2560	3645	5254	6962	9380	11852
Mr ft.-lbs.	1144	2450	4327	6978	10847	14650	20819	27903	37162	47414
O.T.R.	4.642	4.339	4.360	4.374	4.237	4.019	3.962	4.008	3.962	4.001
$\overline{X}$ ft.	0.851	1.073	1.311	1.549	1.774	1.915	2.137	2.376	2.597	2.836
e ft.	0.315	0.427	0.522	0.617	0.726	0.835	0.946	1.040	1.153	1.248
Me ft.-lbs.	332	749	1328	2144	3393	4801	6888	9166	12328	15645
S ft.^{3}	0.907	1.500	2.241	3.130	4.167	5.042	6.338	7.782	9.375	11.116
S.P.t p.s.f.	818	1085	1286	1487	1749	1997	2268	2467	2741	2943
S.P.h p.s.f.	86	86	101	117	120	93	94	112	111	128
Friction	0.191	0.200	0.201	0.202	0.205	0.211	0.213	0.212	0.213	0.213

TABLE 3.7a	PROPERTY LINE RETAINING WALLS — CONCRETE STEM — SOIL OVER FOOTING NO SURCHARGE LOAD								
Stem Height	3'- 0"	4'- 0"	5'- 0"	6'- 0"	7'- 0"	8'- 0"	9'- 0"	10'- 0"	11'- 0"	12'- 0"
W lbs.	1188	1931	2647	3594	4814	6074	7640	9017	10922	12796
F lbs.	202	350	510	700	960	1215	1550	1870	2282	2667
Mo ft.-lbs.	246	565	992	1595	2560	3645	5254	6962	9380	11852
Mr ft.-lbs.	1478	3004	4718	7511	11568	16459	23088	29276	38824	49373
O.T.R.	5.995	5.321	4.753	4.708	4.519	4.516	4.394	4.205	4.139	4.166
$\overline{X}$ ft.	1.037	1.263	1.407	1.646	1.871	2.110	2.334	2.474	2.696	2.932
e ft.	0.297	0.404	0.510	0.604	0.712	0.807	0.916	1.026	1.138	1.234
Me ft.-lbs.	352	780	1349	2172	3429	4902	6996	9247	12425	15795
S ft.^{3}	1.185	1.852	2.449	3.375	4.449	5.671	7.042	8.167	9.796	11.574
S.P.t p.s.f.	743	1000	1242	1442	1702	1906	2169	2421	2693	2900
S.P.h p.s.f.	148	158	140	155	161	177	182	156	156	171
Friction	0.170	0.181	0.193	0.195	0.199	0.200	0.203	0.207	0.209	0.208

TABLE 3.8a	PROPERTY LINE RETAINING WALLS — CONCRETE STEM — SOIL OVER FOOTING NO SURCHARGE LOAD								
Stem Height	3'- 0"	4'- 0"	5'- 0"	6'- 0"	7'- 0"	8'- 0"	9'- 0"	10'- 0"	11'- 0"	12'- 0"
W lbs.	1467	2194	3064	4207	5249	6717	8197	10014	11811	13518
F lbs.	220	350	510	735	960	1260	1550	1927	2282	2667
Mo ft.-lbs.	282	565	992	1715	2560	3851	5254	7279	9380	11852
Mr ft.-lbs.	2186	3944	6454	10181	13884	19866	26750	35785	45790	55416
O.T.R.	7.762	6.987	6.503	5.936	5.423	5.158	5.091	4.917	4.882	4.676
$\overline{X}$ ft.	1.298	1.541	1.782	2.012	2.157	2.384	2.622	2.847	3.083	3.223
e ft.	0.285	0.376	0.468	0.571	0.676	0.783	0.878	0.987	1.084	1.194
Me ft.-lbs.	418	825	1433	2402	3548	5257	7194	9882	12803	16139
S ft.^{3}	1.671	2.449	3.375	4.449	5.352	6.685	8.167	9.796	11.574	13.005
S.P.t p.s.f.	713	909	1106	1354	1589	1847	2052	2315	2523	2771
S.P.h p.s.f.	213	235	256	274	263	274	290	297	311	289
Friction	0.150	0.160	0.167	0.175	0.183	0.188	0.189	0.192	0.193	0.197

TABLE 4.2a, 3a 4a, 5a	PROPERTY LINE RETAINING WALLS — CONCRETE STEM — SOIL NOT OVER FOOTING NO AXIAL LOAD									
Stem Height	3'- 0''	4'- 0''	5'- 0''	6'- 0''	7'- 0''	8'- 0''	9'- 0''	10'- 0''	11'- 0''	12'- 0''
W lbs.	601	759	1108	1346	1615	2130	2450	3078	3448	4189
F lbs.	202	327	510	700	920	1215	1500	1870	2220	2667
Mo ft.-lbs.	246	508	992	1595	2403	3645	5000	6962	9005	11852
Mr ft.-lbs.	706	1034	1800	2796	4264	6446	8812	12213	15647	20551
O.T.R.	2.866	2.036	1.814	1.753	1.774	1.769	1.763	1.754	1.738	1.734
$\overline{X}$ ft.	0.766	0.693	0.729	0.892	1.152	1.315	1.556	1.706	1.926	2.077
e ft.	0.067	0.223	0.354	0.441	0.515	0.602	0.694	0.794	0.907	1.007
Me ft.-lbs.	40	169	392	594	831	1281	1700	2443	3129	4217
S ft.^{3}	0.463	0.560	0.782	1.185	1.852	2.449	3.375	4.167	5.352	6.338
S.P.h p.s.f.	448	716	1013	1006	933	1079	1048	1202	1193	1345
S.P.t p.s.f.	273	111	10	4	36	32	41	29	24	14
Friction	0.336	0.430	0.461	0.520	0.570	0.570	0.612	0.608	0.644	0.637

TABLE 4.6a	PROPERTY LINE RETAINING WALLS — CONCRETE STEM — SOIL NOT OVER FOOTING NO SURCHARGE LOAD									
Stem Height	3'- 0''	4'- 0''	5'- 0''	6'- 0''	7'- 0''	8'- 0''	9'- 0''	10'- 0''	11'- 0''	12'- 0''
W lbs.	601	942	1170	1408	1895	2205	2804	3165	3877	4289
F lbs.	202	350	510	700	960	1215	1550	1870	2282	2667
Mo ft.-lbs.	246	565	992	1595	2560	3645	5254	6962	9380	11852
Mr ft.-lbs.	706	1483	2370	3467	5375	7505	10586	13740	18305	22626
O.T.R.	2.866	2.627	2.387	2.173	2.100	2.059	2.015	1.974	1.951	1.909
$\overline{X}$ ft.	0.766	0.975	1.177	1.329	1.486	1.751	1.901	2.141	2.302	2.512
e ft.	0.067	0.108	0.157	0.254	0.348	0.416	0.516	0.609	0.698	0.821
Me ft.-lbs.	40	102	183	358	659	917	1446	1926	2707	3522
S ft.^{3}	0.463	0.782	1.185	1.671	2.241	3.130	3.894	5.042	6.000	7.407
S.P.h p.s.f.	448	565	593	659	811	802	952	958	1097	1119
S.P.t p.s.f.	273	304	284	231	223	216	209	193	195	168
Friction	0.336	0.372	0.436	0.497	0.507	0.551	0.553	0.591	0.588	0.622

TABLE 4.7a PROPERTY LINE RETAINING WALLS — CONCRETE STEM — SOIL NOT OVER FOOTING NO SURCHARGE LOAD

Stem Height	3'- 0"	4'- 0"	5'- 0"	6'- 0"	7'- 0"	8'- 0"	9'- 0"	10'- 0"	11'- 0"	12'- 0"
W lbs.	634	983	1212	1450	1945	2255	2863	3253	3977	4389
F lbs.	202	350	510	700	960	1215	1550	1870	2282	2667
Mo ft.-lbs.	246	565	992	1595	2560	3645	5254	6962	9380	11852
Mr ft.-lbs.	912	1804	2749	3927	6015	8223	11530	15310	20224	24751
O.T.R.	3.701	3.195	2.770	2.461	2.350	2.256	2.194	2.199	2.156	2.088
$\overline{X}$ ft.	1.050	1.260	1.450	1.608	1.777	2.030	2.192	2.567	2.727	2.939
e ft.	−0.050	−0.010	0.050	0.142	0.223	0.303	0.391	0.433	0.523	0.644
Me ft.-lbs.	−32	−10	61	207	435	683	1119	1410	2082	2828
S ft.^{3}	0.667	1.042	1.500	2.042	2.667	3.630	4.449	6.000	7.042	8.560
S.P.h p.s.f.	269	384	445	516	649	671	806	777	908	943
S.P.t p.s.f.	365	403	363	313	323	295	302	307	316	282
Friction	0.318	0.356	0.421	0.483	0.494	0.539	0.542	0.575	0.574	0.608

TABLE 4.8a PROPERTY LINE RETAINING WALLS — CONCRETE STEM — SOIL NOT OVER FOOTING NO SURCHARGE LOAD

Stem Height	3'- 0"	4'- 0"	5'- 0"	6'- 0"	7'- 0"	8'- 0"	9'- 0"	10'- 0"	11'- 0"	12'- 0"
W lbs.	848	1088	1316	1795	2095	2687	3038	3742	4177	4589
F lbs.	220	350	510	735	960	1260	1550	1927	2282	2667
Mo ft.-lbs.	282	565	992	1715	2560	3851	5254	7279	9380	11852
Mr ft.-lbs.	1596	2632	3768	5832	7985	11235	14412	19131	24212	29151
O.T.R.	5.667	4.662	3.796	3.401	3.119	2.917	2.743	2.628	2.581	2.460
$\overline{X}$ ft.	1.549	1.901	2.109	2.294	2.590	2.748	3.015	3.167	3.551	3.770
e ft.	−0.216	−0.234	−0.192	−0.127	−0.090	0.002	0.069	0.166	0.199	0.314
Me ft.-lbs.	−183	−255	−253	−228	−188	5	208	621	832	1439
S ft.^{3}	1.185	1.852	2.449	3.130	4.167	5.042	6.338	7.407	9.375	11.116
S.P.h p.s.f.	164	189	240	341	374	489	525	645	646	691
S.P.t p.s.f.	473	464	447	487	464	488	460	477	468	432
Friction	0.260	0.322	0.388	0.409	0.458	0.469	0.510	0.515	0.546	0.581

TABLE 5.2a, 3a PROPERTY LINE RETAINING WALLS — MASONRY STEM — SOIL NOT OVER FOOTING NO AXIAL LOAD

Stem Height	3'-0"	4'-0"	5'-0"	6'-0"	7'-0"	8'-0"	9'-0"	10'-0"	11'-0"	12'-0"
W lbs.	577	727	1026	1265	1513	1941	2260	2803	3193	3843
F lbs.	202	327	510	700	920	1215	1500	1870	2220	2667
Mo ft.-lbs.	246	508	992	1595	2403	3645	5000	6962	9005	11852
Mr ft.-lbs.	674	986	1805	2935	4364	6532	8878	12334	15825	20883
O.T.R.	2.736	1.941	1.819	1.839	1.816	1.792	1.776	1.772	1.757	1.762
$\overline{X}$ ft.	0.742	0.658	0.792	1.059	1.295	1.488	1.716	1.916	2.136	2.350
e ft.	0.091	0.259	0.375	0.441	0.538	0.679	0.784	0.917	1.031	1.150
Me ft.-lbs.	52	188	385	558	814	1318	1772	2570	3290	4419
S ft.^{3}	0.463	0.560	0.907	1.500	2.241	3.130	4.167	5.352	6.685	8.167
S.P.h p.s.f.	459	732	864	793	776	869	877	975	996	1090
S.P.t p.s.f.	233	61	16	50	49	27	27	14	12	8
Friction	0.350	0.449	0.497	0.554	0.608	0.626	0.664	0.667	0.695	0.694

TABLE 5.4a, 5a PROPERTY LINE RETAINING WALLS — MASONRY STEM — SOIL NOT OVER FOOTING NO AXIAL LOAD

Stem Height	3'-0"	4'-0"	5'-0"	6'-0"	7'-0"	8'-0"	9'-0"	10'-0"	11'-0"	12'-0"
W lbs.	577	727	1026	1265	1513	1941	2260	2835	3193	3875
F lbs.	202	327	510	700	920	1215	1500	1870	2220	2667
Mo ft.-lbs.	246	508	992	1595	2403	3645	5000	6962	9005	11852
Mr ft.-lbs.	674	986	1805	2935	4364	6532	8878	12493	15825	21083
O.T.R.	2.736	1.941	1.819	1.839	1.816	1.792	1.776	1.794	1.757	1.779
$\overline{X}$ ft.	0.742	0.658	0.792	1.059	1.295	1.488	1.716	1.951	2.136	2.382
e ft.	0.091	0.259	0.375	0.441	0.538	0.679	0.784	0.883	1.031	1.118
Me ft.-lbs.	52	188	385	558	814	1318	1772	2502	3290	4331
S ft.^{3}	0.463	0.560	0.907	1.500	2.241	3.130	4.167	5.352	6.685	8.167
S.P.h p.s.f.	459	732	864	793	776	869	877	968	996	1084
S.P.t p.s.f.	233	61	16	50	49	27	27	33	12	23
Friction	0.350	0.449	0.497	0.554	0.608	0.626	0.664	0.660	0.695	0.688

TABLE 5.6a	PROPERTY LINE RETAINING WALLS — MASONRY STEM — SOIL NOT OVER FOOTING NO SURCHARGE LOAD									
Stem Height	3'- 0"	4'- 0"	5'- 0"	6'- 0"	7'- 0"	8'- 0"	9'- 0"	10'- 0"	11'- 0"	12'- 0"
W lbs.	577	860	1088	1306	1739	2016	2575	2923	3585	3998
F lbs.	202	350	510	700	960	1215	1550	1870	2282	2667
Mo ft.-lbs.	246	565	992	1595	2560	3645	5254	6962	9380	11852
Mr ft.-lbs.	674	1351	2316	3346	5504	7497	10808	13899	18745	23160
O.T.R.	2.736	2.394	2.334	2.097	2.150	2.057	2.057	1.996	1.998	1.954
$\overline{X}$ ft.	0.742	0.915	1.216	1.340	1.693	1.911	2.156	2.373	2.612	2.828
e ft.	0.091	0.168	0.201	0.327	0.390	0.506	0.594	0.710	0.804	0.922
Me ft.-lbs.	52	145	218	427	679	1020	1529	2075	2884	3685
S ft.^{3}	0.463	0.782	1.338	1.852	2.894	3.894	5.042	6.338	7.782	9.375
S.P.h p.s.f.	459	582	547	622	652	679	772	801	895	926
S.P.t p.s.f.	233	212	221	161	183	155	165	147	154	140
Friction	0.350	0.408	0.469	0.536	0.552	0.603	0.602	0.640	0.636	0.667

TABLE 5.7a	PROPERTY LINE RETAINING WALLS — MASONRY STEM — SOIL NOT OVER FOOTING NO SURCHARGE LOAD									
Stem Height	3'- 0"	4'- 0"	5'- 0"	6'- 0"	7'- 0"	8'- 0"	9'- 0"	10'- 0"	11'- 0"	12'- 0"
W lbs.	610	922	1130	1369	1789	2098	2663	3013	3684	4120
F lbs.	202	350	510	700	960	1215	1550	1870	2282	2667
Mo ft.-lbs.	246	565	992	1595	2560	3645	5254	6962	9380	11852
Mr ft.-lbs.	872	1797	2686	3997	6092	8296	12083	15033	20111	24852
O.T.R.	3.539	3.183	2.706	2.505	2.380	2.276	2.300	2.159	2.144	2.097
$\overline{X}$ ft.	1.026	1.336	1.498	1.755	1.974	2.217	2.564	2.679	2.913	3.155
e ft.	−0.026	−0.003	0.085	0.162	0.276	0.366	0.436	0.571	0.670	0.762
Me ft.-lbs.	−16	−3	96	222	493	769	1160	1722	2468	3138
S ft.^{3}	0.667	1.185	1.671	2.449	3.375	4.449	6.000	7.042	8.560	10.227
S.P.h p.s.f.	281	344	414	448	544	579	637	708	802	833
S.P.t p.s.f.	329	348	299	267	251	233	250	219	226	219
Friction	0.331	0.380	0.452	0.512	0.537	0.579	0.582	0.621	0.619	0.647

TABLE 5.8a PROPERTY LINE RETAINING WALLS — MASONRY STEM — SOIL NOT OVER FOOTING NO SURCHARGE LOAD										
Stem Height	3'- 0"	4'- 0"	5'- 0"	6'- 0"	7'- 0"	8'- 0"	9'- 0"	10'- 0"	11'- 0"	12'- 0"
W lbs.	808	1026	1255	1672	1971	2503	2873	3512	3939	4376
F lbs.	220	350	510	735	960	1260	1550	1927	2282	2667
Mo ft.-lbs.	282	565	992	1715	2560	3851	5254	7279	9380	11852
Mr ft.-lbs.	1582	2574	3844	5895	8061	11396	14686	19615	24216	29429
O.T.R.	5.618	4.560	3.873	3.437	3.149	2.959	2.795	2.695	2.582	2.483
$\overline{X}$ ft.	1.610	1.958	2.272	2.500	2.791	3.014	3.283	3.512	3.767	4.017
e ft.	-0.193	-0.208	-0.188	-0.083	-0.041	0.069	0.134	0.238	0.317	0.400
Me ft.-lbs.	-156	-213	-236	-140	-80	174	384	835	1248	1748
S ft.^{3}	1.338	2.042	2.894	3.894	5.042	6.338	7.782	9.375	11.116	13.005
S.P.h p.s.f.	168	189	220	310	342	433	470	557	595	630
S.P.t p.s.f.	402	398	383	382	374	379	371	379	370	361
Friction	0.273	0.341	0.407	0.440	0.487	0.503	0.540	0.549	0.579	0.609

TABLE 6.2a, 3a 4a, 5a UNDERCUT FOOTING RETAINING WALLS — CONCRETE STEM — SOIL AT TOE SIDE NO AXIAL LOAD										
Stem Height	3'- 0"	4'- 0"	5'- 0"	6'- 0"	7'- 0"	8'- 0"	9'- 0"	10'- 0"	11'- 0"	12'- 0"
W lbs.	778	1013	1521	1813	2137	2845	3230	4092	4538	5553
F lbs.	202	327	510	700	920	1215	1500	1870	2220	2667
Mo ft.-lbs.	246	508	992	1595	2403	3645	5000	6962	9005	11852
Mr ft.-lbs.	706	1229	2359	3349	4899	7467	9915	13877	17412	23132
O.T.R.	2.865	2.418	2.376	2.099	2.038	2.049	1.983	1.993	1.934	1.952
$\overline{X}$ ft.	0.591	0.711	0.898	0.967	1.168	1.343	1.522	1.690	1.852	2.031
e ft.	0.243	0.289	0.352	0.449	0.499	0.573	0.645	0.727	0.814	0.885
Me ft.-lbs.	189	293	535	815	1066	1631	2084	2974	3695	4917
S ft.^{3}	0.463	0.667	1.042	1.338	1.852	2.449	3.130	3.894	4.741	5.671
S.P.h p.s.f.	875	946	1122	1249	1216	1408	1411	1610	1630	1819
S.P.t p.s.f.	59	68	95	31	66	76	80	83	71	85
Friction	0.259	0.322	0.336	0.386	0.431	0.427	0.464	0.457	0.489	0.480

TABLE 6.6a	UNDERCUT FOOTING RETAINING WALLS — CONCRETE STEM — SOIL AT TOE SIDE NO SURCHARGE LOAD									
Stem Height	3'- 0"	4'- 0"	5'- 0"	6'- 0"	7'- 0"	8'- 0"	9'- 0"	10'- 0"	11'- 0"	12'- 0"
W lbs.	795	1259	1542	1855	2520	2870	3684	4121	5088	5587
F lbs.	202	350	510	700	960	1215	1550	1870	2282	2667
Mo ft.-lbs.	246	565	992	1595	2560	3645	5254	6962	9380	11852
Mr ft.-lbs.	837	1753	2614	3961	6197	7943	11388	14561	19662	24060
O.T.R.	3.397	3.105	2.634	2.483	2.421	2.179	2.167	2.091	2.096	2.030
$\overline{X}$ ft.	0.743	0.944	1.052	1.275	1.443	1.498	1.665	1.844	2.021	2.185
e ft.	0.174	0.223	0.282	0.308	0.390	0.502	0.585	0.656	0.729	0.815
Me ft.-lbs.	138	281	434	572	983	1442	2156	2703	3711	4552
S ft.^{3}	0.560	0.907	1.185	1.671	2.241	2.667	3.375	4.167	5.042	6.000
S.P.h p.s.f.	680	849	945	928	1126	1258	1457	1473	1661	1690
S.P.t p.s.f.	187	230	212	244	249	177	180	175	189	172
Friction	0.254	0.278	0.331	0.378	0.381	0.423	0.421	0.454	0.448	0.477

TABLE 6.7a	UNDERCUT FOOTING RETAINING WALLS — CONCRETE STEM — SOIL AT TOE SIDE NO SURCHARGE LOAD									
Stem Height	3'- 0"	4'- 0"	5'- 0"	6'- 0"	7'- 0"	8'- 0"	9'- 0"	10'- 0"	11'- 0"	12'- 0"
W lbs.	812	1280	1583	1876	2545	2920	3742	4179	5155	5653
F lbs.	202	350	510	700	960	1215	1550	1870	2282	2667
Mo ft.-lbs.	246	565	992	1595	2560	3645	5254	6962	9380	11852
Mr ft.-lbs.	971	1965	3117	4254	6619	8883	12591	15910	21325	25889
O.T.R.	3.940	3.480	3.141	2.667	2.586	2.437	2.396	2.285	2.273	2.184
$\overline{X}$ ft.	0.893	1.094	1.342	1.417	1.595	1.794	1.961	2.141	2.317	2.483
e ft.	0.107	0.156	0.158	0.249	0.322	0.373	0.456	0.525	0.599	0.684
Me ft.-lbs.	87	200	250	468	819	1088	1707	2196	3090	3865
S ft.^{3}	0.667	1.042	1.500	1.852	2.449	3.130	3.894	4.741	5.671	6.685
S.P.h p.s.f.	536	704	695	815	998	1022	1213	1247	1429	1471
S.P.t p.s.f.	275	320	361	310	330	326	336	320	339	314
Friction	0.248	0.274	0.322	0.373	0.377	0.416	0.414	0.448	0.443	0.472

TABLE 6.8a		UNDERCUT FOOTING RETAINING WALLS — CONCRETE STEM — SOIL AT TOE SIDE NO SURCHARGE LOAD								
Stem Height	3'- 0"	4'- 0"	5'- 0"	6'- 0"	7'- 0"	8'- 0"	9'- 0"	10'- 0"	11'- 0"	12'- 0"
W lbs.	1070	1342	1646	2280	2645	3432	3859	4800	5288	5787
F lbs.	220	350	510	735	960	1260	1550	1927	2282	2667
Mo ft.-lbs.	282	565	992	1715	2560	3851	5254	7279	9380	11852
Mr ft.-lbs.	1764	2603	3907	6177	8324	11912	15091	20361	24762	29658
O.T.R.	6.263	4.610	3.937	3.602	3.252	3.093	2.872	2.797	2.640	2.502
$\overline{X}$ ft.	1.385	1.518	1.771	1.957	2.179	2.348	2.549	2.726	2.909	3.077
e ft.	-0.052	-0.018	-0.021	0.043	0.071	0.152	0.201	0.274	0.341	0.423
Me ft.-lbs.	-56	-25	-34	98	187	520	776	1317	1805	2447
S ft.^{3}	1.185	1.500	2.042	2.667	3.375	4.167	5.042	6.000	7.042	8.167
S.P.h p.s.f.	354	431	453	607	643	811	856	1020	1070	1126
S.P.t p.s.f.	448	464	487	533	532	562	548	580	557	527
Friction	0.206	0.261	0.310	0.322	0.363	0.367	0.402	0.401	0.431	0.461

TABLE 7.2a, 3a		UNDERCUT FOOTING RETAINING WALLS — MASONRY STEM — SOIL AT TOE SIDE NO AXIAL LOAD								
Stem Height	3'- 0"	4'- 0"	5'- 0"	6'- 0"	7'- 0"	8'- 0"	9'- 0"	10'- 0"	11'- 0"	12'- 0"
W lbs.	754	981	1418	1711	1993	2606	2990	3730	4195	5074
F lbs.	202	327	510	700	920	1215	1500	1870	2220	2667
Mo ft.-lbs.	246	508	992	1595	2403	3645	5000	6962	9005	11852
Mr ft.-lbs.	690	1197	2245	3456	4621	7323	9699	13343	16824	22073
O.T.R.	2.800	2.355	2.262	2.166	1.923	2.009	1.940	1.916	1.868	1.862
$\overline{X}$ ft.	0.588	0.702	0.883	1.087	1.112	1.411	1.572	1.711	1.864	2.014
e ft.	0.245	0.298	0.367	0.413	0.554	0.589	0.678	0.789	0.886	0.986
Me ft.-lbs.	185	293	520	706	1105	1534	2029	2944	3717	5001
S ft.^{3}	0.463	0.667	1.042	1.500	1.852	2.667	3.375	4.167	5.042	6.000
S.P.h p.s.f.	852	930	1067	1041	1194	1227	1265	1453	1500	1679
S.P.t p.s.f.	53	52	68	100	1	76	63	39	25	12
Friction	0.267	0.333	0.360	0.409	0.462	0.466	0.502	0.501	0.529	0.526

TABLE 7.4a, 5a	UNDERCUT FOOTING RETAINING WALLS — MASONRY STEM — SOIL AT TOE SIDE NO AXIAL LOAD									
Stem Height	3'- 0"	4'- 0"	5'- 0"	6'- 0"	7'- 0"	8'- 0"	9'- 0"	10'- 0"	11'- 0"	12'- 0"
W lbs.	754	981	1418	1711	1993	2606	2990	3762	4195	5106
F lbs.	202	327	510	700	920	1215	1500	1870	2220	2667
Mo ft.-lbs.	246	508	992	1595	2403	3645	5000	6962	9005	11852
Mr ft.-lbs.	690	1197	2245	3456	4621	7323	9699	13443	16824	22198
O.T.R.	2.800	2.355	2.262	2.166	1.923	2.009	1.940	1.931	1.868	1.873
$\overline{X}$ ft.	0.588	0.702	0.883	1.087	1.112	1.411	1.572	1.723	1.864	2.026
e ft.	0.245	0.298	0.367	0.413	0.554	0.589	0.678	0.777	0.886	0.974
Me ft.-lbs.	185	293	520	706	1105	1534	2029	2924	3717	4972
S ft.^{3}	0.463	0.667	1.042	1.500	1.852	2.667	3.375	4.167	5.042	6.000
S.P.h p.s.f.	852	930	1067	1041	1194	1227	1265	1454	1500	1680
S.P.t p.s.f.	53	52	68	100	1	76	63	51	25	22
Friction	0.267	0.333	0.360	0.409	0.462	0.466	0.502	0.497	0.529	0.522

TABLE 7.6a	UNDERCUT FOOTING RETAINING WALLS — MASONRY STEM — SOIL AT TOE SIDE NO SURCHARGE LOAD									
Stem Height	3'- 0"	4'- 0"	5'- 0"	6'- 0"	7'- 0"	8'- 0"	9'- 0"	10'- 0"	11'- 0"	12'- 0"
W lbs.	771	1177	1439	1732	2289	2656	3397	3820	4696	5196
F lbs.	202	350	510	700	960	1215	1550	1870	2282	2667
Mo ft.-lbs.	246	565	992	1595	2560	3645	5254	6962	9380	11852
Mr ft.-lbs.	817	1676	2483	3743	5731	8175	11572	14672	19625	23968
O.T.R.	3.316	2.969	2.502	2.346	2.239	2.243	2.203	2.107	2.092	2.022
$\overline{X}$ ft.	0.740	0.944	1.036	1.240	1.385	1.705	1.860	2.018	2.182	2.332
e ft.	0.176	0.222	0.297	0.344	0.448	0.461	0.557	0.648	0.735	0.835
Me ft.-lbs.	136	262	428	595	1026	1225	1891	2477	3452	4338
S ft.^{3}	0.560	0.907	1.185	1.671	2.241	3.130	3.894	4.741	5.671	6.685
S.P.h p.s.f.	663	793	901	903	1082	1004	1189	1239	1414	1469
S.P.t p.s.f.	178	216	179	191	167	222	217	194	196	171
Friction	0.262	0.298	0.355	0.404	0.419	0.457	0.456	0.490	0.486	0.513

TABLE 7.7a	UNDERCUT FOOTING RETAINING WALLS — MASONRY STEM — SOIL AT TOE SIDE NO SURCHARGE LOAD									
Stem Height	3'- 0''	4'- 0''	5'- 0''	6'- 0''	7'- 0''	8'- 0''	9'- 0''	10'- 0''	11'- 0''	12'- 0''
W lbs.	788	1198	1481	1774	2339	2713	3455	3910	4795	5318
F lbs.	202	350	510	700	960	1215	1550	1870	2282	2667
Mo ft.-lbs.	246	565	992	1595	2560	3645	5254	6962	9380	11852
Mr ft.-lbs.	947	1874	2953	4310	6502	8710	12714	16076	21287	25920
O.T.R.	3.843	3.319	2.975	2.701	2.540	2.390	2.420	2.309	2.269	2.187
$\overline{X}$ ft.	0.890	1.093	1.324	1.530	1.685	1.867	2.159	2.331	2.483	2.645
e ft.	0.110	0.157	0.176	0.220	0.315	0.383	0.424	0.503	0.600	0.688
Me ft.-lbs.	87	188	261	390	736	1039	1466	1966	2877	3659
S ft.^{3}	0.667	1.042	1.500	2.042	2.667	3.375	4.449	5.352	6.338	7.407
S.P.h p.s.f.	524	660	668	698	861	911	998	1057	1231	1292
S.P.t p.s.f.	263	299	320	316	309	295	339	323	324	304
Friction	0.256	0.293	0.345	0.395	0.410	0.448	0.449	0.478	0.476	0.501

TABLE 7.8a	UNDERCUT FOOTING RETAINING WALLS — MASONRY STEM — SOIL AT TOE SIDE NO SURCHARGE LOAD									
Stem Height	3'- 0''	4'- 0''	5'- 0''	6'- 0''	7'- 0''	8'- 0''	9'- 0''	10'- 0''	11'- 0''	12'- 0''
W lbs.	1008	1281	1564	2107	2471	3162	3607	4437	4983	5507
F lbs.	220	350	510	735	960	1260	1550	1927	2282	2667
Mo ft.-lbs.	282	565	992	1715	2560	3851	5254	7279	9380	11852
Mr ft.-lbs.	1686	2683	3950	6036	8141	11502	14671	19517	24745	29754
O.T.R.	5.985	4.753	3.980	3.520	3.180	2.987	2.792	2.681	2.638	2.511
$\overline{X}$ ft.	1.392	1.653	1.891	2.051	2.259	2.420	2.611	2.758	3.083	3.251
e ft.	-0.059	-0.070	-0.058	0.033	0.075	0.163	0.222	0.325	0.333	0.416
Me ft.-lbs.	-59	-90	-90	69	185	517	802	1441	1661	2289
S ft.^{3}	1.185	1.671	2.241	2.894	3.630	4.449	5.352	6.338	7.782	8.963
S.P.h p.s.f.	328	351	386	529	580	728	786	947	943	1006
S.P.t p.s.f.	428	458	467	482	479	496	487	492	516	496
Friction	0.219	0.273	0.326	0.349	0.389	0.399	0.430	0.434	0.458	0.484

TABLE 8.2a, 3a 4a, 5a	GENERAL RETAINING WALLS — CONCRETE STEM — SOIL AT HEEL SIDE NO AXIAL LOAD									
Stem Height	3'-0''	4'-0''	5'-0''	6'-0''	7'-0''	8'-0''	9'-0''	10'-0''	11'-0''	12'-0''
W lbs.	728	1013	1562	2084	2682	3537	4305	5371	6309	7587
F lbs.	202	327	510	700	920	1215	1500	1870	2220	2667
Mo ft.-lbs.	246	508	992	1595	2403	3645	5000	6962	9005	11852
Mr ft.-lbs.	709	1229	2369	3787	5685	8566	11732	16228	20986	27527
O.T.R.	2.878	2.418	2.387	2.373	2.366	2.350	2.347	2.331	2.331	2.323
$\overline{X}$ ft.	0.636	0.711	0.881	1.051	1.224	1.391	1.564	1.725	1.899	2.066
e ft.	0.198	0.289	0.369	0.449	0.526	0.609	0.686	0.775	0.851	0.934
Me ft.-lbs.	144	293	577	935	1412	2153	2954	4161	5369	7085
S ft.^{3}	0.463	0.667	1.042	1.500	2.042	2.667	3.375	4.167	5.042	6.000
S.P.t p.s.f.	748	946	1179	1318	1458	1691	1832	2073	2212	2445
S.P.h p.s.f.	126	68	71	71	75	77	81	75	82	84
Friction	0.277	0.322	0.327	0.336	0.343	0.344	0.348	0.348	0.352	0.351

TABLE 8.6a	GENERAL RETAINING WALLS — CONCRETE STEM — SOIL AT HEEL SIDE NO SURCHARGE LOAD									
Stem Height	3'-0''	4'-0''	5'-0''	6'-0''	7'-0''	8'-0''	9'-0''	10'-0''	11'-0''	12'-0''
W lbs.	745	1226	1687	2226	3012	3695	4688	5567	6772	7820
F lbs.	202	350	510	700	960	1215	1550	1870	2282	2667
Mo ft.-lbs.	246	565	992	1595	2560	3645	5254	6962	9380	11852
Mr ft.-lbs.	832	1746	2918	4528	7039	9212	13109	17257	22999	28946
O.T.R.	3.376	3.092	2.940	2.838	2.750	2.527	2.495	2.479	2.452	2.442
$\overline{X}$ ft.	0.786	0.964	1.141	1.318	1.487	1.507	1.675	1.849	2.011	2.186
e ft.	0.130	0.203	0.276	0.349	0.429	0.577	0.658	0.734	0.822	0.897
Me ft.-lbs.	97	249	466	777	1293	2131	3085	4085	5568	7017
S ft.^{3}	0.560	0.907	1.338	1.852	2.449	2.894	3.630	4.449	5.352	6.338
S.P.t p.s.f.	580	800	944	1087	1314	1623	1854	1996	2235	2375
S.P.h p.s.f.	233	251	248	248	258	150	155	159	155	161
Friction	0.271	0.286	0.302	0.315	0.319	0.329	0.331	0.336	0.337	0.341

TABLE 8.7a	GENERAL RETAINING WALLS — CONCRETE STEM — SOIL AT HEEL SIDE NO SURCHARGE LOAD									
Stem Height	3'- 0"	4'- 0"	5'- 0"	6'- 0"	7'- 0"	8'- 0"	9'- 0"	10'- 0"	11'- 0"	12'- 0"
W lbs.	812	1313	1792	2347	3153	3878	4897	5792	7022	8087
F lbs.	202	350	510	700	960	1215	1550	1870	2282	2667
Mo ft.-lbs.	246	565	992	1595	2560	3645	5254	6962	9380	11852
Mr ft.-lbs.	971	1975	3239	4959	7619	10554	14774	19248	25457	31794
O.T.R.	3.940	3.498	3.263	3.108	2.976	2.896	2.812	2.765	2.714	2.683
$\overline{X}$ ft.	0.893	1.074	1.254	1.433	1.604	1.782	1.944	2.121	2.290	2.466
e ft.	0.107	0.176	0.246	0.317	0.396	0.468	0.556	0.629	0.710	0.784
Me ft.-lbs.	87	232	441	744	1248	1817	2722	3641	4989	6340
S ft.^{3}	0.667	1.042	1.500	2.042	2.667	3.375	4.167	5.042	6.000	7.042
S.P.t p.s.f.	536	748	891	1035	1256	1400	1633	1775	2002	2144
S.P.h p.s.f.	275	303	303	306	320	323	326	331	339	344
Friction	0.248	0.267	0.285	0.298	0.304	0.313	0.317	0.323	0.325	0.330

TABLE 8.8a	GENERAL RETAINING WALLS — CONCRETE STEM — SOIL AT HEEL SIDE NO SURCHARGE LOAD									
Stem Height	3'- 0"	4'- 0"	5'- 0"	6'- 0"	7'- 0"	8'- 0"	9'- 0"	10'- 0"	11'- 0"	12'- 0"
W lbs.	1145	1509	2021	2780	3462	4432	5284	6467	7488	8587
F lbs.	220	350	510	735	960	1260	1550	1927	2282	2667
Mo ft.-lbs.	282	565	992	1715	2560	3851	5254	7279	9380	11852
Mr ft.-lbs.	1796	2708	4251	6694	9399	13346	17538	23417	29419	36378
O.T.R.	6.378	4.797	4.283	3.903	3.672	3.465	3.338	3.217	3.136	3.069
$\overline{X}$ ft.	1.323	1.421	1.612	1.791	1.976	2.142	2.325	2.496	2.676	2.856
e ft.	0.010	0.079	0.138	0.209	0.274	0.358	0.425	0.504	0.574	0.644
Me ft.-lbs.	12	120	278	581	949	1587	2248	3262	4299	5527
S ft.^{3}	1.185	1.500	2.042	2.667	3.375	4.167	5.042	6.000	7.042	8.167
S.P.t p.s.f.	439	583	714	913	1051	1267	1407	1621	1763	1903
S.P.h p.s.f.	419	423	441	477	488	506	515	534	542	550
Friction	0.193	0.232	0.253	0.264	0.277	0.284	0.293	0.298	0.305	0.311

TABLE 9.2a, 3a	GENERAL RETAINING WALLS — MASONRY STEM — SOIL AT HEEL SIDE NO AXIAL LOAD									
Stem Height	3'- 0"	4'- 0"	5'- 0"	6'- 0"	7'- 0"	8'- 0"	9'- 0"	10'- 0"	11'- 0"	12'- 0"
W lbs.	704	981	1502	2011	2597	3406	4168	5185	6123	7352
F lbs.	202	327	510	700	920	1215	1500	1870	2220	2667
Mo ft.-lbs.	246	508	992	1595	2403	3645	5000	6962	9005	11852
Mr ft.-lbs.	689	1197	2280	3665	5526	8267	11395	15703	20430	26731
O.T.R.	2.797	2.355	2.297	2.297	2.299	2.268	2.279	2.255	2.269	2.255
$\overline{X}$ ft.	0.629	0.702	0.857	1.029	1.202	1.357	1.534	1.686	1.866	2.024
e ft.	0.204	0.298	0.393	0.471	0.548	0.643	0.716	0.814	0.884	0.976
Me ft.-lbs.	144	293	590	947	1422	2190	2982	4223	5413	7177
S ft.^{3}	0.463	0.667	1.042	1.500	2.042	2.667	3.375	4.167	5.042	6.000
S.P.t p.s.f.	734	930	1167	1302	1439	1673	1810	2051	2187	2421
S.P.h p.s.f.	112	52	34	39	45	30	43	24	40	29
Friction	0.286	0.333	0.340	0.348	0.354	0.357	0.360	0.361	0.363	0.363

TABLE 9.4a, 5a	GENERAL RETAINING WALLS — MASONRY STEM — SOIL AT HEEL SIDE NO AXIAL LOAD									
Stem Height	3'- 0"	4'- 0"	5'- 0"	6'- 0"	7'- 0"	8'- 0"	9'- 0"	10'- 0"	11'- 0"	12'- 0"
W lbs.	704	981	1502	2011	2597	3406	4168	5195	6123	7362
F lbs.	202	327	510	700	920	1215	1500	1870	2220	2667
Mo ft.-lbs.	246	508	992	1595	2403	3645	5000	6962	9005	11852
Mr ft.-lbs.	689	1197	2280	3665	5526	8267	11395	15737	20430	26771
O.T.R.	2.797	2.355	2.297	2.297	2.299	2.268	2.279	2.260	2.269	2.259
$\overline{X}$ ft.	0.629	0.702	0.857	1.029	1.202	1.357	1.534	1.689	1.866	2.027
e ft.	0.204	0.298	0.393	0.471	0.548	0.643	0.716	0.811	0.884	0.973
Me ft.-lbs.	144	293	590	947	1422	2190	2982	4213	5413	7166
S ft.^{3}	0.463	0.667	1.042	1.500	2.042	2.667	3.375	4.167	5.042	6.000
S.P.t p.s.f.	734	930	1167	1302	1439	1673	1810	2050	2187	2421
S.P.h p.s.f.	112	52	34	39	45	30	43	28	40	33
Friction	0.286	0.333	0.340	0.348	0.354	0.357	0.360	0.360	0.363	0.362

TABLE 9.6a	GENERAL RETAINING WALLS — MASONRY STEM — SOIL AT HEEL SIDE NO SURCHARGE LOAD								
Stem Height	3'- 0''	4'- 0''	5'- 0''	6'- 0''	7'- 0''	8'- 0''	9'- 0''	10'- 0''	11'- 0''	12'- 0''
W lbs.	721	1177	1627	2153	2897	3589	4553	5420	6596	7640
F lbs.	202	350	510	700	960	1215	1550	1870	2282	2667
Mo ft.-lbs.	246	565	992	1595	2560	3645	5254	6962	9380	11852
Mr ft.-lbs.	808	1676	2818	4394	6783	9509	13402	17633	23432	29498
O.T.R.	3.279	2.969	2.840	2.754	2.650	2.609	2.551	2.533	2.498	2.489
$\overline{X}$ ft.	0.779	0.944	1.122	1.300	1.458	1.634	1.790	1.969	2.130	2.310
e ft.	0.138	0.222	0.294	0.366	0.459	0.533	0.627	0.698	0.786	0.857
Me ft.-lbs.	99	262	479	789	1330	1913	2854	3783	5187	6548
S ft.^{3}	0.560	0.907	1.338	1.852	2.449	3.130	3.894	4.741	5.671	6.685
S.P.t p.s.f.	570	793	932	1072	1299	1439	1675	1814	2045	2186
S.P.h p.s.f.	216	216	216	220	213	217	209	218	216	227
Friction	0.280	0.298	0.314	0.325	0.331	0.339	0.341	0.345	0.346	0.349

TABLE 9.7a	GENERAL RETAINING WALLS — MASONRY STEM — SOIL AT HEEL SIDE NO SURCHARGE LOAD								
Stem Height	3'- 0''	4'- 0''	5'- 0''	6'- 0''	7'- 0''	8'- 0''	9'- 0''	10'- 0''	11'- 0''	12'- 0''
W lbs.	788	1265	1731	2274	3039	3757	4732	5626	6822	7896
F lbs.	202	350	510	700	960	1215	1550	1870	2282	2667
Mo ft.-lbs.	246	565	992	1595	2560	3645	5254	6962	9380	11852
Mr ft.-lbs.	947	1905	3139	4825	7363	10263	14317	18764	24797	31122
O.T.R.	3.843	3.374	3.163	3.024	2.876	2.816	2.725	2.695	2.644	2.626
$\overline{X}$ ft.	0.890	1.060	1.240	1.420	1.581	1.761	1.915	2.098	2.260	2.441
e ft.	0.110	0.190	0.260	0.330	0.419	0.489	0.585	0.652	0.740	0.809
Me ft.-lbs.	87	241	449	750	1275	1836	2766	3669	5051	6391
S ft.^{3}	0.667	1.042	1.500	2.042	2.667	3.375	4.167	5.042	6.000	7.042
S.P.t p.s.f.	524	737	877	1017	1238	1379	1610	1751	1979	2122
S.P.h p.s.f.	263	275	277	282	282	291	282	295	295	307
Friction	0.256	0.277	0.295	0.308	0.316	0.323	0.328	0.332	0.334	0.338

TABLE 9.8a	GENERAL RETAINING WALLS — MASONRY STEM — SOIL AT HEEL SIDE NO SURCHARGE LOAD									
Stem Height	3'- 0"	4'- 0"	5'- 0"	6'- 0"	7'- 0"	8'- 0"	9'- 0"	10'- 0"	11'- 0"	12'- 0"
W lbs.	1108	1548	1960	2682	3357	4288	5139	6281	7311	8418
F lbs.	220	350	510	735	960	1260	1550	1927	2282	2667
Mo ft.-lbs.	282	565	992	1715	2560	3851	5254	7279	9380	11852
Mr ft.-lbs.	1745	2918	4141	6480	9153	12955	17123	22814	28806	35761
O.T.R.	6.197	5.168	4.173	3.778	3.576	3.364	3.259	3.134	3.071	3.017
$\overline{X}$ ft.	1.320	1.520	1.607	1.777	1.964	2.123	2.310	2.473	2.657	2.840
e ft.	0.013	0.063	0.143	0.223	0.286	0.377	0.440	0.527	0.593	0.660
Me ft.-lbs.	14	98	281	599	960	1616	2263	3308	4335	5553
S ft.^{3}	1.185	1.671	2.042	2.667	3.375	4.167	5.042	6.000	7.042	8.167
S.P.t p.s.f.	428	547	698	895	1031	1246	1383	1598	1740	1883
S.P.h p.s.f.	404	430	422	446	462	470	485	496	509	523
Friction	0.199	0.226	0.260	0.274	0.286	0.294	0.302	0.307	0.312	0.317

APPENDIX B

Building Code Requirements for the Design and Construction of Cantilever Retaining Walls

TABLE NO. 29-C — ALLOWABLE SOIL PRESSURE

CLASS OF MATERIAL	MINIMUM DEPTH OF FOOTING BELOW ADJACENT VIRGIN GROUND	VALUE PERMISSIBLE IF FOOTING IS AT MINIMUM DEPTH, POUNDS PER SQUARE FOOT	INCREASE IN VALUE FOR EACH FOOT OF DEPTH THAT FOOTING IS BELOW MINIMUM DEPTH, POUNDS PER SQUARE FOOT	MAXIMUM VALUE POUNDS PER SQUARE FOOT
1	2	3	4	5
Rock	0'	20% of ultimate crushing strength	0	20% of ultimate crushing strength
Compact coarse sand	1'	1500[1]	300[1]	8000
Compact fine sand	1'	1000[1]	200[1]	8000
Loose sand	2'	500[1]	100[1]	3000
Hard clay or sandy clay	1'	4000	800	8000
Medium-stiff clay or sandy clay	1'	2000	200	6000
Soft sandy clay or clay	2'	1000	50	2000
Expansive soils	1'6"[3]	1000[2, 3]	50[3]	
Compact inorganic sand and silt mixtures	1'	1000	200	4000
Loose inorganic sand silt mixtures	2'	500	100	1000
Loose organic sand and silt mixtures and muck or bay mud	0'	0	0	0

[1]These values are for footings 1 foot in width and may be increased in direct proportion to the width of the footing to a maximum of three times the designated value.
[2]For depths greater than 8 feet use values given for clay of comparable consistency.
[3]Also, see Section 2903 (d).

Sec. 2418. (j)

3. **Reinforcement.** All walls using stress permitted for reinforced masonry shall be reinforced with both vertical and horizontal bars.

The area of horizontal reinforcement shall be not less than 0.0013 and that of vertical reinforcement shall be not less than 0.0007 times the gross cross-sectional area of the wall. Wall steel shall be limited to a maximum of 4 feet on center. The minimum diameter shall be ⅜ inch except that joint reinforcement may be considered as part of the required reinforcement.

Horizontal reinforcement shall be provided in the top of footings, at the top of wall openings, at roof and floor levels, and at the top of parapet walls. Only horizontal reinforcement which is continuous in the wall shall be considered in computing the minimum area of reinforcement.

If the wall is constructed of more than two units in thickness, the minimum area of required reinforcement shall be equally divided into two layers, except where designed as retaining walls. Where reinforcement is added above the minimum requirements such additional reinforcement need not be so divided.

In bearing walls of every type of reinforced masonry there shall be not less than one ½-inch bar or two ⅜-inch bars on all sides of, and adjacent to, every opening which exceeds 24 inches in either direction, and such bars shall extend not less than 40 diameters, but in no case less than 24 inches beyond the corners of the opening. The bars required by this paragraph shall be in addition to the minimum reinforcement elsewhere required.

When the reinforcement in bearing walls is designed, placed and anchored in position as for columns, the allowable stresses shall be as for columns. The length of the wall to be considered effective shall not exceed the center-to-center distance between loads nor shall it exceed the width of the bearing plus four times the wall thickness.

Sec. 2310. Retaining walls shall be designed to resist the lateral pressure of the retained material in accordance with accepted engineering practice. Walls retaining drained earth may be designed for pressure equivalent to that exerted by a fluid weighing not less than 30 pounds per cubic foot and having a depth equal to that of the retained earth. Any surcharge shall be in addition to the equivalent fluid pressure.

TABLE NO. 29-B — ALLOWABLE LATERAL SOIL PRESSURE

CLASS OF MATERIAL	ALLOWABLE VALUES PER FOOT OF DEPTH BELOW NATURAL GRADE[1] (Pounds per Square Foot)	MAXIMUM ALLOWABLE VALUES (Pounds per Square Foot)
Good — compact well-graded sand and gravel Hard Clay Well-graded fine and coarse sand (All drained so water will not stand)	400	8000
Average—compact fine sand Medium Clay Compact sandy loam Loose coarse sand and gravel (All drained so water will not stand)	200	2500
Poor—Soft Clay Clay Loam Poorly compacted sand Clays containing large amounts of silt (Water stands during wet season)	100	1500

[1]Isolated poles, such as flagpoles, or signs, may be designed using lateral bearing values equal to two times tabulated values.

TABLE NO. 24-A — MORTAR PROPORTIONS
(Parts by Volume)

MORTAR TYPE	MINIMUM COMPRESSIVE STRENGTH AT 28 DAYS (p.s.i.)	PORT-LAND CEMENT	HYDRATED LIMES OR LIME PUTTY[1] MIN.	HYDRATED LIMES OR LIME PUTTY[1] MAX.	MASONRY CEMENTS	DAMP LOOSE AGGREGATE
M	2500	1	—	¼	—	Not less than 2¼ and not more than 3 times the sum of the volumes of the cement and lime used
M	2500	1	—	—	1	
S	1800	1	¼	½	—	
S	1800	½	—	—	1	
N	750	1	½	1¼	—	
N	750	—	—	—	1	
O	350	1	1¼	2½	—	

[1]When plastic or waterproof cement is used as specified in Section 2403 (p), hydrated lime or putty may be added but not in excess of one-tenth the volume of cement.

Reinforced Grouted Masonry

Sec. 2414. (a) **General.** Reinforced grouted masonry shall conform to all of the requirements for grouted masonry specified in Section 2413 and also the requirements of this Section.

(b) **Construction.** The thickness of grout or mortar between masonry units and reinforcement shall be not less than one-fourth inch (¼″), except that one-fourth-inch (¼″) bars may be laid in horizontal mortar joints at least one-half inch (½″) thick and steel wire reinforcement may be laid in horizontal mortar joints at least twice the thickness of the wire diameter.

(c) **Stresses.** See Section 2418 (a).

Sec. 2415. (a) **General.** Reinforced hollow unit masonry is that type of construction made with hollow masonry units in which certain cells are continuously filled with concrete or grout, and in which reinforcement is embedded. Only Type M or Type S mortar consisting of a mixture of portland cement, hydrated lime and aggregate shall be used.

(b) **Construction.** Requirements for construction shall be as follows:

1. All reinforced hollow unit masonry shall be built to preserve the unobstructed vertical continuity of the cells to be filled. Walls and cross webs forming such cells to be filled shall be full-bedded in mortar to prevent leakage of grout. All head (or end) joints shall be solidly filled with mortar for a distance in from the face of the wall or unit not less than the thickness of the longitudinal face shells. Bond shall be provided by lapping units in successive vertical courses or by equivalent mechanical anchorage.

2. Vertical cells to be filled shall have vertical alignment sufficient to maintain a clear, unobstructed continuous vertical cell measuring not less than two inches by three inches (2″ x 3″).

3. Cleanout openings shall be provided at the bottom of all cells to be filled at each pour of grout where such grout pour is in excess of four feet (4′) in height. Any overhanging mortar or other obstruction or debris shall be removed from the insides of such cell walls. The cleanouts shall be sealed before grouting, after inspection.

4. Vertical reinforcement shall be held in position at top and bottom and at intervals not exceeding 192 diameters of the reinforcement.

5. All cells containing reinforcement shall be filled solidly with grout. Grout shall be poured in lifts of eight feet (8′) maximum height. All grout shall be consolidated at time of pouring by puddling or vibrating and then reconsolidated by again puddling later, before plasticity is lost.

When total grout pour exceeds eight feet (8′) in height the grout shall be placed in four-foot (4′) lifts and special inspection during grouting shall be required. Minimum cell dimension shall be three inches (3″). Special inspection at time of grouting shall not be considered as special inspection under Table No. 24-H.

6. When the grouting is stopped for one hour or longer, horizontal construction joints shall be formed by stopping the pour of grout one and one-half inches (1½″) below the top of the uppermost unit.

(c) **Stresses.** See Section 2418 (a).

TABLE NO. 24-H — MAXIMUM WORKING STRESSES IN POUNDS PER SQUARE INCH FOR REINFORCED SOLID AND HOLLOW UNIT MASONRY[1]

TYPE OF STRESS	SPECIAL INSPECTION REQUIRED — Yes	SPECIAL INSPECTION REQUIRED — No
Compression—Axial, Walls	See Section 2418	One-half of the values permitted under Section 2418
Compression—Axial, Columns	See Section 2418	One-half of the values permitted under Section 2418
Compression—Flexural	$0.33 f'_m$ but not to exceed 900	$0.166 f'_m$ but not to exceed 450
Shear: No shear reinforcement[2]	$0.02 f'_m$ but not to exceed 50	15
Reinforcement taking entire shear: Flexural members	$0.05 f'_m$ but not to exceed 120	50
Shear Walls	$0.04 f'_m$ but not to exceed 75	30
Modulus of Elasticity[3]	$1000 f'_m$ but not to exceed 3,000,000	$500 f'_m$ but not to exceed 1,500,000
Modulus of Rigidity[3]	$400 f'_m$ but not to exceed 1,200,000	$200 f'_m$ but not to exceed 600,000
Bearing on full Area[4]	$0.25 f'_m$ but not to exceed 900	$0.125 f'_m$ but not to exceed 450
Bearing on ⅓ or less of area[4]	$0.30 f'_m$ but not to exceed 1200	$0.15 f'_m$ but not to exceed 600
Bond—Plain bars	60	30
Bond—Deformed	140	100

[1]Stresses for hollow unit masonry are based on net section.

[2]Web reinforcement shall be provided to carry the entire shear in excess of 20 pounds per square inch whenever there is required negative reinforcement and for a distance of one-sixteenth the clear span beyond the point of inflection.

[3]Where determinations involve rigidity considerations in combination with other materials or where deflections are involved, the moduli of elasticity and rigidity under columns entitled "yes" for special inspection shall be used.

[4]This increase shall be permitted only when the least distance between the edges of the loaded and unloaded areas is a minimum of one-fourth of the parallel side dimension of the loaded area. The allowable bearing stress on a reasonably concentric area greater than one-third, but less than the full area, shall be interpolated between the values given.

SEC. 91.2309 — RETAINING WALLS

(a) **Design.** Retaining walls shall be designed to resist the lateral pressure of the retained material determined in accordance with accepted engineering principles.

The soil characteristics and design criteria necessary for such a determination shall be obtained from a special foundation investigation performed by an agency acceptable to the Department. The Department shall approve such characteristics and criteria only after receiving a written opinion from the investigation agency together with substantiating evidence.

EXCEPTION: Freestanding walls which are not over 15' in height or basement walls which have spans of 15' or less between supports may be designed in accordance with Subsection (b) of this Section.

(b) **Arbitrary Design Method.** Walls which retain drained earth and come within the limits of the exception to Subsection (a) of this section may be designed for an assumed earth pressure equivalent to that exerted by a fluid weighing not less than shown in Table 23-E. A vertical component equal to one-third of the horizontal force so obtained may be assumed at the plane of application of the force.

The depth of the retained earth shall be the vertical distance below the ground surface measured at the wall face for stem design or measured at the heel of the footing for overturning and sliding.

(c) **Surcharge.** Any superimposed loading, except retained earth, shall be considered as surcharge and provided for in the design. Uniformly distributed loads may be considered as equivalent added depth of retained earth. Surcharge loading due to continuous or isolated footings shall be determined by the following formulas or by an equivalent method approved by the Superintendent of Building.

TABLE NO. 23-E

Surface Slope of Retained Material* Horiz. to Vert.	Equivalent Fluid Weight lb/ft³
LEVEL	30
5 to 1	32
4 to 1	35
3 to 1	38
2 to 1	43
1½ to 1	55
1 to 1	80

* Where the surface slope of the retained earth varies, the design slope shall be obtained by connecting a line from the top of the wall to the highest point on the slope, whose limits are within the horizontal distance from the stem equal to the stem height of the wall.

Resultant Lateral Force

$$R = \frac{0.3 \, Ph^2}{x^2 + h^2}$$

Location of Lateral Resultant

$$d = x \left[\left(\frac{x^2}{h^2} + 1 \right) \left(\tan^{-1} \frac{h}{x} \right) - \left(\frac{x}{h} \right) \right]$$

Where:

R = Resultant lateral force measured in pounds per foot of wall width.

P = Resultant surcharge load of continuous or isolated footings measured in pounds per foot of length parallel to the wall.

x = Distance of resultant load from back face of wall measured in feet.

h = Depth below point of application of surcharge loading to top of wall footing measured in feet.

d = Depth of lateral resultant below point of application of surcharge loading measured in feet.

$\left(\tan^{-1} \frac{h}{x} \right)$ = The angle in radians whose tangent is equal

to $\left(\frac{h}{x} \right)$

Loads applied within a horizontal distance equal to the wall stem height, measured from the back face of the wall, shall be considered as surcharge.

For isolated footings having a width parallel to the wall less than three feet, "R" may be reduced to 1/6 the calculated value.

The resultant lateral force "R" shall be assumed to be uniform for the length of footing parallel to the wall, and to diminish uniformly to zero at the distance "x" beyond the ends of the footing.

Vertical pressure due to surcharge applied to the top of the wall footing may be considered to spread uniformly within the limits of the stem and planes making an angle of 45° with the vertical.

(d) **Bearing Pressure and Overturning.** The maximum vertical bearing pressure under any retaining wall shall not exceed that allowed in Division 28 of this Article except as provided for by a special foundation investigation. The resultant of vertical loads and lateral pressures shall pass through the middle one-third of the base.

(e) **Friction and Lateral Soil Pressures.** Retaining walls shall be restrained against sliding by friction of the base against the earth, by lateral resistance of the soil, or by a combination of the two. Allowable friction and lateral soil values shall not exceed those allowed in Division 28 of this Article except as provided by a special foundation investigation.

When used, keys shall be assumed to lower the plane of frictional resistance and the depth of lateral bearing to the level of the bottom of the key. Lateral bearing pressures shall be assumed to act on a vertical plane located at the toe of the footing.

(f) **Construction.** No retaining wall shall be constructed of wood.

(g) **Special Conditions.** Whenever, in the opinion of the Superintendent of Building, the adequacy of the foundation material to support a wall is questionable, an unusual surcharge condition exists, or whenever the retained earth is so stratified or of such a character as to invalidate normal design assumptions, he may require a special foundation investigation before approving any permit for such a wall.

SEC. 91.2802 — FOUNDATION ANALYSIS

(a) **General.** The classification of the foundation material under every building shall be based upon the examination of test borings or excavations made at the site. The extent and number of the test borings or excavations shall be sufficient to provide the data necessary to classify the foundation materials under the entire building. The location of the test borings or excavations and the nature of the subsurface materials shall be indicated on the plans.

EXCEPTION: The requirements of this Subsection shall not apply to any building constructed in accordance with the arbitrary requirements of Division 48 (Wood Frame Dwellings).

(b) **Foundation Materials.** The foundation of every structure shall be a uniform natural deposit of rock, gravel, sand, clay, silt or combination thereof which does not contain and which does not overlie strata containing more than 10% by dry weight of organic matter.

EXCEPTION: Foundations may be artificial fill or nonuniform areas of dis-similar materials, provided due allowance is made for the effect of differential settlement.

(c) **Effect of Change in Moisture Content.** Due allowance shall be made in determining the capacity of foundations for the effect of possible change in moisture content.

(d) **Effect of Pressure on Foundations.** Where footings are to be placed at varying elevations or at different elevations from existing footings, the effect of adjacent loads shall be included in the foundation analysis.

(e) **Load Distribution.** A load upon a foundation stratum shall be assumed as distributed uniformly over an area subtended by planes extending downward from the edges of the footing and making an angle of 60 degrees with the horizontal.

(f) **Arbitrary Design Specification.** Certain buildings of Type V construction may have footings and foundations designed in accordance with the provisions of Section 91.1708 (Type V Buildings) and Section 91.4807 (Wood Frame Dwellings).

SEC. 91.2803 — FOUNDATION CLASSIFICATION

(a) **Foundation Classification.** Foundation materials shall be grouped in classes having the designations set forth in Table No. 28-A.

(b) **Variation in Soil Strata.** The classification of foundation material shall be that of the weakest stratum within a depth below the footing equal to twice the least width of the footing.

(c) **Allowable Foundation Pressures.** The design unit pressure upon every foundation shall not exceed the arbitrary values exhibited in Table No. 28-A.

EXCEPTION: The tabulated values may be modified as prescribed in Section 91.2804.

(d) **Friction and Lateral Soil Pressures.** The design unit values for friction and lateral soil pressures shall not exceed the arbitrary values exhibited in Table No. 28-B.

EXCEPTION: The tabulated values may be modified as prescribed in Section 91.2804.

TABLE NO. 28-A — ALLOWABLE FOUNDATION PRESSURE
(Kips per Square Foot — 1 Kip = 1,000 pounds)

CLASS OF MATERIAL

Rock—Depth of Embedment shall be to a Fresh Unweathered Surface Except as Noted	Value at Min. Depth	Increase for Depth	Maximum Value
*Massive crystalline bedrock; basalt, granite and diorite in sound condition......	20		20
*Foliated rocks; schist and slate, in sound condition	8		8
*Sedimentary rocks; hard shales, dense siltstones and sandstones, thoroughly cemented conglomerates	6		6
Soft, or broken bedrocks; soft shales, shattered slates, diatomaceous shales; other badly jointed (fractured) or weathered rock. 12" minimum embedment	2		2

*NOTE: The above values apply only where the strata are level or nearly so, and/or where the area has ample lateral support. Tilted strata, and the relationship to nearby slopes should receive special consideration. These values may be increased one-third to a maximum of two times the assigned value, for each foot of penetration below fresh, unweathered surface.

Soils—Minimum Depth of Embedment shall be one foot below the adjacent undisturbed ground surface*	Loose	Compact	Soft	Stiff	Increase for Depth	Maximum Value
Gravel, well graded. Well graded gravels or gravel-sand mixtures, little or no fines......	1.33	2.0			20	8
Gravel, poorly graded. Poorly graded gravels or gravel-sand mixtures, little or no fines......	1.33	2.0			20	8
Gravel, silty. Silty gravels or poorly graded gravel sand silt mixtures......	1.0	2.0			20	8
Gravel, clayey. Clayey gravels or gravel-sand clay mixtures......	1.0	2.0			20	8
Sand, well graded. Well graded sands or gravelly sands, little or no fines....	1.0	2.0			20	6
Sand, poorly graded. Poorly graded sands or gravelly sands, little or no fines	1.0	2.0			20	6
Sand, silty. Silty sand, or poorly graded sand-silt mixtures......	0.5	1.5			20	4
Sand, clayey. Clayey sands or sand-clay mixtures......	1.0	2.0			20	4
Silt. Inorganic silts and very fine sands, rock flour, silty or clayey fine sands with slight plasticity......	0.5	1.0			20	3
Silt, organic. Organic silts and organic silt-clays of low plasticity......	0.5	1.0	0.5	1.0	10	2
Silt, elastic. Very compressible silts, micaceous or diatomaceous fine sandy or silty soils......	0.5	1.0			10	1.5
Clay, lean. Inorganic clays of low to medium plasticity, silty clays, lean clays	1.0	2.0	1.0	2.0	20	3
Clay, fat. Very compressible clays, inorganic clays of high plasticity......			0.5	1.0	10	1.5
Clay, organic. Organic clays of medium to high plasticity, very compressible......			0.5			0.5
Peat. Peat and other highly organic swamp soils			0			0

NOTES:

1. Values for gravels and sand given are for footings one foot in width and may be increased in direct proportion to footing width to maximum of three times the maximum value, or to the designated maximum value, whichever is the least.

2. Where the bearing values in the above table are used, it should be noted that increased width or unit load will cause increase in settlement.

3. Special attention should be given to the effect of increase in moisture in establishing soil classifications.

*4. Minimum depth for highly expansive soils to be one and one-half feet.

5. Increases for depth are given in percentage of minimum value for each additional foot below the minimum required depth.

TABLE NO. 28-B — ALLOWABLE FRICTIONAL & BEARING VALUES FOR ROCK[1]

Type	Friction Coefficient	Allowable Lateral Bearing lbs. per sq.	per ft. Max. Value
Massive Crystalline Bedrock	1.0	4,000	20,000
Foliated Rocks	.8	1,600	8,000
Sedimentary Rocks	.6	1,200	6,000
Soft or Broken Bedrocks	.4	400	2,000

ALLOWABLE FRICTIONAL & LATERAL BEARING VALUES FOR SOILS
Frictional Resistance — Gravels and Sands[1]

Soil Type	Friction Coefficient
Gravel, Well Graded	0.6
Gravel, Poorly Graded	0.6
Gravel, Silty	0.5
Gravel, Clayey	0.5
Sand, Well Graded	0.4
Sand, Poorly Graded	0.4
Sand, Silty	0.4
Sand, Clayey	0.4

1. Coefficient to be multiplied by the Dead Load.

ALLOWABLE FRICTIONAL RESISTANCE
(lbs. per sq. ft.) — Clay and Silt[2]

Soil Type	Loose or Soft	Compact or Stiff
Silt, Inorganic	250	500
Silt, Organic	250	500
Silt, Elastic	200	400
Clay, Lean	500	1000
Clay, Fat	200	400
Clay, Organic	150	300
Peat	0	0

2. Frictional values to be multiplied by the width of footing subjected to positive soil pressure. In no case shall the frictional resistance exceed ½ the dead load on the area under consideration.

ALLOWABLE LATERAL BEARING PER FT. OF DEPTH BELOW NATURAL GROUND SURFACE
(lbs. per sq. ft.) (Natural Soils or approved compacted fill)

Soil Type	Loose or Soft	Compact or Stiff	Max. Values
Gravel, Well Graded	200	400	8000
Gravel, Poorly Graded	200	400	8000
Gravel, Silty	167	333	8000
Gravel, Clayey	167	333	8000
Sand, Well Graded	183	367	6000
Sand, Poorly Graded	77	200	6000
Sand, Silty	100	233	4000
Sand, Clayey	133	300	4000
Silt, Inorganic	67	133	3000
Silt, Organic	33	67	2000
Silt, Elastic	33	67	1500
Clay, Lean	267	667	3000
Clay, Fat	33	167	1500
Clay, Organic	33	------	500
Peat	0	0	0

GENERAL CONDITIONS OF USE

1. Frictional and lateral resistance of soils may be combined, provided the lateral bearing resistance does not exceed ⅔ of allowable lateral bearing.

2. A ⅓ increase in frictional and lateral bearing values will be permitted to resist loads caused by wind pressure or earthquake forces.

3. Isolated poles such as flag poles or signs may be designed using lateral bearing values equal to two times the tabulated values.

4. Lateral bearing values are permitted only when concrete is deposited against natural ground or compacted fill, approved by the Superintendent of Building.

SEC. 91.2804 — SPECIAL FOUNDATION INVESTIGATION

(a) **Requirements.** Whenever, in the opinion of the Superintendent of Building, the adequacy and class of a foundation cannot be determined by the test borings or excavations required by the provisions of Section 91.2802 (a), he may require a special foundation investigation before approving the use of the foundation.

➤ When the Department of Building and Safety considers a geological exploration necessary, such exploration shall be made by a Registered Certified Engineering Geologist. ◄

(b) **Deviations.** Values in excess of the arbitrary allowable foundation values exhibited in Tables No. 28-A and 28-B shall be permitted only after performance of a special foundation investigation by an agency acceptable to the Department. The Department shall approve such deviations only after receiving a written opinion from the investigating agency together with substantiating evidence.

(c) **Stresses.** Stresses and deformations within the foundations shall be determined by the general principles of soil mechanics.

APPENDIX C

Retaining Wall Design Equations and Design Tables

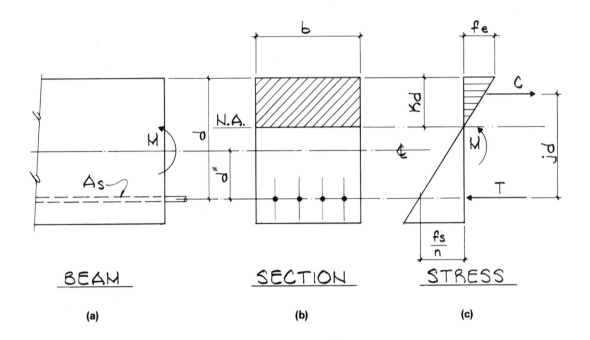

Fig. C.1

RETAINING WALL DESIGN EQUATIONS
REINFORCED CONCRETE (Ref. 22)

Section Coefficients

$$n = \frac{E_s}{E_c} = \frac{29,000,000}{W^{1.5}\ 33\sqrt{f'_c}} \qquad\qquad (C.1)$$

$$K = f_s p j \qquad\qquad (C.2)$$

$$K = \tfrac{1}{2} f_c k j \qquad\qquad (C.3)$$

$$K = \frac{M}{b d^2} \qquad\qquad (C.4)$$

where M = resisting moment of the concrete stresses

$$k = \sqrt{2np + (np)^2} - np \qquad\qquad (C.5)$$

$$k = \frac{1}{1 + \dfrac{f_s}{n f_c}} \qquad\qquad (C.6)$$

$$j = 1 - \frac{k}{3} \qquad\qquad (C.7)$$

Tensile Reinforcement

$$A_s = \frac{M \text{ (in.-lbs.)}}{f_s j d} \tag{C.8}$$

$$A_s = \frac{M \text{ (ft.-kips)}}{\dfrac{f_s j d}{12,000}} \tag{C.9}$$

$$a = \frac{f_s j}{12,000} \tag{C.10}$$

$$A_s = \frac{M}{a \, d} \tag{C.11}$$

Shear Stress

$$v = \frac{V}{b j d} \tag{C.12}$$

Bond Stress

$$\Sigma 0 = \frac{V}{d j u} \tag{C.13}$$

REINFORCED GROUTED MASONRY (Ref. 22)

Section Coefficients

$$n = \frac{E_s}{E_c} \tag{C.14}$$

$$k = \frac{1}{1 + \dfrac{f_s}{n f_c}} \tag{C.15}$$

$$j = 1 - \frac{k}{3} \tag{C.16}$$

Tensile Reinforcement

$$A_s = \frac{M \text{ (in.-lbs.)}}{f_s j d} \tag{C.17}$$

$$A_s = \frac{M \text{ (ft.-kips)}}{a \, d} \tag{C.18}$$

$$p = \frac{A_s}{b d} \tag{C.19}$$

$$k = \sqrt{2np + (np)^2} - np \tag{C.20}$$

$$j = 1 - \frac{k}{3} \tag{C.21}$$

from np calculate $\dfrac{2}{jk}$ $\tag{C.22}$

Note that $n = 20$ with continuous field inspection

$\qquad\qquad n = 40$ without continuous field inspection

$\qquad\qquad$ for tabulated values of np, j, and $\dfrac{2}{jk}$ see (Ref. 1)

$$f_m = \frac{12 M (2/jk)}{b d^2} \quad (M = \text{ft.-lbs.}) \tag{C.23}$$

check with $K = \frac{1}{2} f_m\, j\, k = \dfrac{M}{b d^2}$ $\tag{C.24}$

Shear Stress

$$v = \frac{V}{b j d} \tag{C.25}$$

Bond Stress

$$\Sigma 0 = \frac{V}{d j u} \tag{C.26}$$

$$L = \frac{f_s A_s}{\Sigma 0 u} \tag{C.27}$$

Table C-1. Allowable Stresses for Concrete[6]: f'c = 2000 psi

Description			For Any Strength of Concrete		f'_c = 2000 psi
Modular ratio: n For concrete weighing 145 lbs. per cu. ft.	n	=	$\dfrac{29,000,000}{w^{1.5}\ 33\sqrt{f'_c}}$	=	11
Flexure: f_c Extreme fiber stress in compression	f_c	=	$0.45f'_c$	=	900
Extreme fiber stress in tension in plain concrete footings and walls	f_c	=	$1.6\sqrt{f'_c}$	=	71
Shear: v (as a measure of diagonal tension at a distance "d" from the face of the support) Beams with no web reinforcement[1]	v_c	=	$1.1\sqrt{f'_c}$	=	49
Joists with no web reinforcement	v_c	=	$1.2\sqrt{f'_c}$	=	54
Members with vertical or inclined web reinforcement or properly combined bent bars and vertical stirrups	v	=	$5\sqrt{f'_c}$	=	223
Slabs and footings[1]	v_c	=	$2\sqrt{f'_c}$	=	89[1]

Shear in Walls Resisting other than Seismic Forces:[2]
  Shear carried by concrete[3]
  H/D ≤ 1
	v_c	=	$3\sqrt{f'_c}$	=	134

H/D ≥ 2.7	v_c	=	$1.1\sqrt{f'_c}$	=	49
Shear carried by concrete and reinforcement[4] H/D ≤ 1	v	=	$3\sqrt{f'_c}$	=	134
H/D ≥ 2	v	=	$5\sqrt{f'_c}$	=	223
Bearing: f_c On full area			$0.25f'_c$	=	500
One one-third area or less[5]			$0.375\ f'_c$	=	750

Shear in walls resisting seismic forces in buildings without a 100% moment resisting space frame:[2]
  Shear carried by concrete[3]
  H/D ≤ 1
	v_c	=	$1.25\sqrt{f'_c}$	=	56

H/D ≥ 2.7	v_c	=	$.45\sqrt{f'_c}$	=	20
Shear carried by concrete and reinforcement[4] H/D ≤ 1	v	=	$1.25\sqrt{f'_c}$	=	56
H/D ≥ 2	v	=	$2.30\sqrt{f'_c}$	=	103

1 For shear values for lightweight aggregate concrete see UBC Section 2612 (i). (1970 UBC)
2 The stresses indicated may be increased one-third when caused by seismic or wind forces.
3 For values between "H/D" of 1.0 and 2.7, the allowable shear varies linearly. For lightweight concrete multiply tabulated values by .15 Fsp.
4 For values between "H/D" 1.0 and 2.0, the allowable shear varies linearly.
5 This increase shall be permitted only when the least distance between the edges of the loaded and unloaded areas is a minimum of one-fourth of the parallel side dimension of the loaded area. The allowable bearing stress on a reasonably concentric area greater than one-third, but less than the full area, shall be interpolated between the values given.
6 UBC Table No. 26-D. (1970 UBC)

Table C-2. Maximum Allowable Working Stresses¹ psi Reinforced Solid and Hollow² Unit Masonry

Type of Stress	Allowable Stress or Stress Coefficients (No)	Allowable Stress or Stress Coefficients (Yes)	1350 No	1350 Yes	1500 No	1500 Yes	1800 No	1800 Yes	2000 Yes	2500⁴ Yes	2700 Yes	3000 Yes	3500 Yes	4000 Yes	4500 Yes	5000 Yes	5300⁵ Yes	6000 Yes
Ultimate Compressive Strength	f'_m	f'_m	1350	1350	1500	1500	1800	1800	2000	2500	2700	3000	3500	4000	4500	5000	5300	6000
Special Inspection Required	No	Yes	No	Yes	No	Yes	No	Yes	Yes	Yes	Yes	Yes	Yes	Yes	Yes	Yes	Yes	Yes
Compression—Axial, Walls⁶	$\frac{1}{2}(0.2 f'_m)$	$0.2 f'_m$	135	270	150	300	180	360	400	500	540	600	700	800	900	1000	1060	1200
Compression—Axial, Columns⁷	$\frac{1}{2}(0.18 f'_m)$	$0.18 f'_m$	122	244	135	270	162	324	360	450	486	540	630	720	810	900	954	1080
Compression—Flexural	$\frac{1}{2}(0.33 f'_m)$ 450 psi max.	$0.33 f'_m$ 900 psi max.	225	450	250	500	300	600	667	833	900	900¹²	900	900	900	900	900	900
Shear — No Shear Reinforcement, Flexural	25	$1.1\sqrt{f'_m}$ 50 psi max	25 (Net Section)	40	25	42	25	47	49	50	50	50	50	50	50	50	50	50
Shear Walls⁸ M/vd ≥ 1⁹	17	$.9\sqrt{f'_m}$ 34 psi max.	17	33	17	34	17	34	34	34	34	34	34	34	34	34	34	34
M/vd = 0	25	$2.0\sqrt{f'_m}$ 50 psi max.	25	50	25	50	25	50	50	50	50	50	50	50	50	50	50	50
Reinforcing taking all shear, Flexural	75	$3.0\sqrt{f'_m}$ 150 psi max.	75 (Net Section)	110	75	115	75	127	134	150	150	150	150	150	150	150	150	150
Shear Walls⁸ M/vd ≥ 1⁹	35	$1.5\sqrt{f'_m}$ 75 psi max.	35	55	35	58	35	64	67	75	75	75	75	75	75	75	75	75
M/vd = 0	60	$2.0\sqrt{f'_m}$ 120 psi max.	60	73	60	77	60	85	89	100	104	110	118	120	120	120	120	120
Modulus of Elasticity³	$\frac{1}{2}(1000 f'_m)$ 1,500,000 psi max	$1,000 f'_m$ 3,000,000 psi max	675,000	1,350,000	750,000	1,500,000	900,000	1,800,000	2,000,000	2,500,000	2,700,000	3,000,000	3,000,000	3,000,000	3,000,000	3,000,000	3,000,000	3,000,000
Modular Ratio—n = E_s/E_m	$15,000/f'_m$	$30,000/f'_m$	44	22	40	20	33	16.7	15	12	11	10	10	10	10	10	10	10
Modulus of Rigidity¹⁰	$\frac{1}{2}(400 f'_m)$ 600,000 psi max	$400 f'_m$ 1,200,000 psi max	270,000	540,000	300,000	600,000	360,000	720,000	800,000	1,000,000	1,080,000	1,200,000	1,200,000	1,200,000	1,200,000	1,200,000	1,200,000	1,200,000
Bearing on full area¹¹	$\frac{1}{2}(.25 f'_m)$ 450 psi max.	$0.25 f'_m$ 900 psi max.	170	340	187	375	225	450	500	633	675	750	875	900	900	900	900	900
Bearing on 1/3 or less of area¹¹	$\frac{1}{2}(.30 f'_m)$ 600 psi max.	$0.30 f'_m$ 1200 psi max.	200	400	225	450	270	540	600	750	810	900	1050	1200	1200	1200	1200	1200
Bond – Plain bars	30 psi	60 psi	30	60	30	60	30	60	60	60	60	60	60	60	60	60	60	60
Bond – Deformed	100 psi	140 psi	100	140	100	140	100	140	140	140	140	140	140	140	140	140	140	140

Column group notes:
- 1350: Hollow Clay Units² Grade LB or Hollow Concrete² Units Grade A or N
- 1500: Grouted Solid Hollow Units: Concrete, Grade A or N; Clay, Grade LB; or Solid Units 2500 psi on Gross
- 1800: Solid Units 3000 psi on Gross
- 2000–6000: Special Testing³ f'_m Established by Prism Tests

Table C-2. Footnotes

1. Extension of U B C Table No. 24-H.

2. Stresses for hollow unit masonry are based on net section.

3. Special testing shall include preliminary tests conducted as specified in Section 2404 (c) to establish f′$_m$ and at least one field test during construction of walls per each 5000 square feet of wall but not less than three such field tests for any building.

4. Ultimate compressive strength may be assumed up to f′$_m$ = 2600 psi if solid clay units have an ultimate strength of at least 6000 psi on gross area. No prism test would be required but the strength of the brick would have to be established in accordance with the U.B.C. Standard No. 24-25.

5. Ultimate compressive strength may be assumed up to f′$_m$ = 5300 psi in accordance with Section 2404 (c) 3, but field prism tests would have to be made to verify strength at least one prism per 5000 sq. ft. of wall nor less than 3 prisms per building.

6. The axial stress in reinforced masonry bearing walls shall not exceed the value determined by the following formula:

$$F_a = 0.20 f'_m \left[1 - \left(\frac{h}{40t} \right)^3 \right]$$

Where

F$_a$ = Compressive unit axial stress in masonry wall.
f′$_m$ = Ultimate compressive masonry stress. The value of f′$_m$ shall not exceed 6000 psi.
t = Thickness of wall in inches.
h = Clear height in inches.*

7. The axial load on columns shall not exceed:

$$P = A_g \left(0.18 f'_m + 0.65 p_g f_s \right) \left[1 - \left(\frac{h}{40t} \right)^3 \right]$$

Where

P = Maximum concentric column axial load.
A$_s$ = The gross area of the column.
f′$_m$ = Ultimate compressive masonry strength. The value of f′$_m$ shall not exceed 6000 psi.
p$_g$ = Ratio of the effective cross-sectional area of vertical reinforcement to A$_g$.
f$_s$ = 40% min. yield strength; 24,000 psi max.
t = Least thickness of column in inches.
h = Clear height in inches.*

8. When clacluating shear stresses in shear walls which resist seismic forces, use twice the force required by Sec. 2314(d) 1. e.g. 2V = 2 ZKCW.

9. Interpolate by straight line for M/V$_d$ values between 0 and 1. See Diagram A-2
Where M = Moment on wall = Vh
V = Shear force
d = Length of wall
h = Height of wall

10. Where determinations involve rigidity considerations in combination with other materials or where deflections are involved, the modulii of elasticity and rigidity under columns entitled "yes" for special inspection shall be used.
From tests it appears that the maximum modulus of elasticity for masonry is approximately 3,000,000 psi, and the minimum modular ratio "n" is approximately 10.

11. This increase shall be permitted only when the least distance between the edges of the loaded and unloaded areas is a minimum of one-fourth of the parallel side dimension of the loaded area. The allowable bearing stress on a reasonably concentric area greater than one-third, but less than the full area, shall be interpolated between the values given.

12. Heavy line indicates limit of allowable stresses by U.B.C. values may be extended if local jurisdictions permit.

*EXCEPTION: The height or length to thickness ratio may be increased and the minimum thickness may be decreased when data is submitted which justifies a reduction in the requirements.

Table C-3. Concrete Reinforcing Steel Bar Sizes and Dimensions

Deformed-bar designation	Weight, lb. per ft.	Diameter, in.	Cross-section area, sq. in.	Perimeter, in.	Max. outside dia., in.
#2	0.167	0.250	0.05	0.786	
#3	0.376	0.375	0.11	1.178	7/16
#4	0.668	0.500	0.20	1.571	9/16
#5	1.043	0.625	0.31	1.963	11/16
#6	1.502	0.750	0.44	2.356	7/8
#7	2.044	0.875	0.60	2.749	1
#8	2.670	1.000	0.79	3.142	1 1/8
#9	3.400	1.128	1.00	3.544	1 1/4
#10	4.303	1.270	1.27	3.990	1 7/16
#11	5.313	1.410	1.56	4.430	1 5/8
#14S	7.65	1.693	2.25	5.32	1 15/16
#18S	13.60	2.257	4.00	7.09	2 1/2

Table C-4. Areas and Perimeters of Bars in Sections 1 Ft. Wide

Areas A_s (or A'_s) (top) sq. in.; Perimeters Σo, (bottom) in.

Enter table with values of A_s (or A'_s) and $\Sigma o = \dfrac{V}{7/8\,du}$ (V:lb; d:in.; u:psi)

Coefficients $a = \dfrac{f_s}{12,000} \times j$ inserted in table are for use in $A_s = \dfrac{M}{ad}$ or $A_s = \dfrac{NE}{adi}$

Spacing	#2	#3	#4	#5	#6	#7	#8	#9	#10	#11	Spacing
2	0.30	0.66	1.20	1.86	2.64						2
	4.7	7.1	9.4	11.8	14.2						
2-1/2	0.24	0.53	0.96	1.49	2.11	2.88	3.79				2-1/2
	3.8	5.7	7.5	9.4	11.3	13.2	15.1				
3	0.20	0.44	0.80	1.24	1.76	2.40	3.16	4.00			3
	3.1	4.7	6.3	7.8	9.4	11.0	12.6	14.2			
3-1/2	0.17	0.38	0.69	1.06	1.51	2.06	2.71	3.43	4.36		3-1/2
	2.7	4.0	5.4	6.7	8.1	9.4	10.8	12.2	13.7		
4	0.15	0.33	0.60	0.93	1.32	1.80	2.37	3.00	3.81	4.68	4
	2.4	3.5	4.7	5.9	7.1	8.3	9.4	10.6	12.0	13.3	
4-1/2	0.13	0.29	0.53	0.83	1.17	1.60	2.11	2.67	3.39	4.16	4-1/2
	2.1	3.1	4.2	5.2	6.3	7.3	8.4	9.5	10.6	11.8	
5	0.12	0.26	0.48	0.74	1.06	1.44	1.90	2.40	3.05	3.74	5
	1.9	2.8	3.8	4.7	5.7	6.6	7.5	8.5	9.6	10.6	
5-1/2	0.11	0.24	0.44	0.68	0.96	1.31	1.72	2.18	2.77	3.40	5-1/2
	1.7	2.6	3.4	4.3	5.1	6.0	6.9	7.7	8.7	9.7	
6	0.10	0.22	0.40	0.62	0.88	1.20	1.58	2.00	2.54	3.12	6
	1.6	2.4	3.1	3.9	4.7	5.5	6.3	7.1	8.0	8.9	
6-1/2	0.09	0.20	0.37	0.57	0.81	1.11	1.46	1.85	2.35	2.88	6-1/2
	1.4	2.2	2.9	3.6	4.4	5.1	5.8	6.5	7.4	8.2	
7	0.09	0.19	0.34	0.53	0.75	1.03	1.35	1.71	2.18	2.67	7
	1.3	2.0	2.7	3.4	4.0	4.7	5.4	6.1	6.8	7.6	
7-1/2	0.08	0.18	0.32	0.50	0.70	0.96	1.26	1.60	2.03	2.50	7-1/2
	1.3	1.9	2.5	3.1	3.8	4.4	5.0	5.7	6.4	7.1	
8	0.08	0.17	0.30	0.47	0.66	0.90	1.19	1.50	1.91	2.34	8
	1.2	1.8	2.4	2.9	3.5	4.1	4.7	5.3	6.0	6.6	
8-1/2	0.07	0.16	0.28	0.44	0.62	0.85	1.12	1.41	1.79	2.20	8-1/2
	1.1	1.7	2.2	2.8	3.3	3.9	4.4	5.0	5.6	6.2	
9	0.07	0.15	0.27	0.41	0.59	0.80	1.05	1.33	1.69	2.08	9
	1.0	1.6	2.1	2.6	3.1	3.7	4.2	4.7	5.3	5.9	
9-1/2	0.06	0.14	0.25	0.39	0.56	0.76	1.00	1.26	1.60	1.97	9-1/2
	1.0	1.5	2.0	2.5	3.0	3.5	4.0	4.5	5.0	5.6	
10	0.06	0.13	0.24	0.37	0.53	0.72	0.95	1.20	1.52	1.87	10
	0.9	1.4	1.9	2.4	2.8	3.3	3.8	4.3	4.8	5.3	
10-1/2	0.06	0.13	0.23	0.35	0.50	0.69	0.90	1.14	1.45	1.78	10-1/2
	0.9	1.3	1.8	2.2	2.7	3.1	3.6	4.0	4.6	5.1	
11	0.05	0.12	0.22	0.34	0.48	0.65	0.86	1.09	1.39	1.70	11
	0.9	1.3	1.7	2.2	2.6	3.0	3.4	3.9	4.4	4.8	
11-1/2	0.05	0.11	0.21	0.32	0.46	0.63	0.82	1.04	1.33	1.63	11-1/2
	0.8	1.2	1.6	2.0	2.5	2.9	3.3	3.7	4.2	4.6	
12	0.05	0.11	0.20	0.31	0.44	0.60	0.79	1.00	1.27	1.56	12
	0.8	1.2	1.6	2.0	2.4	2.8	3.1	3.5	4.0	4.4	
13	f_s	a	0.18	0.29	0.41	0.55	0.73	0.92	1.17	1.44	13
			1.4	1.8	2.2	2.5	2.9	3.3	3.7	4.1	
14	16,000	1.13	0.17	0.27	0.38	0.51	0.68	0.86	1.09	1.34	14
	18,000	1.29	1.3	1.7	2.0	2.4	2.7	3.0	3.4	3.8	
15	20,000	1.44	0.16	0.25	0.35	0.48	0.63	0.80	1.02	1.25	15
	22,000	1.60	1.3	1.6	1.9	2.2	2.5	2.8	3.2	3.5	
16	24,000	1.76	0.15	0.23	0.33	0.45	0.59	0.75	0.95	1.17	16
	27,000	2.00	1.2	1.5	1.8	2.1	2.4	2.7	3.0	3.3	
17	30,000	2.24	0.14	0.22	0.31	0.42	0.56	0.71	0.90	1.10	17
	33,000	2.48	1.1	1.4	1.7	1.9	2.2	2.5	2.8	3.1	
18			0.13	0.21	0.29	0.40	0.53	0.67	0.85	1.04	18
			1.0	1.3	1.6	1.8	2.1	2.4	2.7	3.0	

Table C-5. Minimum Clear Cover of Concrete

Location of reinforcement in concrete	Clear distance
Reinforcement in footings and other structural members in which the concrete is poured directly against the ground	3″
Formed concrete surfaces to be exposed to weather or in contact with the ground for bar sizes greater than #5	2″
Formed concrete surfaces to be exposed to weather or in contact with the ground for bar sizes #5 or less	1½″
Slabs and walls not exposed to weather or in contact with the ground	¾″
Beams and girders not exposed to weather or in contact with the ground	1½″
Floor joists with a maximum clear spacing of 30″	¾″
Column spirals or ties (not less than 1½ times the maximum size of the coarse aggregate or . . .)	1½″

Note: Except for concrete slabs or joists, the concrete cover protection shall not be less than the nominal diameter of the reinforcing bar.

Table C-6. Minimum Clear Spacing Distance

Space between bars	Clear distance
Clear distance between parallel bars except in columns and layers of bars in beams and girders	Not less than the nominal diameter of the bars, or 1⅓ times the size of the coarse aggregate, or not less than 1″
Clear distance between layers of reinforcement in beams or girders; the bars in each layer shall be directly above and below the bars in the adjacent layer	Not less than 1″
Clear distance between bars in walls and slabs	Not more than three times the wall or slab thickness or more than 18″
Clear distance between bars in spirally reinforced and tied columns	1½ times the nominal bar diameter, 1½ times the maximum size of the coarse aggregate, or not less than 1½″

Note: The clear distances above also apply for the clear distances between contact splices and adjacent splices of reinforcing bars.

Table C-7. Minimum Reinforcing Steel Spacing and Cover in Grouted Masonry

Location of reinforcment in masonry	Distance
Maximum spacing between reinforcing bars in walls	48″
Minimum spacing between parallel bars	1 bar diameter or not less than 1″
Minimum cover of reinforcing bars at the bottom of foundations	3″
Minimum cover of reinforcing bars in vertical members exposed to weather or soil	2″
Minimum cover of reinforcing bars in columns and at the bottoms or sides of girders or beams	1½″
Minimum cover of reinforcing bars in interior walls	¾″

Note: Reinforcing bars that are perpendicular to each other are permitted point contact at their intersection.

Table C-8. Allowable Bond Stress on Bars in Concrete: f'c =2000 psi

Top Bars $u = \dfrac{3.4 \sqrt{f'_s}}{D}$		Other Bars $u = \dfrac{4.8 \sqrt{f'_s}}{D}$	
Bar No.	u psi	Bar No.	u psi
3	350	3	500
4	304	4	430
5	244	5	344
6	203	6	286
7	174	7	246
8	152	8	215
9	135	9	191
10	122	10	172
11	110	11	156

Table C-9. Length of Embedment with Hook or Bend— 100% stress development

Total Length of Embedment with Hook or Bend

Bar No.	$f_s = 20,000$ psi		$f_s = 24,000$ psi	
	No Special Inspection $u = 100$ psi	Special Inspection $u = 140$ psi	No Special Inspection $u = 100$ psi	Special Inspection $u = 140$ psi
3	14″	10″	19″	13″
4	18″	13″	24″	17″
5	23″	16″	30″	21″
6	26″	19″	34″	25″
7	31″	22″	41″	29″
8	36″	25″	48″	33″
9	40″	29″	53″	38″
10	45″	32″	59″	42″
11	49″	35″	65″	46″

Table C-10a. Length of 12-Bar Diameters

Minimum length, 6 inches

Bar No.	12D	Bar No.	12D
¼″	6.0″	7	10.5″
3	6.0″	8	12.0″
4	6.0″	9	13.5″
5	7.5″	10	15.0″
6	9.0″	11	16.5″

Table C-10. Tension Capacity of Bend or Hook—pounds

Bar No.	90° Bend or 180° Hook
3	825
4	1500
5	2330
6	3300
7	4500
8	5920
9	7500
10	9530
11	11800

Table C-11. Length of Lap of Tension Splice

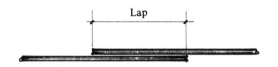

Minimum lap 12″

Bar No.	UBC Masonry Columns	UBC Masonry Walls	ACI Committee 531	
	30 dia.	40 dia.	50 dia.	60 dia.
¼″	12	12	12	15
3	12	15	19	23
4	15	20	25	30
5	19	25	31	38
6	23	30	38	45
7	27	35	44	53
8	30	40	50	60
9	34	45	56	68
10**	38	50	63	75
11**	42	55	69	83

PER ACI COMMITTEE 531

$$L = \frac{f_s A_s}{\Sigma_0 \mu} \times \frac{4}{3} \times 1.2^* = \frac{0.4 f_s D}{\mu}$$

$\mu = 160$ psi
$f_s = 20$ ksi $\qquad f_s = 24$ ksi
$L = 50\ D$ $\qquad\quad L = 60\ D$

* The 1.20 factor is included as satisfying the need for special precautions where all bars are spliced at the same location.

** For #10 and #11 bars, the length of lap required is so long that the cost difference for welding the splice is small and justifies welding for these sizes. Staggering of welds is desirable.

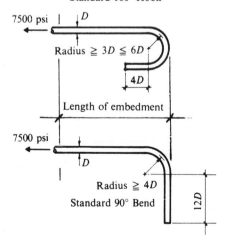

Standard 180° Hook

7500 psi

Radius $\geq 3D \leq 6D$

4D

Length of embedment

7500 psi

Radius $\geq 4D$

Standard 90° Bend

12D

Figure C.2 Standard hook and bends

References

1. Amrhein, James E.: *Reinforced Masonry Engineering Handbook*, Masonry Institute of America (1972).

2. American Society for Testing and Materials, No. C 404-70.

3. American Society for Testing and Materials, No. 615-Grade 40.

4. Building Code of the City of New York, Chap. 26 of Administrative Code, Sec. 1102.3(*c*) (amended 1970), p. 273.

5. Building Code of the City of New York, Chap. 26 of Administrative Code, Sec. C26-1111.2*b* (amended 1970).

6. Building Code of the City of New York, Chap. 26 of Administrative Code, (amended 1970), pp. 272–288.

   Los Angeles City Building Code, Secs., 91.2802, 91.2803 (1972), pp. 296–299.

   Uniform Building Code by I.C.B.O., Secs. 2904, 2905, 2907, Table 29-B (1973).

7. Building Code of the City of New York, Sec. C26-1102.4 (amended 1970), p. 273.

8. Building Code of the City of New York, Administrative Code (amended 1970), R.S. 10-2, p. A62.

9. *Concrete Masonry Design Manual*, Concrete Masonry Association of California (1964), pp. 1.3 to 1.6.

10. Coulomb, Charles Augustin: *Essai sur une application des regles de maximus et minimus a quelques problemes de statique relatifs a l'architecture*, Mem. Div. Savants, Academie Science, Paris, Vol. 7 (1776).

11. Dunham, Clarence W.: *The Theory and Practice of Reinforced Concrete*, 3d ed., McGraw-Hill (1953), pp. 265–267.

12. Dunham, Clarence W.: *Foundations of Structures*, McGraw-Hill (1950), pp. 133–157.

References

13. Dunham, Clarence W.: *Foundations of Structures*, McGraw-Hill (1950), pp. 256–266.

14. Dunham, Clarence W.: *Foundations of Structures*, McGraw-Hill (1950), pp. 7–23.

15. Gaylord, E. H., and C. N. Gaylord: *Structural Engineering Handbook*, McGraw-Hill (1968), pp. 5–48.

16. Los Angeles City Building Code, Sec. 91.2803, Table 28-B (1973).

17. Los Angeles City Building Code, Sec. 91.2309(*b*) (1972), pp. 138–139.

18. Los Angeles City Building Code, Sec. 91.2309(*c*) (1972), pp. 138–139.

19. Los Angeles City Building Code, Table No. 26-H (1972), p. 242.

20. Concrete Reinforcing Steel Institute: *Manual of Standard Practice*, 3d ed. (1973), Chap. 3.

21. Rankine, W. J. M.: *On the Stability of Loose Earth*, Trans. Royal Society of London, Vol. 147 (1857).

22. American Concrete Institute: *Reinforced Concrete Design Handbook*, 3d ed., Pub. SP-3.

23. American Concrete Institute: *Reinforced Concrete Design Handbook*, 3d ed., Pub. SP-3, Table 3a, p. 83.

24. Taylor, Donald W.: *Fundamentals of Soil Mechanics*, John Wiley & Sons (1948), pp. 480–502.

25. Terzaghi, Karl, and Ralph B. Peck: *Soil Mechanics in Engineering Practice*, 2d ed., John Wiley & Sons (1967), Chap. VIII, Art. 46.

26. Tschebotarioff, Gregory P.: *Foundations, Retaining and Earth Structures*, McGraw-Hill (1973), pp. 365–388.

27. Tschebotarioff, Gregory P.: *Foundations, Retaining and Earth Structures*, McGraw-Hill (1973), Chap. 4.

28. Tschebotarioff, Gregory P.: *Foundations, Retaining and Earth Structures*, McGraw-Hill (1973).

29. Uniform Building Code of the I.C.B.O., Table 29-B (1973).

30. Uniform Building Code of the I.C.B.O., Table 23-A (1973).

    Los Angeles City Building Code, Table 23-A (1972).

    Building Code of the City of New York (amended 1970); Chap. 26, R.S. 9-2.

31. Uniform Building Code of the I.C.B.O., Table 24-H (1973).

    Building Code of the City of New York (amended 1970), Chap. 26, R.S. 10-2, R.S. 10-1.

32. Uniform Building Code of the I.C.B.O., Table 24-A (1973).

    American Society for Testing and Materials, No. C144-70.

33. Uniform Building Code of the I.C.B.O., Table 24-H (1973).

    American Society for Testing and Materials, No. C90-70.

34. Uniform Building Code of the I.C.B.O., Sec. 2603(*a*) (1973), pp. 289, 290.

35. Uniform Building Code of the I.C.B.O., Table 24-H (1973).

    Los Angeles City Building Code, Table 24-H (1972).

36. Uniform Building Code of the I.C.B.O., Sec. 2418(*i*)2 (1973).

37. U.S. Army: *Retaining Walls—Engineering and Design*, EM1110-2-2502 (1961).

Index

Index